More Than Just a Textbook

Log on to **algconcepts.com** *to...*

- access your Online Student Edition from home so you don't need to bring your textbook home each night.
- link to Student Workbooks and Online Study Tools.

See mathematical concepts come to life

- Personal Tutor
- Concepts in Motion: BrainPOP® Movies
- Concepts in Motion: Interactive Labs
- Concepts in Motion: Animations

Practice what you've learned

- Extra Examples
- Self-Check Quizzes
- Data Updates
- Career Data
- Vocabulary Review
- Chapter Tests
- Standardized Test Practice

Try these other fun activities

- Scavenger Hunts

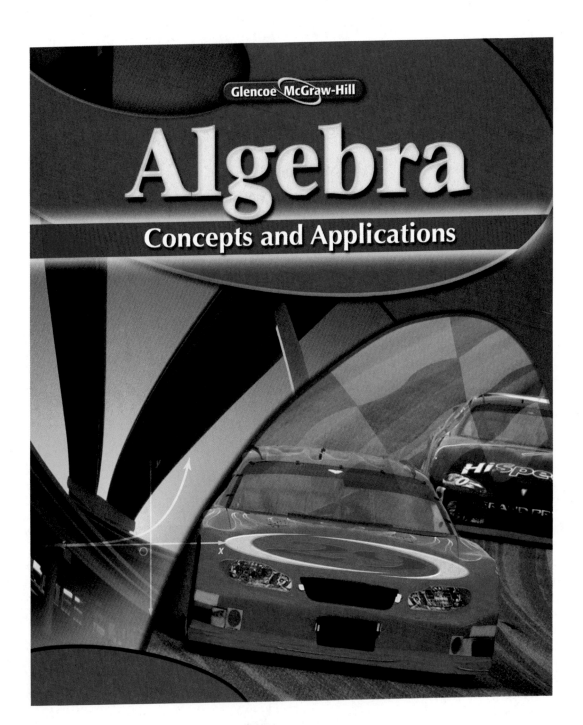

Glencoe McGraw-Hill

Algebra

Concepts and Applications

McGraw Hill Glencoe

New York, New York   Columbus, Ohio   Chicago, Illinois   Woodland Hills, California

Visit the Glencoe Mathematics Internet Site for
Algebra: Concepts and Applications at

www.algconcepts.com

You'll find:

- Concepts in Motion
- Personal Tutor
- Self-Check Quizzes
- Research Helps
- Data Updates
- Career Data
- Review Activities
- Test Practice
- Multilingual Glossary

 links to Web sites relevant to Problem-Solving
Workshops, Investigations, Math In the Workplace
features, exercises, and much more!

 Glencoe

The McGraw-Hill Companies

Send all inquiries to:
Glencoe/McGraw-Hill
8787 Orion Place
Columbus, OH 43240-4027

ISBN: 978-0-07-879912-9
MHID: 0-07-879912-0

Printed in the United States of America.

3 4 5 6 7 8 9 10 055/027 14 13 12 11 10 09

Dear Students, Teachers, and Parents,

Algebra: Concepts and Applications is designed to help you learn algebra and apply it to the real world. Throughout the text, you will be given opportunities to make connections from concrete models to abstract concepts. The real-world photographs and realistic data will help you see algebra in your world. You will also have plenty of opportunities to review and use arithmetic and geometry concepts as you study algebra. And for those of you who love a good debate, you will find plenty of opportunities to communicate your understanding of algebra.

We know that most of you haven't yet decided which careers you would like to pursue, so we've also included a little career guidance. This text offers real examples of how mathematics is used in many types of careers.

You may have to take an end-of-course exam for algebra, a proficiency test for graduation, the SAT, and/or the ACT. When you enter the workforce, you may also have to take job placement tests that include a section on mathematics. All of these tests include algebra problems. Because all Algebra 1 concepts are covered in this text, this program will prepare you for all of those tests.

Each day, as you use **Algebra: Concepts and Applications,** you will see the practical value of algebra. You will grow to appreciate how often algebra is used in ways that relate directly to your life. You will have meaningful experiences that will prepare you for the future. If you don't already see the importance of algebra in your life, you soon will!

Sincerely,
The Authors

P.S. To help you learn how to use your math book, use the Scavenger Hunt at algconcepts.com. The Scavenger Hunt will help you learn where things are located in each chapter.

Contents in Brief

Authors

Jerry Cummins
Staff Development Specialist
Bureau of Education and
   Research
State of Illinois
President, National Council of
   Supervisors of Mathematics
Western Springs, IL

Carol Malloy
Assistant Professor of
   Mathematics Education
University of North Carolina
   at Chapel Hill
Chapel Hill, NC

Kay McClain
Lecturer
George Peabody College
Vanderbilt University
Nashville, TN

Yvonne Mojica
Mathematics Teacher and
   Mathematics Department
   Chairperson
Verdugo Hills High School
Tujunga, CA

Jack Price
Professor, Mathematics
   Education
California State Polytechnic
   University
Pomona, CA

**Contributing Author
Dinah Zike**
Educational Consultant
Dinah-Might Activities, Inc.
San Antonio, TX

Academic Consultants and Teacher Reviewers

Each of the Academic Consultants read all 15 chapters, while each Teacher Reviewer read two chapters. The Consultants and Reviewers gave suggestions for improving the Student Editions and the Teacher's Wraparound Editions.

Academic Consultants

Judith Cubillo
Mathematics Teacher &
  Department Chairperson
Northgate High School
Walnut Creek, California

Mary C. Enderson
Faculty
Middle Tennessee State
  University
Murfreesboro, Tennessee

Alan G. Foster
Former Mathematics Teacher
  & Department Chairperson
Addison Trail High School
Addison, Illinois

Deborah A. Hutchens, Ed.D.
Assistant Principal
Great Bridge Middle School
Chesapeake, Virginia

Nicki Hudson
Mathematics Teacher
West Linn High School
West Linn, Oregon

Daniel Marks, Ph.D.
Associate Professor of
  Mathematics
Auburn University at
  Montgomery
Montgomery, Alabama

Donald McGurrin
Senior Administrator for
  Secondary Mathematics
Wake County Public Schools
Raleigh, North Carolina

C. Vincent Pané, Ed.D.
Associate Professor of
  Education
Molloy College
Rockville Centre, New York

Gary Shannon, Ph.D.
Professor of Mathematics
California State University
Sacramento, California

Marianne Weber
National Mathematics
  Consultant
St. Louis, Missouri

Teacher Reviewers

Breta J. Brown
Mathematics Teacher
Hillsboro High School
Hillsboro, North Dakota

Kimberly A. Brown
Mathematics Teacher
McGehee High School
McGehee, Arkansas

Helen Carpini
Mathematics Department
  Chairperson
Middletown High School
Middletown, Connecticut

Terry Cepaitis
Mathematics Resource
  Teacher
Anne Arundel County
  Public Schools
Annapolis, Maryland

Kindra R. Cerfoglio
Mathematics Teacher
Reed High School
Sparks, Nevada

Tom Cook
Mathematics Department
  Chairperson
Carlisle High School
Carlisle, Pennsylvania

Donna L. Cooper
Mathematics Department
  Chairperson
Walter E. Stebbins High
  School
Riverside, Ohio

James D. Crawford
Instructional Coordinator—
  Mathematics
Manchester Memorial
  High School
Manchester, New Hampshire

David A. Crine
Mathematics Department
  Chairperson
Basic High School
Henderson, Nevada

Table Of Contents

Standardized Test Practice

Hands-On Algebra

*inter*NET
CONNECTION

Graphing Calculator Exploration

Info Graphic

Lesson 1–1, page 7

Chapter ❷ Integers

Standardized Test Practice

Hands-On Algebra

*inter*NET
CONNECTION

Lesson 2–2, page 61

x

Lesson 3–4, page 115

Lesson 4–2, page 150

Chapter 5 · Proportional Reasoning and Probability

Math In the Workplace

Standardized Test Practice

Hands-On Algebra

Lesson 5–5, page 217

*inter*NET CONNECTION

Lesson 6–3, page 254

Chapter 6 Functions and Graphs

xiv

Lesson 7–4, page 302

Chapter 8 Powers and Roots 334

Math
In the Workplace

Standardized Test Practice

Hands-On Algebra

*inter*NET
CONNECTION

Graphing Calculator Exploration

Info Graphic

Photo Graphic

Lesson 8–2, page 345

Investigation, page 411

Lesson 10–3, page 434

Chapter ⑩ Factoring

Lesson 11-2, page 467

Math In the Workplace

Standardized Test Practice

Hands-On Algebra

interNET CONNECTION

Photo Graphic

Lesson 12–3, page 517

Chapter 12 Inequalities

Chapter 13   Systems of Equations and Inequalities

Lesson 13–4, page 567

Chapter 14 Radical Expressions

Standardized Test Practice

Hands-On Algebra

interNET CONNECTION

Graphing Calculator Exploration

Lesson 14–2, page 608

Lesson 15–5, page 666

Chapter 15   Rational Expressions and Equations

Preparing for Standardized Test Success

The **Preparing for Standardized Tests** pages at the end of each chapter and at the end of the book have been created to help you get ready for the mathematics portions of your standardized tests. On these pages, you will find strategies for solving problems and test-taking advice to help you maximize your score.

It is important to remember that there are many different standardized tests given by schools and states across the country. Find out as much as you can about your test. Start by asking your teacher and counselor for any information, including practice materials, that may be available to help you prepare.

To help you get ready for these tests, do the Standardized Test Practice question in each lesson. Also review the concepts and techniques contained in the Preparing for Standardized Tests pages at the end of each chapter listed below. This will help you become familiar with the types of math questions that are asked on various standardized tests.

The **Preparing for Standardized Tests** pages in the Student Handbook review common types of problems on standardized tests and provide practice for each format.

The **Preparing for Standardized Tests** pages are part of a complete test preparation course offered in this text. The test items on these pages were written in the same style as those in state proficiency tests and standardized tests like ACT and SAT. The 15 topics are closely aligned with those tests, the algebra curriculum, and this text. These topics cover all of the types of problems you will see on these tests.

Chapter	Mathematics Topic	Pages
1	Number Concept Problems	48–49
2	Data Analysis Problems	90–91
3	Number Concept Problems	136–137
4	Statistics Problems	184–185
5	Expression and Equation Problems	234–235
6	Probability and Counting Problems	280–281
7	Algebra Word Problems	332–333
8	Pythagorean Theorem Problems	378–379
9	Percent Problems	416–417
10	Function and Graph Problems	454–455
11	Polynomial and Factoring Problems	500–501
12	Angle, Triangle, and Quadrilateral Problems	546–547
13	Perimeter, Area, and Volume Problems	596–597
14	Systems of Equations Problems	634–635
15	Right Triangle Problems	680–681

With some practice, studying, and review, you will be ready for standardized test success. Good luck from Glencoe/McGraw-Hill!

Graphing Calculator Quick Reference Guide

Throughout this text, Graphing Calculator Explorations have been included so you can use technology to solve problems. These activities use the TI–83 Plus or T1–84 Plus graphing calculator. Graphing calculators have a wide variety of applications and features. If you are just beginning to use a graphing calculator, you will not need to use all of its features. This page is designed to be a quick reference for the features you will need to use as you study from this text.

To darken or lighten the screen: 2nd   ▲   or   2nd   ▼

To clear an entry: CLEAR

To get to the home screen: 2nd   [QUIT]

To recall an entry: 2nd   [ENTRY]

To recall an answer: 2nd   [ANS]

To turn the calculator off: 2nd   [OFF]

Task	Keystrokes
Using tables	2nd [TABLE]
Use lists	STAT ENTER
Find the mean and median of listed data	STAT ▶ ENTER ENTER
Solve equations	MATH 0
Set the viewing window	WINDOW
Plot points	2nd [DRAW] ▶ ENTER
Enter an equation	Y=
Enter an inequality symbol	2nd [TEST]
Graph an equation	GRAPH
Zoom in	ZOOM 2
Zoom out	ZOOM 3
Graph in the viewing window x: $[-10, 10]$ and y: $[-10, 10]$	ZOOM 6
Graph an equation with integer coordinates	ZOOM 8
Place a statistical graph in a good viewing window	ZOOM 9
Trace a graph	TRACE
Find the intersection of two graphs	2nd [CALC] 5
Shade an inequality	2nd [DRAW] 7
Enter a program	PRGM ▶ ▶ ENTER

CHAPTER
1
The Language of Algebra

What You'll Learn

Key Ideas

- Write algebraic expressions and equations. *(Lesson 1–1)*
- Simplify and evaluate algebraic expressions. *(Lessons 1–2 to 1–4)*
- Collect, organize, and display data. *(Lessons 1–6 and 1–7)*

Key Vocabulary

algebraic expression *(p. 4)*

data *(p. 32)*

equation *(p. 5)*

variable *(p. 4)*

Why It's Important

Biology The feared great white shark is the largest predatory fish in the ocean. But great white sharks are very rare. It is just one of about 368 species of sharks. These sharks range in size from the size of a person's hand to bigger than a bus.

Algebraic expressions and equations are used to model real-life situations. You will write an algebraic expression for the number of shark species in Lesson 1–1.

Prerequisite Skill, p. 684

✔ Check Your Readiness

Find each product or quotient.

1. 2×9
2. 4×7
3. 9×0
4. 8×3
5. 8×8
6. 12×3
7. 5×1.2
8. 12×1.59
9. 160×0.75
10. $15 \div 3$
11. $35 \div 5$
12. $18 \div 3$
13. $36 \div 6$
14. $24 \div 2$
15. $54 \div 9$
16. $16 \div 2.5$
17. $1.98 \div 2$
18. $150 \div 4.8$

Prerequisite Skill, p. 689

Write each percent as a decimal.

19. 15%
20. 30%
21. 55%
22. 4%
23. 100%
24. 12%

Prerequisite Skill, p. 691

Write the numbers in each set in order from least to greatest.

25. 19, 17.5, 18, 18.5
26. 2.3, 1.2, 1.8, 4.8, 3.0
27. 32, 30, 18, 29, 14, 20, 26, 21, 37
28. 3.1, 2.6, 4.9, 2.9, 3.3, 2.8, 3.4, 4.4, 3.0, 2.7

FOLDABLES™
Study Organizer

Make this Foldable to help you organize your Chapter 1 notes. Begin with four sheets of notebook paper.

❶ **Stack** the sheets with the edges $\frac{3}{4}$ inch apart.

❷ **Fold** up the bottom edges. All tabs should be the same size.

❸ **Staple** along the fold.

❹ **Label** the tabs with topics from the chapter.

Reading and Writing As you read and study the chapter, use each page to write notes, graphs, and examples for each lesson.

Writing Expressions and Equations

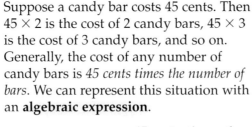

What You'll Learn
You'll learn to translate words into algebraic expressions and equations.

Why It's Important
Communication You can use expressions to represent the cost of your long-distance phone calls.
See Exercise 43.

Suppose a candy bar costs 45 cents. Then 45×2 is the cost of 2 candy bars, 45×3 is the cost of 3 candy bars, and so on. Generally, the cost of any number of candy bars is *45 cents times the number of bars*. We can represent this situation with an **algebraic expression**.

$$\underbrace{45 \text{ cents}}_{45} \quad \underbrace{\text{times}}_{\times} \quad \underbrace{\text{the number of bars}}_{n}$$

The letter n stands for an unknown number, in this case, candy bars. The unknown n is called a **variable** because its value *varies*. An algebraic expression contains at least one variable and at least one mathematical operation, as shown in the examples below.

$$h \div 3 \qquad 5n + 1 \qquad \frac{r}{t} - 1 \qquad xy \qquad 4 \times a$$

A **numerical expression** contains only numbers and mathematical operations. For example, $6 + 2 \div 1$ is a numerical expression.

In an expression involving multiplication, the quantities being multiplied are called **factors**, and the result is the **product**.

$$4 \times 5 \times 8 = 160$$

factors   product

Reading Algebra

When one factor in a product is a variable, the multiplication sign is usually omitted. Read $4a$ as *four a*.

To write a multiplication expression such as $4 \times a$, a raised dot or parentheses can be used. A fraction bar can be used to represent division.

$$\left. \begin{array}{l} 4 \cdot a \\ 4(a) \\ (4)(a) \\ 4a \end{array} \right\} \quad \text{means} \implies 4 \times a \qquad \bigg| \qquad \frac{t}{2} \quad \text{means} \implies t \div 2$$

The result of a division expression is called a **quotient**.

To solve verbal problems in mathematics, you may have to translate words into algebraic expressions. The chart below shows some of the words and phrases used to indicate mathematical operations.

Addition	Subtraction	Multiplication	Division
plus	minus	times	divided by
the sum of	the difference of	the product of	the quotient of
increased by	decreased by	multiplied by	the ratio of
more than	less than	at	per
added to	fewer than	of	
the total of	subtracted from		

Examples

Life Science Link

Write an algebraic expression for each verbal expression.

1 the sum of m and 18

$m + 18$

2 g divided by y

$g \div y$ or $\dfrac{g}{y}$

Your Turn

a. 26 decreased by w

b. 4 more than 8 times k

Some kangaroos can travel 30 feet in a single leap.

3 Write a numerical expression to represent the distance a kangaroo can travel if it leaps 4 times.

$30 \cdot 4$ or $30(4)$

4 Write an algebraic expression to represent the distance a kangaroo can travel if it leaps x times.

$30 \cdot x$ or $30x$

 Online Personal Tutor at algconcepts.com

You can also translate algebraic expressions into verbal expressions.

Examples

Write a verbal expression for each algebraic expression.

5 $32 - b$

32 less b
b less than 32
the difference of 32 and b
b subtracted from 32
32 decreased by b

6 $(y \div 4) + 9$

y divided by 4, plus 9
the quotient of y and 4, increased by 9
9 added to the ratio of y and 4

Your Turn

c. $15v$

d. $r - \dfrac{t}{d}$

An **equation** is a mathematical sentence that contains an equals sign (=). Some words used to indicate the equals sign are in the chart at the right. An equation may contain numbers, variables, or algebraic expressions.

Equality	
equals	is equal to
is	is the same as
is equivalent to	is as much as
	is identical to

Examples

Write an equation for each sentence.

7 Three times g equals 21.

$3g = 21$

8 Five more than twice n is 15.

$2n + 5 = 15$

Your Turn

e. A number k divided by 4 is equal to 18.

Write a sentence for each equation.

9 $x - 2 = 14$
Two less than x is equivalent to 14.

10 $7y + 6 = 34$
Seven times y increased by 6 is 34.

Your Turn

f. $4b - 5 = 3$

Check for Understanding

Communicating Mathematics

1. **Write** three examples of numerical expressions and three examples of algebraic expressions.

2. **Write** three examples of equations.

3. Writing Math Write about a real-life application that can be expressed using an algebraic expression or an equation.

Vocabulary

algebraic expression
variable
numerical expression
factors
product
quotient
equation

Guided Practice

Examples 1 & 2

Write an algebraic expression for each verbal expression.

4. t more than s

5. the product of 7 and m

6. 11 decreased by the quotient of x and 2

Examples 5 & 6

Write a verbal expression for each algebraic expression.

7. $\dfrac{7}{q}$

8. $3\ell - 9$

Examples 7 & 8

Write an equation for each sentence.

9. A number m added to 6 equals 17.

10. Ten is the same as four times r minus 6.

Examples 9 & 10

Write a sentence for each equation.

11. $5 + r = 15$

12. $\dfrac{4p}{3} = 12$

Example 4

13. **Biology** The family of great white sharks has w different species. The blue shark family has nine times w plus three different species. Write an algebraic expression to represent the number of species in the blue shark family.

Exercises •

Practice

Homework Help

For Exercises	See Examples
14–22	1, 2
23–28	5, 6
29–34	7, 8
35–40	9, 10
43	3, 4

Extra Practice
See page 692.

Write an algebraic expression for each verbal expression.

14. twelve less than y **15.** the product of r and s

16. the quotient of t and 5   **17.** three more than five times a

18. p plus the quotient of 9 and 5   **19.** the difference of 1 and n

20. f divided by y **21.** ten plus the product of h and 1

22. seven less than the quotient of j and p

Write a verbal expression for each algebraic expression.

23. $9x$ **24.** $11 + b$ **25.** $6 - y$

26. $2m + 1$ **27.** $\frac{3}{r} - 8$ **28.** $16 - rt$

Write an equation for each sentence.

29. Three plus w equals 15.   **30.** Five times r equals 7.

31. Two is equal to seven divided by x.

32. Five less than the product of two and g equals nine.

33. Three minus the product of five and y is the same as two times z.

34. The quotient of 19 and j is equal to the total of a and b and c.

Write a sentence for each equation.

35. $3r = 18$ **36.** $g + 7 = 3$ **37.** $h = 10 - i$

38. $6v - 2 = 8$ **39.** $\frac{t}{4} = 16$ **40.** $10z + 7 = \frac{6}{r}$

41. Choose a variable and write an equation for *Four times a number minus seven equals the sum of 15 and c and two times the number.*

Applications and Problem Solving

42. Biology A smile requires 26 fewer muscles than a frown. Let f represent the number of muscles it takes for a frown and let s represent the number of muscles for a smile. Write an equation to represent the number of muscles a person uses to smile.

43. Communication A long-distance telephone call costs 20¢ for the first minute plus 10¢ for each additional minute.

   a. Write an expression for the total cost of a call that lasts 15 minutes.

   b. Write an expression for the total cost of a call that lasts m minutes.

44. Critical Thinking The ancient Hindus enjoyed number puzzles like the one below. **Source:** *Mathematical History*

> If 4 is added to a certain number, the result divided by 2, that result multiplied by 5, and then 6 subtracted from that result, the answer is 29. Can you find the number?

   a. Choose a variable and write an algebraic equation to represent the puzzle.

   b. Is your answer in part a the only correct way to write the equation? Explain.

What You'll Learn

You'll learn to use the order of operations to evaluate expressions.

Why It's Important

Business

Businesspeople use the order of operations to determine the cost of renting a car.
See Exercise 18.

Some expressions have more than one operation. The value of the expression depends on the order in which the operations are evaluated. What is the value of $9 \cdot 5 + 4$?

Method 1	Method 2
$9 \cdot 5 + 4 = 45 + 4$ *Multiply 9 and 5.*	$9 \cdot 5 + 4 = 9 \cdot 9$ *Add 5 and 4.*
$ = 49$ *Add 45 and 4.*	$ = 81$ *Multiply 9 and 9.*

Is the answer 49 or 81? The values are different because we multiplied and added in different orders in the two methods. To find the correct value of the expression, follow the **order of operations**.

Order of Operations	1. Find the values of expressions inside grouping symbols, such as parentheses ( ), brackets [ ], and as indicated by fraction bars. 2. Do all multiplications and/or divisions from left to right. 3. Do all additions and/or subtractions from left to right.

According to the order of operations, do multiplication and then addition. So, the value of the expression in Method 1 is correct. The value of the expression is 49.

Examples

Find the value of each expression.

1 $38 - 5 \cdot 6$

$38 - 5 \cdot 6 = 38 - 30$ *Multiply 5 and 6.*
$ = 8$ *Subtract 30 from 38.*

2 $\dfrac{4 \times 9}{26 - 8}$

$\dfrac{4 \times 9}{26 - 8} = \dfrac{36}{18}$ *Evaluate the numerator and the denominator separately.*
$\phantom{\dfrac{4 \times 9}{26 - 8}} = 2$ *Divide 36 by 18.*

Your Turn

a. $7 \cdot 4 + 7 \cdot 3$ b. $12 \div 3 \cdot 5 - 4$ c. $\dfrac{6 + 12}{5(3) - 13}$

The order of operations is useful in solving problems in everyday life.

Example ③

Finance Link

As a 16-year old, Trent Eisenberg ran his own consulting company called *F1 Computer*. Suppose he charged a flat fee of $50, plus $25 per hour. One day he worked 2 hours for one customer and the next day he worked 3 hours for the same customer. Find the value of the expression 50 + 25(2 + 3) to find the total amount of money he earned.

Source: *Scholastic Math*

$$50 + 25(2 + 3) = 50 + 25(5) \quad \text{\textit{Do the operation in parentheses first.}}$$
$$= 50 + 125 \quad \text{\textit{Multiply 25 and 5.}}$$
$$= 175 \quad \text{\textit{Add 50 and 125.}}$$

Trent earned $175.

 Personal Tutor at algconcepts.com

In algebra, statements that are true for any number are called **properties**. Four properties of equality are listed in the table below.

Property of Equality	Symbols	Numbers
Substitution	If $a = b$, then a may be replaced by b.	If $9 + 2 = 11$, then $9 + 2$ may be replaced by 11.
Reflexive	$a = a$	$21 = 21$
Symmetric	If $a = b$, then $b = a$.	If $10 = 4 + 6$, then $4 + 6 = 10$.
Transitive	If $a = b$ and $b = c$, then $a = c$.	If $3 + 5 = 8$ and $8 = 2(4)$, then $3 + 5 = 2(4)$.

Examples

Name the property of equality shown by each statement.

④ **If $9 + 3 = 12$, then $12 = 9 + 3$.**

Symmetric Property of Equality

⑤ **If $z = 8$, then $z \div 4 = 8 \div 4$.**

Substitution Property of Equality   *z is replaced by 8.*

Your Turn

d. $7 - c = 7 - c$
e. If $10 - 3 = 4 + 3$ and $4 + 3 = 7$, then $10 - 3 = 7$.

These properties of numbers may help to find the value of expressions.

Property	Words	Symbols	Numbers
Additive Identity	When 0 is added to any number a, the sum is a.	For any number a, $a + 0 = 0 + a = a$.	$45 + 0 = 45$ $0 + 6 = 6$ *0 is the identity.*
Multiplicative Identity	When a number a is multiplied by 1, the product is a.	For any number a, $a \cdot 1 = 1 \cdot a = a$.	$12 \cdot 1 = 12$ $1 \cdot 5 = 5$ *1 is the identity.*
Multiplicative Property of Zero	If 0 is a factor, the product is 0.	For any number a, $a \cdot 0 = 0 \cdot a = 0$.	$7 \cdot 0 = 0$ $0 \cdot 23 = 0$

When two or more sets of grouping symbols are used, simplify within the innermost grouping symbols first.

Example

6 **Find the value of $5[3 - (6 \div 2)] + 14$. Identify the properties used.**

$5[3 - (6 \div 2)] + 14$
$= 5[3 - 3] + 14$ *Substitution Property of Equality*
$= 5(0) + 14$ *Substitution Property of Equality*
$= 0 + 14$ *Multiplicative Property of Zero*
$= 14$ *Additive Identity*

Your Turn   **f.** $(22 - 15) \div 7 \cdot 9$ **g.** $8 \div 4 \cdot 6(5 - 4)$

You can also apply the properties of numbers to find the value of an algebraic expression. This is called **evaluating** an expression. Replace the variables with known values and then use the order of operations.

Examples

Evaluate each expression if $a = 9$ and $b = 1$.

7 $7 + \left(\dfrac{a}{b} - 9\right)$

$7 + \left(\dfrac{a}{b} - 9\right) = 7 + \left(\dfrac{9}{1} - 9\right)$ *Replace a with 9 and b with 1.*
$= 7 + (9 - 9)$ *Substitution Property of Equality*
$= 7 + 0$ *Substitution Property of Equality*
$= 7$ *Additive Identity*

8 $(a + 4) - 3 \cdot b$

$(a + 4) - 3 \cdot b = (9 + 4) - 3 \cdot 1$ *Replace a with 9 and b with 1.*
$= (13) - 3 \cdot 1$ *Substitution Property of Equality*
$= 13 - 3$ *Multiplicative Identity*
$= 10$ *Substitution Property of Equality*

Your Turn   Evaluate each expression if $m = 8$ and $p = 2$.

h. $6 \cdot p - m \div p$ **i.** $[m + 2(3 + p)] \div 2$

Communicating Mathematics

1. **Name** two of the three types of grouping symbols discussed in this lesson.

2. **Translate** the verbal expression *six plus twelve divided by three* and *the sum of six and twelve divided by three* into numerical expressions. Use grouping symbols. Evaluate the expressions and explain why they are different.

3. Writing Math Label a section of your math journal "Toolbox." Record all properties given in this course, beginning with this lesson.

> **Vocabulary**
> order of operations
> properties
> evaluating

Guided Practice

> **⏱ Getting Ready** State which operation to perform first.
>
> **Sample:** $3 + 2 \cdot 4$ **Solution:** Multiply 2 and 4.

4. $8 \div 4 \cdot 2$ 5. $12 - 6 \cdot 2$

6. $5(7 + 7)$ 7. $(10 - 4) \div 3$

Examples 1–3

Find the value of each expression.

8. $7 \cdot 4 + 3$ 9. $4(1 + 5) \div 8$ 10. $18 \div [3(11 - 8)]$

Examples 4 & 5

Name the property of equality shown by each statement.

11. If $5 + 2n = 5 + 3$ and $5 + 3 = 2 \cdot 4$, then $5 + 2n = 2 \cdot 4$.

12. If $\frac{y}{2} = 19$, then $19 = \frac{y}{2}$.

Example 6

Find the value of each expression. Identify the property used in each step.

13. $8(4 - 8 \div 2)$ 14. $5(2) \cdot (15 \div 15)$

Examples 7 & 8

Evaluate each algebraic expression if $q = 4$ and $r = 1$.

15. $4(q - 2r)$ 16. $\dfrac{7q}{r + 3}$ 17. $r + \dfrac{q}{2} \cdot 6$

Examples 7 & 8

18. **Car Rental** The cost to rent a car is given by the expression $25d + 0.10m$, where d is the number of days and m is the number of miles. If Teresa rents the car for five days and drives 300 miles, what is the cost?

Exercises

Practice

Find the value of each expression.

19. $36 \div 4 + 5$ 20. $16 - 4 \cdot 4$ 21. $4 + 7 \cdot 2 + 8$

22. $42 - (24 \div 2) + 10$ 23. $42 - 24 \div (2 + 10)$ 24. $24 \div 12 \div 2 \cdot 5$

25. $\dfrac{7(3 + 6)}{3}$ 26. $\dfrac{4(8 - 2)}{2 \times 2}$ 27. $38 - [3(9 + 1)]$

Homework Help

For Exercises	See Examples
19–27	1–3
28–33	4, 5
34–39	6
40–49, 51, 53	7, 8

Extra Practice

See page 692.

Name the property of equality shown by each statement.

28. If $x + 3 = 5$ and $x = 2$, then $2 + 3 = 5$.

29. $8t - 1 = 8t - 1$

30. If $6 = 3 + 3$, then $3 + 3 = 6$.

31. $\dfrac{20 - 2}{9} = \dfrac{18}{9}$

32. If $4 \cdot (7 - 7) = 4 \cdot 0$ and $4 \cdot 0 = 0$, then $4 \cdot (7 - 7) = 0$.

33. $a + 1 = 15x$ and $15x = 30$, so $a + 1 = 30$.

Find the value of each expression. Identify the property used in each step.

34. $7(10 - 1 \cdot 3)$

35. $8(9 - 3 \cdot 2)$

36. $19 - 15 \div 5 \cdot 2$

37. $10(6 - 5) - (20 \div 2)$

38. $\dfrac{9 \cdot 9 - 1}{3(1 + 2) - 1}$

39. $6(12 - 48 \div 4) + 7 \cdot 1$

Evaluate each algebraic expression if $j = 5$ and $s = 2$.

40. $7j - 3s$

41. $j(3s + 4)$

42. $j + 5s - 7$

43. $\dfrac{9 \cdot 4 + 5 \cdot s}{7 - j}$

44. $\dfrac{14 + s}{2(j - 1)}$

45. $\dfrac{4js}{s - 1}$

46. $50 \div js + 6$

47. $(3s - j)(5s - j)$

48. $[3j - s(4 + s)] \div 3$

49. $2[16 - (j - s)]$

50. a. Write an algebraic expression for *nine added to the quantity three times the difference of a and b.*

   b. Let $a = 4$ and $b = 1$. Evaluate the expression in part a.

Applications and Problem Solving

51. Real Estate The Phams own a $150,000 home in Rochester, New York, and plan to move to San Diego, California. How much will a similar home in San Diego cost? Evaluate the expression $150,000 \div a \times b$ for $a = 79$ and $b = 164$ to find the answer to the nearest dollar.
Source: *USA TODAY*

52. Sports A person's handicap in bowling is usually found by subtracting the person's average a from 200, multiplying by 2, and dividing by 3.

   a. Write an algebraic expression for a handicap in bowling.

   b. Find a person's handicap whose average is 170.

12 ft

10 ft

53. Gardening Mr. Martin is building a fence around a rectangular garden, as shown at the left. Evaluate the expression $2\ell + 2w$, where ℓ represents the length and w represents the width, to find how much fencing he needs.

54. Critical Thinking The symbol < means "is less than." Are the following properties of equality true for statements containing this symbol? Give examples to explain.

    a. Reflexive   **b.** Symmetric   **c.** Transitive

Mixed Review

Write a sentence for each equation. *(Lesson 1–1)*

55. $x + 8 = 12$ **56.** $2y = 16$ **57.** $25 \div n = 5$

Write an equation for each sentence. *(Lesson 1–1)*

58. Six more than g is 22.

59. Three times c equals 27.

60. Two is the same as the quotient of 8 and x.

61. b increased by 10 and then decreased by 1 is equivalent to 18.

62. Time How many seconds are there in a day? *(Lesson 1–1)*

    a. Write an expression to answer this question.

    b. Evaluate the expression.

Standardized Test Practice

Ⓐ Ⓑ Ⓒ Ⓓ

63. Extended Response Lincoln's Gettysburg Address began "Four score and seven years ago, . . ." *(Lesson 1–1)*

    a. A score is 20. Write a numerical expression for the phrase.

    b. Evaluate the expression to find the number of years.

64. Multiple Choice At the movie theater, the price for an adult ticket a is $1.50 less than two times the price of a student ticket s. Choose the algebraic expression that represents the price of an adult ticket in terms of the price of a student ticket. *(Lesson 1–2)*

    Ⓐ $1.50 - 2s$ Ⓑ $2(s - 1.50)$

    Ⓒ $2s + 2(1.50)$ Ⓓ $2s - 1.50$

Quiz 1   Lessons 1–1 and 1–2

1. Write an algebraic expression for *five less than the product of two and v.* *(Lesson 1–1)*

2. Write an equation for *the sum of nine and y equals 16.* *(Lesson 1–1)*

Evaluate each algebraic expression if $j = 5$ and $s = 2$. *(Lesson 1–2)*

3. $3(j - s) + 4$ **4.** $[(11 - j \div 1) + 8] \cdot s$

5. Carpentry Ana Martinez is putting molding around the ceiling of her family room. The room measures 12 feet by 16 feet. Evaluate the expression $2\ell + 2w$, where ℓ is the length and w is the width, to find how much molding Ana needs. *(Lesson 1–2)*

What You'll Learn
You'll learn to use the commutative and associative properties to simplify expressions.

Why It's Important
Construction
Lumber yards use the commutative and associative properties to determine the amount of wood to order. *See Exercise 25.*

The **Commutative Property of Addition** states that the sum of two numbers does not depend on the order in which they are added. In the example below, adding 35 and 50 in either order does not change the sum.

$$35 + 50 = 85$$
$$50 + 35 = 85$$

This example illustrates the Commutative Property of Addition.

Commutative Property of Addition	Words:	The order in which two numbers are added does not change their sum.
	Symbols:	For any numbers a and b, $a + b = b + a$.
	Numbers:	$5 + 7 = 7 + 5$

COncepts in MOtion
BrainPOP®
algconcepts.com

Likewise, the order in which you multiply numbers does not matter.

Commutative Property of Multiplication	Words:	The order in which two numbers are multiplied does not change their product.
	Symbols:	For any numbers a and b, $a \cdot b = b \cdot a$.
	Numbers:	$3 \cdot 10 = 10 \cdot 3$

Some expressions are easier to evaluate if you group or *associate* certain numbers. Look at the expression below.

$$16 + 7 + 3 = 16 + (7 + 3) \quad \textit{Group 7 and 3.}$$
$$= 16 + 10 \quad \textit{Add 7 and 3.}$$
$$= 26 \quad \textit{Add 16 and 10.}$$

This is an application of the **Associative Property of Addition**.

Associative Property of Addition	Words:	The way in which three numbers are grouped when they are added does not change their sum.
	Symbols:	For any numbers a, b, and c, $(a + b) + c = a + (b + c)$.
	Numbers:	$(24 + 8) + 2 = 24 + (8 + 2)$

The Associative Property also holds true for multiplication.

	Words:	The way in which three numbers are grouped when they are multiplied does not change their product.
Associative Property of Multiplication	**Symbols:**	For any numbers a, b, and c, $(a \cdot b) \cdot c = a \cdot (b \cdot c)$.
	Numbers:	$(9 \cdot 4) \cdot 25 = 9 \cdot (4 \cdot 25)$

Examples

Name the property shown by each statement.

1 $4 \cdot 11 \cdot 2 = 11 \cdot 4 \cdot 2$ Commutative Property of Multiplication

2 $(n + 12) + 5 = n + (12 + 5)$ Associative Property of Addition

Your Turn

a. $(5 \cdot 4) \cdot 3 = 5 \cdot (4 \cdot 3)$ **b.** $16 + t + 1 = 16 + 1 + t$

To **simplify** an expression, eliminate all parentheses first and then add, subtract, multiply, or divide.

Example

3 **Simplify the expression $15 + (3x + 8)$. Identify the properties used in each step.**

$$15 + (3x + 8) = 15 + (8 + 3x) \quad \textit{Commutative Property of Addition}$$
$$= (15 + 8) + 3x \quad \textit{Associative Property of Addition}$$
$$= 23 + 3x \quad \textit{Substitution Property}$$

Your Turn

Simplify each expression. Identify the properties used in each step.

c. $7 + 2a + 6 + 9$ **d.** $(x \cdot 5) \cdot 20$

 nline **Personal Tutor at** algconcepts.com

Example
Geometry Link

4 The volume of a box can be found using the expression $\ell \times w \times h$, where ℓ is the length, w is the width, and h is the height. Find the volume of a box whose length is 30 inches, width is 6 inches, and height is 5 inches.

$$\ell \times w \times h = 30 \times 6 \times 5 \quad \textit{Replace } \ell \textit{ with 30, } w \textit{ with 6, and } h \textit{ with 5.}$$
$$= 30 \times (6 \times 5) \quad \textit{Associative Property of Multiplication}$$
$$= 30 \times 30 \quad \textit{Substitution Property}$$
$$= 900 \quad \textit{Substitution Property}$$

The volume of the box is 900 cubic inches.

Whole numbers are the numbers 0, 1, 2, 3, 4, and so on. When you add whole numbers, the sum is always a whole number. Likewise, when you multiply whole numbers, the product is a whole number. This is an example of the **Closure Property**. We say that the whole numbers are *closed* under addition and multiplication.

Closure Property of Whole Numbers	**Words:**	Because the sum or product of two whole numbers is also a whole number, the set of whole numbers is closed under addition and multiplication.
	Numbers:	$2 + 5 = 7$, and 7 is a whole number. $2 \cdot 5 = 10$, and 10 is a whole number.

Are the whole numbers closed under division? Study these examples.

$$2 \div 1 = 2 \quad \textit{whole number}$$
$$28 \div 4 = 7 \quad \textit{whole number}$$
$$5 \div 3 = \frac{5}{3} \quad \textit{fraction}$$

It is impossible to list every possible division expression to prove that the Closure Property holds true. However, we can easily show that the statement is false by finding one **counterexample**. A counterexample is an example that shows the statement is not true. Consider $5 \div 3$ or $\frac{5}{3}$. While 5 and 3 are whole numbers, $\frac{5}{3}$ is not. So, the statement *The whole numbers are closed under division* is false.

Example ⑤ **State whether the statement *Division of whole numbers is commutative* is *true* or *false*. If false, provide a counterexample.**

Write two division expressions using the Commutative Property and check to see whether they are equal.

$6 \div 3 \overset{?}{=} 3 \div 6$ *Evaluate each expression separately.*

$2 \neq \frac{1}{2}$ $6 \div 3 = 2 \text{ and } 3 \div 6 = \frac{1}{2}$

We found a counterexample, so the statement is false. Division of whole numbers is *not* commutative.

Your Turn

e. State whether the statement *Subtraction of whole numbers is associative* is *true* or *false*. If false, provide a counterexample.

Check for Understanding

Communicating
Mathematics

1. **Describe** what is meant by the statement *The whole numbers are closed under multiplication.*

Vocabulary
simplify
whole numbers
counterexample

2. **Write** an equation that illustrates the Commutative Property of Addition.

3. **You Decide?** Abeque says that the expression $(7 \cdot 2) + 5$ equals $7 \cdot (2 + 5)$ because of the Associative Properties of Addition and Multiplication. Jessie disagrees with her. Who is correct? Explain.

Guided Practice

Examples 1 & 2

Name the property shown by each statement.

4. $27 + 59 = 59 + 27$

5. $(8 + 7) + 3 = 8 + (7 + 3)$

Example 3

Simplify each expression. Identify the properties used in each step.

6. $(n \cdot 2) \cdot 10$

7. $(3 + p + 47)(7 - 6)$

Example 5

8. State whether the statement *Whole numbers are closed under subtraction* is *true* or *false*. If false, provide a counterexample.

Example 4

9. **Geology** The table shows the number of volcanoes in the United States and Mexico.
   a. Find the total number of volcanoes in these two countries mentally.
   b. Describe the properties you used to add the numbers.

Location	Number of Volcanoes
U.S. Mainland	69
Alaska	80
Hawaii	8
Mexico	31

Source: *Kids Discover Volcanoes*

Active volcano in Hawaii

Exercises

Practice

Name the property shown by each statement.

10. $(9 \cdot 5) \cdot 20 = 9 \cdot (5 \cdot 20)$

11. $a + 14 = 14 + a$

12. $r \cdot s = s \cdot r$

13. $4 \times 15 \times 25 = 4 \times 25 \times 15$

14. $(7 + 5) + 5 = 7 + (5 + 5)$

15. $c \cdot (d \cdot 10) = (c \cdot d) \cdot 10$

Simplify each expression. Identify the properties used in each step.

16. $h + 1 + 9$

17. $(r \cdot 30) \cdot 5$

18. $17 + k + 23$

19. $6 + (3 + y)$

20. $2 \cdot (19p)$

21. $2 \cdot j \cdot 7$

Homework Help	
For Exercises	See Examples
10–15	1, 2
16–21	3
22, 23	5
24, 25	4
Extra Practice	
See page 692.	

State whether each statement is *true* or *false*. If false, provide a counterexample.

22. Subtraction of whole numbers is commutative.

23. Division of whole numbers is associative.

Applications and Problem Solving

24. **Sports** The table shows the point values for different plays in football. The expression below represents the total possible points for a team in a game.

$$6t + 1x + 3f + 2c + 2s$$

If a team scores 3 touchdowns, 2 extra points, 2 field goals, and 2 safeties, how many total points are scored?

Type of Score	Number of Points
touchdown, t	6
extra point, x	1
two-point conversion, c	2
field goal, f	3
safety, s	2

25. **Construction** Lumber mills sell wood to lumberyards in *board feet*. The expression shown below represents the number of board feet in a stack of wood.

$$\frac{\text{inches thick} \times \text{inches wide} \times \text{feet long}}{12}$$

Find the number of board feet if the stack of wood is 10 inches thick, 12 inches wide, and 10 feet long.

26. **Geography** The Chattahoochee and Savannah rivers form natural boundaries for the state of Georgia.

   a. Write an expression to approximate the total length of Georgia's borders using the map at the right.

   b. Evaluate the expression that you wrote in part a. Identify any properties that you used.

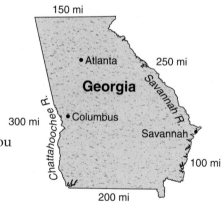

27. **Critical Thinking** Use a counterexample to show that subtraction of whole numbers is not associative.

Mixed Review

Find the value of each expression. *(Lesson 1–2)*

28. $16 \div 2 \cdot 5 \times 3$

29. $48 \div [2(3 + 1)]$

30. $25 - \frac{1}{3}(18 - 9)$

Evaluate each expression if $a = 4$ and $b = 11$. *(Lesson 1–2)*

31. $196 \div [a(b - a)]$

32. $\frac{ab}{a - 2}$

Standardized Test Practice
Ⓐ Ⓑ Ⓒ Ⓓ

33. **Multiple Choice** Which of the following is the value of $3t - 5q(r + 1)$, if $q = 2$, $r = 0$, and $t = 11$? *(Lesson 1–2)*

   Ⓐ 23   Ⓑ 52   Ⓒ 53   Ⓓ 33

www.algconcepts.com/self_check_quiz

What You'll Learn

You'll learn to use the Distributive Property to evaluate expressions.

Why It's Important

Shopping Cashiers use the Distributive Property when they total customers' groceries.
See Exercise 40.

The **Distributive Property** can be applied to simplify expressions. For example, the expression $2 \times (128 + 12)$ can be solved using two different methods.

Method 1
$$2 \times (128 + 12) = 2(140) \quad \textit{First, add.}$$
$$= 280 \quad \textit{Then, multiply.}$$

Method 2
$$2 \times (128 + 12) = (2 \times 128) + (2 \times 12) \quad \textit{First, distribute.}$$
$$= 256 + 24 \quad \textit{Multiply.}$$
$$= 280 \quad \textit{Add.}$$

The Distributive Property is used in Method 2. Using both methods, the value of the expression is 280.

	Symbols:	For any numbers *a*, *b*, and *c*, $a(b + c) = ab + ac$ and $a(b - c) = ab - ac$.
Distributive Property	Numbers:	$2(5 + 3) = (2 \cdot 5) + (2 \cdot 3)$ $2(5 - 3) = (2 \cdot 5) - (2 \cdot 3)$

In the expression $a(b + c)$, it does not matter whether *a* is placed to the left or to the right of the expression in parentheses. So, $(b + c)a = ba + ca$ and $(b - c)a = ba - ca$.

Examples

Simplify each expression.

1 $3(x + 7)$
$$3(x + 7) = (3 \cdot x) + (3 \cdot 7) \quad \textit{Distributive Property}$$
$$= 3x + 21 \quad \textit{Substitution Property}$$

2 $5(2n + 8)$
$$5(2n + 8) = (5 \cdot 2n) + (5 \cdot 8) \quad \textit{Distributive Property}$$
$$= 10n + 40 \quad \textit{Substitution Property}$$

Your Turn

a. $6(a + b)$

b. $(1 + 3t)9$

A **term** is a number, variable, or product or quotient of numbers and variables.

Examples of Terms		Not Terms	
7	7 is a number.	$7 + x$	$7 + x$ is the sum of two terms.
t	t is a variable.	$8rs + 7y + 6$	$8rs + 7y + 6$ is the sum of three terms.
$5x$	$5x$ is a product.	$x - y$	$x - y$ is the difference of two terms.

The numerical part of a term that contains a variable is called the **coefficient**. For example, the coefficient of $2a$ is 2. **Like terms** are terms that contain the same variables, such as $2a$ and $5a$ or $7xy$ and $3xy$.

Reading Algebra

1 is the *coefficient* of x because $x = 1x$.

Consider the expression $5b + 3b + x + 12x$.
- There are four terms.
- The like terms are $5b$ and $3b$, x and $12x$.
- The coefficients are shown in the table.

Term	Coefficient
$5b$	5
$3b$	3
x	1
$12x$	12

The Distributive Property allows us to combine like terms. If $a(b + c) = ab + ac$, then $ab + ac = a(b + c)$ by the Symmetric Property of Equality.

$$2n + 7n = (2 + 7)n \quad \textit{Distributive Property}$$
$$= 9n \quad \textit{Substitution Property}$$

The expressions $2n + 7n$ and $9n$ are called **equivalent expressions** because their values are the same for any value of n. An algebraic expression is in **simplest form** when it has no like terms and no parentheses.

Examples

Simplify each expression.

3 $4x + 9x$
$$4x + 9x = (4 + 9)x \quad \textit{Distributive Property}$$
$$= 13x \quad \textit{Substitution Property}$$

4 $a + 7b + 3a - 2b$
$$a + 7b + 3a - 2b = a + 3a + 7b - 2b \quad \textit{Commutative Property (+)}$$
$$= (a + 3a) + (7b - 2b) \quad \textit{Associative Property (+)}$$
$$= (1 + 3)a + (7 - 2)b \quad \textit{Distributive Property}$$
$$= 4a + 5b \quad \textit{Substitution Property}$$

Your Turn

c. $5st + 2st$

d. $6 + y + 3z + 4y$

 www.algconcepts.com/extra_examples

You can use the Distributive Property to solve problems in different, and possibly simpler, ways.

Example **5**

Sports Link

A soccer game

Write an equation representing the area A of a soccer field given its width w and length ℓ as shown in the diagram. Then simplify the expression and find the area if w is 54 yards and ℓ is 60 yards.

Method 1

$$\begin{aligned} A &= w(\ell + \ell) && \textit{Multiply the total length by the width.} \\ &= 54(60 + 60) && \textit{Replace } w \textit{ with 54 and } \ell \textit{ with 60.} \\ &= 54(120) && \textit{Substitution Property} \\ &= 6480 && \textit{Substitution Property} \end{aligned}$$

Method 2

$$\begin{aligned} A &= w\ell + w\ell && \textit{Add the areas of the smaller rectangles.} \\ &= 54(60) + 54(60) && \textit{Replace } w \textit{ with 54 and } \ell \textit{ with 60.} \\ &= 3240 + 3240 && \textit{Substitution Property} \\ &= 6480 && \textit{Substitution Property} \end{aligned}$$

Using either method, the area of the soccer field is 6480 square yards.

 Online **Personal Tutor at** algconcepts.com

Check for Understanding

Communicating Mathematics

1. **Write** an algebraic expression with five terms. One term should have a coefficient of three. Also, include two pairs of like terms.

2. **Explain** why $3xy$ is a term but $3x + y$ is not a term.

3. **Determine** which two expressions are equivalent. Explain how you determined your answer.

   a. $20n + 3p$
   b. $16n + p - 4n + 2p$
   c. $16n + 4p + 4n$
   d. $12p + 3n$
   e. $12n + 3p$
   f. $20n - p - 16n + 2p$

> *Vocabulary*
>
> term
> coefficient
> like terms
> equivalent expressions
> simplest form

Guided Practice

 Getting Ready **Name the like terms in each list of terms.**

Sample: $3c$, a, ab, 5, $2c$ **Solution:** $3c$, $2c$

4. $5m$, $2n$, $7n$
5. 8, $8p$, $9p$, $9q$
6. $4h$, $10gh$, 8, $2h$
7. $6b$, $6bc$, bc

Simplify each expression.

8. $5x + 9x$

9. $4y + 2 - 3y$

10. $2(5g + 3g)$

11. $3(4 - 6m)$

12. $8(2s + 7)$

13. $(3a + 5t) + (4a + 2t)$

Example 5

14. School Every student at Miller High School must wear a uniform. Suppose shirts or blouses cost $18 and skirts or pants cost $25.

a. If 250 students buy a uniform consisting of a shirt or blouse and a skirt or pants, write an expression representing the total cost.

b. Find the total cost.

Exercises

Practice

Homework Help

For Exercises	See Examples
15–33	1, 4
39, 41	5

Extra Practice

See page 693.

Simplify each expression.

15. $16f + 5f$

16. $9a + 6a$

17. $4r - r$

18. $3 + 7 - 2st$

19. $4g - 2g + 6$

20. $5a + 7a + 8b + 5b$

21. $14x + 6y - y - 8x$

22. $3(2n + 10)$

23. $3(5am - 4)$

24. $6(5q + 3w - 2w)$

25. $4a + 7b + (3a - 2b)$

26. $13x - 1 - 8x + 6$

27. $2y + y + y$

28. $bp + 25bp + p$

29. $2(15xy + 8xy)$

30. $5(n + 2r) + 3n$

31. $(r + 2s)3 - 2s$

32. $3(2v + 5m) + 2(3v - 2m)$

33. Write $5(2n + 3r) + 4n + 3(r + 2)$ in simplest form. Indicate the property that justifies each step.

34. Is the statement $2 + (s \cdot t) = (2 + s) \cdot (2 + t)$ *true* or *false*? Find values for s and t to show that the statement may be true. Otherwise, find a counterexample to show that the statement is false.

35. What is the value of $6y$ decreased by the quantity $2y$ plus 1 if y is equivalent to 3?

36. What is the sum of $14xy$, xy, and $5xy$ if x equals 1 and y equals 4?

Applications and Problem Solving

37. Sports Rich bought two baseballs for $4 each and two basketballs for $22 each. What is the total cost? Use the Distributive Property to solve the problem in two different ways.

38. Retail Marie and Mark work at a local department store. Each earns $6.25 per hour. Maria works 24 hours per week, and Mark works 32 hours per week. How much do the two of them earn together each week?

39. Health If an adult male's height is h inches over five feet, his approximate normal weight is given by the expression $6.2(20 + h)$.

a. What should the normal weight of a 5'9" male be?

b. How many more pounds should the normal weight of a man that is 6'2" tall be than a man that is 5'9" tall?

40. Shopping   Luanda went to the grocery store and bought the items in the table below.

Item	Cost per Item	Quantity
can of soup	$0.99	4
can of corn	$0.49	4
bag of apples	$2.29	4
box of crackers	$3.29	2
jar of jelly	$2.69	2

a. Use the Distributive Property to write an expression representing the cost of the items.

b. Find the change Luanda will receive if she gives the clerk $30.

41. Theater   A school's drama club is creating a stage backdrop with a city theme for a performance. The students sketched a model of buildings as shown at the right.

a. How many square feet of cardboard will they need to make the buildings? Use the expression ℓw, where ℓ is the length and w is the width of each rectangle, to find the area of each rectangle. Then add to find the total area.

8 ft   10 ft   6 ft

4 ft   4 ft   4 ft

b. Show how to use the Distributive Property as another method in finding the total area of the buildings.

42. Critical Thinking   Use the Distributive Property to write an expression that is equivalent to $3ax + 6ay$.

Mixed Review

Name the property shown by each statement. *(Lesson 1–3)*

43. $8(2 \cdot 6) = (2 \cdot 6)8$

44. $(7 + 4) + 3 = 7 + (4 + 3)$

45. If $19 - 3 = 16$, then $16 = 19 - 3$. *(Lesson 1–2)*

Find the value of each expression. *(Lesson 1–2)*

46. $8 + 6 \div 2 + 2$

47. $3(6 - 32 \div 8)$

48. Write an equation for the sentence *Eighteen decreased by d is equal to f.* *(Lesson 1–1)*

Standardized Test Practice
Ⓐ Ⓑ Ⓒ Ⓓ

49. Extended Response   Some toys are 30 decibels louder than jets during takeoff. Suppose jets produce *d* decibels of noise during takeoff. *(Lesson 1–1)*

a. Write an expression to represent the number of decibels produced by the loud toys.

b. Evaluate the expression if $d = 140$.

50. Multiple Choice   Which of the following is an algebraic expression for *six times a number decreased by 17*? *(Lesson 1–1)*

Ⓐ $6n + 17$ Ⓑ $6n - 17$ Ⓒ $17 + 6n$ Ⓓ $17 - 6n$

Lesson 1-4 Distributive Property   **23**

1-5 A Plan for Problem Solving

What You'll Learn
You'll learn to use a four-step plan to solve problems.

Why It's Important
Savings The four-step plan is useful for finding the amount of interest earned in a bank account.
See Exercise 18.

In mathematics, solving problems is an important activity. Any problem can be solved using a problem-solving plan like the one below.

1. **Explore** Read the problem carefully. Identify the information that is given and determine what you need to find.

2. **Plan** Select a strategy for solving the problem. Some strategies are shown at the right. If possible, estimate what you think the answer should be before solving the problem.

3. **Solve** Use your strategy to solve the problem. You may have to choose a variable for the unknown, and then write an expression. Be sure to answer the question.

4. **Examine** Check your answer. Does it make sense? Is it reasonably close to your estimate?

Problem-Solving Strategies

Look for a pattern.

Draw a diagram.

Make a table.

Work backward.

Use an equation or formula.

Make a graph.

Guess and check.

One important problem-solving strategy is *using an equation*. An equation that states a rule for the ralationship between quantities is called a **formula**.

Money in a bank account earns *interest*. You find simple interest by using the formula $I = prt$.

I = interest
p = principal, or amount deposited
r = interest rate, written as a decimal
t = time in years

Example
Savings Link

1 **Suppose you deposit $220 into an account that pays 3% simple interest. How much money would you have in the account after five years?**

Explore What do you know?
- The amount of money deposited is $220.
- The interest rate is 3% or 0.03.
- The time is 5 years.

**Prerequisite
Skills Review**
Decimals and
Percents, p. 689

What do you need to find?

- the amount of money, including interest, at the end of five years

Plan   What is the best strategy to use?

Use the formula $I = prt$ and substitute the known values. Add this amount to the original deposit.

Estimate: 1% of $220 is $2.20. So, 3% of $220 is about $3 \times \$2$ or $6 per year. This will be $30 in five years. You should have approximately $220 + 30$ or $250 in five years.

Solve $I = prt$ *Interest Formula*
$I = 220 \cdot 0.03 \cdot 5$ or 33   *$p = 220, r = 0.03,$ and $t = 5$*

You will earn $33 in interest, so the total amount after five years is $220 + \$33$ or $253.

Examine   Is your answer close to your estimate?

Yes, $253 is close to $250, so the answer is reasonable.

Your Turn

Science   Use $F = 1.8C + 32$ to change degrees Celsius C to degrees Fahrenheit F. Find the temperature in degrees Fahrenheit if it is 29°C.

Another important problem-solving strategy is *using a model*. In the activity below, you will use a model to find a formula for the surface area of a rectangular box.

Hands-On Algebra

Materials: rectangular box
                    ruler

Step 1   Label the edges of a rectangular box ℓ, w, or h to represent the length, width, and height of the box.

Step 2   Take the box apart so that it lies flat on the table with the labels face up.

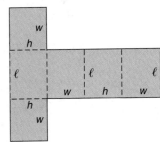

Try These

1. Find the area of each rectangular side of the box in terms of the variables ℓ, w, and h. Be sure to include the top or lid.
2. The sum of the areas is equal to S, the total surface area of a rectangular solid. Express this as a formula in simplest form.
3. Measure the lengths of the sides in centimeters or inches to find the values of ℓ, w, and h.
4. Use the formula in Exercise 2 to find the total surface area of your box.

Example

Money Link

2 **How many ways can you make 25¢ using dimes, nickels, and pennies?**

Explore   A quarter is worth 25¢. How many ways can you make 25¢ without using a quarter?

Plan   Make a chart listing every possible combination.

Solve

Coin	Number											
Dimes	2	2	1	1	1	1	0	0	0	0	0	0
Nickels	1	0	3	2	1	0	5	4	3	2	1	0
Pennies	0	5	0	5	10	15	0	5	10	15	20	25

There are 12 ways to make 25¢.

Examine   Check that each combination totals 25¢ and that there are no other possible combinations. *The solution checks.*

You can use a graphing calculator to solve problems involving formulas.

Graphing Calculator Tutorial
See pp. 750–753.

Graphing Calculator Exploration

The area of a trapezoid is $A = \frac{1}{2}h(a + b)$, where h is the height and a and b are the lengths of the bases. Use a graphing calculator to find the area of trapezoid *JKLM*.

$A = \frac{1}{2} \cdot 2.5(3.2 + 6)$ *Replace each variable with its value.*

Enter: 1 ÷ 2 × 2.5 (3.2 + 6) ENTER *11.5*

Try These

1. Find the area of trapezoid *JKLM* if the height is 15 centimeters and the bases remain the same.

2. Find the area of a trapezoid if base *a* is 14 inches long, base *b* is 10 inches long, and the height is 7 inches.

The chart below summarizes the properties of numbers. The properties are useful when you are solving problems.

The following properties are true for any numbers *a*, *b*, and *c*.		
Property	**Addition**	**Multiplication**
Commutative	$a + b = b + a$	$ab = ba$
Associative	$(a + b) + c = a + (b + c)$	$(ab)c = a(bc)$
Identity	$a + 0 = 0 + a = a$ 0 is the identity.	$a \cdot 1 = 1 \cdot a = a$ 1 is the identity.
Zero		$a \cdot 0 = 0 \cdot a = 0$
Distributive	$a(b + c) = ab + ac$ and $a(b - c) = ab - ac$	
Substitution	If $a = b$, then *a* may be substituted for *b*.	

Check for Understanding

Communicating Mathematics

1. **List** three reasons for "looking back" when examining the answer to a problem.

2. **Write** a problem in which you need to find the surface area of a rectangular solid. Then solve the problem.

> **Vocabulary**
> formula

Guided Practice

🕐 **Getting Ready** For each situation, answer the related questions.

Phoebe and Hai picked fourteen pints of raspberries in three hours. Hai picked five more pints than Phoebe.	
Samples:	**Solutions:**
How many pints were picked in all?	14
How long did Phoebe and Hai work?	3 hours
Who picked more pints?	Hai
If Phoebe picked *x* pints, how many did Hai pick?	$x + 5$

Carlos bought 2 more rock CDs than jazz CDs and 3 fewer country CDs than rock CDs. He bought eight CDs, including 1 classical CD.

3. Did Carlos buy more country than rock?

4. Which type of CD did he buy the most of?

5. If he bought *n* jazz CDs, how many rock CDs did he buy?

Example 1

6. **Geometry** The perimeter P of a rectangle is the sum of two times the length ℓ and two times the width w.

   a. Write a formula for the perimeter of a rectangle.

   b. What is the perimeter of the rectangle?

6 cm

14 cm

Example 2

7. **Money** Nate has $267 in bills. None of the bills is greater than $10. He has eleven $10 bills. He has seven fewer $5 bills than $1 bills.

   a. How many $5 and $1 bills does he have?

   b. Describe the problem-solving strategy that you used to solve this problem.

Example 2

8. **Shopping** Two cans of vegetables together cost $1.08. One of them costs 10¢ more than the other.

   a. Would 2 cans of the less expensive vegetable cost more or less than $1.08?

   b. How much would it cost to buy 3 cans of each?

Practice

Homework Help	
For Exercises	See Examples
10, 15–18	1
12, 13	2
Extra Practice	
See page 693.	

Solve each problem. Use any strategy.

9. Craig is 24 years younger than his mother. Together their ages total 56 years. How old is each person? Explain how you found your answer.

10. Moira has $500 in the bank at an annual interest rate of 4%. How much money will she have in her account after two years?

11. Joanne has 20 books on crafts and cooking. She has 6 more cookbooks than craft books. How many of each does she have?

12. How many ways are there to make 20¢ using dimes, nickels, and pennies?

13. Six Explorer Scouts from different packs met for the first time. They all shook hands with each other when they met.
    a. Make a chart or draw a diagram to represent the problem.
    b. How many handshakes were there in all?
    c. The number of handshakes h can also be found by using the formula $h = \dfrac{p(p - 1)}{2}$, where p represents the number of people. How many handshakes would there be among 12 people?
    d. Which strategy would you prefer to use to solve the problem: make a table, draw a diagram, or use a formula?

Applications and Problem Solving

14. **Savings** The table shows the cost of leasing a car for 36 months. Which option is a better deal? Explain.

Type of Fee	Option A Cost ($)	Option B Cost ($)
monthly payment	99	168
monthly tax	6	7
bank fee	495	0
down payment	1956	0
license plates	75	0

15. **Weather** Meteorologists can predict when a storm will hit their area by examining the travel time of the storm system. To do this, they use the following formula.

distance from storm (miles) ÷ speed of storm (miles per hour) = travel time of storm (hours)

At 4:00 P.M. a storm is heading toward the coast at a speed of 30 miles per hour. The storm is about 150 miles from the coast. What time will the storm hit the coast?

A storm blows in.

16. **Geometry** The area of a triangle is one-half times the product of the base b and the height h.

    a. Write a formula for the area of a triangle.

    b. Find the area of the triangle.

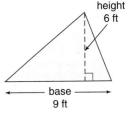

height
6 ft

base
9 ft

17. **History** Distance traveled d equals the product of rate r and time t.

    a. Write the formula for distance.

    b. In 1936, the *Douglas DC-3* became the first commercial airliner to transport passengers. It flew nonstop from New York to Chicago at an average of 190 miles per hour and the flight lasted approximately 3.7 hours. Find the distance it flew.

18. **Savings** Refer to the table at the right.

    a. Suppose you saved 50¢ each school day (180 days) while you were in eighth grade. How much would you have at the end of the school year?

Saving Just 50 Cents a School Day . . .	
If you start saving in this grade . . .	. . . you'll have this amount at high school graduation*
6th grade	$630
7th grade	$540
8th grade	$450
9th grade	$360

*Based on 180 school days in each year × 50 cents a day = $90 a school year.

Source: *Zillions*

    b. Suppose you deposited your "school year savings" in an account with a 4% annual interest rate. How much money would you have in your account after a year?

    c. Suppose you saved the 50¢ each school day (180 days) while you were a freshman in high school. Find the sum of this amount and the money already in your account from the eighth grade.

    d. How much money would you have after the second year?

    e. Repeat steps c and d for your junior and senior years. How much money would you have in your account by the time you graduated from high school? Compare this amount to that listed in the table.

19. **Critical Thinking** Refer to Exercise 13. In a meeting, there were exactly 190 handshakes. How many people were at the meeting?

Mixed Review

Simplify each expression. *(Lesson 1–4)*

20. $21x - 10x$ 21. $5b + 3b$ 22. $3(x + 2y)$ 23. $9a + 15(a + 3)$

24. **Short Response** State the property shown by $4(ab) = (4a)b$. *(Lesson 1–3)*

25. **Multiple Choice** The top of a volleyball net is 7 feet 11 inches from the floor. The bottom of the net is 4 feet 8 inches from the floor. How wide is the volleyball net? *(Lesson 1–3)*

    Ⓐ 3 feet 3 inches   Ⓑ 2.31 feet

    Ⓒ 7 feet 1 inch   Ⓓ 6 feet

Dot-to-Dot

Logical Reasoning

Materials

 5 sheets of different-colored paper

 paper punch

 4 three-inch square slips of paper

 4 two-inch squares of paper

Mathematicians use logical reasoning to discover new ideas and solve problems. *Inductive* and *deductive* reasoning are two forms of reasoning. Let's investigate them to find out how they differ.

Investigate

1. Use the colored paper and the paper punch to make at least 25 dots of each color. You will use these dots to explore patterns.

   a. *Triangular numbers* are represented by the number of dots needed to form different-sized triangles. Use your colored dots to form the first four triangular numbers shown below.

   1st number = 1   2nd number = 3   3rd number = 6   4th number = 10

   b. Draw the fifth triangular number. Do you see a pattern? Use the pattern to write the next five triangular numbers.

   c. In Step 1b, you used **inductive reasoning**, where a conclusion is made based on a pattern or past events.

2. **Deductive reasoning** is the process of using facts, rules, definitions, or properties in a logical order. You use deductive reasoning to reach valid conclusions.

   If-then statements, called **conditionals**, are commonly used in deductive reasoning. Consider the following conditional.

   If <u>I visit the island of Kauai</u>, then <u>I am in Hawaii</u>.

   The portion of the sentence following *if* is called the **hypothesis**, and the part following *then* is called the **conclusion**. This conditional is true since Kauai is a Hawaiian island.

   a. Use the following information to reach a valid conclusion.
   Conditional: If I visit the island of Kauai, then I am in Hawaii.
   Given: I visit the island of Kauai.

The island of Kauai

b. On a two-inch square, write the hypothesis "I visit the island of Kauai." On a three-inch square, write the conclusion "I am in Hawaii." Place the two-inch square inside the three-inch square.

c. Place your pencil on the given statement, "I visit the island of Kauai." Since the pencil is also contained within the square with the conclusion, "I am in Hawaii," the conclusion is valid.

d. Repeat Step 2c, but exchange the hypothesis and conclusion. The new conditional is *If I am in Hawaii, then I visit the island of Kauai.* You can place your pencil in any region marked by the ✕'s in the diagram. You may be in Hawaii, visiting Maui, not the island of Kauai. You cannot reach the conclusion using the conditional and the given information.

I am in Hawaii.

I visit the island of Kauai. ✕

✕   ✕

Extending the Investigation

In this extension, you will continue to investigate inductive and deductive reasoning.

1. The first four *square numbers* are 1, 4, 9, and 16. Use the colored dots to make the first five square numbers. Use inductive reasoning to list the first ten square numbers.

2. The first four *pentagonal numbers* are 1, 5, 12, and 22. Use the colored dots to make the first five pentagonal numbers. Use inductive reasoning to list the first ten pentagonal numbers.

3. For each problem, identify the hypothesis and conclusion. Then use squares as shown above to determine whether a valid conclusion can be made from the conditional and given information.
   a. Conditional: If the living organism is a grizzly bear, then it is a mammal.
      Given: The living organism is a grizzly bear.
   b. Conditional: If Aislyn is in the Sears Tower, then she is in Chicago.
      Given: Aislyn is in Chicago.

4. Write a paragraph explaining the difference between inductive and deductive reasoning. Include an example of each type of reasoning.

Presenting Your Investigation

Here are some ideas to help you present your conclusions to the class.

• Make a poster showing the triangular, square, and pentagonal numbers.

• Include a description of the patterns you observed.

Investigation For more information on logical reasoning, visit: www.algconcepts.com

What You'll Learn
You'll learn to collect and organize data using sampling and frequency tables.

Why It's Important
Marketing
Businesses use surveys to collect data in order to test new ideas. *See Example 3.*

Researchers gather **data,** or information, using **experiments** and **observational studies.** In an experiment, a test is made under controlled conditions. Subjects in an observational study are studied without action by the investigator.

In either an experiment or an observational study, **sampling** is a convenient way to gather data to make predictions about a **population.** A **sample** is a small group that is used to represent a much larger population. Three important characteristics of a good sample are listed below.

Sampling Criteria	A good sample is: • representative of the larger population, • selected at random, and • large enough to provide accurate data.

A survey can be biased and give false results if these criteria are not followed. Note that there is no given number to make the sample large enough. You must consider each survey individually to see if it is based on a good sample.

Real World

Example
Health Link

1 **One hundred people in Lafayette, Colorado, were asked to eat a bowl of oatmeal every day for a month to see whether eating a healthy breakfast daily could help reduce cholesterol. After 30 days, 98 of those in the sample had lower cholesterol. Is this a good sample? Explain.** **Source:** Quaker Oats

If the people were randomly chosen, then this is a good sample. Also, the sample appears to be large enough to be representative of the population. For example, the results of two or three people would not have been enough to make any conclusions.

Your Turn

Determine whether each is a good sample. Explain.

a. Two hundred students at a school basketball game are surveyed to find the students' favorite sport.

b. Every other person leaving a supermarket is asked to name their favorite soap.

After the survey is complete, the gathered data is organized into different types of tables and charts. One way to organize data is by using a **frequency table**. In a frequency table, you use **tally marks** to record and display the frequency of events.

Example ② 2
Science Link

In an experiment, students "charged" balloons by rubbing them with wool. Then the students placed the balloons on a wall and counted the number of seconds they remained. The class results are shown in the chart at the right. Make a frequency table to organize the data.

Static Electricity Time (s)				
15	52	26	22	25
26	29	33	36	20
43	21	30	39	34
35	27	29	42	35
16	18	21	21	40

Step 1 Make a table with three columns: Time (s), Tally, and Frequency. Add a title.

Step 2 It is sometimes helpful to use *intervals* so there are fewer categories. In this case, we are using intervals of size 10.

Step 3 Use tally marks to record the times in each interval.

Step 4 Count the tally marks in each row and record this number in the Frequency column.

Static Electricity		
Time (s)	Tally	Frequency
15–24	⊪ III	8
25–34	⊪ IIII	9
35–44	⊪ II	7
45–54	I	1

Your Turn

c. Make a frequency table to organize the data in the chart at the right.

Noon Temperature (°C)					
32	30	18	29	20	14
21	32	36	15	19	10
16	22	25	30	26	21

In Example 2, suppose the science teacher wanted to know how many balloons stayed on the wall *no more than 44 seconds*. To answer this question, use a **cumulative frequency table** in which the frequencies are accumulated for each item.

Static Electricity		
Time (s)	Frequency	Cumulative Frequency
15–24	8	8
25–34	9	17
35–44	7	**24**
45–54	1	25

From the cumulative frequency table, we see that 24 balloons stayed on the wall for 44 seconds or less. Or, 24 balloons stayed on the wall for no more than 44 seconds.

Once you have summarized data in a frequency table or in a cumulative frequency table, you can analyze the information and make conclusions.

Real World

Example ❸

Marketing Link

The right location?

Owners of a restaurant are looking for a new location. They counted the number of people who passed by the proposed location one day during lunchtime. The frequency table at the right shows the results of their sampling.

Age of People	Tally	Frequency
under 13	JHT II	7
teens	JHT JHT	10
20s	JHT JHT JHT III	18
30s	JHT JHT JHT JHT JHT JHT JHT JHT II	42
40s	JHT JHT JHT JHT JHT JHT JHT I	36
50s	JHT JHT JHT IIII	19
60s	JHT JHT I	11

A. Which two groups of people passed by the location most frequently?

adults in their 30s and 40s

B. If the restaurant is an ice cream shop aimed at teens during their lunchtimes, is this a good location for the restaurant? Explain.

Since very few teens pass by the location compared to adults, the owners should probably look for another location.

Check for Understanding

Communicating Mathematics

1. **Explain** the difference between a frequency table and a cumulative frequency table.

2. **List** some examples of how a survey might be biased.

Vocabulary

data
experiments
observational studies
sampling
population
sample
frequency table
tally marks
cumulative frequency table

Guided Practice

Example 1

Determine whether each is a good sample. Explain.

3. Four people out of 500 are randomly chosen at a senior assembly and surveyed to find the percent of seniors who drive to school.

4. Six hundred randomly chosen pea seeds are used to determine whether wrinkled seeds or round seeds are the more common type of seed.

Refer to the chart at the right.

Number of Soccer Goals Scored This Season		
1	2	5
1	6	2
6	8	4
2	4	5
5	1	3
4	7	2
2	6	4

Example 2   **5.** Make a frequency table to organize the data.

Example 3   **6.** What number of goals was scored most frequently?

Example 3   **7.** How many times did the team score 8 goals?

Example 3   **8.** How many more times did the soccer team score six goals than three goals?

Examples 2 & 3   **9. Technology**   When lines of cars get too long at some traffic lights, computers override the signals to turn the lights green and allow the cars to move. A cycle is the number of seconds it takes a light to change from red back to red. The frequency table below shows different traffic light cycles during one afternoon.

Cycle (s)	Tally	Frequency
80	ЖЖ ЖЖ ЖЖ ЖЖ ЖЖ ЖЖ III	33
90	ЖЖ ЖЖ ЖЖ ЖЖ ЖЖ ЖЖ ЖЖ ЖЖ II	42
100	ЖЖ ЖЖ ЖЖ ЖЖ ЖЖ ЖЖ ЖЖ ЖЖ ЖЖ ЖЖ ЖЖ ЖЖ	60
110	ЖЖ ЖЖ ЖЖ ЖЖ ЖЖ	25

a. Which cycle occurred the most?

b. Make a cumulative frequency table of the data.

c. If the standard cycle for a traffic light is 100 seconds, how many times during this period was the cycle less than the standard?

Exercises

Practice

Determine whether each is a good sample. Describe what caused the bias in each poor sample. Explain.

10. Thirty people standing in a movie line are asked to name their favorite actor.

11. Police stop every fifth car at a sobriety checkpoint.

12. Every other household in a neighborhood of 240 homes is surveyed to determine how many people in the area recycle.

13. Every other household in a neighborhood of 20 homes is surveyed to determine the country's favorite presidential candidate.

14. Every third student on a class roster is surveyed to determine the average number of hours students in the class spend on a computer.

15. All people leaving a sporting goods store are asked to name their favorite golfer.

Homework Help

For Exercises	See Examples
10–15, 20, 21	1
16–18	2, 3
22	3

Extra Practice
See page 693.

16. Refer to the chart at the right.

　a. Make a frequency table to organize the data.

　b. How many fewer sausage pizzas were ordered than cheese pizzas?

　c. Suppose x mushroom pizzas were also ordered. Write an expression representing the total number of mushroom, vegetable, and pepperoni pizzas ordered. Write the expression in simplest form.

Pizzas Ordered				
C	C	C	P	C
C	P	P	V	S
S	P	C	P	C
S	S	C	P	P
C	V	P	C	C
V	S	S	C	C

C = cheese, P = pepperoni,
S = sausage, V = vegetable

17. Refer to the chart at the right.

　a. Make a frequency table to organize the data.

　b. What was the most common score?

　c. Suppose each S represents the score and each F represents the frequency for that score. Explain why the formula below determines the class *average* A for this quiz.

$$A = \frac{(S_1 \cdot F_1) + (S_2 \cdot F_2) + \ldots + (S_8 \cdot F_8)}{30}$$

　d. Find the class average for the quiz.

Quiz Scores (out of 10 points)				
9	8	8	9	8
8	10	6	7	8
10	8	8	9	9
7	9	7	8	10
9	6	7	8	9
9	10	8	9	8

Reading Algebra

Read S_1 as *S sub 1*. The 1 is called a *subscript*.

18. Refer to the chart at the right.

　a. Make a cumulative frequency table to organize the data.

　b. In how many games were there at least three home runs?

　c. In how many games were there no more than four home runs?

Number of Home Runs in a Game per Month				
2	2	3	5	1
0	1	2	6	3
3	1	1	4	5
2	3	4	3	0
1	0	1	1	2

19. Why do you suppose a coffee and bagel shop would want to locate where a lot of people walk past the store between 7:00 A.M. and 10:30 A.M.?

Applications and Problem Solving

20. Health When you have a blood test taken for your health, why does the technician only take a few vials of your blood? Use the terms you learned in this lesson to explain your answer.

21. Marketing A new cola drink is out on the market. Name three places where the cola company could set up taste tests to determine interest in the drink.

22. **Entertainment**   The frequency table at the right shows students' favorite types of movies in one class.

Favorite Type of Movie		
Movie	**Tally**	**Frequency**
Adventure	ЖІІ II	7
Comedy	ЖІІ IIII	9
Horror	IIII	4
Drama	III	3

   a. Suppose you invite students in this class to a party. What type of movie would you show? Explain.

   b. In another class, three times more students favored drama, and two fewer favored comedy. Write an expression to find the total number of people in that class who favored drama and comedy. Then find the number.

23. **Critical Thinking**   Suppose someone takes a phone survey from a large random sample of people. Do you think that the wording of a question or the surveyor's tone of voice can affect the responses and cause biased results? Explain.

Mixed Review

24. An adult bus ticket and a child's bus ticket together cost $2.40. The adult fare is twice the child's fare. What is the adult's fare? Use any strategy to solve the problem.   *(Lesson 1–5)*

25. **Travel**   What distance can a car travel in 5 hours at a constant rate of 55 miles per hour? Use a diagram or the formula $d = rt$ to solve the problem.   *(Lesson 1–5)*

26. Simplify the expression $16a + 21a + 30b - 7b$.   *(Lesson 1–4)*

Standardized Test Practice

Ⓐ Ⓑ Ⓒ Ⓓ

27. **Short Response**   Write a verbal expression for $x + 9$.   *(Lesson 1–1)*

28. **Short Response**   Write an algebraic expression for *4 times n less 3*.   *(Lesson 1–1)*

Quiz 2   Lessons 1–3 through 1–6

Simplify each expression. Identify the properties used in each step.
(Lesson 1–3)

1. $11 + 2a + 6$

2. $4 \cdot (8t)$

3. Health   Your optimum exercise heart rate per minute is given by the expression $0.7(220 - a)$, where a is your age. Use your age for a and find your optimum exercise heart rate.
(Lesson 1–4)

4. Fitness   Lorena runs for 30 minutes each day. Find the distance she runs if she averages 660 feet per minute. Use the formula $d = rt$.   *(Lesson 1–5)*

5. Biology   Make a frequency table to organize the data in the chart at the right. Which eye color occurs the least?   *(Lesson 1–6)*

Eye Color				
H	B	B	G	U
H	G	G	B	B
B	G	U	G	H
H	U	B	B	G
H	B	U	B	B

B = brown, U = blue,
G = green, H = hazel

Graphs are a good way to display and analyze data. The graph at the right is a **line graph**. It shows trends or changes over time. There are no holes in the graph and every point on the graph has meaning. To construct a line graph, include the following items.

1. a title
2. a label on each axis describing the variable that it represents
3. equal intervals on each axis

Note that the graph at the right contains all three items.

Yearly Attendance at Movie Theaters

Source: *USA TODAY*

Example ❶

Travel Link

The number of annual visitors to the Grand Canyon is given in the table at the right. Construct a line graph of the data. Then use the graph to predict the number of annual visitors to the Grand Canyon in the year 2010.

Step 1 Draw a horizontal axis and a vertical axis and label them as shown below. Include a title.

Step 2 Plot the points.

Step 3 Draw a line by connecting the points.

You can see from the graph that the general trend is that the number of visitors to the Grand Canyon increases steadily every ten years. A good prediction for the year 2010 might be about 6 or 6.5 million people.

Year	Grand Canyon Visitors (millions)
1960	1.2
1970	2.3
1980	2.6
1990	3.8
2000	4.8

Source: National Park Service

Grand Canyon Visitors

Grand Canyon

Your Turn

a. The table at the right shows the approximate U.S. consumption of bottled water per person. Construct a line graph of the data. Then use it to predict the amount of bottled water each person will drink in the year 2005.

Year	Bottled Water (gallons)
1991	9
1993	10.5
1995	12
1997	14
1999	17
2001	19.5

Source: International Bottled Water Association

Another type of graph that is used to display data is a **histogram**. A histogram uses data from a frequency table and displays it over equal intervals. To make a histogram, include the same three items as the line graph: title, axes labels, and equal intervals. In a histogram, all bars should be the same width with no space between them.

Example **2**

Physical Science Link

The frequency table is from Example 2 in Lesson 1–6. It shows the various time intervals that "charged" balloons remained stuck to the wall. Construct a histogram of the data.

Static Electricity		
Time (s)	Tally	Frequency
15–24	JHT III	8
25–34	JHT IIII	9
35–44	JHT II	7
45–54	I	1

Step 1 Draw a horizontal axis and a vertical axis and label them as shown below. Include the title.

Step 2 Label equal intervals given in the frequency table on the horizontal axis. Label equal intervals of 1 on the vertical axis.

Step 3 For each time interval, draw a bar whose height is given by the frequency.

The histogram gives a better visual display of the data than the frequency table. In Lesson 1–6, we used cumulative frequency tables to organize data. Likewise, we can construct **cumulative frequency histograms**.

 www.algconcepts.com/extra_examples

Lesson 1–7 Displaying and Interpreting Data **39**

Example **3**

The ages of people who participated in a recent survey are shown in the table at the right. Construct a cumulative frequency histogram to display the data.

Survey		
Age	Tally	Frequency
1–10	JHT III	8
11–20	JHT IIII	9
21–30	JHT	5
31–40	JHT JHT II	12
41–50	JHT III	8

First, make a cumulative frequency table. Then construct a histogram using the cumulative frequencies for the bar heights. Remember to label the axes and include the title.

Survey		
Age	Frequency	Cumulative Frequency
1–10	8	8
11–20	9	17
21–30	5	22
31–40	12	34
41–50	8	42

Survey Participants

Your Turn

b. Construct a cumulative frequency histogram of the data in Example 2.

Another way to display data is a **stem-and-leaf plot**.

The greatest common place value for each data item is used to form the *stem*.

Stem	Leaf
1	1 6
2	1 3 9
3	
4	5 5

$2|3 = 23$

The *leaves* are formed by the next greatest place value.

In this case, the tens digits are the stems. The ones digits are the leaves. Write the leaves in order from least to greatest.

A *key* is always included. This shows how the digits are related.

Info Graphic

In the stem-and-leaf plot at the right, the data are represented by three-digit numbers. In this case, use the digits in the first two place values to form the stems. For example, the values for 102, 108, 114, 115, 125, 127, 131, and 139 are shown in the stem-and-leaf plot at the right.

Stem	Leaf
10	2 8
11	4 5
12	5 7
13	1 9

$11|5 = 115$

Example **4**

School Link

The table shows the class results on a 50-question test. Make a stem-and-leaf plot of the grades.

Class Scores					
29	37	48	40	17	34
28	43	37	35	49	29
13	29	42	45	37	46

The tens digits are the stems, so the stems are 1, 2, 3, and 4. The ones digits are the leaves.

Stem	Leaf
1	7 3
2	9 8 9 9
3	7 4 7 5 7
4	8 0 3 9 2 5 6

$3\,|\,7 = 37$

Now arrange the leaves in numerical order to make the results easier to observe and analyze.

Stem	Leaf
1	3 7
2	8 9 9 9
3	4 5 7 7 7
4	0 2 3 5 6 8 9

$3\,|\,7 = 37$

What were the highest and lowest scores?
49 and 13

Which score occurred most frequently?
29 and 37, three times each

How many students received a score of 35 or better?
11 students

Your Turn

c. Make a stem-and-leaf plot of the quiz grades below.
54, 55, 60, 42, 41, 75, 50, 68, 62, 54, 70, 50

Check for Understanding

Communicating Mathematics

1. **Explain** the differences between the use of line graphs and histograms.

2. **Identify** each essential part of a correctly drawn line graph or histogram.

3. **You Decide?** Marcia says that a histogram works as well as a line graph to show trends over time. Manuel says that a histogram shows intervals, not trends. Who is correct? Explain.

Vocabulary
line graph
histogram
cumulative frequency histograms
stem-and-leaf plot

The table at the right shows the percent of homes in California with internet access.

Example 1

4. Make a line graph of the data.

5. Between which two years was the growth of on-line access the greatest?

6. Predict the percent of homes with on-line access in the year 2001.

Year	Percent of Homes On-Line
1997	19
1998	25
1999	28
2000*	31

Source: Pacific Telesis
*estimated

Refer to the histogram at the right.

Example 2

7. Determine the length of most maple leaves.

Example 2

8. How many leaves were sampled?

Example 3

9. Construct a cumulative frequency histogram of the data.

Example 3

10. How many of the leaves were no more than 15 centimeters long?

Leaf Lengths in a Maple Tree Population

Example 4

11. **Weather** The stem-and-leaf plot at the right shows the daily high temperatures in McComb, Mississippi, in March.
    Source: *The Weather Underground*

    a. What was the highest temperature?

    b. On how many days was the high temperature in the 70s?

    c. What temperature occurred most frequently?

Daily High Temperatures (°F)

Stem	Leaf
4	8
5	4 9 9
6	3 3 3 3 4 4 8 8 8
7	0 0 2 2 2 3 3 5 5
	9 9 9
8	1 1

$4 \mid 8 = 48°F$

Math Online Data Update For the latest information on weather forecasts, visit: algconcepts.com

Exercises

Practice

The percent of unemployment among workers ages 16 to 19 is shown at the right.

12. Make a line graph of the data.

13. When was unemployment at its highest?

14. Describe the general trend in unemployment among teens ages 16 to 19.

Homework Help

For Exercises	See Examples
12–14	1
15–18	2
19–23	4

Extra Practice
See page 694.

Year	Percent of Working Teens Unemployed
1992	17
1993	19
1994	18
1995	17.5
1996	17
1997	15
1998	13

Source: U.S. Labor Dept.

In a survey, men and women were asked how long they were willing to stay on hold when calling a customer service representative about a product they purchased. The results are shown in the table at the right.

How Long on Hold?		
Time (min)	Percent of Men	Percent of Women
0–1	28	18
2–3	32	36
4–5	23	27
6–7	7	10
8+	10	9

Source: Bruskin/Goldring for Inference

15. Make a histogram showing the men's responses.

16. Make a histogram showing the women's responses.

17. How do your histograms compare?

18. Who do you think would hang up the phone sooner, men or women?

The stem-and-leaf plot at the right gives the number of catches of the NFL's leading pass receiver for the first 39 seasons.

Stem	Leaf
6	0 1 2 6 7
7	1 1 1 2 3 3 3 5 7 8
8	2 5 8 8 9
9	0 0 1 2 2 2 3 5
10	0 0 0 1 4 6 8 8
11	2
12	2 3 12|3 = 123

19. What was the greatest number of catches during a season?

20. How many seasons are represented?

21. What number of catches occurred most frequently?

22. How many leading pass receivers have at least 90 catches?

23. **Critical Thinking** *Back-to-back stem-and-leaf plots* are used to compare two sets of data. The back-to-back stem-and-leaf plot below compares the performance of two algebra classes on their first test. Which class do you believe did better on the test? Why do you think so?

First Period	Stem	Second Period
9 8	5	
9 9 8 7	6	7 8
7 7 6 5 3	7	2 4 4 4 5 5 8
5 4 1	8	1 3 4 5 7
8 8 6 5 2	9	0 1 1 3 8

8|9 = 89

Mixed Review

Determine whether each is a good sample. *(Lesson 1–6)*

24. A survey is taken in Alaska to determine how much money an average family in the United States spends on heating their home.

25. In a survey, every third name in the phone book is called and the person answering is interviewed.

26. Write a formula for the perimeter P of a square with side s in simplest form. *(Lesson 1–5)*

27. **Short Response** Write $5x + 3(x - y)$ in simplest form. *(Lesson 1–4)*

28. **Multiple Choice** Evaluate $12 \cdot 6 + 3 \cdot 2 - 8$. *(Lesson 1–2)*

Ⓐ 142   Ⓑ −144   Ⓒ 70   Ⓓ 120

Understanding and Using the Vocabulary

After completing this chapter, you should be able to define each term, property, or phrase and give an example of each.

inter NET CONNECTION **Review Activities**
For more review activities, visit:
www.algconcepts.com

Algebra

algebraic expression (p. 4)
coefficient (p. 20)
equation (p. 5)
equivalent expressions (p. 20)
evaluating (p. 10)
factors (p. 4)
formula (p. 24)
like terms (p. 20)
numerical expression (p. 4)
order of operations (p. 8)
product (p. 4)
quotient (p. 4)
simplest form (p. 20)
simplify (p. 15)

term (p. 20)
variable (p. 4)
whole numbers (p. 16)

Statistics

cumulative frequency
    histogram (p. 39)
cumulative frequency
    table (p. 33)
data (p. 32)
experiment (p. 32)
frequency table (p. 32)
histogram (p. 39)
line graph (p. 38)
observational study (p. 32)

population (p. 32)
sample (p. 32)
sampling (p. 32)
stem-and-leaf plot (p. 40)
tally marks (p. 32)

Logic

conclusion (p. 30)
conditional (p. 30)
counterexample (p. 16)
deductive reasoning (p. 30)
hypothesis (p. 30)
if-then statement (p. 30)
inductive reasoning (p. 30)

Choose the correct term to complete each sentence.

1. A (coefficient, term) is a number, a variable, or a product or quotient of numbers and variables.
2. The result of two numbers multiplied together is the (factor, product).
3. A(n) (numerical expression, algebraic expression) contains variables.
4. According to the (order of operations, like terms), you do multiplication before addition.
5. A (counterexample, hypothesis) shows that a statement is not always true.
6. Some examples of (like terms, whole numbers) are $2x$, $10x$, and $-6x$.
7. A (sample, variable) is a group used to represent a much larger population.
8. Any sentence that contains an equals sign is a(n) (equation, formula).
9. Using (sampling, frequency tables) is a way to organize data.
10. A (histogram, stem-and-leaf plot) makes it easier to identify specific data items.

Skills and Concepts

Objectives and Examples

• **Lesson 1–1**   Translate words into algebraic expressions and equations.

Write an algebraic expression for *7 decreased by the quantity x divided by 2.*

$7 - (x \div 2)$ or $7 - \dfrac{x}{2}$

Review Exercises

Write an algebraic expression.

11. the product of 5 and n
12. the sum of 2 and three times x

Write an equation for each sentence.

13. Six less than two times y equals 14.
14. The quotient of 20 and x is 4.

Objectives and Examples

Review Exercises

• **Lesson 1–2** Use the order of operations to evaluate expressions.

$$2 \cdot 7 + 2 \cdot 3 = 14 + 6 \quad \textit{Multiply.}$$
$$= 20 \qquad \textit{Add.}$$

Find the value of each expression.

15. $3 + 8 \div 2$

16. $12 \div 4 + 15 \cdot 3$

17. $29 - 3(9 - 4)$

18. $4(11 + 7) - 9 \cdot 8$

19. Find the value of $3ac - b$ if $a = 6$, $b = 9$, and $c = 1$.

• **Lesson 1–3** Use the commutative and associative properties to simplify expressions.

Name the property shown by
$3 + x + 2 = 3 + 2 + x$. Then simplify.

$$3 + x + 2 = 3 + 2 + x \quad \textit{Commutative (+)}$$
$$= 5 + x \qquad \textit{Substitution}$$

Name the property shown by each statement. Then simplify.

20. $6 + (7 + b) = (6 + 7) + b$

21. $2 \cdot c \cdot 10 = 2 \cdot 10 \cdot c$

22. $9 \cdot (5 \cdot f) = (9 \cdot 5) \cdot f$

23. $x(5 + 4) = (5 + 4)x$

24. $3 + a + 8 = 3 + 8 + a$

25. $(g + 1) + 2 = g + (1 + 2)$

• **Lesson 1–4** Use the Distributive Property to evaluate expressions.

$$5b + 3(b + 2) = 5b + 3 \cdot b + 3 \cdot 2$$
$$= 5b + 3b + 6 \quad \textit{Multiply.}$$
$$= (5 + 3)b + 6 \quad \textit{Distribute.}$$
$$= 8b + 6 \qquad \textit{Simplify.}$$

Simplify each expression.

26. $4(8 + y)$

27. $7(v - 1)$

28. $10x + x$

29. $h(2 + a)$

30. $5z + 2z - 6$

31. $10 + 3(4 - d)$

• **Lesson 1–5** Use the four-step plan to solve problems.

Explore What do you know? What are you trying to find?

Plan How will you go about solving this? What problem-solving strategy could you use?

Solve Carry out your plan. Does it work? Do you need another plan? If necessary, choose a variable for an unknown and write an expression.

Examine Check your answer. Does it make sense? Is it reasonably close to your estimate?

Use the four-step plan to solve each problem.

32. Finance Mr. Rockwell deposited $1000 in an account that pays 2% interest. How much money would he have in the account after ten years?

33. School Jamal is typing a three-page report with approximately 400 words per page for school. He thought he could finish typing the report in 2 hours. After $1\frac{1}{2}$ hours, he had finished 2 pages.

a. How many words are in his paper?

b. About how many words had Jamal typed in $1\frac{1}{2}$ hours?

Mixed Problem Solving
See pages 724–731.

Objectives and Examples

- **Lesson 1–6** Collect and organize data using sampling and frequency tables.

Make a frequency table for the data
{1, 4, 3, 4, 0, 2, 3, 1, 0, 2, 0, 2, 0, 4, 0, 0, 4, 1, 2}.

Number	Tally	Frequency
0	JHT I	6
1	III	3
2	IIII	4
3	II	2
4	IIII	4

Review Exercises

Use the frequency table at the left to answer each question.

34. How many numbers are in the sample?
35. Which number occurs most frequently?
36. How many times does the number 2 occur?
37. Make a cumulative frequency table from the data.
38. How many times does a number less than 2 occur?
39. How many times does a number greater than or equal to 2 occur?

- **Lesson 1–7** Construct and interpret line graphs, histograms, and stem-and-leaf plots.

Construct a histogram for the data {10, 10, 10, 10, 11, 11, 12, 13, 14, 14, 14, 15, 15}.

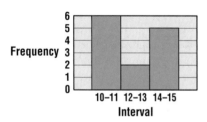

Use the histogram at the left to answer each question.

40. How large is each interval?
41. Which interval has the most data?
42. How many numbers have a value greater than 11?
43. Make a cumulative histogram from the data.
44. How many numbers are in the sample?
45. How many numbers have a value less than 14?

Applications and Problem Solving

46. **Geometry** Write an equation to represent the perimeter P of the figure below. Then solve for P if $x = 9$ and $y = 5$. *(Lesson 1–5)*

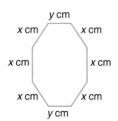

47. **Testing** The stem-and-leaf plot below shows the scores from a driver's test. *(Lesson 1–7)*

   a. What were the highest and lowest scores?
   b. Which score occurred most frequently?
   c. How many people received a score of 76 or better?

Stem	Leaf
6	2 8
7	4 5 5 6
8	0 4 8 8 8
9	2 4 7
10	0

$7|4 = 74$

1. **Explain** why we use the order of operations in mathematics.

2. **List** three like terms with the variable k.

Write an algebraic expression for each verbal expression.

3. x increased by 12

4. the quotient of 5 and y

5. 1 less than 8 times p

Use the order of operations to find the value of each expression.

6. $13 + 4 \cdot 5$

7. $12 + 6 \div 3 - 4$

8. $3(8 + 2) - 7$

Evaluate each expression if $h = 8$, $j = 3$, and $k = 2$.

9. $k(4 + j) + 6$

10. $\dfrac{h + k}{h - j}$

Name the property shown by each statement.

11. If $11 = 7 + x$, then $7 + x = 11$.

12. $28 \cdot 1 = 28$

13. $(r \cdot 9) \cdot 3 = r \cdot (9 \cdot 3)$

14. $10 + b = b + 10$

15. $6(m + 2) = 6 \cdot m + 6 \cdot 2$

Simplify each expression.

16. $n + 5n$

17. $6x - 4x + 9y - 4y$

18. $4(2s + 8t - 1)$

19. **Sports** Danny stayed late after every basketball practice to shoot 5 free throws. The chart shows how many free throws he made out of 5 for each night of practice.
    a. Make a frequency table to organize the data.
    b. If Danny has basketball practice 5 days a week, how many weeks did he stay late, shooting free throws?
    c. What number of free throws did he make most often?
    d. How many times did he not make any free throws?
    e. How many times did he make all 5 free throws?

Free Throws (out of 5)			
1	4	0	2
1	3	3	2
4	3	3	2
5	3	2	2
1	0	4	3

20. **Communication** The line graph shows the growth in sales of prepaid calling cards.
    a. Between which two years was growth in sales the greatest?
    b. Predict the number of sales for the year 2002.

Sales of Prepaid Calling Cards

Sales (billions)

Year (estimated)

Source: Atlantic ACM

Number Concept Problems

Standardized tests include many questions written with realistic settings. Read each question carefully. Be sure you understand the situation and what the question asks.

A calculator can help, but you can often find the answer faster with a pencil and your own math skills. Since standardized tests are timed, you will want to find the correct answers as quickly as possible.

Test-Taking Tip
Translate words into arithmetic symbols.
is $=$
of $\times$
per $\div$

Example 1

Mrs. Lopez estimates that $\frac{2}{3}$ of the families in her neighborhood will participate in the annual garage sale. If there are 225 families in her neighborhood, how many families does she expect to participate?

- (A) 75
- (B) 150
- (C) 175
- (D) 220

Hint Estimate the answer before making any calculations.

Solution First estimate. Since $\frac{2}{3}$ is greater than $\frac{1}{2}$, more than one half of the 225 families will participate. One half of 225 is about 112. So, choice A is not possible.

The word *of* ($\frac{2}{3}$ *of* the families) tells you to use multiplication.

$$\frac{2}{3} \cdot 225 = \frac{2(225)}{3} \qquad \textit{Multiply.}$$

$$= \frac{2(\overset{75}{\cancel{225}})}{\underset{1}{\cancel{3}}} \qquad \textit{Simplify.}$$

$$= 2(75) \text{ or } 150 \qquad \textit{Multiply.}$$

If you use your calculator, multiply 2 by 225, and then divide the answer by 3.

Two-thirds of the 225 families is 150 families. So, the answer is B.

Example 2

Jan drove 144 miles between 10:00 A.M. and 12:40 P.M. What was her average speed in miles per hour?

Hint Pay attention to the units of measure.

Solution From 10:00 A.M. to 12:40 P.M. is 2 hours and 40 minutes. You need time in *hours*. Convert minutes to hours.

2 hours 40 minutes $= 2\frac{40}{60}$ hours or $2\frac{2}{3}$ hours

$$\frac{144}{2\frac{2}{3}} = \frac{144}{\frac{8}{3}} \qquad \textit{Divide the miles by time. Rename } 2\frac{2}{3} \textit{ as } \frac{8}{3}.$$

$$= 144 \cdot \frac{3}{8} \qquad \textit{Multiply by the reciprocal of } \frac{8}{3}.$$

$$= \overset{18}{\cancel{144}} \cdot \frac{3}{\underset{1}{\cancel{8}}} \qquad \textit{Divide by the GCF, 8.}$$

$$= 18(3) \text{ or } 54 \qquad \textit{Multiply.}$$

The answer is 54 miles per hour. Record it on the grid.

- Start with the *left* column.
- Write the answer in the boxes at the top. Write one digit in each column.
- Mark the corresponding oval in each column.
- *Never* grid a mixed number; change it to a fraction or a decimal.

After you work each problem, record your answer on the answer sheet provided or on a sheet of paper.

Multiple Choice

1. Grant purchased a shirt for $29.95 and 2 pairs of socks for $2.95 a pair. The sales tax on these purchases was $2.42. What was the total amount Grant spent? *(Prerequisite Skill)*
   - Ⓐ $35.32
   - Ⓑ $35.85
   - Ⓒ $38.27
   - Ⓓ $39.37

2. Ariel adds $\frac{1}{2}$ cup of flour to a bowl that already has $3\frac{2}{3}$ cups of flour. How many total cups of flour will be in the bowl? *(Prerequisite Skill)*
   - Ⓐ $7\frac{1}{3}$
   - Ⓑ $4\frac{1}{6}$
   - Ⓒ 4
   - Ⓓ $3\frac{1}{6}$

3. The number 1134 is divisible by all of the following except— *(Basic Skill)*
   - Ⓐ 3.
   - Ⓑ 6.
   - Ⓒ 9.
   - Ⓓ 12.
   - Ⓔ 14.

4. Dr. Hewson has 758 milliliters of a solution to use for a class lab experiment. She divides the solution evenly among 32 students. If 22 milliliters are left after the experiment, how much of the solution did she give each student? *(Lesson 1–5)*
   - Ⓐ 23.0 mL
   - Ⓑ 24.2 mL
   - Ⓒ 33.0 mL
   - Ⓓ 35.9 mL

5. For shipping, a company charges $2.75 in addition to $1.25 for each $10 ordered. Which equation represents the cost c for shipping an order worth $50? *(Lesson 1–1)*
   - Ⓐ $\frac{c}{50} = 2.75 + 1.25$
   - Ⓑ $c + 2.75 = 1.25 + \frac{50}{10}$
   - Ⓒ $c = 2.75 + 1.25\left(\frac{50}{10}\right)$
   - Ⓓ $c = \frac{50}{10}(2.75) + 1.25$

6. Baseballs are packed one dozen per box. There are 208 baseballs to be packed. How many more baseballs will be needed to fill the last, partially filled box? *(Lesson 1–5)*
   - Ⓐ 0
   - Ⓑ 4
   - Ⓒ 8
   - Ⓓ 12
   - Ⓔ 18

7. Franco is making a casserole. The recipe uses 8 cups of macaroni and serves 12 people. How many cups of macaroni does the recipe use per person? *(Lesson 1–5)*
   - Ⓐ $\frac{2}{3}$
   - Ⓑ 1
   - Ⓒ $\frac{1}{2}$
   - Ⓓ $\frac{1}{3}$

8. Use the commutative and associative properties to compute the product. *(Lesson 1–3)*
$$2 \cdot 4 \cdot 2.5 \cdot 15 \cdot 5 \cdot 10$$
   - Ⓐ 1500
   - Ⓑ 12,000
   - Ⓒ 15,000
   - Ⓓ 120,000

Grid In

9. The daily newspaper always follows a particular format. Each even-numbered page contains 6 articles, and each odd-numbered page contains 7 articles. If today's paper has 36 pages, how many articles does it contain? *(Lesson 1–5)*

Extended Response

10. The average annual snowfall in Denver, Colorado, is 59.8 inches. How many feet of snow can Denver residents expect in the next 4 years? *(Lesson 1–5)*

    Part A   List the operations you use to solve this problem. Calculate the answer to the nearest hundredth of a foot. Show your work.

    Part B   Round your answer to the nearest foot.

www.algconcepts.com/standardized_test

Integers

What You'll Learn

Key Ideas

- Graph, compare, and order integers. *(Lesson 2–1)*
- Graph points on a coordinate plane. *(Lesson 2–2)*
- Add, subtract, multiply, and divide integers. *(Lessons 2–3 to 2–6)*

Key Vocabulary

coordinate plane *(p. 58)*

graph *(p. 53)*

integers *(p. 52)*

ordered pair *(p. 58)*

origin *(p. 58)*

Why It's Important

Meteorology Visitors to the arctic enjoy beautiful views of rugged landscape by day and aurora borealis, or the northern lights, by night. But travelers must be prepared for the extreme weather. Monitoring the windchill factor is one way to ensure safety in cold weather.

Integers are used to represent temperatures and other real-life quantities. You will compare integers and find windchill factors in Lesson 2–1.

✓ Check Your Readiness

Study these lessons to improve your skills.

Lesson 1–4,
pp. 19–23

Simplify each expression.

1. $6x + 4x$
2. $15a + 2a$
3. $11d + 8d$
4. $22t + 9t$
5. $8y - 2y$
6. $14b - 6b$
7. $18r - 12r$
8. $24c - 11c$
9. $32p - 7p$

Lesson 1–2,
pp. 8–13

Find the value of each expression if $a = 2$, $b = 7$, $c = 6$, and $d = 12$.

10. $a + b$
11. $d + b$
12. $d - c$
13. $b - c$
14. $5a$
15. $9c$
16. $c \div a$
17. $\dfrac{d}{c}$
18. $3a + b$
19. $4d - c$
20. $9a + 2c$
21. $\dfrac{3d}{c}$

Lesson 1–3,
pp. 14–18

Simplify each expression.

22. $5(3v)$
23. $4(8q)$
24. $6(6a)$
25. $3(12x)$
26. $7(6r)$
27. $13(2c)$
28. $4a(3b)$
29. $9s(4t)$
30. $8p(8q)$

FOLDABLES™
Study Organizer

Make this Foldable to help you organize your Chapter 2 notes. Begin with four sheets of plain $8\frac{1}{2}$" by 11" paper.

❶ **Stack** the sheets with the edges $\frac{3}{4}$ inch apart.

❷ **Fold** up the bottom edges. All tabs should be the same size.

❸ **Staple** along the fold.

❹ **Label** the tabs as shown.

Reading and Writing As you read and study the chapter, use each page to write notes and examples for each lesson.

---

www.algconcepts.com/chapter_readiness

Chapter 2 Integers **51**

2-1 Graphing Integers on a Number Line

What You'll Learn
You'll learn to graph integers on a number line and to compare and order integers.

Why It's Important
Meteorology
Weather forecasters use integers to determine wind chill. *See Exercise 49.*

There are many ways to represent numbers. One way to represent numbers is with a **number line**. The number line also shows the order of numbers; 2 is to the left of 3, so 2 is smaller than three.

This number line represents the set of whole numbers.

+—•—•—•—•—•—•—•—•—→
   0 1 2 3 4 5 6 7

A number line is drawn by choosing a starting position on a line and marking off equal distances from that point.

The arrowhead indicates that the line and the set of numbers continue indefinitely.

A **negative number** is a number less than zero. To include negative numbers on a number line, extend the line to the left of zero and mark off equal distances. Negative whole numbers are members of the set of **integers**. So, integers can also be represented on a number line.

Reading Algebra

Read −3 as *negative 3*. Read +4 as *positive 4*. Positive integers usually are written without the + sign. So, +4 and 4 are the same number.

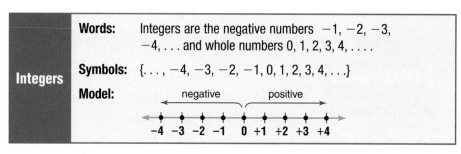

Integers

Words: Integers are the negative numbers −1, −2, −3, −4, . . . and whole numbers 0, 1, 2, 3, 4, ...

Symbols: {. . . , −4, −3, −2, −1, 0, 1, 2, 3, 4, . . .}

Model:

negative   positive

←—•—•—•—•—•—•—•—•—•—→
−4 −3 −2 −1 0 +1 +2 +3 +4

Zero is neither negative nor positive.

Sets of numbers can also be represented by **Venn diagrams**.

natural numbers   1, 2, 3, 4, . . .

whole numbers   0, 1, 2, 3, . . .

integers   . . . −3, −2, −1, 0, 1, 2, 3, . . .

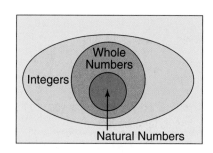

The Venn diagram shows that every natural number is also a whole number. Natural numbers are a *subset* of whole numbers. Similarly, whole numbers are a subset of integers.

To **graph** a set of integers, locate the points named by those numbers on a number line and place a dot on the number line. The number that corresponds to a point is called the **coordinate** of that point.

Examples

1 Name the coordinates of *A*, *B*, and *C*.

The coordinate of *A* is −4, *B* is 2, and *C* is −1.

2 Graph points *X*, *Y*, and *Z* on a number line if *X* has coordinate 4, *Y* has coordinate 0, and *Z* has coordinate −3.

Find each number on a number line. Place a dot on the mark above the number. Then write the letter above the dot.

Your Turn

a. Name the coordinates of *D*, *E*, and *F*.

b. Graph points *M*, *N*, and *P* on a number line if *M* has coordinate −3, *N* has coordinate −4, and *P* has coordinate 1.

The numbers on a number line increase as you move to the right and decrease as you move to the left. When graphing two integers on a number line, the number to the right is always greater.

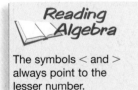

Reading Algebra

The symbols < and > always point to the lesser number.

Words: 3 is greater than −2.   **Words:**   −2 is less than 3.

Symbols: $3 > -2$ **Symbols:** $-2 < 3$

Replace each ● with < or > to make a true sentence.

③ 4 ● −1

4 is to the right of −1 on the number line. So, 4 > −1.

④ −5 ● −3

−5 is to the left of −3 on the number line. So, −5 < −3.

Your Turn

c. −1 ● −2   d. 2 ● −2   e. 0 ● 1

Integers are used to compare numbers in many everyday applications.

Example

Meteorology Link

⑤ **The table shows the average high temperatures for January in selected cities. Order the temperatures from least to greatest.**

Graph each integer on a number line. Use the first letter of each city name to label the points.

City	Temperature (°C)
Boston, MA	2
Chicago, IL	−2
Detroit, MI	−1
Juneau, AK	3
New York, NY	4
St. Louis, MO	−3

Write the integers as they appear on the number line from left to right.

−3°, −2°, −1°, 2°, 3°, 4° are in order from least to greatest.

Online **Personal Tutor at** algconcepts.com

Looking at the graphs of 4 and −4 on a number line, you can see that they are the same number of units from 0. We say that they have the same **absolute value**.

Reading Algebra

The symbol for absolute value is two vertical bars on either side of the number. Read |−4| = 4 as *the absolute value of negative 4 is 4.*

Absolute Value	**Words:**	The absolute value of a number is the distance it is from 0 on the number line.
	Model:	
	Numbers:	|−4| = 4, |4| = 4

Examples

Evaluate each expression.

6 $|-3|$

$|-3| = 3$ *The graph of −3 is 3 units away from 0.*

7 $|-5| - |2|$

$|-5| - |2| = 5 - 2$ *The absolute value of −5 is 5.*

$= 3$ *The absolute value of 2 is 2.*

Your Turn

f. $|9|$ **g.** $|-2| + |-6|$ **h.** $|15| - |-4|$

Check for Understanding

Communicating Mathematics

1. **Describe** a situation in the real world where negative integers are used.

2. **Draw** a number line from −6 to 6. Graph two points whose coordinates have the same absolute value.

3. Tiffany says that 0 is a negative number. Ramon says that 0 is a positive number. Who is correct? Explain.

> **Vocabulary**
>
> number line
> negative number
> integer
> Venn diagram
> natural number
> graph
> coordinate
> absolute value

Guided Practice

 Write an integer for each situation.

Sample: 10 feet below sea level   **Solution:** −10

4. 4 degrees above zero

5. a loss of 6 pounds

6. 3 inches less rain than normal

7. a salary increase of $150

Example 1   **Name the coordinates of each point.**

8. A 9. B

Example 2   **Graph each set of numbers on a number line.**

10. $\{-3, 1, 4\}$ 11. $\{5, 0, -4\}$

Examples 3 & 4

Replace each ● **with < or > to make a true sentence.**

12. -8 ● -5 **13.** -4 ● 2 **14.** 9 ● -7

Examples 6 & 7

Evaluate each expression.

15. $|-8| + |-2|$ **16.** $|-7| - |4|$

Example 5

17. Meteorology The table gives the record low temperatures for each month at the Grand Canyon Airport in Arizona. Order the temperatures from least to greatest.

Month	J	F	M	A	M	J	J	A	S	O	N	D
Temperature (°F)	-22	-17	-7	9	10	26	35	35	22	13	-1	-14

Source: *The Weather Almanac*

Exercises

Practice

Name the coordinate of each point.

18. C **19.** D **20.** E

21. F **22.** G **23.** H

Homework Help

For Exercises	See Examples
18–23	1
24–29, 46, 47	2
30–38	3, 4
39–45	6, 7
48, 49	5

Extra Practice
See page 694.

Graph each set of numbers on a number line.

24. $\{2, 3, 5\}$ **25.** $\{-1, -3, 4\}$ **26.** $\{-2, 4, 0\}$

27. $\{-3, -2, 1\}$ **28.** $\{-2, -1, 0, 1\}$ **29.** $\{-4, -3, -2, -1\}$

Replace each ● **with < or > to make a true sentence.**

30. 4 ● -4 **31.** 0 ● -2 **32.** -2 ● -1

33. 2 ● -3 **34.** -10 ● 1 **35.** -15 ● -10

36. -5 ● $|-5|$ **37.** $|4|$ ● -4 **38.** $|-6|$ ● $|-3|$

Evaluate each expression.

39. $|-6|$ **40.** $|10|$ **41.** $|-5| - |3|$

42. $|-7| + |-2|$ **43.** $|14| - |-5|$ **44.** $|-13| + |-17|$

45. Is $|5| = |-5|$ *sometimes*, *always* or *never* true? Explain.

46. Order $-3, -4, 0, 1, -5,$ and 3 from least to greatest.

47. Order $-25, 78, -36, 14,$ and -14 from greatest to least.

Applications and Problem Solving

48. **Population** In 2002, the population of Texas was 16 million greater than the average of all 50 state populations. The population of Arkansas was 3 million less than the average state population. Write an integer for each situation.

49. **Meteorology** *Windchill factor* is an estimate of the cooling effect the wind has on a person in cold weather.

Windchill Factor (°F)

Wind Speed (mph)	Actual Temperature (F°)						
	30	20	10	0	−10	−20	−30
0	30	20	10	0	−10	−20	−30
5	27	16	6	−5	−15	−26	−36
10	16	4	−9	−21	−33	−46	−58
15	9	−5	−18	−36	−45	−58	−72

Source: BMFA, 1999

a. Find the windchill factor when the actual temperature is 0° with a wind speed of 15 mph.

b. What is the windchill factor when the actual temperature is −20° with a wind speed of 5 mph?

c. Which is less: the windchill factor in part a or part b?

50. **Critical Thinking** Determine whether each statement is *true* or *false*. If *false*, give a counterexample.

a. Every integer is a whole number.

b. Every whole number is an integer.

Mixed Review

History Refer to the table for Exercises 51–53.

Heights of United States Presidents

Height (in.)	63–65	66–68	69–71	72–74	75–77
Number	1	9	14	18	1

51. Make a histogram of the data. *(Lesson 1–7)*

52. Make a cumulative frequency table for the data. *(Lesson 1–6)*

53. How many presidents were at least six feet tall? *(Lesson 1–6)*

Simplify each expression. *(Lesson 1–4)*

54. $5x + 6x$

55. $9a − 3a$

56. $9x − x + 7x$

57. $3m + 2n + 4m$

58. $3r + 2s + 2r + s$

59. $5x + 12y − y + 2x$

Standardized Test Practice
Ⓐ Ⓑ Ⓒ Ⓓ

60. **Multiple Choice** You have two more sisters than brothers. If you have s sisters, which equation could be used to find b, the number of brothers you have? *(Lesson 1–1)*

Ⓐ $s = b − 2$ Ⓑ $s − 2 = b$ Ⓒ $b = s + 2$ Ⓓ $b = 2s$

Why It's Important

Meteorology
Weather forecasters at the National Hurricane Center use a coordinate system to track hurricanes. *See Exercise 36.*

In mathematics, you locate a point on a **coordinate system** that is similar to a grid. The coordinate system is formed by the intersection of two number lines that meet at right angles at their zero points.

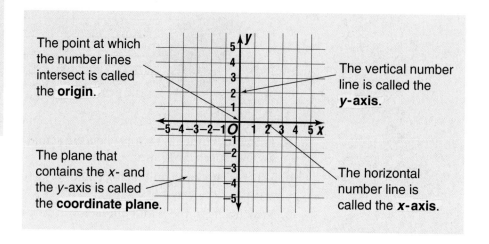

The point at which the number lines intersect is called the **origin**.

The vertical number line is called the **y-axis**.

The plane that contains the x- and the y-axis is called the **coordinate plane**.

The horizontal number line is called the **x-axis**.

The directions *east* and *north* tell you how to locate a point on a map. In mathematics, an **ordered pair** of numbers is used to locate any point on a coordinate plane.

The first number in an ordered pair is called the **x-coordinate**. It corresponds to a number on the x-axis. The second number is called the **y-coordinate**. It corresponds to a number on the y-axis.

(1, 5)

x-coordinate⎯⎯⎯↑ ↑⎯⎯⎯y-coordinate

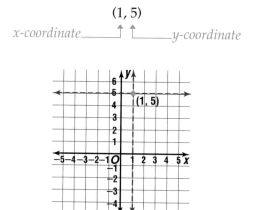

Notice that (1, 5) and (5, 1) are not the same points on the coordinate system.

Write the ordered pair that names each point.

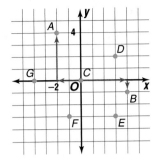

1 *A*

- Start at the origin. Move left on the *x*-axis to find the *x*-coordinate of point *A*. The *x*-coordinate is −2.
- Move up along the grid lines to find the *y*-coordinate. The *y*-coordinate is 4.

The ordered pair for point *A* is (−2, 4).

2 *B*

The *x*-coordinate is 4, and the *y*-coordinate is −1.
The ordered pair for point *B* is (4, −1).

3 *C*

Point *C* is the origin. The ordered pair for the origin is (0, 0).

Your Turn

a. *D*   **b.** *E*   **c.** *F*   **d.** *G*

Online Personal Tutor at algconcepts.com

A point can be named by both a letter and its ordered pair. For example, *P*(2, 3) means point *P* has an *x*-coordinate of 2 and a *y*-coordinate of 3. To graph an ordered pair on a coordinate plane, draw a dot at the point that corresponds to the ordered pair. This is called *plotting* the point.

Example

4 **Graph *P*(2, 3) on a coordinate plane.**

- Start at the origin, *O*.
- The *x*-coordinate is 2. So, move 2 units to the right.
- The *y*-coordinate is 3. Move 3 units up and draw a dot.
- Label the dot with the letter *P*.

You can assume that each unit on the axes represents 1 unit.

Example **5** Graph $Q(-3, 0)$ on a coordinate plane.

- Start at the origin, O.
- The x-coordinate is -3. So, move 3 units to the left.
- The y-coordinate is 0. So the dot is placed on the axis.

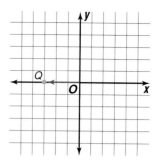

Your Turn

Graph each point on a coordinate plane.

  e. $R(2, -4)$ **f.** $S(-1, 4)$ **g.** $T(0, -3)$

The x-axis and the y-axis separate the coordinate plane into four regions, called **quadrants**. The quadrants are numbered as shown at the right. Note that the axes are not located in any of the quadrants.

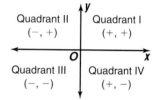

Examples Name the quadrant in which each point is located.

6 $A(5, -4)$

The x-coordinate is positive, and the y coordinate is negative. So, point A is located in Quadrant IV.

7 $B(2, 0)$

Point B lies on the x-axis. It is not located in a quadrant.

Your Turn

  h. $C(-2, -7)$ **i.** $D(-4, 9)$ **j.** $E(0, -3)$

Example **8**

Biology Link

You can use ordered pairs to show how data are related.

Dolphins can swim at 30 mph over long distances. Let x represent the number of hours. Then, $30x$ represents the total distance traveled. Evaluate the expression to find the distances traveled in 1, 2, and 3 hours. Then graph the ordered pairs (time, distance).

The data are graphed in the first quadrant because both values are positive.

Time (hours)	Distance (miles)
x	$30x$
1	30
2	60
3	90

Graphing
Calculator Tutorial
See pp. 750–753.

Graphing Calculator Exploration

You can plot points on a graphing calculator.

Step 1 Press ZOOM 8 ENTER
to display a coordinate grid.

Step 2 Press 2nd [DRAW] ▶
ENTER . Use the arrow keys
to move the cursor to each
desired location. Press
ENTER to plot the point.

Try These

1. Choose four ordered pairs such that the sum of their x- and y-coordinates is 5. Graph them.

2. What do you notice about the graphs of the points?

Check for Understanding

Communicating Mathematics

1. **Explain** how to graph $(-5, 1)$ on a coordinate plane.

2. **Name** an ordered pair whose graph satisfies each condition.
   a. located in Quadrant IV
   b. not located in any quadrant

3. Writing Math Draw a coordinate system and label the origin, x-axis, y-axis, and quadrants.

Vocabulary
coordinate system
y-axis
x-axis
origin
coordinate plane
ordered pair
x-coordinate
y-coordinate
quadrant

Guided Practice

Examples 1–3

Examples 4 & 5

Examples 6 & 7

Example 8

Write the ordered pair that names each point.

4. F 5. G

Graph each point on a coordinate plane.

6. $R(-5, 2)$ 7. $S(0, -2)$

Name the quadrant in which each point is located.

8. $D(-9, 1)$ 9. $E(0, -6)$

10. **Biology** A tortoise is one of the slowest animals on land. It travels at an average speed of only 20 feet per minute.
   a. Find the distance traveled in 2, 4, and 6 minutes.
   b. Graph the ordered pairs (time, distance).

Exercises ● ● ● ● ● ● ● ● ● ● ● ● ● ● ● ●

Practice

Homework Help

For Exercises	See Examples
11–16, 29, 36–37	1–3
17–22, 30, 34	4, 5
23–28, 31–33	6, 7
35	8

Extra Practice
See page 694.

Write the ordered pair that names each point.

11. A **12.** B **13.** C

14. D **15.** E **16.** F

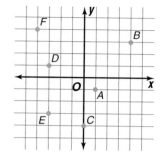

Graph each point on a coordinate plane.

17. $J(-3, 3)$ **18.** $K(4, 0)$ **19.** $L(4, -3)$

20. $M(3, 1)$ **21.** $N(-1, -2)$ **22.** $P(0, 5)$

Name the quadrant in which each point is located.

23. $(-3, -4)$ **24.** $(6, -2)$ **25.** $(0, 4)$

26. $(11, 15)$ **27.** $(-15, 25)$ **28.** $(-18, 0)$

29. What point lies on both the x-axis and the y-axis?

30. Graph three ordered pairs in which the x- and y-coordinates are equal. Describe the graph.

If the graph of $A(x, y)$ satisfies the given conditions, name the quadrant in which point A is located.

31. $x > 0, y > 0$ **32.** $x < 0, y < 0$ **33.** $x > 0, y < 0$

Applications and Problem Solving

34. Geometry   Graph the points $A(-1, 1)$, $B(4, 1)$, $C(4, 0)$, and $D(-1, 0)$ on the same coordinate plane. Connect the points in alphabetical order and then connect A and D. Describe the figure.

35. Entertainment   It costs $3 to rent a video for a day.

    a. Find the total cost of renting 1, 3, and 5 videos for a day.

    b. Graph the ordered pairs (number of videos, cost).

    c. Make a prediction about the location of the graph of (4, 12). Check your prediction by graphing (4, 12).

36. Meteorology   Weather forecasters use a coordinate system composed of latitude (horizontal) and longitude (vertical) lines to locate hurricanes. For example, the position of Hurricane Dennis is 35°N latitude and 74°W longitude, or (35°N, 74°W). Write the position of each hurricane as an ordered pair.

    a. Bret   **b.** Cindy

37. **Geometry** A *vertex* of a triangle is a point where two sides of the triangle meet.

   a. Identify the coordinates of the vertices in the triangle at the right.

   b. Multiply each *x*- and *y*-coordinate of the vertices by 2 and graph the new ordered pairs. Connect the points.

   c. Compare the two figures. Write a sentence that tells how the figures are the same and how they are different.

38. **Critical Thinking** Where are all of the possible locations for the graph of (x, y) if $x = y$?

Mixed Review

39. **Geography** The Caribbean Sea has an average depth of 8685 feet below sea level. Use an integer to express this depth. *(Lesson 2–1)*

40. **Compare and contrast** a histogram and a cumulative frequency histogram. *(Lesson 1–7)*

Name the property shown by each statement. *(Lesson 1–3)*

41. $15 + 4 = 4 + 15$

42. $a(bc) = (ab)c$

43. $3 + 4 = 7, 7$ is a whole number

44. $4 + (5 + 6) = 4 + (6 + 5)$

Standardized Test Practice
Ⓐ Ⓑ Ⓒ Ⓓ

45. **Multiple Choice** Evaluate $8x - 3y$ if $x = 2$ and $y = 3$. *(Lesson 1–2)*

   Ⓐ 7   Ⓑ 25   Ⓒ 39   Ⓓ 57

Quiz 1   Lessons 2–1 and 2–2

Replace each ● with < or > to make a true sentence. *(Lesson 2–1)*

1. $3 ● -2$ 2. $0 ● -5$ 3. $-6 ● -2$

4. Order $-6, -10, 10, 5, -7$, and 0 from least to greatest. *(Lesson 2–1)*

5. Evaluate $|-3| + |-8|$. *(Lesson 2–1)*

Name the quadrant in which each point is located. *(Lesson 2–2)*

6. $A(4, -2)$ 7. $B(-5, -5)$

8. $C(0, -4)$ 9. $D(-8, 6)$

10. **Entertainment** It costs \$4 to buy a student ticket to the movies.

   a. Find the cost of 2, 4, and 5 tickets.

   b. Graph the ordered pairs (number of tickets, cost). *(Lesson 2–2)*

There are several ways to add integers. One way is to use the 1-tiles from a set of algebra tiles.

Find 3 + 2.

Combine 3 positive tiles with 2 positive tiles on a mat.

There are 5 positive tiles on the mat. Therefore, 3 + 2 = 5.

Find −3 + (−2).

Combine 3 negative tiles with 2 negative tiles.

There are 5 negative tiles on the mat. Therefore, −3 + (−2) = −5.

You can also add integers on a number line. Start at 0. Positive integers are represented by arrows pointing *right*. Negative integers are represented by arrows pointing *left*. Start at 0. Move 3 units to the right. From there, move another 2 units to the right.

Start at 0. Move 3 units to the left. From there, move another 2 units to the left.

$$-3 + (-2) = -5$$

These and other similar examples suggest the following rule for adding integers with the same sign.

Adding Integers with the Same Sign	**Words:** To add integers with the same sign, add their absolute values. Give the result the same sign as the integers.
	Numbers: $3 + 2 = 5$, $-3 + (-2) = -5$

Examples

Find each sum.

① $4 + 5$

$4 + 5 = 9$ *Both numbers are positive, so the sum is positive.*

② $-6 + (-2)$

$-6 + (-2) = -8$ *Both numbers are negative, so the sum is negative.*

Your Turn

a. $8 + 9$ **b.** $-2 + (-4)$ **c.** $-5 + (-10)$ **d.** $11 + 6$

What is the result when you add two numbers that differ only in sign, like 3 and -3?

Start at zero. Move 3 units to the right. From there, move 3 units to the left.

$$3 + (-3) = 0$$

You can also use tiles. When one positive tile is paired with one negative tile, the result is a **zero pair**. You can remove zero pairs from the mat because removing zero does not change the value.

The models above show $3 + (-3) = 0$. If the sum of two numbers is 0, the numbers are called **opposites** or **additive inverses**.

-3 is the additive inverse, or opposite, of 3. $3 + (-3) = 0$
7 is the additive inverse, or opposite, of -7. $-7 + 7 = 0$

Additive Inverse Property	**Words:**	The sum of any number and its additive inverse is 0.
	Symbols:	$a + (-a) = 0$
	Numbers:	$3 + (-3) = 0, -7 + 7 = 0$

In the following activity, you'll use tiles to find a rule for adding two integers with different signs.

Materials: algebra tiles integer mat

Find the sum $3 + (-2)$ using 1-tiles.

Step 1 Place 3 positive tiles and 2 negative tiles on the mat.

Step 2 Make as many zero pairs as you can. Remove them from the mat. The remaining tiles represent the sum.

$3 + (-2)$ = 1

Try These

1. Is the sum $3 + (-2)$ positive or negative?
2. Which number, 3 or -2, has the greater absolute value?
3. Use tiles to find the sum $-3 + 2$. Compare the sign of the sum with the sign of the number with the greater absolute value.
4. **Make a conjecture** about the sign of each sum. Verify using tiles.
   a. $4 + (-6)$ **b.** $-7 + 1$ **c.** $8 + (-2)$ **d.** $-5 + 9$

The results of the activity suggest this rule.

Adding Integers with Different Signs	**Words:**	To add integers with different signs, find the difference of their absolute values. Give the result the same sign as the integer with the greater absolute value.
	Numbers:	$3 + (-2) = 1, -3 + 2 = -1$

Examples

Find each sum.

3 $5 + (-3)$
$|5| - |3| = 5 - 3$ or 2
$|5| > |-3|$, so the sum is positive.
Therefore, $5 + (-3) = 2$.

④ $4 + (-6)$

$|-6| - |4| = 6 - 4$ or 2
$|-6| > |4|$, so the sum is negative.
Therefore, $4 + (-6) = -2$.

 Your Turn

  e. $-7 + 5$ f. $6 + (-8)$ g. $-4 + 9$ h. $11 + (-8)$

nline **Personal Tutor at** algconcepts.com

Real World

Example **⑤**

Banking Link

Talisa opened a checking account with a deposit of $25. During the next two weeks, she wrote checks for $20 and $15 and made a deposit of $30. Find the balance in her account.

Explore You know that Talisa made deposits of $25 and $30. She wrote checks for $20 and $15. You want to find the balance in her account.

Plan Deposits are represented by positive integers (+25 and +30). Checks are represented by negative integers (−20 and −15). Write an addition sentence and solve.

Solve Let x represent the balance in her account.
$x = 25 + (-20) + (-15) + 30$
$x = 5 + (-15) + 30$ *$25 + (-20) = 5$*
$x = -10 + 30$ *$5 + (-15) = -10$*
$x = 10$ *$-10 + 30 = 20$*
The balance in Talisa's account is $20.

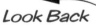
Look Back

Commutative
Property:
Lesson 1–3

Examine Addition of integers is commutative. So, you can check the solution by adding the integers in a different order. One way is to group all of the positive numbers and all of the negative numbers.
$x = 25 + 30 + (-20) + (-15)$
$x = 55 + (-35)$ *$25 + 30 = 55; -20 + (-15) = -35$*
$x = 20$ ✓

You can use the rules for adding integers to simplify expressions.

Example **⑥** **Simplify $5x + (-3x)$.**
$5x + (-3x) = [5 + (-3)]x$ *Use the Distributive Property.*
$\qquad\qquad = 2x$ *$5 + (-3) = 2$*

 Your Turn

Simplify each expression.

  i. $-8y + 3y$ j. $6m + 4m + (-2m)$ k. $-5x + 4x$

Check for Understanding

Communicating Mathematics

1. **Show** how to find the sum of -5 and -3 on a number line.
2. **Explain** why -10 and 10 are additive inverses.
3. **Draw** a diagram that shows how to find the sum of -4 and 6 using tiles.
4. **Writing Math** Write a paragraph that describes how to add two integers. Be sure to include examples with your description.

Vocabulary
zero pair
opposites
additive inverse

Guided Practice

Getting Ready Tell whether each sum is *positive* or *negative*.

Sample 1: $-4 + (-3)$

Solution: Both integers are negative, so the sum is negative.

Sample 2: $-9 + 11$

Solution: $|11| > |-9|$, so the sum is positive.

5. $5 + 12$
6. $12 + (-15)$
7. $-3 + (-7)$
8. $-3 + 9$
9. $-5 + (-2)$
10. $-8 + 12$

Examples 1–4

Find each sum.

11. $7 + 9$
12. $-2 + (-8)$
13. $8 + (-9)$
14. $-12 + 15$
15. $-10 + 5$
16. $11 + (-2)$

Example 6

Simplify each expression.

17. $4x + (-2x)$
18. $-9y + (-2y)$
19. $3a + (-4a) + 3a$

Example 5

20. **Games** On a famous TV game show, contestants earn money for each correct answer and lose money for each incorrect answer. Suppose a contestant answered questions worth $100, $200, and $400 correctly, but answered questions worth $300, $300, and $400 incorrectly. What was the contestant's final score?

Exercises

Practice

Find each sum.

Homework Help	
For Exercises	See Examples
21–26	1, 2
27–38	3, 4
39–44	5
45–47	1–4
48–56	6
Extra Practice	
See page 695.	

21. $3 + 9$
22. $8 + 6$
23. $5 + 16$
24. $-3 + (-10)$
25. $-5 + (-6)$
26. $-11 + (-7)$
27. $-13 + 5$
28. $12 + (-7)$
29. $-6 + 15$
30. $6 + (-6)$
31. $5 + (-18)$
32. $-9 + (-9)$
33. $-15 + 7$
34. $16 + (-11)$
35. $-10 + (-11)$
36. $30 + (-15)$
37. $-20 + (-35)$
38. $-40 + 26$
39. $8 + (-5) + 10$
40. $3 + 15 + (-6)$
41. $-10 + (-4) + (-8)$
42. $15 + 7 + (-7) + (-13)$
43. $-6 + 12 + (-11) + 1$
44. $17 + (-21) + 10 + (-17)$

45. Find the value of y if $y = -3 + 2$.
46. What is the value of w if $-7 + (-2) = w$?
47. Find the value of b if $b = 3 + (-6)$.

Simplify each expression.

48. $-9a + 3a$
49. $-5x + (-10x)$
50. $-16y + 15y$
51. $-11m + 14m$
52. $4z + (-3z)$
53. $8c + (-8c)$
54. $-8b + 4b + (-2b)$
55. $3y + 8y + (-3y)$
56. $-2n + (-4n) + 3n$

Evaluate each expression if $x = -4$, $y = -5$, and $z = 4$.

57. $x + 4 + (-9)$
58. $-7 + y + z$
59. $|x| + y$

Applications and Problem Solving

60. **Sports** In golf, a score of 0 is called *even par*. One over par is represented by $+1$, and one under par is represented by -1. In the 1999 U.S. Open, Tiger Woods had scores represented by -2, $+1$, $+2$, and 0. What was his final score?

61. **Geometry** The points $A(2, 3)$, $B(3, -3)$, and $C(-3, -2)$ are connected with line segments to form a triangle.
    a. Add 2 to each y-coordinate and draw another triangle.
    b. How did the position of the triangle change?

62. **Critical Thinking** Refer to Exercise 61. What change would you make to the ordered pairs so that the triangle would move to the right?

Mixed Review

Name the quadrant in which each point is located. *(Lesson 2–2)*

63. $A(6, -5)$
64. $B(-2, -2)$
65. $C(-5, 3)$
66. $(0, 4)$

Write an integer for each situation. *(Lesson 2–1)*

67. a debt of $5
68. 2 inches more rain than normal
69. a loss of 10 yards
70. a deposit of $17
71. maintaining your present weight

Standardized Test Practice
Ⓐ Ⓑ Ⓒ Ⓓ

72. **Extended Response** One hundred people were surveyed outside a movie theater to determine the favorite leisure-time activity for a large population. Is this a good sample? Explain your reasoning. *(Lesson 1–6)*

73. **Multiple Choice** Use the pattern in the perimeter P of each rectangle to determine the perimeter of a rectangle made up of ten unit squares. *(Lesson 1–5)*

$P = 4$ $P = 6$ $P = 8$

Ⓐ 14   Ⓑ 30   Ⓒ 22   Ⓓ 28

What You'll Learn
You'll learn to subtract integers.

Why It's Important
Budgeting Families often need to find the difference between the amount of money in a budget and the actual amount spent.
See Exercise 48.

When you add or subtract two integers, the sum or the difference is also an integer. Algebra tiles can be used as a model for subtraction of integers. In the examples below, you can see how addition and subtraction of integers are related.

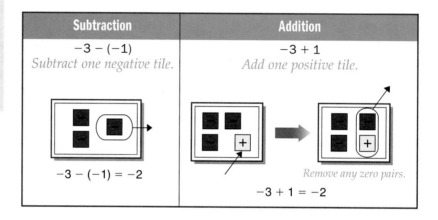

In the examples above, notice that $-3 - (-1) = -3 + 1$.

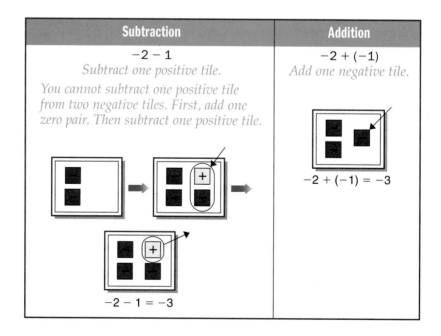

In the examples above, notice that $-2 - 1 = -2 + (-1)$. This example shows that subtracting 1 from -2 is the same as adding -1 to -2.

The examples on the previous page suggest that subtracting an integer is the same as adding the additive inverse or opposite of the integer.

Subtraction	Addition	Subtraction	Addition
additive inverses		*additive inverses*	
$-3 - (-1) = -2$	$-3 + 1 = -2$	$-2 - 1 = -3$	$-2 + (-1) = -3$
same result		*same result*	

COncepts in MOtion
BrainPOP®
algconcepts.com

Subtracting Integers	**Words:**	To subtract an integer, add its additive inverse.
	Model:	$a - b = a + (-b)$
	Numbers:	$9 - 7 = 9 + (-7)$ or 2

Examples

Find each difference.

❶ $6 - 4$
$6 - 4 = 6 + (-4)$ *To subtract 4, add −4.*
$ = 2$ *Simplify.*

❷ $-5 - (-3)$
$-5 - (-3) = -5 + 3$ *To subtract −3, add 3.*
$ = -2$ *Simplify.*

❸ $-3 - 2$
$-3 - 2 = -3 + (-2)$ *To subtract 2, add −2.*
$ = -5$ *Simplify.*

❹ $4 - (-1)$
$4 - (-1) = 4 + 1$ *To subtract −1, add 1.*
$ = 5$ *Simplify.*

❺ $2 - 5$
$2 - 5 = 2 + (-5)$ *To subtract 5, add −5.*
$ = -3$ *Simplify.*

❻ $-3 - (-6)$
$-3 - (-6) = -3 + 6$ *To subtract −6, add 6.*
$ = 3$ *Simplify.*

Your Turn

a. $9 - 3$ **b.** $-7 - (-2)$ **c.** $-8 - 3$
d. $5 - (-1)$ **e.** $-4 - 6$ **f.** $-7 - (-11)$

When you evaluate expressions, it is helpful to write any subtraction expressions as addition expressions first.

Examples

7 **Evaluate $x - y$ if $x = -2$ and $y = 1$.**

$$
\begin{aligned}
x - y &= -2 - 1 && \textit{Replace x with } -2 \textit{ and y with 1.} \\
&= -2 + (-1) && \textit{Write } -2 - 1 \textit{ as } -2 + (-1). \\
&= -3 && -2 + (-1) = -3
\end{aligned}
$$

8 **Evaluate $a - b + c$ if $a = 6$, $b = -2$, and $c = -6$.**

$$
\begin{aligned}
a - b + c &= 6 - (-2) + (-6) && \textit{Replace a with 6, b with } -2, \textit{ and c with } -6. \\
&= 6 + 2 + (-6) && \textit{Write } 6 - (-2) \textit{ as } 6 + 2. \\
&= 8 + (-6) && 6 + 2 = 8 \\
&= 2 && 8 + (-6) = 2
\end{aligned}
$$

Your Turn

g. Evaluate $m - n$ if $m = 5$ and $n = -3$.

h. Evaluate $w - x + y - z$ if $w = -5$, $x = -7$, $y = 10$, and $z = -5$.

Integers are often used to show how data has changed for a given time.

Example

Population Link

9 **The map shows the number of people who moved to Ohio from Indiana in a recent year. It also shows the number of people who left Ohio for Indiana. The change in Ohio's population p can be found by using the formula $p = m - l$, where m is the number of people moving to Ohio and l is the number of people leaving Ohio.**

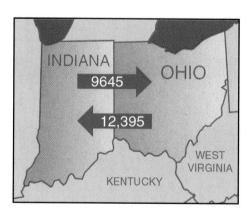

Find the net change in Ohio's population resulting from people moving to and from Indiana.

$$p = m - l$$
$$p = 9645 - 12{,}395 \quad \textit{Replace m with 9645 and l with 12,395.}$$

9645 $\boxed{-}$ 12395 $\boxed{\text{ENTER}}$ -2750

$$p = -2750 \qquad \text{Ohio's population decreased by 2750 people.}$$

Technology Tip

The $\boxed{-}$ key is used for subtraction. The $\boxed{(-)}$ key is used to enter negative integers.

Check for Understanding

Communicating Mathematics

1. **Explain** how additive inverses are used in subtraction.
2. **Draw** a diagram using algebra tiles that shows how $2 - 5$ and $2 + (-5)$ have the same result.

Guided Practice

 Getting Ready Write each expression as an addition expression.

Sample: $4 - (-3)$	**Solution:** $4 + 3$

3. $10 - 3$ 4. $-2 - (-5)$ 5. $-4 - 8$

Examples 1–6

Find each difference.

6. $8 - 2$ 7. $-6 - (-4)$ 8. $-5 - 4$
9. $7 - (-4)$ 10. $8 - 11$ 11. $-4 - (-9)$

Examples 7 & 8

Evaluate each expression if $a = -2$, $b = 6$, $c = -3$, and $d = -1$.

12. $a - b$ 13. $b - c + d$

Example 9

14. **Population** In a recent year, 5899 people moved to Ohio from West Virginia, and 5394 people left Ohio for West Virginia. Find the net change in Ohio's population.

Exercises

Practice

Find each difference.

15. $15 - 2$ 16. $11 - 6$ 17. $14 - 7$
18. $-9 - (-3)$ 19. $-10 - (-2)$ 20. $-15 - (-4)$
21. $-10 - 3$ 22. $-8 - 4$ 23. $-9 - 2$
24. $5 - (-2)$ 25. $5 - (-11)$ 26. $9 - (-8)$
27. $4 - 10$ 28. $9 - 16$ 29. $0 - 9$
30. $-4 - (-10)$ 31. $0 - (-12)$ 32. $-8 - (-14)$

33. Find the value of x if $3 - (-4) = x$.

34. What is the value of y if $y = -3 - (-12)$?

35. Find the value of v if $v = 2 - 19$.

Evaluate each expression if $x = 10$, $y = -7$, $z = -10$, and $w = 12$.

36. $x - y$ 37. $y - z$ 38. $15 - w$
39. $7 - x + y$ 40. $x - z - w$ 41. $x + z - w$

Simplify each expression.

42. $5y - 2y$ 43. $20n - (-5n)$ 44. $4a - 9a + 3a$

45. What is the difference of 25 and -25?

46. Write $a - (-b)$ as an addition expression.

Homework Help

For Exercises	See Examples
15–35, 45–48	1–6
36–44	7, 8

Extra Practice

See page 695.

47. Meteorology The record high temperature in Minneapolis-St. Paul, Minnesota, is 108°F. The record low temperature is 142°F lower. What is the record low temperature?

Expenditures for July

Expenses	Amount Budgeted (dollars)	Amount Spent (dollars)
Food	160	175
Electric	45	44
Telephone	35	41
Heating Fuel	50	15
Water	25	32
Cable TV	25	25

48. Budgeting The table shows the Thomas family's budget and expense summary for food and household utilities for July.

   a. For each item, find the difference between the budgeted amount and the amount spent.

   b. What does a negative difference indicate?

   c. Was the total amount spent for these items more or less than the amount budgeted? by how much?

49. Critical Thinking Determine whether each statement is *true* or *false*. If *false*, give a counterexample.

   a. Subtraction of integers is commutative.

   b. Subtraction of integers is associative.

   c. The set of integers is closed under the operation of subtraction.

Mixed Review

Find each sum. *(Lesson 2–3)*

50. $16 + (-5)$

51. $-12 + (-8)$

52. $9 + (-15)$

53. $-24 + (-3)$

54. $18 + 6$

55. $-12 + 4$

56. Communications A new long-distance plan charges a flat rate of 5¢ per minute. *(Lesson 2–2)*

   a. Find the amount spent for calls of 5, 8, and 10 minutes.

   b. Graph the ordered pairs (time, cost).

Replace each ● with < or > to make a true sentence. *(Lesson 2–1)*

57. $2 ● -3$

58. $-4 ● -8$

59. $-15 ● -14$

Standardized Test Practice
Ⓐ Ⓑ Ⓒ Ⓓ

60. Short Response The table shows the record high temperatures for each state in the United States. Make a histogram of the data. Use 100–104, 105–109, 110–114, 115–119, 120–124, 125–129, and 130–134 as categories for the histogram. *(Lesson 1–7)*

Record High Temperatures (°F)									
112	100	128	120	134	118	106	110	109	112
100	118	117	116	118	121	114	114	105	109
107	112	114	115	118	117	118	125	106	110
122	108	110	121	113	120	119	111	104	111
120	113	120	117	105	110	118	112	114	114

Source: *World Almanac*

 www.algconcepts.com/self_check_quiz

Multiplying integers can be modeled by repeat addition. The multiplication of integers can be represented on a number line.

$$3(-2) = (-2) + (-2) + (-2)$$
$$= -6$$

Therefore, $3(-2) = -6$.

The number line below models the product $2(-3)$.

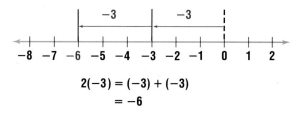

$$2(-3) = (-3) + (-3)$$
$$= -6$$

Look Back

Commutative Property of Multiplication
Lesson 1–3

What happens if the order of the factors is changed to $(-2)3$? The Commutative Property of Multiplication guarantees that $3(-2) = (-2)3$. Therefore, $-2(3) = -6$.

In $3(-2) = -6$ and $-2(3) = -6$, one factor is positive, one factor is negative, and the product is negative. These examples suggest the following rule for multiplying two integers with different signs.

Multiplying Two Integers with Different Signs	**Words:**	The product of two integers with different signs is negative.
	Numbers:	$3(-2) = -6$, $-2(3) = -6$

Examples

Find each product.

1 **$6(-8)$**

$6(-8) = -48$ *The factors have different signs. The product is negative.*

2 **$-5(9)$**

$-5(9) = -45$ *The factors have different signs. The product is negative.*

 Your Turn

a. $10(-3)$ **b.** $-7(7)$ **c.** $15(-3)$

You already know that the product of two positive numbers is positive. What is the sign of the product of two negative numbers? Consider the product $-2(-3)$.

Look Back

Multiplicative Property of Zero: Lesson 1–4

$0 = -2(0)$	*Multiplicative Property of Zero*
$0 = -2[3 + (-3)]$	*Replace 0 with 3 + (−3) or any zero pair.*
$0 = -2(3) + (-2)(-3)$	*Distributive Property*
$0 = \quad -6 \quad + \quad ?$	*−2(3) = −6*

By the Additive Inverse Property, $-6 + 6 = 0$. Therefore, $-2(-3)$ must be equal to 6. This example suggests the following rule for multiplying two integers with the same sign.

COncepts in MOtion

Animation

algconcepts.com

Multiplying Two Integers with the Same Sign		
	Words:	The product of two integers with the same sign is positive.
	Numbers:	$2(3) = 6$, $-2(-3) = 6$

Examples

Find each product.

3 **15(2)**

$15(2) = 30$ *The factors have the same sign. The product is positive.*

4 **−5(−6)**

$-5(-6) = 30$ *The factors have the same sign. The product is positive.*

Your Turn

d. 11(9)   e. −6(−7)   f. −10(−8)

To find the product of three or more numbers, multiply the first two numbers. Then multiply the result by the next number, until you come to the end of the expression.

Examples

Find each product.

5 **8(−10)(−4)**

$8(-10)(-4) = -80(-4)$ *8(−10) = −80*
$\qquad\qquad\quad = 320$ *−80(−4) = 320*

6 **5(−3)(−2)(−2)**

$5(-3)(-2)(-2) = -15(-2)(-2)$ *5(−3) = −15*
$\qquad\qquad\qquad = 30(-2)$ *−15(−2) = 30*
$\qquad\qquad\qquad = -60$ *30(−2) = −60*

Your Turn

g. −2(−3)(4)   h. 6(−2)(3)   i. (−1)(−5)(−2)(−3)

www.algconcepts.com/extra_examples

You can use the rules for multiplying integers to evaluate algebraic expressions and to simplify expressions.

Examples

7 **Evaluate $2xy$ if $x = -4$ and $y = -2$.**

$2xy = 2(-4)(-2)$ *Replace x with -4 and y with -2.*

$\quad\quad = -8(-2)$ *$2(-4) = -8$*

$\quad\quad = 16$ *$-8(-2) = 16$*

8 **Simplify $(2a)(-5b)$.**

$(2a)(-5b) = (2)(a)(-5)(b)$ *$2a = (2)(a); -5b = (-5)(b)$*

$\quad\quad\quad\quad = (2)(-5)(a)(b)$ *Commutative Property*

$\quad\quad\quad\quad = -10ab$ *$(2)(-5) = -10; (a)(b) = ab$*

Your Turn

j. Evaluate $-5n$ if $n = -7$.   **k.** Simplify $12(-3z)$.

Example

Geometry Link

9 **The graphs of $A(3, 5)$, $B(1, 2)$, and $C(5, -1)$ are connected with line segments to form a triangle. Multiply each x-coordinate by -1 and redraw the triangle. Describe how the position of the triangle changed.**

Read A' as A prime. A' corresponds to point A on the original triangle.

$A(3, 5) \rightarrow (3 \times -1, 5) \rightarrow A'(-3, 5)$
$B(1, 2) \rightarrow (1 \times -1, 2) \rightarrow B'(-1, 2)$
$C(5, -1) \rightarrow (5 \times -1, -1) \rightarrow C'(-5, -1)$

Triangle $A'B'C'$ is shown in green. It is the same size and shape as triangle ABC, but it is reflected, or flipped, over the y-axis.

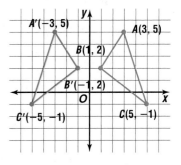

Check for Understanding

Communicating Mathematics

1. **Write** the multiplication sentence represented by the model.

2. **Name** the property that allows you to write $-5(6)$ as $6(-5)$.

Find each product.

3. $2(-6)$ **4.** $-4(9)$ **5.** $10(8)$

6. $-7(-11)$ **7.** $2(-6)(-3)$ **8.** $4(-1)(-5)(-2)$

Example 7

Evaluate each expression if $a = -4$ and $b = -6$.

9. $-7a$ **10.** $-3ab$

Example 8

Simplify each expression.

11. $9(-2x)$ **12.** $(-3m)(-2n)$

Example 9

13. Geometry The graphs of $A(4, 2)$, $B(-3, 4)$, and $C(-1, 1)$ are connected with line segments to form a triangle.

   a. Multiply each y-coordinate by -1 and redraw the triangle.

   b. Describe how the position of the triangle changed.

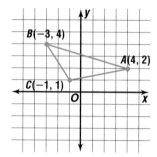

Exercises • • • • • • • • • • • • • • • • •

Practice

Homework Help	
For Exercises	See Examples
14–25, 32–34	1–4
26–31	5, 6
35–40, 44	7
41–43, 45	8

Extra Practice
See page 695.

Find each product.

14. $5(8)$ **15.** $12(-4)$ **16.** $-1(-1)$

17. $9(-1)$ **18.** $-6(5)$ **19.** $3(15)$

20. $5(-15)$ **21.** $13(0)$ **22.** $-8(-9)$

23. $-3(8)$ **24.** $-12(-5)$ **25.** $-13(3)$

26. $3(-2)(4)$ **27.** $-1(-3)(9)$ **28.** $-2(-2)(-2)$

29. $3(4)(-7)$ **30.** $-2(4)(-5)(2)$ **31.** $-1(-1)(1)(-1)$

32. Find the value of a if $a = -3(14)$.

33. What is the value of n if $n = (-11)(-9)$?

34. Find the value of p if $12(-10) = p$.

Evaluate each expression if $x = 2$, $y = -3$, and $z = -5$.

35. $-4x$ **36.** $7xy$ **37.** xyz

38. $2y + z$ **39.** $5x - y$ **40.** $3y + 4z$

Simplify each expression.

41. $4(-2a)$ **42.** $-8(5m)$ **43.** $(-4m)(-8n)$

44. What is the product of -3, -4, and -5?

45. Evaluate $8a - 2b$ if $a = -2$ and $b = 3$.

Applications and Problem Solving

46. Patterns Find the next term in the pattern $-1, 2, -4, 8, \ldots$

47. Oceanography A research submarine descends to the ocean floor at a rate of 100 feet per minute. Write a multiplication equation that tells how far the submarine moves in 5 minutes.

48. Health From 1995 through 1998, deaths from AIDS decreased by an average of about 11,000 per year. If 49,351 people died in 1995, about how many died in 1998?

49. Geometry $A(-5, 0)$, $B(-3, -5)$, and $C(-1, -2)$ are connected with line segments to form a triangle.
   a. Multiply each x- and y-coordinate by -1 and draw another triangle.
   b. Describe how the position of the triangle changed.

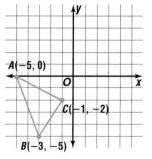

50. Critical Thinking If the product of three integers is negative, what can you conclude about the signs of the integers? Write a rule for determining the sign of the product of three nonzero integers.

Mixed Review

Evaluate each expression if $a = -3$, $b = 7$, $c = -8$, and $d = -15$.
(Lessons 2–3 & 2–4)

51. $a + b$ **52.** $b - (-1)$ **53.** $c - (-3)$

54. $c + d$ **55.** $d + b$ **56.** $d - b$

57. $5 - b$ **58.** $a - b$ **59.** $c + 8$

60. Short Response The melting point of several common elements are shown. Which element has the lowest melting point? *(Lesson 2–1)*

Element	Melting Point (°F)
Helium	−458
Hydrogen	−435
Mercury	−38
Oxygen	−361

Exercise 60

61. Multiple Choice Which verbal expression represents the algebraic expression $5x - 3$? *(Lesson 1–1)*

  Ⓐ three minus five times a number x

  Ⓑ a number x decreased by three

  Ⓒ three less than five times a number x

  Ⓓ five more than a number x minus three

Quiz 2   Lessons 2–3 through 2–5

▶ **Find each sum, difference, or product.** *(Lessons 2–3, 2–4, & 2–5)*

1. $-5 + (-2)$ **2.** $3 - 8$ **3.** $-4(-8)$

4. $6(-9)$ **5.** $9 - (-4)$ **6.** $-10 + 5$

7. $-15(3)$ **8.** $18 + (-2)$ **9.** $-11 - (-2)$

10. Meteorology The temperature between the ground and 11 kilometers above the ground drops about 7°C for each kilometer higher in altitude. Suppose the ground temperature is 0°C. Find the temperature 2 kilometers above the ground. *(Lesson 2–5)*

Chapter 2 Investigation

Bits, Bytes, and BUGS!

Materials

 calculator

Matrices

Manufacturer	Year 1	Year 2
C	3,417,360	5,035,118
P	3,030,398	2,776,144
I	2,196,318	2,738,588
D	1,790,755	2,930,235
A	1,687,161	1,276,249
G	1,666,706	2,219,395

Shipments of Personal Computers

Source: *The Wall Street Journal Almanac*

Investigate

1. A **matrix** is a rectangular arrangement of numbers in rows and columns. Each number in a matrix is called an **element**. A matrix is an **ordered array** because the order of the elements matters. The **dimensions** of a matrix tell how many rows and columns it has. The data about the computers shipped in Year 1 could be organized in a 6 × 1 matrix as shown. Write the data for Year 2 as a matrix.

$$\begin{bmatrix} 3,417,360 \\ 3,030,398 \\ 2,196,318 \\ 1,790,755 \\ 1,687,161 \\ 1,666,706 \end{bmatrix}$$

> **Reading Algebra**
>
> The plural of *matrix* is *matrices*.

2. Two matrices can be added as shown below.

$$\begin{bmatrix} 2 & -3 \\ -1 & 8 \\ 0 & 5 \end{bmatrix} + \begin{bmatrix} -5 & 3 \\ 7 & -7 \\ -10 & 3 \end{bmatrix} = \begin{bmatrix} 2+(-5) & -3+3 \\ -1+7 & 8+(-7) \\ 0+(-10) & 5+3 \end{bmatrix} = \begin{bmatrix} -3 & 0 \\ 6 & 1 \\ -10 & 8 \end{bmatrix}$$

   a. Write your own rule for adding two matrices.

   b. Use matrix addition to find the total number of personal computers shipped by the manufacturers in both years.

3. Two matrices can be subtracted as shown below.

$$\begin{bmatrix} -1 & 0 \\ 4 & -2 \end{bmatrix} - \begin{bmatrix} 1 & 5 \\ -3 & 6 \end{bmatrix} = \begin{bmatrix} -1-1 & 0-5 \\ 4-(-3) & -2-6 \end{bmatrix} = \begin{bmatrix} -2 & -5 \\ 7 & -8 \end{bmatrix}$$

   a. Write your own rule for subtracting two matrices.

   b. Use matrix subtraction to find how many more personal computers were shipped by each manufacturer in Year 2 than in Year 1.

   c. What do negative elements indicate?

4. You can multiply any matrix by a number called a **scalar**. When **scalar multiplication** is performed, each element is multiplied by the scalar, and a new matrix is formed.

$$6\begin{bmatrix} 8 & -2 & 10 \\ -5 & 4 & 6 \end{bmatrix} = \begin{bmatrix} 6(8) & 6(-2) & 6(10) \\ 6(-5) & 6(4) & 6(6) \end{bmatrix} = \begin{bmatrix} 48 & -12 & 60 \\ -30 & 24 & 36 \end{bmatrix}$$

Suppose the computer industry predicted a 20% increase in shipments compared to the number of shipments in Year 2. Use scalar multiplication to find the predicted number of computer shipments. (*Hint:* Multiply the matrix by 1.2 to show an increase of 20%.)

Extending the Investigation

In this extension, you will investigate how matrices are used in the real world. Here are some suggestions.

- The matrices below show the sales and expenses for two different companies for 2003 and 2004. Use the information in the matrices to find a matrix that shows each company's profits in 2003 and 2004. (*Hint:* Profits = Sales − Expenses)

	Sales (million dollars)		Expenses (million dollars)	
	2003	2004	2003	2004
Company A →	4761	6471	4362	5917
Company B →	5061	3483	4904	4838

- Find some data that can be organized using matrices. Then write a problem using the data that can be solved by adding, subtracting, or using scalar multiplication.

Presenting Your Conclusions

Here are some ideas to help you present your conclusions to the class.

- Prepare a poster presenting matrices in a creative manner. Show how you solved problems using matrices.
- Make a booklet of your problems, matrices, and solutions.

*inter*NET
CONNECTION   **Investigation**   For more information on matrix addition, visit: www.algconcepts.com

What You'll Learn

You'll learn to divide integers.

Why It's Important

Farming The average change in populations can be found by dividing integers. *See Exercise 42.*

Dividing two integers can be modeled by separating objects into new groups. In the example below, algebra tiles are used to represent the division of integers.

Find $-6 \div 2$. *The expression $-6 \div 2$ means to separate six negative tiles into 2 groups.*

$$-6 \div 2 = -3$$

Therefore, $-6 \div 2 = -3$. Is a negative integer divided by a positive integer always negative? Recall that division is related to multiplication.

$$-6 \div 2 = -3 \qquad 2 \times (-3) = -6$$

What if a negative integer is divided by a negative integer?

$$-9 \div (-3) = 3 \qquad -3 \times 3 = -9$$

Study the pairs of related sentences in the table below. Look for a pattern in the signs.

Related Sentences	
Multiplication	**Division**
$2 \times (-3) = -6$	$-6 \div 2 = -3$
$-2 \times (-3) = 6$	$6 \div (-2) = -3$
$-2 \times 3 = -6$	$-6 \div (-2) = 3$
$2 \times 3 = 6$	$6 \div 2 = 3$

Signs are different.
  The quotients are negative.
Signs are the same.
  The quotients are positive.

The pattern suggests the following rule for dividing integers.

Dividing Integers		
	Words:	The quotient of two integers with the same sign is positive.
	Numbers:	$6 \div 2 = 3, -6 \div (-2) = 3$
	Words:	The quotient of two integers with different signs is negative.
	Numbers:	$-6 \div 2 = -3, 6 \div (-2) = -3$

Examples

Find each quotient.

1 $-10 \div 2$
The signs are different.
The quotient is negative.
$-10 \div 2 = -5$

2 $-32 \div (-8)$
The signs are the same.
The quotient is positive.
$-32 \div (-8) = 4$

Your Turn

a. $-9 \div 3$ **b.** $-20 \div (-4)$ **c.** $16 \div (-2)$

Recall that fractions are another way of showing division.

Example

3 Evaluate $\dfrac{6x}{y}$ if $x = -4$ and $y = 8$.

$\dfrac{6x}{y} = \dfrac{6(-4)}{8}$ *Replace x with −4 and y with 8.*

$= \dfrac{-24}{8}$ $6(-4) = -24$

$= -3$ $\dfrac{-24}{8}$ *means* $-24 \div 8$.

Your Turn

Evaluate each expression if $x = -3$ and $y = 6$.

d. $-12 \div x$ **e.** $\dfrac{xy}{-2}$ **f.** $\dfrac{36}{3x}$

Example
Media Link

4 The table shows the number of CDs and cassettes that were shipped in 1990 and 2000. What was the average change in the number of cassettes that were shipped for each of those ten years?

Recording Media Shipped (millions)

Medium	1990	2000
CDs	287	942
Cassettes	442	76

Source: *Statistical Abstract of the United States*

First, find the change in the number of cassettes that were shipped.
$76 - 442 = -366$ *There were 366 million fewer cassettes shipped.*

To find the average change, divide -366 by 10.
$-366 \div 10 = -36.6$

The average change in the number of cassettes that were shipped was -36.6 per year. This means that each year there were about 36,600,000 fewer cassettes shipped than the year before.

 Personal Tutor at algconcepts.com

**Communicating
Mathematics**

1. **Write** two division sentences related to the multiplication sentence $-5 \times 2 = -10$.

2. **You Decide?** Joel claims that a positive number divided by a negative number is a positive number. Abbey claims that a negative number divided by a negative number is a negative number. Who is correct? Explain.

Guided Practice

Examples 1 & 2

Find each quotient.

3. $-55 \div 11$
4. $-14 \div (-2)$
5. $15 \div (-3)$
6. $16 \div 4$
7. $-20 \div (-5)$
8. $\frac{-8}{2}$

Example 3

Evaluate each expression if $a = 3$, $b = -12$, and $c = -6$.

9. $-24 \div a$
10. $\frac{ab}{9}$
11. $\frac{6b}{c}$

Example 4

12. **Economy** In July, 1998, about 6,200,000 people were unemployed in the United States. Twelve months later, this figure dropped to 5,900,000. What was the average change in unemployment for each of the last twelve months?

Exercises ● ● ● ● ● ● ● ● ● ● ● ● ● ● ● ● ● ● ●

Practice

Find each quotient.

13. $-12 \div (-12)$
14. $-18 \div 3$
15. $36 \div 6$
16. $-10 \div (-2)$
17. $30 \div (-5)$
18. $15 \div 5$
19. $-25 \div (-5)$
20. $-21 \div 7$
21. $45 \div (-5)$
22. $24 \div (-24)$
23. $-20 \div (-2)$
24. $-72 \div 9$
25. $64 \div (-8)$
26. $-48 \div (-4)$
27. $-40 \div 8$
28. $\frac{-49}{-7}$
29. $\frac{60}{-5}$
30. $\frac{-26}{2}$

For Exercises	See Examples
13–33, 40, 41	1, 2
34–39, 45	3
42–44	4

Homework Help

Extra Practice
See page 696.

31. Find the value of a if $-42 \div 7 = a$.
32. What is the value of m if $m = -81 \div (-9)$?
33. Find the value of w if $w = 85 \div (-17)$.

Evaluate each expression if $x = 5$, $y = -6$, $z = 2$, and $w = -3$.

34. $18 \div y$
35. $y \div z$
36. $\frac{x}{5}$
37. $\frac{-4w}{2}$
38. $\frac{x - z}{w}$
39. $\frac{y - 8}{z}$

40. What is the quotient of -42 and -7?
41. Divide 100 by -50.

42. **Animals** Experts estimate that there were about 100,000 tigers living 100 years ago. Today, there are only about 6000. What was the average change in tiger population for each of the last 100 years?

43. **Farming** The table shows the number of thousands of farms in the U.S. according to their size.

    a. Find the average yearly change in the number of farms that are between 50 and 179 acres in size.

    b. For farms of which size(s) is the average change a positive number?

U.S. Farms (thousands)

Acres	1997	2002
1–9	205	197
10–49	531	563
50–179	694	659
180–499	428	389
500–999	179	162
1000–1999	103	99
2000 or more	74	78

Source: Census of Agriculture

44. **Media** Refer to the table on page 83.

    a. What was the average change in the number of CDs that were shipped for each of the seven years from 1990 to 1997?

    b. If this trend continues, estimate the number of CDs that will be shipped in 2005.

45. **Energy** A measure called *degree days* is used to estimate the energy needed for heating on cold days. The formula $d = 65 - \frac{h + l}{2}$ can be used to find degree days. In the formula, d represents degree days, h represents the high temperature of a given day, and l represents the low temperature of that day. Find the degree days for a day in which the high temperature was $-2°F$ and the low temperature was $-16°F$.

46. **Critical Thinking** Explain why division by zero is not possible.

Find each sum, difference, or product. *(Lessons 2–3, 2–4, & 2–5)*

47. $9(-6)$
48. $-11 + (-4)$
49. $9 - (-7)$
50. $15 + (-25)$
51. $-10(-8)$
52. $8 - 10$
53. $-7 - (-5)$
54. $-8(9)$
55. $-16 + 20$

56. **Short Response** A competition swimming pool is 75 feet long and 72 feet wide. It is filled to a depth of 6 feet. Use the formula $V = \ell wh$, where ℓ is the length, w is the width, and h is the depth, to find the volume V in cubic feet of water in the pool. *(Lesson 1–5)*

57. **Multiple Choice** Which property of real numbers allows you to conclude that if $2t + 4t = 36$, then $4t + 2t = 36$? *(Lesson 1–3)*
    - (A) Distributive Property
    - (B) Commutative Property
    - (C) Associative Property
    - (D) Additive Inverse Property

Understanding and Using the Vocabulary

After completing this chapter, you should be able to define each term, property, or phrase and give an example or two of each.

inter NET CONNECTION Review Activities
For more review activities, visit:
www.algconcepts.com

absolute value (p. 54)
additive inverse (p. 65)
coordinate (p. 53)
coordinate plane (p. 58)
coordinate system (p. 58)
dimensions (p. 80)
element (p. 80)
graph (p. 53)
integers (p. 52)

matrix (p. 80)
natural numbers (p. 52)
negative numbers (p. 52)
number line (p. 52)
opposites (p. 65)
ordered array (p. 80)
ordered pair (p. 58)
origin (p. 58)
quadrants (p. 60)

scalar (p. 81)
scalar multiplication (p. 81)
Venn diagrams (p. 52)
x-axis (p. 58)
x-coordinate (p. 58)
y-axis (p. 58)
y-coordinate (p. 58)
zero pair (p. 65)

Complete each sentence using a term from the vocabulary list.

1. On a number line, the numbers to the left of zero are ___?___.
2. The ___?___ is the plane that contains the x-axis and the y-axis.
3. If the sum of two numbers is 0, the numbers are called ___?___.
4. The ___?___ is the first number in an ordered pair.
5. The distance a number is from 0 on the number line is the ___?___.
6. The numbers 1, 2, 3, 4, . . . are ___?___.
7. Whole numbers are a subset of ___?___.
8. The ___?___ of a point is the number corresponding to that point on a number line.
9. A(n) ___?___ is used to locate any point on a coordinate plane.
10. The x-axis and y-axis separate the coordinate plane into four regions called ___?___.

Skills and Concepts

Objectives and Examples

- **Lesson 2–1** Graph integers on a number line and compare and order integers.

  Replace the ● with $<$ or $>$ to make a true sentence.

  $$7 ● -3$$

  7 is to the right of -3 on the number line, so $7 > -3$.

Review Exercises

Replace each ● with $<$ or $>$ to make a true sentence.

11. $0 ● -5$
12. $-3 ● 3$
13. $-9 ● -7$
14. $|-12| ● -12$
15. Order $-4, 7, 4, -2, -3$, and 0 from least to greatest.
16. Order $-15, -23, -18$, and -20 from greatest to least.

 www.algconcepts.com/vocabulary_review

Objectives and Examples

Review Exercises

• **Lesson 2–2** Graph points on a coordinate plane.

Write the ordered pair that names each point and name the quadrant in which each point is located.

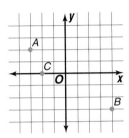

$A(-3, 2)$, II
$B(4, -3)$, IV
$C(-2, 0)$, none

Write the ordered pair that names each point.

17. P

18. Q

19. N

20. M

Name the quadrant in which each point is located.

21. $(6, 10)$

22. $(-4, 8)$

23. $(0, -12)$

24. $(13, -7)$

• **Lesson 2–3** Add integers.

Find $-2 + (-3)$.
Both numbers are negative, so the sum is negative.
$-2 + (-3) = -5$

Find $4 + (-12)$.
$|-12| > |4|$, so the sum is negative.
$4 + (-12) = -8$

Find each sum.

25. $8 + (-14)$

26. $-7 + 5$

27. $-8 + (-2)$

28. $-8 + 8$

29. $23 + (-18)$

30. $-14 + (-12)$

31. $-10 + 3 + (-6) + 8$

32. $7 + (-5) + (-7) + 15$

Simplify each expression.

33. $7x + (-5x)$

34. $-4y + (-y)$

35. $14m + (-10m)$

36. $-31x + 27x$

• **Lesson 2–4** Subtract integers.

Find $7 - (-3)$.
$7 - (-3) = 7 + 3$ *To subtract -3, add 3.*
$\qquad = 10$

Find $-4 - 8$.
$-4 - 8 = -4 + (-8)$ *To subtract 8, add -8.*
$\qquad = -12$

Find each difference.

37. $6 - 14$

38. $-11 - (-5)$

39. $4 - (-5)$

40. $-3 - 5$

41. $-6 - (-2)$

42. $10 - (-10)$

Evaluate each expression if $x = 3$, $y = -5$, and $z = -1$.

43. $2 - x$

44. $y - z$

45. $x + y - z$

46. $x - y + z$

Mixed Problem Solving
See pages 724–731.

Objectives and Examples

Review Exercises

• **Lesson 2–5** Multiply integers.

$-6(-4) = 24$ *The integers have the same sign, so the product is positive.*

$3(-5) = -15$ *The integers have different signs, so the product is negative.*

Find each product.

47. $-7(-5)$ 48. $8(-4)$

49. $-3(-2)(6)$ 50. $-1(-4)(-5)$

Evaluate each expression if $a = -3$ and $b = -6$.

51. $-9a$ 52. $-7ab$

Simplify each expression.

53. $-7(6m)$ 54. $(-3x)(-15y)$

• **Lesson 2–6** Divide integers.

$-9 \div (-3) = 3$ *The integers have the same sign, so the quotient is positive.*

$\frac{8}{-2} = -4$ *The integers have different signs, so the quotient is negative.*

Find each quotient.

55. $42 \div (-6)$ 56. $-63 \div -7$

57. $-24 \div (4)$ 58. $-40 \div (-5)$

Evaluate each expression if $a = -4$, $b = -2$, and $c = 3$.

59. $\frac{6a}{c}$ 60. $\frac{a + b}{2c}$

Applications and Problem Solving

61. **Banking** Mikaela opened a checking account by depositing $250. She later wrote a check for $25 for the phone bill and $32 for a magazine subscription. Then Mikaela received $20 for her birthday and deposited it into her account. What was her balance after her birthday deposit? *(Lesson 2–3)*

63. **Scuba Diving** A scuba diver descends at a rate of 40 feet per minute. Write a multiplication equation that tells how far the scuba diver moves in 2 minutes. *(Lesson 2–5)*

62. **Geometry** The graphs of $M(5, 4)$, $N(-4, 3)$, and $P(0, 1)$ are connected with line segments to form a triangle. *(Lesson 2–5)*

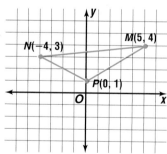

a. Multiply each y-coordinate by -1 and draw another triangle.

b. Describe how the position of the triangle changed.

1. **Write** two division sentences related to the multiplication sentence $6 \times (-7) = -42$.
2. **Graph** $\{4, -2, 1\}$ on a number line.

Replace each ● with < or > to make a true sentence.

3. $-5 \ ● \ -8$

4. $9 \ ● \ -2$

Graph each point on a coordinate plane and name the quadrant in which each point is located.

5. $X(4, 3)$

6. $M(-3, 2)$

7. $A(0, -4)$

Find each sum, difference, product, or quotient.

8. $-16 + 9$

9. $5 - (-2)$

10. $-3 - 5$

11. $4(-6)$

12. $-14 \div (-7)$

13. $-8(-7)$

14. $\dfrac{-32}{8}$

15. $\dfrac{-25}{-5}$

16. $-8 - 2$

17. $-7 + 5 + (-12)$

18. $8 + (-14) + (-6)$

19. $5(-2)(3)$

Evaluate each expression if $m = 5$, $n = -8$, and $p = -3$.

20. $n - p$

21. $m + n$

22. $2(n)$

23. $p - m + n$

24. $\dfrac{n}{-2}$

25. $\dfrac{m + n}{p}$

Simplify each expression.

26. $3x - 8x$

27. $10y - (-3y)$

28. $9(-4x)$

29. $-2(-5y)$

30. $(-7m)(3n)$

31. $3x + 4y + x - 2y$

32. **Weather** The table shows record high and low temperatures for six cities for the month of November.
    a. Find the differences in temperatures for each city.
    b. Which city had the greatest difference in its record temperatures?

33. **Sports** The Tigers football team had a gain of 7 yards on their first run. They lost 3 yards on their second run and gained 12 yards on their third run. What was the total gain or loss of yardage in the three runs?

Record Temperatures in November (°F)		
City	**High**	**Low**
Atlanta, GA	84	3
Boston, MA	78	15
Columbus, OH	80	5
Duluth, MN	70	−23
Juneau, AK	56	−5
Kahului, HI	93	55

Exercise 32

Data Analysis Problems

You will need to create and interpret frequency tables as well as data graphs. This includes bar graphs, histograms, line graphs, and stem-and-leaf plots.

Test-Taking Tip
Read the graph or table before you read the question.

Example 1

Use the information on movie-making costs in the table. Make two line graphs on one grid, one for average production costs and the other for average marketing costs. Title the graph, label the axes, use appropriate scales, and accurately graph the data.

Movie-Making Costs ($ millions)

Year	Average Production Costs	Average Marketing Costs
1980	9.4	4.3
1985	16.8	6.5
1990	26.8	12.0
1995	36.3	17.7

Hint In open-ended questions, you may need to construct a graph, draw a diagram, or explain your answer.

Solution The x-axis shows the years. Decide on a scale for the y-axis, which represents costs. Since the lowest cost is 4.3 and the highest is 36.6, use a scale of $0 - 40$ with intervals of 5. Mark each point (year, cost). Connect the points with line segments.

Example 2

The graph below represents the amount of money each person earns per day. How many days must Andy work to earn as much as Jill would earn in four days?

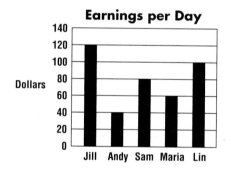

Earnings per Day

Ⓐ 480 Ⓑ 120 Ⓒ 80 Ⓓ 12 Ⓔ 3

Hint Look carefully at the graph. Read the title and labels. The range of the y-axis scale is 140.

Solution Calculate the amount that Jill earns in four days. The graph shows that she earns $120 per day. In four days, she will earn $4 \times \$120$ or $480.

Now calculate how many days it will take Andy to earn the amount of $480. The graph shows that Andy earns $40 per day.

Divide 480 by 40.

$$\frac{\$480}{\$40 \text{ per day}} = 12 \text{ days}$$

Andy needs to work 12 days to earn $480. So, the answer is D.

After you work each problem, record your answer on the answer sheet provided or on a sheet of paper.

Multiple Choice

1. One winter night the temperature dropped 3° every hour. If the temperature was 0° at midnight, what was the temperature at 4:00 A.M.? *(Lesson 2–5)*
   Ⓐ −12° Ⓑ −15° Ⓒ 12° Ⓓ 32°

2. Which of the following numbers, when subtracted from −8, gives a result greater than −8? *(Lesson 2–4)*
   Ⓐ −2 Ⓑ 0 Ⓒ 2 Ⓓ 3

3. How many even integers are there between −4 and 4? *(Lesson 2–1)*
   Ⓐ 2 Ⓑ 3 Ⓒ 4 Ⓓ 6 Ⓔ 8

4. If hot dogs are sold in packs of 10 and buns are sold in packs of 12, what is the smallest number of each you can buy to have no extra hot dogs or buns? *(Basic Skill)*
   Ⓐ 30 Ⓑ 60 Ⓒ 90 Ⓓ 120

5. Find the greatest number of fat grams of any hamburger shown. *(Lesson 1–7)*

Hamburgers	Stem	Chicken
	0	9
9 8 0	1	0 1 2 5 5 7 9
7 2	2	
5 2 0	3	*1 | 2 = 12 g*

   Ⓐ 53 Ⓑ 91 Ⓒ 35 Ⓓ 21

6. When was income closest to $20,000? *(Lesson 1–7)*

   Texas's per Capita Personal Income

   Ⓐ 2000 Ⓑ 1995 Ⓒ 1990 Ⓓ 1980

7. The frequency table shows how many books each student read over the summer. Which statement is correct? *(Lesson 1–6)*

Number of Books	Tally	Frequency
0	III	3
1	ЖII	7
2	ЖIIII	9
3	III	3
4	I	1

   Ⓐ The students read 4 different books.
   Ⓑ The most students read only 2 books.
   Ⓒ Two students read 9 books.
   Ⓓ There are a total of 21 students.

8. Which has the greatest value? *(Lesson 2–5)*
   Ⓐ −38 × (−10) Ⓑ −38 × 10
   Ⓒ 38 × (−10) Ⓓ 1 × 38

Short Response

9. Evaluate $\dfrac{-2 \times 4}{36 \div 2 - 5 \times 2}$. *(Lesson 1–2)*

Extended Response

10. The table below shows the winners of the first 17 World Cup soccer competitions. *(Lesson 1–6)*

Winners of the First 17 World Cup Soccer Competitions

Year	Champion	Year	Champion
1930	Uruguay	1974	West Germany
1934	Italy	1978	Argentina
1938	Italy	1982	Italy
1950	Uruguay	1986	Argentina
1954	West Germany	1990	West Germany
1958	Brazil	1994	Brazil
1962	Brazil	1998	France
1966	England	2002	Brazil
1970	Brazil		

Part A Construct a frequency table of the World Cup soccer champions.

Part B Use the table to make a bar graph.

CHAPTER 3

Addition and Subtraction Equations

What You'll Learn

Key Ideas

- Compare, order, add, and subtract rational numbers. *(Lessons 3–1 and 3–2)*

- Find the mean, median, mode, and range of a set of data. *(Lesson 3–3)*

- Solve addition and subtraction equations. *(Lessons 3–4 to 3–6)*

- Solve absolute value equations. *(Lesson 3–7)*

Key Vocabulary

equivalent equations *(p. 122)*

inequality *(p. 95)*

rational numbers *(p. 94)*

solution *(p. 112)*

Why It's Important

Lighthouses The Outer Banks of North Carolina are home to two of the ten United States National Seashores — Cape Hatteras and Cape Lookout. Tourists enjoy swimming, bird watching, and other activities on their beaches, sand dunes, marshes, and woodlands.

Addition and subtraction equations are used to solve real-life problems like comparing the heights of lighthouses. You will use an equation to compare the heights of the Cape Hatteras and Cape Lookout lighthouses in Lesson 3–6.

✓ Check Your Readiness

Study these lessons to improve your skills.

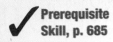
Prerequisite Skill, p. 685

Write each fraction in simplest form.

1. $\frac{4}{12}$

2. $\frac{3}{18}$

3. $\frac{5}{40}$

4. $\frac{6}{15}$

5. $\frac{10}{25}$

6. $\frac{16}{28}$

Prerequisite Skill, p. 684

Find each sum or difference.

7. $3.5 + 2.1$

8. $6.6 + 9.8$

9. $12.1 + 11.6$

10. $11.8 - 3.7$

11. $45.8 - 12.7$

12. $23.1 - 15.9$

13. $1.5 + 2.1 + 3.2 + 1.8 + 2.6 + 3.1 + 2.9$

14. $8.9 + 11.1 + 10.7 + 9.8 + 11.0 + 9.3 + 8.7$

Prerequisite Skill, p. 684

Find each quotient.

15. $15.6 \div 3$

16. $18.4 \div 4$

17. $48.6 \div 12$

18. $54.9 \div 6$

19. $188 \div 5$

20. $225 \div 18$

Prerequisite Skill, p. 687

Find each sum or difference.

21. $\frac{3}{8} + \frac{1}{8}$

22. $\frac{1}{6} + \frac{2}{3}$

23. $\frac{7}{10} + \frac{3}{5}$

24. $\frac{5}{12} - \frac{1}{12}$

25. $\frac{5}{6} - \frac{2}{3}$

26. $\frac{7}{8} - \frac{3}{4}$

FOLDABLES™ Study Organizer

Make this Foldable to help you organize your Chapter 3 notes. Begin with a sheet of 11" by 17" paper.

❶ **Fold** the short sides to meet in the middle.

❷ **Fold** the top to the bottom.

❸ **Open.** Cut along the second fold to make four tabs.

❹ **Label** the tabs as shown.

Reading and Writing As you read and study the chapter, write notes and examples under each tab.

Why It's Important

Shopping Knowing how to compare rational numbers will help you determine better buys.
See Example 6.

Rational numbers are frequently used in daily life. They consist of all positive and negative fractions. All integers are also rational numbers because integers can be expressed as fractions with 1 in the denominator.

For example, 2 equals $\frac{2}{1}$ and -3 equals $-\frac{3}{1}$.

Rational Number	Words:	A rational number is any number that can be expressed as a fraction where the numerator and denominator are integers and the denominator is not zero.
	Symbols:	$\frac{a}{b}$, where a and b are integers and $b \neq 0$

Some examples of rational numbers and their form as a fraction are listed in the table below.

Rational Number	-2	$3\frac{1}{4}$	0.625	0	$-0.66\overline{6}$
Form $\frac{a}{b}$	$-\frac{2}{1}$	$\frac{13}{4}$	$\frac{5}{8}$	$\frac{0}{1}$	$-\frac{2}{3}$

Rational numbers can be graphed on a number line in the same manner as integers.

Graphing rational numbers on a number line helps you to compare them. Just as with integers, the numbers on a number line increase in value as you move to the right and decrease in value as you move to the left.

The following statements can be made about the number line above.

Words	Symbols
The graph of -1 is to the left of the graph of $-\frac{1}{2}$. -1 is less than $-\frac{1}{2}$.	$-1 < -\frac{1}{2}$
The graph of 1 is to the right of -0.25. 1 is greater than -0.25.	$1 > -0.25$
The graph of $\frac{1}{4}$ is to the left of 0.5. $\frac{1}{4}$ is less than 0.5.	$\frac{1}{4} < 0.5$

A mathematical sentence that uses $<$ and $>$ to compare two expressions is called an **inequality**. When you compare two numbers, the following property applies.

Comparison Property	**Symbols:** For any two numbers a and b, exactly one of the following sentences is true.
	$\quad\quad a < b \quad\quad\quad a > b \quad\quad\quad a = b$
	Numbers: If $a = 2$ and $b = 3$, then only $2 < 3$ is true.

Examples

Replace each ● with $<$, $>$, or $=$ to make a true sentence.

① 1.5 ● -8

Since any positive number is greater than any negative number, $1.5 > -8$.

② $-14 + 8$ ● $2(-3)(5)$

$-14 + 8$ ● $2(-3)(5)$
$\quad\quad -6$ ● -30 *Find the value of each side.*

-6 is to the right of -30 on a number line and $-6 > -30$.
So, $-14 + 8 > 2(-3)(5)$.

③ $-\dfrac{4}{5}$ ● $\dfrac{2}{3}$

Since any negative number is less than any positive number, $-\dfrac{4}{5} < \dfrac{2}{3}$.

Your Turn

a. -1 ● -14.2 b. $-5(0)$ ● $16 + (-16)$ c. $-\dfrac{3}{8}$ ● $\dfrac{5}{8}$

To compare two fractions with different denominators, you can use **cross products**. Cross products are the products of the diagonal terms of two fractions.

$$\dfrac{2}{7} \ \bullet \ \dfrac{3}{8}$$

$8(2)$ ● $7(3)$ *Find the cross products.*

$16 < 21$

If $16 < 21$, then $\dfrac{2}{7} < \dfrac{3}{8}$.

Lesson 3–1 Rational Numbers   **95**

This example illustrates the following property.

Comparison Property for Rational Numbers	**Symbols:** For any rational numbers $\frac{a}{b}$ and $\frac{c}{d}$, with $b > 0$ and $d > 0$: **1.** If $\frac{a}{b} < \frac{c}{d}$, then $ad < bc$. **2.** If $ad < bc$, then $\frac{a}{b} < \frac{c}{d}$. **Numbers:** **1.** If $\frac{2}{3} < \frac{3}{4}$, then $8 < 9$. **2.** If $8 < 9$, then $\frac{2}{3} < \frac{3}{4}$.

This property also holds true if $<$ is replaced by $>$ or $=$.

Examples

Replace each ● with $<$, $>$, or $=$ to make a true sentence.

4
$$\frac{3}{8} \bullet \frac{4}{10}$$

$$\frac{3}{8} \bullet \frac{4}{10}$$
$$10(3) \bullet 8(4) \qquad \textit{Find the}$$
$$30 < 32 \qquad \textit{cross products.}$$
So, $\frac{3}{8} < \frac{4}{10}$.

5
$$-\frac{5}{6} \bullet -\frac{3}{4}$$
$$-\frac{5}{6} \bullet -\frac{3}{4}$$
$$4(-5) \bullet 6(-3)$$
$$-20 < -18$$
So, $-\frac{5}{6} < -\frac{3}{4}$.

Technology Tip

You can use a calculator to express rational numbers as decimals so they can be easily compared.

Your Turn

d. $\frac{4}{5} \bullet \frac{7}{8}$

e. $-\frac{5}{10} \bullet -\frac{3}{5}$

Another way to compare rational numbers is to express them as terminating or repeating decimals.

Example

6 Write $\frac{3}{8}$, $\frac{1}{3}$, and $\frac{2}{5}$ in order from least to greatest.

$$\frac{3}{8} = 0.375 \qquad \textit{This is a terminating decimal.}$$
$$\frac{1}{3} = 0.3333 \ldots \text{ or } 0.\overline{3} \qquad \textit{This is a repeating decimal.}$$
$$\frac{2}{5} = 0.4 \qquad \textit{This is a terminating decimal.}$$

In order from least to greatest, the decimals are $0.\overline{3}$, 0.375, 0.4. So, the fractions in order from least to greatest are $\frac{1}{3}$, $\frac{3}{8}$, $\frac{2}{5}$.

Reading Algebra

A bar over a number indicates that it repeats indefinitely. The number $0.\overline{3}$ is read as *zero point three repeating*.

Your Turn **Write the numbers in each set from least to greatest.**

f. $\frac{4}{8}$, $\frac{9}{10}$, $\frac{2}{3}$

g. $-\frac{1}{3}$, $-\frac{5}{8}$, $-\frac{1}{6}$

You can use the cost per unit, or **unit cost**, to compare the costs of similar items. Many people use unit cost to comparison shop. The least unit cost is the best buy.

$$\text{unit cost} = \text{total cost} \div \text{number of units}$$

Example — **7**

Shopping Link

Rolando needs to buy colored pencils. The cost of a package of 12 pencils is $6.39. A package of 24 pencils costs $12.89. Which is the better buy? Explain.

Find the unit cost of each package. In each case, the unit cost is expressed in cents per pencil.

unit cost of package of 12:
$6.39 \div 12 = 0.5325$ or about $0.53 per pencil

unit cost of package of 24:
$12.89 \div 24 \approx 0.5371$ or about $0.54 per pencil

Since $0.53 < $0.54, the package of 12 pencils is the better buy.

Smooth Bright Colors!

Check for Understanding

Communicating Mathematics

1. **Write** three examples of rational numbers. Include one in decimal form and one negative number.

2. **Give an example** of two different fractions whose cross products are equal.

3. **Writing Math** Create a list of at least five commonly used fraction-decimal equivalents such as $\frac{1}{2} = 0.5$. Explain why memorizing them is useful.

Vocabulary

rational number
inequality
cross products
unit cost

Guided Practice

Getting Ready Express each fraction as a decimal.

Sample 1: $\frac{1}{4}$

Solution: $1 \div 4 = 0.25$

Sample 2: $\frac{2}{3}$

Solution: $2 \div 3 = 0.\overline{6}$

4. $\frac{8}{10}$ 5. $\frac{3}{4}$ 6. $\frac{2}{6}$ 7. $\frac{5}{8}$ 8. $\frac{7}{9}$

9. Graph the set of numbers $\left\{-3, 1.5, \frac{1}{4}, 0, -2.\overline{6}\right\}$ on a number line. Use it to explain why $-3 < -2.\overline{6}$.

Examples 1–4

Replace each ● with <, >, or = to make a true sentence.

10. 0 ● -4.3

11. $15 - 24$ ● $-3(2)(-6)$

12. $\frac{1}{2}$ ● $\frac{1}{10}$

13. $\frac{2}{8}$ ● $\frac{1}{4}$

Example 5

Write the numbers in each set from least to greatest.

14. $-\frac{2}{3}, -\frac{7}{8}, -\frac{3}{5}$

15. $\frac{4}{5}, \frac{7}{10}, \frac{6}{8}$

Example 6

16. **Shopping** Thompson's Market sells a 25-pound bag of Happy Chow dog food for $15.49 and a 30-pound bag of Super Chow dog food for $17.39. Which brand is the better buy? Explain.

Exercises

Practice

Replace each ● with <, >, or = to make a true sentence.

17. -12 ● -2

18. 8.2 ● -7

19. -0.88 ● -0.86

20. -3 ● $3(0)$

21. $-5 - 3$ ● 9

22. $-8 + 3$ ● $7(2)(-3)$

23. $\frac{1}{8}$ ● 0.124

24. $\frac{4}{5}$ ● 0.8

25. $-\frac{3}{8}$ ● $\frac{3}{4}$

26. $-\frac{1}{5}$ ● $-\frac{2}{10}$

27. $\frac{4}{6}$ ● $\frac{2}{5}$

28. $\frac{5}{8}$ ● $\frac{5}{6}$

29. Compare the numbers -2.002 and -2.02 using an inequality.

30. Write an inequality that compares $\frac{5}{8}$ and $\frac{2}{3}$.

31. Using a number line, explain why $-\frac{8}{10} > -\frac{5}{6}$.

Write the numbers in each set from least to greatest.

32. $0.2, -\frac{5}{6}, \frac{4}{5}$

33. $\frac{1}{2}, \frac{5}{8}, 0.6$

34. $-\frac{3}{5}, -\frac{8}{10}, \frac{4}{6}$

35. $-\frac{1}{4}, -\frac{2}{6}, -\frac{1}{8}$

36. $\frac{3}{10}, -\frac{2}{8}, -\frac{1}{3}$

37. $\frac{3}{5}, 0.\overline{6}, \frac{3}{8}$

Homework Help

For Exercises	See Examples
17–19, 29	1
20–22	2
23–28, 30	3, 4, 5
32–37, 42, 43	6
38–41	7

Extra Practice
See page 696.

Which is the better buy? Explain.

38. a 16-ounce bottle of juice for $0.89 or a 20-ounce bottle for $1.09

39. a dozen eggs for $1.59 or 18 eggs for $2.49

40. a 25-pack of computer diskettes for $9.95 or a 30-pack for $12.97

41. an $8\frac{1}{3}$-yd by 12-in. roll of aluminum foil for $1.29, a $16\frac{2}{3}$-yd by 12-in. roll for $2.89, or a 25-yd by 12-in. roll for $2.99

Applications and Problem Solving

42. Tools Julie is using a socket wrench to tighten a bolt on her car. She needs to use the next size smaller than the $\frac{7}{16}$-inch socket. Which socket should she use, $\frac{1}{2}$-inch, $\frac{3}{8}$-inch, or $\frac{9}{16}$-inch?

43. Life Science Scientists have identified over 360,000 species of beetles. The table below lists the average lengths of six beetles.

Species	Firefly	Confused Flour Beetle	Elm Leaf Beetle	Mealworm	Lady Beetle	Japanese Beetle
Length (in.)	$\frac{3}{4}$	$\frac{1}{8}$	$\frac{1}{4}$	$\frac{5}{8}$	$\frac{1}{3}$	$\frac{1}{2}$

a. Which beetles are larger than the Lady Beetle?

b. Name the beetle whose length is half that of the Japanese Beetle.

44. Critical Thinking Does $\frac{8}{6.4}$ name a rational number? Explain.

Mixed Review

Find each quotient. *(Lesson 2–6)*

45. $-42 \div 6$ **46.** $-16 \div (-4)$ **47.** $-\frac{12}{3}$

Evaluate each expression if a = 4, b = −5, and c = 2. *(Lesson 2–5)*

48. $-3b$ **49.** $6ac$ **50.** $a + bc$

Standardized Test Practice
Ⓐ Ⓑ Ⓒ Ⓓ

51. Grid In The deepest lake in the world is Lake Baikal, in Siberia. Its deepest point is 5315 feet deep. If its surface is 1493 feet above sea level, how many feet below sea level is the deepest point? *(Lesson 2–3)*

52. Multiple Choice Use the graph to name the coordinates of point *F*. *(Lesson 2–2)*

Ⓐ $(-2, 3)$ Ⓑ $(3, 2)$

Ⓒ $(-2, -3)$ Ⓓ $(2, 3)$

To add rational numbers, you can use the same rules you used to add integers.

Examples

1 Find each sum.

$-4.8 + (-8.7)$

$$-4.8 + (-8.7) = -(|-4.8| + |-8.7|) \quad \textit{The numbers have the same sign.}$$
$$= -(4.8 + 8.7) \quad \textit{Add their absolute values.}$$
$$= -13.5 \quad \textit{Simplify.}$$

2 $3\frac{2}{3} + \left(-4\frac{1}{6}\right)$

$$3\frac{2}{3} + \left(-4\frac{1}{6}\right) = 3\frac{4}{6} + \left(-4\frac{1}{6}\right) \quad \textit{The LCD is 6. Replace } 3\frac{2}{3} \textit{ with } 3\frac{4}{6}.$$
$$= -\left(\left|-4\frac{1}{6}\right| - \left|3\frac{4}{6}\right|\right) \quad \textit{The signs differ. Subtract the lesser absolute value from the greater.}$$
$$= -\left(4\frac{1}{6} - 3\frac{4}{6}\right) \quad \textit{The sign is negative. Why?}$$
$$= -\left(3\frac{7}{6} - 3\frac{4}{6}\right) \quad \textit{Replace } 4\frac{1}{6} \textit{ with } 3\frac{7}{6}. \textit{ Then subtract.}$$
$$= -\frac{3}{6} \text{ or } -\frac{1}{2} \quad \textit{Simplify.}$$

Prerequisite Skills Review

Operations with Decimals, p. 684

Your Turn

a. $-0.76 + (-1.34)$

b. $-1\frac{3}{4} + \left(-4\frac{1}{8}\right)$

When adding three or more rational numbers, you can use the Commutative and Associative Properties to rearrange the addends.

Example

3 Find $14.8 + (-7.2) + 30.7$.

$$14.8 + (-7.2) + 30.7$$
$$= (14.8 + 30.7) + (-7.2) \quad \textit{Comm. \& Assoc. Properties } (+)$$
$$= 45.5 + (-7.2) \quad \textit{Add.}$$
$$= 38.3 \quad \textit{Simplify.}$$

Your Turn Find each sum.

c. $28.3 + (-56.1) + 32.4 + (-75.1)$ **d.** $-\frac{4}{3} + \frac{5}{8} + \left(-\frac{7}{3}\right)$

Example ④

Agriculture Link

Prerequisite Skills Review
Adding and Subtracting Fractions, p. 687

Jason Wilson has kept track of the level of water in his farm's pond over the past five years. He compared the water level to the overall average. Find the net change in the water level of the pond.

Year	Level Above or Below Average (feet)
2002	2
2003	$1\frac{1}{3}$
2004	$-\frac{3}{4}$
2005	$-2\frac{1}{2}$
2006	$1\frac{1}{4}$

To find the net change in the water level, add.

$$2 + 1\frac{1}{3} + \left(-\frac{3}{4}\right) + \left(-2\frac{1}{2}\right) + 1\frac{1}{4}$$

Group positive numbers together and negative numbers together.

$$2 + 1\frac{1}{3} + \left(-\frac{3}{4}\right) + \left(-2\frac{1}{2}\right) + 1\frac{1}{4}$$

$$= 2 + 1\frac{1}{3} + 1\frac{1}{4} + \left(-\frac{3}{4}\right) + \left(-2\frac{1}{2}\right) \qquad \textit{Commutative Property}$$

$$= 2 + 1\frac{4}{12} + 1\frac{3}{12} + \left(-\frac{9}{12}\right) + \left(-2\frac{6}{12}\right) \qquad \textit{The LCD is 12.}$$

$$= \frac{24}{12} + \frac{16}{12} + \frac{15}{12} + \left(-\frac{9}{12}\right) + \left(-\frac{30}{12}\right) \qquad \textit{Rewrite using the LCD.}$$

$$= \frac{55}{12} + \left(-\frac{39}{12}\right) \qquad \textit{Add positives and negatives.}$$

$$= \frac{16}{12} \qquad \textit{Add.}$$

$$= \frac{16 \div 4}{12 \div 4} \qquad \textit{Divide 16 and 12 by their GCF, 4. Then simplify.}$$

$$= \frac{4}{3} \text{ or } 1\frac{1}{3} \qquad \textit{Simplify.}$$

◉nline **Personal Tutor at algconcepts.com**

You subtract rational numbers the same way that you subtract integers.

Example ⑤

Find −4.5 − 6.8.

$$-4.5 - 6.8 = -4.5 + (-6.8) \qquad \textit{To subtract 6.8, add −6.8.}$$

$$= -11.3 \qquad \textit{The numbers have the same sign. Add their absolute values.}$$

Your Turn Find each difference.

e. $-72.5 - 81.3$

f. $7\frac{3}{8} - \left(-4\frac{1}{3}\right)$

Evaluate $a - b$ if $a = -\dfrac{7}{8}$ and $b = -\dfrac{3}{5}$.

$$a - b = -\frac{7}{8} - \left(-\frac{3}{5}\right) \quad \text{\textit{Replace a with }} -\frac{7}{8} \text{ \textit{and b with }} -\frac{3}{5}.$$

$$= -\frac{7}{8} + \frac{3}{5} \quad \text{\textit{To subtract }} -\frac{3}{5}, \text{ \textit{add }} \frac{3}{5}.$$

$$= -\frac{35}{40} + \frac{24}{40} \quad \text{\textit{The LCD is 40.}}$$

$$= -\frac{11}{40} \quad \text{\textit{Simplify.}}$$

Your Turn

g. Evaluate $x - y$ if $x = 25.8$ and $y = -13.9$.

h. If $m = -\dfrac{3}{4}$ and $n = -\dfrac{2}{5}$, what is the value of $m - n$?

Check for Understanding

Communicating Mathematics

1. **Explain** how the sum $-4.78 + 10.23 + (-7.04) + 0.92$ can be solved in more than one way.

2. **Describe** an example of adding rational numbers from a newspaper.

Guided Practice

Examples 1–5

Find each sum or difference.

3. $4.5 + (-3.6)$ **4.** $-1.6 - 3.8$ **5.** $-5.6 - 9.45$

6. $-5\dfrac{7}{8} - 2\dfrac{3}{4}$ **7.** $4\dfrac{1}{2} + \left(-7\dfrac{3}{4}\right) + 3\dfrac{1}{4}$

Example 6

Evaluate each expression if $a = -\dfrac{1}{2}$, $b = -0.85$, $c = 1.36$, and $d = -\dfrac{5}{6}$.

8. $b - c$ **9.** $a - d$

Example 6

10. Stock Market On Monday, stock in JAB Corporation rose $3\dfrac{1}{8}$ points. The next day, the stock dropped $1\dfrac{3}{4}$ points. What was the net change in the price of the stock?

Exercises ● ● ● ● ● ● ● ● ● ● ● ● ● ● ●

Practice

Find each sum or difference.

11. $-4.7 + (-8.6)$ **12.** $89.3 + (-14.4)$

13. $-2.68 - 3.14$ **14.** $-18.7 + 12.0 + (-9.2)$

15. $-0.23 - 0.13 + (-0.9)$ **16.** $-3.12 + 4.33 + (-1.89)$

17. $-18.9 + (-3.15) - 7.43$ **18.** $-0.8 + 3.5 + (-7.6) + 2.8$

19. $-\dfrac{1}{2} + \dfrac{1}{8}$ **20.** $-\dfrac{3}{5} - \left(-\dfrac{5}{8}\right)$ **21.** $3\dfrac{4}{5} - 9\dfrac{1}{10}$

22. $-\dfrac{5}{6} + \dfrac{5}{8} - \left(-\dfrac{1}{2}\right)$

23. $-\dfrac{2}{7} + \dfrac{2}{5} + \dfrac{3}{7}$

24. $\dfrac{1}{4} + 2 + \left(-\dfrac{3}{4}\right)$

25. $-2\dfrac{3}{4} + 5\dfrac{1}{2} - \left(-\dfrac{3}{8}\right)$

Homework Help

For Exercises	See Examples
11, 12	1
13	5
14–18	3
19–25	2
26–32	6
33, 35	3, 4

Extra Practice
See page 696.

Evaluate each expression if $a = -\dfrac{2}{5}$, $b = 1\dfrac{3}{4}$, $c = 14.6$, and $d = -5.9$.

26. $a + b$

27. $a - b$

28. $c - d$

29. $1 - d$

30. Find the value of $y - 0.5$ if $y = -0.8$.

31. Evaluate $x - 3\dfrac{1}{3}$ if $x = -2\dfrac{5}{6}$.

32. Find the value of a if $a = \dfrac{3}{4} + \left(-\dfrac{4}{5}\right) - \dfrac{2}{5}$.

Applications and Problem Solving

33. Banking   On Wednesday, Lakeesha wrote checks for $289.53, $312.41, and $76.89. On Thursday, she deposited $210.08 and $315.17 into her checking account. What is the net increase or decrease in Lakeesha's account?

34. Sports   The game of *quoits* is played with metal rings. As in horseshoes, the object of the game is to throw the quoit onto or as close as possible to a vertical metal pole. The dimensions of the quoit are shown at the right. What is the diameter of the hole in the center of the quoit?

$1\dfrac{5}{8}$ in. $1\dfrac{5}{8}$ in.

6 in.

35. Critical Thinking   The net snow pack level in the Sierra Nevadas over the past five years has been $1\dfrac{1}{4}$ feet above normal. The following snow pack levels were recorded for the first four years: $1\dfrac{1}{2}$ feet above normal, $1\dfrac{7}{8}$ feet below normal, $\dfrac{3}{4}$ foot below normal, and $2\dfrac{1}{8}$ feet above normal. What was the snow pack level for the fifth year?

Mixed Review

36. Shopping   Which is a better buy: a 184-gram can of peanuts for $1.09 or a 340-gram can for $1.99? Explain.   *(Lesson 3–1)*

Evaluate each expression if $g = -4$, $h = 5$, and $j = -6$.   *(Lesson 2–6)*

37. $-20 \div (h + g)$

38. $\dfrac{j - 9}{h}$

39. Simplify $(4b)(-3c)$.   *(Lesson 2–5)*

40. Music   Monica has 16 CDs that are either country or jazz. She has 8 more jazz CDs than country CDs. How many of each does Monica have?   *(Lesson 1–5)*

Standardized Test Practice
Ⓐ Ⓑ Ⓒ Ⓓ

41. Short Response   Simplify $5y + 2(7 + 3y)$.   *(Lesson 1–4)*

42. Grid In   Find the value of $16 \div [2(13 - 11)]$.   *(Lesson 1–2)*

What You'll Learn
You'll learn to find the mean, median, mode, and range of a set of data.

Why It's Important
Meteorology A meteorologist can compare climates by comparing the mean, median, mode, and range of a set of temperature data.
See Example 5.

A set of data can contain many numbers. To help understand the data, you can let one number describe the set of data. This number is called a **measure of central tendency** because it represents the center, or middle, of the data. The most commonly used measures of central tendency are the **mean**, **median**, and **mode**.

Mean	The mean, or *average*, of a set of data is the sum of the data divided by the number of pieces of data.

The table below shows how much snack food Americans consume.

Favorite Snack Foods

Amount Consumed (lb per person per year)

Potato Chips 6.7, Tortilla Chips 4.9, Pretzels 2.5, Microwave Popcorn 1.5, Snack Nuts 1.5, Cheese Puffs 1.2, Corn Chips 0.8, Ready-to-eat Popcorn 0.6, Party Mix 0.4, Homemade Popcorn 0.4

Source: Snack Food Association   **Snack Food**

Example ①
Food Link

Find the mean of the snack food data.

First, find the sum of the amounts consumed. Then divide by the number of items of data. In this case there are 10 items of data.

$$\text{mean} = \frac{6.7 + 4.9 + 2.5 + 1.5 + 1.5 + 1.2 + 0.8 + 0.6 + 0.4 + 0.4}{10}$$

$$= \frac{20.5}{10} \text{ or } 2.05 \quad \text{The mean of the data is 2.05 pounds.}$$

Prerequisite Skills Review
Operations with Decimals, p. 684

Your Turn   Find the mean of each set of data.

a. 19, 21, 18, 17, 18, 22, 46

b.
Stem	Leaf
7	3 5 6
8	2 2 4
9	0 4 7 9
10	5 8
11	4 6 $9|4 = 94$

Notice that the amount of potato chips consumed is far greater than the amounts of other snacks consumed. Because the mean is an average of several numbers, a single number that is so much greater or less than the others can affect the mean a great deal. In such cases, the mean becomes less representative of the values in a set of data.

The median is another measure of central tendency.

| **Median** | The median of a set of data is the middle number when the data in the set are arranged in numerical order. |

Real World

Example — ② **Find the median of the snack food data.**

Food Link

Arrange the numbers in order from least to greatest.

0.4   0.4   0.6   0.8   ⌊1.2 ↑ 1.5⌋   1.5   2.5   4.9   6.7
                                    median

Since there is an even number of data items, the median is the mean of the two middle values, 1.5 and 1.2. *If there were an odd number of data items, the middle one would be the median.*

$$\text{median} = \frac{1.5 + 1.2}{2} = \frac{2.7}{2} \text{ or } 1.35$$

The median is 1.35 pounds. *The number of values that are greater than the median is the same as the number of values that are less than the median.*

Your Turn   **Find the median of each set of data.**

c. 4, 6, 12, 5, 8   **d.** 10, 3, 17, 1, 8, 6, 12, 15

Graphing Calculator Tutorial
See pp. 750–753.

Graphing Calculator Exploration

You can use a graphing calculator to find the mean and median of the snack food data. First press [STAT] [ENTER]. Under L1, enter in the numerical data. Once all of the data are entered, press [STAT] [▶] [ENTER] [ENTER]. Statistics will appear on the screen. The first statistic, $\bar{x}$ the mean. Scroll down, and you will find the median given by the label "Med."

Try These

1. Use a graphing calculator to find the mean and median of the snack food data. Did you get the same answers?

2. An *outlier* is an item that is much greater or much less than the other data. In the snack food data, potato chips, at 6.7 pounds per person per year, is an outlier. Find the mean and median of the data without potato chips. Is there a significant difference in your answers? Explain.

A third measure of central tendency is the mode.

Mode	The mode of a set of data is the number that occurs most often in the set.

Sometimes a set of data has only one mode. In other cases, every item in a data set occurs the same number of times. When this happens, the set has no mode.

Example

Food Link

3 **Find the mode of the snack food data.**

Look for the number that occurs most often.

0.4   0.4   0.6   0.8   1.2   1.5   1.5   2.5   4.9   6.7

In this set, 0.4 and 1.5 each appear twice. So, the set of data has two modes, 0.4 pound and 1.5 pounds.

 Find the mode of each set of data.

e. 7, 19, 9, 4, 7, 2   f. 300, 34, 40, 50, 60

Online Personal Tutor at algconcepts.com

The snack food data has a mean of 2.05 pounds, a median of 1.35 pounds, and two modes of 0.4 pound and 1.5 pounds. As you can see, the mean, median, and mode may not be the same value.

Measures of central tendency may not give an accurate description of a set of data. Often, **measures of variation** are used to describe the *distribution* of the data. A common measure of variation is the **range**.

Range	The range of a set of data is the difference between the greatest and the least values of the set.

To find the range of the snack food data, subtract the least value of the data set from the greatest value.

The greatest value is 6.7. The least value is 0.4.
So, the range of the data is 6.7 − 0.4 or 6.3.

Example

4 **Find the range of the data set {19, 21, 18, 17, 18, 22, 46}.**

The greatest value is 46. The least value is 17.
So, the range is 46 − 17 or 29.

Your Turn Find the range of each set of data.

g. 4, 6, 12, 5, 8   h.

$$
\begin{array}{cccc}
 & & \times & \\
 & \times & \times & \times \\
 & \times & \times & \times \\
\times & \times & \times & \times \\
\hline
10 & 11 & 12 & 13
\end{array}
$$

Example

⑤

Meteorology Link

Suppose the temperature in a building is always 10°F cooler than the temperature outside. The average outside temperature for two weeks is given below. Find the mean, median, mode, and range for both the inside and outside temperatures. Compare these values.

Temperature	Mean	Median	Mode	Range
Outside	84°F	84°F	83°F, 85°F	6°F
Inside	74°F	74°F	73°F, 75°F	6°F

Note that the mean, mode, and median of the outside and inside temperatures differ by 10°F, while the range is the same.

Data can be classified in several ways. Suppose a person tracks the number of men and women who walk into a theater. These data are **univariate** because there is only one variable being measured, gender. The data are also **categorical** because they fit into one of two categories, man or woman. If the person also asks each theatergoer's age, the data would be **bivariate** because there are two quantities being measured, gender and age. The age data are classified as **measurement** because there are several values for age rather than just a few categories.

Check for Understanding

Communicating Mathematics

1. **Illustrate** how the mean of a set of data is affected by an extremely high or low value.
2. **Describe** the steps you would take to find the mean, median, mode, and range of the data below.

Vocabulary
measure of central tendency
mean, median, mode
measure of variation
range
univariate
catagorical
bivariate

3. **You Decide** Eric says his mean test score is a good measure of how he is doing in class. Sonia disagrees. Who is correct? Explain.

Guided Practice

⏱ Getting Ready Find each quotient without using a calculator.

Sample 1: 156 ÷ 8

Solution:
```
        19.5
    8)156
     -8
      76
     -72
       40
      -40
        0
```

Sample 2: 98.7 ÷ 3

Solution:
```
        32.9
    3)98.7
     -9
      8
     -6
      2 7
     -2 7
        0
```

(continued on the next page)

4. $87 \div 6$ **5.** $126 \div 3$ **6.** $2.04 \div 8$

7. $0.453 \div 3$ **8.** $35.6 \div 2$ **9.** $189 \div 4$

Examples 1–4

Find the mean, median, mode, and range of each set of data.

10. 26, 30, 45, 61, 68   **11.** 12.5, 11.2, 12.4, 12.1, 12.5, 32.3

12. 3.8, 6, 4.5, 9, 6.2, 4.9, 7, 7.6, 3.8, 4, 7

13.

Stem	Leaf
5	5 6 7 8
6	0 0 1
7	2 2 2 $7\,|\,2 = 72$

14.

Example 5

15. Sales The district manager of a cellular phone company wants to promote an employee to sales manager. The quarterly sales for the two employees for the last two years are listed below.

Name	Quarterly Sales (thousands of dollars)							
Ms. Diaz	200.8	190.9	200.0	210.0	200.1	200.5	200.7	210.0
Mr. Cruz	210.0	180.1	191.2	239.2	210.8	190.8	180.9	210.0

a. Find the mean, median, mode, and range of the quarterly sales data for each employee.

b. Based on sales, who should be promoted to sales manager? Explain.

Exercises • • • • • • • • • • • • • • • • • • •

Practice

Find the mean, median, mode, and range of each set of data.

16. 7, 9, 7, 9, 8, 9, 7   **17.** 10, 3, 14, 1, 8, 7, 20

18. 1.5, 1.2, 1.1, 1.3, 1.6, 1.1   **19.** 0.3, 0.3, 0.3, 0.1

20. 45, 32, 17, 65, 80, 55   **21.** 200, 24, 20, 40, 20

22. 95, 76, 88, 82, 73, 65, 76, 76, 84, 90

23. 10.2, 8.8, 10.0, 9.4, 12.6, 10.2, 9.8, 12.0

Homework Help	
For Exercises	See Examples
16–27	1–4
28	2
29	3
30–34	1
35, 36	1-3
Extra Practice	
See page 697.	

24.

25.

26.

Stem	Leaf
5	6 6 7
6	0 1 1 8
7	0 2 4 4 4
8	0 9 $7\,|\,2 = 72$

27.

Stem	Leaf
18	3 3 4
19	0 0 2 5
20	0 8
21	0 0 $20\,|\,8 = 20.8$

28. Find the median of $12.99, $5.89, $6.75, $2.45, $9.25, and $5.88.

29. Give an example of a set of data that has no mode.

Write a set of data with six numbers that satisfies each set of conditions.

30. The mean is greater than all but one of the numbers.

31. The mean is smaller than all but one of the numbers.

Applications and Problem Solving

32. Sports The batting averages for 10 players on a baseball team are 0.234, 0.253, 0.312, 0.333, 0.281, 0.240, 0.183, 0.222, 0.297, and 0.275. Find the mean batting average for these players.

33. School Jack's last four test scores in French were 88, 77, 81, and 83. What must he score on the next test so that his average is exactly 85?

34. Zoos The table shows the number of acres for the five largest zoos in the United States.

Zoo	San Diego Wild Animal Park	Minnesota Zoo	Miami Metrozoo	Bronx Zoo	Albuquerque Biological Park
Acres	2200	500	300	265	240

Source: *World Almanac*

a. What was the mean size of the zoos?

b. Does the mean size accurately describe the values in the set of data? Explain.

35. Business The salaries of the 12 employees at the Hanson Company are $28,500, $32,000, $29,500, $31,200, $28,600, $38,500, $20,100, $85,000, $36,000, $25,350, $26,500, and $19,850.

a. Find the mean, median, and mode of the data.

b. Suppose the president is interviewing an applicant. Should the president quote the mean, median, or mode as the "average" salary? Explain.

c. If the employees are asking for a pay raise, should they quote the mean, median, or mode as their "average" salary? Explain.

36. Critical Thinking List six numbers whose mean is 50, median is 40, and mode is 20.

Mixed Review

Find each sum or difference. *(Lesson 3–2)*

37. $-2.5 - 7.4$

38. $6\frac{1}{3} + \left(-3\frac{3}{4}\right)$

39. $\frac{3}{4} + \left(-\frac{5}{4}\right) - \frac{1}{2}$

40. Write an inequality that compares $-\frac{8}{17}$ and $\frac{1}{9}$. *(Lesson 3–1)*

Standardized Test Practice
Ⓐ Ⓑ Ⓒ Ⓓ

41. Short Response Graph the set of points $\{-3, 4, 2, -2, 0, -1\}$ on a number line. *(Lesson 2–1)*

42. Grid In You have 40 pieces of fencing that are each 1 meter long. What is the largest rectangular area in square meters you can enclose with these pieces of fencing? (*Hint*: The formula for perimeter is $P = 2(\ell + w)$, and the formula for area is $A = \ell w$, where ℓ is the length and w is the width.) *(Lesson 1–5)*

 www.algconcepts.com/self_check_quiz

Sugar and Pizza and Everything Nice!

Materials

 sugar cubes or wooden cubes

Arithmetic Sequences

A **sequence** is a set of numbers in a specific order.

Investigate

1. Use cubes to form the four figures shown. Examine how Figure 2 is different from Figure 1, Figure 3 from Figure 2, and so on.

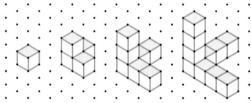

Figure 1   Figure 2   Figure 3   Figure 4

a. Make a table like the one below. Write the numbers 1–10 in columns 1 and 2. Record the number of cubes in each of the four figures. Build a fifth figure and record the number of cubes used. Look for a pattern and complete column 3.

Figure Number	Term Number	Number of Cubes	New Number of Cubes– Previous Number of Cubes
1	1	1	—
2	2	4	4 − 1 = ?

b. Find the difference between the new number of cubes and the number of cubes from the previous row to complete column 4.

c. List the numbers in column 3 from least to greatest, with commas separating the numbers. This list is a sequence. Each number in the sequence is called a **term**.

d. The sequence in part c is called an **arithmetic sequence** because the difference between any two consecutive terms is always the same. What is this difference?

2. An arithmetic sequence can be written as a formula in several ways.

a. Copy and complete a table like the one on the next page for terms 1–10.

Term Number	Term Value	Form 1	Form 2	Form 3
1	1	1	1	$1 + (0 \cdot 3)$
2	4	$1 + 3$	$1 + 3$	$1 + (1 \cdot 3)$
3	7	$4 + 3$	$1 + 3 + 3$	$1 + (2 \cdot 3)$

b. A **recursive formula** for a sequence describes the new term as it relates to the previous term. Look at the column labeled "Form 1." For this arithmetic sequence, you can describe a recursive formula as *term value = previous term value + 3*. Another way to write this formula is $f_{n+1} = f_n + 3$. f_{n+1} is the value you are trying to find. f_n is the value of the previous term. Describe Forms 2 and 3 using a formula like $f_{n+1} = f_n + 3$.

3. A pizza chain offers the deal *Buy one pizza at regular price and get additional pizzas for only $4 each*. The first pizza costs $7.99.

a. Copy and complete a table like the one at the left for terms 1–10.

Pizza Number	Term Number	Cost of Pizzas	New Cost of Pizza – Previous Cost of Pizza
1	1	$7.99	—
2	2	$11.99	$11.99 − $7.99 = ?

b. Write a sequence representing the cost of the pizzas. Then write a recursive formula representing the sequence.

Extending the Investigation

In this extension, you will examine other sequences and their formulas.

1. Use cubes to make the figures shown.

2. Make a table similar to the one in Exercise 1.

3. Write a recursive formula for the sequence.

Figure 1   Figure 2   Figure 3

Presenting Your Investigation

Here are some ideas to help you present your conclusions to the class.

- Make a brochure showing the sequences you wrote in this investigation. Include your tables, formulas, and any diagrams.

- Describe a situation that results in a sequence similar to those in this investigation. Make a poster showing tables, formulas, and diagrams for this sequence.

*inter*NET CONNECTION **Investigation** For more information on sequences, visit: www.algconcepts.com

Why It's Important

Health Knowing how to find the solution of an equation can help you determine your normal blood pressure. *See Example 3.*

A **statement** is any sentence that is either true or false, but not both. Look at the sentences below. Are they true or false?

Statement	True or False?
The sum of 4 and 5 is 9.	true
A number divided by 0 is a rational number.	false
A square has 4 sides of equal lengths.	true
The sum of an even and odd number is even.	false
The sum of angles measures in a triangle equals 180.	true

You can choose symbols to express mathematical statements.

Words	Symbols
A number m minus 5 is equal to 12.	$m - 5 = 12$
20 equals a number y plus 4.	$20 = y + 4$
Eight increased by a number d is 6.	$8 + d = 6$
Seven minus a number equals 5.	$7 - x = 5$
The sum of a number and 11 equals 20.	$z + 11 = 20$

Mathematical sentences like these are called **open sentences**. An open sentence is neither true nor false until the variable is replaced by a value. A set of numbers from which replacements for a variable may be chosen is called a **replacement set**. Finding the replacements for the variable that results in a true sentence is called **solving** the open sentence. The replacement values are called a **solution** of the open sentence.

Examples **1** Find the solution of $m - 5 = 12$ if the replacement set is {15, 16, 17, 18}.

Value for m	$m - 5 = 12$	True or False?
15	$15 - 5 \stackrel{?}{=} 12$	false
16	$16 - 5 \stackrel{?}{=} 12$	false
17	$17 - 5 \stackrel{?}{=} 12$	true ✓
18	$18 - 5 \stackrel{?}{=} 12$	false

Since 17 makes the sentence $m - 5 = 12$ true, the solution is 17.

② Find the solution of $2n + 3 = 9$ if the replacement set is {2, 3, 4, 5}.

Value for n	$2n + 3 = 9$	True or False?
2	$2(2) + 3 \overset{?}{=} 9$	false
3	$2(3) + 3 \overset{?}{=} 9$	true ✓
4	$2(4) + 3 \overset{?}{=} 9$	false
5	$2(5) + 3 \overset{?}{=} 9$	false

Therefore, the solution of $2n + 3 = 9$ is 3.

Your Turn

a. Find the solution of $3n - 2 = 13$ if the replacement set is {4, 5, 6, 7}.
b. Find the solution of $-2x + 1 = 7$ if the replacement set is {−3, −2, 2, 3}.

Open sentences and replacement sets are used frequently in real-life situations.

Example ③
Health Link

The normal systolic blood pressure p of a woman who is A years old is given by the formula $p = \dfrac{A(A + 5)}{100} + 107$. Ms. Adams recently had her annual physical. Her blood pressure was normal. If Ms. Adams' blood pressure was 121, how old is she: 34, 35, 36, or 37?

You need to find the replacement for the variable that results in a true sentence. First, replace p with 121.

$$121 = \frac{A(A + 5)}{100} + 107$$

Then replace A with each of the replacements.

Value for A	$121 = \dfrac{A(A + 5)}{100} + 107$	True or False?
34	$121 \overset{?}{=} \dfrac{34(34 + 5)}{100} + 107$	false
35	$121 \overset{?}{=} \dfrac{35(35 + 5)}{100} + 107$	true ✓
36	$121 \overset{?}{=} \dfrac{36(36 + 5)}{100} + 107$	false
37	$121 \overset{?}{=} \dfrac{37(37 + 5)}{100} + 107$	false

The solution is 35. So, Ms. Adams is 35 years old.

Blood pressure cuff

Look Back

Equations:
Lesson 1–1

Recall that an open sentence containing an equals sign, =, is an *equation*. Sometimes, you can solve an equation by simply applying the order of operations.

Examples

Solve each equation.

4 $4(8) + 6 = a$

$4(8) + 6 = a$ *Original equation*

$32 + 6 = a$ *Multiply 4 and 8.*

$38 = a$ *Add 32 and 6.*

The solution is 38.

**Prerequisite
Skills Review**

Simplifying Fractions,
p. 685

5 $m = \dfrac{3 + 4 - 2}{28 - 3}$

$m = \dfrac{3 + 4 - 2}{28 - 3}$ *Evaluate the numerator and denominator separately.*

$m = \dfrac{5}{25}$ $3 + 4 - 5 = 7 - 2 \text{ or } 5$
$28 - 3 = 25$

$m = \dfrac{1}{5}$ *Simplify.*

The solution is $\dfrac{1}{5}$.

**CONcepts
in MOtion**

Interactive Lab
algconcepts.com

Your Turn

c. $t = -3(4 \cdot 3 + 6)$

d. $\dfrac{5 \cdot 3 + 3}{4 \cdot 2 - 2} = h$

Check for Understanding

**Communicating
Mathematics**

1. **Describe** the difference between a statement and an open sentence.

2. **Define** the term *solution*.

3. **Explain** how to find the solution of $2n - 3 = 25$ if the replacement set is {12, 13, 14, 15}.

Vocabulary

statement
open sentences
replacement set
solving
solution

Guided Practice

Examples 1 & 2

Find the solution of each equation if the replacement sets are a = {5, 6, 7}, b = {−1, 0, 1}, and c = {−2, −1, 0, 1}.

4. $6 = c + 8$

5. $2a + 5 = 17$

6. $\dfrac{3 + 15}{6} = 3b$

Examples 4 & 5

Solve each equation.

7. $12 - 0.3 = y$ 8. $p = 5(10) + 2 - 4$ 9. $n = \dfrac{24 \div 8 + 1}{15 \div 3}$

Example 3

10. **Health**   The normal systolic blood pressure p of a man who is A years old is given by the formula $p = \dfrac{A(3A - 10)}{500} + 120$. How old is Mr. Nichols if his normal blood pressure is 122: 18, 19, 20, or 21?

Exercises • • • • • • • • • • • • • • • • • • •

Practice

Find the solution of each equation if the replacement sets are $x = \{2, 3, 4, 5\}$, $m = \{-5, -4, -3\}$, and $d = \{-1, 0, 1, 2\}$.

11. $x + 5 = 9$ 12. $2d = -2$ 13. $8 - m = 11$

14. $-11 = -7 + m$ 15. $4x + (-2) = 10$ 16. $9d - 20 = -20$

17. $6x + 7 = 31$ 18. $-23 = 3d - 4(5)$ 19. $3 - m = 8$

20. $\dfrac{3 + 15}{m} - 6 = -12$ 21. $\dfrac{6d}{4} + 3 = 3d$ 22. $\dfrac{2(x - 2)}{3} = \dfrac{4}{7 - 5}$

Solve each equation.

23. $6 \cdot 7 = y$ 24. $g = 14.8 - 3.75$ 25. $h = 5 + 2 \cdot 3$

26. $-7.7 - 5.2 = k$ 27. $3 + 18 \div 6 = w$ 28. $m = -2(5) + 4 \cdot 5$

29. $n = (18 \div 6) - 4$ 30. $20 - 3 \cdot 3 - 14 = m$ 31. $7 \cdot 3 - 15 \div 5 = b$

32. $\dfrac{7 \cdot 5 + 5}{(9 \cdot 3) - 7} = p$ 33. $a = \dfrac{14 + 7}{2 \cdot 2}$ 34. $\dfrac{4 \cdot 7 - 4}{2 \div 2 + 5} = d$

35. Find the solution of $2(7) - 12 = v$.

36. What is the value of g if $g = \dfrac{24 \div 2}{6}$?

Homework Help

For Exercises	See Examples
11–22	1, 2
23–34	4, 5
35	4
36	5
37–39	3

Extra Practice
See page 697.

Applications and Problem Solving

37. **Recreation**   At Sinclair's Balloon Rental, the cost of renting a hot air balloon is given by the formula $C = 85 + 36h$, where C is the cost in dollars and h is the number of hours. Michael and Keisha paid a total of $301 for a balloon ride. Did they rent the balloon for 5, 6, 7, or 8 hours?

38. **Geometry**   To find the area of a trapezoid, you can use the formula $A = \frac{1}{2}h(a + b)$, where h is the height of the trapezoid and a and b are the lengths of the bases. If a trapezoid has an area of 35 square feet and bases of 4 feet and 10 feet, is the height 3, 4, 5, or 6 feet?

39. **Life Science** At birth, a baby blue whale weighs 4000 pounds and gains 200 pounds each day while nursing. The formula $W = 4000 + 200d$ represents this situation, where W is the weight of the baby and d is the number of days it has nursed. How many days has a baby blue whale been nursing if it weighs 17,000 pounds: 63, 64, or 65?

40. **Critical Thinking** Write four different open sentences that each have 3 as a solution.

Blue Whales

Mixed Review

Find the mean, median, mode, and range of each set of data. *(Lesson 3–3)*

41. 38, 42, 46, 40, 40

42. 3.1, 3.2, 3.3, 3.4, 3.5, 3.6, 3.7, 3.8

Evaluate each expression if $x = -\frac{1}{3}$, $y = \frac{1}{2}$, and $z = 3\frac{1}{4}$. *(Lesson 3–2)*

43. $2 - x$

44. $y + z$

45. $x - y - z$

Standardized Test Practice
Ⓐ Ⓑ Ⓒ Ⓓ

46. **Short Response** Thomas buys $\frac{2}{3}$ yard of ribbon, $\frac{3}{8}$ yard of plaid fabric, and $\frac{1}{2}$ yard of striped fabric. Write the lengths in order from least to greatest. *(Lesson 3–1)*

47. **Multiple Choice** Which of the following is equal to the expression $[(8 - 7) - (6 - 5)] - [(8 - 7 - 6) - 5]$? *(Lesson 2–4)*

　Ⓐ -31　　　Ⓑ -11　　　Ⓒ 0　　　Ⓓ 10

Quiz 1   Lessons 3–1 through 3–4

Replace each ● with <, >, or = to make a true sentence. *(Lesson 3–1)*

1. $-16.3 \bullet -16.03$

2. $3(-4) + 7 \bullet -12 + 6$

3. $\frac{2}{3} \bullet \frac{4}{6}$

4. Which is the better buy, a 12-pack of soda for $2.79 or a 24-pack of soda for $5.69? *(Lesson 3–1)*

5. Find $3.54 - (-4.78) + 1.02 + (-7.65)$. *(Lesson 3–2)*

Find the mean, median, mode, and range for each set of data. *(Lesson 3–3)*

6. 45, 12, 67, 22, 34

7.

Stem	Leaf
1	1 3 7
2	1 6 7
3	0 0 4 6
4	4 7 $1|1 = 1.1$

8.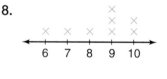

9. Find the solution of $3d - 7 = -13$ if the replacement set is $\{-4, -3, -2, -1\}$. *(Lesson 3–4)*

10. Solve $m = \dfrac{2 + 4 \cdot 6}{3 + 10} - 8$. *(Lesson 3–4)*

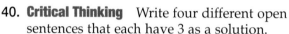 www.algconcepts.com/self_check_quiz

What You'll Learn

You'll learn to solve addition and subtraction equations by using models.

Why It's Important

Zoology Knowing how to solve equations can help in finding unknown values. *See Example 3.*

Many real-world situations can be approached by developing a model, writing an equation, and solving it. You can use algebra tiles to solve equations. The green tile represents the variable, x.

1 | **Examples**

Use algebra tiles to solve each equation.

$x + (-2) = -7$

Step 1 Model $x + (-2) = -7$ by placing 1 green tile and 2 red square tiles on one side of the mat to represent $x + (-2)$. Place 7 red square tiles on the other side to represent -7.

$x + (-2) = -7$

Step 2 To get the green tile by itself, remove 2 red square tiles from each side.

$x + (-2) - (-2) = -7 - (-2)$

Step 3 The green tile on the left side of the mat is matched with 5 red square tiles. Therefore, $x = -5$.

$x = -5$

2 $x - 3 = 5$

Step 1 Write the equation in the form $x + (-3) = 5$. Place 1 green tile and 3 red square tiles on one side of the mat to represent $x + (-3)$. Place 5 yellow square tiles on the other side of the mat to represent $+5$.

(continued on the next page)

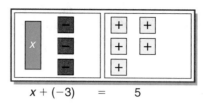

$$x + (-3) = 5$$

Step 2 To get the green tile by itself, you need to remove 3 red square tiles from each side. Since there are no red square tiles on the right side of the mat, you will need to add 3 yellow square tiles to each side to make 3 zero pairs on the left side of the mat.

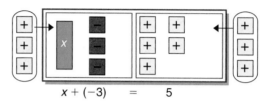

$$x + (-3) = 5$$

Step 3 Remove the zero pairs.

$$x + (-3) + 3 = 5 + 3$$

Step 4 The green tile is matched with 8 yellow square tiles. Therefore, $x = 8$.

$$x = 8$$

Your Turn Use algebra tiles to solve each equation.

a. $x - 4 = -8$ **b.** $2 + x = -7$ **c.** $x + 6 = 5$

In Examples 1 and 2, you solved each equation for the variable by isolating the variable on one side of the equation.

Example
Zoology Link

3

The largest bear is the Alaskan brown bear. When fully grown, it has a height of 9 feet. The smallest bear is the sun bear. The Alaskan brown bear is 6 feet longer than the sun bear. Find the length of the sun bear.

Explore   You know that the Alaskan brown bear is 6 feet longer than the sun bear. You need to find the length of the sun bear.

Plan   Let x represent the length of the sun bear. Translate the problem into an equation using the variable x. Model the equation and then solve.

Solve

length of sun bear	plus	6 feet	equals	length of Alaskan brown bear
x	$+$	6	$=$	9

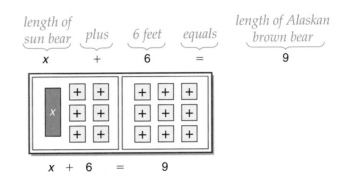

$$x + 6 = 9$$

To solve for x, subtract 6 from each side.

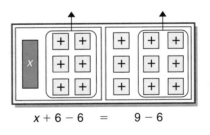

$$x + 6 - 6 = 9 - 6$$

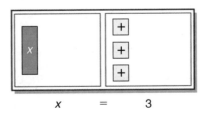

$$x = 3$$

The solution is 3. So, the length of the sun bear is 3 feet.

Examine   Check the solution.

The length of the Alaskan brown bear is equal to the length of the sun bear plus 6 feet. The length of the sun bear is 3 feet, and 3 feet plus 6 feet is 9 feet. √

The answer is correct.

Alaskan Brown Bear

Sun Bear

Check for Understanding

1. **Show** how to model $6 + x = -2$ using algebra tiles.

2. **Explain** why the model below represents $x - (-8) = -6$.

Guided Practice

🕐 **Getting Ready** **Write an equation for each model.**

Sample 1:

Solution: $x + 7 = -5$

Sample 2:

Solution: $-4 = x + (-8)$

3.

4.

5.

6.

Examples 1 & 2

Solve each equation. Use algebra tiles if necessary.

7. $m + 2 = -2$ 8. $3 = n - 6$ 9. $t - 4 = -8$

10. $y + (-3) = -1$ 11. $4 = x - 2$ 12. $-5 + c = -5$

13. Find the value of h if $-4 = h - 7$.

Example 3

14. **Sales** Jeff Simons sold 27 cars last month. This is 36 fewer cars than he sold during the same time period one year ago. What were his sales one year ago?

    a. Write an equation that can be used to find Mr. Simons' sales one year ago.

    b. What were Mr. Simons' sales one year ago?

Exercises

• • • • • • • • • • • • • • • • • •

Practice

Homework Help

For Exercises	See Examples
15–35	1, 2
36–37	3

Extra Practice
See page 697.

Solve each equation. Use algebra tiles if necessary.

15. $-2 + y = -8$ **16.** $w - 1 = 3$ **17.** $n + 5 = -2$
18. $7 = q - 4$ **19.** $p - 6 = 2$ **20.** $3 = z + 10$
21. $-3 + k = -3$ **22.** $4 = g + 6$ **23.** $2 + f = -1$
24. $10 = 8 + b$ **25.** $-2 = d + (-3)$ **26.** $x + 6 = -23$
27. $-7 = a + (-9)$ **28.** $k - 7 = -8$ **29.** $p - 9 = 7$
30. $-12 + r = -2$ **31.** $v + 9 = 2$ **32.** $-13 = c + (-4)$

33. What is the value of m if $8 + m = -13$?

34. If $a + (-14) = -16$, what is the value of a?

35. When 6 is subtracted from d, the result is 5. Find the value of d.

Applications and Problem Solving

36. Sports The 1998 NASCAR season consisted of 33 races. Out of the 33 races, driver Jeff Gorden finished in one of the top five positions 26 times. This number is 15 more than the number of times driver Bobby Labonte finished in the top five positions. **Source:** NASCAR

  a. Write an equation that represents this situation.

  b. How many times did Bobby Labonte finish in one of the top five positions?

37. Weather Jaclyn needs to record the outside temperature for her science project. At 6:00 P.M., the temperature was 8°F. The difference between the temperature she recorded in the morning and the evening temperature was −2°F.

  a. Write an equation that Jaclyn can use to determine the morning temperature.

  b. What was the morning temperature?

38. Critical Thinking Explain how algebra tiles can be used to solve the equation $2n + 1 = 7$.

Mixed Review

Find the solution of each equation if the replacement set is $x = \{-3, -2, -1, 0, 1, 2, 3\}$. *(Lesson 3–4)*

39. $18 + x = 20$ **40.** $x \div 9 = 0$ **41.** $5x - 7 = -2$

42. Sports The stem-and-leaf plot shows Tanya's bowling scores for the past eight games. Find Tanya's mean bowling score. *(Lesson 3–3)*

Stem	Leaf
9	0 1
10	4 5 8
11	5 6 9

$10|5 = 105$

43. Find the product of −9 and −6. *(Lesson 2–5)*

44. Simplify $(4m - 5n) + (3m + 3n)$. *(Lesson 1–4)*

Standardized Test Practice
(A) (B) (C) (D)

45. Short Response Identify the property shown by $24 + 7 = 7 + 24$. *(Lesson 1–3)*

46. Multiple Choice Choose the expression that represents *5 subtracted from the sum of a number and 2.* *(Lesson 1–1)*

  (A) $5 - x + 2$ (B) $x + 2 - 5$
  (C) $2x - 5$ (D) $5 - 2x$

What You'll Learn

You'll learn to solve addition and subtraction equations by using the properties of equality.

Why It's Important

History Knowing how to solve equations can help you find important dates in history.
See Exercise 17.

The Addition Property of Equality and the Subtraction Property of Equality can be used to solve equations. The example below illustrates the **Addition Property of Equality.**

$$14 = 14$$

$$14 + 3 = 14 + 3 \quad \textit{Add 3 to each side.}$$

$$17 = 17$$

Addition Property of Equality	**Words:**	If you add the same number to each side of an equation, the two sides remain equal.
	Symbols:	For any numbers a, b, and c, if $a = b$, then $a + c = b + c$.
	Numbers:	If $x = 2$, then $x + 3 = 2 + 3$.

Note that c can be positive, negative, or 0. Look at the equations below. In the equation on the left, $c = 3$. In the equation on the right, $c = -3$.

$$a + c = b + c \qquad\qquad\qquad a + c = b + c$$
$$15 + 3 = 15 + 3 \qquad\qquad 15 + (-3) = 15 + (-3)$$

When the same number is added to each side of an equation, the result is an **equivalent equation**. Equivalent equations have the same solution.

$$x + 2 = 6 \qquad \textit{The solution of the equation is 4.}$$

$$x + 2 + 3 = 6 + 3 \quad \textit{Using the Addition Property of Equality,}$$
$$\textit{add 3 to each side.}$$

$$x + 5 = 9 \qquad \textit{The solution of this equation is also 4.}$$

The Addition Property of Equality can be modeled with algebra tiles. Model the equation $x - 3 = 5$. Each side of the mat represents a side of the equation. When you add 3 tiles to each side, the result is an equivalent equation, $x = 8$.

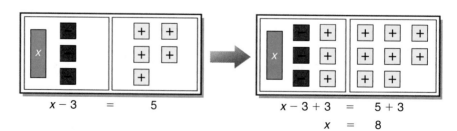

$x - 3$	$=$	5
$x - 3 + 3$	$=$	$5 + 3$
x	$=$	8

1 **Solve each equation. Check your solution.**

$m - 10 = 14$

$$m - 10 = 14 \qquad \textit{Original equation}$$
$$m - 10 + 10 = 14 + 10 \qquad \textit{Add 10 to each side.}$$
$$m + 0 = 24 \qquad \textit{-10 + 10 = 0}$$
$$m = 24 \qquad \textit{Simplify.}$$

To check that 24 is the solution, replace m in the original equation with 24.

Check: $\quad m - 10 = 14 \qquad \textit{Original equation}$
$$24 - 10 \stackrel{?}{=} 14 \qquad \textit{Replace m with 24.}$$
$$14 = 14 \quad \checkmark \quad \textit{24 - 10 = 14}$$

The solution is 24.

2 $-6.2 + x = 15.5$

$$-6.2 + x = 15.5 \qquad \textit{Original equation}$$
$$-6.2 + x + 6.2 = 15.5 + 6.2 \qquad \textit{Add 6.2 to each side.}$$
$$x + 0 = 21.7 \qquad \textit{Simplify.}$$
$$x = 21.7 \qquad \textit{Simplify.}$$

To check that 21.7 is the solution, replace x with 21.7.

Check: $\quad -6.2 + x = 15.5 \qquad \textit{Original equation}$
$$-6.2 + 21.7 \stackrel{?}{=} 15.5 \qquad \textit{Replace x with 21.7.}$$
$$15.5 = 15.5 \qquad \checkmark \quad \textit{-6.2 + 21.7 = 15.5}$$

The solution is 21.7.

Prerequisite Skills Review
Operations with Decimals, p. 684

Your Turn

a. $a - 5 = -14$ $\qquad\qquad$ **b.** $-20.4 = k + 7.8$

In addition to the Addition Property of Equality, there is also a **Subtraction Property of Equality**.

Subtraction Property of Equality	**Words:**	If you subtract the same number from each side of an equation, the two sides remain equal.
	Symbols:	For any numbers a, b, and c, if $a = b$, then $a - c = b - c$.
	Numbers:	If $x = 5$, then $x - 3 = 5 - 3$.

Solve 9 + *a* = −3. Check your solution.

$$9 + a = -3 \qquad \textit{Original equation}$$
$$9 + a - 9 = -3 - 9 \qquad \textit{Subtract 9 from each side.}$$
$$a = -12 \qquad \textit{Simplify.}$$

Check:
$$9 + a = -3 \qquad \textit{Original equation}$$
$$9 + (-12) \overset{?}{=} -3 \qquad \textit{Replace a with }-12.$$
$$-3 = -3 \quad \checkmark \quad \textit{Add }-12\textit{ to 9.}$$

The solution is −12.

Sometimes the Subtraction Property of Equality can be used instead of the Addition Property of Equality. Recall that subtracting a number is the same as adding its inverse.

Romanian Constantin Brancusi and American David Smith are successful twentieth-century sculptors. Brancusi's *Bird in Space* and Smith's *Cubi XVIII* were created in 1918 and 1964, respectively. Use the equation 1918 + *y* = 1964 to find the number of years between these two sculptures.

Method 1: Use the Subtraction Property of Equality.

$$1918 + y = 1964 \qquad \textit{Original equation}$$
$$1918 + y - 1918 = 1964 - 1918 \qquad \textit{Subtract 1918 from each side.}$$
$$y = 46$$

Check:
$$1918 + y = 1964 \qquad \textit{Original equation}$$
$$1918 + 46 \overset{?}{=} 1964 \qquad \textit{Replace y with 46.}$$
$$1964 = 1964 \quad \checkmark$$

Bird in Space

Method 2: Use the Addition Property of Equality.

$$1918 + y = 1964 \qquad \textit{Original equation}$$
$$1918 + y + (-1918) = 1964 + (-1918) \qquad \textit{Add }-1918\textit{ to each side.}$$
$$y = 46 \qquad \textit{The answers are the same.}$$

The solution is 46. So, the sculptures were created 46 years apart.

nline **Personal Tutor at** algconcepts.com

5 Solve $\frac{5}{6} = y - \left(-\frac{2}{3}\right)$. **Check your solution.**

$$\frac{5}{6} = y - \left(-\frac{2}{3}\right) \qquad \text{\textit{Original equation}}$$

$$\frac{5}{6} = y + \frac{2}{3} \qquad \text{\textit{Rewrite the equation. Why?}}$$

$$\frac{5}{6} - \frac{2}{3} = y + \frac{2}{3} - \frac{2}{3} \qquad \text{\textit{Subtract $\frac{2}{3}$ from each side.}}$$

$$\frac{5}{6} - \frac{4}{6} = y \qquad \text{\textit{Rewrite $\frac{2}{3}$ as $\frac{4}{6}$.}}$$

$$\frac{1}{6} = y \qquad \text{\textit{Simplify.}}$$

Check: $\quad \frac{5}{6} = y - \left(-\frac{2}{3}\right) \qquad \text{\textit{Original equation}}$

$$\frac{5}{6} \stackrel{?}{=} \frac{1}{6} - \left(-\frac{2}{3}\right) \qquad \text{\textit{Replace y with $\frac{1}{6}$.}}$$

$$\frac{5}{6} \stackrel{?}{=} \frac{1}{6} + \frac{2}{3} \qquad \text{\textit{Rewrite the equation.}}$$

$$\frac{5}{6} \stackrel{?}{=} \frac{1}{6} + \frac{4}{6} \qquad \text{\textit{Rewrite $\frac{2}{3}$ as $\frac{4}{6}$.}}$$

$$\frac{5}{6} = \frac{5}{6} \quad \checkmark \qquad \text{\textit{Add $\frac{1}{6}$ and $\frac{4}{6}$.}}$$

The solution is $\frac{1}{6}$.

Prerequisite Skills Review
Adding and Subtracting Fractions, p. 687

Your Turn Solve each equation. Check your solution.

c. $h + 8.5 = 3.4$ \qquad\qquad **d.** $-\frac{4}{7} = b - \left(-\frac{3}{7}\right)$

Check for Understanding

Communicating Mathematics

1. **Explain** how you can show that -4 is a solution of $12 + m = 8$.

2. **Write** two equivalent equations.

3. **Complete** the following statement. Justify your answer.
   If $x - 5 = 12$, then $x - 2 = $ ___?___ .

4. **You Decide** Malcom thinks that $y - (-24) = -36$ is equivalent to $y + (-24) = -36$. Adita disagrees. Who is correct, and why?

Vocabulary
equivalent equations

Guided Practice

Getting Ready State the number you would add to or subtract from each side of the equation to solve it.

Sample 1: $c + 16 = 14$ \qquad **Solution:** Subtract 16 from each side.

Sample 2: $-3 + h = -13$ \qquad **Solution:** Add 3 to each side.

5. $y + 21 = -7$ \qquad 6. $-10 + p = 25$ \qquad 7. $34 = d + (-9)$

8. $14 = 16 + s$ \qquad 9. $u - (-9.3) = -53.1$ \quad 10. $8 = a + (-5)$

Solve each equation. Check your solution.

11. $h + 8 = -7$

12. $4 = -5 + d$

13. $2.75 + x = 3.85$

14. $x + 12 = -32$

15. $h - \dfrac{1}{2} = -\dfrac{2}{3}$

16. $8 = x - (-12)$

Example 3

17. History The state of Georgia entered the union in 1788. New Mexico entered the union 124 years later.

  a. Write an equation that could be used to find the year New Mexico entered the union.

  b. When did New Mexico enter the union?

Exercises • • • • • • • • • • • • • • • • •

Practice

Solve each equation. Check your solution.

18. $k + (-8) = 29$

19. $y - (-7) = 3$

20. $-9 = 8 + q$

21. $b - 14 = -11$

22. $-21 = 5 + c$

23. $-3 + m = 10$

24. $27 + y = -49$

25. $7 + w = -6$

26. $11 = r + (-9)$

27. $-15 + y = -63$

28. $-39 = n - 24$

29. $52 = 18 + j$

30. $k - 14.6 = 9.2$

31. $3.4 = m + 7.2$

32. $-3.41 + n = 0.23$

33. $\dfrac{5}{6} = g - \left(-\dfrac{1}{6}\right)$

34. $\dfrac{3}{4} + y = \dfrac{7}{8}$

35. $g + \left(-\dfrac{4}{9}\right) = \dfrac{2}{3}$

Homework Help	
For Exercises	See Examples
18–32, 36, 37	1–3
33–35	5
38–41	4

Extra Practice
See page 698.

36. Solve for s if $7 + s = -10$.

37. What is the solution of the equation $g + (-2) = 11$?

Write an equation and solve.

38. A number increased by -22 is 45. Find the number.

39. What number decreased by -15 is 20?

40. The sum of a number and 8 is -47. What is the number?

Applications and Problem Solving

41. Inflation Today, the cost of a 50-pound box of nails is about $32.50. This is $27.16 more than the cost of the same size box of nails sold in 1800.

Source: *Family Handyman*

  a. Write an equation that could be used to find the cost of a 50-pound box of nails sold in 1800.

  b. In 1800, how much did a 50-pound box of nails cost?

42. **Lighthouses**   There are an estimated 850 lighthouses in the United States. The tallest is located in Cape Hatteras, North Carolina. It is 27 feet taller than the one located in Cape Lookout, North Carolina. The height of the Cape Lookout lighthouse is 169 feet.
   a. Write an equation that could be used to find the height of the lighthouse located in Cape Hatteras, North Carolina.
   b. What is the height of the lighthouse?

43. **Critical Thinking**   What value of x makes $x + x = x$ true?

Mixed Review

44. **Sports**   On the first day of an LPGA golf tournament, professional golfer Pearl Sinn's score was $+4$ or 4 over par. Her score on the second day was added to the first day's score, and the total was $+1$ or 1 over par. What was Sinn's score on the second day?   *(Lesson 3–5)*

   a. Let s represent Sinn's score on the second day. Then translate the problem into an equation using the variable s.
   b. Solve for s to find Sinn's score on the second day.

Pearl Sinn

Solve each equation.   *(Lesson 3–4)*

45. $3.75 - 0.5 = m$
46. $x = -4(8) + 6$
47. $(-15 \div -3) + 1 = y$
48. $t = [3 - (-5)] \cdot 7$

49. **Statistics**   Find the range of the set of data $5.25, $4.39, $1.36, $8.12, $2.90.   *(Lesson 3–3)*

Standardized Test Practice
Ⓐ Ⓑ Ⓒ Ⓓ

50. **Short Response**   Find the quotient of 36 and -4.   *(Lesson 2–6)*

51. **Multiple Choice**   Choose the set of numbers that is correctly ordered from greatest to least.   *(Lesson 2–1)*
   Ⓐ $\{-3, 2, 1, 0\}$ Ⓑ $\{5, -4, 0, -1\}$
   Ⓒ $\{-4, -3, -2, 0\}$ Ⓓ $\{7, 4, -1, -3\}$

Quiz 2   Lessons 3–5 and 3–6

Solve each equation. Check your solution.   *(Lessons 3–5 & 3–6)*

1. $y - 3 = -7$
2. $6 = x + 9$
3. $17.8 = s + (-5.2)$
4. $-\dfrac{5}{6} = p - \dfrac{1}{2}$

5. **Dining**   In Sacramento, California, the average cost of a fast-food meal was $3.89 in a recent year. This was 62 cents less than the average cost of a fast-food meal in Chicago, Illinois. What was the average price of a fast-food meal in Chicago?   **Source:** Runzheimer International, 1999
   a. Choose a variable and write an equation that can be used to find the average cost of a fast-food meal in Chicago.
   b. Find the average cost of a fast-food meal in Chicago.

What You'll Learn

You'll learn to solve equations involving absolute value.

Why It's Important

Science Knowing how to solve equations involving absolute value can help you determine greatest and least values.
See Example 4.

Look Back

Absolute Value: Lesson 2–1

Some equations involve absolute value. Consider the graph of the equation $|a| = 4$.

$$|a| = 4$$

The distance from 0 to a is 4 units. Therefore, $a = -4$ or $a = 4$. When an equation has more than one solution, the solutions are often written as a set $\{a, b\}$. The solution set is $\{-4, 4\}$.

Equations that involve absolute value can be solved by using a number line or by writing them as compound sentences and solving them.

Example ❶ Solve $|x - 3| = 5$. Check your solution.

Method 1: Use a number line.

$|x - 3| = 5$ means the distance between x and 3 is 5 units. So, to find x on the number line, start at 3 and move 5 units in either direction.

$$x = -2 \qquad\qquad x = 8$$

Method 2: Write and solve a compound sentence.
$|x - 3| = 5$ also means $x - 3 = 5$ or $x - 3 = -5$.

$$
\begin{array}{lcl}
x - 3 = 5 & \text{or} & x - 3 = -5 \\
x - 3 + 3 = 5 + 3 & \text{\textit{Add 3 to each side.}} & x - 3 + 3 = -5 + 3 \\
x = 8 & & x = -2
\end{array}
$$

Check: *Replace x with 8.*

$$|x - 3| = 5$$
$$|8 - 3| \stackrel{?}{=} 5$$
$$|5| \stackrel{?}{=} 5$$
$$5 = 5 \quad \checkmark$$

Replace x with −2.

$$|x - 3| = 5$$
$$|-2 - 3| \stackrel{?}{=} 5$$
$$|-5| \stackrel{?}{=} 5$$
$$5 = 5 \quad \checkmark$$

The solution set is $\{-2, 8\}$.

Your Turn **Solve each equation. Check your solution.**

a. $|c - 4| = 2$

b. $6 = |5 + h|$

When solving an equation involving an absolute value symbol, you can think of the absolute value bars as grouping symbols.

Example **Solve $|a + 6| + 5 = 12$. Check your solution.**

To solve the equation, first simplify the expression.

$$|a + 6| + 5 = 12 \qquad \textit{Original equation}$$
$$|a + 6| + 5 - 5 = 12 - 5 \quad \textit{Subtract 5 from each side.}$$
$$|a + 6| = 7 \qquad \textit{Simplify.}$$

Next, write a compound sentence and solve it.

$$a + 6 = 7 \qquad\qquad \text{or} \qquad\qquad a + 6 = -7$$
$$a + 6 - 6 = 7 - 6 \quad \textit{Subtract 6 from each side.} \quad a + 6 - 6 = -7 - 6$$
$$a = 1 \qquad\qquad\qquad\qquad\qquad a = -13$$

Check: *Replace a with 1.*

$$|a + 6| + 5 = 12$$
$$|1 + 6| + 5 \stackrel{?}{=} 12$$
$$|7| + 5 \stackrel{?}{=} 12$$
$$7 + 5 \stackrel{?}{=} 12$$
$$12 = 12 \quad \checkmark$$

Replace a with −13.

$$|a + 6| + 5 = 12$$
$$|-13 + 6| + 5 \stackrel{?}{=} 12$$
$$|-7| + 5 \stackrel{?}{=} 12$$
$$7 + 5 \stackrel{?}{=} 12$$
$$12 = 12 \quad \checkmark$$

The solution set is $\{1, -13\}$.

Your Turn **Solve each equation. Check your solution.**

c. $|m + 5| - 4 = 18$

d. $13 = |-8 + d| + 2$

Sometimes an equation has no solution. For example, $|x| = -6$ is never true. The absolute value of a number is always positive or zero. So, there is no replacement for x that will make the sentence true. The solution set has no members. It is called the **empty set,** and its symbols are { } or $\varnothing$.

Example ──●3 **Solve** $|d| + 7 = 2$. **Check your solution.**

First, simplify the expression.

$$|d| + 7 = 2 \qquad \textit{Original equation}$$
$$|d| + 7 - 7 = 2 - 7 \qquad \textit{Subtract 7 from each side.}$$
$$|d| = -5 \qquad \textit{Simplify.}$$

This sentence can never be true. The solution is the empty set or $\emptyset$.

Your Turn

e. $3 = |s - 4| + 9$ **f.** $|w| - 18 = -6$

Science Link

Real World

●4 **Hydrogen can exist as a solid, liquid, or gas. For hydrogen to be a liquid, its temperature must be within 2° of −257°C. Write and then solve an equation that can be used to find the least and greatest temperatures at which hydrogen is a liquid.**

Let t represent temperature. t differs from -257 by exactly 2 degrees. Write an equation that represents the least and greatest temperatures.

The difference between t and −257   *is*   *2 degrees.*

$$|t - (-257)| \qquad\qquad = \qquad 2$$

$t + 257 = 2$ *Rewrite the equation.* $t + 257 = -2$

$t + 257 - 257 = 2 - 257$ *Subtract 257.* $t + 257 - 257 = -2 - 257$

$t = -255$ *Simplify.* $t = -259$

The solutions are -255 and -259. So, the least temperature for hydrogen to remain a liquid is $-259°C$, and the greatest temperature is $-255°C$.

Check for Understanding

Communicating Mathematics

1. **Write** an absolute value equation that has no solution.

2. **Draw** a number line to show the solution of $|d - 2| = 1$.

3. **Writing Math** Explain why an absolute value equation can have two solutions.

> **Vocabulary**
>
> empty set

Guided Practice

Examples 1 & 3

Solve each equation. Check your solution.

4. $|a| = 6$ 5. $|w| - 2 = 1$ 6. $|x + 1| = 4$

7. $-5 = |w - 9|$ 8. $|3 + g| = 7$ 9. $|h - 3| + 6 = 16$

Example 3   **10.** The absolute value of the sum of 6 and a number is 2. What is the number?

Example 4   **11. Sports** Ricardo's bowling score was within 6 points of his average score of 145. Write and solve an equation that can be used to find the least and greatest bowling score Ricardo could have received in this game.

Exercises

Practice

Homework Help	
For Exercises	See Examples
12, 13, 16–20, 22–25, 30–33	1
14, 15, 21	3
26–29, 34	2
Extra Practice	
See page 698.	

Solve each equation. Check your solution.

12. $|\ell| = -8$ **13.** $|b| = 2$ **14.** $4 + |x| = 7$

15. $|y| - 10 = -4$ **16.** $|x - 6| = 8$ **17.** $|10 + n| = -2$

18. $1 = |2 + p|$ **19.** $|-1 + y| = 0$ **20.** $8 = |-9 + q|$

21. $|s| + 12 = 6$ **22.** $-9 = |2 + h|$ **23.** $11 = |10 + d|$

24. $17 = |y + (-3)|$ **25.** $|z - (-6)| = 13$ **26.** $9 = 4 + |k + 1|$

27. $|b - 2| + 5 = 16$ **28.** $6 + |-5 + f| = 18$ **29.** $|2 + r| + 4 = 1$

30. How many solutions exist for $|a + 3| = 6$?

31. How many solutions exist for $|4 + b| = 0$?

32. Is $-2 = |3 + c|$ sometimes, always, or never true? Explain.

Applications and Problem Solving

For Exercises 33 and 34, write and solve an equation.

33. Real Estate Tamika Samuel is going to place an offer on a home. If she comes within $10,000 of the asking price of $165,000, she has a good chance of getting the home. Write and solve an equation to find out what her greatest and least offers should be.

34. Shipping The Jones Packaging Company needs to ship a crate of books to a book store. The crate must weigh within 6 kilograms of 40 kilograms. If the crate alone weighs 8 kilograms, what are the greatest and least amounts the books can weigh?

35. Critical Thinking Which number line represents the solution set of the equation $1 = -2 + |c + 1|$?

A   ⟵|–♦|–|–|–|–|–|–♦|–|–|→
    −5−4−3−2−1 0 1 2 3 4 5

B   ⟵|–|–|–|–|–|–|–|–♦|–|–|→
    −5−4−3−2−1 0 1 2 3 4 5

C   ⟵|–|–♦|–|–|–♦|–|–|–|→
    −5−4−3−2−1 0 1 2 3 4 5

D   ⟵|–|–|–♦|–|–|–|–|–♦|–|→
    −5−4−3−2−1 0 1 2 3 4 5

Mixed Review

Solve each equation. Check your solution. *(Lesson 3–6)*

36. $13 + w = -6$ **37.** $t - 3 = 45$ **38.** $11.3 = h - 5.7$

39. What is the value of x if $9 + x = -5$? *(Lesson 3–5)*

Standardized Test Practice

40. Grid In If $y - (-8) = 12$, what is the value of y? *(Lesson 3–5)*

41. Multiple Choice A store stocks five different brands of cookies priced at $2.25, $2.50, $2.00, $2.25, and $1.85. What is the difference between the mode price and the mean price? *(Lesson 3–3)*

Ⓐ $0.00   Ⓑ $0.08   Ⓒ $2.25   Ⓓ $4.42

Understanding and Using the Vocabulary

After completing this chapter, you should be able to define each term, property, or phrase and give an example of each.

*inter*NET CONNECTION **Review Activities**
For more review activities, visit:
www.algconcepts.com

arithmetic sequence *(p. 110)*
bivariate *(p. 107)*
categorical *(p. 107)*
cross products *(p. 95)*
empty set *(p. 129)*
equivalent equations *(p. 122)*
inequality *(p. 95)*
mean *(p. 104)*
measurement *(p. 107)*

measure of central tendency *(p. 104)*
measure of variation *(p. 106)*
median *(p. 104)*
mode *(p. 104)*
open sentences *(p. 112)*
range *(p. 106)*
rational numbers *(p. 94)*
recursive formula *(p. 111)*

replacement set *(p. 112)*
sequence *(p. 110)*
solution *(p. 112)*
solving *(p. 112)*
statement *(p. 112)*
term *(p. 110)*
unit cost *(p. 97)*
univariate *(p. 107)*

Choose the correct term to complete each sentence.

1. The (median, mean) is the middle number when data are arranged in numerical order.
2. The (range, unit cost) is used to compare the price of similar items.
3. The difference between the greatest and least values of a set is the (mode, range).
4. When the same number is subtracted from each side of an equation, the result is a(n) (replacement set, equivalent equation).
5. A (rational number, mean) can be expressed in the form $\frac{a}{b}$, where $b \neq 0$.
6. A solution set with no members is the (replacement set, empty set).
7. An (inequality, equivalent equation) uses the symbols $<$ and $>$ to compare two expressions.
8. (Measures of central tendency, measures of variation) describe the distribution of data.
9. A (solution, rational number) can always be expressed as a fraction.
10. You can use (cross products, open sentences) to compare two fractions with different denominators.

Skills and Concepts

Objectives and Examples

• **Lesson 3–1** Compare and order rational numbers.

Replace the ● with $<$, $>$, or $=$ to make a true sentence.

$$\frac{2}{3} \;\bullet\; \frac{4}{5}$$

$$5 \cdot 2 \;\bullet\; 3 \cdot 4$$

If $10 < 12$, then $\frac{2}{3} < \frac{4}{5}$.

Review Exercises

Replace each ● with $<$, $>$, or $=$ to make a true sentence.

11. $-9 \;\bullet\; -11$
12. $-13 \;\bullet\; 13$
13. $0.35 \;\bullet\; -3.5$
14. $2.2 \;\bullet\; 2.20$
15. $\frac{3}{8} \;\bullet\; \frac{3}{4}$
16. $-\frac{7}{4} \;\bullet\; -1.75$
17. Write $-1.1, \frac{1}{8}, 0.25, 0$ in order from least to greatest.

 www.algconcepts.com/vocabulary_review

Objectives and Examples

Review Exercises

• **Lesson 3–2** Add and subtract rational numbers.

Find the difference.
$$-0.3 - (-1.8) = -0.3 + 1.8$$
$$= 1.5$$

Find each sum or difference.

18. $-4.5 + (-8.1)$ 19. $3.7 - (-6.8)$

20. $\frac{8}{9} - \frac{1}{3}$ 21. $-\frac{13}{7} + \frac{6}{7}$

22. $-\frac{4}{3} + \frac{5}{6} + \left(-\frac{7}{3}\right)$

• **Lesson 3–3** Find the mean, median, mode, and range of a set of data.

Find the mean, median, mode, and range of 2, 4, 5, 6, 6, 13, 14, 14, 14, 16, and 16.

mean: $\frac{110}{11} = 10$

median: The 6th or middle value is 13.
mode: The most frequent number is 14.
range: $16 - 2$ or 14

Find the mean, median, mode, and range of each set of data.

23. 6, 8, 6, 5, 5, 7, 9, 2, 7, 5, 3

24. 8.6, 7.5, 9.9, 5.1, 7.1

25.

Stem	Leaf
0	3 6 7 7 7 7 9
1	0 0 0 0 0 0 3 4 4 6 6 7 7 9 9
2	0 1
3	1 $3\,|\,1 = 31$

• **Lesson 3–4** Determine whether a given number is a solution of an equation.

Find the solution of $x + (-3) = 2$ if the replacement set is $x = \{4, 5, 6\}$.

The solution is 5, since $5 + (-3) = 2$.

Find the solution of each equation if the replacement sets are $x = \{-3, -2, -1\}$, $y = \{0, 2, 4\}$, and $z = \{1, 3, 5\}$.

26. $3x = -6$ 27. $5y - 4 = -4$

28. $\frac{3}{5}z = 3$ 29. $\frac{2 + 8}{x} + 9 = -1$

Solve each equation.

30. $a = 21 - 5$ 31. $5(6) - 3(-5) = w$

• **Lesson 3–5** Solve addition and subtraction equations using algebra tiles.

$x + 3 = 2$ $x + 3 + (-3) = 2 + (-3)$

$x = -1$

Solve each equation. Use algebra tiles if necessary.

32. $x + (-3) = 5$ 33. $4 = 6 + y$

34. $-7 + z = 3$ 35. $p + 2 = -2$

36. $s - (-6) = 1$ 37. $-1 = g + (-1)$

38. What is the value of h if $h + 4 = 2$?

39. A number increased by 4 is 11. Write an equation. Then solve for the number.

Mixed Problem Solving
See pages 724–731.

Objectives and Examples

Review Exercises

- **Lesson 3–6** Solve addition and subtraction equations by using the properties of equality.

Solve $m + 16 = 8$. Check your solution.

$$m + 16 = 8$$
$$m + 16 - 16 = 8 - 16 \quad \textit{Subtract 16 from}$$
$$m = -8 \quad \textit{each side.}$$

Check:
$$m + 16 = 8$$
$$-8 + 16 \overset{?}{=} 8 \quad \textit{Replace m with } -8.$$
$$8 = 8 \quad \checkmark$$

Solve each equation. Check your solution.

40. $z + 15 = -9$ **41.** $19 = y + 7$

42. $p + (-7) = 31$ **43.** $m - (-4) = 21$

44. $r - (-1.2) = -7.3$ **45.** $j + (-2.4) = 6.6$

46. $\frac{1}{2} + t = \frac{1}{2}$ **47.** $-\frac{1}{6} = v - \frac{2}{3}$

Write an equation and solve.

48. Some number added to -16 is equal to 39. What is the number?

49. Twelve more than a number is -108. What is the number?

- **Lesson 3–7** Solve equations involving absolute value.

Solve $|x + 6| = 14$. Check your solution.

$$\begin{array}{ccc} x + 6 = 14 & \text{or} & x + 6 = -14 \\ x + 6 - 6 = 14 - 6 & & x + 6 - 6 = -14 - 6 \\ x = 8 & & x = -20 \end{array}$$

Check: $\quad |8 + 6| = 14 \qquad |-20 + 6| = 14$
$$|14| \overset{?}{=} 14 \qquad\quad |-14| \overset{?}{=} 14$$
$$14 = 14 \ \checkmark \qquad\quad 14 = 14 \ \checkmark$$

Solve each equation. Check your solution.

50. $|x| = 22$

51. $|c| - 13 = -5$

52. $7 = 10 + |y|$

53. $6 = |-2 + f|$

54. $|h - 1| = 0$

55. $|a - 3| = -8$

56. Write an equation to represent the situation *The distance between a number x and -4 is 10 units.*

Applications and Problem Solving

57. Shopping The approximate prices for 16 different types of athletic shoes are shown on the line plot below. *(Lesson 3–3)*

Dollars

Find the mean, median, mode, and range for the prices.

58. Manufacturing A certain bolt used in lawn mowers will work properly only if its diameter differs from 2 centimeters by no more than 0.04 centimeter. Write and then solve an equation that could be used to find the least and greatest diameter of the bolt. *(Lesson 3–7)*

1. **List** at least two methods you can use to compare fractions with unlike denominators.

2. **Describe** the process of finding the mean, median, mode, and range of a set of data.

Replace each ● with <, >, or = to make a true sentence.

3. $-4 \bullet -3$

4. $-1.01 \bullet -1.1$

5. $\dfrac{3}{5} \bullet \dfrac{8}{15}$

6. Write $-\dfrac{1}{2}, -\dfrac{2}{9},$ and $-\dfrac{3}{8}$ in order from least to greatest.

Find each sum or difference.

7. $-4.0 - 2.8$

8. $0.32 - (-0.45)$

9. $6.1 + (-3.2)$

10. $\dfrac{7}{10} + \left(-\dfrac{3}{10}\right)$

11. $\dfrac{1}{5} - \left(-\dfrac{3}{4}\right)$

12. $\dfrac{3}{8} + 2\dfrac{1}{2}$

13. **Finance** Sally recorded the amount of money she spent on lunch for five school days in the table at the right.
    a. Find Sally's mean, median, mode, and range of lunch money.
    b. Explain which measure of central tendency is the least useful in describing Sally's lunch costs.

Day	Cost
Monday	$0.99
Tuesday	$1.50
Wednesday	$3.99
Thursday	$0.99
Friday	$2.28

Solve each equation.

14. $10 \div 2 + 6 = x$

15. $8(0.3) - 0.5 = y$

16. $a = \dfrac{2 \cdot 6 + 4}{8 \div 2}$

Solve each equation. Use algebra tiles if necessary.

17. $g + 7 = 2$

18. $-6 + h = -1$

Solve each equation. Check your solution.

19. $a - (-76) = 44$

20. $5.2 + f = 16.4$

21. $\dfrac{1}{2} = b - \dfrac{7}{8}$

22. $3 + |c| = 6$

23. $|m - 4| = 1$

24. $-12 = |t - (-7)|$

25. Taryn visited the "Guess Your Age" booth at the state fair. If the person in the booth could not guess Taryn's age within three years, Taryn would win a stuffed animal. Taryn is 16 years old. Write and solve an equation that can be used to find the person's greatest and least guesses, without Taryn's winning a prize.

Number Concept Problems

Standardized tests include many questions that use fractions and decimals. You'll need to convert between fractions and decimals. You'll also need to add, subtract, multiply, and divide fractions and decimals. Here are some terms to remember.

reciprocal   repeating decimal

Example 1

Which of the following numbers, when divided by $\frac{1}{2}$, gives a result less than $\frac{1}{2}$?

(A) $\frac{2}{8}$ (B) $\frac{7}{12}$ (C) $\frac{2}{3}$ (D) $\frac{5}{24}$

Hint In some multiple-choice questions, you can check each answer choice to see if it is correct.

Solution Recall that dividing by a fraction is equivalent to multiplying by its reciprocal. Dividing by $\frac{1}{2}$ is the same as multiplying by 2. Remember that 2 is also written as $\frac{2}{1}$.

A $\frac{2}{8} \times \frac{2}{1} = \frac{4}{8}$ or $\frac{1}{2}$ **B** $\frac{7}{12} \times \frac{2}{1} = \frac{7}{6}$ or $1\frac{1}{6}$

C $\frac{2}{3} \times \frac{2}{1} = \frac{4}{3}$ or $1\frac{1}{3}$ **D** $\frac{5}{24} \times \frac{2}{1} = \frac{10}{24}$ or $\frac{5}{12}$

Decide which of the four resulting numbers is less than $\frac{1}{2}$. Answer choice A is equal to $\frac{1}{2}$, choices B and C are greater than $\frac{1}{2}$, but choice D is less than $\frac{1}{2}$.

$$\frac{5}{12} < \frac{6}{12} = \frac{1}{2}$$

The answer is D.

Example 2

If $\frac{2}{11}$ is written as a decimal carried out to 100 places, what is the sum of the first 50 digits to the right of the decimal point?

(A) 100 (B) 225 (C) 350 (D) 450 (E) 900

Hint Your calculator can be useful for some types of questions.

Solution First write the fraction as a decimal. Use either long division or your calculator.

$$\frac{2}{11} = 0.1818181818\ldots \text{ or } 0.\overline{18}$$

Since this is a repeating decimal, the first 50 digits are 25 pairs of the digits 1 and 8.

The question asks for the *sum* of the first 50 digits. The sum of each pair of 1 and 8 is 9. The sum of 25 pairs is 25×9 or 225.

The answer is B.

After you work each problem, record your answer on the answer sheet provided or on a sheet of paper.

Multiple Choice

1. In Store X, which color button costs the most per individual unit? *(Lesson 3–1)*

Price of Buttons in Store X	
Color	**Cost**
Black	$2 per 5 buttons
Blue	$2 per 6 buttons
Brown	$3 per 8 buttons
Orange	$4 per 12 buttons
Red	$4 per 7 buttons

   Ⓐ black   Ⓑ blue   Ⓒ brown
   Ⓓ orange   Ⓔ red

2. You burn 15 additional Calories each hour by standing instead of sitting. Suppose you burn x Calories per hour while sitting. How many Calories would you burn in one hour of standing? *(Lesson 1–1)*
   Ⓐ $15x$ Ⓑ $x + 15$
   Ⓒ $x - 15$ Ⓓ $15 - x$

3. The total number of students in a class is represented by $5x - 6$. If $2x + 6$ represents the number of boys, which expression represents the number of girls?
   (Lesson 1–3)
   Ⓐ $3x$ Ⓑ $12 - 3x$
   Ⓒ $3x - 12$ Ⓓ $3x + 12$

4. Order $-\dfrac{3}{4}, \dfrac{3}{4}$, and $-\dfrac{4}{3}$ from least to greatest. *(Lesson 3–1)*
   Ⓐ $\dfrac{3}{4}, -\dfrac{4}{3}, -\dfrac{3}{4}$ Ⓑ $-\dfrac{4}{3}, \dfrac{3}{4}, -\dfrac{3}{4}$
   Ⓒ $-\dfrac{3}{4}, -\dfrac{4}{3}, \dfrac{3}{4}$ Ⓓ $-\dfrac{4}{3}, -\dfrac{3}{4}, \dfrac{3}{4}$

5. What is the GCF of the numbers in the sequence 6, 12, 18, 24, 30, . . . ? *(Basic Skill)*
   Ⓐ 3   Ⓑ 6   Ⓒ 12   Ⓓ 18

6. Two years ago, Luna was ℓ years of age. How old will Luna be 2 years from now? *(Lesson 1–1)*
   Ⓐ $\ell + 4$ Ⓑ $\ell + 2$ Ⓒ 2ℓ Ⓓ ℓ

7. If you evaluate the following expression according to the order of operations, in what order should the operations be performed? *(Lesson 1–2)*
   $$12.5 \times 3 + 5 + 20 - 12$$
   Ⓐ $\times, +, +, -$ Ⓑ $+, -, \times, +$
   Ⓒ $+, \times, -, +$ Ⓓ $-, +, +, \times$

8. Which expression finds multiples of 5? *(Lesson 1–1)*
   Ⓐ $x + 5$ Ⓑ $x - 5$ Ⓒ $5x$ Ⓓ $5 - x$

Grid In

9. Yvette's goal is to run 30 miles a week. So far this week, she has run 6.5, 5.2, 7.8, 3, and 6.9 miles. How many more miles does she need to run to reach her goal? *(Lesson 3–2)*

Extended Response

10. Band members Scott and Ben are taking the bus to a football game. The bus leaves at 5:45 P.M. They have to arrive 15 minutes before the bus leaves to load instruments. It takes Scott 10 minutes to drive to Ben's house and 25 minutes to drive from there to the school. What is the latest time that Scott can leave his house? *(Lesson 3–2)*

    Part A   Explain the steps you use in solving this problem.

    Part B   Calculate the latest time Scott can leave home.

CHAPTER

4

Multiplication and Division Equations

What You'll Learn

Key Ideas

- Multiply and divide rational numbers. *(Lessons 4–1 and 4–3)*

- Use tree diagrams or the Fundamental Counting Principle to count outcomes. *(Lesson 4–2)*

- Solve equations involving multiplication and division, more than one operation, variables on both sides, or grouping symbols. *(Lessons 4-4 to 4-7)*

Key Vocabulary

combination *(p. 152)*

event *(p. 147)*

multiplicative inverses *(p. 154)*

permutation *(p. 152)*

tree diagram *(p. 146)*

Why It's Important

The Tower of Pisa is one of the most-visited tourist attractions in Europe. Soon after its construction in the twelfth century, the foundation sank, causing the tower to lean.

Equations involving multiplication and division are used to model real-life situations. You will multiply rational numbers to analyze an experiment Galileo performed at the Tower of Pisa in Lesson 4–1.

Study these lessons to improve your skills.

 Lesson 2-5, pp. 75-79

Find each product.

1. $6(9)$ **2.** $-4(2)$ **3.** $-5(0)$

4. $-11(-2)$ **5.** $-15(-4)$ **6.** $8(-11)$

7. $4(-1)(-5)$ **8.** $4(2)(3)$ **9.** $-3(3)(-6)$

 Lesson 2-6, pp. 82-85

Find each quotient.

10. $18 \div 2$ **11.** $-14 \div 2$ **12.** $-36 \div (-9)$

13. $42 \div (-6)$ **14.** $24 \div 6$ **15.** $-56 \div 8$

16. $\dfrac{-16}{-4}$ **17.** $\dfrac{56}{-7}$ **18.** $\dfrac{-81}{3}$

 Prerequisite Skill, p. 685

Write each fraction in simplest form.

19. $\dfrac{2}{8}$ **20.** $\dfrac{9}{12}$ **21.** $\dfrac{16}{24}$

22. $\dfrac{50}{60}$ **23.** $\dfrac{21}{35}$ **24.** $\dfrac{60}{96}$

 Lesson 1-4, pp. 19-23

Simplify each expression.

25. $8a - a$ **26.** $9x - 3x$ **27.** $7t - 6t$

28. $13y - 5y$ **29.** $4(x - 2)$ **30.** $3(n + 6)$

31. $2(b - 1.6)$ **32.** $5(3b + 6)$ **33.** $4(8m - 3)$

FOLDABLES™ Study Organizer

Make this Foldable to help you organize your Chapter 4 notes. Begin with seven sheets of grid paper.

❶ **Fold** each sheet in half along the width.   ❷ **Unfold** and cut four rows from the left side of each sheet, from the top to the fold.

❸ **Stack** the sheets and staple to form a book.   ❹ **Label** each page with a lesson number and title.

Reading and Writing As you read and study the chapter, unfold each page and fill the journal with notes, graphs, and examples.

Multiplying Rational Numbers

What You'll Learn
You'll learn to multiply rational numbers.

Why It's Important
Health One way to determine safe backpack weights is to multiply rational numbers.
See Exercise 16.

Knowing how to multiply rational numbers is essential in simplifying algebraic expressions and solving equations. The multiplication of rational numbers can be modeled by repeated addition.

Find $4 \times (-0.2)$.

$$4 \times (-0.2) = -0.2 + (-0.2) + (-0.2) + (-0.2)$$
$$= -0.8$$

This example illustrates the rule for multiplying two rational numbers.

Multiplying Two Rational Numbers	Words	Numbers
Different Signs	The product of two rational numbers with different signs is negative.	$1.2(-4.5) = -5.4$ $-1.2(4.5) = -5.4$
Same Sign	The product of two rational numbers with the same sign is positive.	$3(2.8) = 8.4$ $-3(-2.8) = 8.4$

Examples

Find each product.

1 $2.3(-4)$

$2.3(-4) = -9.2$ *The factors have different signs. The product is negative.*

2 $-3.5(-0.8)$

$-3.5(-0.8) = 2.8$ *The factors have the same sign. The product is positive.*

Your Turn

a. $-5(-1.3)$ **b.** $-2.4(7.5)$ **c.** $8 \cdot (-3.2)$

Example

Sports Link

3 A skydiver jumps from 12,000 feet. Solve the equation $h = 12,000 + (0.5)(-32.1)(576)$ to find his height after he free-falls for 24 seconds.

$h = 12,000 + (0.5)(-32.1)(576)$

$= 12,000 + (-16.05)(576)$ *Multiply 0.5 and −32.1.*

$= 12,000 + (-9244.8)$ *Multiply −16.05 and 576.*

$= 2755.2$ *Add 12,000 and −9244.8.*

After 24 seconds, the skydiver's height is 2755.2 feet.

You can use grid paper to model the multiplication of fractions.

Materials: grid paper colored pencils

 straightedge

Multiply $\frac{3}{7}$ and $\frac{1}{5}$ using a model.

Step 1 Use a straightedge to draw a rectangle that has seven rows and five columns.

Step 2 Shade three of the seven rows yellow to represent $\frac{3}{7}$.

Step 3 Shade one column blue to represent $\frac{1}{5}$.

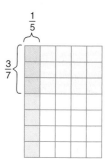

Try These

1. How many of the squares are shaded green?

2. What fraction of the total number of squares in the rectangle is shaded green?

3. The portion of squares that is shaded green represents the product of the fractions. What is the product of $\frac{3}{7}$ and $\frac{1}{5}$?

4. Use grid paper to model $\frac{3}{4}$ and $\frac{5}{8}$. Then use your model to find the product of $\frac{3}{4}$ and $\frac{5}{8}$.

5. What do you notice about the relationship between the numerators and denominators of $\frac{3}{4}$ and $\frac{5}{8}$ and their product?

The Hands-On Algebra activity suggests the following rule for multiplying fractions.

Multiplying Fractions	**Words:**	To multiply fractions, multiply the numerators and multiply the denominators.
	Symbols:	$\frac{a}{b} \cdot \frac{c}{d} = \frac{ac}{bd}$, where $b, d \neq 0$
	Numbers:	$\frac{1}{5} \cdot \frac{3}{4} = \frac{1 \cdot 3}{5 \cdot 4}$ or $\frac{3}{20}$

Find each product.

④ $-\dfrac{4}{5} \cdot \dfrac{1}{3}$

$-\dfrac{4}{5} \cdot \dfrac{1}{3} = -\dfrac{4 \cdot 1}{5 \cdot 3}$ *Multiply the numerators and multiply the denominators.*

$= -\dfrac{4}{15}$ *The factors have different signs.*
The product is negative.

⑤ $6 \cdot \left(\dfrac{2}{5}\right)$

$6 \cdot \left(\dfrac{2}{5}\right) = \dfrac{6}{1} \cdot \dfrac{2}{5}$ *Rewrite 6 as an improper fraction.*

$= \dfrac{6 \cdot 2}{1 \cdot 5}$ *Multiply the numerators and multiply the denominators.*

$= \dfrac{12}{5}$ or $2\dfrac{2}{5}$ *The factors have the same sign.*
The product is positive.

Prerequisite Skills Review
Simplifying Fractions, p. 685

⑥ $-\dfrac{1}{4} \cdot \left(-4\dfrac{1}{2}\right)$

$-\dfrac{1}{4} \cdot \left(-4\dfrac{1}{2}\right) = -\dfrac{1}{4} \cdot \left(-\dfrac{9}{2}\right)$ *Rewrite* $-4\dfrac{1}{2}$ *as an improper fraction.*

$= \left(\dfrac{1 \cdot 9}{4 \cdot 2}\right)$ *Multiply the numerators and multiply the denominators.*

$= \dfrac{9}{8}$ or $1\dfrac{1}{8}$ *The factors have the same sign.*
The product is positive.

Your Turn

d. $\dfrac{3}{4}\left(\dfrac{5}{7}\right)$ **e.** $\dfrac{4}{5} \cdot (-3)$ **f.** $-\dfrac{1}{2}\left(-1\dfrac{3}{8}\right)$

Study these products.

$-1(5) = -5$ *−1 times 5 equals −5.*

$(-1)(-2.4) = 2.4$ *−1 times −2.4 equals 2.4.*

Notice that multiplying a number by −1 results in the opposite of the number. This suggests the **Multiplicative Property of −1**.

Multiplicative Property of −1	**Words:**	The product of −1 and any number is the number's additive inverse.	
	Symbols:	$-1 \cdot a = -a$	**Numbers:** $-1 \cdot \dfrac{2}{7} = -\dfrac{2}{7}$

You can use what you know about multiplying rational numbers to simplify algebraic expressions.

Example

⑦ **Simplify $-7.2t(5u)$.**

$$-7.2t(5u) = (-7.2)(t)(5)(u) \quad \textit{-7.2t = (-7.2)(t); (5u) = (5)(u)}$$
$$= (-7.2)(5)(t)(u) \quad \textit{Commutative Property}$$
$$= (-7.2 \cdot 5)(t \cdot u) \quad \textit{Associative Property}$$
$$= -36tu \quad \textit{Simplify.}$$

⑧ **Simplify $\frac{2}{5}m\left(\frac{3}{5}n\right)$.**

$$\frac{2}{5}m\left(\frac{3}{5}n\right) = \left(\frac{2}{5}\right)(m)\left(\frac{3}{5}\right)(n) \quad \textit{$\frac{2}{5}m = \left(\frac{2}{5}\right)(m)$; $\left(\frac{3}{5}n\right) = \left(\frac{3}{5}\right)(n)$}$$
$$= \left(\frac{2}{5}\right)\left(\frac{3}{5}\right)(m)(n) \quad \textit{Commutative Property}$$
$$= \left(\frac{2}{5} \cdot \frac{3}{5}\right)(m \cdot n) \quad \textit{Associative Property}$$
$$= \frac{6}{25}mn \quad \textit{Simplify.}$$

Your Turn

g. $2x(-4.5y)$

h. $\frac{5}{7}g\left(\frac{2}{3}h\right)$

When you add any two rational numbers, is the sum always a rational number? What happens when you subtract or multiply rational numbers? In all three cases, the result is a rational number.

Closure Property of Rational Numbers	The sum, difference, or product of two rational numbers is always a rational number. Therefore, the set of rational numbers is closed under addition, subtraction, and multiplication.

For example, you can add, subtract, or multiply $\frac{5}{7}$ and $\frac{1}{7}$, and the result will be a rational number. *Check this.*

Check for Understanding

Communicating Mathematics

1. **State** whether $1.2bc$ is *sometimes*, *always*, or *never* a rational number if b and c are rational numbers. Explain.

2. **Determine** whether the product of three negative rational numbers can ever be positive. Explain.

3. **Write** a multiplication equation that can be shown using the model.

Exercise 3

Getting Ready Find each product.

| **Sample:** $(-4)(-6)$ | **Solution:** $(-4)(-6) = 24$ |

4. $-1 \cdot 8$ **5.** $3(-3)$ **6.** $-7 \cdot (-4)$

Examples 1–6   **Find each product.**

7. $2.1(-7)$ **8.** $(-1.8)(-3.5)$ **9.** $0.5(-4.6)$

10. $-\frac{1}{3}\left(-\frac{4}{7}\right)$ **11.** $-3 \cdot \frac{3}{4}$ **12.** $2\frac{3}{5}\left(\frac{1}{4}\right)$

Example 7   **Simplify each expression.**

13. $3\left(\frac{1}{5}y\right)$ **14.** $-\frac{1}{4} \cdot \frac{7}{10}k$ **15.** $-2.5c(-2.4d)$

Example 5   **16. Health** Students who carry backpacks that are too heavy can develop spinal problems. The maximum recommended backpack weight y is given by the equation $y = \frac{3}{20}x$, where x is a student's weight. Would a 12-pound backpack be too heavy for a student who weighs 110 pounds? Explain.
Source: *Zillions*

Exercises

Practice

Find each product.

17. $4 \cdot 6.1$ **18.** $3.8(8)$ **19.** $7.1(-1)$

20. $(-0.5)(-5.6)$ **21.** $-6.0(8.5)$ **22.** $-0.4(1.5)$

23. $-2.9 \cdot 10$ **24.** $-7.3 \cdot 0$ **25.** $(-12.8)(-1)(4.5)$

26. $\frac{1}{5}\left(\frac{2}{3}\right)$ **27.** $-1 \cdot \left(-\frac{5}{8}\right)$ **28.** $-\frac{4}{7} \cdot \frac{3}{5}$

29. $(0)\left(-\frac{6}{7}\right)$ **30.** $-\frac{3}{5}\left(-\frac{9}{10}\right)$ **31.** $\frac{7}{3}\left(-\frac{3}{7}\right)$

32. $\left(-\frac{3}{5}\right)\left(\frac{-2}{7}\right)(-1)$ **33.** $3 \cdot \frac{7}{8}$ **34.** $\frac{6}{7}\left(-2\frac{1}{3}\right)$

Homework Help	
For Exercises	See Examples
17–25, 48	1, 2
26–34	4, 5, 6
35–42	3, 7, 8
43, 45–47	3
Extra Practice	
See page 698.	

Simplify each expression.

35. $2(-1.5x)$ **36.** $0.6(-15y)$

37. $(4r)(2.2s)$ **38.** $\frac{5}{6}r(-18s)$

39. $\left(\frac{3}{5}a\right)\left(\frac{2}{5}b\right)$ **40.** $\frac{1}{3}y\left(-\frac{1}{4}z\right)$

41. $-\frac{1}{2}s\left(-\frac{1}{2}\right) + \frac{2}{3}\left(\frac{3}{2}s\right)$ **42.** $\frac{5}{6}\left(\frac{3}{4}a\right) - \left(\frac{1}{8}a\right)\left(-\frac{1}{3}\right)$

43. Solve $x = 2(-3.2) - 0.5(-4.8)$ to find the value of x.

44. Evaluate $7a(-4.5b) + (-2ab)$ if $a = 2$ and $b = -1$.

Applications and Problem Solving

45. Shopping Teri wants to buy a bicycle with the money she earns from her after-school job. If she can save $22.50 a week, how much money will she have after six weeks?

46. Space A day on Jupiter is much shorter than a day on Earth because Jupiter rotates on its axis about $2\frac{2}{5}$ times as fast as Earth rotates on its axis. During our month of June, how many days will Jupiter have?

47. History Galileo (1564–1642) was a mathematician and scientist who is said to have experimented with falling objects from the Leaning Tower of Pisa. Suppose Galileo dropped a stone from the tower and it took $3\frac{1}{2}$ seconds to reach the ground. Solve $d = -\frac{1}{2}\left(9\frac{4}{5}\right)\left(3\frac{1}{2}\right)\left(3\frac{1}{2}\right)$ to find the distance d in meters that the stone fell. *A negative answer means that the final position is lower than the initial position.*

48. Critical Thinking Describe the values of a and b for each statement to be true.

a. ab is positive.   **b.** ab is negative.   **c.** $ab = 0$

Mixed Review

Solve each equation. Check your solution. *(Lesson 3–7)*

49. $|t| = -5$ **50.** $2 + |x| = 6$ **51.** $3 = |-9 - a|$

52. Aviation A traffic helicopter descended 160 meters to observe road conditions. It leveled off at 225 meters. *(Lesson 3–6)*

a. Let a represent the original altitude. Write an equation to represent this problem.

b. What was the helicopter's original altitude?

Find each sum or difference. *(Lesson 3–2)*

53. $3.5 + (-2.1)$ **54.** $-1.7 - 4.0$ **55.** $-\frac{3}{4} + \frac{1}{8}$

56. Find $24 \div (-6)$. *(Lesson 2–6)*

Standardized Test Practice
(A) (B) (C) (D)

57. Extended Response The stem-and-leaf plot shows the heights of 30 students. *(Lesson 1–7)*

a. What is the height of the tallest student?

b. Which height occurs most frequently?

Class Heights

Stem	Leaf
5	7 8 8 9 9 9
6	0 0 1 2 2 2 3 3 4 4 4 4 4
	5 5 7 8 9 9
7	0 0 2 3 3 $7\,|\,2 = 72$ *in.*

58. Multiple Choice The expression $(220 - y) \times 0.8 \div 4$ gives the target 15-second heart rate for an athlete y years old during a workout. Find the target 15-second heart rate for a 17-year-old athlete during a workout. *(Lesson 1–2)*

(A) 27   (B) 40.6   (C) 216.6   (D) 219.7

What You'll Learn

You'll learn to use tree diagrams and the Fundamental Counting Principle to count outcomes.

Why It's Important

Manufacturing Car manufacturers count outcomes to determine the number of different key combinations that are possible.
See Exercise 23.

Baroness Martine de Beausoleil was a French scientist in the 17th century who spent 30 years studying geology and mathematics. She determined which rocks were valuable by the minerals they contained. Some other ways to classify rocks are by their texture and by their color.

How many different rocks are possible having the characteristics shown in the table? We can represent this situation by using a **tree diagram**.

Rock Characteristics		
Mineral	**Texture**	**Color**
gold quartz iron silver	glassy dull	brown gray

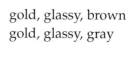

Quartz

Mineral	Texture	Color	Outcome
gold	glassy	brown	gold, glassy, brown
		gray	gold, glassy, gray
	dull	brown	gold, dull, brown
		gray	gold, dull, gray
quartz	glassy	brown	quartz, glassy, brown
		gray	quartz, glassy, gray
	dull	brown	quartz, dull, brown
		gray	quartz, dull, gray
iron	glassy	brown	iron, glassy, brown
		gray	iron, glassy, gray
	dull	brown	iron, dull, brown
		gray	iron, dull, gray
silver	glassy	brown	silver, glassy, brown
		gray	silver, glassy, gray
	dull	brown	silver, dull, brown
		gray	silver, dull, gray

Gold

By using the tree diagram, we find that there are 16 different rocks possible. These results are called **outcomes**. For example, the first outcome is a rock that contains gold, has a glassy texture, and is brown. The list of all the possible outcomes is called the **sample space**.

Example ─① **Dance Link**

The Greenwood Dance Company has three lead female dancers, Teresa, Kendra, and Angelica, and two lead male dancers, Keith and Jamal. How many different ways can the director choose one female dancer and one male dancer to lead the next production?

Make a tree diagram to find the number of combinations.

Female	**Male**	**Outcome**
Teresa	Keith	Teresa, Keith
	Jamal	Teresa, Jamal
Kendra	Keith	Kendra, Keith
	Jamal	Kendra, Jamal
Angelica	Keith	Angelica, Keith
	Jamal	Angelica, Jamal

The sample space contains six outcomes. So, there are six different ways that the director can choose the leads for the next production.

Your Turn

a. The Ice Cream Parlor offers the choices below for making sundaes. Draw a tree diagram to find the number of different sundaes that can be made.

Ice Cream	Topping	Whipped Cream
chocolate vanilla rocky road	chocolate butterscotch	yes no

 Personal Tutor at algconcepts.com

An **event** is a subset of the possible outcomes, or sample space. In Example 1, the choice of female dancers is one event, and the choice of male dancers is another event. Notice that the product of the number of choices in each event is 3 · 2 or 6. This method of finding the number of possible outcomes is called the **Fundamental Counting Principle**.

Fundamental Counting Principle	If event *M* can occur in *m* ways and is followed by event *N* that can occur in *n* ways, then the event *M* followed by event *N* can occur in *m* × *n* ways.

Example ──➋ **How many different kinds of school sweatshirts are possible?**

Clothing Link

There are 3 styles, 2 colors, and 5 sizes, so the number of different sweatshirts is $3 \times 2 \times 5$ or 30.

Style	Color	Size
school name	red	small
school logo	tan	medium
team graphic		large
		1X
		2X

Your Turn

b. Catina wants to buy a red sweatshirt in size large or 1X in any style. How many choices does she have?

Check for Understanding

Communicating Mathematics

1. **Explain** the Fundamental Counting Principle in your own words.

2. **Represent** a real-life situation by drawing a tree diagram. Include a description of each event and a list of all the outcomes in the sample space.

3. **You Decide** Ling says that if three coins are tossed, then the number of outcomes with 2 heads and 1 tail is the same as the number with 1 head and 2 tails. Lorena thinks he is wrong. Who is correct and why?

Vocabulary

tree diagram
outcome
sample space
event

Guided Practice

⟳ **Getting Ready** **Suppose you spin the spinner twice. Determine whether each is an *outcome* or a *sample space*.**

Sample 1: (red, blue)

Solution: Spinning a red and then a blue (red, blue) is an outcome because it is only one possible result.

Sample 2: (red, red), (red, blue), (blue, red), (blue, blue)

Solution: This is a sample space because it is a list of all possible outcomes.

Determine whether each is an *outcome* or a *sample space* for the given experiment.

4. (5, 2, 2); rolling a number cube three times

5. (H, T); tossing a coin once

6. (H, H), (H, T), (T, H), (T, T); tossing a coin twice

7. (4, 10, J, 3, 7); choosing five cards from a standard deck

Example 1

8. Suppose you can order a burrito or a taco with beef, chicken, or bean filling. Find the number of possible outcomes by drawing a tree diagram.

Example 2

9. Suppose you roll a die twice. Find the number of possible outcomes by using the Fundamental Counting Principle.

Example 1

10. **Shopping** Enrique wants to buy a bicycle, but he is having trouble deciding what kind to buy.

Type	Speed	Handlebars
road	21	touring
mountain	16	racing
	14	
	10	

   a. Draw a tree diagram to represent his choices.
   b. How many different kinds of bicycles are possible?
   c. How many different kinds of bicycles have racing handlebars and 16 or greater gear speeds?

Exercises

Practice

Find the number of possible outcomes by drawing a tree diagram.

11. three tosses of a coin
12. choosing one marble from each box shown in the table at the right

Box A	Box B	Box C
blue	green	green
green	red	white

13. Anita purchased five T-shirts and three pairs of jeans for school, as shown in the table at the right. How many different T-shirt and jeans outfits are possible?

T-Shirt	Jeans
red	black
orange	white
green	blue
white	
striped	

Find the number of possible outcomes by using the Fundamental Counting Principle.

14. different cars with options shown at the right
15. possible sequences of answers on a 5-question true-false quiz

Color	Engine	Transmission
red	4-cylinder	manual
blue	6-cylinder	automatic
white		
green		

16. different 1-topping pizzas with the choices shown at the right

Crust	Size	Topping	
thin	individual	pepperoni	onion
regular	small	mushroom	olive
thick	medium	sausage	pepper
	large		

17. Refer to Example 1. How many different ways can the director choose one female dancer and one male dancer if there are five female dancers and six male dancers?

18. Suppose the spinner at the right is spun twice. Assume that the spinner will not land on a border.

 a. Let R represent red, B represent blue, and Y represent yellow. Write all possible outcomes in the sample space as ordered pairs.

 b. How many outcomes have at least one blue?

19. Write a situation that fits the tree diagram below.

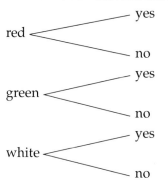

red — yes
red — no
green — yes
green — no
white — yes
white — no

Applications and Problem Solving

20. **Dining** A free continental breakfast offers one type of bread and one beverage.

 a. How many different breakfasts are possible?

 b. How many of the breakfasts include a muffin?

Bread	Beverage
bagel	tea
bran muffin	coffee
English muffin	juice
white toast	milk
wheat toast	
raisin toast	

21. **History** In 1869, Fanny Jackson Coppin became the first African-American school principal. One of her favorite teaching poems began:

> A noun is the name of anything,
> As _school_, or _garden_, _hoop_, or _swing_.
> Adjectives tell the kind of noun,
> As _great_, _small_, _pretty_, _white_, or _brown_.

How many two-word phrases can you make using one of the underlined nouns and one of the underlined adjectives from the poem?

22. **Family** Valerie and Jessie just got married. They hope to eventually have two girls and a boy, in any order.

 a. How many combinations of three children are possible?

 b. How many outcomes will give them two girls and a boy?

 c. In how many of the outcomes from part b will the girls be born in consecutive order?

23. **Keys**   A car manufacturer makes keys with six sections.

   a. Until the 1960s, there were only two patterns for each section. How many different keys were possible?

   b. After the 1960s, the car manufacturer made keys having three different patterns for each section. How many different keys were possible after the 1960s?

24. **Critical Thinking**   The president, vice president, secretary, and treasurer of the Drama Club pose for a yearbook picture. If they sit in four chairs, how many different seating arrangements are possible?

Mixed Review

Find each product.   *(Lesson 4–1)*

25. $-3 \cdot 5.4$

26. $-7.2(-1.5)$

27. $\dfrac{3}{5} \cdot \dfrac{2}{7}$

28. **Manufacturing**   A certain bolt used in lawn mowers will work properly only if its diameter differs from 2 centimeters by exactly 0.04 centimeter.   *(Lesson 3–7)*

   a. Let d represent the diameter. Write an equation to represent this problem.

   b. What are the least and greatest diameters for this bolt?

29. Solve $x - \dfrac{1}{3} = -\dfrac{3}{4}$. Check your solution.   *(Lesson 3–6)*

Standardized Test Practice

Ⓐ Ⓑ Ⓒ Ⓓ

30. **Grid In**   In the graph below, how many degrees did the temperature rise from 1:00 to 5:00?   *(Lesson 2–4)*

31. **Multiple Choice**   Ayani has $600 in the bank at an annual interest rate of 3%. How much money will he have in his account after four years? (*Hint:* Use the formula for simple interest $I = prt$.)   *(Lesson 1–5)*

   Ⓐ $72   Ⓑ $672   Ⓒ $607   Ⓓ $720

Investigation

Materials

 index cards

 green, blue, pink, and yellow self-adhesive notes

 graphing calculator

Permutations and Combinations

Mandy makes and sells stuffed rabbits at craft shows. There are three sets of pieces in the rabbit pattern, and each set is cut from a different color fabric. One set of pieces is for the body, one is for the arms, and the other is for the legs. How many different rabbits can she make?

Investigate

1. Label the index cards Set 1, Set 2, and Set 3.

   a. Suppose Mandy has three different fabrics. Let a green, a blue, and a pink self-adhesive note represent the three fabrics.

   b. Lay the cards on a table and place a different color note on each.

This represents one arrangement of colors.

   c. Find and record all of the different arrangements that are possible using each color only once. Two different arrangements are shown.

Arrangement	Set 1 (body)	Set 2 (arms)	Set 3 (legs)
A	blue	green	pink
B	blue	pink	green

   d. An arrangement of objects in which order is important is called a **permutation**. How many permutations of the three colors are there?

   e. Suppose the order of the three colors does not matter. For example, a rabbit with a green body, blue arms, and pink legs is considered the same as a rabbit with a blue body, pink arms, and green legs. An arrangement of objects in which order is not important is called a **combination**. How many combinations are possible using green, blue, and pink fabrics?

2. Suppose Mandy has four different fabrics. Use a yellow self-adhesive note to represent the fourth fabric.

a. Use the index cards to find the number of different rabbits that are possible when there are four colors from which to choose. Record your results in a new table. Remember that you can use only three colors at a time. Two possible arrangements are shown below.

Arrangement	Set 1 (body)	Set 2 (arms)	Set 3 (legs)
A	green	blue	pink
B	yellow	blue	pink

How many permutations are possible?

b. How many different rabbits are possible if the order of the colors is not important?

Extending the Investigation

In this extension, you will use formulas to find permutations and combinations. The mathematical notation 4! means $4 \cdot 3 \cdot 2 \cdot 1$. The symbol 4! is read *four **factorial***. *n*! means the product of all counting numbers beginning with *n* and counting backward to 1.

- The number of permutations of *n* objects taken *r* at a time is defined as $_nP_r = \dfrac{n!}{(n-r)!}$.

  The number of permutations of 4 colors taken 3 at a time can be written $_4P_3$. Find $_4P_3$ and compare your answer to the answer in Exercise 2a.

- The number of combinations of *n* objects taken *r* at a time is defined $_nC_r = \dfrac{n!}{(n-r)!r!}$.

  The number of combinations of 4 colors taken 3 at a time can be written $_4C_3$. Find $_4C_3$ and compare your answer to the answer in Exercise 2b.

- Use the formulas to find the number of permutations and combinations of 5 colors taken 3 at a time and 6 colors taken 3 at a time.

- We define 0! as 1. **Make a conjecture** comparing $_nP_n$ and $_nC_n$.

Presenting Your Conclusions

Here are some ideas to help you present your conclusions to the class.

- Make a pamphlet showing the permutations and combinations for the rabbit pieces and fabrics.

- Include diagrams and tables as needed. Show how the formulas can be used to make the calculations.

*inter*NET
CONNECTION **Investigation** For more information on permutations and combinations, visit: www.algconcepts.com

What You'll Learn
You'll learn to divide rational numbers.

Why It's Important
Cooking To adjust recipes for different serving sizes, cooks must know how to divide rational numbers.
See Example 5.

When you divide two rational numbers, the quotient is also a rational number. You can use the rules below to find the sign of the quotient. The sign rules are the same as the ones used for dividing integers.

Dividing Rational Numbers	Words	Numbers
Different Signs	The quotient of two numbers with different signs is negative.	$-2.1 \div 7 = -0.3$ $2.1 \div (-7) = -0.3$
Same Sign	The quotient of two numbers with the same sign is positive.	$4.5 \div 0.9 = 5$ $-4.5 \div (-0.9) = 5$

Examples

Prerequisite Skills Review
Operations with Decimals, p. 684

Find each quotient.

1 $-6 \div 0.5$
$-6 \div 0.5 = -12$ *Numbers have different signs. The quotient is negative.*

2 $-7.4 \div (-2)$
$-7.4 \div (-2) = 3.7$ *Numbers have the same sign. The quotient is positive.*

Your Turn

a. $16 \div (-2.5)$ **b.** $-3.9 \div 3$ **c.** $-8.4 \div (-1.2)$

Two numbers whose product is 1 are called **multiplicative inverses** or **reciprocals**.

$\dfrac{7}{8}$ and $\dfrac{8}{7}$ are reciprocals because $\dfrac{7}{8} \cdot \dfrac{8}{7} = 1$.

-3 and $-\dfrac{1}{3}$ are reciprocals because $-3 \cdot \left(-\dfrac{1}{3}\right) = 1$.

a and $\dfrac{1}{a}$, where $a \neq 0$, are reciprocals because $a \cdot \dfrac{1}{a} = 1$.

These examples demonstrate the **Multiplicative Inverse Property**.

Multiplicative Inverse Property		
	Words:	The product of a number and its multiplicative inverse is 1.
	Symbols:	For every number $\dfrac{a}{b}$, where $a, b \neq 0$, there is exactly one number $\dfrac{b}{a}$ such that $\dfrac{a}{b} \cdot \dfrac{b}{a} = 1$.
	Numbers:	$-\dfrac{4}{9}$ and $-\dfrac{9}{4}$ are multiplicative inverses.

In Lesson 2–4, you learned that subtracting a number is the same as adding its additive inverse. In the same way, dividing by a number is the same as multiplying by its multiplicative inverse or reciprocal.

Dividing Fractions	**Words:**	To divide a fraction by any nonzero number, multiply by the reciprocal of the number.
	Symbols:	$\dfrac{a}{b} \div \dfrac{c}{d} = \dfrac{a}{b} \cdot \dfrac{d}{c}$, where $b, c, d \neq 0$
	Numbers:	$\dfrac{3}{4} \div \dfrac{2}{5} = \dfrac{3}{4} \cdot \dfrac{5}{2}$ or $\dfrac{15}{8}$

Use the rules for dividing rational numbers to determine the sign of the quotient.

Examples

Find each quotient.

3 $\dfrac{4}{7} \div (-8)$

$\dfrac{4}{7} \div (-8) = \dfrac{4}{7} \cdot \left(-\dfrac{1}{8}\right)$ *To divide by -8, multiply by its reciprocal, $-\dfrac{1}{8}$.*

$\qquad\qquad = -\dfrac{4}{56}$ *The numbers have different signs. The product is negative.*

$\qquad\qquad = -\dfrac{1}{14}$ *Write the fraction in simplest form.*

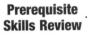

Prerequisite Skills Review
Simplifying Fractions, p. 685

4 $\dfrac{5}{6} \div \left(1\dfrac{2}{5}\right)$

$\dfrac{5}{6} \div \left(1\dfrac{2}{5}\right) = \dfrac{5}{6} \div \dfrac{7}{5}$ *Rewrite $1\dfrac{2}{5}$ as an improper fraction.*

$\qquad\qquad = \dfrac{5}{6} \cdot \dfrac{5}{7}$ *To divide by $\dfrac{7}{5}$, multiply by its reciprocal, $\dfrac{5}{7}$.*

$\qquad\qquad = \dfrac{25}{42}$ *The numbers have the same sign. The product is positive.*

Real World

Example
Cooking Link

5 **How much sugar do you need to make one dozen cookies?**

Divide the recipe by 2. The recipe calls for $\dfrac{2}{3}$ cup of sugar.

$\left(\begin{matrix}\text{amount}\\\text{of sugar}\end{matrix}\right) \div 2 = \dfrac{2}{3} \div 2$

$\qquad\qquad\qquad = \dfrac{2}{3} \cdot \dfrac{1}{2}$

$\qquad\qquad\qquad = \dfrac{2}{6}$ or $\dfrac{1}{3}$ *Write the fraction in simplest form.*

> **Raspberry Almond Cookies**
>
> | 1 cup butter | 2 cups flour |
> |---|---|
> | $\dfrac{2}{3}$ cup sugar | $\dfrac{1}{2}$ cup raspberry jam |
> | $\dfrac{1}{2}$ tsp. almond extract | |
> | Makes two dozen cookies. | |

To make one dozen cookies, you need $\dfrac{1}{3}$ cup of sugar.

Your Turn   Find each quotient.

   d. $14 \div \left(-\dfrac{2}{3}\right)$ **e.** $-\dfrac{2}{7} \div \left(-\dfrac{5}{9}\right)$ **f.** $\dfrac{1}{6} \div 2\dfrac{4}{5}$

You can use what you know about dividing rational numbers to evaluate algebraic expressions.

Example

6   Evaluate $\dfrac{x}{4}$ if $x = \dfrac{3}{5}$.

$\dfrac{x}{4} = \dfrac{\frac{3}{5}}{4}$ *Replace x with $\dfrac{3}{5}$.*

$= \dfrac{3}{5} \div 4$ *Rewrite the fraction as a division sentence.*

$= \dfrac{3}{5} \cdot \dfrac{1}{4}$ *To divide by 4, multiply by its reciprocal, $\dfrac{1}{4}$.*

$= \dfrac{3}{20}$ *Multiply the numerators and multiply the denominators.*

Your Turn   Evaluate if $x = \dfrac{3}{5}$.

   g. $\dfrac{6}{x}$ **h.** $-\dfrac{x}{2}$ **i.** $\dfrac{4}{7x}$

Check for Understanding

Communicating Mathematics

1. **Determine** whether the reciprocal of a negative rational number can ever be positive. Explain.

2. **Compare** a positive rational number n with the value of $n \div \dfrac{1}{2}$ and with the value of $n \cdot \dfrac{1}{2}$. Which is the greatest? the least?

3. **Writing Math** Describe real-life situations in which you would divide fractions to solve problems.

Vocabulary

multiplicative inverse
reciprocal

Guided Practice

Getting Ready   Name the reciprocal of each number.

Sample: $-\dfrac{1}{4}$ **Solution:** $-\dfrac{1}{4}\left(-\dfrac{4}{1}\right) = 1$, so the reciprocal is $-\dfrac{4}{1}$ or -4.

   4. 3 **5.** $-\dfrac{2}{3}$ **6.** $2\dfrac{3}{4}$ **7.** $-\dfrac{1}{x}, x \neq 0$

Examples 1–5

Find each quotient.

8. $-6 \div 1.5$

9. $3.8 \div 19$

10. $-4.7 \div (-0.5)$

11. $-\frac{1}{2} \div \left(\frac{7}{3}\right)$

12. $4 \div \left(\frac{7}{8}\right)$

13. $2\frac{2}{5} \div \frac{1}{3}$

Example 6

Evaluate each expression if $a = \frac{1}{4}$ and $b = -\frac{2}{3}$.

14. $\frac{a}{3}$

15. $\frac{5}{b}$

16. $\frac{a}{b}$

Example 4

17. **Lunch** Students working on decorations for homecoming ordered pizza for lunch. They ordered five 8-slice pizzas, and each person ate $2\frac{1}{2}$ slices. There was no pizza left over. How many students were there?

Exercises • • • • • • • • • • • • • • • • •

Practice

Find each quotient.

18. $8 \div (-1.6)$

19. $-7.5 \div 3$

20. $9.6 \div 3.2$

21. $2.7 \div -27$

22. $0.4 \div -0.4$

23. $15.6 \div 1.3$

24. $\frac{1}{8} \div \frac{1}{7}$

25. $0 \div \frac{3}{8}$

26. $\frac{3}{4} \div \left(-\frac{4}{5}\right)$

27. $-\frac{5}{6} \div \frac{5}{6}$

28. $\frac{1}{-7} \div \frac{4}{9}$

29. $-\frac{2}{3} \div \left(-\frac{5}{7}\right)$

30. $-\frac{7}{9} \div \left(-\frac{9}{7}\right)$

31. $5 \div \frac{1}{9}$

32. $-\frac{1}{5} \div 10$

33. $\frac{-3}{4} \div 8$

34. $-14 \div \left(-2\frac{4}{5}\right)$

35. $-3\frac{1}{2} \div \frac{1}{6}$

Homework Help	
For Exercises	See Examples
18–23	1, 2
24–35	3, 4
36–49	5, 6
Extra Practice	
See page 699.	

Evaluate each expression if $j = -\frac{1}{2}$, $k = \frac{3}{4}$, and $n = \frac{1}{5}$.

36. $\frac{3}{n}$

37. $\frac{9}{k}$

38. $\frac{k}{2}$

39. $\frac{-j}{5}$

40. $\frac{1}{2n}$

41. $\frac{n}{j}$

42. $\frac{k}{j}$

43. $\frac{2n}{j}$

44. $\frac{jk}{n}$

45. Find the quotient of -12 and $-\frac{2}{3}$.

46. Solve the equation $t = -\frac{7}{18} \div 3\frac{1}{2}$ to find the value of t.

**Applications and
Problem Solving**

47. Track Mr. Vance is training for a 10-kilometer race in a senior citizen track meet. Suppose $3\frac{1}{4}$ laps equals 1 kilometer and he wants to finish the race in 65 minutes. In how many minutes will he have to run each lap?

48. Nutrition A 20-ounce bottle of soda contains $16\frac{2}{3}$ teaspoons of sugar. A 12-ounce can of soda contains $16\frac{2}{3} \div \frac{5}{3}$ teaspoons of sugar. How much sugar is in a 12-ounce can of soda?

49. Sales The performing arts series at Central State University offers a package ticket to students for $62.50. If there are five performances, how much do students pay for each performance?

50. Critical Thinking

   a. Is the division of two rational numbers always a rational number? If not, give a counterexample.

   b. Is the set of rational numbers closed under division? Explain.

   c. What number has no reciprocal? Explain.

Mixed Review

51. Uniforms The school dress for female students is a uniform, a blouse, and socks. A sweater may be added, if necessary. Find the number of different outfits that female students can wear.
(Lesson 4–2)

Uniform	Blouse	Socks	Sweater
skirt	white	white	yes
jumper	yellow	blue	no
	blue	gray	
	pink		

52. Sales Ms. Ortiz wants to buy a washer and dryer from Bargain Appliances. If she buys on the "90 days same as cash" plan, she will have to make 5 equal payments of $146.12 each. If she writes a check for each payment, what is the net effect on her checking account?
(Lesson 4–1)

Find the mean, median, mode, and range for each set of data.
(Lesson 3–3)

53. 6, 8, 8, 8, 6, 4, 3, 7, 4

54. 9, 4, 13, 11, 7, 6, 20

55. 20, 11, 20, 23, 20, 29

Replace each ● with <, >, or = to make a true sentence. *(Lesson 3–1)*

56. -3.6 ● 0

57. $\frac{3}{10}$ ● 0.3

58. $-7(-4)$ ● $7 - 15$

Which is the better buy? Explain. *(Lesson 3–1)*

59. a $\frac{1}{2}$-pound bag of cashews for $3.15 or $\frac{3}{4}$-pound bag for $5.19

60. three liters of soda for $2.25 or two liters for $1.98

61. a 48-ounce bottle of dishwashing liquid for $2.69 or a 22-ounce bottle for $1.09

62. Grid In Evaluate $16 + k$ if $k = -11$. *(Lesson 2–3)*

63. Multiple Choice Write the ordered pair that names point Q. *(Lesson 2–2)*

Ⓐ $(2, -3)$ Ⓑ $(-2, -3)$

Ⓒ $(-3, 2)$ Ⓓ $(-3, -2)$

Quiz 1   Lessons 4–1 through 4–3

Find each product. *(Lesson 4–1)*

1. $9.4 \cdot (-2)$

2. $-\frac{3}{8}\left(-\frac{1}{7}\right)$

3. $4 \cdot \frac{5}{8}$

Find the number of possible outcomes. *(Lesson 4–2)*

4. choosing a sandwich, a side, and a beverage from the table

5. different three-digit security codes using the numbers 1, 2, and 3 (You can use the numbers more than once.)

6. rolling a number cube three times

Sandwich	Side	Beverage
ham	French fries	milk
turkey	cole slaw	tea
egg salad	fruit	soda
		juice

Exercise 4

Find each quotient. *(Lesson 4–3)*

7. $-8.1 \div (-3)$

8. $\frac{1}{2} \div \left(-\frac{3}{4}\right)$

9. $2\frac{1}{3} \div \frac{3}{4}$

10. World Records The longest loaf of bread ever baked was 2132 feet $2\frac{1}{2}$ inches. If this loaf were cut into $\frac{1}{2}$-inch wide slices, how many slices of bread would there have been? *(Lesson 4–3)*

Solving Multiplication and Division Equations

In Lesson 3-6 you learned about the Addition Property of Equality and the Subtraction Property of Equality. There are properties of equality for multiplication and division as well. One way to solve algebraic equations is to use the **Division Property of Equality**.

Division Property of Equality	Words:	If you divide each side of an equation by the same nonzero number, the two sides remain equal.
	Symbols:	For any numbers a, b, and c, with $c \neq 0$, if $a = b$, then $\dfrac{a}{c} = \dfrac{b}{c}$.
	Numbers:	If $3x = 12$, then $\dfrac{3x}{3} = \dfrac{12}{3}$.

Examples

Solve each equation. Check your solution.

1 $3y = 45$

$3y = 45$ *Original equation*

$\dfrac{3y}{3} = \dfrac{45}{3}$ *Divide each side by 3.*

$y = 15$ *Simplify.*

Check: $3y = 45$ *Original equation*

$3(15) \stackrel{?}{=} 45$ *$y = 15$*

$45 = 45$ ✓

2 $16 = -2h$

$16 = -2h$ *Original equation*

$\dfrac{16}{-2} = \dfrac{-2h}{-2}$ *Divide each side by −2.*

$-8 = h$ *Simplify.*

Check: $16 = -2h$ *Original equation*

$16 \stackrel{?}{=} -2(-8)$ *$h = -8$*

$16 = 16$ ✓

3 $7.4a = -37$

$7.4a = -37$ *Original equation*

$\dfrac{7.4a}{7.4} = \dfrac{-37}{7.4}$ *Divide each side by 7.4.*

$a = -5$ *Check by substituting into the original equation.*

Your Turn   Solve each equation. Check your solution.

  a. $4t = 28$ **b.** $19.2 = -6d$ **c.** $-0.1m = -7$

Sometimes, you will have to write and solve an equation to solve a problem.

Example **4**
Consumer Link

Xue Wu is in charge of buying sandwiches for a hiking trip. She has $24, and sandwiches cost $3.80 each. How many sandwiches can she buy?

Explore You know that the sandwiches cost $3.80 and Xue Wu has $24.

Plan Let *s* represent the number of sandwiches and write an equation to represent the problem.

price per sandwich	times	number of sandwiches	equals	total cost
3.80	$\times$	*s*	=	24

Solve Solve the equation for *s*.

$$3.80s = 24 \quad \textit{Original equation}$$

$$\frac{3.80s}{3.80} = \frac{24}{3.80} \quad \textit{Divide each side by 3.80.}$$

$$s \approx 6.3 \quad \textit{Simplify.}$$

Since Xue Wu can't buy part of a sandwich, she has enough money to buy 6 sandwiches.

Examine Does the answer make sense?
Round 3.80 to 4
and multiply by 6.

$4 \times 6 = 24$

So 6 sandwiches
would cost
about $24. The
answer makes sense.

Another useful property in solving equations is the **Multiplication Property of Equality**.

Multiplication Property of Equality	**Words:**	If you multiply each side of an equation by the same number, the two sides remain equal.
	Symbols:	For any numbers *a*, *b*, and *c*, if $a = b$, then $ac = bc$.
	Numbers:	If $\frac{1}{5}x = 8$, then $5\left(\frac{1}{5}x\right) = 5(8)$.

Solve each equation. Check your solution.

5 $\frac{g}{4} = 9$

$\frac{g}{4} = 9$ *Original equation*   **Check:** $\frac{g}{4} = 9$

$4\left(\frac{g}{4}\right) = 4(9)$ *Multiply each side by 4.* $\frac{36}{4} \stackrel{?}{=} 9$ $g = 36$

$g = 36$ *Simplify.* $9 = 9$ ✓

6 $-5 = -\frac{1}{6}t$

$-5 = -\frac{1}{6}t$ *Original equation*   **Check:** $-5 = -\frac{1}{6}t$

$-6(-5) = -6\left(-\frac{1}{6}t\right)$ *Multiply each side by -6.* $-5 \stackrel{?}{=} -\frac{1}{6}(30)$

$30 = t$ *Simplify.* $-5 = -5$ ✓

7 $-\frac{2}{3}x = 12$

$-\frac{2}{3}x = 12$ *Original equation*

$-\frac{3}{2}\left(-\frac{2}{3}x\right) = -\frac{3}{2}(12)$ *Multiply each side by $-\frac{3}{2}$.*

$x = -18$ *Check by substituting into the original equation.*

Your Turn

d. $\frac{t}{4} = 8$ e. $\frac{5}{6}a = 25$ f. $36 = -\frac{3}{4}x$

Example
Geography Link

8 In 1999, Nunavut (NU-na-voot) became Canada's newest territory in fifty years. There are 15,000 people in Nunavut under age 25. This is about $\frac{3}{5}$ of the total population. What is the population of Nunavut?

Let n represent the total population. Then $\frac{3}{5}n = 15{,}000$ represents the problem.

$\frac{3}{5}n = 15{,}000$ *Original equation*

$\frac{5}{3}\left(\frac{3}{5}n\right) = \frac{5}{3}(15{,}000)$ *Multiply each side by $\frac{5}{3}$. Why?*

$n = 25{,}000$ *Simplify.*

The total population of Nunavut is 25,000.

 Personal Tutor at algconcepts.com

Check for Understanding

Communicating Mathematics

1. **Describe** how to solve $4x = 36$.

2. **List** the steps you take when you check the solution of $4x = 36$.

3. **Write** a multiplication equation whose solution is -7.

Guided Practice

Examples 1–8

Solve each equation. Check your solution.

4. $8x = -24$

5. $-5v = -40$

6. $6 = 0.3y$

7. $\frac{r}{15} = 3$

8. $-7 = \frac{f}{8}$

9. $-9 = -\frac{9}{4}n$

Example 3

10. **Geometry** The area of the rectangle is 51.2 square centimeters. Find the length.

6.4 cm

x cm

Exercises

Practice

Solve each equation. Check your solution.

11. $4c = 24$

12. $9p = 63$

13. $30 = 2x$

14. $-6a = 60$

15. $-8t = 56$

16. $-3 = -5n$

17. $14 = 0.5x$

18. $0 = 3.9c$

19. $2.2r = 11$

20. $\frac{9}{2}t = 1$

21. $\frac{1}{2}y = 32$

22. $\frac{b}{3} = -5$

23. $10 = \frac{a}{-7}$

24. $-\frac{2}{7}x = -6$

25. $-18 = -\frac{2}{3}p$

26. $-\frac{8}{5}p = 40$

27. $-\frac{24}{5}k = 12$

28. $-\frac{63}{10} = -\frac{21}{10}b$

29. What is the solution of $-18.4 = -9.2n$?

30. Solve $\frac{k}{6} = -6$. Then check your solution.

31. Find the value of r in the equation $\left(-4\frac{1}{2}\right)r = 36$.

32. If $7h - 5 = 4$, then $21h - 15 = \underline{\quad?\quad}$.

Write an equation and solve.

33. Eight times a number x is 112. What is the number?

34. Negative 154 equals the product of negative 7 and a number p. What is p?

35. Three fifths of a number y is negative 9. What is the number?

Homework Help	
For Exercises	See Examples
11–13	1
14–16	2
17–19	3
20–22	5
23–28	6, 7
29–31	5–7
36–39	4
Extra Practice	
See page 699.	

36. Plumbing Two meters of copper tubing weigh 0.25 kilogram. How
much do 50 meters of the same tubing weigh?

37. Entertainment James is in charge of purchasing student tickets for an
Atlanta Braves baseball game. Each student paid $6 per ticket, and he
collected $288.

    a. Write an equation to represent the number of students *n* who
bought tickets.

    b. How many students are going to the game?

38. Geometry Find the length
of the base *b* of the triangle
if its area *A* is 182 square
centimeters. Use the formula
$A = \frac{1}{2}bh$.

39. Technology Engineering students at The Ohio State University built
an experimental electric race car that went around a 1.1-mile lap in
36 seconds. How fast did it go? Find the rate *r* in miles per hour by
using the formula $d = rt$, where *d* is the distance and *t* is the time.
(*Hint:* Convert seconds to hours.)

40. Critical Thinking Explain why the Multiplication and Division
Properties of Equality for fractions can be thought of as one property.

Mixed Review

Find each quotient. *(Lesson 4–3)*

41. $4 \div 0.2$ **42.** $-8.1 \div 9$ **43.** $\frac{1}{4} \div \frac{3}{7}$

Solve each equation. *(Lesson 3–4)*

44. $n = 5 + 16 \div 4$ **45.** $2.6(5) - 13 = x$ **46.** $\frac{5 + 3}{3 \cdot 3} = t$

47. Animals Giant deep-sea squids can be up to 19 yards long. How
many feet long can they be? *(Lesson 2–5)*

**Standardized
Test Practice**

Ⓐ   Ⓑ   Ⓒ   Ⓓ

48. Extended Response A field
hockey playing field is shown
at the right. *(Lesson 1–4)*

    a. Write an equation to represent
the area *A* of the field.

    b. Simplify the expression and
find the area if *a* is 60 yards
and *b* is 25 yards.

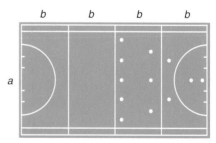

49. Multiple Choice Write an equation for the following sentence. Three
less than six times *y* equals 14. *(Lesson 1–1)*

    Ⓐ $3 + 6y = 14$ Ⓑ $3 - 6y = 14$

    Ⓒ $6y = 14 - 3$ Ⓓ $6y - 3 = 14$

  www.algconcepts.com/self_check_quiz

What You'll Learn
You'll learn to solve equations involving more than one operation.

Why It's Important
Shopping Multi-step equations are used to calculate purchase orders.
See Exercise 40.

Equations that have more than one operation require more than one step to solve. To solve this type of problem, the best strategy is to undo each operation in reverse order. In other words, work backward.

Examples

❶ Solve each equation. Check your solution.

$\frac{x}{3} + 5 = 14$

$\frac{x}{3} + 5 = 14$	*Original equation*
$\frac{x}{3} + 5 - 5 = 14 - 5$	*Subtract 5 from each side.*
$\frac{x}{3} = 9$	*Simplify.*
$3\left(\frac{x}{3}\right) = 3(9)$	*Multiply each side by 3.*
$x = 27$	*Simplify.*

Check:	$\frac{x}{3} + 5 = 14$	*Original equation*
	$\frac{27}{3} + 5 \overset{?}{=} 14$	*Replace x with 27.*
	$9 + 5 \overset{?}{=} 14$	*Simplify $\frac{27}{3}$.*
	$14 = 14 \quad \checkmark$	*Add 9 and 5.*

❷ $2k - 7 = 23$

$2k - 7 = 23$	*Original equation*
$2k - 7 + 7 = 23 + 7$	*Add 7 to each side.*
$2k = 30$	*Simplify.*
$\frac{2k}{2} = \frac{30}{2}$	*Divide each side by 2.*
$k = 15$	*Check the solution.*

❸ Solve $\frac{h - 7}{2.5} = -10$. Check your solution.

$\frac{h - 7}{2.5} = -10$	*Original equation*
$2.5\left(\frac{h - 7}{2.5}\right) = 2.5(-10)$	*Multiply each side by 2.5.*
$h - 7 = -25$	*Simplify.*
$h - 7 + 7 = -25 + 7$	*Add 7 to each side.*
$h = -18$	*Simplify.*

(continued on the next page)

Check: $\dfrac{h-7}{2.5} = -10$ *Original equation*

$\dfrac{-18-7}{2.5} \overset{?}{=} -10$ *Replace h with −18.*

$\dfrac{-25}{2.5} \overset{?}{=} -10$ *Simplify.*

$-10 = -10$ ✓

Your Turn   Solve each equation. Check your solution.

a. $11 + 9v = 119$ **b.** $\dfrac{a}{8} - 5.2 = 3$ **c.** $\dfrac{b+4}{7} = 6$

Technology Link

❹ **Trains that float on magnets above rails have been developed in Japan. They are designed to run at a maximum speed of 341 miles per hour. This is 71 miles per hour faster than twice the speed of Japan's bullet trains. Find the speed of the bullet trains in Japan and graph the solution.**

Explore   You know that the speed of the floating trains is 341 mph. This is 71 mph faster than twice Japan's bullet trains.

Plan   Let x represent the speed of the bullet trains. Then translate the given information into an equation and solve.

Solve

speed of floating train	*equals*	*71 miles per hour*	*plus*	*twice the speed of bullet trains*
341	=	71	+	2x

$341 = 71 + 2x$ *Original equation*

$341 - 71 = 71 + 2x - 71$ *Subtract 71 from each side.*

$270 = 2x$ *Simplify.*

$\dfrac{270}{2} = \dfrac{2x}{2}$ *Divide each side by 2.*

$135 = x$ *Simplify.*

The bullet trains in Japan travel at 135 miles per hour.

Examine   Check by substituting 135 into the original equation.

$341 = 71 + 2x$ *Original equation*

$341 \overset{?}{=} 71 + 2(135)$ *Replace x with 135.*

$341 \overset{?}{=} 71 + 270$ *Multiply 2 and 135.*

$341 = 341$ ✓   *Add 71 and 270.*

www.algconcepts.com/extra_examples

You can use a graphing calculator to solve multi-step equations.

Graphing Calculator Tutorial

See pp. 750–753.

To edit or replace an equation, press ▲ until the equation editor is displayed. Then edit or replace the equation.

 Graphing Calculator Exploration

Solve $2 = \dfrac{x - 5}{4}$ using a graphing calculator.

Step 1 Rewrite the equation so that one side is equal to 0.

$$2 = \frac{x - 5}{4} \qquad \textit{Original equation}$$

$$2 - 2 = \frac{x - 5}{4} - 2 \quad \textit{Subtract 2 from each side.}$$

$$0 = \frac{x - 5}{4} - 2 \quad \textit{Simplify.}$$

Step 2 Press [MATH] and select 0:Solver.

Step 3 Enter the equation after the 0 =. Press
([X,T,θ,n] [−] 5)
[÷] 4 [−] 2 [ENTER].

Step 4 Press [ALPHA] [SOLVE].

The solution is 13.

Try These

1. What would you enter into the calculator to solve $3x - 4 = 6$?

2. Work through the examples in this lesson with a calculator.

Consecutive integers are integers in counting order, such as 3, 4, and 5. Beginning with an even integer and counting by two will result in *consecutive even integers*. For example, $-6, -4, -2, 0,$ and 2 are consecutive even integers. Beginning with an odd integer and counting by two will result in *consecutive odd integers*. For example, $-1, 1, 3,$ and 5 are consecutive odd integers.

Consecutive Integers		
Integers	**Even Integers**	**Odd Integers**
$n = 1$	$n = 2$	$n = 3$
$n + 1 = 2$	$n + 2 = 4$	$n + 2 = 5$
$n + 2 = 3$	$n + 4 = 6$	$n + 4 = 7$
⋮	⋮	⋮

Example ❺

Number Theory Link

Find three consecutive even integers whose sum is -18.

Explore   You know that the sum of three consecutive even integers is -18. You need to find the three numbers.

Plan   Let n represent the first even integer. Then $n + 2$ represents the second even integer, and $n + 4$ represents the third. *Why?*

(continued on the next page)

Solve	$n + (n + 2) + (n + 4) = -18$	Write an equation.
	$3n + 6 = -18$	Add like terms.
	$3n + 6 - 6 = -18 - 6$	Subtract 6 from each side.
	$3n = -24$	Simplify.
	$\dfrac{3n}{3} = \dfrac{-24}{3}$	Divide each side by 3.
	$n = -8$	Simplify.

$$n = -8 \qquad n + 2 = -8 + 2 \qquad n + 4 = -8 + 4$$
$$= -6 \qquad\qquad = -4$$

The numbers are -8, -6, and -4.

Examine Check by adding the three numbers.
$$-8 + (-6) + (-4) \overset{?}{=} -18$$
$$-18 = -18 \quad \checkmark$$

Your Turn

d. Find three consecutive integers whose sum is 27.

 Personal Tutor at algconcepts.com

Check for Understanding

Communicating Mathematics

1. **Describe** how you work backward to solve multi-step equations.

2. **Write** an equation that requires more than one operation to solve.

3. Soto solved the equation $\dfrac{t + 16}{4} = 1$ using the steps at the right. Jean does not agree with his answer. Who is correct? Explain.

Vocabulary

consecutive integers

$$\dfrac{t + 16}{4} = 1$$
$$t + 16 = 1$$
$$t + 16 - 16 = 1 - 16$$
$$t = -15$$

Guided Practice

 State the first step in solving each equation.

Sample: $3x - 4 = -10$ **Solution:** Add 4 to each side.

4. $7 + 6v = 43$ **5.** $11 = \dfrac{g}{8} - 5$ **6.** $\dfrac{-y - 2}{3} = 15$

Examples 1–4

Solve each equation. Check your solution.

7. $6t - 1 = 11$ **8.** $7 = 4 - \dfrac{n}{3}$ **9.** $-5r + 2 = 27$

10. $-5.6 + 4d = 2$ **11.** $2 = \dfrac{f + 8}{-6}$ **12.** $\dfrac{7n - 1}{8} = 6$

Example 1

13. **Animals** A teacher in Wisconsin came up with a formula that more accurately gives a cat's or dog's age in people years.

> Take your cat's/dog's age, subtract 1, multiply by 6, add 21, and that will give you the equivalent in people years.

Source: *The Mathematics Teacher*

a. Write an equation to represent the age of a dog or cat in people years y if the dog or cat is x years old.

b. How old is a dog if he is 17 in people years?

Exercises • • • • • • • • • • • • • • • • • •

Practice

Solve each equation. Check your solution.

14. $6 = 4n + 2$
15. $8 + 3k = 5$
16. $3b - 7 = 2$
17. $15 = 1 - 2t$
18. $12 = -5h + 2$
19. $-13 - 9y = -13$
20. $8 + 1.6a = 0$
21. $0.3x + 3 = 4.8$
22. $6.5 = 2x + 4.1$
23. $27 = 3 + 2.5t$
24. $-4 - 0.7m = 2.3$
25. $-6z + 8 = -40.6$
26. $7 = \frac{x}{2} + 5$
27. $\frac{y}{3} - 6 = 4$
28. $8 + \frac{c}{-4} = 12$
29. $28 = 7 - \frac{3}{2}t$
30. $\frac{3+n}{7} = -5$
31. $-10 = \frac{s-8}{-6}$

Homework Help

For Exercises	See Examples
14–25, 33	2
26–29	1
30–32	3
34–42	4, 5

Extra Practice
See page 700.

32. What is the solution of $6 = \frac{s-8}{-7}$?

33. Find the value of y in the equation $-2.5y - 4 = 14$.

For Exercises 34–36, write an equation and solve each problem. Graph the solution.

34. Five more than three-fourths of a number c is 26. Find the number.

35. The steps below are applied to a number n to get $\frac{1}{3}$. What is the number?
   - subtract 2
   - divide by -3

36. Start with the number x. If you multiply the number by 2, divide by 5, and add 20, you get 10. What is the value of x?

Applications and Problem Solving

37. **Geometry** Find the value of x if the perimeter of the square is 20 inches.

38. **Number Theory** Four consecutive odd integers have a sum of 56.
   a. Define a variable and write an equation to represent the problem.
   b. Find the numbers.

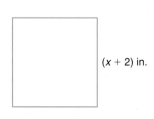

$(x + 2)$ in.

Exercise 37

39. Consumer Spending Dwayne was checking his expenses and found that in eight visits to the barbershop, he had spent $143 for haircuts. Of that amount, $15 was tips.

   a. Write an equation to represent the cost c of the haircuts.

   b. How much does Dwayne pay for each haircut before the tip?

40. Shopping Tanisa and her sister ordered a total of 5 pairs of shoes from a catalog. Each pair cost the same amount. The sisters had to pay $10 for shipping and handling. If their total purchase was $170, how much did each pair of shoes cost?

41. Gardening Mr. Green purchased 39 plants for his garden. His purchase included an equal number of ferns, irises, and gladiolas, as well as 12 impatiens.

   a. Write an equation representing the number of items he purchased.

   b. How many ferns did he buy?

Impatiens

42. Critical Thinking According to legend, Islamic mathematician al-Khowarizmi (810 A.D. – ?) stated in his will that if he had a son, the son should receive twice as much of the estate as al-Khowarizmi's wife when he died. If he had a daughter, the daughter should receive half as much as al-Khowarizmi's wife. However, al-Khowarizmi had twins, a boy and a girl. What fraction of his estate should each one get? Explain how you found your answer.

Mixed Review

Geometry **Find each missing measure.** *(Lesson 4–4)*

43. $A = 15$ in^2

44. $A = 49$ cm^2

45. $A = 27$ m^2

x in.

5 in.

7 cm

x cm

4.5 m

x m

Solve each equation. *(Lesson 3–5)*

46. $-1 + x = -5$

47. $n - 3 = 1$

48. $t + 4 = -2$

Standardized Test Practice
Ⓐ Ⓑ Ⓒ Ⓓ

49. Short Response Find the next term in the pattern $-1, 3, -9, 27, \ldots$ *(Lesson 2–5)*

50. Short Response Graph two positive integers and two negative integers on a number line. *(Lesson 2–1)*

 www.algconcepts.com/self_check_quiz

4-6 Variables on Both Sides

What You'll Learn

You'll learn to solve equations with variables on both sides.

Why It's Important

Sports Equations comparing athletes' performances have variables on both sides.
See Example 3.

Many equations contain variables on both sides. To solve these equations, first use the Addition or Subtraction Property of Equality to write an equivalent equation that has all of the variables on one side. Then solve.

Examples

Solve each equation. Check your solution.

❶ $5x = x - 12$

$$5x = x - 12 \qquad \textit{Original equation}$$
$$5x - x = x - 12 - x \qquad \textit{Subtract x from each side.}$$
$$4x = -12 \qquad \textit{Simplify.}$$
$$\frac{4x}{4} = \frac{-12}{4} \qquad \textit{Divide each side by 4.}$$
$$x = -3 \qquad \textit{Simplify.}$$

Check:
$$5x = x - 12 \qquad \textit{Original equation}$$
$$5(-3) \overset{?}{=} -3 - 12 \qquad \textit{Replace x with } -3.$$
$$-15 = -15 \quad \checkmark$$

❷ $16 + \frac{1}{4}n = \frac{3}{4}n$

$$16 + \frac{1}{4}n = \frac{3}{4}n \qquad \textit{Original equation}$$
$$16 + \frac{1}{4}n - \frac{1}{4}n = \frac{3}{4}n - \frac{1}{4}n \qquad \textit{Subtract } \frac{1}{4}n \textit{ from each side.}$$
$$16 = \frac{1}{2}n \qquad \qquad \frac{3}{4}n - \frac{1}{4}n = \frac{2}{4}n \textit{ or } \frac{1}{2}n$$
$$2(16) = 2\left(\frac{1}{2}n\right) \qquad \textit{Multiply each side by 2.}$$
$$32 = n \qquad \textit{Check the solution.}$$

Your Turn

a. $a + 9 = 4a$

b. $\frac{2}{3}n = \frac{1}{3}n - 2$

c. $5k - 2.4 = 3k + 1$

You can solve an equation with a variable on both sides to determine when a quantity that is increasing and one that is decreasing will be the same.

Example ③
Sports Link

The 2003 World Championship 100-meter dash times were 10.07 seconds for men and 10.85 seconds for women. If each year the men's times decrease 0.08 second and the women's times decrease 0.14 second, when would they have the same winning times?

Let y represent the number of years. Write an equation.

Words	*Men's times decrease 0.08 second per year.*	*Women's times decrease 0.14 second per year.*
Variables	$10.07 - 0.08y$	$10.85 - 0.14y$

Equation

$$10.07 - 0.08y = 10.85 - 0.14y \quad \textit{Write an equation.}$$

$$10.07 - 0.08y + 0.14y = 10.85 - 0.14y + 0.14y \quad \textit{Add 0.14y to each side.}$$

$$10.07 + 0.06y = 10.85 \quad \textit{Simplify.}$$

$$10.07 - 10.07 + 0.06y = 10.85 - 10.07 \quad \textit{Subtract 10.07 from each side.}$$

$$0.06y = 0.78 \quad \textit{Simplify.}$$

$$\frac{0.06y}{0.06} = \frac{0.78}{0.06} \quad \textit{Divide each side by 0.2.}$$

$$y = 13 \quad \textit{Simplify.}$$

At these rates, men and women have the same times 13 years after 2003, or in 2016.

 Personal Tutor at algconcepts.com

Some equations have no solution. This means that there is no value for the variable that will make the equation true. Other equations may have every number as the solution. An equation that is true for every value of the variable is called an **identity**.

Examples ④

Solve each equation.

$5x + 3 = 8 + 5x$

$$5x + 3 = 8 + 5x \quad \textit{Original equation}$$

$$5x + 3 - 5x = 8 + 5x - 5x \quad \textit{Subtract 5x from each side.}$$

$$3 = 8 \quad \textit{False statement}$$

The equation has no solution.   *3 = 8 is never true.*

⑤ $3 + 3t - 5 = 3t - 2$

$$3 + 3t - 5 = 3t - 2 \quad \textit{Original equation}$$

$$3t - 2 = 3t - 2 \quad \textit{Simplify.}$$

The equation is an identity.   *3t − 2 = 3t − 2 is true for all values of t.*

Your Turn

d. $2t + 4 - t = 4 + t$

e. $16h + 7 = 16 + 16h$

www.algconcepts.com/extra_examples

Communicating Mathematics

1. **List** the steps you would take to solve the equation $3x + 5 = -2x - 16$.

2. **Choose** the correct term and justify your answer. Equations with variables on both sides (*sometimes, always, never*) have a solution.

3. **Explain** the difference between an equation that is an identity and an equation with no solution.

Guided Practice

⏱ Getting Ready State the first step in solving each equation.

Sample: $9r = -10 + 7r$ **Solution:** Subtract $7r$ from each side.

4. $2y = 4 - 3y$ 5. $0.5p + 4 = 1.5p$ 6. $\dfrac{-y - 6}{3} = 7y$

Examples 1–5

Solve each equation. Check your solution.

7. $10x = 3x + 14$

8. $2r - 21 = 3 - 4r$

9. $8y + 1 = 1 + 8y$

10. $7 - 6x = 41 - 6x$

11. $3.4a + 3 = 2a - 4$

12. $\dfrac{1}{3}m - 10 = \dfrac{2}{3}m + 4$

Example 1

13. **Number Theory**   Three times a number n is 24 less than five times the number.

   a. Write an equation to represent the problem.

   b. Find the value of n.

Exercises

Practice

Solve each equation. Check your solution.

14. $10h = 8h + 6$

15. $7u - 19 = 6u$

16. $6x + 2 = 3x - 1$

17. $5t - 9 = -3t + 7$

18. $2n + 15 = 2n + 20 - 5$

19. $3x + 6 = 7x - 4$

20. $8y - 5 = -y + 2$

21. $18 + 3k = 22 + 3k$

22. $12 + 2.5a = 5a$

23. $a + 3.1 = 6.6 - 4a$

24. $6.2y + 7 = 3y - 1$

25. $3.2x - 1.8 = 6 + 11x$

26. $4.5 - x = 2x - 8.1$

27. $1.6 - 4.3y = -5.2 - 2.3y$

28. $\dfrac{1}{2}d = 2 - \dfrac{1}{2}d$

29. $\dfrac{1}{5}p - 4 = \dfrac{2}{5}p$

30. $\dfrac{1}{3}z - 4 = -\dfrac{1}{3}z + 18$

31. $\dfrac{1}{9}p - 9 = \dfrac{4}{9}p - 29$

Homework Help

For Exercises	See Examples
14–21	1, 4, 5
22–33	2, 4, 5
34–37	3

Extra Practice
See page 700.

32. Solve $\frac{3}{4}n + 16 = 2 - \frac{1}{8}n$.

33. Find the solution of $18 - 3.8x = 7.5 - 1.3x$.

34. Eight times a number n is 51 more than five times the number.

  a. Write an equation to represent the problem.

  b. Find the number.

Applications and Problem Solving

35. Geometry The length of the rectangle is twice the width.

  a. Write an equation to represent the relationship between the length and width.

$(3w - 18)$ cm

w cm

  b. What are the dimensions of the rectangle?

36. Temperature On the Fahrenheit scale, water freezes at 32°. On the Celsius scale, water freezes at 0°. The formula $F = \frac{9}{5}C + 32$ gives the temperature in degrees Fahrenheit when the temperature in degrees Celsius is known. Is the temperature ever the same on both scales? To find a temperature when both are the same, let $F = C$ in the equation $F = \frac{9}{5}C + 32$ and solve $C = \frac{9}{5}C + 32$.

37. Critical Thinking Write an equation with variables on both sides that has a solution of 6.

Mixed Review

38. Solve $3 + 6k = 45$. Check your solution. *(Lesson 4–5)*

Solve each equation. Check your solution. *(Lesson 4–4)*

39. $-7t = 63$ **40.** $0 = 4.5c$ **41.** $-\frac{2}{3}x = -8$

42. School Supplies How many different kinds of transparent tape are possible? *(Lesson 4–2)*

Transparent Tape		
Type	**Size**	**Color**
permanent	$\frac{1}{2}$ in.	clear
removable		lime
double-sided	$\frac{3}{4}$ in.	purple
	1 in.	pink

43. Entertainment To raise money for new uniforms, the marching band set up a lottery using three numbers. The first number is between 1 and 4, inclusive. The second number is between 3 and 8, inclusive. The third number is between 5 and 15, inclusive. How many different lottery numbers are possible?
(Lesson 4–2)

Solve each equation. Check your solution. *(Lesson 3–6)*

44. $8 + g = 17$ **45.** $b + (-5) = 26$ **46.** $k - 6 = -11$

47. Grid In Most people drink more than the recommended eight glasses of beverage each day. However, some beverages actually make people thirsty. The graph shows the daily average glasses of beverages per person. Find the difference between the glasses of water and the glasses of caffeine soda that a person drinks on average per day. *(Lesson 3–2)*

Some Drinks Leave You Thirsty

Quenches Your Thirst

Water	4.6
Juice	1.4
Milk	1.3
No-caffeine soda	0.6

Leaves You Thirsty

Coffee/tea	2.8
Caffeine soda	1.3
Alcohol	0.8

Source: Yankelovich for New York Hospital Nutrition Information Center

48. Multiple Choice The frequency table shows the results of a survey that asked students how many times a week they buy lunch at school. Determine which statement is correct. *(Lesson 1–6)*

Ⓐ Most students buy lunch four times a week.

Ⓑ The greatest number of students buys lunch three times a week.

Ⓒ The least number of students buys lunch every day.

Ⓓ Five students buy lunch twice a week.

Buying Lunch		
Number of Days	**Tally**	**Frequency**
0	III	3
1	JHT	5
2	JHT I	6
3	JHT JHT	10
4	JHT III	8
5	JHT	5

Quiz 2   Lessons 4–4 through 4–6

Solve each equation. Check your solution.

1. $-3t = 18$

2. $\dfrac{c}{-10} = -5$

3. $\dfrac{5}{7}d = 45$ *(Lesson 4–4)*

4. $2g + 11 = -7$

5. $\dfrac{b}{4} - 11 = 7$

6. $-9 = \dfrac{d + 5}{3}$ *(Lesson 4–5)*

7. $5n + 4 = 7n - 20$

8. $8t + 15 = -9t - 2$

9. $6a - 4 = 15 + 6a$ *(Lesson 4–6)*

10. Sports According to NBA standards, a basketball should bounce back $\dfrac{2}{3}$ the distance from which it is dropped. How high should a basketball bounce when it is dropped from a height of $3\dfrac{3}{4}$ feet? *(Lesson 4–4)*

What You'll Learn

You'll learn to solve equations with grouping symbols.

Why It's Important

Painting Painters can solve equations by grouping symbols to determine how much paint to buy.

A gallon of paint covers about 350 square feet. Figure the room's square footage by multiplying the combined wall lengths by wall height and subtracting 15 square feet for each window and door.

Source: *The Family Handyman Magazine Presents Handy Hints for Home, Yard, and Workshop*

Suppose a room measures 16 feet long by 12 feet wide by 7 feet high and has two windows and a door. How many gallons of paint are needed to paint the room? To solve this problem, first find the square footage of the room. *The sum of the wall lengths is the perimeter of the room.*

$$\underbrace{x}_{\substack{square \\ footage}} \quad = \quad \underbrace{2(16 + 12)}_{\substack{perimeter \\ of\ room}} \quad \cdot \quad \underbrace{7}_{\substack{wall \\ height}} \quad - \quad \underbrace{(15 + 15 + 15)}_{window\ +\ window\ +\ door}$$

Now solve the equation to find x.

$x = 2(16 + 12) \cdot 7 - (15 + 15 + 15)$	*Original equation*
$x = 2(28) \cdot 7 - 45$	*Simplify inside the parentheses.*
$x = 56 \cdot 7 - 45$	*Multiply 2 and 28.*
$x = 392 - 45$	*Multiply 56 and 7.*
$x = 347$	*Subtract 45 from 392.*

The room has an area of 347 square feet. Therefore, only one gallon of paint is needed to paint the room.

In solving the equation above, the first step was to remove the parentheses. This is often the first step in solving any equation that contains grouping symbols.

Examples

Solve each equation. Check your solution.

1 $8 = 4(3x + 5)$

$8 = 4(3x + 5)$	*Original equation*
$8 = 12x + 20$	*Distributive Property*
$8 - 20 = 12x + 20 - 20$	*Subtract 20 from each side.*
$-12 = 12x$	*Simplify.*
$\dfrac{-12}{12} = \dfrac{12x}{12}$	*Divide each side by 12.*
$-1 = x$	*Simplify.*

COncepts in MOtion

BrainPOP®

algconcepts.com

Check: $8 = 4(3x + 5)$ *Original equation*

$8 \overset{?}{=} 4[3(-1) + 5]$ *Replace x with −1.*

$8 \overset{?}{=} 4(-3 + 5)$ *Multiply.*

$8 \overset{?}{=} 4(2)$ *Add.*

$8 = 8$ ✓   *Multiply.*

2 $2(3n - 5) - 3(1 - n) = 5$

$2(3n - 5) - 3(1 - n) = 5$ *Original equation*

$6n - 10 - 3 + 3n = 5$ *Distributive Property*

$9n - 13 = 5$ *Add like terms.*

$9n - 13 + 13 = 5 + 13$ *Add 13 to each side.*

$9n = 18$ *Simplify.*

$\dfrac{9n}{9} = \dfrac{18}{9}$ *Divide each side by 9.*

$n = 2$ *Simplify.*

Check: $2(3n - 5) - 3(1 - n) = 5$ *Original equation*

$2[3(2) - 5) - 3(1 - 2)] \overset{?}{=} 5$ *Replace n with 2.*

$2(6 - 5) - 3(-1) \overset{?}{=} 5$ *Simplify.*

$2(1) + 3 \overset{?}{=} 5$ *Simplify.*

$5 = 5$ ✓

Your Turn

a. $7 = 3(x + 1) - 2$ **b.** $4(t + 5) + 6(2t - 3) = 12$

Building Link

Real World

3   **The Rock and Roll Hall of Fame in Cleveland, Ohio, has a trapezoid-shaped cinema with an area of about 4675 square feet. Find x, the approximate width of the movie screen in the cinema.**

80 ft

x

90 ft

Rock and Roll Hall of Fame

$A = \dfrac{1}{2}(b_1 + b_2)h$ *Formula for the area of a trapezoid*

$4675 = \dfrac{1}{2}(90 + 80)x$ *Replace A with 4675, b_1 with 90, b_2 with 80, and h with x.*

$4675 = \dfrac{1}{2}(170)x$ *Add 90 and 80.*

$4675 = 85x$ *Multiply $\dfrac{1}{2}$ and 170.*

$\dfrac{4675}{85} = \dfrac{85x}{85}$ *Divide each side by 85.*

$55 = x$ *Simplify.*

The width of the movie screen is 55 feet.

 Online Personal Tutor at algconcepts.com

Check for Understanding

Communicating Mathematics

1. **State** the first step in solving $w - 8(w + 15) = -9$.

2. **Describe** a way to remove grouping symbols from an equation.

3. Writing Math   Write a summary of the techniques you learned in this chapter for solving equations.

Guided Practice

Examples 1–3

Solve each equation. Check your solution.

4. $12 = 2(k + 1)$

5. $3n - 2 = 3(n + 1) - 5$

6. $7 + 6x = 2(10 + 3x)$

7. $3p - 8(1 + p) = 12$

8. $5(2k - 3) + 2(k + 4) = 17$

9. $5 - \frac{1}{2}(b - 6) = 4$

Example 3

10. **Geometry**   Find the value of x in the trapezoid if its area is 39 square centimeters.

Exercises

Practice

Solve each equation. Check your solution.

11. $3(a + 2) = 24$

12. $14 = 2(4n - 1)$

13. $7 = 5(x - 4) + 2$

14. $6(2 + y) - 4 = -10$

15. $-2(6 + p) = 8$

16. $5(z - 1) = 5z$

17. $t + 4(1 - t) = 9t$

18. $3(b - 2) - 4(2b + 5) = 9$

19. $18 = 2(g + 4) + 3(2g - 6)$

20. $7(y - 2) + 8 = 3(y - 4) - 2$

21. $2.4h + 1 = 7(2h - 4)$

22. $2(f + 3) - 2.7 = 3(f + 1.1)$

23. $6(3.5 + r) - 1 = 7r - r$

24. $4(5 - x) - 2.1(6x) = -4.9$

25. $4(2a - 8) = 7a + 10$

26. $3 + \frac{5}{8}x = 2(x - 4)$

27. $3(1 - 4k) = \frac{1}{2}(-24k + 6)$

28. $\frac{5(n - 1)}{3} = n + 1$

29. Find the solution of $2[3 + 4(g + 1)] = -2$. Check your solution.

30. What is the value of y in $6(3 + 5y) + 5(y - 8) = 13$?

Applications and Problem Solving

31. **Geometry**   The area of a trapezoid is 70 square inches. Find the lengths of the bases if base 1 is four more than twice the length of base 2, and the altitude is 5 inches.

Homework Help	
For Exercises	See Examples
11–17, 29	1
18–28, 30	2
31–34	3
Extra Practice	
See page 700.	

32. Gardening Carlos is building a fence around his garden to keep rabbits out. One side of the garden is against the garage, as shown in the diagram at the right. He has 88 feet of fencing to fit around the three other sides.

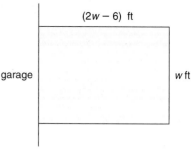

garage

a. Write an equation to find the width w of the garden.

b. What are the dimensions of the garden?

33. Number Theory Twice the greater of two consecutive odd integers is 13 less than three times the lesser. Find the integers.

34. Critical Thinking *Complementary angles* are two angles whose measures have a sum of 90°. Each angle is said to be the *complement* of the other angle. The measure of angle A is one-half the measure of its complement. Find the measure of angle A and the measure of its complement.

A

Mixed Review

Solve each equation. Check your solution. *(Lesson 4–6)*

35. $2n + 3 = 6 - n$

36. $5t - 12 = 8t$

37. Number Theory Four less than one-half of a number x is 7. *(Lesson 4–5)*

a. Write an equation to represent the problem.

b. What is the number?

38. Evaluate $\dfrac{ab}{-3}$ if $a = 6$ and $b = -5$. *(Lesson 2–6)*

Standardized Test Practice
Ⓐ Ⓑ Ⓒ Ⓓ

39. Short Response The table gives the average temperatures for six planets. Order the temperatures from least to greatest. *(Lesson 2–1)*

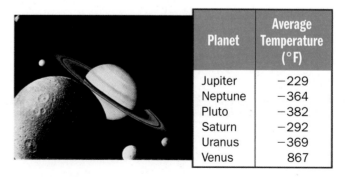

Planet	Average Temperature (°F)
Jupiter	−229
Neptune	−364
Pluto	−382
Saturn	−292
Uranus	−369
Venus	867

40. Multiple Choice A pitcher's earned-run average (ERA) is given by the formula $ERA = 9\left(\dfrac{a}{b}\right)$, where a is the number of earned runs the pitcher has allowed and b is the number of innings pitched. Find a pitcher's ERA to the nearest hundredth if he has allowed 23 earned runs in 180 innings. *(Lesson 1–5)*

Ⓐ 0.01　　Ⓑ 0.13　　Ⓒ 1.15　　Ⓓ 7.83

Understanding and Using the Vocabulary

After completing this chapter, you should be able to define each term, property, or phrase and give an example or two of each.

*inter*NET
CONNECTION **Review Activities**
For more review activities, visit:
www.algconcepts.com

Algebra

consecutive integers *(p. 167)*
identity *(p. 172)*
multiplicative inverses *(p. 154)*
reciprocals *(p. 154)*

Counting

combination *(p. 152)*
factorial *(p. 153)*
event *(p. 147)*
outcomes *(p. 146)*

permutation *(p. 152)*
sample space *(p. 146)*
tree diagram *(p. 146)*

Complete each sentence using a term from the vocabulary list.

1. A subset of the possible outcomes of a sample space is called a(n) ___?___.
2. Two numbers whose product is 1 are called ___?___.
3. To find the sample space of a particular situation, a(n) ___?___ may be used.
4. Integers in counting order are called ___?___.
5. An equation that is true for every value of the variable is called a(n) ___?___.
6. The ___?___ is the list of all the possible outcomes.
7. The results shown by a tree diagram are called ___?___.
8. The numbers $-\frac{4}{3}$ and $-\frac{3}{4}$ are called ___?___.
9. To find the possible number of ___?___ by multiplying, the Fundamental Counting Principle is used.
10. The equation $\frac{1}{2}x + 1 = \frac{1}{2}x + 1$ is an example of a(n) ___?___.

Skills and Concepts

Objectives and Examples	Review Exercises

• **Lesson 4–1** Multiply rational numbers.

$-3.2(-4) = 12.8$ *The signs are the same.*
The product is positive.

$\frac{4}{5}\left(-\frac{2}{9}\right) = -\frac{4 \cdot 2}{5 \cdot 9}$

$= -\frac{8}{45}$ *The signs are different.*
The product is negative.

Find each product.

11. $6(3.8)$
12. $-9.2 \cdot 0$
13. $-1\left(\frac{3}{4}\right)$
14. $-\frac{5}{6}\left(-\frac{1}{3}\right)$

Simplify each expression.

15. $-2.7(4y)$
16. $5a(3.4b)$
17. $6s\left(-\frac{1}{7}\right)$
18. $\left(-\frac{3}{8}m\right)\left(-\frac{3}{5}n\right)$

 www. algconcepts.com/vocabulary_review

Objectives and Examples

Review Exercises

- **Lesson 4-2** Use tree diagrams or the Fundamental Counting Principle to count outcomes.

Shirt	Pants	Socks
red	black	red
green	blue	green
white		white

There are 3 shirts, 2 pants, and 3 socks, so the number of different combinations is $3 \times 2 \times 3$ or 18.

19. A sandwich can be made with wheat or white bread. It can have ham, turkey, or beef. Find the number of possible outcomes by drawing a tree diagram.

20. Find the number of different team jerseys.

Shirt	Stripe	Numbers
white	green	green
blue	yellow	black
	gold	gold
	red	red

- **Lesson 4-3** Divide rational numbers.

Evaluate $\frac{t}{3}$ if $t = \frac{1}{5}$.

$$\frac{t}{3} = \frac{\frac{1}{5}}{3} \qquad \textit{Replace t with } \frac{1}{5}.$$

$$= \frac{1}{5} \div 3 \qquad \textit{Rewrite as division.}$$

$$= \frac{1}{5} \cdot \frac{1}{3} \text{ or } \frac{1}{15} \qquad \textit{Multiply by reciprocal.}$$

Find each quotient.

21. $5 \div (-2.5)$

22. $-7.2 \div (-1.6)$

23. $\frac{4}{5} \div \frac{3}{2}$

24. $3\frac{1}{2} \div \left(-\frac{1}{2}\right)$

Evaluate each expression if $x = \frac{2}{3}$ and $y = -\frac{1}{2}$.

25. $\frac{y}{4}$

26. $\frac{5}{x}$

27. $\frac{y}{x}$

- **Lesson 4-4** Solve multiplication and division equations by using the properties of equality.

$$\frac{x}{3} = -12$$

$$3\left(\frac{x}{3}\right) = 3(-12) \qquad \textit{Multiply each side by 3.}$$

$$x = -36 \qquad \textit{Simplify.}$$

Solve each equation. Check your solution.

28. $5n = 30$

29. $16 = 2a$

30. $-k = 5$

31. $-8y = 32$

32. $3t = 4.2$

33. $-2x = -24.6$

34. $\frac{m}{4} = -6$

35. $\frac{5}{9}r = 5$

- **Lesson 4-5** Solve equations involving more than one operation.

$$\frac{x}{4} + 6 = 18$$

$$\frac{x}{4} + 6 - 6 = 18 - 6 \qquad \textit{Subtract 6 from each side.}$$

$$4\left(\frac{x}{4}\right) = 4(12) \qquad \textit{Multiply each side by 4.}$$

$$x = 48 \qquad \textit{Simplify.}$$

Solve each equation. Check your solution.

36. $7t + 3 = 24$

37. $1 + 6y = -11$

38. $8 - 2b = 10$

39. $-4y + 1 = 1$

40. $2.1 - 3r = 4.8$

41. $-1.8 = 5.1w - 12$

42. $14 = 6 + \frac{x}{5}$

43. $\frac{4 + x}{7} = -3$

Mixed Problem Solving
See pages 724–731.

Objectives and Examples

Review Exercises

• **Lesson 4–6** Solve equations with variables on both sides.

$$7x = 4x - 3$$
$$7x - 4x = 4x - 3 - 4x \quad \textit{Subtract 4x from}$$
$$3x = -3 \qquad\qquad \textit{each side.}$$
$$\frac{3x}{3} = \frac{-3}{3} \qquad\qquad \textit{Divide each side by 3.}$$
$$x = -1 \qquad\qquad \textit{Simplify.}$$

Solve each equation. Check your solution.

44. $3a = a + 4$

45. $2t + 4 = 10 + t$

46. $11 + 4m = 4m + 8$

47. $5x - 5.2 = 9.4 + 3x$

48. $6 + \frac{1}{3}t = -2 + \frac{2}{3}t$

• **Lesson 4–7** Solve equations with grouping symbols.

$$5(x - 7) = 10$$
$$5x - 35 = 10 \qquad\qquad \textit{Distributive Property}$$
$$5x - 35 + 35 = 10 + 35 \quad \textit{Add 35 to each side.}$$
$$5x = 45 \qquad\qquad \textit{Simplify.}$$
$$\frac{5x}{5} = \frac{45}{5} \qquad\qquad \textit{Divide each side by 5.}$$
$$x = 9 \qquad\qquad \textit{Simplify.}$$

Solve each equation. Check your solution.

49. $6a = 2(4 - a)$

50. $10 = 5(n + 3)$

51. $4(3h - 2) = 7h + 2$

52. $2(1 + 2x) + 3(x - 4) = 11$

53. $\frac{1}{3}(12 - 6t) = 4 - 2t$

Applications and Problem Solving

54. Dining With each shrimp, salmon, or perch dinner at the Seafood Palace, you may have soup or salad. You may also have broccoli, baked potato, or rice. How many different dinners are possible? *(Lesson 4–2)*

56. Number Theory Four more than two-thirds of a number n is zero. *(Lesson 4–5)*
   a. Write an equation to represent the problem.
   b. What is the value of n?

55. Space Halley's Comet flashes through the sky every 76.3 years. How many times will it appear in 381.5 years? *(Lesson 4–3)*

57. Geometry The width of the rectangle is equal to one-half the length. What is the width of the rectangle? *(Lesson 4–6)*

$(x + 4)$ in.

$6x$ in.

1. **Write** an equation using grouping symbols that results in $x = -5$.

Brand	Type	Power
A	scientific	solar
B	graphing	battery
C		

Exercise 2

2. **Draw** a tree diagram to show the different calculators that are possible from the choices given in the table.

3. **Explain** whether the equation $3(2x - 3) - 1 = 6x - 10$ is an identity or has no solution.

Simplify each expression.

4. $7.2(-4x)$

5. $-3.9(-10t)$

6. $-\frac{4}{5}a\left(\frac{2}{3}b\right)$

7. $\frac{2}{3}\left(\frac{1}{2}g\right) + \frac{1}{3}g$

8. Find the number of possible outfits by drawing a tree diagram.

Shirt	Pants
red	blue
yellow	black
green	gray

Find the number of possible outcomes by using the Fundamental Counting Principle.

9. rolling an eight-sided die twice

10. spinning the spinner three times, and it never lands on a border

Evaluate each expression if $f = \frac{1}{2}$, $g = \frac{2}{5}$, and $h = -\frac{1}{3}$.

11. $\frac{f}{5}$

12. $\frac{3}{h}$

13. $\frac{g}{f}$

14. $\frac{h}{g}$

Solve each equation.

15. $5n = 35$

16. $18.4 = -2m$

17. $\frac{3}{4}x = 6$

18. $20 = -\frac{5}{4}a$

19. $3r + 1 = 7$

20. $13 - 4y = 29$

21. $10 = 7 + \frac{5}{8}w$

22. $7b + 8 = 5b - 4$

23. $2(h - 3) = 5 + 13h$

24. **Number Theory** Find three consecutive even integers whose sum is -12.

25. **Geometry** Find the dimensions of the rectangle if the perimeter is 220 feet.

$2w - 40$

w

Preparing for Standardized Tests

Statistics Problems

Many standardized test questions provide a table or a list of numerical data and ask you to find the *mean*, *median*, *mode*, and *range*.

$$\text{mean} = \frac{\text{sum of the numbers}}{\text{number of numbers}}$$

Example 1

The average speeds of the winners of the Indianapolis 500 from 1994 to 2003 are listed in the table. What is the median of the speeds?

Indianapolis 500 Winners			
Year	Average Speed (mph)	Year	Average Speed (mph)
1994	160.9	1999	153.2
1995	153.6	2000	167.6
1996	148.0	2001	141.6
1997	145.8	2002	166.5
1998	145.2	2003	156.3

(A) 153.2 (B) 153.4
(C) 153.6 (D) 153.87

Hint The median of an even number of values is the average (mean) of the middle two values when they are arranged in numeric order.

Solution Don't be confused by the phrase *average speed*; think of it as the winning speed. Write the winning speeds in order from least to greatest.

141.6, 145.2, 145.8, 148.0, 153.2, 153.6, 156.3, 160.9, 165.5, 167.6

There are ten values, so the median is the mean of the two middle values.

$$\frac{153.2 + 153.6}{2} = \frac{306.8}{2} \text{ or } 153.4$$

The median is 153.4 mph. The answer is B.

Example 2

A survey of the town of Niceville found a mean of 3.2 persons per household and a mean of 1.2 televisions per household. If 48,000 people live in Niceville, how many televisions are there in Niceville?

(A) 15,000 (B) 16,000 (C) 18,000
(D) 40,000 (E) 57,600

Hint Use your calculator for large numbers.

Solution Solve this problem one step at a time.

You need to find the number of televisions. The survey information is based on households, so first calculate the number of households.

number of people	divided by	people per household	equals	number of households
48,000	÷	3.2	=	15,000

The survey information says there is a mean (average) of 1.2 televisions per household. Multiply the number of households by the televisions per household.

number of households	times	televisions per household	equals	number of televisions
15,000	×	1.2	=	18,000

The answer is C.

After you work each problem, record your answer on the answer sheet provided or on a sheet of paper.

Multiple Choice

1. A company plans to open a new fitness center. It conducts a survey of the number of hours people exercise weekly. The results for 12 people chosen at random are 7, 2, 4, 8, 3, 0, 3, 1, 0, 5, 3, and 4 hours. What is the mode of the data? *(Lesson 3–3)*

   A 3   **B** 3.3   **C** 4
   D There is no mode.

2. For their science fiction book reports, students must choose one book from a list of eight. The books have the following numbers of pages: 272, 188, 164, 380, 442, 216, 360, and 262. What is the median number of pages? *(Lesson 3–3)*

   A 267   **B** 285.5   **C** 303   **D** 442

3. $\dfrac{0.5 + 0.5 + 0.5 + 0.5}{4} =$ *(Lesson 4–3)*

   A 0.05   **B** 0.125   **C** 0.5
   D 1   **E** 2.0

4. The graph shows the percent of men and women who held multiple jobs. In what year was the total percent of men and women holding multiple jobs greatest? *(Lesson 3–2)*

Having Multiple Jobs

Source: U.S. Bureau of Labor Statistics

   A 1970   **B** 1980
   C 1990   **D** 2000

5. If $\dfrac{12}{4} = x$, then $4x + 2 =$ *(Lesson 4–3)*

   A 2.   **B** 3.   **C** 4.   **D** 12.   **E** 14.

6. Over nine games, a baseball team had an average of 8 runs per game. In the first seven games, they had an average of 6 runs per game. They scored the same number of runs in each of the last two games. How many runs did they score in the last game? *(Lesson 3–3)*

   A 5   **B** 15   **C** 26   **D** 30   **E** 46

7. The stem-and-leaf plot shows the populations of ten cities in Oklahoma. Which statement is correct? *(Lesson 3–3)*

Stem	Leaf
35	6
36	2 4
37	5 6 7
38	1 2 3 6

   $35|6 = 35,600$

   A The mode is 38,000.
   B The range of the data is 35,600.
   C Most of these cities have fewer than 36,000 people.
   D The median population is 37,650.

8. If $5 + \dfrac{n}{4} = 9$, then what is the value of n?
   (Lesson 4–5)

   A 3.5   **B** 4   **C** 16   **D** 56

Grid In

9. If $3x = 12$, then $8 \div x = ?$ *(Lesson 4–4)*

Extended Response

10. The number of daily visitors to the local history museum during the month of June is shown below.   *(Lesson 1–7)*

    11 19 61 18 43 22 18 14 21 8
    19 41 36 16 16 14 24 31 64 7
    29 24 27 33 31 71 89 61 41 34

    Part A Construct a frequency table that represents the data.

    Part B Construct a histogram from the frequency table.

CHAPTER
5

Proportional Reasoning and Probability

What You'll Learn

Key Ideas

- Solve problems involving proportions. *(Lessons 5–1 and 5–2)*

- Solve problems involving percents. *(Lessons 5–3 to 5–5)*

- Find the probability of mutually exclusive events and inclusive events. *(Lesson 5–7)*

Key Vocabulary

odds *(p. 221)*

percent *(p. 198)*

probability *(p. 219)*

proportion *(p. 188)*

ratio *(p. 188)*

Why It's Important

Agriculture Corn is America's number one crop and one of the most versatile crops grown today. It is not just used for food. Products made from corn are all around us. The raw materials from processed corn are used to make antibiotics, glue, paint, fuel, and many other products.

Proportional reasoning is used in many branches of mathematics and other fields such as business and navigation. You will use a proportion to predict the amount of corn a farmer can expect from a crop in Lesson 5–1.

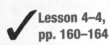
Study these lessons to improve your skills.

✓Check Your Readiness

Lesson 4–4,
pp. 160–164

Solve each equation. Check your solution.

1. $6a = 18$ **2.** $5d = 100$ **3.** $9x = 81$

4. $8c = 32$ **5.** $4n = 60$ **6.** $7k = 98$

Prerequisite
Skill, p. 689

Write each percent as a decimal.

7. 25%   **8.** 60%   **9.** 45%

10. 8%   **11.** 125%   **12.** 33.3%

Prerequisite
Skill, p. 687

Find each sum or difference.

13. $\frac{1}{6} + \frac{1}{6}$ **14.** $\frac{1}{10} + \frac{3}{10}$ **15.** $\frac{3}{10} + \frac{3}{5}$

16. $\frac{5}{6} + \frac{1}{12}$ **17.** $\frac{3}{4} + \frac{1}{12}$ **18.** $1 - \frac{3}{5}$

Prerequisite
Skill, p. 686

Find each product.

19. $\frac{1}{10} \cdot \frac{1}{10}$ **20.** $\frac{3}{8} \cdot \frac{1}{8}$ **21.** $\frac{3}{10} \cdot \frac{1}{2}$

22. $\frac{1}{6} \cdot \frac{2}{3}$ **23.** $\frac{7}{10} \cdot \frac{2}{5}$ **24.** $\frac{4}{13} \cdot \frac{1}{2}$

Prerequisite
Skill, p. 684

Find each sum or product.

25. $0.65 + 0.12$ **26.** $0.15 + 0.06$ **27.** $0.34 + 0.28$

28. $0.5 \cdot 0.5$ **29.** $0.4 \cdot 0.6$ **30.** $0.125 \cdot 0.125$

FOLDABLES™
Study Organizer

Make this Foldable to help you organize your Chapter 5 notes. Begin with a sheet of notebook paper.

❶ Fold lengthwise to the holes.

❷ Cut four tabs.

❸ Label the tabs as shown.

Reading and Writing Store your foldable in a 3-ring binder. As you read and study the chapter, write notes and examples under each tab.

In mathematics, a **ratio** is a comparison of two numbers by division. For example, the ratio of 15 and 10 can be expressed in the following three ways.

$$15 \text{ to } 10 \qquad 15\!:\!10 \qquad \frac{15}{10}$$

You can express $\frac{15}{10}$ in simplest form as $\frac{3}{2}$.

$$\frac{15}{10} \underset{\div 5}{\overset{\div 5}{=}} \frac{3}{2}$$

An equation stating that two ratios are equal is called a **proportion**. So, $\frac{15}{10} = \frac{3}{2}$ is a proportion. Every proportion has two equal cross products.

15(2) is one cross product.   10(3) is another cross product.

$$\frac{15}{10} = \frac{3}{2}$$

$$15(2) = 10(3)$$

$$30 = 30$$

The cross products are equal.

Info Graphic

In the proportion, $b \neq 0$ and $d \neq 0$.

Property of Proportions		
Words:	The cross products of a proportion are equal.	
Symbols:	If $\frac{a}{b} = \frac{c}{d}$, then $ad = bc$. If $ad = bc$, then $\frac{a}{b} = \frac{c}{d}$.	
Numbers:	If $\frac{15}{10} = \frac{3}{2}$, then $15(2) = 10(3)$.	
	If $15(2) = 10(3)$, then $\frac{15}{10} = \frac{3}{2}$.	

You can use cross products to solve proportions. Sometimes you'll use the Distributive Property.

Solve each proportion.

1 $\dfrac{5}{3} = \dfrac{75}{n}$

$\dfrac{5}{3} = \dfrac{75}{n}$	*Original equation*
$5n = 3(75)$	*Find the cross products.*
$5n = 225$	*Multiply.*
$\dfrac{5n}{5} = \dfrac{225}{5}$	*Divide each side by 5.*
$n = 45$	*Simplify.*

Check: $\quad \dfrac{5}{3} = \dfrac{75}{n} \qquad$ *Original equation*

$\qquad\qquad \dfrac{5}{3} \stackrel{?}{=} \dfrac{75}{45} \qquad$ *Replace n with 45.*

$\qquad 5(45) \stackrel{?}{=} 3(75) \quad$ *Find the cross products.*

$\qquad\qquad 225 = 225 \quad \checkmark$

The solution is 45.

Technology Tip

A graphing calculator follows the order of operations. So, it is not necessary to press ENTER after multiplying 3 and 75 below.

3 ✕ 75 ÷ 5

ENTER 45

2 $\dfrac{y}{8} = \dfrac{y+4}{10}$

$\dfrac{y}{8} = \dfrac{y+4}{10}$	*Original equation*
$10y = 8(y+4)$	*Find the cross products.*
$10y = 8y + 32$	*Distributive Property*
$10y - 8y = 8y + 32 - 8y$	*Subtract 8y from each side.*
$2y = 32$	*Simplify.*
$\dfrac{2y}{2} = \dfrac{32}{2}$	*Divide each side by 2.*
$y = 16$	*Simplify.*

Check: $\quad \dfrac{y}{8} = \dfrac{y+4}{10} \qquad$ *Original equation*

$\qquad\qquad \dfrac{16}{8} \stackrel{?}{=} \dfrac{16+4}{10} \qquad$ *Replace y with 16.*

$\qquad\qquad \dfrac{16}{8} \stackrel{?}{=} \dfrac{20}{10} \qquad$ *Add 16 and 4.*

$\qquad\qquad 2 = 2 \quad \checkmark \qquad$ *Simplify.*

The solution is 16.

Your Turn

a. $\dfrac{3}{a} = \dfrac{18}{24}$ $\qquad$ **b.** $\dfrac{r}{100} = \dfrac{7}{8}$ $\qquad$ **c.** $\dfrac{5}{c} = \dfrac{13}{c+8}$

You can use proportions to find equivalent measurements within the English or metric systems.

Examples

Convert each measurement as indicated.

3 **72 ounces to pounds**

16 ounces = 1 pound

Let x represent the number of pounds. Write a proportion.

$$\frac{72 \text{ ounces}}{x \text{ pounds}} = \frac{16 \text{ ounces}}{1 \text{ pound}}$$

$$\frac{72}{x} = \frac{16}{1}$$

$$72(1) = 16x$$

$$72 = 16x$$

$$\frac{72}{16} = \frac{16x}{16}$$

$$4.5 = x$$

So, 72 ounces = 4.5 pounds.

4 **15 meters to centimeters**

1 meter = 100 centimeters

Let x represent the number of centimeters. Write a proportion.

$$\frac{1 \text{ meter}}{100 \text{ centimeters}} = \frac{15 \text{ meters}}{x \text{ centimeters}}$$

$$\frac{1}{100} = \frac{15}{x}$$

$$1x = 100(15)$$

$$x = 1500$$

So, 15 meters = 1500 centimeters.

Your Turn

d. 96 centimeters to meters

e. 2 gallons to quarts

A **rate** is a ratio of two measurements having different units of measure. For example, 120 miles in 2 hours is a rate. When a rate is simplified so it has a denominator of 1, it is called a **unit rate**. To find the unit rate of 120 miles in 2 hours, divide by 2. The result is a unit rate of 60 miles in 1 hour, or 60 miles per hour.

To solve problems involving rates, you can use **dimensional analysis**. This is the process of carrying units throughout a computation.

Example **5**

Science Link

Density is a measure of the amount of matter that occupies a given volume. The density of wood is 0.71 gram per cubic centimeter. Suppose you have a piece of wood whose volume is 50 cubic centimeters. How many grams of wood does it contain?

$$0.71 \text{ gram per cubic centimeter} = \frac{0.71 \text{ gram}}{1 \text{ cubic centimeter}}$$

Multiply the unit rate by the number of cubic centimeters of wood.

Note that the unit of measure, cubic centimeters, can be canceled.

$$\frac{0.71 \text{ gram}}{1 \text{ cubic centimeter}} \cdot \frac{50 \text{ cubic centimeters}}{1} = \frac{35.5 \text{ grams}}{1} \text{ or } 35.5 \text{ grams}$$

So, the piece of wood contains 35.5 grams of wood.

Example — **6**

Travel Link

Juanita's family drove 400 miles in 8 hours. The next day, they need to drive another 700 miles. At the same rate, how long will it take them to drive the 700 miles?

You can also solve this problem by solving the proportion
$$\frac{400 \text{ miles}}{8 \text{ hours}} = \frac{700 \text{ miles}}{h \text{ hours}}.$$

Explore   You know that the family drove 400 miles in 8 hours. You need to find how long it will take to drive 700 miles farther.

Plan   Write the rate 400 miles in 8 hours as a unit rate. Then divide 700 miles by the unit rate.

Solve $\frac{400 \text{ miles}}{8 \text{ hours}} = \frac{50 \text{ miles}}{1 \text{ hour}}$ *The unit rate is 50 miles per hour.*

$$700 \text{ miles} \div \frac{50 \text{ miles}}{1 \text{ hour}} = \frac{700 \text{ miles}}{1} \cdot \frac{1 \text{ hour}}{50 \text{ miles}}$$

$$= \frac{\overset{14}{\cancel{700 \text{ miles}}}}{1} \cdot \frac{1 \text{ hour}}{\underset{1}{\cancel{50 \text{ miles}}}} \quad \textit{Note that the units cancel.}$$

$$= 14 \text{ hours}$$

Examine   It takes 8 hours to travel 400 miles. A trip of 700 miles would take nearly twice as long. So, 14 hours is reasonable.

Online Personal Tutor at algconcepts.com

Check for Understanding

Communicating Mathematics

1. **Explain** how to use the Property of Proportions to solve $\frac{x}{12} = \frac{5}{6}$.

2. **Write** two ratios that form a proportion and two other ratios that do *not* form a proportion.

Guided Practice

 Getting Ready   **Find the cross products for each proportion.**

Sample: $\frac{2}{8} = \frac{4}{16}$	**Solution:** 2(16) = 8(4)

3. $\frac{75}{100} = \frac{3}{4}$ 4. $\frac{16}{12} = \frac{12}{9}$ 5. $\frac{8}{12} = \frac{10}{15}$

Examples 1 & 2   **Solve each proportion.**

6. $\frac{6}{2} = \frac{9}{a}$ 7. $\frac{3}{5} = \frac{b}{100}$ 8. $\frac{5}{c} = \frac{25}{35}$

9. $\frac{2}{1.9} = \frac{4}{x}$ 10. $\frac{y}{4} = \frac{y+2}{5}$ 11. $\frac{z}{5} = \frac{z+2}{7}$

Examples 3 & 4   **Convert each measurement as indicated.**

12. 3 pounds to ounces   13. 2000 grams to kilograms

Example 5

14. **Agriculture**   A farmer gets 300 bushels of corn for each acre of land that he plants. At this rate, how many bushels will he expect to get if he plants 90 acres of corn?

Example 6

15. **Travel**   Katie knows that she can drive her car an average of 40 miles while using one gallon of gasoline. How many gallons of gasoline will she need for a trip of 180 miles?

Exercises · · · · · • • • • • • • • • • • • • • •

Practice

Solve each proportion.

16. $\dfrac{14}{6} = \dfrac{21}{m}$ 17. $\dfrac{5}{4} = \dfrac{y}{12}$ 18. $\dfrac{5}{8} = \dfrac{x}{100}$

Homework Help	
For Exercises	**See Examples**
16–38	1, 2
39–46, 49	3, 4
47, 48	5, 6
Extra Practice	
See page 701.	

19. $\dfrac{3}{a} = \dfrac{18}{24}$ 20. $\dfrac{18}{27} = \dfrac{t}{3}$ 21. $\dfrac{8}{20} = \dfrac{30}{x}$

22. $\dfrac{6}{1} = \dfrac{1200}{r}$ 23. $\dfrac{45}{5} = \dfrac{81}{p}$ 24. $\dfrac{5}{8} = \dfrac{6}{3}$

25. $\dfrac{2}{3} = \dfrac{7}{y}$ 26. $\dfrac{a}{3.5} = \dfrac{14}{7}$ 27. $\dfrac{n}{2} = \dfrac{0.7}{0.4}$

28. $\dfrac{10}{y} = \dfrac{2.5}{4}$ 29. $\dfrac{0.35}{3} = \dfrac{c}{18}$ 30. $\dfrac{y+2}{y} = \dfrac{8}{6}$

31. $\dfrac{8}{x+3} = \dfrac{5}{x}$ 32. $\dfrac{x+6}{5} = \dfrac{x}{4}$ 33. $\dfrac{x}{x+4} = \dfrac{3}{5}$

34. $\dfrac{5}{x} = \dfrac{9}{x+4}$ 35. $\dfrac{x-7}{x+8} = \dfrac{2}{3}$ 36. $\dfrac{4}{x-3} = \dfrac{5}{x+5}$

37. Are $\dfrac{15}{2}$ and $\dfrac{7.5}{1}$ equivalent ratios? Explain your reasoning.

38. Find the value of x that makes $\dfrac{18}{9} = \dfrac{28}{x}$ a proportion.

Convert each measurement as indicated.

39. 5 feet to inches
40. 3 quarts to pints
41. 3 liters to milliliters
42. 2.4 grams to milligrams
43. 16 quarts to gallons
44. 4200 pounds to tons
45. 520 meters to kilometers
46. 1600 milliliters to liters

Applications and Problem Solving

47. **Health**   An average adult's body contains about 5 quarts of blood. If an adult donates 1 pint of blood, how many pints are left?

48. **City Planning**   A water treatment plant can filter about 350,000 gallons of water in seven days. At this rate, about how many gallons of water can be filtered in 30 days?

49. Measurement   Is a 2-quart pitcher large enough to hold 1 batch of fruit punch?

Fruit Punch

Mix together:
$2\frac{1}{2}$ cups cherry juice
2 cups orange juice
$1\frac{1}{2}$ cups pineapple juice
3 cups ginger ale

Exercise 49

50. Cooking   A recipe for 72 chocolate chip cookies needs $4\frac{1}{2}$ cups of flour. How much flour is needed to make 48 cookies?

51. Construction   A batch of concrete is made by mixing 1 part cement, $2\frac{1}{2}$ parts sand, and $3\frac{1}{2}$ parts gravel. How many pounds of sand and gravel are needed to make 300 pounds of cement?

52. Education   A student-to-teacher ratio is used to compare the number of students at a university with the number of faculty.

University	Students	Faculty
Texas	48,857	2517
Ohio State	48,278	4151
UCLA	35,558	3276
Georgia	29,693	3075
Tennessee	25,397	1453

Source: *The World Almanac*

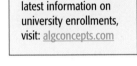

Math Data Online Update For the latest information on university enrollments, visit: algconcepts.com

  a. To the nearest tenth, express the student-to-teacher ratio for each school as a decimal.
  b. Rank the universities in the chart in order from greatest student-to-teacher ratio to least.

53. Critical Thinking   If $\frac{6}{a} = \frac{b}{8}$, what happens to the value of b as the value of a increases?

Mixed Review

Solve each equation.   *(Lessons 4–6 and 4–7)*

54. $5a - 3 = 8a + 12$

55. $10n + 6 = 8n - 6$

56. $6x + 7 = 8x - 9$

57. $3y - 4 = -2y - 14$

58. $4(m - 2) = 6m$

59. $-4(a - 2) = 10$

60. $6(z + 2) - 4 = -10$

61. $3x - 2(x + 3) = 6$

62. Food   The Pasta Palace offers a choice of six types of pasta with five types of sauce. Each dinner comes with or without a meatball. How many different dinners are possible? *(Lesson 4–2)*

Standardized Test Practice
Ⓐ Ⓑ Ⓒ Ⓓ

63. Short Response   Find the product of $\frac{2}{3}$ and $-\frac{3}{5}$. *(Lesson 4–1)*

64. Multiple Choice   For what value(s) of x is the equation $|x - 2| = 5$ true? *(Lesson 3–7)*
  Ⓐ -3 only   Ⓑ 7 only   Ⓒ -3 and 7 Ⓓ -7 and 3

What You'll Learn

You'll learn to solve problems involving scale drawings and models.

Why It's Important

Movies Movie directors use models for special effects.
See Example 2.

Leonardo da Vinci was commissioned to construct a huge bronze horse for the Duke of Milan. Unfortunately, he died before he could complete it.

Five hundred years later, master sculptor Nina Akamu created an 8-foot clay model from da Vinci's sketches. The model was then enlarged by sections to become the 24-foot *Il Cavallo*—the largest bronze horse in the world.

A **scale drawing** or **scale model** is used to represent an object that is too large or too small to be drawn or built at actual size. The **scale** is the ratio of a length on the drawing or model to the corresponding length of the real object.

Il Cavallo

$$\begin{array}{l} \textit{length of clay model} \quad \rightarrow \\ \textit{length of bronze statue} \quad \rightarrow \end{array} \quad \frac{8 \text{ ft}}{24 \text{ ft}} = \frac{1 \text{ ft}}{3 \text{ ft}}$$

The scale is 1 foot = 3 feet or 1:3.
If the scale uses the same units, it is not necessary to include them.

In the following activity, you will investigate scales with rectangles.

Hands-On Algebra

Materials: grid paper

Step 1 Draw a rectangle that is 4 units wide and 6 units long on your grid paper.

Step 2 Draw a second rectangle so that the ratio of the sides of the first rectangle to the second rectangle is 1:2.

Try These

1. What are the lengths of the sides of the second rectangle?
2. Draw two rectangles so that the scale is 1:3.
3. Draw two rectangles so that the scale is 2:3.
4. Draw a rectangle that is 5 units by 12 units. Then draw another rectangle that is 15 units by 36 units. What is the scale?

COncepts in MOtion

Interactive Lab
algconcepts.com

Example ❶

Map Link

One of the most common types of scale drawings is a map.

The scale on a map of Texas is 1 inch = 50 miles. Find the actual distance between Dallas and Houston if the distance between them on the map is 4.5 inches.

Use the scale and the distance given on the map to write a proportion.

$$\frac{1 \text{ inch}}{50 \text{ miles}} = \frac{4.5 \text{ inches}}{x \text{ miles}} \quad \begin{array}{l} \leftarrow \textit{map distance} \\ \leftarrow \textit{actual distance} \end{array}$$

$$1x = 50(4.5) \qquad \textit{Find the cross products.}$$
$$x = 225 \qquad \textit{Simplify.}$$

The distance between Dallas and Houston is about 225 miles.

Your Turn

Find the actual distance between San Antonio and Houston if the distance between them on the map is $3\frac{3}{4}$ inches.

Example ❷

Movie Link

You can determine the scale if you know the dimensions of the model and the dimensions of the actual object.

The ship *Titanic*, which sank in 1912, was 880 feet long. When a movie was made about it more than 80 years later, a 44-foot-long model of the ship was created for special effects shots. Find the scale that was needed to design other parts of the model.

Write the ratio of the length of the model to the length of the ship. Then solve a proportion in which the length of the model is 1 foot and the length of the ship is x feet.

$$\frac{44 \text{ feet}}{880 \text{ feet}} = \frac{1 \text{ foot}}{x \text{ feet}} \quad \begin{array}{l} \leftarrow \textit{model length} \\ \leftarrow \textit{actual length} \end{array}$$

$$44x = 880(1) \qquad \textit{Find the cross products.}$$
$$44x = 880 \qquad \textit{Simplify.}$$
$$\frac{44x}{44} = \frac{880}{44} \qquad \textit{Divide each side by 44.}$$
$$x = 20 \qquad \textit{Simplify.}$$

The scale is 1 foot = 20 feet or 1:20.

880 ft

Photo Graphic

Check for Understanding

Communicating Mathematics

1. **Sketch** two rectangles in which the ratio of the sides of the first rectangle to the sides of the second rectangle is 1:2.

2. **Name** two careers in which you would use scale drawings or scale models.

Guided Practice

Example 1

On a map, the scale is 1 inch = 50 miles. Find the actual distance for each map distance.

	From	To	Map Distance
3.	Sacramento, CA	Fresno, CA	3 inches
4.	Kansas City, MO	St. Louis, MO	$4\frac{3}{4}$ inches

Example 2

5. **Transportation** A scale model of a Boeing 747 jumbo jet is 2 meters long. If an actual jet is 70.4 meters long, find the scale of the model.

Exercises

Practice

On a map, the scale is 1 inch = 40 miles. Find the actual distance for each map distance.

	From	To	Map Distance
6.	Tampa, FL	Orlando, FL	2 inches
7.	Albany, NY	Syracuse, NY	3 inches
8.	Jacksonville, FL	Pensacola, FL	8 inches
9.	Washington, DC	Richmond, VA	2.5 inches
10.	Cleveland, OH	Pittsburgh, PA	$2\frac{3}{4}$ inches
11.	Hartford, CT	New Haven, CT	$\frac{3}{4}$ inch

Homework Help

For Exercises	See Examples
6–11, 16	1
12–15	2

Extra Practice
See page 701.

Applications and Problem Solving

12. **Bridges** The 1500-foot Natchez Trace Bridge in Natchez, Mississippi, is the longest precast segmental arch bridge in North America. If a model of the bridge is 12 inches long, find the scale of the model.

13. **Model Cars** In a scale model of a race car, the wheels have a diameter of 2 inches. If the actual wheels have a diameter of 30 inches, find the scale of the model.

14. **Architecture** On a blueprint of a house, one bedroom is 4 inches long. When the house is built, the room will be 16 feet long. Find the scale of the blueprint.

15. **Life Science** In a photograph in a science textbook, the length of a ladybug is 2.5 centimeters. The scale of the photograph is 1 centimeter = 0.2 centimeter. What is the actual length of the ladybug?

16. **Travel** On a map of Colorado, the scale is 1 centimeter = 40 kilometers. The distance on the map between the northern and southern boundaries of Colorado is 11.5 centimeters.

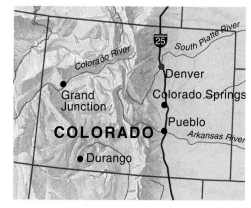

   a. How many kilometers is the northern boundary from the southern boundary?

   b. Suppose you travel at an average speed of 80 kilometers per hour. How long will it take you to travel through the state along Route 25? Round your answer to the nearest hour.

17. **Critical Thinking** The state of Pennsylvania is 332 miles long and 179 miles wide. Suppose you want to draw a map of Pennsylvania on an $8\frac{1}{2}$-by-11 inch piece of paper. Find the scale that will allow you to draw the map as large as possible.

Mixed Review

18. **Monuments** A monument of Crazy Horse, the famous Oglala chief, has been under construction for more than 50 years. The sculptors use a model to calculate how to carve the statue. The length of the model is 18.8 feet, the height of the model is 16.6 feet, and the height of the monument is 563 feet. To the nearest foot, what is the length of the monument?
(Lesson 5–1)

Crazy Horse Monument

Solve each equation. *(Lesson 4–7)*

19. $3(-2x + 5) = 7$ 20. $-2(3y - 4) = 3 - y$ 21. $2(3x + 1) - 8x = 10$

22. **Multiple Choice** What relationship exists between the x- and y-coordinates of each of the points shown on the graph? *(Lesson 2–2)*

   Ⓐ They are opposites.

   Ⓑ Their sum is 2.

   Ⓒ The y-coordinate is 1 more than the x-coordinate.

   Ⓓ Their sum is 1.

What You'll Learn
You'll learn to solve problems by using the percent proportion.

Why It's Important
Health Nurses use percents to calculate the correct dosage of medicines.
See Exercise 42.

Quantities are often expressed as percents. A **percent** is a ratio that compares a number to 100. Percent also means *per hundred* or *hundredths*. For example, a 5% glucose solution contains 5 grams of glucose in 100 milliliters of solution.

				Model:
Percent	**Words:**	five percent		
	Symbols:	5%		
	Numbers:	$\frac{5}{100}$ or 5 to 100		

You can express a fraction or a ratio as a percent by using a proportion in which one denominator is 100.

Examples

Express each fraction or ratio as a percent.

1 $\frac{2}{5}$ of the circle is shaded.

$$\frac{2}{5} = \frac{r}{100} \quad r \text{ is the percent.}$$

$2(100) = 5r \quad$ *Find the cross products.*

$200 = 5r \quad$ *Multiply 2 and 100.*

$\frac{200}{5} = \frac{5r}{5} \quad$ *Divide each side by 5.*

$40 = r \quad$ *Simplify.*

So, $\frac{2}{5} = 40\%$.

2 A study found that 3 out of 8 people switch channels when a commercial is on television. **Source:** *American Demographics*

$$\frac{3}{8} = \frac{r}{100} \quad \text{Write a proportion.}$$

$3(100) = 8r \quad$ *Find the cross products.*

$300 = 8r \quad$ *Multiply 3 and 100.*

$\frac{300}{8} = \frac{8r}{8} \quad$ *Divide each side by 8.*

$37.5 = r \quad$ 37.5% of people switch channels.

Your Turn

CONcepts in MOtion
Animation
algconcepts.com

a. $\frac{73}{50}$

b. 2 out of 3

In Example 1, the number of parts shaded, 2, is called the **percentage** (*P*). It is being compared to the total number of parts, 5, which is called the **base** (*B*).

$$\text{percentage} \rightarrow \frac{2}{5} = \frac{40}{100} \leftarrow \text{percent}$$
$$\text{base} \rightarrow$$

This is an example of a **percent proportion**.

Percent Proportion	If *P* is the percentage, *B* is the base, and *r* is the percent, the percent proportion is $\frac{P}{B} = \frac{r}{100}$.

Examples ③ **Forty is what percent of 160?**

$\dfrac{P}{B} = \dfrac{r}{100}$ *Use the percent proportion. Write "what percent" as $\dfrac{r}{100}$.*

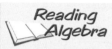

Reading Algebra

In percent problems, the base follows the word *of*. It is *not* always the larger number.

$\dfrac{40}{160} = \dfrac{r}{100}$ *Replace P with 40 and B with 160.*

$40(100) = 160r$ *Find the cross products.*

$4000 = 160r$ *Multiply 40 and 100.*

$\dfrac{4000}{160} = \dfrac{160r}{160}$ *Divide each side by 160.*

$25 = r$ *Simplify.*

So, 40 is 25% of 160.

④ **50% of what number is 42.5?**

$\dfrac{P}{B} = \dfrac{r}{100}$ *Use the percent proportion.*

$\dfrac{42.5}{B} = \dfrac{50}{100}$ *Replace P with 42.5 and r with 50.*

$42.5(100) = 50B$ *Find the cross products.*

$4250 = 50B$ *Multiply 425 and 100.*

$\dfrac{4250}{50} = \dfrac{50B}{50}$ *Divide each side by 50.*

$85 = B$ *Simplify.*

So, 50% of 85 is 42.5.

Your Turn

c. 7 is what percent of 20?   **d.** 60 is 15% of what number?

e. Find 25% of 66.   **f.** What number is 10% of 88?

Businesses can use a time study to determine what percent of an employee's time is spent on various activities.

Real World

Example ⑤

Business Link

The manager of The Furniture Store conducted a time study of the employees in its warehouse. The chart at the right shows the average number of hours spent on each activity during the workday. What percent of the time do the employees spend on each activity?

Activity	Time (hours)
Loading furniture	3
Delivering furniture	4
Lunch	1

The employees work 3 + 4 + 1 or 8 hours each day. This is the base. To find each percent, write and solve the percent proportion for each activity.

Loading: $\dfrac{3}{8} = \dfrac{r}{100}$ 3 ☒ 100 ÷ 8 ENTER *37.5*

Delivering: $\dfrac{4}{8} = \dfrac{r}{100}$ 4 ☒ 100 ÷ 8 ENTER *50*

Lunch: $\dfrac{1}{8} = \dfrac{r}{100}$ 1 ☒ 100 ÷ 8 ENTER *12.5*

The results are summarized in the chart.

Note that the total of the percent column is 100%.

Activity	Time (hours)	Percent
Loading furniture	3	37.5
Delivering furniture	4	50.0
Lunch	1	12.5

Data are often displayed using circle graphs. A **circle graph** is a graph that shows the relationship between parts of the data and the whole.

Real World

Example ⑥

Statistics Link

Make a circle graph of the data in Example 5.

Loading furniture accounts for 37.5% of the time. A circle is composed of 360°. To find the number of degrees for this section of the graph, find 37.5% of 360. Repeat this process for the other sections. *Replace B with 360 and r with each percent.*

Loading: $\dfrac{P}{360} = \dfrac{37.5}{100}$ 360 ☒ 37.5 ÷ 100 ENTER *135*

Delivering: $\dfrac{P}{360} = \dfrac{50}{100}$ 360 ☒ 50 ÷ 100 ENTER *180*

Lunch: $\dfrac{P}{360} = \dfrac{12.5}{100}$ 360 ☒ 12.5 ÷ 100 ENTER *45*

Use a compass to draw a circle. Then draw a radius. Use a protractor to draw a 135° angle. *You can start with any of the angles.*

From the new radius, draw a 180° angle. The remaining section should be 45°. Label each section of the graph with the category and percent. Give the graph a title.

The Furniture Store Employee Time Study

Loading Furniture 37.5%

Delivering Furniture 50%

Lunch 12.5%

Check for Understanding

Communicating Mathematics

1. **Explain** what P, B, and r represent in the percent proportion.

2. Brian and Julia are trying to find what percent 30 is of 20. Brian uses the proportion $\frac{30}{20} = \frac{r}{100}$. Julia uses the proportion $\frac{20}{30} = \frac{r}{100}$. Who is correct? Explain your reasoning.

> **Vocabulary**
> percent
> percentage
> base
> percent proportion
> circle graph

Guided Practice

> ⏱ **Getting Ready** Write a proportion that can be used to find each number.
>
> **Sample:** 5 is what percent of 40? **Solution:** $\frac{5}{40} = \frac{r}{100}$

3. What percent of 24 is 12?

4. 80 is 75% of what number?

5. Find 45% of 60.

6. What percent of 50 is 100?

Examples 1 & 2

Express each fraction or ratio as a percent.

7. 3 out of 5

8. $\frac{1}{3}$

9. 7 to 10

10. In a recent year, 17 out of 20 students used the Internet as a reference source.

Use the percent proportion to find each number.

11. 30 is what percent of 120? **12.** 16 is 40% of what number?

13. Find 20% of 82. **14.** Find 250% of 8.

15. Government In the 108th Congress, there were 48 Democratic Senators, 51 Republican Senators, and 1 Independent Senator.

Example 5

   a. What percent of the Senators were Democrats? What percent were Republicans? What percent were Independents?

Example 6

   b. Make a circle graph of the data.

Exercises • • • • • • • • • • • • • • • •

Practice

Express each fraction or ratio as a percent.

16. $\frac{1}{4}$ **17.** 5 out of 8 **18.** 9 to 10

19. 4 to 5 **20.** $\frac{5}{4}$ **21.** $\frac{9}{12}$

22. 9 out of 20 **23.** 3 to 2 **24.** 1 out of 6

25. Three out of 5 people have eaten in an Italian restaurant during the past six months.

26. Two-thirds of all households own their homes.

27. About 3 of every 50 people in the United States are between the ages of 14 and 17.

Homework Help	
For Exercises	**See Examples**
16–27	1–2
28–39	3–4
40–41	3
42	5
43	6
Extra Practice	
See page 701.	

Use the percent proportion to find each number.

28. 60 is what percent of 150? **29.** 9 is what percent of 25?

30. 15 is 20% of what number? **31.** 50% of what number is 95?

32. What number is 60% of 5? **33.** Find 10% of 125.

34. What percent of 50 is 75? **35.** What number is 125% of 48?

36. 5% of what number is 6.5? **37.** 1 is what percent of 200?

38. Find 0.1% of 450. **39.** 18.6 is 200% of what number?

Applications and Problem Solving

40. Education If you answer 35 problems correctly on a 40-problem test, what percent did you answer correctly?

41. Pharmacy A peroxide solution contains 5 milliliters of peroxide in 10 milliliters of solution. What is the percent of peroxide in the solution?

42. Nursing A nurse prepares a 5% glucose solution by dissolving 5 grams of glucose in 100 milliliters of solution. How many grams of glucose are in 20 milliliters of the solution?

43. Sales About 21 million people give floral gifts for Mother's Day. The chart shows the percent of fresh-cut flowers that are typically purchased. Make a circle graph of the data.

Mother's Day Flowers
(percent)

Mixed Bouquets	47
Roses	23
Carnations	16
Single-Stem Flowers	14

Source: California Cut Flower Commission

44. Critical Thinking Describe a real-life situation in which the percentage is greater than the base.

Mixed Review

45. Maps The distance from St. Louis, Missouri, to Dallas, Texas, is 630 miles. On a map, the distance is 9 inches. What is the scale of the map? *(Lesson 5–2)*

Convert each measurement as indicated. *(Lesson 5–1)*

46. 54 inches to yards

47. 2.5 kilometers to meters

48. Health An ulcer medication has 300 milligrams in 2 tablets. How many milligrams are in 3 tablets? *(Lesson 5–1)*

Add or subtract. *(Lesson 3–2)*

49. $-1.5 - 3.7$

50. $\frac{3}{8} - \left(-\frac{1}{8}\right)$

51. $\frac{3}{5} + \left(-\frac{1}{5}\right)$

Standardized Test Practice
Ⓐ Ⓑ Ⓒ Ⓓ

52. Multiple Choice Which set of numbers is *not* closed under multiplication? *(Lesson 1–3)*

- Ⓐ real numbers
- Ⓑ negative integers
- Ⓒ positive integers
- Ⓓ whole numbers

Quiz 1   Lessons 5–1 through 5–3

Solve each proportion. *(Lesson 5–1)*

1. $\frac{9}{25} = \frac{x}{75}$

2. $\frac{z}{9} = \frac{z + 4}{21}$

3. Surveying A civil engineer uses the scale 1 inch = 40 feet to draw a parcel of land. How many linear feet of land are represented by a 2.5-inch line on the drawing? *(Lesson 5–2)*

Use the percent proportion to find each number. *(Lesson 5–3)*

4. Six is what percent of 15?

5. 95 is 10% of what number?

What You'll Learn

You'll learn to solve problems by using the percent equation.

Why It's Important

Business Employers use the percent equation to calculate taxes.
See Example 3.

Interest and tax payments involve finding percents. You can use the percent proportion to solve these kinds of problems. However, it is usually easier to write the percent proportion as an equation.

$$\frac{P}{B} = \frac{r}{100} \qquad \textit{Start with the percent proportion.}$$

$$\frac{P}{B} = R \qquad \textit{Replace } \frac{r}{100} \textit{ with R.}$$

$$\frac{P}{B}B = RB \qquad \textit{Multiply each side of the equation by B.}$$

$$P = RB$$

The equation $P = RB$ is called the **percent equation**. In this equation, R is the **rate**. The rate is the decimal form of the percent.

Percent Equation	**Words:**	The percentage is equal to the rate times the base.
	Symbols:	$P = RB$, where P is the percentage, B is the base, and R is the rate.

The percent equation is easier to use when the rate and base are known. However, the percent equation can be used to solve any percent problem.

Examples

1 Find 4% of $160.

$$P = RB \qquad \textit{Use the percent equation.}$$
$$ = 0.04(160) \quad \textit{Replace R with 0.04 and B with 160.}$$
$$ = 6.4 \qquad \textit{0.04} \boxed{\times} \textit{160} \boxed{\text{ENTER}} \text{ 6.4}$$

So, 4% of $160 is $6.40.

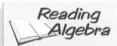
Reading Algebra

To write a percent as a decimal, place or move the decimal point two places to the left and drop the percent sign.

4% = 04% = 0.04

60% = 60% = 0.6

1.8% = 01.8% = 0.018

2 12 is 60% of what number?

$$P = RB \qquad \textit{Use the percent equation.}$$
$$12 = 0.6B \qquad \textit{Replace P with 12 and R with 0.6.}$$
$$\frac{12}{0.6} = \frac{0.6B}{0.6} \qquad \textit{Divide each side by 0.6.}$$
$$20 = B \qquad \textit{12} \boxed{\div} \textit{0.6} \boxed{\text{ENTER}} \text{ 20}$$

So, 12 is 60% of 20.

Your Turn

a. Find 62% of 120.

b. 75% of what number is 12?

Many real-world problems can be solved by using the percent equation.

Example ❸

Tax Link

The Federal Insurance Contributions Act (FICA) requires employers to deduct **6.2% of your income for social security taxes. Suppose your weekly pay is $140. What amount would be deducted from your pay for social security taxes?**

To find the amount deducted, find 6.2% of 140.

$P = RB$ *Use the percent equation.*
$\quad = 0.062(140)$ *Replace R with 0.062 and B with 140.*
$\quad = 8.68$ *0.062* ☓ *140* ENTER *8.68*

So, $8.68 would be deducted from your pay.

Many scientific calculators have a ☐%☐ key.

6.2 ☐%☐ ☐×☐ 140

☐=☐ *8.68*

Percents are also used in simple interest problems. **Simple interest** is the amount paid or earned for the use of money. If you have a savings account, you earn interest. If you borrow money through a loan or with a credit card, you pay interest.

The formula $I = prt$ is used to solve problems involving interest.

- I represents the interest,
- p represents the amount of money invested or borrowed, which is called the *principal*,
- r represents the annual interest rate, and
- t represents the time in years.

Example ❹

Banking Link

Rodney Turner is opening a savings account that earns **4% annual interest. He wants to earn at least $50 in interest after 2 years. How much money should he save in order to earn $50 in interest?**

$I = prt$
$50 = p(0.04)(2)$ *Replace I with 50, r with 0.04, and t with 2.*
$50 = 0.08p$ *0.04 × 2 = 0.08*
$\dfrac{50}{0.08} = \dfrac{0.08p}{0.08}$ *Divide each side by 0.08.*
$625 = p$ *50* ☐÷☐ *0.08* ENTER *625*

Rodney should invest at least $625 to earn $50 in interest.

Your Turn

c. Jessica deposited $3000 in a savings account that pays an interest rate of 6%. How long should she leave the money in the account if she wants to earn $90 in interest?

Mixture problems involve combining two or more parts into a whole. The parts that are combined usually have a different price or a different percent of something.

Example 5

Sales Link

Crystal sold tickets to the Drama Club's spring play. Adult tickets cost $8.00, and student tickets cost $5.00. Crystal sold 35 more student tickets than adult tickets. She collected a total of $1475. How many of each type of ticket did she sell?

Explore   Let a be the number of adult tickets that Crystal sold. Since there were 35 more student tickets sold than adult tickets, $a + 35$ is the number of student tickets sold.

Plan   Make a chart of the information.

	Number Sold	Price	Sales
Adult Tickets	a	$8	$8a$
Student Tickets	$a + 35$	$5	$5(a + 35)$

$$\underbrace{\text{Sales of adult tickets}}_{8a} \quad \underset{+}{\text{plus}} \quad \underbrace{\text{sales of student tickets}}_{5(a + 35)} \quad \underset{=}{\text{equals}} \quad \underbrace{\text{total sales.}}_{1475}$$

Solve

$$8a + 5(a + 35) = 1475 \qquad \textit{Original equation}$$
$$8a + 5a + 175 = 1475 \qquad \textit{Distributive Property}$$
$$13a + 175 = 1475 \qquad \textit{8a + 5a = 13a}$$
$$13a + 175 - 175 = 1475 - 175 \qquad \textit{Subtract 175 from each side.}$$
$$13a = 1300 \qquad \textit{Simplify.}$$
$$\frac{13a}{13} = \frac{1300}{13} \qquad \textit{Divide each side by 13.}$$
$$a = 100 \qquad \textit{Simplify.}$$

Crystal sold 100 adult tickets and $100 + 35$ or 135 student tickets.

Examine   If 100 adult tickets were sold, the total amount of money collected for them would be 100×8 or $800.
If 135 student tickets were sold, the total amount of money collected for them would be 135×5 or $675.
The total sales would be $800 + 675$ or $1475. ✓

Look Back

Solving Multi-Step Equations: Lesson 4–5

Mixture problems occur often in chemistry.

Example — **6**

Chemistry Link

Kelsey is doing a chemistry experiment that calls for a 30% solution of copper sulfate. She has 40 milliliters of a 25% solution. How many milliliters of a 60% solution should she add to make the required 30% solution?

Let x represent the amount of 60% solution to be added. Since she starts with 40 milliliters of solution, the final solution will have $40 + x$ milliliters.

	Amount of Solution (mL)	Amount of Copper Sulfate
25% solution	40	0.25(40)
60% solution	x	0.60x
30% solution	$40 + x$	0.30(40 + x)

Amount of copper sulfate in 25% solution	plus	amount of copper sulfate in 60% solution	equals	amount of copper sulfate in mixture.
0.25(40)	+	0.60x	=	0.30(40 + x)

$0.25(40) + 0.60x = 0.30(40 + x)$	*Original equation*
$10 \quad + 0.6x = 12 + 0.3x$	*Distributive Property*
$10 + 0.6x - 0.3x = 12 + 0.3x - 0.3x$	*Subtract 0.3x from each side.*
$10 + 0.3x = 12$	*Simplify.*
$10 + 0.3x - 10 = 12 - 10$	*Subtract 10 from each side.*
$0.3x = 2$	*Simplify.*
$\dfrac{0.3x}{0.3} = \dfrac{2}{0.3}$	*Divide each side by 0.3.*
$x \approx 6.7$	2 ÷ 0.3 ENTER 6.666666667

Kelsey should add about 6.7 milliliters of the 60% solution.

 Personal Tutor at algconcepts.com

Check for Understanding

Communicating Mathematics

1. **Explain** how r and R are different in the percent proportion and the percent equation.

2. **Write** a percent equation to find the following. 20% of what amount is $500?

3. **YOU Decide** Akira and Nikki are using the percent equation to find 5% of 17.99. Akira multiplies 17.99 by 5. Nikki multiplies 17.99 by 0.05. Who is correct? Explain.

Vocabulary

percent equation
rate
simple interest
mixture problem

Use the percent equation to find each number.

4. Find 90% of 200.

5. What number is 38% of 110?

6. 60 is 80% of what number?

7. 220 is 40% of what number?

8. **Business** Employers are required to deduct 1.45% of your income for Medicare taxes. If your weekly pay is $100, what amount would be deducted from your pay?

9. **Banking** Miguel bought a $500 certificate of deposit that paid an annual interest rate of 5%. If the certificate was issued for three years, how much simple interest will he earn?

10. **Business** Marian works at The Daily Grind coffee shop. She makes a blend of coffee by mixing hazelnut, which sells for $4 a pound, and chocolate almond, which sells for $7 a pound. How many pounds of chocolate almond should she mix with 10 pounds of hazelnut if she wants to sell the mixture for $6 a pound?

Exercises

Practice

Homework Help	
For Exercises	See Examples
11–24	1–2
25–26, 28	4
29	5
30	6
Extra Practice	
See page 702.	

Applications and Problem Solving

Use the percent equation to find each number.

11. Find 40% of 280.

12. What number is 5% of 120?

13. What number is 65% of 180?

14. Find 200% of 90.

15. 18 is 75% of what number?

16. 110 is 20% of what number?

17. 250 is 40% of what number?

18. 60 is 150% of what number?

19. Find 110% of 80.

20. 90 is 200% of what number?

21. What number is 400% of 16?

22. Find 5% of 3200.

23. 25 is 40% of what number?

24. Find 0.5% of 240.

25. **Banking** How much interest will Marissa earn if she invests $250 at an annual rate of 4% for 5 years?

26. **Banking** How long will it take Mr. Albany to earn $75 if he invests $3000 at an annual rate of 5%?

27. **Statistics** The graph shows ways that teenagers ages 12 to 17 get spending money.

   a. Suppose you survey a group of 150 15-year-olds. Write a percent equation to predict how many students get a regular allowance.

   b. Solve the equation. Round to the nearest whole number.

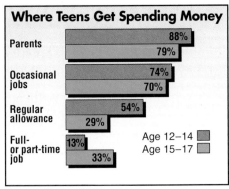

Where Teens Get Spending Money

Parents 88% / 79%
Occasional jobs 74% / 70%
Regular allowance 54% / 29%
Full- or part-time job 13% / 33%
Age 12–14 / Age 15–17

Source: *ICR TeenEXCEL*

28. Credit Cards   A credit card company charges an annual rate of 18% on the unpaid balance on credit card accounts. Suppose you have an unpaid balance of $800. How much monthly interest will you be charged? (*Hint:* First find the monthly interest rate.)

29. Sales   Great Smoky Mountains Tours conducts hiking trips along the Appalachian Trail. The owner charges $250 for adults and $175 for children. He takes groups of up to 15 people on a single trip. If he wants to earn $3150 for his next trip to pay for some new equipment, how many of the 15 people on the trip should be adults?

30. Chemistry   Roberto needs a solution that is 30% silver nitrate. He has 12 ounces of a 25% silver nitrate solution. How many ounces of a 40% silver nitrate solution should he add to make a solution that is 30% silver nitrate?

31. Critical Thinking   If $a\%$ of $b = x$ and $b\%$ of $a = y$, what is the relationship between x and y?

Mixed Review

32. Health   Nutritionists recommend a diet in which less than 30% of your total Calories come from fat. Which subs listed in the chart have less than 30% of their Calories from fat? (*Lesson 5–3*)

The Sub Shop

Kind of Sub	Total Calories	Calories from Fat
Italian	522	288
Meatball	459	198
Turkey	322	90
Roast Beef	345	108
Ham & Cheese	322	81

33. On a map, the scale is 1 inch = 60 miles. Find the actual distance for each map distance.   (*Lesson 5–2*)

	From	To	Map Distance
a.	Memphis, TN	Nashville, TN	3 inches
b.	Shreveport, LA	Jackson, MS	$3\frac{1}{2}$ inches
c.	Little Rock, AR	Memphis, TN	$2\frac{1}{4}$ inches

Solve each equation.   (*Lesson 4–6*)

34. $12x + 15 = 35 + 2x$

35. $3y + 10 = 2y - 21$

36. $6 - 8a = 20a + 20$

37. $7n - 13 = 3n + 7$

38. Extended Response   List a set of data that includes at least five numbers for which the mean is greater than the median.   (*Lesson 3–3*)

Investigation

BUILDING BRIDGES

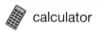 calculator

Box-and-Whisker Plots

The table shows the lengths of the longest suspension bridges.

Name of Bridge (inside U.S.)	Length (ft)	Name of Bridge (outside U.S.)	Length (ft)
Verrazano-Narrows	4260	Akashi Kaikyo	6529
Golden Gate	4200	Izmit Bay	5472
Mackinac Straits	3800	Storebaelt	5328
George Washington	3500	Humber	4626
Tacoma Narrows II	2800	Jiangyin Yangtze	4543
San Francisco-Oakland	2310	Tsing Ma	4518
Bronx-Whitestone	2300	Hardanger Fjord	4347
Delaware Memorial	2150	Hoga Kusten	3970
Seaway Skyway	2150	High Coast	3969
Walt Whitman	2000	Minami Bisan-Seto	3668
Ambassador International	1850	Second Bosporus	3576

Source: *Information Please Almanac*

Let's use a **box-and-whisker plot** to investigate these data.

Investigate

1. To make box-and-whisker plots for these data, you need to find five important values for each set of data: the median (M), the two extremes, the upper quartile (UQ), and the lower quartile (LQ).

   a. First find the median of each list.

   b. Consider only the upper half of each list. Find the medians. These numbers are the **upper quartiles**.

   c. Consider only the bottom half of each list. Find the medians. These numbers are the **lower quartiles**.

   d. The greatest (GV) and least values (LV) in each list are called the **extremes**. Find the extremes.

Look Back

Median: Lesson 3–3

2. Draw a number line.

   a. Make a box-and-whisker plot for the U.S. bridges using the five values from Exercise 1a–1d. Mark dots above their coordinates.

   b. Complete the box-and-whisker plot as shown below.

The lines outside the box are called whiskers.

3. Make a box-and-whisker plot for the bridges outside the U.S.

Extending the Investigation

In this extension, you will compare the box-and-whisker plots for the two sets of data. You will also investigate how changes in the data affect a box-and-whisker plot.

1. Use the box-and-whisker plots to answer these questions.

   a. What percent of the data lies below the median?

   b. What percent of the data lies inside the box?

   c. What percent of the data lies in one whisker?

   d. What percent of bridges outside the U.S. are longer than the longest U.S. bridge? Explain your reasoning.

2. When data are arranged in order from least to greatest, you can describe the data with percentiles. A **percentile** is the point below which a given percent of the data lies. For example, 50% of the data falls below the median. So, the median is the 50th percentile for the data.

   a. Which U.S. bridge is at the 50th percentile?

   b. Which bridge outside the U.S. is at the 25th percentile?

   c. Which U.S. bridge is at the 75th percentile?

3. Suppose a 7000-foot bridge is built in the United States.

   a. Redraw the box-and-whisker plot, including this value.

   b. Describe the change in the box-and-whisker plot.

   c. How do these data change your comparisons of the bridges?

Presenting Your Conclusions

Here are some ideas to help you present your conclusions to the class.

- Prepare a poster displaying your box-and-whisker plots. Write a comparison of the data.

- Find two related sets of data that you would like to compare with box-and-whisker plots. Draw the plots. Then prepare an oral presentation using the plots as visual aids.

*inter*NET CONNECTION **Investigation** For more information on box-and-whisker plots, visit: www.algconcepts.com

In 1990, the average cost of one dozen large eggs was $1.00. By 2000, the cost had decreased to $0.96. The decrease in the cost of goods and services is called *deflation*. **Source:** *Statistical Abstract of the United States*

To find the deflation rate, write a ratio that compares the amount of the decrease to the original price.

amount of decrease: $1.00 − $0.96 = $0.04 or 4¢
original price: $1.00 or 100¢

amount of decrease → $\dfrac{4}{100} = 4\%$ ← *deflation rate*
original price →

So, the cost of one dozen eggs decreased 4% from 1990 to 2000.

When an increase or decrease is expressed as a percent, the percent is called the **percent of increase** or the **percent of decrease**. The earlier amount is always used as the base in the percent equation.

Examples

Find the percent of increase or decrease. Round to the nearest percent.

① original: 25
new: 29

Find the amount of increase.
29 − 25 = 4

Use the percent proportion.

$\dfrac{P}{B} = \dfrac{r}{100}$ *Percent proportion*

$\dfrac{4}{25} = \dfrac{r}{100}$ *P = 4, B = 25*

$4(100) = 25r$ *Cross products*
$400 = 25r$ *Multiply.*
$\dfrac{400}{25} = \dfrac{25r}{25}$ *Divide each side by 25.*
$16 = r$ *Simplify.*

The percent of increase is 16%.

② original: 18
new: 12

Find the amount of decrease.
18 − 12 = 6

Use the percent proportion.

$\dfrac{P}{B} = \dfrac{r}{100}$ *Percent proportion*

$\dfrac{6}{18} = \dfrac{r}{100}$ *P = 6, B = 18*

$6(100) = 18r$ *Cross products*
$600 = 18r$ *Multiply.*
$\dfrac{600}{18} = \dfrac{18r}{18}$ *Divide each side by 18.*
$33 \approx r$ *Simplify.*

The percent of decrease is about 33%.

Your Turn

a. original: 12
new: 20

b. original: 50
new: 49

Two applications of percent of change are sales tax and discounts. **Sales tax** is a tax that is added to the cost of the item. It is an example of a percent of increase. **Discount** is the amount by which the regular price of an item is reduced. It is an example of a percent of decrease.

Examples ❸ **Ms. Cruz bought a car for $12,500. A state sales tax of 4% is then added to the price of the car. What was the total price?**

Sales Link

Method 1	**Method 2**
First, use the percent equation to find the sales tax.	A sales tax of 4% means that Ms. Cruz will pay 100% + 4% or 104% of the price of the car.

Method 1

$$P = RB$$
$$= 0.04(12{,}500)$$
$$= 500$$

Then, add the $500 sales tax to $12,500.

$$12{,}500 + 500 = 13{,}000$$

The total price was $13,000.

Method 2

A sales tax of 4% means that Ms. Cruz will pay 100% + 4% or 104% of the price of the car.

Use the percent equation to find the total price.

$$P = RB$$
$$= 1.04(12{,}500)$$
$$= 13{,}000$$

❹ **All shoes at The Runner's Place are on sale at a 25% discount. If a pair of running shoes originally cost $65, what is the sale price?**

Method 1

First, use the percent equation to find the discount.

$$P = RB \qquad \textit{Percent equation}$$
$$= 0.25(65) \quad \textit{R = 0.25, B = 65}$$
$$= 16.25 \qquad \textit{Multiply.}$$

Then, subtract the $16.25 discount from $65.

$$65 - 16.25 = 48.75$$

Method 2

A discount of 25% means that the buyer will pay 100% − 25% or 75% of the selling price. Use the percent equation to find the sale price.

$$P = RB \qquad \textit{Percent equation}$$
$$= 0.75(65) \quad \textit{R = 0.75, B = 65}$$
$$= 48.75 \qquad \textit{Multiply.}$$

The sale price of the running shoes is $48.75.

Your Turn

c. What is the total cost of a basketball that sells for $45 if the sales tax rate is 7%?

d. All long-sleeve T-shirts are on sale for 40% off. If the original price was $19.95, what is the discount price?

Sometimes discount stores advertise an additional discount. For example, suppose all merchandise in an outlet store is 50% off retail prices every day. If the store has a sale where you can take another 25% off, there are two ways you might interpret this.

- The discounts can be successive. That is, 50% is taken off the original price and then 25% is taken off the resulting price.
- The discounts can be combined. That is, the discount is 50% + 25% or 75% of the original price.

You can use a graphing calculator to examine these two interpretations.

BLOW-OUT SALE

25% off already low prices

Graphing Calculator Tutorial

See pp. 750–753.

 Graphing Calculator Exploration

The graphing calculator program at the right finds the cost of an item for successive discounts and combined discounts.

Copy the program into your calculator. Run the program and enter the original price and each discount (as a decimal) when prompted.

```
PROGRAM:DISCOUNT
:Disp "ORIG. PRICE"
:Input P
:Disp "1ST DISCOUNT?"
:Input A
:Disp "2ND DISCOUNT?"
:Input B
:P(1-A)(1-B) → C
:P(1-(A+B)) → D
:Disp "SUCCESSIVE
 DISCOUNT", C
:Disp "COMBINED
DISCOUNT", D
```

Try These

Copy and complete the table.

	Price	First Discount	Second Discount	Sale Price Successive Discount	Sale Price Combined Discount
1.	$ 49.00	20%	10%		
2.	$185.00	25%	10%		
3.	$ 12.50	30%	20%		
4.	$156.95	30%	15%		

5. What is the relationship between the sale price using successive discounts and the sale price using combined discounts?
6. Which type of discount is usually used in stores?

Percent of change is often used to describe how data change from one year to another. Sometimes the percent of increase is greater than 100%.

Example **5** **Statistics Link**

The graph shows on-line music sales for 2001 and projected sales for 2006. To the nearest percent, find the percent of increase.

The amount of increase is 5500 − 900 or 4600.

$$\frac{P}{B} = \frac{r}{100}$$

$$\frac{4600}{900} = \frac{r}{100}$$

$$4600(100) = 900r$$

$$460{,}000 = 900r$$

$$\frac{460{,}000}{900} = \frac{900r}{900}$$

460,000 ÷ 900 **ENTER** *511.111111*

$$r \approx 511$$

On-line music sales are expected to increase by about 511% from 2001 to 2006.

On-line Music Sales

Sales (millions)

$8000
7000
6000
5000 — $5500
4000
3000
2000
1000
0 — $900
        2001   2006
          Year

Source: Jupiter Communications

 Personal Tutor at algconcepts.com

Check for Understanding

Communicating Mathematics

1. **Describe** two different methods for finding the total price of an item if the sales tax rate and the cost of the item are given.

2. **Explain** how a percent of increase can be greater than 100%.

3. Writing Math Find an example of percent of change in a newspaper or magazine. Then write and solve a problem using your example.

Vocabulary

percent of increase
percent of decrease
sales tax
discount

Guided Practice

⊙ **Getting Ready** **Write a percent proportion to find the percent of change.**

Sample: original: 15, new: 20   **Solution:** $\frac{5}{15} = \frac{r}{100}$

4. original: 25   5. original: 14   6. original: 30
   new: 23   new: 18   new: 70

Find the percent of increase or decrease. Round to the nearest percent.

7. original: 20
   new: 16

8. original: 8
   new: 20

Example 3

The cost of an item and a sales tax rate are given. Find the total price of each item to the nearest cent.

9. in-line skates: $90; 5%

10. make-up: $14.95; 4%

Example 4

The original cost of an item and a discount rate are given. Find the sale price of each item to the nearest cent.

11. computer: $1200; 10%

12. tennis balls: $4.50; 25%

Example 5

13. **Sales**   In one year, online retail sales increased from $6 to $11 billion. Find the percent of increase to the nearest percent.

Exercises • • • • • • • • • • • • • • • • • • •

Practice

Find the percent of increase or decrease. Round to the nearest percent.

14. original: 10
    new: 12

15. original: 50
    new: 56

16. original: 30
    new: 12

17. original: 500
    new: 420

18. original: 15
    new: 36

19. original: 48
    new: 112

Homework Help	
For Exercises	See Examples
14–19	1, 2
20–25	3
26–34	4

Extra Practice
See page 702.

The cost of an item and a sales tax rate are given. Find the total price of each item to the nearest cent.

20. television: $500; 6%

21. CD player: $40; 7%

22. dress: $55; 6%

23. jeans: $38; 4%

24. CD: $17.99; 5%

25. book: $25.99; 2.5%

The original cost of an item and a discount rate are given. Find the sale price of each item to the nearest cent.

26. shoes: $40; 25%

27. tires: $280; 5%

28. earrings: $12; 40%

29. watch: $35; 15%

30. video: $24.99; 30%

31. radio: $18.95; 25%

32. What number is 30% less than 40?

33. Find the percent of increase from $18 to $36.

Applications and Problem Solving

34. **Retail Sales**   Georgia bought a new belt that was on sale for 15% off. If the original cost was $24, what was the sale price?

35. Travel The table shows the number of new passports issued in the United States. Between which years was the percent of increase greater: 1999 to 2001, or 2001 to 2003? Explain your reasoning.

New Passports (millions)	
1999	6.7
2001	7.1
2003	7.3

Source: U.S. State Department

36. Biology Over a period of years, the number of breeding pairs of bald eagles increased from 2.2 thousand pairs to 5.7 thousand pairs. Find the percent of increase to the nearest percent.
Source: Fish and Wildlife Service

37. Critical Thinking An amount is increased by 10%. The result is decreased by 10%. Is the final result less than, greater than, or equal to the original amount?

Bald Eagle

Mixed Review

38. Real Estate A *commission* is a percent of sales that is paid for selling a product. Tom Augustus sells real estate at a 7% commission. Last week, his sales totaled $90,000. What was his commission? *(Lesson 5–4)*

Express each fraction or ratio as a percent. *(Lesson 5–3)*

39. 5 out of 8   **40.** 2 to 5   **41.** $\frac{7}{20}$

Convert each measurement as indicated. *(Lesson 5–1)*

42. 15 inches to feet   **43.** 1500 grams to kilograms

Standardized Test Practice
Ⓐ Ⓑ Ⓒ Ⓓ

44. Multiple Choice Which point corresponds to a number that *cannot* be obtained by subtracting two integers? *(Lesson 2–4)*

Ⓐ A   Ⓑ B   Ⓒ C   Ⓓ D

Quiz 2   Lessons 5–4 and 5–5

Use the percent equation to find each number. *(Lesson 5–4)*

1. Find 4% of 625.   **2.** What percent of 50 is 6?

3. 15 is 75% of what number?   **4.** Find 105% of 80.

5. Economics The inflation rate for groceries is 5%. How much will a cart of groceries cost next year if it costs $134.10 this year? Round to the nearest cent. *(Lesson 5–5)*

In the Workplace

Accountant

Accountants prepare and analyze financial reports. Some accountants track the depreciation of their company's property and inventory. *Depreciation* is the decrease in value of an item due to its age. There are different ways to calculate depreciation.

- *Straight-Line Method:* The depreciation is the same each year.
- *Double-Declining-Balance Method:* The depreciation is determined by a percent of decrease.

Suppose a media production company purchased equipment for $240,000. The company estimates that the equipment will last for 5 years. At that time, the value of the equipment will be $20,000. The table and graph show the value each year for a straight-line (SL) depreciation of $44,000 and a double-declining-balance (DDB) depreciation of 40%.

Value (dollars)		
Year	**SL**	**DDB**
0	240,000	240,000
1	196,000	144,000
2	152,000	86,400
3	108,000	51,840
4	64,000	31,104
5	20,000	20,000

Key:
— SL
— DDB

1. Why do you think the first method is called the straight-line method?
2. For which year is the difference between the straight-line value and the double-declining-balance value the greatest? How is this shown on the graph?

FAST FACTS About Accountants

Working Conditions
- usually work in a comfortable environment
- generally work a 40-hour week, but some work 50 hours a week or more

Education
- most have a college degree in accounting
- knowledge of computers a necessity

Earnings

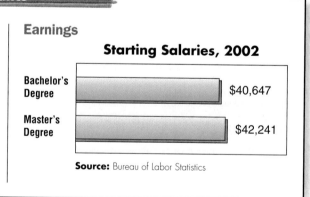

Source: Bureau of Labor Statistics

Career Data For the latest information on accountants, visit:
www.algconcepts.com

What You'll Learn

You'll learn to find the probability and odds of a simple event.

Why It's Important

Manufacturing
Inspectors use experimental probability when they find defective items. *See Examples 2 & 3.*

The table shows all of the possible outcomes when you roll a pair of dice. The highlighted outcomes are doubles.

	1	**2**	**3**	**4**	**5**	**6**
1	(1, 1)	(1, 2)	(1, 3)	(1, 4)	(1, 5)	(1, 6)
2	(2, 1)	(2, 2)	(2, 3)	(2, 4)	(2, 5)	(2, 6)
3	(3, 1)	(3, 2)	(3, 3)	(3, 4)	(3, 5)	(3, 6)
4	(4, 1)	(4, 2)	(4, 3)	(4, 4)	(4, 5)	(4, 6)
5	(5, 1)	(5, 2)	(5, 3)	(5, 4)	(5, 5)	(5, 6)
6	(6, 1)	(6, 2)	(6, 3)	(6, 4)	(6, 5)	(6, 6)

There are 36 possible outcomes. If the dice are fair, each outcome is equally likely to occur. Of those 36 outcomes, 6 are doubles. You can measure the chances of an event happening with **probability**.

Reading Algebra

P(doubles) is read as the *probability of rolling doubles*.

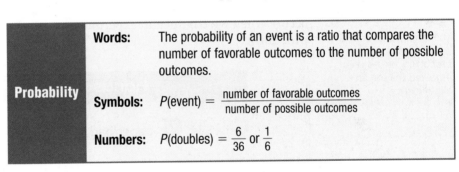

Probability

Words: The probability of an event is a ratio that compares the number of favorable outcomes to the number of possible outcomes.

Symbols: $P(\text{event}) = \dfrac{\text{number of favorable outcomes}}{\text{number of possible outcomes}}$

Numbers: $P(\text{doubles}) = \dfrac{6}{36}$ or $\dfrac{1}{6}$

The probability that an event will happen is between 0 and 1 inclusive.

- A probability of 0 means that the event is impossible.
- A probability of 1 means the event is certain to happen.
- The closer a probability is to 1, the more likely it is to happen.

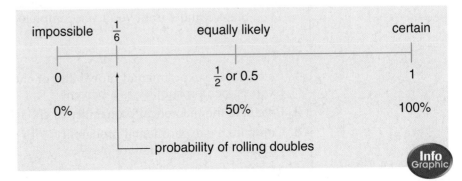

impossible $\frac{1}{6}$ equally likely   certain

0      $\frac{1}{2}$ or 0.5   1

0%   50%   100%

— probability of rolling doubles

COncepts in MOtion

Interactive Lab
algconcepts.com

Info Graphic

When all possible outcomes have an equally likely chance of happening, the outcomes are said to be **random**.

Real World

Example ①
Population Link

A population distribution for California is shown. If a person is chosen at random, what is the probability that the person is age 65 or older?

There are 4 million people who are age 65 or older. The total population is $3 + 6 + 3 + 5 + 5 + 4 + 2 + 4$ or 32 million.

P(65 or older)

$= \dfrac{\text{number of people age 65 or older}}{\text{total population}}$

$= \dfrac{4}{32}$ or $\dfrac{1}{8}$

The probability of choosing a person age 65 or older is $\dfrac{1}{8}$ or 12.5%.

Population of California

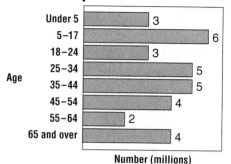

Age	Number (millions)
Under 5	3
5–17	6
18–24	3
25–34	5
35–44	5
45–54	4
55–64	2
65 and over	4

Number (millions)
Source: U.S. Census Bureau

Reading Algebra

The **empirical probability** is the most accurate probability based upon repeated trials in an experiment.

The probability that was found in Example 1 is called the **theoretical probability**. Theoretical probability is what *should* occur. What *actually* occurs when we conduct an experiment is called the **experimental probability**.

Hands-On Algebra

Materials: 📕 paper bag   📄 20 two-inch pieces of paper

Work with a partner.

Step 1 Mark 4 slips of paper with an X, 7 slips of paper with a Y, and 9 slips of paper with a Z.

Step 2 Put the slips of paper in the bag and mix well.

Step 3 Draw one slip of paper from the bag and record its letter.

Step 4 Return the slip of paper to the bag and mix well. Repeat Steps 3 and 4 until you have completed 20 trials.

Try These

1. Calculate the experimental probability of choosing each letter. Express each probability as a percent.
2. Calculate the theoretical probability of choosing each letter.
3. Compare the experimental probability with the theoretical probability. How similar are they?

Experimental probability is often used by quality-control inspectors.

2 A quality-control inspector for Office Suppliers checked a sample of 250 marking pens and found that 14 of them were defective. Find the experimental probability of choosing a defective pen.

$$P(\text{defective pen}) = \frac{\text{number of defective pens}}{\text{total number of pens}}$$

$$= \frac{14}{250} \text{ or } \frac{7}{125}$$

3 If the percent of defective pens in Example 2 is greater than 6%, production will be stopped. Should the quality-control inspector stop production?

First, change $\frac{7}{125}$ to a percent. Then compare it to 6%.

$\dfrac{P}{B} = \dfrac{r}{100}$ *Use the percent proportion.*

$\dfrac{7}{125} = \dfrac{r}{100}$ *Replace P with 7 and B with 125.*

$7(100) = 125r$ *Find the cross products.*

$700 = 125r$ *Multiply 7 and 100.*

$\dfrac{700}{125} = \dfrac{125r}{125}$ *Divide each side by 125.*

$5.6 = r$ $700 \boxed{\div} 125 \boxed{\text{ENTER}} \; 5.6$

The sample contains 5.6% defective pens. Since this is less than 6%, the quality-control inspector should not stop production.

Another way to measure the chance of an event occurring is with **odds**.

Odds	**Words:**	The odds of an event occurring is a ratio that compares the number of favorable outcomes to the number of unfavorable outcomes.
	Symbols:	Odds $= \dfrac{\text{number of favorable outcomes}}{\text{number of unfavorable outcomes}}$

Example

4 A bag contains 6 red marbles, 3 blue marbles, and 1 yellow marble. Find the odds of choosing a red marble.

There are 6 red marbles. So, there are 6 favorable outcomes.
There are 3 + 1 or 4 marbles that are *not* red. So, there are 4 unfavorable outcomes.

odds of choosing a red marble = 6:4 or 3:2

Your Turn

Find the odds of choosing a blue marble.

Check for Understanding

Communicating Mathematics

1. **Compare and contrast** theoretical probability and experimental probability.

2. **Describe** how to find the odds of an event happening.

3. Can the odds of an event happening be greater than 1? **Explain** why or why not.

Vocabulary

probability
random
theoretical probability
experimental probability
empirical probability
odds

Guided Practice

Example 1

Refer to the application at the beginning of the lesson. Find the probability of each outcome if a pair of dice is rolled.

4. an even number on the first die
5. a sum of 2

Example 4

Find the odds of each outcome if a die is rolled.

6. a number greater than 4
7. an even number

Examples 2 & 3

8. **Sports** Michael made 7 free throws out of 8 attempts during the last basketball game.
   a. What was the experimental probability of making a free throw?
   b. Express the probability as a percent.

Exercises

• • • • • • • • • • • • • • • • • • • •

Practice

Refer to the application at the beginning of the lesson. Find the probability of each outcome if a pair of dice are rolled.

9. an odd number on the first die
10. a sum of 12
11. a sum of 5
12. an even sum
13. a sum greater than 12
14. a sum less than 13

Homework Help	
For Exercises	**See Examples**
9–14, 21	1
15–20	4
23	2, 3
Extra Practice	
See page 702.	

Find the odds of each outcome if the spinner at the right is spun.

15. an even number
16. greater than 2
17. *not* a 5
18. red
19. red or blue
20. *not* red

21. What is the probability that a month picked at random starts with the letter *J*?

22. If the probability that an event occurs is $\frac{1}{2}$, find the odds that it occurs.

23. Manufacturing A quality-control inspector checked three samples of sweaters during the day. Each sample contained 40 sweaters. The chart lists the number of sweaters that were defective.

Quality Control Results	
Sample (time)	**Number Defective**
9:00	2
1:00	1
4:00	5

   a. For each sample, find the experimental probability of choosing a defective sweater.

   b. Production is stopped if the percent of defective sweaters is greater than 5%. For which sample(s) should production be stopped?

24. Marketing The local video store advertises that 1 out of 4 customers will receive a free box of popcorn when they rent a video.

   a. What are the odds of receiving free popcorn?

   b. What are the odds *against* receiving free popcorn?

   c. At the end of the first day, 15 customers out of 75 had received free popcorn. Find the experimental probability.

   d. Did the store give away as much free popcorn as it had advertised?

25. Allowance The graph shows the weekly allowance for students in grades 6 through 12.

   a. If a student is chosen at random, what is the probability the student receives $5 or more as a weekly allowance?

   b. What are the odds that a student chosen at random receives no allowance?

Weekly Allowance

	Number
No allowance	45
Less than $5	5
$5–$9.99	16
$10–$14.99	13
$15–$19.99	4
$20 or more	17

Amount (dollars)

Source: *USA WEEKEND*

26. Critical Thinking The **complement** of an event's occurring is the event's *not* occurring. If the probability that an event will occur is $\frac{3}{5}$, what is the probability of its complement?

The original cost of an item and a discount rate are given. Find the sale price of each item to the nearest cent. *(Lesson 5–5)*

27. basketball: $45, 20%   **28.** television: $399, 10%

29. Banking How much interest will Ben earn if he invests $1000 at a rate of 6% for 3 years? *(Lesson 5–4)*

Ⓐ Ⓑ Ⓒ Ⓓ

30. Grid In What percent of 25 is 30? *(Lesson 5–3)*

31. Multiple Choice Which expression is equivalent to $-2(x + 5) + 6(x + 5)$? *(Lesson 2–5)*

   Ⓐ $4(x + 5)$ Ⓑ $4x + 40$ Ⓒ $8x + 40$ Ⓓ $4x - 20$

What You'll Learn

You'll learn to find the probability of mutually exclusive and inclusive events.

Why It's Important

Marketing
Restaurant owners can use probability to make predictions about their business.
See Exercise 21.

Compound events consist of two or more simple events that are connected by the words *and* or *or*. Let's investigate a case where simple events are connected by the word *and*.

Hands-On Algebra

Materials: 2 paper bags red and yellow counters

Work with a partner.

Step 1   Place a red counter and a yellow counter in each bag.

Step 2   Without looking, remove one counter from each bag. Record the color combination in the order that you drew the counters. Return the counters to their respective bags.

Step 3   Repeat 99 times. Count and record the number of red/red, red/yellow, yellow/red, and yellow/yellow combinations.

Try These

Estimate the probability of each outcome.

1. P(red *and* red)
2. P(red *and* yellow)
3. P(yellow *and* red)
4. P(yellow *and* yellow)

Reading Algebra

P(red *and* yellow) means the probability of choosing red from the first bag and yellow from the second bag.

Choosing a counter from bag 1 did not affect choosing a counter from bag 2. These events are called **independent events** because the outcome of one event does not affect the outcome of the other event.

You can analyze the experiment with a tree diagram.

Look Back

Tree diagrams:
Lesson 4–2

Bag 1	Bag 2	Outcome
	R	RR
R		
	W	RW
	R	WR
W		
	W	WW

There are four equally-likely outcomes. So, the probability of choosing white on the first draw *and* white on the second draw is $\frac{1}{4}$. You can also multiply to find the probability of two independent events.

$$\underbrace{P(\text{white from bag 1})}_{\frac{1}{2}} \times \underbrace{P(\text{white from bag 2})}_{\frac{1}{2}} = \underbrace{P(\text{white and white})}_{\frac{1}{4}}$$

<table>
<tr><td rowspan="3">**Probability of Independent Events**</td><td>**Words:**</td><td>The probability of two independent events is found by multiplying the probability of the first event by the probability of the second event.</td></tr>
<tr><td>**Symbols:**</td><td>$P(A \text{ and } B) = P(A) \cdot P(B)$</td></tr>
<tr><td>**Model:**</td><td>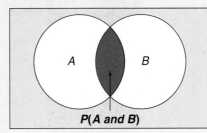</td></tr>
</table>

$P(A \text{ and } B)$

Example **1** **Two dice are rolled. Find the probability that an odd number is rolled on the first die and the number 4 is rolled on the second.**

$P(\text{odd number}) = \dfrac{3}{6} \text{ or } \dfrac{1}{2}$ *There are three ways to roll an odd number.*

$P(4) = \dfrac{1}{6}$ *There is 1 way to roll a 4.*

$P(\text{odd number and } 4) = \dfrac{1}{2} \cdot \dfrac{1}{6} \text{ or } \dfrac{1}{12}$

Your Turn

a. Two dice are rolled. Find the probability that an even number is rolled on the first die and a number greater than 4 on the second die.

Two events can also be connected by the word *or*. For example, consider the probability of drawing a jack or a queen from a standard deck of 52 cards. Since a card cannot be both a jack and a queen, the events are **mutually exclusive**. That is, both events cannot occur at the same time.

The probability of two mutually exclusive events is found by adding.
$P(\text{jack or queen}) = P(\text{jack}) + P(\text{queen})$

$$= \dfrac{4}{52} + \dfrac{4}{52}$$

$$= \dfrac{8}{52} \text{ or } \dfrac{2}{13}$$

The probability of drawing a jack or a queen is $\dfrac{2}{13}$.

<table>
<tr>
<td rowspan="3">**Probability of Mutually Exclusive Events**</td>
<td>**Words:**</td>
<td>The probability of two mutually exclusive events is found by adding the probability of the first event and the probability of the second event.</td>
</tr>
<tr>
<td>**Symbols:**</td>
<td>$P(A \text{ or } B) = P(A) + P(B)$</td>
</tr>
<tr>
<td>**Model:**</td>
<td>

A   B

P(A or B)
</td>
</tr>
</table>

Example ❷ **Jamal has 4 quarters, 2 dimes, and 4 nickels in his pocket. He takes one coin from his pocket at random. What is the probability that the coin is either a quarter or a dime?**

A coin cannot be both a quarter and a dime, so the events are mutually exclusive. Find the sum of the individual probabilities.

$P(\text{quarter } or \text{ dime}) = P(\text{quarter}) + P(\text{dime})$

$$= \frac{4}{10} + \frac{2}{10} \quad \textit{4 of the coins are quarters and 2 are dimes.}$$

$$= \frac{6}{10} \quad \textit{Add.}$$

$$= \frac{3}{5} \quad \textit{Simplify.}$$

The probability of choosing a quarter or a dime is $\frac{3}{5}$.

Your Turn

b. Find the probability of Jamal's choosing a quarter or a nickel.

Sometimes events are connected by the word *or,* but they are not mutually exclusive. For example, suppose there is a chance of rain on Saturday and there is a chance of rain on Sunday. You want to find the chance of rain over the weekend. Because it could rain on *both* Saturday *and* Sunday, rainfall on Saturday and Sunday are not mutually exclusive events. The two events are called **inclusive events.**

	Words:	The probability of two inclusive events is found by adding the probabilities of the events, then subtracting the probability of both events.
Probability of Inclusive Events	Symbols:	$P(A \text{ or } B) = P(A) + P(B) - P(A \text{ and } B)$
	Model:	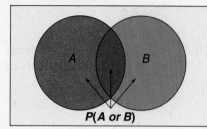

Example ③

Meteorology Link

If there is a 40% chance of rain on Saturday and a 60% chance of rain on Sunday, find the probability that it will rain on either Saturday or Sunday.

Since it is possible to rain on both days, these events are inclusive.

$P(\text{Saturday}) = 0.4$ $P(\text{Sunday}) = 0.6$

These events are independent since the weather on Saturday does not affect the weather on Sunday.

$P(\text{Saturday } or \text{ Sunday})$
$= P(\text{Saturday}) + P(\text{Sunday}) - P(\text{Saturday } and \text{ Sunday})$
$= \quad 0.4 \quad + \quad 0.6 \quad - \quad (0.4)(0.6)$
$= 1.0 - 0.24$ *Simplify.*
$= 0.76 \text{ or } 76\%$ *Write as a percent.*

The probability that it will rain on the weekend is 76%.

	Probability	$P(A \text{ and } B)$	$P(A \text{ or } B)$	$P(A \text{ or } B)$
Concept Summary	**Type of Events**	A and B are independent.	A and B are inclusive (sample spaces overlap).	A and B are mutually exclusive (sample spaces do not overlap).
	Formula	$P(A) \cdot P(B)$	$P(A) + P(B) - P(A \text{ and } B)$	$P(A) + P(B)$

Check for Understanding

Communicating Mathematics

1. **Explain** the difference between $P(A \text{ and } B)$ and $P(A \text{ or } B)$.

2. Compare and contrast independent events, mutually exclusive events, and inclusive events. Give an example of each.

> *Vocabulary*
>
> compound events
> independent events
> mutually exclusive
> inclusive

Guided Practice

Example 1

A die is rolled and the spinner is spun. Find the probability of each event.

3. P(3 and blue)

4. P(even and red)

Examples 2 & 3

A card is drawn from a standard deck of cards. Determine whether each event is *mutually exclusive* or *inclusive*. Then find each probability.

5. P(jack or king)

6. P(ace or red card)

7. P(jack or diamond)

8. P(diamond or club)

Example 3

9. **Savings** The chart shows the probability that students in grades 6 through 12 are saving for certain items. What is the probability that a student picked at random is saving for a CD player or a computer?

What Students Save For	
Item	**Probability**
Car	31%
College	27%
Clothing	19%
CDs or CD player	15%
Trip or vacation	6%
Computer or software	6%

Source: *USA WEEKEND*

Exercises

Practice

A card is drawn from a deck of ten cards numbered 1 through 10. The card is replaced in the deck and another card is drawn. Find the probability of each outcome.

10. P(5 and then a 3)

11. P(two even numbers)

12. P(two numbers greater than 4)

13. P(6 and then an odd number)

14. P(an odd number and then an even number)

15. P(a number greater than 7 and then a number less than 6)

16. What is the probability of tossing a coin three times and getting heads each time?

Homework Help

For Exercises	See Examples
10–16	1
17–21	2, 3

Extra Practice

See page 703.

Determine whether each event is *mutually exclusive* or *inclusive*. Then find each probability.

17. There are 3 books about the American Revolution, 2 books about the Civil War, and 4 books about World War II on a shelf. If a book is selected at random, what is the probability of choosing a book about the Civil War or World War II?

18. A card is drawn from a deck of cards. What is the probability that it is a black card or a face card?

19. **Sports**   The probability that a player hits a home run during an official time at bat during the 2003 season was 0.115. Find the probability that he hit a home run on two consecutive times at bat. Express your answer as a decimal rounded to the nearest thousandth.

20. **Games**   A radio station sponsored The Birthday Game, in which it would award $1 million to any caller whose birthday matched a certain month, day, and year. If there were 62 possible years, find the probability that a caller wins $1 million. Assume 30 days in each month.

21. **Marketing**   The graph shows the percent of adults who say they've eaten at certain types of restaurants during the last six months.

Restaurant Choices	
Kind	**Percent**
American	70
Chinese	60
French	10
Italian	61
Japanese	18
Mexican	57
Thai	11

Source: *American Demographics*

  a. What is the probability that a person chosen at random has eaten in both a Chinese and Mexican restaurant?

  b. Suppose no one has eaten at both a French and a Japanese restaurant. What was the probability that a person ate at either a French or a Japanese restaurant?

  c. What is the probability that a person chosen at random has eaten at either an American or Italian restaurant?

22. **Critical Thinking**   A standard domino set has 28 tiles. Seven of these tiles have the same number of dots on each side and are called *doubles*. As the game begins, each player draws one tile. What is the probability that the first and second players each draw doubles?

Determine the probability of each outcome.   *(Lesson 5–6)*

23. A month is chosen at random that ends in *-ber*.

24. A die is rolled and shows a number greater than 1.

25. A brown-haired student is chosen at random from a class of 25 students. There are 15 brown-haired students in the class.

26. **Short Response**   In 2003, the population of North America was about 323 million. It is expected to increase to about 390 million by 2050. By what percent is the population expected to increase?   *(Lesson 5–5)*

27. **Multiple Choice**   Evaluate the expression $|3| + |-5|$.   *(Lesson 2–1)*
    Ⓐ 2   Ⓑ −2   Ⓒ −8   Ⓓ 8

Understanding and Using the Vocabulary

After completing this chapter, you should be able to define each term, property, or phrase and give an example or two of each.

*inter*NET
CONNECTION **Review Activities**
For more review activities, visit:
www.algconcepts.com

Algebra
base *(p. 199)*
dimensional analysis *(p. 190)*
discount *(p. 213)*
mixture problems *(p. 206)*
percent *(p. 198)*
percent equation *(p. 204)*
percent of decrease *(p. 212)*
percent of increase *(p. 212)*
percent proportion *(p. 199)*
percentage *(p. 199)*
proportion *(p. 188)*
rate *(pp. 190, 204)*
ratio *(p. 188)*
sales tax *(p. 213)*
scale *(p. 194)*
scale drawing *(p. 194)*
scale model *(p. 194)*
simple interest *(p. 205)*
unit rate *(p. 190)*

Statistics
box-and-whisker plot *(p. 210)*
circle graph *(p. 200)*
extremes *(p. 210)*
lower quartile *(p. 210)*
percentile *(p. 211)*
upper quartile *(p. 210)*

Probability
complement *(p. 223)*
compound event *(p. 224)*
empirical probability *(p. 220)*
experimental probability
    (p. 220)
inclusive events *(p. 226)*
independent events *(p. 224)*
mutually exclusive *(p. 225)*
odds *(p. 221)*
probability *(p. 219)*
random *(p. 220)*
theoretical probability *(p. 220)*

Choose the correct term to complete each sentence.

1. The equation $\frac{2}{3} = \frac{10}{15}$ is a (proportion , ratio).

2. A ratio is a comparison of two numbers by (multiplication, division).

3. A probability of 0 means that an event is (certain to happen, impossible).

4. The probability that occurs when you conduct an experiment is called the (empirical probability , theoretical probability).

5. The (odds , probability) that an event will occur is the ratio of the number of favorable outcomes to the number of unfavorable outcomes.

6. In the percent proportion $\frac{3}{8} = \frac{r}{100}$, 8 is the (base , percentage).

7. A (unit rate , ratio) always has a denominator of 1.

8. In a percent equation, the base is (always, not always) the largest number.

9. (Discount , Sales tax) is an example of a percent of decrease.

10. If two events cannot occur at the same time, they are (mutually exclusive , inclusive).

Skills and Concepts

Objectives and Examples	Review Exercises

• Lesson 5–1 Solve proportions.

$$\frac{7}{4} = \frac{n}{2}$$

$14 = 4n$ *Find the cross products.*

$3.5 = n$ *Divide each side by 4.*

Solve each proportion.

11. $\frac{5}{8} = \frac{n}{72}$ **12.** $\frac{r}{11} = \frac{35}{55}$

13. $\frac{x}{x-1} = \frac{4}{3}$ **14.** $\frac{z-7}{6} = \frac{z+3}{7}$

• Lesson 5–2 Solve problems involving scale drawings and models.

In a drawing, 3 cm = 45 m. Find the length of a line representing 75 meters.

$$\frac{3 \text{ centimeters}}{45 \text{ meters}} = \frac{x \text{ centimeters}}{75 \text{ meters}}$$

$225 = 45x$ *Find the cross products.*

$5 = x$ *Divide each side by 45.*

On a map, 2 inches = 5 miles. Find the actual distance for each map distance.

15. 4 inches   **16.** 5 inches

17. 12 inches   **18.** 9 inches

19. A scale model of the Statue of Liberty is 10 inches high. If the Statue of Liberty is 305 feet tall, find the scale of the model.

• Lesson 5–3 Solve problems by using the percent proportion.

75 is what percent of 250?

$$\frac{75}{250} = \frac{r}{100}$$ *Use the percent proportion.*

$7500 = 250r$ *Find the cross products.*

$30 = r$ *Divide each side by 250.*

So, 75 is 30% of 250.

Use the percent proportion to find each number.

20. What number is 60% of 80?

21. 21 is 35% of what number?

22. 7 is what percent of 56?

23. 60 is what percent of 40?

24. Find 12% of 5200.

25. 15 is 30% of what number?

• Lesson 5–4 Solve problems by using the percent equation.

Find 15% of 82.

$P = RB$ *Use the percent equation.*

$P = 0.15(82)$ *Replace R with 0.15 and*

$P = 12.3$ *B with 82.*

So, 15% of 82 is 12.3.

Use the percent equation to find each number.

26. 54 is 150% of what number?

27. Find 45% of 18.

28. What is 8% of 80?

29. 21 is 14% of what number?

30. What percent of 34 is 17?

Mixed Problem Solving
See pages 724–731.

Objectives and Examples

Review Exercises

• **Lesson 5–5** Solve problems involving percent of increase or decrease.

original: $120   new: $108
amount of decrease: $120 − $108 = $12

$\frac{12}{120} = \frac{r}{100}$ *Use the percent proportion.*

$1200 = 120r$ *Find the cross products.*

$10 = r$ *Divide each side by 120.*

The percent of decrease is 10%.

Find the percent of increase or decrease. Round to the nearest percent.

31. original: 8
new: 10

32. original: 10
new: 15

33. original: 18
new: 12

34. original: 800
new: 300

35. Find the percent of increase from $25 to $50.

36. What number is 20% less than 80?

• **Lesson 5–6** Find the probability of a simple event.

Find the probability of randomly choosing the letter I in the word PITTSBURGH.

$\frac{\text{number of favorable outcomes}}{\text{number of possible outcomes}} = \frac{1}{10}$

One letter from the word MISSISSIPPI is chosen at random. Find each probability.

37. $P(M)$

38. $P(I)$

39. $P(\text{consonant})$

40. $P(\text{vowel})$

41. Find the probability that a letter chosen at random from the word OHIO is *not* an O.

• **Lesson 5–7** Find the probability of mutually exclusive and inclusive events.

mutually exclusive:

$P(A \text{ or } B) = P(A) + P(B)$

inclusive:

$P(A \text{ or } B) = P(A) + P(B) - P(A \text{ and } B)$

A six-sided die is rolled. Determine whether each event is *mutually exclusive* or *inclusive*. Then find each probability.

42. $P(\text{even or less than 5})$

43. $P(6 \text{ or odd})$

Applications and Problem Solving

44. Food Of the students surveyed, 40% chose pizza as their favorite lunch. If there are 1250 students in the school, how many would you expect to order pizza? *(Lesson 5–4)*

45. Shopping Inez is buying a sound system that costs $399. Since she is an employee of the store, she receives a 15% discount. How much will Inez pay for the sound system? *(Lesson 5–5)*

1. **Write** the percent proportion and the percent equation. Explain the meaning of P, B, r, and R.

2. **Compare and contrast** experimental probability and theoretical probability.

Solve each proportion.

3. $\dfrac{3}{x} = \dfrac{12}{16}$

4. $\dfrac{9}{y} = \dfrac{15}{10}$

5. $\dfrac{x}{5} = \dfrac{x + 2}{6}$

6. $\dfrac{3}{3a + 2} = \dfrac{3}{8}$

Find each number.

7. 20% of 40 is what number?

8. 12 is what percent of 60?

9. 20 is what percent of 16?

10. 23 is 25% of what number?

11. Find 120% of 32.

12. What is 35% of 60?

13. **Measurement**   Convert 42 inches to feet.

14. **Recycling**   When 2000 pounds of paper are recycled or reused, 17 trees are saved. How many trees would be saved if 8000 pounds of paper are recycled?

15. **Hobbies**   Model railroads are scaled-down models of real trains. The scale on an HO model train is 1 inch = 87 inches. An HO model of a modern diesel locomotive is 8 inches long. How long is the real locomotive?

16. **Banking**   How long will it take Mr. Roberts to earn $1500 if he invests $5000 at a rate of 6%?

17. **Shopping**   The Just Skates sporting goods store advertises that all in-line skates are on sale for 20% off the regular price. Find the sale price of a pair of skates that cost $160.

18. **Taxes**   What is the cost of a pair of jeans that sells for $49 if the sales tax rate is 6%?

19. **Sports**   A quarterback threw 18 completed passes out of 30 attempts. Find the experimental probability of making a completed pass. Express the probability as a percent.

20. **Probability**   A die is rolled. What is the probability of rolling a 5 or a number greater than 3?

 www.algconcepts.com/chapter_test

Preparing for Standardized Tests

Expression and Equation Problems

All standardized tests include questions that ask you to evaluate expressions and solve equations. You'll need to calculate with positive and negative integers, as well as with fractions and decimals. Be sure that you know and can apply the properties of equality.

Test-Taking Tip

You can use a strategy called "backsolving" on some multiple-choice questions. To use this strategy, substitute each answer choice into the expression or equation to determine which is correct.

Example 1

Solve $y = -14x - 5$ if $x = -1$.

(A) -19 (B) -9 (C) 9 (D) 19

Hint Work carefully with negative numbers. Apply the rules for adding and subtracting integers.

Solution Substitute -1 for x and evaluate the right side of the equation.

$y = -14x - 5$	*Original equation*
$y = -14(-1) - 5$	*Substitution*
$y = 14 - 5$	*$-14(-1) = 14$*
$y = 9$	*$14 - 5 = 9$*

The answer is C.

You can check your answer by replacing y with 9 and x with -1 in the original equation. If the statement is true, then your answer is correct.

$y = -14x - 5$	*Original equation*
$9 \stackrel{?}{=} -14(-1) - 5$	*Replace x with -1 and y with 9.*
$9 \stackrel{?}{=} 14 - 5$	*Multiply.*
$9 = 9$ ✓	*Subtract.*

Example 2

If $2 + a = 2 - a$, then $a =$

(A) -1 (B) 0 (C) 1 (D) 2 (E) 4

Hint Use the properties of equality to solve an equation.

Solution

$2 + a = 2 - a$	*Original equation*
$2 + a + (-2) = 2 - a + (-2)$	*Add -2 to each side.*
$a = -a$	*$2 + (-2) = 0$*
$a + a = -a + a$	*Add a to each side.*
$2a = 0$	*$-a + a = 0$*
$a = 0$	*Divide each side by 2.*

The answer is B.

Alternate Solution Use the strategy known as "backsolving." Substitute each of the answer choices for a and see which value makes the equation true.

A $2 + (-1) = 2 - (-1)$	$1 = 3$	*not true*
B $2 + 0 = 2 - 0$	$2 = 2$	*true*
C $2 + 1 = 2 - 1$	$3 = 1$	*not true*
D $2 + 2 = 2 - 2$	$4 = 0$	*not true*
E $2 + 4 = 2 - 4$	$6 = -2$	*not true*

After you work each problem, record your answer on the answer sheet provided or on a sheet of paper.

Multiple Choice

1. The time that a traffic light remains yellow is given by the formula $t = \frac{1}{8}s + 1$, where t is the time in seconds and s is the speed limit (mph). If the speed limit is 40 mph, how long will a light stay yellow? *(Lesson 4–5)*

- **A** 5 seconds
- **B** 6 seconds
- **C** 7 seconds
- **D** 8 seconds

2. The heights of Monica's sunflowers are listed below. What is the median height, in inches, of the sunflowers? *(Lesson 3–3)*

Height of Sunflowers	
2 ft 4 in.	3 ft 11 in.
2 ft 9 in.	4 ft 5 in.
4 ft 3 in.	3 ft 5 in.

- **A** 41
- **B** 42.2
- **C** 44
- **D** 49

3. If $9b = 81$, then $3 \times 3b =$ *(Lesson 4–4)*
- **A** 9.
- **B** 27.
- **C** 81.
- **D** 243.
- **E** 729.

4. The Tigers baseball team scored four more runs than the Bears. The number of runs the Bears scored is represented by n. Which expression represents the number of runs the Tigers scored? *(Lesson 1–1)*
- **A** $n - 4$
- **B** $n + 4$
- **C** $4 - n$
- **D** $4n$

5. The formula $F = \frac{9}{5}C + 32$ is used to convert between degrees Celsius and degrees Fahrenheit. If the temperature is $82°$ F, what is the temperature in degrees Celsius? *(Lesson 1–5)*
- **A** 27.8
- **B** 45.6
- **C** 63.3
- **D** 90

6. Which expression is equivalent to $\frac{4 + 8x}{12x}$? *(Lesson 4–1)*
- **A** $\frac{1 + 2x}{3x}$
- **B** $\frac{1 + 8x}{3x}$
- **C** 1
- **D** $\frac{8}{3}$
- **E** $\frac{1 + 2x}{3}$

7. The Huang family had weekly grocery bills of $105, $115, $120, and $98 last month. What was their mean (average) weekly grocery bill last month? *(Lesson 3–3)*
- **A** $101.50
- **B** $102.00
- **C** $109.50
- **D** $117.50

8. The equation $y = \frac{1}{3}x - \frac{3}{8}$ relates x and y. When y is $-\frac{1}{2}$, what is the value of x? *(Lesson 4–5)*
- **A** $-\frac{1}{24}$
- **B** $-\frac{13}{24}$
- **C** $-2\frac{5}{8}$
- **D** $-\frac{3}{8}$

Grid In

9. If $\frac{x + 2x + 3x}{2} = 6$, find the value of x.
(Lesson 4–5)

Extended Response

10. Consider $3(x - 2) + 4(2x + 1) - 2(1 - x)$. *(Lesson 1–4)*

Part A Simplify the expression completely. Show your work.

Part B List the properties you used.

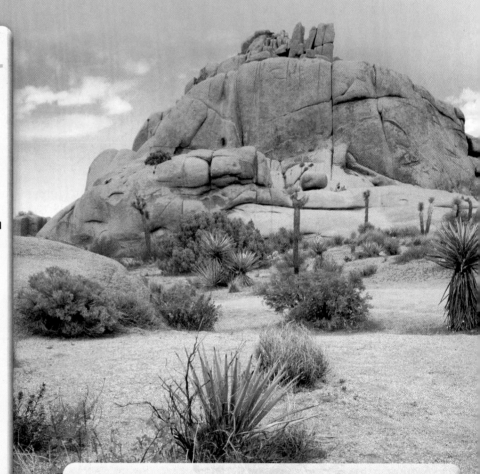

CHAPTER 6

Functions and Graphs

What You'll Learn

Key Ideas

- Show relations as ordered pairs, tables, and graphs. *(Lesson 6–1)*

- Solve and graph linear equations. *(Lessons 6–2 and 6–3)*

- Determine whether a relation is a function. *(Lesson 6–4)*

- Solve problems involving direct variation and inverse variation. *(Lessons 6–5 and 6–6)*

Key Vocabulary

domain *(p. 238)*

function *(p. 256)*

linear equation *(p. 250)*

range *(p. 238)*

relation *(p. 238)*

Why It's Important

Plants Even deserts get a little rain each year. When it rains, a cactus plant's leaves, stems, or roots take in as much water as possible to live on during the dry season. A cactus's spines also help it survive. They discourage hungry animal and also slow down wind, making less water evaporate from the cactus.

Functions and relations are used to represent relationships between real-life quantities. You will use a linear relation to predict the number of blooms on a cactus in Lesson 6–3.

Check Your Readiness

Graph each point on a coordinate plane.

1. $A(2, 4)$ **2.** $B(3, 1)$ **3.** $C(0, 4)$

4. $D(-1, 3)$ **5.** $E(-2, 2)$ **6.** $F(4, -2)$

7. $G(3, -4)$ **8.** $H(-4, -5)$ **9.** $I(-1, -3)$

Evaluate each algebraic expression if *x* = 2.

10. $2x$ **11.** $-3x$ **12.** $x + 5$

13. $8 - x$ **14.** $2x + 3$ **15.** $6x - 2$

16. $-3x + 4$ **17.** $-4x + 2$ **18.** $7 - 3x$

Solve each proportion.

19. $\dfrac{5}{6} = \dfrac{y}{12}$ **20.** $\dfrac{3}{4} = \dfrac{y}{16}$ **21.** $\dfrac{y}{5} = \dfrac{12}{15}$

22. $\dfrac{4}{24} = \dfrac{9}{x}$ **23.** $\dfrac{-3}{-9} = \dfrac{-5}{x}$ **24.** $\dfrac{0.5}{x} = \dfrac{2}{20}$

25. $\dfrac{12}{1} = \dfrac{x}{2.5}$ **26.** $\dfrac{150}{2.50} = \dfrac{225}{x}$ **27.** $\dfrac{-2}{3} = \dfrac{y}{-6}$

FOLDABLES™
Study Organizer

Make this Foldable to help you organize your Chapter 6 notes. Begin with a sheet of notebook paper.

❶ **Fold** lengthwise to the holes.

❷ **Cut** along the top line and then cut 10 tabs.

❸ **Label** the tabs using the vocabulary words as shown.

Reading and Writing Store your foldable in a 3-ring binder. As you read and study the chapter, write notes and examples under each tab.

What You'll Learn
You'll learn to show relations as sets of ordered pairs, as tables, and as graphs.

Why It's Important
Demographics
Population growth data can be expressed as a relation.
See Example 3.

In Lesson 1-6, you learned that data can be represented in a table. The table at the right shows the estimated number of sea turtles found stranded on the Texas Gulf Coast over several months in a recent year.

The data can also be represented by a set of ordered pairs, as shown below. Each first coordinate represents the month, and the second coordinate is the number of stranded sea turtles.

Sea Turtles Stranded	
Month	**Number**
January	22
February	29
March	43
April	76
May	41
June	23

Source: HEART

$$\{(1, 22), (2, 29), (3, 43), (4, 76), (5, 41), (6, 23)\}$$

The first coordinate in an ordered pair is called the **x-coordinate**. The second coordinate is called the **y-coordinate**.

Each ordered pair can be graphed.

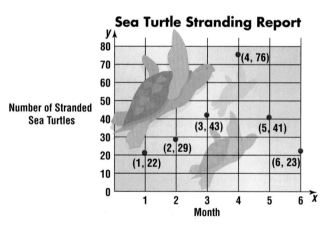

Sea Turtle Stranding Report

A set of ordered pairs, such as the one for the sea turtle data, is a **relation**. The set of all first coordinates of the ordered pairs, or x-coordinates, is called the **domain** of the relation. The set of all second coordinates, or y-coordinates, is called the **range**.

Domain and Range of a Relation	The domain of a relation is the set of all first coordinates from the ordered pairs of the relation.
	The range of a relation is the set of all second coordinates from the ordered pairs of the relation.

For the relation of the sea turtle data, the domain is {1, 2, 3, 4, 5, 6}, and the range is {22, 29, 43, 76, 41, 23}.

A relation can be shown in several ways. Three ways of representing the relation {(−1, −1), (0, 2), (1, 3)} are shown below.

Concept Summary		
Ordered Pairs	Table	Graph
(−1, −1) (0, 2) (1, 3)		

x	y
−1	−1
0	2
1	3

Examples

1 Express the relation {(−2, 4), (2, 5), (3, 3), (4, 2), (5, −1)} as a table and as a graph. Then determine the domain and range.

x	y
−2	4
2	5
3	3
4	2
5	−1

The domain is {−2, 2, 3, 4, 5}, and the range is {4, 5, 3, 2, −1}.

2 Express the relation shown on the graph as a set of ordered pairs and in a table. Then determine the domain and range.

The set of ordered pairs for the relation is {(−3, −1), (−2, 5), (1, 2), (2, −3), (5, 4)}.

x	y
−3	−1
−2	5
1	2
2	−3
5	4

The domain is {−3, −2, 1, 2, 5}, and the range is {−1, 5, 2, −3, 4}.

(continued on the next page)

Your Turn

a. Express the relation $\{(-4, -1), (-3, 1), (0, 3), (2, -6), (5, 5)\}$ as a table and as a graph. Then determine the domain and range.

b. Express the relation shown on the graph as a set of ordered pairs and in a table. Then find the domain and range.

In real-life situations, you may need to select a range of values for the x- or y-axis that does not begin with 0.

The table shows the population of Kentucky since 1930.

A. Determine the domain and range of the relation.

B. Graph the relation.

C. During which decade was there the greatest increase in population?

Kentucky State Capitol Building

Year	1930	1940	1950	1960	1970	1980	1990	2000
Population (millions)	2.6	2.8	2.9	3.0	3.2	3.7	3.7	4.0

Source: U.S. Census

A. The domain is {1930, 1940, 1950, 1960, 1970, 1980, 1990, 2000}. The range is {2.6, 2.8, 2.9, 3.0, 3.2, 3.7, 4.0}.

B. The x-coordinates go from 1930 to 2000. The scale does not have to begin at 0. The y-coordinates include values from 2.6 to 4.0. You can include 0 and use units of 0.2.

C. The graph shows that the greatest increase in population was during the 1970s.

nline **Personal Tutor at** algconcepts.com

Check for Understanding

Communicating Mathematics

1. **Describe** three ways to express a relation.
2. **Describe** the following relation.
   {(1, Washington), (2, Adams), (3, Jefferson), . . .}
3. **Writing Math** Write a sentence explaining the difference between the domain and the range of a relation.

Guided Practice

Example 1

Express each relation as a table and as a graph. Then determine the domain and the range.

4. {(−4, 2), (−2, 0), (0, 2), (2, 4)}
5. {(−3, −3.5), (−1, 3.9), (0, 2), (5, 2.5)}

Example 2

Express each relation as a set of ordered pairs and in a table. Then determine the domain and the range.

6.

7.

Example 3

8. **Economics** For his economics class, Alonso is researching the growth of the national debt since 1975. He organized his results in a table.
   a. Determine the domain and range of the relation.
   b. Graph the relation.
   c. Predict the national debt in 2005. Check your answer on the Internet.

National Debt	
Year	Amount (billions)
1975	$577
1980	$930
1985	$1946
1990	$3233
1995	$4974
2000	$5674

Source: Bureau of the Public Debt

Exercises • • • • • • • • • • • • • • • • • •

Practice

Express each relation as a table and as a graph. Then determine the domain and the range.

9. {(4, 3), (−2, 3), (−2, 4), (4, 4)}
10. {(2, 3.5), (2.9, 1), (4.5, 7.5)}
11. {(−2, 0), (3, −7), (2, −5), (−6, 3), (1, 5)}
12. {(3.1, 0), (4, 1.5), (−1, −1), (−5.5, 1.9)}
13. $\left\{\left(-\frac{1}{2}, \frac{1}{2}\right), (1, 0), \left(-\frac{1}{2}, -5\right)\right\}$
14. $\left\{\left(\frac{1}{2}, 3\right), \left(\frac{1}{2}, 1\right), (0, -1), \left(\frac{3}{4}, -2\right)\right\}$

Homework Help

For Exercises	See Examples
9–14, 25	1
15–20, 24, 27	2
26	1, 2
Extra Practice	
See page 703.	

Express each relation as a set of ordered pairs and in a table. Then determine the domain and the range.

15.

16.

17.

18.

19.

20.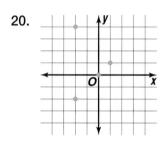

Express each relation as a set of ordered pairs.

21.

x	y
−1	1
0	2
1	3
2	4

22.

x	y
6	−12
5	−10
4	−8
3	−6

23.

x	y
0	0
−1	−0.5
1	0.5
−2	−1

Applications and Problem Solving

24. **Counting**   A die is rolled, and the spinner shown is spun.

   a. Write a relation that shows the possible outcomes. Describe the data as the set of ordered pairs (number, color).

   b. Determine the domain and range of the relation.

   c. Use this relation to find the number of ways to roll an even number and land on blue.

25. **Food**   Kelly buys a dozen bagels to share with her co-workers. She knows that her co-workers prefer blueberry bagels to plain bagels. So, she buys at least twice as many blueberry bagels as plain bagels.

   a. Write a relation to show the different possibilities. (*Hint:* Let the domain represent the number of blueberry bagels.)

   b. Express the relation in a table.

26. Entertainment The size of a movie on a screen and the distance from the projector to the screen can be expressed as a relation. The diagram below shows that if the distance between the screen and the projector is 3 units, the size of the picture is 9 squares.

0   1   2   3   4

   a. Express the relation as a set of ordered pairs.

   b. Describe the pattern between the distance from the projector to the screen and the size of the picture.

   c. Find the size of the picture if the projector is 12 units from the screen.

   d. Graph the relation. Describe the graph.

27. Critical Thinking There are 8 counters numbered 1 through 8 in a bag. Two counters are chosen at random without replacement.

   a. What possible sums can you get?

   b. Describe the data as the set of ordered pairs (sum, number of ways to occur).

Mixed Review

28. Probability A card is drawn from a standard deck of cards. Find the probability of drawing an ace or a jack. *(Lesson 5–7)*

29. Odds A bag contains 4 yellow marbles, 3 green marbles, and 3 red marbles. Find the odds of drawing a yellow marble. *(Lesson 5–6)*

30. Food The Dessert Factory offers the choices shown for ordering a piece of cheesecake. Draw a tree diagram to find the number of different cheesecakes that can be ordered. *(Lesson 4–2)*

Cheesecake	Topping
original	blueberry
chocolate chip	strawberry
fudge swirl	raspberry
	cherry

Evaluate each expression. *(Lesson 2–1)*

31. $-|-24|$

32. $|8| - |-12|$

33. Multiple Choice The stem-and-leaf plot shows the scores for Neshawn's bowling team. What are the highest and lowest scores? *(Lesson 1–7)*

Stem	Leaf
15	3 5 9
16	5 5 8
17	2 3 5 6

$17 | 3 = 173$

   Ⓐ 15, 17

   Ⓑ 153, 176

   Ⓒ 153, 172

   Ⓓ 150, 170

What You'll Learn

You'll learn to solve linear equations for a given domain.

Why It's Important

Sales Knowing how to solve linear equations can help salespeople determine their monthly income. *See Exercise 38.*

The table below shows a relationship between two sets of numbers, x and y.

x	4.2x	y	(x, y)
0.1	4.2(0.1)	0.42	(0.1, 0.42)
0.25	4.2(0.25)	1.05	(0.25, 1.05)
0.5	4.2(0.5)	2.1	(0.5, 2.1)
0.75	4.2(0.75)	3.15	(0.75, 3.15)
1.0	4.2(1.0)	4.2	(1.0, 4.2)

Each ordered pair (x, y) is a *solution* of the equation $y = 4.2x$.

The set of solutions of the problem is called the **solution set**. In this example, you would write the solution set as {(0.1, 0.42), (0.25, 1.05), (0.5, 2.1), (0.75, 3.15), (1.0, 4.2)}.

The equation $y = 4.2x$ is an example of an **equation in two variables**.

Solution of an Equation in Two Variables	If a true statement results when the numbers in an ordered pair are substituted into an equation in two variables, then the ordered pair is a solution of the equation.

Example ❶ **Which of the ordered pairs (0, 1), (2, 3), (−1, 1), or (3, 5) are solutions of $y = 2x - 1$?**

Make a table. Substitute the x and y values of each ordered pair into the equation.

x	y	y = 2x − 1	True or False?
0	1	1 = 2(0) − 1 1 = −1	false
2	3	3 = 2(2) − 1 3 = 3	true ✓
−1	1	1 = 2(−1) − 1 1 = −3	false
3	5	5 = 2(3) − 1 5 = 5	true ✓

A true statement results when the ordered pairs (2, 3) and (3, 5) are substituted into the equation. Therefore, ordered pairs (2, 3) and (3, 5) are solutions of the equation $y = 2x - 1$.

a. Which of the ordered pairs (1, 1), (0, 2), (−2, 8), or (−1, −5) are solutions of $y = -3x + 2$?

Since the solutions of an equation in two variables are ordered pairs, this type of equation describes a relation. The set of values of x is the domain of the relation. The set of corresponding values of y is the range of the relation.

Example ② Solve $y = 3x$ if the domain is {−2, −1, 0, 1, 3}. Graph the solution set.

Make a table. Substitute each value of x into the equation to determine the corresponding values of y.

x	$3x$	y	(x, y)
−2	3(−2)	−6	(−2, −6)
−1	3(−1)	−3	(−1, −3)
0	3(0)	0	(0, 0)
1	3(1)	3	(1, 3)
3	3(3)	9	(3, 9)

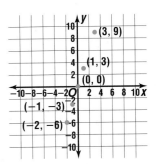

The solution set is {(−2, −6), (−1, −3), (0, 0), (1, 3), (3, 9)}.

Your Turn

b. Solve $y = -4x$ if the domain is {−2, −1, 0, 1, 2}. Graph the solution set.

Sometimes you can solve an equation for y before substituting each domain value into the equation. This makes creating a table of values easier.

Example ③ Solve $4x + 2y = 8$ if the domain is {−2, −1, 0, 1, 2}. Graph the solution set.

First, solve the equation for y in terms of x.

$4x + 2y = 8$ *Original equation*

$2y = 8 - 4x$ *Subtract 4x from each side.*

$\dfrac{2y}{2} = \dfrac{8 - 4x}{2}$ *Divide each side by 2.*

$y = \dfrac{8}{2} - \dfrac{4x}{2}$ *Write $\dfrac{8 - 4x}{2}$ as the difference of fractions.*

$y = 4 - 2x$ *Simplify.*

(continued on the next page)

 www.algconcepts.com/extra_examples

Next, substitute each value of x from the domain to determine the corresponding values of y.

x	$4 - 2x$	y	(x, y)
-2	$4 - 2(-2)$	8	$(-2, 8)$
-1	$4 - 2(-1)$	6	$(-1, 6)$
0	$4 - 2(0)$	4	$(0, 4)$
1	$4 - 2(1)$	2	$(1, 2)$
2	$4 - 2(2)$	0	$(2, 0)$

The solution set is $\{(-2, 8), (-1, 6), (0, 4), (1, 2), (2, 0)\}$.

Your Turn

c. Solve $3x + 6y = 12$ if the domain is $\{-3, -2, -1, 5, 7\}$.

You can also solve an equation for given range values.

Example ④ **Find the domain of $y = 2x - 5$ if the range is $\{-3, -1, 1, 3\}$.**

Make a table. Substitute each value of y into the equation. Then solve each equation to determine the corresponding values of x.

y	$y = 2x - 5$	x	(x, y)
-3	$-3 = 2x - 5$	1	$(1, -3)$
-1	$-1 = 2x - 5$	2	$(2, -1)$
1	$1 = 2x - 5$	3	$(3, 1)$
3	$3 = 2x - 5$	4	$(4, 3)$

The domain is $\{1, 2, 3, 4\}$.

Your Turn

d. Find the domain of $y = -3x$ if the range is $\{-6, -3, 0, 3, 6\}$.

Sometimes variables other than x and y are used in an equation. In this text, the values of the variable that comes first alphabetically are from the domain.

$$m = 2n - 5 \qquad 3d = c \qquad 2a + 3b = 6$$

domain   range   *range   domain*   *domain   range*

5 The equation $2w + 2\ell = P$ can be used to find the perimeter P of a rectangle. Suppose a rectangle has a perimeter of 48 inches. Find the possible dimensions of the rectangle given the domain values {3, 4, 8, 12}.

Assume that the values of ℓ come from the domain. Therefore, the equation should be solved for w in terms of ℓ.

$2w + 2\ell = P$	*Perimeter formula*
$2w + 2\ell = 48$	*Replace P with 48.*
$2w = 48 - 2\ell$	*Subtract 2ℓ from each side.*
$\dfrac{2w}{2} = \dfrac{48 - 2\ell}{2}$	*Divide each side by 2.*
$w = \dfrac{48}{2} - \dfrac{2\ell}{2}$	*Write $\dfrac{48 - 2\ell}{2}$ as the difference of fractions.*
$w = 24 - \ell$	*Simplify.*

Now, substitute each value of ℓ from the domain into the equation to find the corresponding values of w.

ℓ	$24 - \ell$	w	(ℓ, w)
3	24 − 3	21	(3, 21)
4	24 − 4	20	(4, 20)
8	24 − 8	16	(8, 16)
12	24 − 12	12	(12, 12)

The dimensions of the rectangle could be (3, 21), (4, 20), (8, 16), or (12, 12).

Online Personal Tutor at algconcepts.com

Check for Understanding

1. **Explain** why (2, 5) is a member of the solution set of $y = 2x + 1$.

2. **Identify** the set of numbers that make up the range of $y = x + 2$ if the domain is {0, 1, 2, 4}.

Vocabulary

equation in two variables
solution set

3. Lorena and Dan are solving the equation $3n + 2m = 11$ for the domain {−1, 3, 6, 7}. Lorena says to replace n with each of these values. Dan disagrees. He says to replace m with each value. Who is correct? Explain your reasoning.

Which ordered pairs are solutions of each equation?

4. $3a + b = 8$ **a.** (4, −4)   **b.** (2, 2)   **c.** (8, 0)   **d.** (3, 1)

5. $2c + 3d = 11$ **a.** (3, 1)   **b.** (4, −1)   **c.** (1, 3)   **d.** (−2, 5)

Solve each equation if the domain is {−2, −1, 0, 1, 2}. Graph the solution set.

6. $4x = y$

7. $y = -5x - 3$

8. $2x + 2 = y$

9. $5 + 2b = 3a$

Example 4

10. Find the domain of $y = x + 3$ if the range is {−4, −1, 0, 1, 2}.

Example 5

11. **Geometry** Refer to Example 5. Suppose a rectangle has a perimeter of 56 inches. Find the possible dimensions of the rectangle given the domain values {3, 4, 8, 12}.

Exercises • • • • • • • • • • • • • • • • • • •

Practice

Which ordered pairs are solutions of each equation?

12. $2p - 5q = 1$ **a.** (7, 3)   **b.** (2, 1)   **c.** (−7, −3) **d.** (−2, −1)

13. $3g = h + 7$ **a.** (2, 4)   **b.** (2, 3)   **c.** (−1, 2)   **d.** (2, −1)

14. $3x + 3y = 0$ **a.** (2, −2)   **b.** (−2, 2)   **c.** (1, −1)   **d.** (−1, 1)

15. $8 - 2n = 4m$ **a.** (0, 2)   **b.** (2, 0)   **c.** (1, −2)   **d.** (0.5, −3)

16. $2c + 4d = 8$ **a.** (0, 2)   **b.** (−2, 1)   **c.** (2, 0)   **d.** (−3, 0.5)

17. $8u - 4 = 3v$ **a.** (4, 2)   **b.** $\left(0, \frac{1}{2}\right)$ **c.** (2, 4)   **d.** $\left(\frac{2}{3}, \frac{3}{4}\right)$

Homework Help

For Exercises	See Examples
12–17	1
18–29	2, 3
31–33, 37	4
36–38	3, 5

Extra Practice
See page 703.

Solve each equation if the domain is {−1, 0, 1, 2, 3}. Graph the solution set.

18. $y = 2x$

19. $y = -3x$

20. $y = 7 - x$

21. $x - y = 6$

22. $y = 2x + 1$

23. $y = -5x + 5$

24. $3y = 9x + 3$

25. $2m - 2n = 4$

26. $3 - 5r = 2s$

27. $12 = 3b + 6a$

28. $6w - 3u = 36$

29. $4 = 3a - b$

30. Name an ordered pair that is a solution of $y = 6x - 2$.

Find the domain of each equation if the range is {−2, 0, 2, 4}.

31. $y = 2x - 2$

32. $y = 6 - 2x$

33. $2y = 4x$

34. Find the value of d if a solution of $y - dx = 1$ is (2, 5).

35. A solution of $4 = ay + 2x$ is (−1, 2). What is the value of a?

Applications and Problem Solving

36. **Shopping** Cedric has 90¢ to spend on school supplies. Erasers cost 10¢ each, and pencils cost 15¢ each. Suppose r represents the number of erasers and p represents the number of pencils. The equation $10r + 15p = 90$ represents this situation.

   a. Determine the ordered pairs (p, r) that satisfy the equation if the domain is {0, 2, 4, 6}.

   b. Express the relation in a table.

   c. Graph the relation.

37. Life Science The equation $y = 0.025w$ gives the weight of the brain of an infant at birth, where w is the birth weight in pounds. The table shows the brain weights of five newborns.

w	y
?	0.15
?	0.17
?	0.18
?	0.2
?	0.22

    a. Find the domain value for each range value given in the table.

    b. Express the relation as a set of ordered pairs.

    c. Graph the relation.

38. Sales Mr. Richardson is a sales representative for an electronics store. Each month, he receives a salary of $1800 plus a 6% commission on sales. Suppose s represents his sales. The equation $t = 1800 + 0.06s$ can be used to determine his total monthly income. In July, his sales will either be $800, $1300, or $2000. Find the possible total incomes for July.

39. Critical Thinking The sum of the measures of the two acute angles in a right triangle is 90°. In a given triangle, one acute angle is twice as large as the other acute angle.

    a. Write two equations that represent these two relations.

    b. Determine the ordered pair that is a solution of both equations.

Mixed Review

40. Express the relation shown in the table as a set of ordered pairs and as a graph. Then find the domain and range. *(Lesson 6–1)*

x	y
−4	−4
−3	0
2	5
5	−3

Exercise 40

41. Shopping At Crafter's Corner, all paint supplies are on sale at a 30% discount. A set of paint brushes normally sells for $15.70. What is the sale price? *(Lesson 5–5)*

42. Banking McKenna invests $580 at a rate of 6% for 5 years. How much interest will she earn? *(Lesson 5–4)*

Solve each equation.

43. $5(4 - 2c) + 3(c + 9) = 12$ *(Lesson 4–7)*

44. $4n - 9 = 5 + 3n$ *(Lesson 4–6)*

45. $3x - 7 = -10$ *(Lesson 4–5)*

Standardized Test Practice
Ⓐ Ⓑ Ⓒ Ⓓ

46. Short Response Simplify $5(7ab - 4ab)$. *(Lesson 1–4)*

47. Multiple Choice Name the property of equality shown by the statement *If $6 + 2 = 8$, then $8 = 6 + 2$. (Lesson 1–2)*

    Ⓐ Reflexive

    Ⓑ Transitive

    Ⓒ Symmetric

    Ⓓ Commutative

Why It's Important

Earth Science
Geologists can use the graph of a relation to predict the time at which Old Faithful will erupt.
See Example 8.

The relation at the right is a solution set for the equation $y = x + 2$. If you graph the relation, the points lie on a straight line, and the graph of the line includes all possible solutions of the equation of the graph.

x	y
−2	0
−1	1
0	2
1	3
2	4

$y = x - 2$

The line contains an infinite number of points. The ordered pair for every point on the line is a solution of the equation $y = x + 2$. An equation whose graph is a straight line is called a **linear equation**.

Linear Equation in Standard Form	A linear equation is an equation that can be written in the form $Ax + By = C$, where A, B, and C are any numbers, and A and B are not both zero. $Ax + By = C$, where $A \geq 0$, is called the *standard form* if A, B, and C are integers whose greatest common factor is 1.

Linear equations can contain one or two variables. However, not all equations in one or two variables are linear equations.

Linear equations	Nonlinear equations
$3x + 4y = 6$	$2x + 4y = 3z^2$
$y = -2$	$2x + 2xy = 3$
$8b = -2a - 5$	$2x^2 = 7$

Examples

Determine whether each equation is a linear equation. Explain. If an equation is linear, identify A, B, and C when written in standard form.

1 $3x = 5 + y$

First, rewrite the equation so that both variables are on the same side of the equation.

$3x = 5 + y$

$3x - y = 5$ *Subtract y from each side.*

The equation is now in the form $Ax + By = C$, where $A = 3$, $B = -1$, and $C = 5$. Therefore, this is a linear equation.

2 $-2x + 3xy = 6$

Since the term $3xy$ has two variables, the equation cannot be written in the form $Ax + By = C$. So, this is not a linear equation.

3 $x = -2$

This equation can be written as $x + 0y = -2$. Therefore, it is a linear equation in the form $Ax + By = C$, where $A = 1$, $B = 0$, and $C = -2$.

Your Turn

a. $8 + y = 2x$ **b.** $-5z = 4x + 2y$ **c.** $y^2 = 4$

To graph a linear equation, make a table of ordered pairs that are solutions. Graph the ordered pairs and connect the points with a line. Note that the domain of the equation and its graph includes the values of x and the range includes the values of y.

Examples

Graph each equation.

4 $y = 2x - 3$

Select several values for the domain and make a table. Then graph the ordered pairs and connect them to draw the line.

Usually, values of x such as 0 and integers near 0 are chosen so that calculations are simple.

x	$2x - 3$	y	(x, y)
-2	$2(-2) - 3$	-7	$(-2, -7)$
-1	$2(-1) - 3$	-5	$(-1, -5)$
0	$2(0) - 3$	-3	$(0, -3)$
1	$2(1) - 3$	-1	$(1, -1)$
2	$2(2) - 3$	1	$(2, 1)$

Reading Algebra

When graphing an equation, use arrows to show that the graph continues. Also, label the line with the equation.

CΟncepts in MΟtion

Animation

algconcepts.com

www.algconcepts.com/extra_examples

⑤ $2x + 4y = 8$

In order to find values for y more easily, solve the equation for y.

$$2x + 4y = 8$$
$$4y = 8 - 2x \quad \textit{Subtract 2x from each side.}$$
$$y = \frac{8 - 2x}{4} \quad \textit{Divide each side by 4.}$$

Now make a table and draw the graph.

x	$\dfrac{8 - 2x}{4}$	y	(x, y)
-4	$\dfrac{8 - 2(-4)}{4}$	4	$(-4, 4)$
-2	$\dfrac{8 - 2(-2)}{4}$	3	$(-2, 3)$
0	$\dfrac{8 - 2(0)}{4}$	2	$(0, 2)$
2	$\dfrac{8 - 2(2)}{4}$	1	$(2, 1)$
4	$\dfrac{8 - 2(4)}{4}$	0	$(4, 0)$

Your Turn Graph each equation.

d. $y = 2x - 1$

e. $3x + 2y = 4$

Sometimes, the graph of an equation is a vertical or horizontal line.

Example

⑥ **Graph $y = 3$. Describe the graph.**

In standard form, this equation is written as $0x + y = 3$. So, for any value of x, $y = 3$. For example, if $x = 0$, $y = 3$; if $x = 1$, $y = 3$; if $x = 3$, $y = 3$. By graphing the ordered pairs $(0, 3)$, $(1, 3)$, and $(3, 3)$, you find that the graph of $y = 3$ is a horizontal line.

Your Turn

f. Graph the equation $x = -2$. Describe the graph.

Recall that the *domain* of a linear function is all of the x-coordinates of the ordered pairs for points on the line. The *range* of a linear function is all of the y-coordinates of the ordered pairs for points on the line. Since a line extends forever in both directions, if it is not vertical and not horizontal, the domain and the range are both the set of all real numbers.

Graph $y = 2x$.

Make a table and draw the graph.

x	$2x$	y	(x, y)
-2	$2(-2)$	-4	$(-2, -4)$
-1	$2(-1)$	-2	$(-1, -2)$
0	$2(0)$	0	$(0, 0)$
1	$2(1)$	2	$(1, 2)$
2	$2(2)$	4	$(2, 4)$

Write $y = 2x$ in standard form.

$y = 2x$	*Original equation*
$0 = 2x - y$	*Subtract y from each side.*
$2x - y = 0$	*Symmetric Property of Equality*

In any equation where $C = 0$, the graph passes through the origin.

Your Turn

g. Graph the equation $y = -3x$.

Example ──8

Earth Science

Old Faithful is a famous geyser. The equation $y = 14x + 27$ can be used to predict its eruptions, where y is the time until the next eruption (in minutes) and x is the length of the eruption (in minutes). Suppose Old Faithful erupts at 9:46 A.M. for 3.4 minutes. At about what time will the next eruption occur?

Explore   You need to find the time of the next eruption.

Plan   Use the graph of $y = 14x + 27$ to solve the problem.

Length of an Eruption (min)

Solve   The graph shows that if an eruption lasts 3.4 minutes, the next eruption will occur in about 75 minutes.

$$9\ h\ 46\ min \rightarrow 9\ h\ 46\ min$$
$$+\quad 75\ min \rightarrow \underline{1\ h\ 15\ min}$$
$$10\ h\ 61\ min \rightarrow 11\ h\ 1\ min$$

The next eruption should take place at 11:01 A.M.

Examine   You can verify the answer by using the equation.

$y = 14x + 27$	*Original equation*
$y \stackrel{?}{=} 14(3.4) + 27$	*Replace x with 3.4.*
$y = 74.6$	*Simplify.*

The answer makes sense because 74.6 minutes added to 9:46 is about 11:01.

Old Faithful,
Yellowstone National Park

Check for Understanding

Communicating Mathematics

1. **Describe** the graph of $x = -8$.

2. **Explain** how you know that the graph of $y = -2x$ passes through the origin.

Vocabulary

linear equation

Guided Practice
Examples 1–3

Determine whether each equation is a linear equation. Explain. If an equation is linear, identify A, B, and C.

3. $xy = 3$

4. $3y + 4x = 9$

5. $2x = 8$

Examples 4–7

Graph each equation.

6. $y = 3x + 2$

7. $y = -\frac{1}{2}x$

8. $x = -5$

9. $-2 = 4x + y$

10. $3x + 2y = 6$

11. $5y = -2x + 10$

Cactus

•12. **Plants**   The number of blooms on a cactus is related to the number of days of sun it gets in a month. This relation is given by the equation $y = 7x - 1$, where x is the number of days of sun and y is the number of blooms.   **Example 8**

  a. Graph the equation.

  b. Use the graph to determine the number of blooms on a cactus if it gets two days of sun a month.

Exercises

Practice

Determine whether each equation is a linear equation. Explain. If an equation is linear, identify A, B, and C.

13. $xy = 4$

14. $6 + 2xy = 4$

15. $2y - 6x = 0$

16. $4x = 8y$

17. $y = x$

18. $x + 3xy = 5$

19. $y = x^2 + 8$

20. $2 - 3x = 8$

21. $4x + x + y = 9$

Homework Help	
For Exercises	See Examples
13–21	1–3
22–36	4, 5, 7
Extra Practice	
See page 704.	

Graph each equation.

22. $x = y$

23. $y = 3x$

24. $-2x = y$

25. $y = -3$

26. $x = -3$

27. $2 = x$

28. $y = 3x + 2$

29. $2x - 5 = y$

30. $y = 4 - 2x$

31. $-1 = y + 4x$

32. $2x + y = -2$

33. $3 = -3x + y$

34. $4x + 2y = 8$

35. $3y - 2x = 6$

36. $-2y = 3x + 1$

Each table lists coordinates of points on a linear graph. Copy and complete each table. (Hint: Graph the ordered pairs.)

37.

x	y
0	?
1	-1
2	0
3	1
4	?

38.

x	y
-4	-2
-2	?
0	0
2	1
4	?

39.

x	y
-2	-4
-1	?
0	2
1	?
2	8

40. **Business** Bongo's Clown Service charges a $50 fee for a clown to perform at a birthday party plus an additional $2 per guest.
    a. Write an equation that represents this situation. Let x represent the number of guests and let y represent the cost of a clown's services.
    b. Graph the equation.
    c. Serena's parents paid $68 for a clown to perform at Serena's party. How many people attended Serena's party?

41. **Geometry** Complementary angles are two angles whose sum is 90°. Suppose two complementary angles measure $(2x + 2y)°$ and $(x + 4y)°$.
    a. Write an equation for the sum of the measures of these two angles.
    b. Determine five ordered pairs that satisfy this equation.
    c. Graph the equation.

42. **Critical Thinking** Graph the equations $y = 2x$, $y = 2x + 1$, $y = 2x + 2$, and $y = 2x + 3$ on the same coordinate plane. Then describe the graphs.

Mixed Review

43. Solve $y = x - 5$ if the domain is $\{-3, 4, 5, 7, 9\}$. *(Lesson 6–2)*

44. Express the relation $\{(-1, -1), (0, 5), (1, -2), (3, 4)\}$ as a table and as a graph. *(Lesson 6–1)*

Solve each proportion. *(Lesson 5–1)*

45. $\dfrac{3}{4} = \dfrac{x}{8}$

46. $\dfrac{a}{45} = \dfrac{3}{15}$

47. $\dfrac{y}{3} = \dfrac{y + 4}{6}$

Standardized Test Practice
Ⓐ Ⓑ Ⓒ Ⓓ

48. **Grid In** Find the range of the data: $\{17, 24, 16, 25, 38, 57\}$. *(Lesson 3–3)*

49. **Short Response** The temperature outside is 67°F. Suppose it rises an average of 2°F every hour for h hours to reach a high of 79°F. Write an equation to represent this situation. *(Lesson 1–1)*

Quiz 1 — Lessons 6–1 through 6–3

Express each relation as a table and as a graph. Then determine the domain and the range. *(Lesson 6–1)*

1. $\{(1, -1), (3, 5), (0, -2), (3, 3)\}$

2. $\{(-0.5, 3), (1.5, 0), (2.5, 4), (3.5, 2.5)\}$

3. Which ordered pairs are solutions to the equation $2y + 4 = 4x$? *(Lesson 6–2)*
   a. $(-1, 3)$
   b. $(0, -2)$
   c. $(-2, -6)$
   d. $(3, -4)$

4. Solve $y = 4x - 3$ if the domain is $\{-2, -1, 2, 3, 5\}$. Graph the solution set. *(Lesson 6–2)*

5. Anne is saving money to buy a $2500 used car. She started with $500 and has been saving $80 per week since then. The equation $y = 500 + 80x$ represents her total savings y after any number of weeks x. *(Lesson 6–3)*
   a. Graph the equation.
   b. Use the graph to estimate when Anne will have enough money.
   c. Verify your estimate using the equation.

What You'll Learn

You'll learn to determine whether a given relation is a function.

Why It's Important

Travel Functions can be used to determine the cost of renting a car.
See Exercise 48.

Study the graph of the line below. Notice that for each value of *x*, there is exactly one value of *y*. This type of relation is called a **function**. Because the graph of the function is a line, this graph represents a **linear function**.

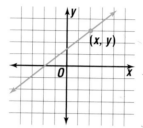

The point at which the graph crosses the *x*-axis is called the **zero** of the function. In the graph above, the zero of the linear function is −2.

Functions	A function is a relation in which each member of the domain is paired with *exactly* one member of the range.

Examples

Determine whether each relation is a function. Explain your answer.

1 {(5, 2), (3, 5), (−2, 3), (−5, 1)}

Since each element of the domain is paired with *exactly* one element of the range, this relation is a function.

2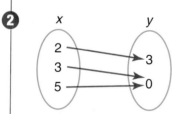

The mapping between *x* and *y* represents a function since there is *only one* corresponding element in the range for each element of the domain. *It does not matter if two elements of the domain are paired with the same element in the range.*

3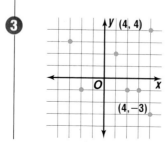

The graph represents a relation that is not a function. Look at the points with the ordered pairs (4, −3) and (4, 4). The member 4 in the domain is paired with both −3 and 4 in the range.

Your Turn

Determine whether each relation is a function. Explain your answer.

a. {(5, −2), (3, 2), (4, −1), (−2, 2)}

b.

x	y
−1	2
0	7
1	4
1	−2
4	5
6	−4

c.
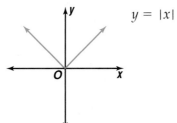

To determine whether an equation is a function, you can use the **vertical line test** on the graph of the equation. Consider the graph below.

COncepts in MOtion
Animation
algconcepts.com

To perform the test, place a pencil at the left of the graph to represent a vertical line. Move it to the right across the graph.

For each value of x, this vertical line passes through exactly one point on the graph. So, the equation is a function.

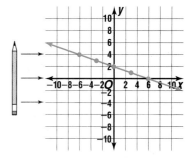

Vertical Line Test for a Function	If any vertical line passes through no more than one point of the graph of a relation, then the relation is a function.

Examples

Use the vertical line test to determine whether each relation is a function.

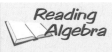

4 $y = |x|$

5 $y = \pm\sqrt{x}$

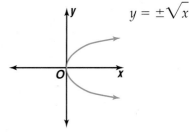

Reading Algebra

When $x = 4$ in $y = \pm\sqrt{x}$, y has two values, 2 and −2. So, $y = \pm\sqrt{x}$ does not represent a function.

The relation in Example 4 is a function since any vertical line passes through no more than one point of the graph of the relation. The graph in Example 4 is the graph of the function $y = |x|$. The relation in Example 5 is not a function since a vertical line can pass through more than one point. While the entire graph is not the graph of a function, it can be separated into two parts: the graph of the function $y = \sqrt{x}$ and the graph of the function $y = -\sqrt{x}$.

(continued on the next page)

Your Turn

Use the vertical line test to determine whether each relation is a function.

d.

e.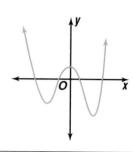

Equations that are functions can be written in a form called **functional notation**. For example, consider the equation $y = 3x + 5$.

equation	functional notation
$y = 3x + 5$	$f(x) = 3x + 5$

Reading Algebra

Read the symbol $f(x)$ as *f at x* or *f of x*. Letters other than *f* are also used for names of functions.

In a function, x represents the elements of the domain, and $f(x)$ represents the elements of the range. For example, $f(2)$ is the element in the range that corresponds to the member 2 in the domain. $f(2)$ is called the **functional value** of f for $x = 2$.

You can find $f(2)$ as shown at the right. The ordered pair (2, 11) is a solution of the function of f.

$f(x) = 3x + 5$
$f(2) = 3 \cdot 2 + 5$ *Replace x with 2.*
$f(2) = 6 + 5$ or 11

You can use functional values to find the zero of a function. The value of x for which $f(x) = 0$ is the zero of the function $f(x)$.

Examples

If $f(x) = 2x + 3$, find each value.

6 $f(4)$

$f(x) = 2x + 3$
$f(4) = 2(4) + 3$ *x = 4*
$\quad\quad = 8 + 3$ *Multiply.*
$\quad\quad = 11$ *Add.*

7 $f\left(-\frac{1}{2}\right)$

$f(x) = 2x + 3$
$f\left(-\frac{1}{2}\right) = 2\left(-\frac{1}{2}\right) + 3$ $x = -\frac{1}{2}$
$\quad\quad\quad = -1 + 3$ *Multiply.*
$\quad\quad\quad = 2$ *Add.*

8 $f(6a)$

$f(x) = 2x + 3$
$f(6a) = 2(6a) + 3$ *Replace x with 6a.*
$\quad\quad = 12a + 3$ *Multiply.*

Your Turn

f. If $g(x) = 3x - 7$, find $g(-2)$, $g(1.5)$, and $g(2h)$.

Functions are often used to solve real-life problems.

Example 9

Anatomy Link

Anthropologists use the length of certain bones of a human skeleton to estimate the height of the living person. One of these bones is the femur, which extends from the hip to the knee. To estimate the height in centimeters of a female with a femur of length x, the function $h(x) = 61.41 + 2.32x$ can be used. What was the height of a female whose femur measures 46 centimeters?

$h(x) = 61.41 + 2.32x$ *Original equation*

$h(46) = 61.41 + 2.32(46)$ *Replace x with 46.*

$\quad\quad = 168.13$ *Simplify.*

The woman was about 168 centimeters tall.

Check for Understanding

Communicating Mathematics

1. **Compare and contrast** relations and functions.

2. **Explain** why $f(3) = -1$ if $f(x) = 2x - 7$.

3. Lisa says that the graph of every straight line represents a function. Zina says that there are some graphs of straight lines that do not represent a function. Who is correct? Explain your reasoning.

Vocabulary

function
linear function
zero
vertical line test
functional notation
functional value

Guided Practice

Examples 1–3

Determine whether each relation is a function.

4. {(2, 3), (3, 5), (4, 6), (2, 5), (1, 4)}   5. {(−1, 6), (1, 4), (−2, 5), (2, 6)}

6.

x	y
−2	4
0	3
5	2
2	4

7.

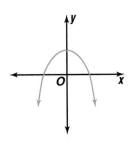

Examples 4 & 5

8. Use the vertical line test to determine whether the relation at the right is a function.

If *f(x)* = 4*x* − 3, find each value.

9. $f(-3)$ **10.** $f\left(\frac{1}{4}\right)$ **11.** $f(b)$ **12.** $f(-2a)$

Example 9

13. Safety The time in seconds that a traffic light remains yellow is given by the function $t(s) = 0.05s + 1$, where *s* represents the speed limit. How long will a light remain yellow if the speed limit is 45 miles per hour?

Exercises • • • • • • • • • • • • • • • • • •

Practice

Determine whether each relation is a function.

14. {(3, 5), (4, 3), (−2, 1), (−1, 4)}   **15.** {(−2, 3), (−3, 2), (5, 2), (7, 2)}

16. {(2, −3), (3, −2), (5, −2), (2, −5)}   **17.** {(0, 2), (−1, 3), (2, 3), (−1, 2)}

18. {(−3, 5), (−2, 4), (−3, 6), (−2, 3)}   **19.** {(1, 2), (2, 1), (−1, 2), (−2, 1)}

Homework Help

For Exercises	See Examples
14–25	1–3
26–28	4, 5
29–46	6–8
47, 48	6–9

Extra Practice
See page 704.

20.

x	y
−2	5
3	−2
4	5
5	5
6	1

21.

x	y
0	1
0	0
−4	0
2	6
0	−4

22.

23.   **24.**   **25.**

Use the vertical line test to determine whether each relation is a function.

26.   **27.**   **28.**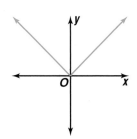

If *f(x)* = 2*x* − 5 and *g(x)* = −3*x* + 4, find each value.

29. $f(2)$ **30.** $g(4)$ **31.** $g(-3)$ **32.** $f(-7)$

33. $g(0.2)$ **34.** $f(0.6)$ **35.** $g(1.9)$ **36.** $f(-2.6)$

37. $g\left(\frac{1}{3}\right)$ **38.** $f\left(-\frac{1}{2}\right)$ **39.** $f\left(\frac{3}{4}\right)$ **40.** $g\left(-\frac{5}{6}\right)$

41. $g(h)$ **42.** $f(-2d)$ **43.** $g(-2a)$ **44.** $f(4.5c)$

45. Find $f(-4)$ if $f(x) = 5x + 8$.

46. What is the value of $m(3) + n(-5)$ if $m(x) = 2x$ and $n(x) = 3x - 1$?

Applications and Problem Solving

47. Health A person's normal systolic blood pressure p depends on the person's age a. To determine the normal systolic blood pressure for an individual, you can use the equation $p = 0.5a + 110$.

   a. Write this relation in functional notation.

   b. Find the normal systolic blood pressure for a person who is 36 years old.

   c. Graph the relation. Describe the graph.

48. Travel The cost of a one-day car rental from City-Wide Rentals is given by the function $C(x) = 0.18x + 35$, where x is the number of miles that the car is driven. Suppose Ms. Burton drove a distance of 95 miles and back in one day. What is the cost of the car rental?

49. Critical Thinking Consider the functions $h(x) = 2x - 5$ and $g(x) = 3x + 2$.

   a. For what value of x will $h(x)$ and $g(x)$ be equal?

   b. Graph $h(x)$ and $g(x)$. Describe what you observe.

Mixed Review

50. Graph $y = -2x - 6$. *(Lesson 6–3)*

51. Geometry The equation $A = \frac{1}{2}bh$ can be used to find the area of a triangle. Suppose a triangle has an area of 28 square meters. Find the possible heights of the triangle given the domain values $\{2, 4, 7\}$. *(Lesson 6–2)*

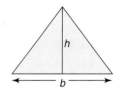

Use the percent proportion to answer each question. *(Lesson 5–3)*

52. What is 145% of 86?

53. What percent of 60 is 18?

54. Life Science A scale model of a sequoia tree is 4 feet tall. If the actual sequoia tree is 311 feet tall, what is the scale of the model? *(Lesson 5–2)*

55. Evaluate $\frac{x}{5}$ if $x = \frac{3}{4}$. *(Lesson 4–3)*

Gentoo Penguin

56. Animals The fastest swimming bird is the gentoo penguin of Antarctica. It moves through the water at about 17 miles per hour. Traveling at this rate, how far can the gentoo penguin swim in 45 minutes? *(Lesson 4–1)*

Standardized Test Practice
Ⓐ Ⓑ Ⓒ Ⓓ

57. Grid In What is the value of w if $w - 13.8 = -12.6$? *(Lesson 3–6)*

58. Multiple Choice Find the only solution of $3x + (-5) = 10$ if the replacement set is $\{-5, 3, 5, 8\}$. *(Lesson 3–4)*

   Ⓐ -5 Ⓑ 3 Ⓒ 5 Ⓓ 8

How Many CENTIMETERS in a Foot?

Materials

 grid paper

 meterstick

 red and blue pens or pencils

 almanac

Functions

Many real-world situations represent functions. Is there a relationship between foot length and shoe size? Does a relationship exist between the month of the year and the average temperature? Let's find out!

Investigate

1. Ask eight to ten females and eight to ten males to participate in your data research.

   a. Make two tables with the headings shown. Use one table to record data for the females and the other to record data for the males.

Name	Foot Length (cm)	Shoe Size	(Foot Length, Shoe Size)

   b. Measure the foot length in centimeters as each participant stands without shoes on. Find out each person's shoe size. Record the results in your tables.

   c. Write an ordered pair (foot length, shoe size) for each person. Then graph the ordered pairs on a coordinate plane. Use a red pencil to graph the ordered pairs that represent the female data. Use a blue pencil to graph the ordered pairs that represent the male data. Connect each set of ordered pairs with the corresponding color.

   d. Is either relation a function? Why or why not? Describe any relationship that exists between foot length and shoe size.

   e. Save your tables to use later in the investigation.

2. Choose a city in the United States. Research temperature data for that city. Find the average high and low temperatures for each month. It may be helpful to use an almanac or the Internet.

a. Make a table like the one shown. Record the temperature data for the selected city.

Month	Average High Temperature (°F)	Average Low Temperature (°F)
January		
February		
⋮		

b. Write two sets of ordered pairs. The first set should contain the ordered pairs (month, high temperature). Assign January the number 1, February the number 2, and so on. The second set should contain the ordered pairs (month, low temperature).

c. Graph both sets of data on one coordinate plane. Connect the ordered pairs in the first set using a red pencil. Use a blue pencil to connect the ordered pairs in the second set.

d. Is either graph a function? Explain. Describe any relationship that exists between the month and the average temperature. Are there any similarities between these two graphs?

Extending the Investigation

In this extension, you will continue to investigate functions.

- Select another city in the United States. Make a table of average high and low temperatures for each month. Then graph the data and determine if the graph is a function. Describe any relationships that exist between the graphs.

- Select a city in Hawaii. Make a table of average high and low temperatures for each month. Then graph the data. How does the graph of the temperature data for Hawaii compare to the graph in Exercise 2?

Presenting Your Investigation

Here are some ideas to help you present your conclusions to the class.

- Make a brochure showing the results of your research. Be sure to include the tables and graphs. List any relationships you have found.

- Discuss any similarities or differences in the graphs.

- Look in an almanac or on the Internet for real-world data whose graph would be a function. Make a poster presenting your data, tables, and graphs.

*inter*NET CONNECTION **Investigation** For more information on temperatures, visit: www.algconcepts.com

What You'll Learn

You'll learn to solve problems involving direct variations.

Why It's Important

Electronics
Electricians can use direct variation to determine the weight of copper wire.
See Exercise 32.

Study the table and graph of the linear function $y = 2x$ shown below.

x	y
-2	-4
-1	-2
0	0
1	2
2	4

The value of y depends directly on the corresponding value of x. That is, as the value of x increases, the value of y increases. Note that when $x = 0$, $y = 0$. Therefore, the line passes through the origin. The equation $y = 2x$ is called a **direct variation.** We say that y *varies directly as x.*

A direct variation describes a linear relationship because $y = 0.4x$ is a linear equation.

Direct Variation	A direct variation is a linear function that can be written in the form $y = kx$, where $k \neq 0$.

Since the value of y depends on the value of x, y is called the **dependent variable.** x is called the **independent variable.**

As stated, a direct variation can be written in the form $y = x$. Notice that when $x = 0$, $y = 0$. The graph of a direct variation passes through the origin.

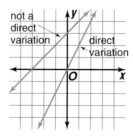

In the equation $y = kx$, k is called the **constant of variation.**

Determine whether the equation is a direct variation.

1 $y = 4x$

Graph the equation.

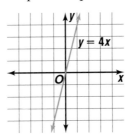

The graph passes through the origin. So, the equation is a direct variation. *The constant of variation is 4.*

2 $y = x - 1$

Graph the equation.

The graph does not pass through the origin. So, the equation is not a direct variation.

Your Turn

Determine whether each equation is a direct variation. Verify your answer with a graph.

a. $y = 2x + 1$

b. $y = -3x$

Many real-world problems involve direct variation.

Real World

Example

3

Astronomy Link

The weight of an object on Venus varies directly as its weight on Earth. An object that weighs 80 pounds on Earth weighs 72 pounds on Venus. How much would an object weigh on Venus if its weight on Earth is 90 pounds?

First, find the constant of variation. Let x = weight on Earth and let y = weight on Venus, since the weight on Venus depends on the weight on Earth.

$y = kx$ *Definition of direct variation*

$72 = k \cdot 80$ *Replace y with 72 and x with 80.*

$\dfrac{72}{80} = \dfrac{k \cdot 80}{80}$ *Divide each side by 80.*

$0.9 = k$ *Simplify.*

Next, use the constant of variation to find y when $x = 90$.

$y = kx$ *Definition of direct variation*

$y = 0.9(90)$ *Replace k with 0.9 and x with 90.*

$y = 81$ *Simplify.*

So, the object would weigh 81 pounds on Venus.

(continued on the next page)

c. An object that weighs 110 pounds on Earth weighs 18.7 pounds on the moon. How much would an object weigh on the moon if its weight on Earth is 150 pounds?

Direct variations can be used to solve **rate problems**.

Real World

Example **4**

Travel Link

The length of a trip varies directly as the amount of gasoline used. Suppose 4 gallons of gasoline were used for a 120-mile trip. At that rate, how many gallons of gasoline are needed for a 345-mile trip?

Let ℓ represent the length of the trip and let g represent the amount of gasoline used for the trip. The statement *the length varies directly as the amount of gasoline* translates into the equation $\ell = kg$ in the same way as *y varies directly as x* translates into $y = kx$. Find the value of k.

$\ell = kg$ *Direct variation*

$120 = k \cdot 4$ *Replace ℓ with 120 and g with 4.*

$\dfrac{120}{4} = \dfrac{4k}{4}$ *Divide each side by 4.*

$30 = k$ *Simplify.*

Next, find the amount of gasoline needed for a 345-mile trip.

$\ell = kg$ *Direct variation*

$345 = 30 \cdot g$ *Replace ℓ with 345 and k with 30.*

$\dfrac{345}{30} = \dfrac{30g}{30}$ *Divide each side by 30.*

$11.5 = g$ *Simplify.*

A 345-mile trip would use 11.5 gallons of gasoline.

Your Turn

d. How many gallons of gasoline would be needed for a 363-mile trip if a 72.6-mile trip used 3 gallons of gasoline?

 Personal Tutor at algconcepts.com

Direct variations are also related to proportions. Using the table on page 264, many proportions can be formed. Two examples are shown.

$$\begin{array}{ll} x_1 \to & \dfrac{5}{2} = \dfrac{10}{4} \quad \leftarrow x_2 \\ y_1 \to & \qquad\qquad\quad \leftarrow y_2 \end{array}$$

$$\begin{array}{ll} x_1 \to & \dfrac{5}{10} = \dfrac{2}{4} \quad \leftarrow y_1 \\ x_2 \to & \qquad\qquad\quad \leftarrow y_2 \end{array}$$

Three general forms for proportions like these are $\dfrac{y_1}{y_2} = \dfrac{x_1}{x_2}$, $\dfrac{x_1}{y_1} = \dfrac{x_2}{y_2}$ and $\dfrac{y_1}{x_1} = \dfrac{y_2}{x_2}$. You can use any of these forms to solve direct variation problems.

Suppose y varies directly as x and $y = 36$ when $x = 9$. Find x when $y = 48$.

Use $\dfrac{y_1}{x_1} = \dfrac{y_2}{x_2}$ to solve the problem.

$\dfrac{36}{9} = \dfrac{48}{x_2}$ *Let $y_1 = 36$, $y_2 = 48$, and $x_1 = 9$.*

$36x_2 = 9(48)$ *Find the cross products.*

$36x_2 = 432$ *Multiply.*

$\dfrac{36x_2}{36} = \dfrac{432}{36}$ *Divide each side by 36.*

$x_2 = 12$ *Simplify.*

So, $x = 12$ when $y = 48$.

Your Turn

e. Suppose y varies directly as x and $y = 27$ when $x = 6$. Find x when $y = 45$.

Example ❻
Measurement Link

You can also use direct variation to convert measurements.

If there are 12 inches in 1 foot, how many inches are in 4.5 feet?

Use $\dfrac{y_1}{y_2} = \dfrac{x_1}{x_2}$ to solve the problem.

$\dfrac{12}{1} = \dfrac{x_1}{4.5}$ *Let $y_1 = 12$, $y_2 = 1$, and $x_2 = 4.5$.*

$12(4.5) = x_1$ *Find the cross products.*

$54 = x_1$ *Simplify.*

There are 54 inches in 4.5 feet.

Your Turn

f. How many pints are in 2.3 gallons if there are 8 pints in 1 gallon?

Check for Understanding

Communicating Mathematics

1. **Explain** how you find the constant of variation in a direct variation.

2. **Give an example** of an equation that is a direct variation.

3. **Writing Math** Refer to Example 2. Write a few sentences describing the dependent variable and independent variable.

Vocabulary

direct variation
independent variable
dependent variable
constant of variation
rate problems

⏱ Getting Ready Find the constant of variation for each direct variation.

Sample 1: $y = 3x$	**Sample 2:** $n = -0.3m$
Solution: 3	**Solution:** -0.3

4. $y = 2x$ **5.** $5c = d$ **6.** $h = -4g$ **7.** $t = \frac{2}{5}r$ **8.** $1.4p = q$

Examples 1 & 2   Determine whether each equation is a direct variation. Verify the answer with a graph.

9. $y = -x$ **10.** $y = 5x + 5$

Example 5   Solve. Assume that y varies directly as x.

11. If $y = 28$ when $x = 7$, find x when $y = 52$.

12. Find y when $x = 10$ if $y = -8$ when $x = 2$.

Example 6   **13.** If there are 3 feet in a yard, how many yards are in 14.1 feet?

Examples 3 & 4   **14.** **Astronomy**   The weight of an object on Jupiter varies directly as its weight on Earth. An object that weighs 32.5 pounds on Earth weighs 81.25 pounds on Jupiter. How much would an object weigh on Jupiter if its weight on Earth is 90.2 pounds?

Exercises • • • • • • • • • • • • • • • • • •

Practice

Determine whether each equation is a direct variation. Verify the answer with a graph.

15. $y = x + 4$ **16.** $y = -2x$ **17.** $y = -3$

18. $y = -0.5x$ **19.** $x = 2$ **20.** $\frac{y}{x} = 6$

Homework Help	
For Exercises	See Examples
15–20	1, 2
21–32	5, 6
33, 34	3, 4
Extra Practice	
See page 704.	

Solve. Assume that y varies directly as x.

21. Find x when $y = 5$ if $y = 3$ when $x = 15$.

22. If $x = 24$ when $y = 8$, find y when $x = 12$.

23. If $y = -7$ when $x = -14$, find x when $y = 10$.

24. Find y when $x = 9$ if $y = -4$ when $x = 6$.

25. If $x = 15$ when $y = 12$, find x when $y = 21$.

26. Find y when $x = 0.5$ if $y = 100$ when $x = 20$.

27. If $y = 12$ when $x = 1\frac{1}{8}$, find y when $x = \frac{1}{4}$.

Solve by using direct variation.

28. How many inches are in 1.25 yards if there are 36 inches in 1 yard?

29. If 1 bushel is 4 pecks, how many bushels are in 2.4 pecks?

30. How many acres are in 0.75 square mile if there are 640 acres in 1 square mile?

Applications and Problem Solving

31. **Shopping**   For $1.89, you can buy 75 square feet of plastic wrap. At this rate, how much does 125 square feet of plastic wrap cost?

32. **Electronics** How much will 25 meters of copper wire weigh if 3.4 meters of the same copper wire weighs 0.442 kilogram?

33. **Health** The height h of an average person varies directly with their foot length f. Suppose the constant of variation is 7.
   a. Write an equation that represents this situation.
   b. Graph the equation.
   c. Use the graph to find the height of a person whose foot length is 9.5 inches.

34. **Critical Thinking** Assume that y varies directly as x.
   a. Suppose the value of x is doubled. What is the constant of variation? What happens to the value of y?
   b. Suppose the value of x is tripled. What is the constant of variation? What happens to the value of y?
   c. Suppose the value of y is doubled. What is the constant of variation? What happens to the value of x?
   d. Suppose the value of y is tripled. What is the constant of variation? What happens to the value of x?

Mixed Review

35. Find $d(-3)$ if $d(x) = -5d - 4$. *(Lesson 6–4)*

36. Graph $x = 4$. *(Lesson 6–3)*

37. **Life Science** A full-grown male seal weighs about 550 pounds. This is 50 pounds less than the weight of a full-grown male sea lion. *(Lesson 3–5)*
   a. Write an equation that represents the weight of a full-grown male sea lion.
   b. How much does a full-grown male sea lion weigh?

Find each quotient. *(Lesson 2–6)*

38. $-36 \div 4$
39. $-64 \div (-8)$
40. $\dfrac{-96}{-16}$

Standardized Test Practice

41. **Multiple Choice** Name the coordinates of U, X, and Y. *(Lesson 2–1)*

$$\begin{array}{c} U \quad\ V\ W \quad\ X \qquad\quad Y \quad\ Z \\ \overset{\longleftarrow}{\underset{-5\,-4\,-3\,-2\,-1\ \ 0\ \ 1\ \ 2\ \ 3\ \ 4\ \ 5}{\rule{0pt}{0pt}}} \end{array}$$

    Ⓐ $-5, -2, 3$ Ⓑ $-5, 0, 3$ Ⓒ $-3, 3, 5$ Ⓓ $-4, -2, 0$

Quiz 2 Lessons 6–4 and 6–5

Determine whether each relation is a function. *(Lesson 6–4)*

1. $\{(1, 2), (2, -1), (0, 3), (-1, 4)\}$
2. $5x = 6 + y$
3. If $g(x) = 4x + 3$, find $g(-2.6)$. *(Lesson 6–4)*
4. Find x when $y = -8$ if $y = 4$ when $x = 20$. Assume that y varies directly as x. *(Lesson 6–5)*
5. **Transportation** The length of a trip varies directly as the amount of gasoline used. Julio's car used 5 gallons of gasoline in the first 145 miles of his trip. How much gasoline should he expect to use in the remaining 232 miles of the trip? *(Lesson 6–5)*

What You'll Learn
You'll learn to solve problems involving inverse variations.

Why It's Important
Music Musicians can use inverse variations to find the wavelength of a musical tone. *See Exercise 12.*

Each of the rectangles shown has an area of 12 square units. Notice that as the length increases, the width decreases. As the length decreases, the width increases. However, their product stays the same.

4 units
3 units

6 units
2 units

12 units
1 units

Length	Width	Area
4	3	12
6	2	12
12	1	12

This is an example of an **inverse variation**. We say that y varies inversely as x. This means that as x increases in value, y decreases in value, or as y decreases in value, x increases in value.

Inverse Variation	An inverse variation is described by an equation of the form $xy = k$, where $k \neq 0$.

The equation may also be written as $y = \dfrac{k}{x}$.

The graphs below show how the graph of a direct variation differs from the graph of an inverse variation.

	Direct Variation	Inverse Variation
Concept Summary	$y = kx$ y varies directly as x. y is directly proportional to x. The constant of variation is $\dfrac{y}{x}$.	$xy = k$ or $y = \dfrac{k}{x}$ y varies inversely as x. y is inversely proportional to x. The constant of variation is xy.

Example ①

Construction Link

The number of carpenters needed to frame a house varies inversely as the number of days needed to complete the project. Suppose 5 carpenters can frame a house in 16 days. How many days will 8 carpenters take to frame the house? Assume that they all work at the same rate.

Explore   You know that it takes 16 days for 5 carpenters to frame the house. You need to know how many days it will take 8 carpenters to frame the house.

Plan   Solve the problem by using inverse variation.

Solve   Let x = the number of carpenters. Let y = the number of days. First, find the value of k.

$xy = k$ *Definition of inverse variation*

$(5)(16) = $ *Replace x with 5 and y with 16.*

$80 = $ *The constant of variation is 80.*

Next, find the number of days for 8 carpenters to frame the house.

$y = \dfrac{k}{x}$ *Divide each side of xy = k by x.*

$= \dfrac{80}{8}$ *Replace k with 80 and x with 8.*

$= 10$ *Simplify.*

A crew of 8 carpenters can frame the house in 10 days.

Examine   It takes 16 days for 5 carpenters to frame a house. It makes sense that 8 carpenters could do the work in less time. 10 days is a reasonable solution.

Your Turn

a. The number of masons needed to build a block basement varies inversely as the number of days needed. If 8 masons can build a block basement in 3 days, how long will 9 masons take to do it?

You can use a graphing calculator to check the answer to Example 1.

Graphing Calculator Tutorial See pp. 750–753.

Graphing Calculator Exploration

Step 1 Set the viewing window for x: [0, 10] by 1 and y: [0, 200] by 20.

Step 2 To graph $xy = 80$, rewrite the equation as $y = \frac{80}{x}$. Press [Y=] 80 [÷] [X,T,θ,n] to enter the equation.

Step 3 Press [TRACE] to estimate the x- and y-coordinates for any point on the graph. Notice that when $x = 5$, $y = 16$. When x is 8, y is 10. Therefore, the answer given in Example 1 is correct.

Try These

1. Why doesn't the viewing window allow for negative values to be shown on the graph?

2. Show how to rewrite $xy = 80$ as $y = \frac{80}{x}$.

3. Solve Your Turn Exercise a by using a graphing calculator.

Just as with direct variation, you can use a proportion to solve problems involving inverse variation. The proportion $\frac{x_1}{x_2} = \frac{y_2}{y_1}$ is only one of several that can be formed. *Can you name others?*

Example

❷ **Suppose y varies inversely as x and $y = -6$ when $x = -2$. Find y when $x = 3$.**

$$\frac{x_1}{x_2} = \frac{y_2}{y_1} \qquad \textit{Inverse variation proportion}$$

$$\frac{-2}{3} = \frac{y_2}{-6} \qquad \textit{Let } x_1 = -2, y_1 = -6, \textit{ and } x_2 = 3.$$

$$-2(-6) = 3y_2 \qquad \textit{Find the cross products.}$$

$$12 = 3y_2 \qquad \textit{Multiply.}$$

$$4 = y_2 \qquad \textit{Divide each side by 3.}$$

Therefore, when $x = 3$, $y = 4$.

Your Turn

b. Suppose y varies inversely as x and $y = 3$ when $x = 12$. Find x when $y = 4$.

 Personal Tutor at algconcepts.com

Inverse variations can be used to solve rate problems.

Example ③

Sports Link

In the formula $d = rt$, the time t varies inversely as the rate r. A race car traveling 125 miles per hour completed one lap around a race track in 1.2 minutes. How fast was the car traveling if it completed the next lap in 0.8 minute?

First, solve for the distance d, the constant of variation.

$d = rt$ *Formula for distance*

  $= (125)(1.2)$ *Replace r with 125 and t with 1.2.*

  $= 150$ *The constant of variation is 150.*

Next, find the rate if one lap was completed at 0.8 minute.

$\dfrac{d}{t} = r$ *Divide each side of d = rt by t.*

$\dfrac{150}{0.8} =$ *Replace d with 150 and t with 0.8.*

$187.5 =$ *Simplify.*

The car was traveling 187.5 miles per hour.

Check for Understanding

Communicating Mathematics

1. **Explain** how to find the constant of variation in an inverse variation.

2. **Complete** the table for the inverse variation $xy = 24$.

x	1	2	3	4	6	8	12	24
y	24	?	?	?	?	?	?	?

> *Vocabulary*
>
> inverse variation

3. **Writing Math** Write a few sentences to explain the difference between inverse variation and direct variation.

Guided Practice

Getting Ready Determine if each equation is an inverse or a direct variation. Find the constant of variation.

Sample 1: $cd = -3$

Solution: This equation is of the form $xy = k$. So, it is an inverse variation. The constant of variation is -3.

Sample 2: $h = \dfrac{1}{2}g$

Solution: This equation is of the form $y = kx$. So, it is a direct variation. The constant of variation is $\dfrac{1}{2}$.

4. $ab = 4$ 5. $s = 5r$ 6. $mn = -7$

7. $y = \dfrac{1}{3}x$ 8. $y = \dfrac{7}{x}$ 9. $t = \dfrac{r}{4}$

Example 2

Solve. Assume that *y* varies inversely as *x*.

10. Suppose $y = 6$ when $x = 2$. Find y when $x = 3$.

11. Find x when $y = 2$ if $y = -4$ when $x = -3$?

Example 1

12. Music The frequency of a musical tone varies inversely as its wavelength. If one tone has a frequency of 440 vibrations per second and a wavelength of 2.4 feet, find the wavelength of a tone that has a frequency of 660 vibrations per second.

Example 3

13. Travel The time it takes to travel a certain distance varies inversely to the speed of travel. Suppose it takes 2.75 hours to drive from one city to another at a rate of 65 miles per hour. How long will the return trip take at a rate of 55 miles per hour?

Exercises • • • • • • • • • • • • • • • • • •

Practice

Homework Help

For Exercises	See Examples
14–19, 28	2
20–23, 29, 30	1, 3

Extra Practice
See page 705.

Solve. Assume that *y* varies inversely as *x*.

14. Find y when $x = 10$ if $y = 15$ when $x = 4$.

15. Suppose $y = 20$ when $x = 6$. Find x when $y = 12$.

16. If $y = 1.2$ when $x = 3$, find y when $x = 6$.

17. Find y when $x = 2.5$ if $y = 0.15$ when $x = 1.5$.

18. If $y = \frac{2}{3}$ when $x = 6$, find x when $y = 1\frac{1}{3}$.

19. Suppose $y = -\frac{1}{2}$ when $x = -\frac{3}{4}$. Find y when $x = \frac{3}{5}$.

Find the constant of variation. Then write an equation for each statement.

20. *y* varies inversely as *x*, and $y = 5$ when $x = 10$.

21. *y* varies directly as *x*, and $x = -18$ when $y = 3$.

22. *y* varies inversely as *x*, and $y = 4$ when $x = -2$.

23. *y* varies directly as *x*, and $x = 7$ when $y = 2.1$.

Match each situation with the equation that represents it. Then tell whether the situation represents a direct variation or an inverse variation.

24. DeShawn purchases a number of pencils for $0.25 each.

25. A winning $125 lottery ticket is shared with a number of people.

26. Twenty-five pieces of candy are shared with several people.

27. Each person who has a winning lottery ticket wins $125.

a. $xy = 125$

b. $y = \frac{25}{x}$

c. $y = 125x$

d. $25x = y$

28. Music The length of a piano string varies inversely as the frequency of its vibrations. A piano string 36 inches long vibrates at a frequency of 480 cycles per second. Find the frequency of a 24-inch string.

29. Geometry The length ℓ of the rectangle varies inversely as its width w.

12 m
3 m

   a. Find the constant of variation.

   b. What is the width of the rectangle when the length is 15 meters?

   c. Write an equation that represents this inverse variation.

   d. Draw the graph of the equation.

30. Critical Thinking Assume that y varies inversely as x.

   a. Suppose the value of x is doubled. What is the constant of variation? What happens to the value of y?

   b. Suppose the value of y is tripled. What is the constant of variation? What happens to the value of x?

Mixed Review

31. Suppose y varies directly as x and $y = 16$ when $x = 12$. Find x when $y = 36$. *(Lesson 6–5)*

32. Is $\{(6, 3), (5, -2), (2, 3), (12, -12)\}$ a function? *(Lesson 6–4)*

Simplify. *(Lesson 3–2)*

33. $-38.9 + 24.2$

34. $12.5 + (-15.6) + 22.7 + (-35.8)$

35. Grid In The surface of Lake Michigan is 176 meters above sea level. Its deepest point is 281 meters below sea level. How many meters below the surface is the deepest part of the lake? *(Lesson 2–4)*

36. Multiple Choice Which of the following shows the graph of the relation $\{(-1, 2), (1, 1), (1, 3), (2, 0)\}$? *(Lesson 2–2)*

Ⓐ

Ⓑ

Ⓒ

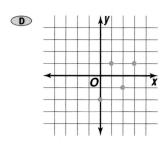
Ⓓ

Understanding and Using the Vocabulary

*inter***NET**
CONNECTION **Review Activities**
For more review activities, visit:
www.algconcepts.com

After completing this chapter, you should be able to define each term, property, or phrase and give an example or two of each.

constant of variation (*p. 264*)
dependent variable (*pp. 238, 264*)
direct variation (*p. 264*)
domain (*p. 238*)
equation in two variables (*p. 244*)
function (*p. 256*)
functional notation (*p. 258*)

functional value (*p. 258*)
independent variable
  (*pp. 238, 264*)
inverse variation (*p. 270*)
linear equation (*p. 250*)
linear function (*p. 258*)
range (*p. 238*)

rate problems (*p. 266*)
relation (*p. 238*)
solution set (*p. 244*)
vertical line test (*p. 257*)
x-coordinate (*p. 238*)
y-coordinate (*p. 238*)
zero (*p. 258*)

Choose the correct term or expression to complete each sentence.

1. A (function, linear equation) is an equation whose graph is a straight line.
2. The set of all second coordinates of a set of ordered pairs is called the (range, domain).
3. Equations that are functions can be written in (functional notation, functional value).
4. The domain of $\{(-2, 3), (0, 4), (1, 5)\}$ is $(\{-2, 4, 1\}, \{-2, 0, 1\})$.
5. As x increases in value, y decreases in value in a(n) (direct, inverse) variation.
6. Any set of ordered pairs is a (function, relation).
7. A direct variation is a linear function that can be written in the form $(xy = k, y = kx)$.
8. In a (relation, function), each member of the domain is paired with one member of the range.
9. The equation $(2 + x = 8, y = 2x + 3)$ is an example of *an equation in two variables*.
10. The (first, second) coordinate in an ordered pair is called the *y*-coordinate.

Skills and Concepts

Objectives and Examples

- **Lesson 6–1**   Show relations as sets of ordered pairs, as tables, and as graphs.

  Express $\{(-1, 0), (0, 3), (1, 2), (2, 1)\}$ as a table and as a graph. Then determine the domain and the range.

x	y
−1	0
0	3
1	2
2	1

  The domain is $\{-1, 0, 1, 2\}$.
  The range is $\{0, 3, 2, 1\}$.

Review Exercises

Express each relation as a table and as a graph. Then determine the domain and the range.

11. $\{(-1, 0), (0, 0), (2, 3), (4, 1)\}$
12. $\{(-3, 2), (-2, -1), (0, 4), (3, 3)\}$
13. $\{(-1.5, 2), (1, 3.5), (4.5, -5.5)\}$
14. $\left\{\left(-\frac{1}{2}, 3\right), (1, 2), \left(\frac{1}{2}, 4\right)\right\}$

15. Express the relation as a set of ordered pairs and in a table. Find the domain and range.

www.algconcepts.com/vocabulary_review

Objectives and Examples	**Review Exercises**

• **Lesson 6–2** Solve linear equations for a given domain.

Solve $y = 2x$ if the domain is $\{-1, 0, 2\}$.

x	2x	y	(x, y)
−1	2(−1)	−2	(−1, −2)
0	2(0)	0	(0, 0)
2	2(2)	4	(2, 4)

The solution is $\{(-1\ -2), (0, 0), (2, 4)\}$.

Solve each equation if the domain is $\{-1, 0, 2, 4\}$. Graph the solution set.

16. $y = -x$ **17.** $y = \frac{1}{2}x$

18. $y = 3x - 4$ **19.** $3x + 2y = 10$

20. Find the domain of $y = x - 2$ if the range is $\{-3, -1, 4, 6\}$.

21. Find the domain of $y = 2x + 2$ if the range is $\{-2, 0, 2, 4\}$.

• **Lesson 6–3** Graph linear relations.

Graph $y = 2x - 4$.

x	2x − 4	y	(x, y)
0	2(0) − 4	−4	(0, −4)
1	2(1) − 4	−2	(1, −2)
2	2(2) − 4	0	(2, 0)
3	2(3) − 4	2	(3, 2)

Determine whether each equation is a linear equation. Explain. If an equation is linear, identify A, B, and C.

22. $yx = -2$ **23.** $-x = y$

24. $7x + y = -2$ **25.** $3x + 4y = 2z$

Graph each equation.

26. $y = 2x$ **27.** $y = x - 3$

28. $-1 = y$ **29.** $x = 5$

30. $y = -3x + 4$ **31.** $4 = 8x - 12$

32. The table lists the coordinates of points on a linear graph. Copy and complete the table.

x	y
?	−2
−2	0
?	2
1	3

• **Lesson 6–4** Determine whether a given relation is a function.

Determine whether $\{(3, 2), (5, 3), (4, 3), (5, 2)\}$ is a function.

Because there are two values of y for one value of x, 5, the relation is *not* a function.

Determine whether each relation is a function.

33. $\{(-2, 6), (3, -2), (3, 0), (4, 6)\}$

34. $\{(2, 8), (9, 3), (-3, 7), (4, 3)\}$

35. $\{(-1, 0), (3, 0), (6, 2)\}$

If $g(x) = -3x - 6$, find each value.

36. $g(2)$ **37.** $g(-3.5)$ **38.** $g(-4a)$

Mixed Problem Solving
See pages 724–731.

Objectives and Examples

- **Lesson 6–5** Solve problems involving direct variation.

Suppose y varies directly as x and $y = 10$ when $x = 15$. Find x when $y = 6$.

Use $\dfrac{y_1}{x_1} = \dfrac{y_2}{x_2}$ to solve the proportion.

$\dfrac{10}{15} = \dfrac{6}{x_2}$ *Substitute.*

$10x_2 = 15(6)$ *Cross products*

$10x_2 = 90$ *Multiply.*

$x_2 = 9$ So, $x = 9$ when $y = 6$.

Review Exercises

Solve. Assume that y varies directly as x.

39. Find y when $x = 7$ if $y = 15$ when $x = 5$.

40. If $y = 12$ when $x = -6$, find y when $x = 2$.

41. Find x when $y = -5$ if $x = -37.5$ when $y = 15$.

Solve by using direct variation.

42. If there are 16 ounces in 1 pound, how many ounces are in 2.8 pounds?

- **Lesson 6–6** Solve problems involving inverse variation.

Suppose y varies inversely as x and $y = 10$ when $x = 72$. Find x when $y = 24$.

$\dfrac{x_1}{x_2} = \dfrac{y_2}{y_1}$ *Inverse variation proportion*

$\dfrac{72}{x_2} = \dfrac{24}{10}$ *Substitute.*

$720 = 24x_2$ *Cross products*

$30 = x_2$ So, when $y = 24$, $x = 30$.

Solve. Assume that y varies inversely as x.

43. If $x = 3$ when $y = 25$, find x when $y = 15$.

44. Find y when $x = 42$ if $y = 21$ when $x = 56$.

Find the constant of variation. Then write an equation for each statement.

45. y varies inversely as x, and $y = 4$ when $x = 2$.

46. y varies directly as x, and $x = -12$ when $y = 3$.

Applications and Problem Solving

47. Science The table shows how far away a lightning strike is, given the number of seconds between when the lightning is seen and when thunder is heard. Express the relation as a set of ordered pairs. Then find the domain and range. *(Lesson 6–1)*

Time (s)	5	10	15	20	25
Distance (mi)	1	2	3	4	5

48. Sales Jennifer earns a weekly salary of $255 plus a bonus of $0.25 for each CD over 150 that she sells each week. The equation $C(r) = 255 + 0.25(r - 150)$ can be used to determine her total weekly income. How much more will she earn by selling 450 CDs than 200 CDs? *(Lesson 6–2)*

49. Electronics In an electrical transformer, voltage varies directly as the number of turns on the coil. What would be the voltage produced by 66 turns, if 110 volts come from 55 turns? *(Lesson 6–5)*

1. **Write** $y = 2x - 3$ in functional notation.
2. **List** three ways to show a relation.

Express each relation as a table and as a graph. Then determine the domain and range.

3. $\{(-3, -2), (-1, 4), (0, -1), (2, 3), (3, 3)\}$
4. $\{(-1, 1.5), (2, -4), (3, 2.5), (4, -1.5)\}$

Solve each equation if the domain is $\{-2, -1, 0, 4, 6\}$.

5. $y = \frac{1}{2}x$
6. $y = 4 - x$
7. $2x - 2y = 6$

Graph each equation.

8. $x = -5$
9. $y = -3x - 1$
10. $6x - 2y = 8$

Determine whether each relation is a function.

11. $\{(4, 1), (2, 4), (-1, 2), (-4, 2)\}$
12. $\{(-3, 2), (0, 4), (1, 5), (3, 0), (-3, -1)\}$

13.

x	y
−1	2
3	−4
−1	0
2	3
1	1

14.

15.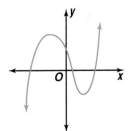

If $f(x) = -3x + 4$ and $g(x) = 2x + 2$, find each value.

16. $f(-1)$
17. $g(3.5)$
18. $g\left(\frac{1}{4}\right)$
19. $f(2d)$

Solve. Assume that y varies directly as x.

20. Find y when $x = 12$ if $y = 18$ when $x = 27$.
21. Suppose $y = -84$ when $x = -12$. Find y when $x = -1$.

Solve. Assume that y varies inversely as x.

22. Suppose $x = -16$ when $y = 4$. Find x when $y = 8$.
23. Find y when $x = 10$ if $y = 2.5$ when $x = 19.2$.

24. **Counting** Refer to the spinners.
   a. Suppose each spinner is spun once. Write a relation that shows the different possible outcomes.
   b. Use the relation to find the number of ways to land on an odd number and green.

25. **Geometry** To determine the perimeter P of a rectangle, the equation $P = 2w + 2\ell$, where w is the width and ℓ is the length, can be used. Suppose a rectangle has a perimeter of 64 inches. Find the possible dimensions of the rectangle given the domain values $\{3, 6, 7, 10\}$.

Probability and Counting Problems

You will need to solve problems involving combinations, permutations, and probability. Familiarize yourself with the terms *combination*, *permutation*, *tree diagram*, and *outcomes*. You will also need to know the definition of probability.

$$P(\text{event}) = \frac{\text{number of favorable outcomes}}{\text{number of possible outcomes}}$$

Test-Taking Tip

Sketching a picture can help you understand and solve many problems.

Example 1

How many different five-digit ZIP codes can be made if no digit is repeated in a code?

(A) 15,120   (B) 30,240

(C) 60,480   (D) 100,000

Hint Be sure you understand the question before you start to solve it.

Solution A ZIP code consists of five digits. The digits can be 0–9. There are ten possible digits. The problem states that no digit can appear more than once in these ZIP codes. Find the number of possible choices for each of the five positions.

——— ——— ——— ——— ———

There are 10 possible choices for the first position. There are 9 choices for the second position, because one digit was used for the first position. Following the same reasoning, there are 8 choices for the third position, 7 choices for the fourth position, and 6 choices for the fifth position. Multiply these numbers to find the total number of different codes. Use a calculator.

$$10 \times 9 \times 8 \times 7 \times 6 = 30{,}240$$

So, there are 30,240 possible ZIP codes. Thus, the answer is B.

Example 2

Five cards are numbered from 0 to 4. Suppose two cards are selected at random without replacement. What is the probability that the sum of the numbers on the two cards is an even number?

(A) $\frac{1}{5}$ (B) $\frac{3}{10}$ (C) $\frac{2}{5}$ (D) $\frac{1}{2}$ (E) $\frac{3}{5}$

Hint Decide whether or not order matters. In this problem order doesn't matter. So, it is a combination problem.

Solution First, write the possible outcomes of selecting two cards at random.

| 0 | 1 | | 0 | 2 | | 0 | 3 | | 0 | 4 |

| 1 | 2 | | 1 | 3 | | 1 | 4 |

| 2 | 3 | | 2 | 4 |

| 3 | 4 |

There are 10 possible combinations. There are 4 ways to get a sum that is an even number.

$$P(\text{even}) = \frac{4}{10} \text{ or } \frac{2}{5}$$

The answer is C.

After you work each problem, record your answer on the answer sheet provided or on a sheet of paper.

Multiple Choice

1. Suppose the spinner shown is spun. What is the probability of landing on an even number? *(Lesson 5–6)*

  (A) $\dfrac{5}{8}$ (B) $\dfrac{1}{2}$

  (C) $\dfrac{3}{8}$ (D) $\dfrac{3}{4}$

2. How many different starting squads of 6 players can be picked from 11 volleyball players? *(Lesson 4–2)*

  (A) 120   (B) 246   (C) 462   (D) 720

3. Diego has a collection of music CDs. Ten are rock, 5 are jazz, and 8 are country. If he chooses a CD at random, what is the probability that he will *not* choose a jazz CD? *(Lesson 5–6)*

  (A) $\dfrac{5}{23}$ (B) $\dfrac{5}{18}$ (C) $\dfrac{15}{23}$

  (D) $\dfrac{13}{18}$ (E) $\dfrac{18}{23}$

4. How many different 3-digit security codes can you make using the numbers 1, 2, and 3? You can use the numbers more than once. *(Lesson 4–2)*

  (A) 6   (B) 9   (C) 18   (D) 27

5. A bag contains only white and blue marbles. The probability of selecting a blue marble is $\dfrac{1}{5}$. The bag contains 200 marbles. If 100 white marbles are added to the bag, what is the probability of selecting a white marble? *(Lesson 5–6)*

  (A) $\dfrac{2}{15}$ (B) $\dfrac{7}{15}$ (C) $\dfrac{8}{15}$

  (D) $\dfrac{4}{5}$ (E) $\dfrac{13}{15}$

6. On Tuesday, the temperature fell 6° in 2 hours. Which of these expresses the temperature change per hour? *(Lesson 2–6)*

  (A) −3°   (B) −6°   (C) 6°   (D) 12°

7. Tomorrow 352 students will take a test in the cafeteria. If each table seats 12 students and there are already 15 tables in the cafeteria, how many additional tables will be needed? *(Lesson 1–5)*

  (A) 14   (B) 15   (C) 23   (D) 24

8. Ellen earns $86.80 for working 14 hours. At this rate, how much would she earn if she worked 20 hours? *(Lesson 5–1)*

  (A) $6.20   (B) $60.76

  (C) $124.00   (D) $1,215.20

Grid In

9. If $5x - 4 = x - 1$, what is the value of x? *(Lesson 4–6)*

Extended Response

10. The table shows the average monthly low temperatures for Detroit, Michigan. *(Lesson 3–2)*

Month	Temperature (°F)
January	44
February	46
March	54
April	60
May	66
June	72
July	74
August	74
September	70
October	61
November	52
December	45

Source: Excite Travel

Part A Organize the data in a stem-and-leaf plot.

Part B Find the mean, median, and mode. Tell which of these best represents the data.

 www.algconcepts.com/standardized_test

CHAPTER

7

Linear Equations

What You'll Learn

Key Ideas

- Find the slope of a line.
  (Lesson 7–1)

- Write linear equations in slope-intercept form and in point-slope form.
  (Lessons 7–2 and 7–3)

- Graph and interpret points on scatter plots. *(Lesson 7–4)*

- Graph and explore linear equations and families of linear equations.
  (Lessons 7–5 and 7–6)

- Write equations of parallel and perpendicular lines.
  (Lesson 7–7)

Key Vocabulary

point-slope form *(p. 290)*

rate of change *(p. 285)*

slope *(p. 284)*

slope-intercept form *(p. 296)*

Why It's Important

Entertainment The Mauch Chunk Switchback Railway, called the first American roller coaster, was built in Pennsylvania in 1827 to carry coal to freight trains. When rail lines were built closer to the mine, the switchback railway became a ride that attracted thousands of tourists every year.

Linear equations represent relationships between real-life quantities. Slope tells you how a quantity changes over time. You will find the slope of the steepest wooden roller coaster in Lesson 7–1.

✓ Check Your Readiness

Study these lessons to improve your skills.

Lesson 2–4, pp. 70–74

Find each difference.

1. $6 - 2$
2. $15 - 7$
3. $-3 - 2$
4. $-5 - (-5)$
5. $8 - 9$
6. $7 - (-2)$
7. $-2 - 4$
8. $-4 - 8$
9. $14 - (-4)$

Lesson 2–6, pp. 82–85

Find the value of each expression.

10. $\dfrac{5 - 1}{6 - 4}$
11. $\dfrac{10 - 1}{7 - 4}$
12. $\dfrac{8 - 5}{9 - 5}$
13. $\dfrac{-2 - (-1)}{5 - 3}$
14. $\dfrac{-3 - 0}{8 - 6}$
15. $\dfrac{-12 - (-9)}{2 - (-1)}$
16. $\dfrac{-8 - (-4)}{-2 - (-1)}$
17. $\dfrac{3 - (-3)}{5 - 10}$
18. $\dfrac{-3 - 2}{0 - 3}$

Lesson 2–2, pp. 58–63

Write the ordered pair for each point.

19. A
20. B
21. C
22. D
23. E
24. F

Study Organizer

Make this Foldable to help you organize your Chapter 7 notes. Begin with four sheets of grid paper.

❶ **Fold** each sheet in half from top to bottom.

❷ **Cut** along each fold. Staple the eight half-sheets together to form a booklet.

❸ **Cut** tabs into the margin. The top tab is 4 lines wide, the next tab is 8 lines wide, and so on.

❹ **Label** each page with a lesson number and title. The last tab is for vocabulary.

Reading and Writing As you read and study the chapter, fill the journal with notes, diagrams, and examples.

What You'll Learn

You'll learn to find the slope of a line given the coordinates of two points on the line.

Why It's Important

Construction
Building codes require that certain slopes be used for stairways.
See Exercise 28.

The rows in some movie theaters are so steep that there are no bad seats. We say that the theater has a steep **slope**. Slope is the ratio of the *rise*, or the vertical change, to the *run*, or the horizontal change.

rise 9 ft
run 10 ft

$$\text{slope} = \frac{\text{rise}}{\text{run}} = \frac{9}{10} \quad \begin{array}{l} \textit{vertical change} \\ \textit{horizontal change} \end{array}$$

The slope of the floor in the theater at the right is $\frac{9}{10}$.

On the graph below, the line passes through points at (4, 0) and (11, 3). The change in y or rise is $3 - 0$ or 3, while the change in x or run is $11 - 4$ or 7. Therefore, the slope of this line is $\frac{3}{7}$.

$$\text{slope} = \frac{\text{rise}}{\text{run}} = \frac{\text{change in } y}{\text{change in } x}$$

Slope

Words: The slope of a line is the ratio of the change in y to the corresponding change in x.

Model:

Symbols:

$$\text{slope} = \frac{\text{change in } y}{\text{change in } x}$$

Examples

Determine the slope of each line.

❶

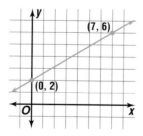
(7, 6)
(0, 2)

$$\text{slope} = \frac{\text{change in } y}{\text{change in } x}$$

$$= \frac{6 - 2}{7 - 0}$$

$$= \frac{4}{7} \qquad \text{The slope is } \frac{4}{7}.$$

❷

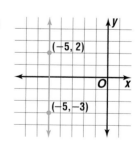
(−5, 2)
(−5, −3)

$$\text{slope} = \frac{\text{change in } y}{\text{change in } x}$$

$$= \frac{-3 - 2}{-5 - (-5)}$$

$$= \frac{-5}{0} \qquad \text{The slope is undefined. } \textit{Why?}$$

a.

b.

In Example 1, suppose $2 - 6$ had been used as the change in y and $0 - 7$ had been used as the change in x. Since $\frac{2-6}{0-7}$ is also equal to $\frac{4}{7}$, it does not matter which order is chosen. However, the coordinates of both points must be used in the same order.

In many real-world applications, the slope is the **rate of change.**

Example ❸

Income Link

The graph at the right shows the hours worked and income for Helena and Steve. Find the slope of each line. To what does the slope refer?

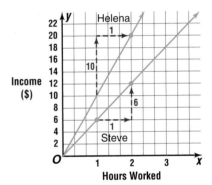

Helena: slope $= \dfrac{\text{change in } y}{\text{change in } x}$

$= \dfrac{10}{1}$ or 10

Steve: slope $= \dfrac{\text{change in } y}{\text{change in } x}$

$= \dfrac{6}{1}$ or 6

The graph represents Helena's income and Steve's income. The steepness depends on their hourly pay rate. So, the slope refers to each person's pay rate. Helena makes $10 per hour and Steve makes $6 per hour. Notice that the domain and range are nonnegative numbers since they cannot work fewer than 0 hours or earn less than $0.

 Personal Tutor at algconcepts.com

Look at the pattern in the graph at the right. Each time x increases 3 units, y decreases 1 unit.

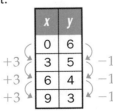

	x	y	
$+3$	0	6	-1
$+3$	3	5	-1
$+3$	6	4	-1
	9	3	

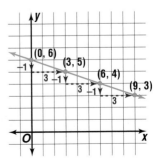

$$\text{slope} = \dfrac{\text{change in } y}{\text{change in } x} = \dfrac{-1}{3} \text{ or } -\dfrac{1}{3}$$

4 A line contains the points whose coordinates are listed in the table. Determine the slope of the line.

x	y
6	−3
4	0
2	3
0	6

-2, -2, -2 (change in x) ; $+3$, $+3$, $+3$ (change in y)

Each time x decreases 2 units, y increases 3 units.

$$\text{slope} = \frac{\text{change in } y}{\text{change in } x} = \frac{3}{-2}$$

The slope of the line containing these points is $-\frac{3}{2}$.

Your Turn

c.

x	y
0	0
1	1
2	2
3	3

d.

x	y
−5	2
−1	3
3	4
7	5

The examples above suggest the following.

Determining Slope Given Two Points

Words: The slope m of a line containing any two points with coordinates (x_1, y_1) and (x_2, y_2) is given by the following formula.

$$\text{slope} = \frac{\text{difference of the } y\text{-coordinates}}{\text{difference of the corresponding } x\text{-coordinates}}$$

Symbols: $m = \dfrac{y_2 - y_1}{x_2 - x_1}$, where $x_2 \neq x_1$

Determine the slope of each line.

5 the line through (2, 9) and (6, 9)

$$m = \frac{y_2 - y_1}{x_2 - x_1}$$

$$= \frac{9 - 9}{6 - 2}$$

$$= \frac{0}{4} \text{ or } 0$$

The slope is 0.

6 the line through (3, −2) and (−4, 7)

$$m = \frac{y_2 - y_1}{x_2 - x_1}$$

$$= \frac{7 - (-2)}{-4 - 3}$$

$$= \frac{9}{-7} \text{ or } -\frac{9}{7}$$

The slope is $-\frac{9}{7}$.

Your Turn

e. the line through (2, 5) and (3, 9)

f. the line through (−8, 1) and (4, 1)

The slopes of lines can be summarized as follows.

Concept Summary

positive slope	negative slope	0 slope	undefined slope
$m > 0$	$m < 0$	$m = 0$	m is undefined

Check for Understanding

Communicating Mathematics

1. **Sketch** the graph of a line having each slope.

   a. negative   **b.** $\frac{1}{2}$

   c. 0   **d.** 1

Vocabulary
rate of change
slope

2. **Choose** the line whose slope is $-\frac{1}{4}$.

   a.   **b.**   **c.**

3. **You Decide** Naomi says that only the coordinates of points A and D can be used to find the slope of the line at the right. Percy says he could use the coordinates of B and D to find the slope. Who is correct? Explain.

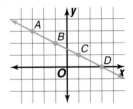

Guided Practice

⏱ **Getting Ready** Simplify each expression.

Sample: $\dfrac{-2 - 1}{4 - (-3)}$ **Solution:** $\dfrac{-2 - 1}{4 - (-3)} = \dfrac{-3}{7}$ or $-\dfrac{3}{7}$

4. $\dfrac{6 - 1}{-2 - 5}$ 5. $\dfrac{4 - (-2)}{7 - 4}$ 6. $\dfrac{-3 - 8}{9 - (-5)}$

Examples 1–3   **Determine the slope of each line.**

7.   8.

Example 4

9. Determine the slope of the line passing through the points whose coordinates are listed in the table.

x	−2	−1	0	1
y	−8	−4	0	4

Determine the slope of each line.

10. the line through
    $(1, 0)$ and $(−2, 9)$

11. the line through
    $(10, −3)$ and $(5, 3)$

Example 1

12. **Entertainment**   The Cyclone roller coaster has the steepest first drop of any wooden coaster in the world. It drops about 5 feet for every 3 feet of horizontal change. What is the slope of the first drop?

Cyclone, Coney Island, N.Y.

Exercises

Practice

Determine the slope of each line.

13.

14.

15.

16.

17.

18.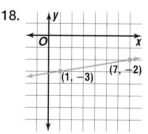

Homework Help	
For Exercises	See Examples
13–18	1, 2
19–21, 27	4
22–26	5, 6
28, 29	3
Extra Practice	
See page 705.	

Determine the slope of the line passing through the points whose coordinates are listed in each table.

19.

x	y
0	12
1	10
2	8
3	6

20.

x	y
0	7
2	6
4	5
6	4

21.

x	y
−2	−7
1	−6
4	−5
7	−4

Determine the slope of each line.

22. the line through (2, 5) and (6, 9)

23. the line through (2, 3) and (−3, 5)

24. the line through (4, −3) and (4, 5)

25. Find the slope of a line that passes through (−8, 9) and (0, 6).

26. What is the slope of a line that passes through $A(-8, 2)$ and $B(5, 8)$?

Applications and Problem Solving

27. Driving   Some roads in the Rocky Mountains have a rise of 7 feet for every 100 horizontal feet. What is the slope of such roads?

28. Construction   Building codes regulate the steepness of stairs. Some homes must have steps that are at least 11 inches wide for each 9 inches that they rise.

   a. What is the slope of the stairs?

   b. Describe how changing the width or the height affects the steepness of the stairs.

29. Money   The linear function $f(x) = 0.25x$ represents the value of x quarters.

   a. Explain how changes in x affect $f(x)$. What does this tell you about the slope of the graph of the equation?

   b. Describe a reasonable domain and range for the function.

Number of Quarters x	Value of Quarters $f(x)$
1	0.25
2	0.50
3	0.75
4	1.00

30. Critical Thinking   Determine whether the table represents a function that is linear or nonlinear. Explain.

x	1	3	5	7
y	2	4	8	16

Mixed Review

31. Assume that y varies inversely as x. Suppose $y = 4$ when $x = 6$. Find y when $x = 2$.   *(Lesson 6–6)*

Determine whether each relation is a function.   *(Lesson 6–4)*

32. {(4, 8), (5, 5), (−1, 3), (0, 6)}

33. {(7, 1), (−4, 3), (0, 1), (−4, 0)}

34. Communication   The coastline of New Zealand is 8161 miles long. This is about 86% of the length of a cable under the ocean that connects New Zealand and Canada. What is the approximate length of the cable?   *(Lesson 5–3)*

Standardized Test Practice
Ⓐ Ⓑ Ⓒ Ⓓ

35. Short Response   Solve $9x = 4x - 10$.   *(Lesson 4–6)*

36. Multiple Choice   Solve $-5 - 0.8y = 7$.   *(Lesson 4–5)*

   Ⓐ 15   Ⓑ −15   Ⓒ 9.6   Ⓓ −2.5

Writing Equations in Point-Slope Form

What You'll Learn

You'll learn to write a linear equation in point-slope form given the coordinates of a point on the line and the slope of the line.

Why It's Important

Population Point-slope equations are useful in making predictions about the population.
See Exercise 40.

You can write an equation of a line if you know its slope and the coordinates of one point on the line.

$$\frac{y_2 - y_1}{x_2 - x_1} = m \qquad \textit{Slope of a line}$$

$$\frac{y - y_1}{x - x_1} = m \qquad \textit{Replace } (x_2, y_2) \textit{ with } (x, y).$$

$$\left(\frac{y - y_1}{x - x_1}\right)(x - x_1) = m(x - x_1) \qquad \textit{Multiply each side by } (x - x_1).$$

$$y - y_1 = m(x - x_1)$$

This equation is said to be in **point-slope form**. Recall that it is called a *linear equation* because its graph is a straight line. *The graph of a nonlinear equation is not a straight line.*

Point-Slope Form	**Words:**	For a nonvertical line through the point at (x_1, y_1) with slope m, the point-slope form of a linear equation is $y - y_1 = m(x - x_1)$.
	Numbers:	$y - 6 = 4(x - 1)$ $(x_1, y_1) = (1, 6)$ and $m = 4$

The equation of a vertical line through the point at (x_1, y_1) is $x = x_1$.

Examples

Write the point-slope form of an equation for each line passing through the given point and having the given slope.

 (4, −5), $m = \dfrac{3}{4}$

$$y - y_1 = m(x - x_1) \qquad \textit{Point-Slope Form}$$

$$y - (-5) = \frac{3}{4}(x - 4) \qquad \textit{Replace } x_1 \textit{ with 4, } y_1$$
$$\qquad\qquad\qquad\qquad\quad \textit{with } -5, \textit{ and } m \textit{ with } \tfrac{3}{4}.$$

$$y + 5 = \frac{3}{4}(x - 4) \qquad \textit{Simplify.}$$

An equation of the line is $y + 5 = \dfrac{3}{4}(x - 4)$.

Check: Look at the graph. Choose some other point on the line and determine whether it is a solution of $y + 5 = \dfrac{3}{4}(x - 4)$. Try $(0, -8)$.

$$y + 5 = \frac{3}{4}(x - 4) \qquad \textit{Original equation}$$

$$-8 + 5 \stackrel{?}{=} \frac{3}{4}(0 - 4) \qquad \textit{Replace x with 0 and y with } -8.$$

$$-3 = -3 \quad \checkmark$$

② **$(-7, 2)$, $m = 0$**

$$y - y_1 = m(x - x_1) \qquad \textit{Point-Slope Form}$$
$$y - 2 = 0[x - (-7)] \qquad \textit{Replace } x_1 \textit{ with } -7, y_1 \textit{ with } 2, \textit{ and } m \textit{ with } 0.$$
$$y - 2 = 0 \qquad \textit{Simplify.}$$
$$y - 2 + 2 = 0 + 2 \qquad \textit{Add 2 to each side.}$$
$$y = 2 \qquad \textit{Simplify.}$$

An equation of the line is $y = 2$.

Your Turn

a. $(-3, 2)$, $m = 2$

b. $(5, 4)$, $m = -\dfrac{2}{3}$

You can also write an equation of a line if you know the coordinates of two points on the line.

Example **③** **Write the point-slope form of an equation of the line at the right.**

First, determine the slope of the line.

$$m = \frac{y_2 - y_1}{x_2 - x_1} = \frac{5 - 1}{-2 - 3} \text{ or } -\frac{4}{5}$$

The slope is $-\dfrac{4}{5}$. Now use the slope and either point to write an equation.

Method 1 Use $(-2, 5)$.

$$y - y_1 = m(x - x_1) \qquad \textit{Point-Slope Form}$$

$$y - 5 = -\frac{4}{5}[x - (-2)] \qquad \begin{array}{l}(x_1, y_1) = \\ (-2, 5)\end{array}$$

$$y - 5 = -\frac{4}{5}(x + 2) \qquad \textit{Simplify.}$$

Check:

$$y - 5 = -\frac{4}{5}(x + 2)$$

$$1 - 5 \stackrel{?}{=} -\frac{4}{5}(3 + 2) \qquad \begin{array}{l}(x, y) = \\ (3, 1)\end{array}$$

$$-4 \stackrel{?}{=} -\frac{4}{5}(5)$$

$$-4 = -4 \quad \checkmark$$

Method 2 Use $(3, 1)$.

$$y - y_1 = m(x - x_1) \qquad \textit{Point-Slope Form}$$

$$y - 1 = -\frac{4}{5}(x - 3) \qquad \begin{array}{l}(x_1, y_1) = \\ (3, 1)\end{array}$$

Check:

$$y - 1 = -\frac{4}{5}(x - 3)$$

$$5 - 1 \stackrel{?}{=} -\frac{4}{5}(-2 - 3) \qquad \begin{array}{l}(x, y) = \\ (-2, 5)\end{array}$$

$$4 \stackrel{?}{=} -\frac{4}{5}(-5)$$

$$4 = 4 \quad \checkmark$$

Both $y - 5 = -\dfrac{4}{5}(x + 2)$ and $y - 1 = -\dfrac{4}{5}(x - 3)$ are point-slope forms of an equation for the line passing through $(3, 1)$ and $(-2, 5)$.

Your Turn

Write the point-slope form of an equation for each line.

c.

(0, 1)

(−3, −1)

d. the line passing through (1, 4) and (3, −5)

 ♁nline **Personal Tutor at** algconcepts.com

Real World

Example ❹

Sports Link

In the 1988 Olympics, the women's winning long jump was about 24.7 feet. It is predicted that by 2020 the record will be about 26.3 feet. Write the point-slope form of an equation for the line in the graph.

Women's Long Jump

Distance (ft)

(2020, 26.3)

(1988, 24.7)

1980 1990 2000 2010 2020

Year

Source: USA Today

First, determine the slope of the line.

Prerequisite Skills Review

Operations with Decimals, p. 684

$$m = \frac{y_2 - y_1}{x_2 - x_1} = \frac{26.3 - 24.7}{2020 - 1988} \quad (x_1, y_1) = (1988, 24.7), (x_2, y_2) = (2020, 26.3)$$

$$= \frac{1.6}{32} \text{ or } \frac{1}{20} \quad \text{The slope is } \frac{1}{20}.$$

Now use the slope and either point to write an equation.

$$y - y_1 = m(x - x_1) \quad \textit{Point-Slope Form}$$

$$y - 24.7 = \frac{1}{20}(x - 1988) \quad \textit{Replace } (x_1, y_1) \textit{ with } (1988, 24.7) \textit{ and m with } \frac{1}{20}.$$

An equation of the line is $y - 24.7 = \frac{1}{20}(x - 1988)$.

Check by substituting (2020, 26.3) into the equation.

Check for Understanding

Communicating Mathematics

1. **Tell** what information is needed to write a linear equation in point-slope form.

2. **Sketch** a line whose slope is 2. Then use a point on the line to write an equation of the line in point-slope form.

Vocabulary

point-slope form

Getting Ready State the slope and the coordinates of a point on the graph of the equation.

Sample: $y - 2 = 3(x - 5)$

Solution: $y - y_1 = m(x - x_1)$
$$y - 2 = 3(x - 5)$$
$$m = 3, (5, 2)$$

3. $y - 6 = 2(x - 1)$ **4.** $y + 4 = \frac{2}{5}(x - 8)$ **5.** $y - (-5) = -3(x + 9)$

Examples 1 & 2

Write the point-slope form of an equation for each line passing through the given point and having the given slope.

6. $(5, 6), m = 2$

7. $(-2, 4), m = \frac{3}{4}$

8. $(3, -4), m = -5$

9. $(-8, 1), m = 0$

Example 3

Write the point-slope form of an equation for each line.

10.

11.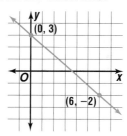

12. the line through $(5, 3)$ and $(-2, 1)$

13. the line through $(-2, 5)$ and $(4, -7)$

Example 4

14. Physical Science The table shows the growth of the Pacific giant kelp plant.

    a. Write the point-slope form of an equation that represents the growth.

    b. How much will the plant grow in 7 days?

    c. Would the plant show faster or slower growth if the slope of the equation was 30?

Time (days)	Growth (cm)
x	y
2	90
3	135

Exercises   • • • •   • • • • • • • • • • • • • •

Practice

Write the point-slope form of an equation for each line passing through the given point and having the given slope.

15. $(-1, 4), m = -3$ **16.** $(7, 2), m = 5$ **17.** $(9, 1), m = 6$

18. $(-3, 6), m = \frac{2}{3}$ **19.** $(-2, -2), m = \frac{4}{5}$ **20.** $(-1, -5), m = -7$

21. $(4, -2),$ m is undefined.   **22.** $(0, 6), m = \frac{3}{5}$ **23.** $\left(\frac{1}{2}, -4\right), m = -\frac{5}{2}$

Write the point-slope form of an equation for each line.

Homework Help	
For Exercises	See Examples
15–23, 36	1, 2
25–35, 37	3
38–40	4
Extra Practice	
See page 705.	

24.

25.

26.

27.

28.

29.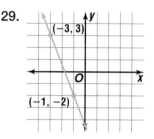

30. the line through (2, 1) and (3, 8)

31. the line through (4, 5) and (0, 3)

32. the line through (4, −2) and (0, −1)

33. the line through (1, −4) and (4, 5)

34. the line through (6, 9) and (−2, 9)

35. the line through (−3, 7) and (5, 2)

36. Write an equation in point-slope form of a line that has a slope of −1.8 and passes through (−5, 11).

37. What is the point-slope form of an equation of the line containing the points $Q(3, 6)$ and $R(−4, 7)$?

Applications and Problem Solving

38. **Transportation** The graph shows the growth in the number of sport-utility vehicles in the United States.

    a. Write the point-slope form of an equation for the graph of the line.

    b. Use the equation to predict the number of sport-utility vehicles in the United States in 2004.

SUVs on the Rise

Source: Polk

39. Savings Lynda adds $5 to her savings account each week, as shown in the graph at the right.

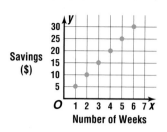

Savings ($)

Number of Weeks

  a. Write the point-slope form of an equation of the line through the points that represent her savings.

  b. Describe a reasonable domain and range for the equation.

Math Data
Online Update
For the latest information on U.S. population, visit: algconcepts.com

40. Population The graph shows the percent of U.S. residents who are born outside the United States.

Residents Not Born in U.S.A.

Percent

(2000, 9.5)

(1970, 4.7)

Year

Source: U.S. Census Bureau

  a. Write the point-slope form of an equation for the line passing through the points.

  b. Use the graph to estimate the percent of residents that were born outside the United States in 1995. Use the equation to check your estimate.

  c. Predict the percent of residents that will be born outside the United States in 2005.

41. Critical Thinking Find the coordinates of a point through which the graph of $y - 11 = \frac{3}{4}(x - 8)$ passes, if the y-coordinate is twice the x-coordinate.

Mixed Review

42. Construction A ramp installed to give handicapped people access to a building has a 3-foot rise and a 36-foot run. What is the slope of the ramp? *(Lesson 7–1)*

Find the domain of each equation if the range is {0, 2, 4}.
(Lesson 6–2)

43. $y = 2x$ **44.** $y = 4 + x$ **45.** $y = 2x - 2$

Find the percent of increase or decrease. *(Lesson 5–5)*

46. original: 15   **47.** original: 6
    new: 18   new: 24

48. Use the percent equation to find 45% of 160. *(Lesson 5–4)*

Standardized Test Practice
Ⓐ Ⓑ Ⓒ Ⓓ

49. Grid In Find $(-1.2)(-7)$. *(Lesson 4–1)*

50. Short Response List six numbers whose median is 12. *(Lesson 3–3)*

What You'll Learn

You'll learn to write a linear equation in slope-intercept form given the slope and y-intercept.

Why It's Important

Catering Caterers can use linear equations to represent the total cost. *See Exercise 47.*

The line crosses the y-axis at $(0, 5)$. The number 5 is called the **y-intercept** of the equation of the line. The line crosses the x-axis at $(2, 0)$. Thus, 2 is the **x-intercept** of the equation of the line.

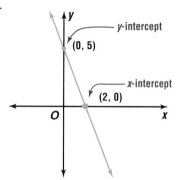

In Lesson 7–2, you learned how to write an equation in point-slope form by using the slope and a point on the line, and two points on the line. You can also write an equation of a line if you know the slope and y-intercept. Consider the graph below, which crosses the y-axis at $(0, b)$. Because $f(x) = 0$ at the x-intercept, the x-intercept is a zero of the function.

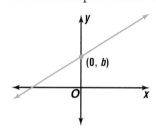

$$y - y_1 = m(x - x_1) \quad \textit{Point-Slope Form}$$
$$y - b = m(x - 0) \quad \textit{Replace } (x_1, y_1) \textit{ with}$$
$$y - b = mx \qquad\qquad (0, b).$$
$$y - b + b = mx + b \quad \textit{Add b to each side.}$$
$$y = mx + b$$
$$\uparrow \qquad \uparrow$$
$$\textit{slope} \quad \textit{y-intercept}$$

Slope-Intercept Form	**Words:** Given the slope m and y-intercept b of a line, the slope-intercept form of an equation of the line is $y = mx + b$.	**Model:**

Examples

Write an equation in slope-intercept form of each line with the given slope and y-intercept.

1 $m = -4, b = 5$

$y = mx + b \quad$ *Slope-Intercept Form*

$ = -4x + 5 \quad$ *Replace m with -4 and b with 5.*

An equation of the line is $y = -4x + 5$.

2 $m = \dfrac{1}{3}, b = -6$

$y = mx + b \qquad$ *Slope-Intercept Form*

$ = \dfrac{1}{3}x + (-6) \quad$ *Replace m with $\dfrac{1}{3}$ and b with -6.*

$ = \dfrac{1}{3}x - 6 \qquad$ *Simplify.*

An equation of the line is $y = \dfrac{1}{3}x - 6$.

③ $m = 0, b = 3$

$y = mx + b$ *Slope-Intercept Form*
$ = 0x + 3$ *Replace m with 0 and b with 3.*
$ = 3$ *Simplify.*

The equation of the line is $y = 3$. Remember that a line with a slope of 0 is a horizontal line.

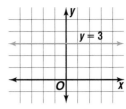

Your Turn

a. $m = 2, b = 1$ **b.** $m = -\dfrac{5}{3}, b = 0$ **c.** $m = 0, b = -8$

Now you can use the methods in Lesson 7–2 to write equations in slope-intercept form.

Examples

Write an equation of the line in slope-intercept form for each situation.

④ **slope 1 and passes through (2, 5)**

$y - y_1 = m(x - x_1)$ *Point-Slope Form*
$y - 5 = 1(x - 2)$ *Replace (x_1, y_1) with $(2, 5)$*
$y - 5 = x - 2$ *and m with 1.*

$y - 5 + 5 = x - 2 + 5$ *Add 5 to each side.*
$ y = x + 3$ *Slope-Intercept Form*

An equation of the line is $y = x + 3$.   *What are the x- and y-intercepts?*

Check: $y = x + 3$ *Original equation*
 $5 \stackrel{?}{=} 2 + 3$ *Replace x with 2 and y with 5.*
 $5 = 5$ $\checkmark$

⑤ **passing through (−4, 4) and (2, 1)**

First, determine the slope of the line.

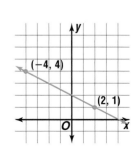

$m = \dfrac{y_2 - y_1}{x_2 - x_1}$ *Slope formula*

$ = \dfrac{1 - 4}{2 - (-4)}$ *Replace (x_1, y_1) and (x_2, y_2) with $(-4, 4)$ and $(2, 1)$.*

$ = \dfrac{-3}{6}$ or $-\dfrac{1}{2}$ *Simplify.*

(continued on next page)

www.algconcepts.com/extra_examples

Now substitute the known values into the point-slope form.

$$y - y_1 = m(x - x_1) \qquad \text{\textit{Point-Slope Form}}$$

$$y - 4 = -\frac{1}{2}[x - (-4)] \qquad \text{\textit{Replace }}(x_1, y_1) \text{ \textit{with }}(-4, 4) \text{ \textit{and m with }} -\frac{1}{2}.$$

$$y - 4 = -\frac{1}{2}x - 2 \qquad \text{\textit{Distributive Property}}$$

Then write in slope-intercept form.

$$y - 4 + 4 = -\frac{1}{2}x - 2 + 4 \qquad \text{\textit{Add 4 to each side.}}$$

$$y = -\frac{1}{2}x + 2 \qquad \text{\textit{Slope-Intercept Form}}$$

An equation of the line is $y = -\frac{1}{2}x + 2$. You can see from the graph that the y-intercept is 2. *You can also check by substituting the coordinates of one of the points into the equation.*

Your Turn

Write an equation in slope-intercept form of each line.

d. the line whose slope is $\frac{3}{4}$ and passes through $(8, -2)$

e. the line passing through $(2, 4)$ and $(0, 5)$

 Personal Tutor at algconcepts.com

Example — **6**

Science Link

A California inventor designed the *Skycar*, a car that can fly. The graph represents its landing from 50 meters off the ground. Write an equation of the line in slope-intercept form.

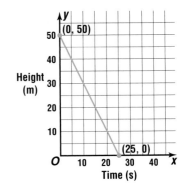

The y-intercept of the line is 50. Determine the slope.

$$m = \frac{y_2 - y_1}{x_2 - x_1} = \frac{0 - 50}{25 - 0} \text{ or } -2$$

Now substitute these values into the slope-intercept form.

$$y = mx + b \qquad \text{\textit{Slope-Intercept Form}}$$

$$= -2x + 50 \qquad \text{\textit{Replace m with }}-2 \text{ \textit{and b with 50.}}$$

An equation of the line is $y = -2x + 50$.

Check for Understanding

Communicating Mathematics

1. **Sketch** a line that has a y-intercept of 4 and a negative slope. Then find the x-intercept.

2. **Determine** whether the graph of $y = -3x + 8$ passes through $(-5, 23)$. Explain how you found your answer.

3. **You Decide** Andrew says that every line has an x-intercept. Jacquie disagrees. Who is correct? Give an example to support your answer.

Vocabulary

y-intercept
x-intercept
slope-intercept form

Guided Practice

Examples 1–3

Write an equation in slope-intercept form of the line with each slope and y-intercept.

4. $m = \dfrac{3}{2}, b = 4$ 5. $m = \dfrac{1}{4}, b = -3$ 6. $m = 0, b = 12$

Example 4

Write an equation in slope-intercept form of the line having the given slope and passing through the given point.

7. $m = 4, (2, 1)$ 8. $m = \dfrac{3}{5}, (0, 5)$ 9. $m = -\dfrac{1}{2}, (4, -5)$

Example 5

Write an equation in slope-intercept form of the line passing through each pair of points.

10. $(2, 3)$ and $(1, 0)$ 11. $(2, -4)$ and $(5, 2)$ 12. $(-4, 9)$ and $(5, 9)$

13. **Recycling** Yoshi collected 100 pounds of cans to recycle. He plans to collect an additional 25 pounds each week. The graph shows the amount of cans he plans to collect.   **Example 6**

   a. Write an equation in slope-intercept form of the line.

   b. What does the slope represent?

   c. Use the equation to predict the total amount of cans Yoshi will have collected after 12 weeks.

Cans (lb) vs. Number of Weeks
(0, 100) (4, 200)

A recycling plant

Exercises

Practice

Write an equation in slope-intercept form of the line with each slope and y-intercept.

14. $m = 3, b = 1$ 15. $m = -2, b = 5$ 16. $m = 4, b = -2$

17. $m = \dfrac{3}{2}, b = 7$ 18. $m = -\dfrac{1}{4}, b = -8$ 19. $m = -\dfrac{2}{5}, b = 3$

20. $m = -3, b = 0$ 21. $m = 0, b = 4$ 22. $m = -1.6, b = -12$

Homework Help

For Exercises	See Examples
14–22, 44–46	1–3
23–31	4
32–40, 47	5
41–43	6

Extra Practice
See page 706.

Write an equation in slope-intercept form of the line having the given slope and passing through the given point.

23. $m = 3, (5, -2)$ **24.** $m = -\frac{1}{4}, (0, 8)$ **25.** $m = 5, (0, 0)$

26. $m = 0, (5, 6)$ **27.** $m = -2, (4, 5)$ **28.** $m = 3, (4, -3)$

29. $m = -5, (6, -4)$ **30.** $m = -\frac{3}{7}, (1, 1)$ **31.** $m = \frac{1}{2}, (-3, -6)$

Write an equation in slope-intercept form of the line passing through each pair of points.

32. $(3, 5)$ and $(2, 1)$ **33.** $(4, 0)$ and $(0, 5)$ **34.** $(4, -3)$ and $(2, 1)$

35. $(3, 0)$ and $(2, -4)$ **36.** $(1, 6)$ and $(-2, 6)$ **37.** $(9, -3)$ and $(-1, -3)$

38. $(7, -2)$ and $(14, -4)$ **39.** $(1, 0)$ and $(6, -4)$ **40.** $(-1, 8)$ and $(3, 6)$

Write an equation in slope-intercept form for each line. Determine the x-intercept and y-intercept.

41.

42.

43.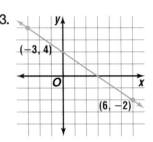

44. Write an equation in slope-intercept form of a line with slope $-\frac{2}{3}$ and y-intercept the same as the line whose equation is $y = 4x - 7$.

45. Write an equation in slope-intercept form of a line with a y-intercept of 11 and slope the same as the line whose equation is $y = -3x + 9$.

Applications and Problem Solving

46. Work For babysitting, Nicole charges a flat fee of $3, plus $5 per hour. The graph shows how much she earns.

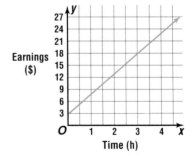

  a. Write an equation in slope-intercept form of the line.

  b. What do you think the slope and y-intercept represent?

  c. How much money will she make if she babysits 5 hours?

  d. What are the effects if Nicole changes her rates so that the equation is $y = 6x + 2$?

47. Catering A caterer charges $120 to cater a party for 15 people and $200 for 25 people. Assume that the cost y is a linear function of the number of people x.

  a. Write an equation in slope-intercept form for this function.

  b. Explain what the slope represents.

  c. How much would a party for 40 people cost?

48. Critical Thinking A line contains points at (5, 5) and (9, 1). Write a convincing argument showing that the x-intercept of the line is 10. Check by sketching the line.

Mixed Review

Write the point-slope form of an equation for each line passing through the given point and having the given slope. *(Lesson 7–2)*

49. $(4, 5)$, $m = -2$

50. $(6, -2)$, $m = \dfrac{1}{2}$

Express each relation as a table and as a graph. Then determine the domain and the range. *(Lesson 6–1)*

51. $\{(-3, 1), (-1, 5), (1, 3), (2, 6)\}$

52. $\{(-3, 0), (-1, 3), (0, -3), (3, 2)\}$

53. Measurement Convert 9144 meters to kilometers. *(Lesson 5–1)*

Find each quotient. *(Lesson 4–3)*

54. $7.2 \div 72$

55. $6 \div (-1.2)$

56. $-0.5 \div (-0.2)$

Standardized Test Practice
Ⓐ Ⓑ Ⓒ Ⓓ

57. Multiple Choice Solve $|c - 3| + 6 = 17$. *(Lesson 3–7)*

Ⓐ $\{-8, 14\}$ Ⓑ $\{8, 14\}$ Ⓒ $\{-14, 8\}$ Ⓓ $\{-14, -8\}$

Quiz 1   Lessons 7–1 through 7–3

Determine the slope of each line. *(Lesson 7–1)*

1.

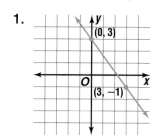

2. the line through points at $(0, -2)$ and $(2, 2)$

Write the point-slope form of an equation for each line passing through the given point and having the given slope. *(Lesson 7–2)*

3. $(5, 4)$, $m = -2$

4. $(-1, 6)$, $m = \dfrac{8}{5}$

5. Education In order to "curve" a set of test scores, a teacher uses the equation $y = 2.5x + 10$, where y is the curved test score and x is the number of problems answered correctly. *(Lesson 7–3)*

a. Find the test score of a student who answers 32 problems correctly.

b. Explain what the slope and y-intercept mean in the equation.

What You'll Learn

You'll learn to graph and interpret points on scatter plots.

Why It's Important

Insurance Insurance companies can use scatter plots to determine policy rates. See Exercise 17.

The Caribbean islands have many different species of birds.

To determine if there is a relationship between area and number of bird species, we can graph the data points in a **scatter plot**. In a scatter plot, two sets of data are plotted as ordered pairs in the coordinate plane.

(area, number of species)
↑ ↑
x-coordinate   *y-coordinate*

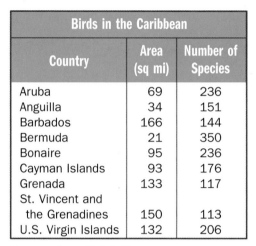

Birds in the Caribbean		
Country	**Area (sq mi)**	**Number of Species**
Aruba	69	236
Anguilla	34	151
Barbados	166	144
Bermuda	21	350
Bonaire	95	236
Cayman Islands	93	176
Grenada	133	117
St. Vincent and the Grenadines	150	113
U.S. Virgin Islands	132	206

Source: *Thayer Birder's Diary*

For example, the point with the box around it is at (95, 236). *The area is the independent quantity, and the number of species is the dependent quantity.*

You can use the scatter plot to draw conclusions and make predictions about the data.

- The country with the least area has 350 species of birds.

- In general, as area increases, the number of species of birds appears to decrease. This shows a relationship between area and the number of species. That is, the number of species is a function of area. *The country of Anguilla is an exception to this generalization.*

- The domain is the set of all *x*-coordinates:

  {21, 34, 69, 93, 95, 132, 133, 150, 166}.

- The range is the set of all corresponding *y*-coordinates:

  {350, 151, 236, 176, 236, 206, 117, 113, 144}.

A gold and blue macaw

- Based on the data in the scatter plot, you could predict that a Caribbean country with an area of 50 square miles would have fewer than 350 species of birds.

Often, when data are displayed in a scatter plot, you can determine whether there is a pattern, trend, or relationship between the variables.

- Data points that appear to go uphill show a relationship that is *positive*.
- Data points that appear to go downhill show a relationship that is *negative*.

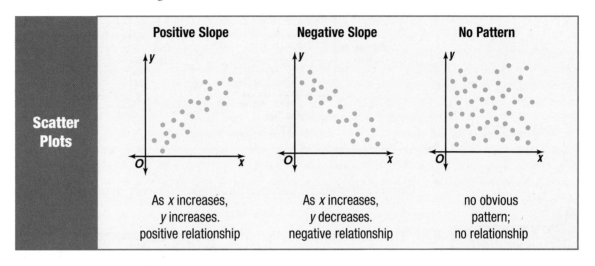

Scatter Plots

Positive Slope	Negative Slope	No Pattern
As *x* increases, *y* increases. positive relationship	As *x* increases, *y* decreases. negative relationship	no obvious pattern; no relationship

In the opening application, suppose a line were drawn close to most of the data points. The line would have a negative slope. That is, as the area increases, the number of species decreases. Thus, area and number of species would have a negative relationship.

Example ❶
School Link

The scatter plot at the right shows the word processing speeds of 12 students and the number of weeks they have studied word processing. Determine whether the scatter plot shows a *positive* relationship, *negative* relationship, or *no* relationship. If there is a relationship, describe it.

The plot indicates that as the number of weeks of experience increases, the word processing speed increases. Thus, there is a positive relationship between experience and word processing speed.

Determine whether each scatter plot shows a *positive* relationship, *negative* relationship, or *no* relationship. If there is a relationship, describe it.

a.

b.

Organizing data in a scatter plot makes it easier to observe patterns.

Example ② **School Link**

The table shows the amount of time students in math class spent studying and the scores they received on the test.

Study Time (min)	Test Score	Study Time (min)	Test Score
60	92	30	77
55	79	95	94
10	65	60	83
75	87	15	68
120	98	35	79
90	95	70	78
110	75	25	97
45	73	115	100

A. Make a scatter plot of the data.

Let the horizontal axis represent study time, and let the vertical axis represent test scores. Then plot the data.

B. Does the scatter plot show a relationship between study time and test scores? Explain.

In general, the scatter plot seems to show that a higher grade is directly related to the amount of time spent studying. There is a positive relationship.

C. Describe the independent and dependent variables. Then state the domain and the range.

Because test scores depend on the amount of study time, test scores is the dependent variable, and study time is the independent variable. The domain is the set of all study times and the range is the set of all test scores.

c. Make a scatter plot of the data. Determine whether there is a relationship between annual rainfall and weight of pumpkins. Explain.

Annual Rainfall (in.)	43	35	38	42	30	29	44	40	41
Weight of Pumpkin (lb)	204	195	198	200	190	190	206	202	200

Check for Understanding

Communicating Mathematics

1. **Describe** how you can use a scatter plot to display two sets of related data. Then explain what the domain, range, independent variable, and dependent variable are.

Vocabulary

scatter plot

2. **Explain** how you know whether there appears to be a linear relationship between the two sets of data plotted on a scatter plot. Include the concept of slope in your explanation.

3. **Writing Math** Describe situations in which data graphed in a scatter plot would have a positive relationship, negative relationship, and no relationship. Tell what the labels on the horizontal and vertical axes for each scatter plot would be.

Guided Practice

Example 1

4. Determine whether the scatter plot has a *positive* relationship, *negative* relationship, or *no* relationship. If there is a relationship, describe it.

Example 2

5. **Weather** The average maximum and minimum monthly temperatures in July of ten cities are given in the table at the right.

  a. Make a scatter plot of the data.

  b. Does the scatter plot show a linear relationship between maximum and minimum temperatures? Explain.

City	Minimum (°F)	Maximum (°F)
Atlanta, GA	70	88
Lexington, KY	66	86
Galveston, TX	79	87
Columbus, OH	63	84
Jackson, MS	71	92
Albany, NY	60	84
Los Angeles, CA	65	84
Miami, FL	76	89
Portland, OR	57	80
Chicago, IL	63	84

Source: *The World Almanac*

Example 2

6. **Health** The scatter plot at the right shows data that compare a person's weekly exercise to that person's resting heart rate. Does there appear to be any relationship between hours of exercise and resting heart rate? If so, describe it.

Exercises

Practice

Determine whether each scatter plot has a *positive* relationship, *negative* relationship, or *no* relationship. If there is a relationship, describe it.

Homework Help	
For Exercises	**See Examples**
7–8	1
9–17	2
Extra Practice	
See page 706.	

7.

CO_2 (g/mi) — Car Speed (mph)

8.

Income ($1000s) — College (years)

Make a scatter plot of each set of data. Then describe any trend shown in the scatter plot.

9.

Dallas Stars Hockey Team, Selected Players			
Goals	**Assists**	**Goals**	**Assists**
34	47	3	9
28	27	2	6
12	33	20	32
13	18	14	34
12	17	17	17
1	3	9	21
13	14	6	23
4	11	6	14
6	11	4	12

Source: *Infoplease.com*

10.

Breakfast Sweets	
Sugar (g)	**Calories**
30	260
36	430
13	420
19	270
15	530
18	430
47	420
52	580
23	520

Source: *Nutrition Action Healthletter*

11. Emission of CO_2 from car exhaust is harmful to the environment. Refer to the scatter plot in Exercise 7. What recommendation would you make about speed limits based on the data? Explain.

Determine whether a scatter plot of the data for the following would show a *positive*, *negative*, or *no* relationship between the variables.

12. playing time and points scored

13. your height and month of birth

14. size of household and pounds of garbage produced

15. age of car and its value

Applications and Problem Solving

16. Insurance The table shows data on the age of drivers and the number of traffic accidents they have for each 100,000 drivers of that age.

Age	Number of Accidents	Age	Number of Accidents
16	168	23	120
17	181	24	117
18	190	25	92
19	195	26	80
20	145	27	69
21	126	28	70
22	114	29	75

a. Make a scatter plot of the data. Determine if there is a *positive*, *negative*, or *no* relationship. If there is a relationship, describe it.

b. Based on the scatter plot, what recommendation would you make to insurance companies about how to determine insurance rates for different age drivers?

17. Earth Science The table below shows the elevation and the average annual precipitation for places around the world.

Elevation (ft)	Average Yearly Precipitation (in.)	Elevation (ft)	Average Yearly Precipitation (in.)
20	49	138	21
66	59	164	47
118	37	79	33
82	45	108	27
79	50	125	59
200	32	10	46
23	44	82	27
118	55	171	21
190	23		

Source: *The World Almanac*

a. Make a scatter plot of the data.

b. Does there seem to be a relationship? If so, describe it.

c. Describe the slope of a line passing through the points.

18. Critical Thinking Refer to the scatter plot in Exercise 6. Would the relationship between the variables change if you relabeled the axes so that resting heart rate was on the horizontal axis? Explain.

Mixed Review

19. Write an equation in slope-intercept form of the line passing through $(-2, -8)$ and $(1, 7)$. *(Lesson 7–3)*

20. Graph $y = x + 4$. *(Lesson 6–3)*

21. Architecture The Chrysler Building in New York is approximately 300 meters high. If a model of the building is 60 centimeters high, find the scale of the model. *(Lesson 5–2)*

22. Solve $-3(5 + x) = 12$. Check your solution. *(Lesson 4–7)*

Standardized Test Practice
Ⓐ Ⓑ Ⓒ Ⓓ

23. Multiple Choice The area of the rectangle is 185.9 square centimeters. Find the length. *(Lesson 4–4)*

Ⓐ 22 cm   Ⓑ 163.9 cm
Ⓒ 174.9 cm   Ⓓ 16.9 cm

11 cm

x cm

Humidity Heats Things Up

Materials

grid paper

straightedge

graphing calculator or graphing software

Best-Fit Lines and Prediction

In the summertime, have you ever noticed that it feels hotter on days that are humid? The *heat index* describes what the temperature feels like to the body when there is humidity.

Investigate

1. The table shows the heat index for various air temperatures and percents of relative humidity. For example, at 100°F and 30% relative humidity, the heat index is 104°F. So, it feels like 104°F.

Air Temperature	Heat Index at Relative Humidities										
	0%	5%	10%	15%	20%	25%	30%	35%	40%	45%	50%
70°F	64	65	65	65	66	66	67	67	68	68	69
75°F	69	69	70	71	72	72	73	73	74	74	75
80°F	73	74	75	76	77	77	78	79	79	80	81
85°F	78	79	80	81	82	83	84	85	86	87	88
90°F	83	84	85	86	87	88	90	91	93	95	96
95°F	87	88	90	91	93	94	96	98	101	104	107
100°F	91	93	95	97	99	101	**104**	107	110	115	120

   a. Look at the row for 75°F. Write ordered pairs of the form (percent humidity, heat index) for the entire row. For example, the first ordered pair is (0, 69).

   b. Make a scatter plot of the data.

   c. Does the scatter plot show any relationship between humidity and heat index? If so, describe it.

   d. Describe the domain and the range using words and using numbers.

2. When there is a relationship between two sets of data, you can draw a **best-fit line** through the points, as you did in Lesson 7–4. This line is close to most of the data points. Use a straightedge to draw a best-fit line on the scatter plot you made in Exercise 1.

Another way to fit a line to data is with a **median-median line.** To make a median-median line, you divide the data set into 3 equal parts and find the median for each part. You then use these three points to draw a line to approximate the data.

3. You can use the best-fit line to write a linear equation that describes a relationship between humidity and heat index.

   a. Select two points on the line that you drew in Exercise 2. Use the points to determine the slope of the line. *The points may or may not be points in the scatter plot.*

   b. Write an equation of the line in slope-intercept form.

   c. Describe the independent and dependent variables.

   d. A best-fit line can help you predict values even when you don't have data for every point. **Interpolation** is used to predict values between the greatest and least data points. **Extrapolation** is used to predict values that are below the least value or above the greatest value. Use your equation to predict the heat index at 75°F when the relative humidity is 60%.

Extending the Investigation

The process of finding a line that best fits a set of data is called **linear regression.** The slope of the best-fit line is called the **correlation coefficient.** The best-fit line will probably not pass through all of the data points. The distances from data points to the best-fit line are called **residuals.**

In this extension, you will use a graphing calculator or graphing software to explore the line of best fit. On a track or football field, collect data for running times for the following distances: 10 yards, 20 yards, 70 yards, and 100 yards.

- Make a scatter plot and draw best-fit lines for the data you collected. Write an equation in slope-intercept form for the best-fit line.

- Use your equations to predict the times for running 50, 80, and 150 yards. Are your predictions reasonable?

- Use a graphing calculator or graphing software to find linear equations for the data. Compare your equation for the best-fit line to the equation given by the calculator or software.

- Use the best-fit line equation to predict the time to run 1760 yards.

- Explain whether the time prediction for running 1760 yards is reasonable.

Presenting Your Conclusions

- Make a poster displaying the data table, your scatter plots, best-fit line, and equation.

 Investigation For more information on windchill factor, visit: www.algconcepts.com

7-5 Graphing Linear Equations

What You'll Learn

You'll learn to graph linear equations by using the *x*- and *y*-intercepts or the slope and *y*-intercept.

Why It's Important

Rates Linear graphs are helpful in showing phone costs.
See Exercise 38.

A simple method of graphing a linear equation is using the points where the line crosses the *x*-axis and the *y*-axis. Consider the equations $x + y = 8$.

To find the *x*-intercept, let $y = 0$.	To find the *y*-intercept, let $x = 0$.
$x + y = 8$ *Original equation*	$x + y = 8$ *Original equation*
$x + 0 = 8$ *$y = 0$*	$0 + y = 8$ *$x = 0$*
$x = 8$ *Simplify.*	$y = 8$ *Simplify.*

The *x*-intercept is 8, and the *y*-intercept is 8. This means that the graph intersects the *x*-axis at (8, 0) and the *y*-axis at (0, 8). Graph these ordered pairs. Then draw the line that passes through these points. Each point on the graph represents two numbers whose sum is 8. Notice that the table of values confirms these intercepts.

x	y
−1	9
0	8
1	7
2	6
3	5
4	4
5	3
6	2
7	1
8	0
9	−1

Recall that the zero of the function is the *x*-intercept of the graph of the related equation.

Examples

Determine the *x*-intercept and *y*-intercept of the graph of each equation. Then graph the equation.

1 $5y - x = 10$

To find the *x*-intercept, let $y = 0$.	To find the *y*-intercept, let $x = 0$.
$5y - x = 10$ *Original equation*	$5y - x = 10$ *Original equation*
$5(0) - x = 10$ *Replace y with 0.*	$5y - 0 = 10$ *Replace x with 0.*
$-x = 10$ *Simplify.*	$5y = 10$ *Simplify.*
$\dfrac{-x}{-1} = \dfrac{10}{-1}$ *Divide each side by 5.*	$\dfrac{5y}{5} = \dfrac{10}{5}$ *Divide each side by 5.*
$x = -10$ *Simplify.*	$y = 2$ *Simplify.*

The *x*-intercept is −10, and the *y*-intercept is 2. This means that the graph intersects the *x*-axis at (−10, 0) and the *y*-axis at (0, 2). Graph these ordered pairs. Then draw the line that passes through these points.

2 $2x - 4y = 8$

Select values from the domain and make a table.

x	−2	0	2	4	6
y	−3	−2	1	0	1

↑
zero

From the table, you can see that the *x*-intercept is 4, and the *y*-intercept is −2. This means that the graph intersects the *x*-axis at (4, 0) and the *y*-axis at (0, −2). Graph these ordered pairs. Then draw the line that passes through these points. The zero of this linear function is 4.

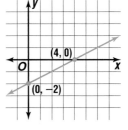

Check: Look at the graph. Choose some other point on the line and determine whether it is a solution of $2x - 4y = 8$. Try (2, −1).

$$2x - 4y = 8 \quad \textit{Original equation}$$
$$2(2) - 4(-1) \stackrel{?}{=} 8 \quad \textit{Replace x with 2 and y with −1.}$$
$$4 - (-4) \stackrel{?}{=} 8 \quad \textit{Multiply.}$$
$$8 = 8 \quad \checkmark$$

Concepts in MOtion
BrainPOP®
algconcepts.com

Your Turn

a. $x + y = 2$ **b.** $3x + y = 3$ **c.** $4x - 5y = 20$

You can easily find the slope and *y*-intercept of the graph of an equation that is written in slope-intercept form.

Example

Mail Link

3 **To mail a letter in 2004, it cost \$0.37 for the first ounce and \$0.23 for each additional ounce. This can be represented by $y = 0.37 + 0.23x$. Determine the slope and *y*-intercept of the graph of the equation.**

$$y = mx + b \quad \textit{Slope-Intercept Form}$$
$$y = 0.23x + 0.37$$

The slope is 0.23, and the *y*-intercept is 0.37. So the slope represents the cost per ounce after the first ounce, and the *y*-intercept represents the cost of the first ounce of mail.

Determine the slope and *y*-intercept of the graph of $10 + 5y = 2x$.

Write the equation in slope-intercept form to find the slope and *y*-intercept.

$$10 + 5y = 2x \qquad \textit{Original equation}$$
$$10 + 5y - 10 = 2x - 10 \qquad \textit{Subtract 10 from each side.}$$
$$5y = 2x - 10 \qquad \textit{Simplify.}$$
$$\frac{5y}{5} = \frac{2x - 10}{5} \qquad \textit{Divide each side by 5.}$$
$$y = \frac{2}{5}x - 2 \qquad \textit{Simplify.}$$

The slope is $\frac{2}{5}$, and the *y*-intercept is -2.

Your Turn

d. $y = 5x + 9$ **e.** $4x + 3y = 6$

You can also graph a linear equation by using the slope and *y*-intercept.

Graph each equation by using the slope and *y*-intercept.

5 $y = \frac{2}{3}x - 5$

$$y = \frac{2}{3}x + (-5)$$

The slope is $\frac{2}{3}$, and the *y*-intercept is -5.
Graph the point at $(0, -5)$. Then go up 2 units and right 3 units. This will be the point at $(3, -3)$. Then draw the line through points at $(0, -5)$ and $(3, -3)$.

Check: The graph appears to go through the point at $(6, -1)$. Substitute $(6, -1)$ into $y = \frac{2}{3}x + (-5)$.

$$y = \frac{2}{3}x + (-5) \qquad \textit{Original equation}$$

$$-1 \stackrel{?}{=} \frac{2}{3}(6) + (-5) \qquad \textit{Replace x with 6 and y with -1.}$$

$$-1 \stackrel{?}{=} 4 - 5 \qquad \textit{Simplify.}$$

$$-1 = -1 \quad \checkmark$$

6 $3x + 2y = 6$

First, write the equation in slope-intercept form.

$$3x + 2y = 6 \qquad \textit{Original equation}$$
$$3x + 2y - 3x = 6 - 3x \qquad \textit{Subtract 3x from each side.}$$
$$2y = 6 - 3x \qquad \textit{Simplify.}$$
$$\frac{2y}{2} = \frac{6 - 3x}{2} \qquad \textit{Divide each side by 2.}$$
$$y = -\frac{3}{2}x + 3 \quad \text{The slope is } -\frac{3}{2}, \text{ and the } y\text{-intercept is 3.}$$

Graph the point at (0, 3). Then go up 3 units and left 2 units. This will be the point at (−2, 6). Then draw the line through (0, 3) and (−2, 6). *Check by substituting the coordinates of another point that appears to lie on the line, such as (2, 0).*

Your Turn

f. $y = \frac{1}{2}x + 3$

g. $x + 4y = -8$

The graph of a horizontal line has a slope of 0 and no x-intercept. The graph of a vertical line has an undefined slope and no y-intercept.

Examples

Graph each equation.

7 $y = 4$

$$y = mx + b$$
$$\qquad \downarrow \qquad\quad \downarrow$$
$$y = 0x + 4 \quad \textit{slope = 0, y-intercept = 4}$$

No matter what the value of x, $y = 4$. So, all ordered pairs are of the form $(x, 4)$. Some examples are (0, 4) and (−3, 4).

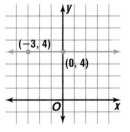

8 $x = -2$

slope: undefined, y-intercept: none

No matter what the value of y, $x = -2$. So, all ordered pairs are of the form $(-2, y)$. Some examples are (−2, −1) and (−2, 3).

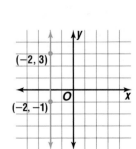

Your Turn

h. $y = -1$

i. $x = 3$

Check for Understanding

Communicating Mathematics

1. **Explain** how many points are needed to draw the graph of a linear equation.

2. Gwen says that $y = 8$ and $x = 8$ are both functions. Darnell says that only one of the equations is a function. Who is correct? Explain.

Guided Practice

Examples 1 & 2

Use a table to determine the *x*-intercept and *y*-intercept of the graph of each equation. Then graph the equation.

3. $x + y = -3$
4. $x + 4y = 4$
5. $5x + 2y = 10$
6. $2x - 6y = 12$

Examples 3–8

Determine the slope and *y*-intercept of the graph of each equation. Then graph the equation.

7. $y = x + 1$
8. $y = -4x$
9. $y = \frac{1}{4}x - 3$
10. $y = -2$

Example 3

11. **Temperature** The function $F = \frac{9}{5}C + 32$ gives the temperature in degrees Fahrenheit if you know the degrees Celsius.

    a. What are the slope and *y*-intercept of the graph of the equation?

    b. What is the zero of the function, and what does it represent?

Exercises

Practice

Use a table to determine the *x*-intercept and *y*-intercept of the graph of each equation. Then graph the equation.

Homework Help	
For Exercises	See Examples
12–23	1, 2
24–35	4–6
36, 37	7–8
38, 39	3
Extra Practice	
See page 706.	

12. $x + y = 4$
13. $x + y = -5$
14. $x + 2y = 4$
15. $x - 3y = 3$
16. $5x + y = 5$
17. $2x + y = -6$
18. $4x + 5y = 20$
19. $-3x + 4y = 12$
20. $6x - 3y = 6$
21. $7x - 2y = 14$
22. $2x + 5y = -10$
23. $x + \frac{1}{2}y = 4$

Determine the slope and *y*-intercept of the graph of each equation. Then graph the equation.

24. $y = x + 3$
25. $y = -x + 2$
26. $y = 2x + 1$
27. $y = 3x - 1$
28. $x = 6$
29. $y = 4$
30. $y = \frac{1}{2}x + 3$
31. $y = \frac{3}{4}x - 4$
32. $y = -\frac{3}{2}x + 5$
33. $-2x + y = 3$
34. $x + 2y = 4$
35. $3x + 5y = 10$

36. a. Graph $3x - 2y = 4$.

    b. What is the slope of the line?

    c. Where does the line intersect the x-axis?

37. a. Graph a line that goes through points at $(-4, 5)$ and $(2, -2)$.

    b. Write an equation of the line in slope-intercept form.

    c. Name the slope and y-intercept.

38. Rates A long-distance phone company charges $5 per month plus $0.10 per minute.

    a. Write a linear equation to represent the total monthly cost $f(x)$ as a function of the number of minutes x.

    b. Graph the equation.

    c. Explain what the slope and y-intercept represent.

39. Physical Science The weight of a bucket of water is a linear function of the depth of the water. When there are 4 centimeters of water in the bucket, it weighs 2 pounds. When there are 12 centimeters of water in the bucket, it weighs 4 pounds.

    a. Write an equation for the weight of the bucket $f(x)$ as a function of the depth of the water x.

    b. Graph the equation.

    c. Estimate how much a bucket of water will weigh if there are 18 centimeters of water in it.

40. Critical Thinking Explain how the pattern in the arithmetic sequence 2, 2.5, 3, 3.5, 4, 4.5, 5, . . . is related to the graph at the right.

A long-distance call

41. Sketch a scatter plot that shows a positive relationship between two sets of related data. *(Lesson 7–4)*

42. Entertainment It costs $60 to rent jet skis for 2.5 hours. At that rate, how much will it cost to rent jet skis for 6 hours? *(Lesson 6–5)*

Find the probability of each outcome if a die is rolled. *(Lesson 5–6)*

43. a 4

44. an even number

45. a number greater than 6

Solve each equation. Check your solution. *(Lesson 3–6)*

46. $-6 + x = 20$

47. $12 = b - 7$

48. $0 = k + 1.5$

49. Multiple Choice Which fractions are ordered from least to greatest? *(Lesson 3–1)*

  Ⓐ $\dfrac{3}{14}, \dfrac{5}{23}, \dfrac{9}{23}$
  Ⓑ $\dfrac{5}{23}, \dfrac{9}{23}, \dfrac{3}{14}$
  Ⓒ $\dfrac{9}{23}, \dfrac{5}{23}, \dfrac{3}{14}$
  Ⓓ $\dfrac{3}{14}, \dfrac{9}{23}, \dfrac{5}{23}$

What You'll Learn

You'll learn to explore the effects of changing the slopes and y-intercepts of linear functions.

Why It's Important

Business Families of graphs can display different fees.
See Exercise 10.

The graph representing a cheetah's speed is much steeper than the graph representing a spider's speed or an elephant's speed. This is because the cheetah runs faster and therefore covers greater distance in each unit of time.

What do the graphs have in common? They have the same y-intercept, 0. They are called a **family of graphs** because they have at least one characteristic in common that makes them different from other groups of graphs.

Maximum Animal Speeds

cheetah 62 mph   elephant 25 mph

Distance (mi)

spider 1 mph

Time (hr)

Families of linear graphs often fall into two categories—those with the same slope or those with the same x- or y-intercept.

Family of Graphs	Family of Graphs	Not a Family of Graphs
Linear Graphs $y = \frac{1}{2}x + 3$ $y = \frac{1}{2}x - 1$	$y = -x + 1$ $y = \frac{1}{3}x + 1$	$y = x + 2$ $y = \frac{1}{3}x$
same slope	same y-intercept	different slopes and different intercepts

Examples

Graph each pair of equations. Describe any similarities or differences. Explain why they are a family of graphs.

 $y = 3x + 4$
$y = 3x - 2$

The graphs have y-intercepts of 4 and −2, respectively.

They are a family of graphs because the slope of each line is 3.

$y = 3x + 4$
$y = 3x - 2$

② $y = x + 3$

$y = -\frac{1}{2}x + 3$

Each graph has a different slope.
Each graph has a y-intercept of 3.
Thus, they are a family of graphs.

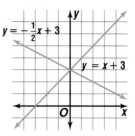

Your Turn

a. $y = 2x - 1$
$y = 2x + 5$

b. $y = x + 1$
$y = 3x + 1$

A graphing calculator is a good tool for exploring families of graphs.

Graphing Calculator Exploration

Graphing Calculator Tutorial
See pp. 750–753.

Step 1 Graph $y = x$ in the standard viewing window.

| Y= | X,T,θ,n | ZOOM | 6

Step 2 Graph $y = 2x$ and $y = \frac{1}{2}x$.

Try These

1. Describe the differences in the graphs of $y = 2x$ and $y = x$ in terms of their slopes and intercepts.

2. Describe the differences in the graph of $y = \frac{1}{2}x$ and the graphs in Exercise 1 in terms of their slopes and intercepts.

3. Clear the Y = list of the previous equations. Then graph $y = |x|$, $y = |x| + 3$, and $y = |x + 2|$. Describe the differences in these graphs in terms of their slopes and intercepts.

Example

Business Link

③ **Matthew and Juan are starting their own pet care business. Juan wants to charge $5 an hour. Matthew thinks they should charge $3 an hour. Suppose x represents the number of hours. Then $y = 5x$ and $y = 3x$ represent how much they would charge, respectively. Compare and contrast the graphs of the equations.**

You can compare graphs of lines by looking at their equations.

The equations have the same y-intercept, but the graph of $y = 5x$ is steeper. This is because its slope, which represents $5 per hour, is greater than the slope of the graph of $y = 3x$.

(continued on the next page)

Your Turn Compare and contrast the graphs of the equations. Verify by graphing the equations.

c. $y = -3x + 4$
$y = -x + 4$

d. $y = \frac{2}{3}x + 3$
$y = \frac{2}{3}x - 1$

A **parent graph** is the simplest of the graphs in a family. Let's summarize how changing the m or b in $y = mx + b$ affects the graph of the equation.

COncepts in MOtion
Animation
algconcepts.com

<table>
<tr><td colspan="3" align="center">**Parent Graphs**</td></tr>
<tr><td align="center">**Parent: $y = x$**</td><td align="center">**Parent: $y = -x$**</td><td align="center">**Parent: $y = 2x$**</td></tr>
<tr><td></td><td></td><td></td></tr>
<tr><td>As the value of m increases, the line gets steeper.</td><td>As the value of m decreases, the line gets steeper.</td><td>As the value of b increases, the graph shifts up on the y-axis. As the value of b decreases, the graph shifts down on the y-axis.</td></tr>
</table>

You can change a graph by changing the slope or y-intercept.

Examples

Change $y = -\frac{1}{2}x + 3$ so that the graph of the new equation fits each description.

 same y-intercept, steeper negative slope

The y-intercept is 3, and the slope is $-\frac{1}{2}$.
The new equation will also have a y-intercept of 3. In order for the slope to be steeper and still be negative, its value must be less than $-\frac{1}{2}$, such as -2. The new equation is $y = -2x + 3$.

 same slope, y-intercept is shifted up 4 units

The slope of the new equation will be $-\frac{1}{2}$. Since the current y-intercept is 3, the new y-intercept will be $3 + 4$ or 7. The new equation is $y = -\frac{1}{2}x + 7$. *Check by graphing.*

Your Turn Change $y = 2x + 1$ so that the graph of the new equation fits each description.

e. same slope, shifted down 1 unit

f. same y-intercept, less steep positive slope

Check for Understanding

Communicating Mathematics

1. **Describe** how the graph of each equation is different from the graph of $y = \frac{1}{4}x + 1$.

Vocabulary

family of graphs
parent graph

a. $y = \frac{1}{4}x - 3$

b. $y = \frac{1}{8}x + 1$

c. $y = 2x + 1$

d. $y = \frac{1}{4}x + 4$

2. **Sketch** a family of graphs. Identify the parent graph and explain how the graphs are similar.

3. a. **Explain** the connection between a rate, such as miles per hour or dollars per hour, and slope.

   b. **Describe** how changing the rate affects the slope of the graph.

   c. **Include** some examples and sketches of graphs.

Guided Practice

Examples 1 & 2

Graph each pair of equations. Describe any similarities or differences and explain why they are a family of graphs.

4. $y = 3x + 3$

   $y = 3x - 2$

5. $y = \frac{1}{2}x - 4$

   $y = \frac{3}{2}x - 4$

Example 3

Compare and contrast the graphs of each pair of equations. Verify by graphing the equations.

6. $2x + 1 = y$
   $2x = y$

7. $y = -\frac{1}{4}x + 3$

   $y = -x + 3$

Examples 4 & 5

Change $y = x - 3$ so that the graph of the new equation fits each description.

8. same y-intercept, steeper positive slope

9. same slope, shifted up 3 units

Example 2

10. **Pet Care** Refer to Example 3. Suppose Juan and Matthew both agree to charge $3 an hour, but Juan wants to charge an additional $5 fee per visit. Then $y = 3x + 5$ represents how much Juan would charge.

    a. Graph $y = 3x$ and $y = 3x + 5$.

    b. Explain how the graphs are similar and how they are different.

    c. What does the slope represent?

Practice

Homework Help

For Exercises	See Examples
11–16	1, 2
17–22, 32	3
23–31	4, 5

Extra Practice
See page 707.

Graph each pair of equations. Describe any similarities or differences. Explain why they are a family of graphs.

11. $y = 5x$
$y = 2x$

12. $y = 3x + 4$
$y = x + 4$

13. $y = 4x$
$y = 4x - 2$

14. $-\frac{2}{3}x + 1 = y$
$\frac{1}{4}x + 1 = y$

15. $y = -\frac{1}{2}x$
$y = -3x$

16. $y = -2x + 2$
$y = -2x - 3$

Compare and contrast the graphs of each pair of equations. Verify by graphing the equations.

17. $y = -x$
$y = x$

18. $\frac{1}{5}x = y$
$5x = y$

19. $y = 3x + 1$
$y = 3x - 8$

20. $y = -\frac{1}{3}x + 4$
$y = -\frac{1}{3}x$

21. $y = -2x - 2$
$y = -\frac{3}{4}x - 2$

22. $y = \frac{1}{2}x - 1$
$x - 3 = 2y$

Change $y = -\frac{4}{5}x + 6$ so that the graph of the new equation fits each description.

23. y-intercept is 0, same slope

24. positive slope, same y-intercept

25. shifted down 4 units, same slope

26. shifted up 2 units, same slope

27. steeper negative slope, same y-intercept

28. less steep negative slope, same y-intercept

29. Write an equation in slope-intercept form of a line passing through points at $(-3, -5)$ and $(3, 7)$. Then write an equation that has the same slope but a different y-intercept.

30. Write an equation of a line whose graph lies between the graphs of $y = 3x + 4$ and $y = 3x + 2$.

Applications and Problem Solving

31. Shipping Most cargo is transported by barge, train, or truck. The graph represents the distances traveled for the same amount of fuel.

    a. Are these graphs a family of graphs? Explain.

    b. What does the slope represent?

    c. Which form of transportation gets the most miles per gallon? Explain.

32. Population

> The U.S. Latino population ... is growing at a rate five times faster than that of the general population, according to the Census Bureau.

Source: *USA TODAY*

Compare and contrast the graphs that represent the growth of the Latino population and the growth of the general population.

33. Critical Thinking Given $A(0, 7)$, $B(2, -3)$, $C(-4, 6)$, and $D(0, 7)$, determine whether $\overline{AB}$ and $\overline{CD}$ are a family of graphs. Explain how you know.

Mixed Review

Determine the slope and y-intercept of the graph of each equation. *(Lesson 7–5)*

34. $y = 5x + 1$ **35.** $y = -2x$ **36.** $y = \frac{3}{4}x - 7$

37. Jewelry Necklace charms can be gold or silver, boy-shaped or girl-shaped, and contain a different birthstone for each month of the year. How many different charms are possible? *(Lesson 4–2)*

Standardized Test Practice

38. Short Response Find $-14 + (-3)$. *(Lesson 3–2)*

39. Multiple Choice What is the property shown by $5(x + 2) = 5x + 10$? *(Lesson 1–4)*

  (A) Commutative ($\times$)   (B) Identity

  (C) Distributive   (D) Associative ($+$)

Quiz 2   Lessons 7–4 through 7–6

1. Animals Determine whether a scatter plot showing the temperature of a glass of water at a given air temperature has a *positive* relationship, *negative* relationship, or *no* relationship. If there is a relationship, describe it. *(Lesson 7–4)*

Determine the x-intercept and y-intercept of the graph of each equation. Then graph the equation. *(Lesson 7–5)*

2. $2x + y = 4$ **3.** $x + 4y = 4$ **4.** $2x - 3y = -12$

5. Videos The graph shows how much two different video stores pay for movies. *(Lesson 7–6)*

  a. Describe any similarities or differences and explain why they are a family of graphs.

  b. Explain what the slope represents.

  c. Which store pays less per movie?

The graphs of the equations shown at the right are a family of graphs because they have the same slope. Because $4x$ is never equal to $4x + 5$, the value of y will never be the same for any given value of x, and the graphs will never intersect. These lines are **parallel**.

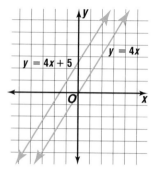

Parallel Lines	**Words:** If two lines have the same slope, then they are parallel.	
	Model:	**Symbols:** $\ell \parallel m$

All vertical lines are parallel.

Example ① **Determine whether the graphs of the equations are parallel.**

$$y = -\frac{3}{4}x - 2$$

$$4y = -3x + 12$$

First, determine the slopes of the lines. Write each equation in slope-intercept form.

$y = -\frac{3}{4}x - 2$ *Slope-Intercept Form*	$4y = -3x + 12$
	$\dfrac{4y}{4} = \dfrac{-3x + 12}{4}$ *Divide each side by 4.*
The slope is $-\frac{3}{4}$.	$y = -\frac{3}{4}x + 3$ The slope is $-\frac{3}{4}$.

The slopes are the same, so the lines are parallel. *Check by graphing.*

Your Turn

a. $y = 2x$
$7 = 2x - y$

b. $y = -3x + 3$
$2y = 6x - 5$

A *parallelogram* is a four-sided figure with two sets of parallel sides.

Example

2

Geometry Link

Determine whether figure $ABCD$ is a parallelogram.

Explore To be a parallelogram, $\overline{AB}$ and $\overline{DC}$ must be parallel. Also, $\overline{AD}$ and $\overline{BC}$ must be parallel.

Plan Find the slope of each side.

Solve $\overline{AB}$: $m = \dfrac{-1-2}{1-(-2)} = \dfrac{-3}{3}$ or -1

$\overline{BC}$: $m = \dfrac{-1-2}{1-7} = \dfrac{-3}{-6}$ or $\dfrac{1}{2}$

$\overline{DC}$: $m = \dfrac{2-5}{7-4} = \dfrac{-3}{3}$ or -1

$\overline{AD}$: $m = \dfrac{5-2}{4-(-2)} = \dfrac{3}{6}$ or $\dfrac{1}{2}$

Prerequisite Skills Review
Simplifying Fractions, p. 685

$\overline{AB}$ is parallel to $\overline{DC}$ because their slopes are both -1.
$\overline{BC}$ is parallel to $\overline{AD}$ because their slopes are both $\dfrac{1}{2}$.
Therefore, $ABCD$ is a parallelogram.

Examine A parallelogram has opposite sides that are equal in length. Use a ruler to check that this appears to be true of $ABCD$.

You can use the slope of a line to write an equation of a line that is parallel to it.

Example

3

Write an equation in slope-intercept form of the line that is parallel to the graph of $y = -4x + 8$ and passes through $(1, 3)$.

The slope of the given line is -4. So, the slope of the new line will also be -4. Find the new equation by using the point-slope form.

$y - y_1 = m(x - x_1)$ *Point-Slope Form*
$y - 3 = -4(x - 1)$ *Replace (x_1, y_1) with $(1, 3)$ and m with -4.*
$y - 3 = -4x + 4$ *Distributive Property*
$y - 3 + 3 = -4x + 4 + 3$ *Add 3 to each side.*
$y = -4x + 7$ *Simplify.*

An equation whose graph is parallel to the graph of $4x + y = 8$ and passes through $(1, 3)$ is $y = -4x + 7$. *Check by substituting $(1, 3)$ into $y = -4x + 7$ or by graphing.*

Your Turn

Write an equation in slope-intercept form of the line that is parallel to the graph of each equation and passes through the given point.

c. $y = 6x - 4$; $(2, 3)$ **d.** $3x + 2y = 9$; $(2, 0)$

Materials: grid paper protractor

Step 1 Draw line ℓ through the origin and $P(5, 3)$.

Step 2 Use a protractor to rotate the line 90°. Label the new line ℓ'.

Try These

1. What are the slopes of ℓ and ℓ'?

2. Compare the slopes. What is their product?

The results of the Hands-On Algebra activity lead to the following definition of **perpendicular lines**.

Perpendicular Lines	**Words:** If the product of the slopes of two lines is -1, then the lines are perpendicular. **Model:** **Symbols:** $\ell \perp m$

In a plane, vertical lines are perpendicular to horizontal lines.

Example ④ **Determine whether the graphs of the equations are perpendicular.**

$y = \frac{2}{3}x + 1$

$y = -\frac{3}{2}x + 2$

The graphs are perpendicular because the product of their slopes is $\frac{2}{3} \cdot \left(-\frac{3}{2}\right)$ or -1.

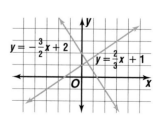

Your Turn

e. $y = \frac{1}{5}x + 2$

   $y = 5x + 1$

f. $y = -4x + 3$

   $4y = x - 5$

Write an equation in slope-intercept form of the line that is perpendicular to the graph of $y = \frac{1}{3}x - 2$ and passes through $(-4, 2)$.

The slope is $\frac{1}{3}$. A line perpendicular to the graph of $y = \frac{1}{3}x - 2$ has slope -3. Find the new equation by using the point-slope form.

$$
\begin{aligned}
y - y_1 &= m(x - x_1) && \textit{Point-Slope Form} \\
y - 2 &= -3[x - (-4)] && \textit{Replace } (x_1, y_1) \textit{ with } (-2, 4) \textit{ and } m \textit{ with } -3. \\
y - 2 &= -3x - 12 && \textit{Distributive Property} \\
y - 2 + 2 &= -3x - 12 + 2 && \textit{Add 2 to each side.} \\
y &= -3x - 10
\end{aligned}
$$

The new equation is $y = -3x - 10$. *Check by substituting $(-4, 2)$ into the equation or by graphing.*

Your Turn Write an equation in slope-intercept form of the line that is perpendicular to the graph of each equation and passes through the given point.

g. $y = 2x + 6; (0, 0)$ **h.** $2x + 3y = 2; (3, 0)$

 Personal Tutor at algconcepts.com

Check for Understanding

Communicating Mathematics

1. **Compare and contrast** the slopes of lines that are parallel and the slopes of lines that are perpendicular.

2. **Choose** the graph that is parallel to the graph of $3x + 4y = -12$. Explain.

a.
b.
c.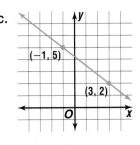

Guided Practice

Getting Ready State the slopes of the lines parallel to and perpendicular to the graph of each equation.

Sample: $y = -3x + 6$

Solution: The slope is -3. Lines parallel to the graph have a slope of -3. Lines perpendicular to the graph have a slope of $\frac{1}{3}$.

3. $y = 4x + 2$ **4.** $5x + y = 8$ **5.** $2x - 3y = 7$

Examples 1 & 4 **Determine whether the graphs of each pair of equations are *parallel*, *perpendicular*, or *neither*.**

6. $y = 4x - 9$
   $y = 4x - 11$

7. $y = -\dfrac{3}{2}x + 5$
   $y = 2x + 1$

8. $y = 5x + 3$
   $x + 5y = 2$

Example 4 **9. Geometry** Determine $m\angle STQ$ if the slope of $\overline{RQ}$ is $\dfrac{2}{5}$ and the slope of $\overline{SP}$ is $-\dfrac{5}{2}$. Explain.

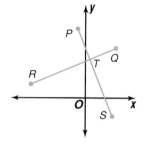

Exercise 9

Example 3 **Write an equation in slope-intercept form of the line that is parallel to the graph of each equation and passes through the given point.**

10. $y = x + 5$; $(7, 2)$

11. $y - \dfrac{3}{4}x = 4$; $(2, 0)$

Example 5 **Write an equation in slope-intercept form of the line that is perpendicular to the graph of each equation and passes through the given point.**

12. $y = 2x + 3$; $(3, -4)$

13. $3x + 8y = 4$; $(0, 4)$

Bicycling near San Francisco

14. Bicycling Joey and Cortez are both riding their bikes at 12 miles per hour. However, Cortez started 5 miles ahead of Joey. The equations $y = 12x$ and $y = 12x + 5$ represent the positions of Joey and Cortez, respectively.
Example 1

a. If they ride for 3 hours, will Joey catch up to Cortez? Explain your answer in terms of slopes of the graphs of the equations.

b. Are the graphs a family of graphs? Explain.

Exercises

• • • •   • • • • • • • • • • • •

Practice

Determine whether the graphs of each pair of equations are *parallel*, *perpendicular*, or *neither*.

Homework Help	
For Exercises	See Examples
15–23, 39	1, 4
24–29, 36	3
30–35, 37	5
36–37, 40	2
Extra Practice	
See page 707.	

15. $y = \dfrac{3}{5}x + 2$
   $y = -\dfrac{5}{3}x - 1$

16. $y = 5x + 4$
   $y = 5x + 8$

17. $x + 3y = 7$
   $y = \dfrac{1}{3}x - 9$

18. $y = -2x + 6$
   $y = 2x$

19. $y = 2x + 3$
   $x + 2y = 4$

20. $y = \dfrac{2}{3}x$
   $3x + 2y = 0$

21. $3y = 8$
   $y = 3$

22. $-\dfrac{1}{4}x + 5 = 0$
   $2y - 9 = 0$

23. $7x + 3y = 4$
   $3x - 7y = 1$

Write an equation in slope-intercept form of the line that is parallel to the graph of each equation and passes through the given point.

24. $y = 4x + 5; (2, -3)$ **25.** $y = -\frac{1}{2}x - 3; (0, 0)$ **26.** $y = 4; (2, 5)$

27. $4x + y = 1; (-2, 1)$ **28.** $3x + y = 3; (3, 5)$ **29.** $x = 3; (2, 6)$

Write an equation in slope-intercept form of the line that is perpendicular to the graph of each equation and passes through the given point.

30. $y = 2x + 3; (3, -4)$ **31.** $y = -4x + 5; (1, 1)$ **32.** $2x - 5y = 3; (-2, 7)$

33. $y = 2x; (2, 0)$ **34.** $x = 6; (4, 2)$ **35.** $3x - 5y = 2; (-1, 0)$

Determine whether $\overline{AB}$ and $\overline{CD}$ are *parallel*, *perpendicular*, or *neither*.

36. $A(4, 3), B(5, 2), C(8, 3), D(9, 2)$

37. $A(-2, 3), B(4, 0), C(2, 5), D(-1, 11)$

38. Write an equation in slope-intercept form of each line.

    a. passes through the point at $(4, -2)$ and is parallel to the graph of $5x - 2y = 6$

    b. perpendicular to the line through points at $(1, 2)$ and $(8, 6)$ and passes through the point at $(0, 5)$

Applications and Problem Solving

39. Surveying Determine whether roads $\overleftrightarrow{AB}$ and $\overleftrightarrow{BC}$ are perpendicular in the survey at the right. Explain.

Exercise 39

40. Geometry Determine whether quadrilateral *JKLM* is a rectangle if its vertices are $J(0, 5)$, $K(9, 2)$, $L(7, -4)$, and $M(-2, -1)$. Explain how you know.

41. Critical Thinking Suppose the line through points at $(-2, 2)$ and $(x, 5)$ is parallel to the graph of $2x - 4y = 9$. What is x?

Mixed Review

Compare and contrast the graphs of each pair of equations. Verify by graphing the equations. *(Lesson 7–6)*

42. $y = 4x$
    $y = 4x - 3$

43. $y = -3x + 2$
    $y = 3x + 2$

44. $y = x + 2$
    $7y = -2x + 14$

A card is drawn from a standard deck of cards. Determine whether each event is *mutually exclusive* or *inclusive*. *(Lesson 5–7)*

45. P(queen or ace)   **46.** P(8 or red card)   **47.** P(heart or spade)

48. Solve $6 = q - 5$. *(Lesson 3–5)*

Standardized Test Practice
Ⓐ Ⓑ Ⓒ Ⓓ

49. Grid In Solve $g = 15.6 - 2.75$. *(Lesson 3–4)*

50. Multiple Choice In which quadrant does the graph of $G(-15, 4)$ lie? *(Lesson 2–2)*

    Ⓐ I   Ⓑ II   Ⓒ III   Ⓓ IV

Understanding and Using the Vocabulary

After completing this chapter, you should be able to define each term, property, or phrase and give an example or two of each.

*inter***NET**
CONNECTION **Review Activities**
For more review activities, visit:
www.algconcepts.com

best-fit line (*p. 308*)
correlation coefficient (*p. 309*)
extrapolation (*p. 309*)
family of graphs (*p. 316*)
interpolation (*p. 309*)
linear regression (*p. 309*)

median-median line (*p. 309*)
parallel lines (*p. 322*)
parent graph (*p. 318*)
perpendicular lines (*p. 324*)
point-slope form (*p. 290*)
rate of change (*p. 285*)

residual (*p. 309*)
scatter plot (*p. 302*)
slope (*p. 284*)
slope-intercept form (*p. 296*)
x-intercept (*p. 296*)
y-intercept (*p. 296*)

State whether each sentence is *true* or *false*. If false, replace the underlined word(s) to make a true statement.

1. In the equation $y = mx + b$, b is the <u>x-intercept</u> of the line.
2. A line that appears to go downhill from left to right has a <u>negative</u> slope.
3. <u>Perpendicular</u> lines have the same slopes.
4. The graph of a <u>linear equation</u> is a straight line.
5. You can write an equation of a line if you know the slope and a <u>point on the line</u>.
6. The <u>point-slope form</u> of a linear equation is $y - y_1 = m(x - x_1)$.
7. The graph of a line having a slope of 0 is a <u>vertical</u> line.
8. To determine the y-intercept of the graph of an equation, let <u>y</u> = 0 and solve.
9. A <u>scatter plot</u> is a graph in which data are plotted as ordered pairs.
10. Data points that appear to go uphill in a scatter plot show a <u>positive</u> relationship.

Skills and Concepts

Objectives and Examples

• **Lesson 7–1**   Find the slope of a line given the coordinates of two points on the line.

Determine the slope of the line through points at $(-3, 1)$ and $(7, 4)$.

$$m = \frac{y_2 - y_1}{x_2 - x_1} \quad \text{\textit{Slope formula}}$$

$$= \frac{4 - 1}{7 - (-3)} \quad \begin{array}{l}(x_1, y_1) = (-3, 1),\\ (x_2, y_2) = (7, 4)\end{array}$$

$$= \frac{3}{10} \quad \text{\textit{Simplify.}}$$

The slope is $\frac{3}{10}$.

Review Exercises

Determine the slope of each line.

11.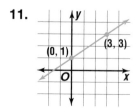

12.

13. the line through $(6, 0)$ and $(2, 8)$
14. the line through $(-4, 9)$ and $(-4, 5)$

Objectives and Examples

- **Lesson 7–2** Write a linear equation in point-slope form given the coordinates of a point on the line and the slope of the line.

$$y - y_1 = m(x - x_1) \quad \textit{Point-Slope Form}$$
$$y - 8 = 2(x - 1) \quad (x_1, y_1) = (1, 8) \text{ and } m = 2$$

- **Lesson 7–3** Write a linear equation in slope-intercept form given the slope and y-intercept.

$$y = mx + b \quad \textit{Slope-Intercept Form}$$
$$= 4x + 12 \quad m = 4 \text{ and } b = 12$$

slope   y-intercept

- **Lesson 7–4** Graph and interpret points on scatter plots.

Height (cm)   170   160

Forearm Length (cm)

There is a positive relationship between forearm length and height of students. As forearm length increases, height increases.

- **Lesson 7–5** Graph linear equations by using the x- and y-intercepts or the slope and y-intercept.

slope $= \dfrac{2}{3}$

y-intercept $= 2$

(0, 2)   2   3

Review Exercises

Write the point-slope form of an equation for each line passing through the given point and having the given slope.

15. $(2, 5)$, $m = 3$ **16.** $(-1, 6)$, $m = \dfrac{1}{4}$

17. $(0, 3)$, $m = -5$ **18.** $(4, 8)$, $m = 0$

19. Write an equation in point-slope form of a line passing through $(10, 4)$ and $(-8, 7)$.

Write an equation in slope-intercept form of the line with each slope and y-intercept.

20. $m = 1$, $b = 4$ **21.** $m = 3$, $b = -9$

22. $m = \dfrac{1}{2}$, $b = 5$ **23.** $m = 2$, $b = 0$

24. $m = 0$, $b = 7$ **25.** $m = 11$, $b = -6$

Determine whether each scatter plot has a positive relationship, negative relationship, or no relationship. If there is a relationship, describe it.

26.

Grades

Watching TV

27.

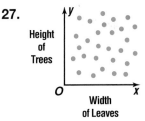

Height of Trees

Width of Leaves

Determine the x-intercept and y-intercept of the graph of each equation. Then graph.

28. $x + 3y = 6$ **29.** $2x - y = 4$

Determine the slope and y-intercept of the graph of each equation. Then graph.

30. $y = x + 5$ **31.** $y = -2x + 3$

Mixed Problem Solving
See pages 724–731.

Objectives and Examples

Review Exercises

• **Lesson 7–6** Explore the effects of changing the slopes and y-intercepts of linear functions.

The graphs have the same y-intercept and different negative slopes. They are a family of graphs.

Compare and contrast the graphs of each pair of equations. Verify by graphing the equations.

32. $y = x + 4$
$y = 3x + 4$

33. $y = \frac{1}{3}x - 1$
$y = x - 1$

34. $y = 2x$
$y = 2x + 5$

35. $\frac{1}{2}x + 3 = y$
$\frac{1}{2}x - 4 = y$

• **Lesson 7–7** Write an equation of a line that is parallel or perpendicular to the graph of a given equation and that passes through a given point.

ℓ is parallel to a because their slopes are each $\frac{1}{2}$.

ℓ is perpendicular to b because the product of their slopes is -1.

Determine whether the graphs of each pair of equations are *parallel, perpendicular,* or *neither.*

36. $y = 2x + 8$
$y = 2x - 4$

37. $y = -3x$
$y = 3x + 1$

38. $y = \frac{1}{7}x - 1$
$y = -7x - 1$

39. $\frac{3}{4}x - 5 = y$
$y = \frac{3}{4}x + 6$

Applications and Problem Solving

40. Building Paved areas are usually slightly inclined so that puddles do not form. The pavement at the right rises 1 inch for every 25 inches of horizontal change. What is the slope? *(Lesson 7–1)*

41. Employment The table shows earnings for Employees A and B. *(Lesson 7–6)*

a. Sketch a graph of each employee's earnings by plotting points with the coordinates (hours, earnings).

b. Suppose Employee C earns more money per hour than Employee A and less money per hour than Employee B. How would a graph of this employee's earnings compare with the graphs in part a?

Number of Hours	Earnings ($)	
	A	B
1	6	11
2	12	22
3	18	33
4	24	44
5	30	55

1. **Describe** how changing each of the following affects the graph of a linear function.
   a. slope   **b.** y-intercept

2. **Compare and contrast** the graphs of $y = 4x - 1$, $y = x - 1$, and $y = 3x - 1$. Verify by graphing the equations.

Determine the slope of each line.

3.

4.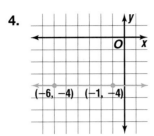

5. the line passing through points at $(1, -2)$ and $(6, 0)$

Write the point-slope form of an equation for each line passing through the given point and having the given slope.

6. $(5, 6)$, $m = 3$ 7. $(-3, 1)$, $m = -2$ 8. $(-4, 8)$, $m = 0$

Write an equation in slope-intercept form of the line passing through each pair of points.

9. $(1, 5)$ and $(2, 8)$ 10. $(3, 1)$ and $(-7, 11)$ 11. $(-4, 0)$ and $(2, 3)$

12. **Music**   A disc jockey notices that as the music gets faster, more people start dancing. Would a scatter plot showing speed of music and number of dancers have a *positive* relationship, *negative* relationship, or *no* relationship?

Graph each equation.

13. $x + y = 2$ 14. $x + 3y = 3$ 15. $2x - 4y = 12$

16. $y = 2x - 5$ 17. $y = \frac{1}{3}x + 2$ 18. $y = -1$

19. Are the graphs of the equations *parallel*, *perpendicular*, or *neither*?
    $y = -4x + 9$
    $y = \frac{1}{4}x - 6$

20. **Construction**   The *pitch* of a roof describes its steepness. Suppose a roof is 40 feet wide and its pitch is $\frac{1}{2}$. Find x, its height above the rafters at its peak. ($Hint$: Use $\frac{\text{rise}}{\text{run}}$.)

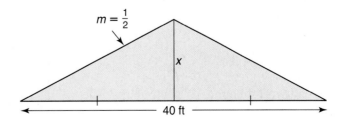

Algebra Word Problems

All standardized tests involve writing and solving equations from realistic settings. You'll need to translate words into equations. You'll need to find patterns in number tables and write equations to represent the patterns.

Example 1

Ami dropped a rubber ball from several heights and measured how high the ball bounced. Her results are shown in the table. How high would the ball bounce if it were dropped from a height of 36 centimeters?

Height of Drop (cm)	Bounce Height (cm)
40	30
60	45
80	60
100	75

Ⓐ 19 cm   Ⓑ 24 cm   Ⓒ 27 cm   Ⓓ 30 cm

Hint Look for a pattern in the table.

Solution The bounce is 30 when the drop height is 40. The bounce is 75 when the drop height is 100. So, the bounce is $\frac{3}{4}$ of the drop height. Check that this pattern is true for each of the other numbers in the table.

Let b represent the bounce height and let d represent the drop height.

$b = \frac{3}{4}d$ *Write an equation.*

$= \frac{3}{4}(36)$ *Replace d with 36.*

$= \frac{108}{4}$ or 27   *Simplify.*

The ball would bounce 27 centimeters. The answer is C.

Example 2

After N chocolate bars are divided equally among 6 children, 3 bars remain. How many would remain if $(N + 4)$ chocolate bars were divided equally among 6 children?

Ⓐ 0   Ⓑ 1   Ⓒ 2   Ⓓ 3   Ⓔ 4

Hint Sketch a diagram to understand the problem. Use the "plug-in" strategy.

Solution One way to solve this kind of problem is to choose a numeric value for N and "plug it in." Suppose each of the 6 children got 2 chocolate bars, and 3 bars remain. Then $N = 6(2) + 3 = 15$.

Now, use the value of 15 for N.

$$N + 4 = 15 + 4 \text{ or } 19 \text{ bars}$$

Divide 19 bars among 6 children.

$$19 = 18 + 1 \text{ or } 3(6) + 1$$

The remainder is 1. The answer is B.

After you work each problem, record your answer on the answer sheet provided or on a sheet of paper.

Multiple Choice

1. Refer to the table and the equation in Example 1 on the opposite page. From what height should Ami drop the ball so that the bounce height is 105 centimeters? *(Lesson 6–1)*
   - Ⓐ 110 cm
   - Ⓑ 140 cm
   - Ⓒ 180 cm
   - Ⓓ 200 cm

2. A cable TV company charges $21.95 per month for basic service. Each premium channel selected costs an additional $5.95 per month. If x represents the number of premium channels selected, which expression can be used to find the monthly cost of cable service? *(Lesson 1–1)*
   - Ⓐ $5.95 + 21.95x$
   - Ⓑ $21.95 + 5.95x$
   - Ⓒ $21.95 + 5.95 + x$
   - Ⓓ $21.95 - 5.95x$

3. Steve ran a 12-mile race at an average speed of 8 miles per hour. If Adam ran the same race at an average speed of 6 miles per hour, how many minutes longer than Steve did Adam take to complete the race? *(Lesson 6–6)*
   - Ⓐ 9
   - Ⓑ 12
   - Ⓒ 16
   - Ⓓ 24
   - Ⓔ 30

4. The bar graph shows the number of people who watch prime time television each day of the week. About how many more people watch on Monday than on Saturday? *(Lesson 1–6)*

   - Ⓐ 5 million
   - Ⓑ 15 million
   - Ⓒ 5 million
   - Ⓓ 50 million

5. Kim earns $7 per hour baby-sitting. If x represents the number of hours Kim baby-sits, which expression could be used to find the amount she earns? *(Lesson 1–1)*
   - Ⓐ $x + 7$
   - Ⓑ $x + 5$
   - Ⓒ $7x$
   - Ⓓ $5x$

6. If Nathan is $\frac{1}{4}$ as old as his father and the sum of their ages is 60, then how old is Nathan? *(Lesson 4–5)*
   - Ⓐ 8
   - Ⓑ 12
   - Ⓒ 15
   - Ⓓ 20
   - Ⓔ 48

7. There are 5 different pizza toppings. How many different 3-topping pizzas are possible? *(Lesson 4–2)*
   - Ⓐ 10
   - Ⓑ 15
   - Ⓒ 60
   - Ⓓ 125

8. A group of 5 adults and 3 children sees a play. A child's ticket costs $6.25, and an adult's ticket costs $9.75. Which equation can be used to find c, the amount of change from $100 after paying for the group's tickets? *(Lesson 1–1)*
   - Ⓐ $5(6.25) + 3(9.75) = 100 - c$
   - Ⓑ $5(9.75) + 3(6.25) = c$
   - Ⓒ $5(9.75) + 3(6.25) + c = 100$
   - Ⓓ $5(9.75) + 3(6.25) = 100 + c$

Grid In

9. What number increased by two is equal to two less than twice the number? *(Lesson 4–6)*

Extended Response

10. The fare charged by a taxi driver is a $3 fixed charge plus $0.35 per mile. Beth pays $10 for a ride of m miles. *(Lesson 7–3)*

    Part A Write an equation that can be used to find m. Show your work.

    Part B Use the equation in Part A to find how many miles Beth rode. Show your work.

CHAPTER

8

Powers and Roots

What You'll Learn

Key Ideas

- Use, multiply, and divide powers. *(Lessons 8–1 and 8–2)*

- Simplify expressions containing negative exponents. *(Lesson 8–3)*

- Express numbers in scientific notation. *(Lesson 8–4)*

- Simplify radicals. *(Lesson 8–5)*

- Estimate square roots. *(Lesson 8–6)*

- Use the Pythagorean Theorem. *(Lesson 8–7)*

Key Vocabulary

exponent *(p. 336)*

irrational numbers *(p. 362)*

power *(p. 336)*

radical expression *(p. 358)*

square root *(p. 357)*

Why It's Important

Firefighting Every year, fires take thousands of lives and destroy property worth billions. Firefighters help protect us from these dangers. They are frequently the first emergency personnel at a traffic accident or medical emergency to put out a fire, treat injuries, or perform other vital functions.

Functions involving powers and roots are used to describe scientific concepts like motion. You will use a square root function to find the velocity of water in a fire hose in Lesson 8–5.

✓ Check Your Readiness

✓ **Lesson 2-5,**
   pp. 75-79

Find each product.

1. $6 \cdot 6$

2. $10 \cdot 10 \cdot 10$

3. $(-3)(-3)$

4. $4 \cdot 4 \cdot 4$

5. $5 \cdot 5 \cdot 5 \cdot 5$

6. $(-2)(-2)(-2)(-2)(-2)$

7. $(-3)(-3)(5 \cdot 5)$

8. $-4(6 \cdot 6)$

9. $3(-4)(2)(2)$

✓ **Prerequisite**
   Skill, p. 685

Write each fraction in simplest form.

10. $\dfrac{2}{6}$

11. $\dfrac{4}{64}$

12. $\dfrac{24}{32}$

13. $\dfrac{24}{4}$

14. $\dfrac{15}{10}$

15. $\dfrac{9}{72}$

16. $\dfrac{12}{28}$

17. $\dfrac{27}{9}$

18. $\dfrac{12}{60}$

✓ **Prerequisite**
   Skill, p. 684

Find each product or quotient.

19. $2.5 \cdot 3$

20. $1.6 \cdot 4$

21. $5 \cdot 1.2$

22. $7.8 \cdot 1.1$

23. $9.4 \cdot 3.2$

24. $\dfrac{4.8}{1.6}$

25. $\dfrac{10.8}{2.4}$

26. $\dfrac{3.6}{0.6}$

27. $\dfrac{9.2}{2}$

FOLDABLES™
Study Organizer

Make this Foldable to help you organize your Chapter 8 notes. Begin with four sheets of grid paper.

❶ Fold each sheet in half along the width. Then cut along the crease.

❷ Staple the eight half-sheets together to form a booklet.

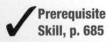

❸ Cut seven lines from the bottom of the top sheet, six lines from the second sheet, and so on.

❹ Label the tabs with lesson topics as shown.

Reading and Writing As you read and study the chapter, use each page to write notes and graph examples for each lesson.

What You'll Learn
You'll learn to use powers in expressions.

Why It's Important
Landscaping
Landscape architects use the formula $A = \pi r^2$ to find the area of a circular region. The 2 in the formula is an exponent. See Exercise 43.

Perfect squares like 1, 4, 9, and 16 can be represented by a square array of dots.

| 1 | 4 | 9 | 16 |

A perfect square is the product of a number and itself. For example, 16 is a perfect square because $16 = 4 \times 4$. The expression 4×4 can be written using exponents. An **exponent** tells how many times a number, called the **base**, is used as a factor. Numbers that are expressed using exponents are called **powers**. The expression 4×4 can be written as 4^2.

$$base \to 4^2 \leftarrow exponent$$

Symbols	Words	Meaning
4^1	4 to the first power	4
4^2	4 to the second power or 4 squared	$4 \cdot 4$
4^3	4 to the third power or 4 cubed	$4 \cdot 4 \cdot 4$
4^4	4 to the fourth power	$4 \cdot 4 \cdot 4 \cdot 4$
4^n	4 to the nth power	$4 \cdot 4 \cdot 4 \cdot \ldots \cdot 4$
		n factors

Examples

Write each expression using exponents.

1 $2 \cdot 2 \cdot 2 \cdot 2 \cdot 2$

The base is 2. It is a factor 5 times.

$2 \cdot 2 \cdot 2 \cdot 2 \cdot 2 = 2^5$

2 $m \cdot m \cdot m \cdot m$

The base is m. It is a factor 4 times..

$m \cdot m \cdot m \cdot m = m^4$

Reading Algebra

When no exponent is shown, it is understood to be 1. For example, $10 = 10^1$.

3 7

The base is 7. It is a factor 1 time.

$7 = 7^1$

Your Turn

a. $4 \cdot 4 \cdot 4 \cdot 4$ **b.** $x \cdot x \cdot x$ **c.** 10

Write (2)(2)(2)(−5)(−5) using exponents.

Use the Associative Property to group the factors with like bases.

$(2)(2)(2)(-5)(-5) = [(2)(2)(2)][(-5)(-5)]$
$= (2)^3(-5)^2$

Your Turn

d. Write $(-1)(-1)(-1)(-1)(3)(3)$ using exponents.

You can use the definition of exponent to write a power as a multiplication expression.

Examples

Write each power as a multiplication expression.

⑤ 10^2

The base is 10. The exponent 2 means that 10 is a factor 2 times.
$10^2 = 10 \cdot 10$

⑥ b^3

The base is b. The exponent 3 means that b is a factor 3 times.
$b^3 = b \cdot b \cdot b$

⑦ $10x^2y^4$

10 is used as a factor once, x is used twice, and y is used 4 times.
$10x^2y^4 = 10 \cdot x \cdot x \cdot y \cdot y \cdot y \cdot y$

Your Turn

e. 2^3 **f.** y^4 **g.** $5a^3b^2$

You can also use the definition of exponent to evaluate expressions.

Real World

Example ―⑧

Science Link

The distance between the Earth and the Sun is about 10^8 kilometers. Write this number as a multiplication expression and then evaluate the expression.

$10^8 = 10 \cdot 10 \cdot 10 \cdot 10 \cdot 10 \cdot 10 \cdot 10 \cdot 10$
$= 100,000,000$

The distance between Earth and the sun is about 100 million kilometers.

When an expression contains an exponent, simplify the expression using the rules for order of operations.

Look Back

Order of
Operations:
Lesson 1–2

Order of Operations	1. Do all operations within grouping symbols first; start with the innermost grouping symbols. 2. Evaluate all powers in order from left to right. 3. Do all multiplications and divisions from left to right. 4. Do all additions and subtractions from left to right.

In any expression, an exponent goes with the number or the quantity in parentheses that immediately precedes it.

$2 \cdot 5^3$ means $2 \cdot 5 \cdot 5 \cdot 5$ *The exponent 3 goes with the 5.*

$(2 \cdot 5)^3$ means $(2 \cdot 5)(2 \cdot 5)(2 \cdot 5)$ *The exponent goes with $(2 \cdot 5)$.*

Examples

Evaluate each expression.

9 $4m^3$ if $m = 2$

$4m^3 = 4(2)^3$ *Replace m with 2.*
$\quad\;\; = 4(8)$ *Evaluate the power: $2 \cdot 2 \cdot 2 = 8$.*
$\quad\;\; = 32$ *Multiply.*

10 $3x + y^2$ if $x = -2$ and $y = -3$

$3x + y^2 = 3(-2) + (-3)^2$ *Replace x with -2 and y with -3.*
$\qquad\quad\; = 3(-2) + (9)$ *$(-3)^2 = (-3)(-3)$ or 9*
$\qquad\quad\; = (-6) + (9)$ *Multiply.*
$\qquad\quad\; = 3$ *Add.*

Your Turn

h. $3a^3$ if $a = -2$ **i.** $-5(m + n)^2$ if $m = 4$ and $n = 2$

 Personal Tutor at algconcepts.com

You can use exponents to find perimeters and areas of squares.

Graphing Calculator Exploration

Graphing Calculator Tutorial
See pp. 750–753.

The pattern at the right contains perfect squares.

Step 1 Draw the next three squares in the pattern.

Step 2 Find the perimeter and area of each square. Organize your data in a table.

Length of Side	Perimeter	Area

Try These

1. Write an equation that shows the relationship between the length of a side of a square x and its perimeter y.

2. Write an equation that shows the relationship between the length of a side of a square x and its area y.

3. Graph the two equations. Compare and contrast the graphs.

4. When is the value of the perimeter greater than the value of the area? When is the value of the perimeter equal to the area?

5. If the length of each side of a square is doubled, how does its perimeter change? How does its area change?

Check for Understanding

Communicating Mathematics

1. **Write** a definition of *perfect square*.

2. **Explain** what the 2 represents in 10^2.

3. **You Decide** Jan thinks that $(6n)^3$ is equal to $6n^3$. Becky thinks they are not equal. Who is correct? Explain your reasoning.

<table>
<tr><td>**Vocabulary**</td></tr>
<tr><td>perfect squares
exponent
base
powers</td></tr>
</table>

Guided Practice

Examples 1–3

Write each expression using exponents.

4. $9 \cdot 9 \cdot 9 \cdot 9$ 5. $a \cdot a \cdot a \cdot a \cdot a$ 6. 3

Examples 5–7

Write each power as a multiplication expression.

7. 12^4 8. x^5 9. $m^4 n^3$

Examples 8–10

Evaluate each expression if $a = 3$, $b = -2$, and $c = 4$.

10. c^3 11. $2a^4$ 12. $3a^2b$

Example 4

13. **Number Theory**   The prime factorization of 360 is $2 \cdot 2 \cdot 2 \cdot 3 \cdot 3 \cdot 5$. Write the prime factorization using exponents.

Exercises

Practice

Write each expression using exponents.

14. $10 \cdot 10 \cdot 10$ 15. $(-2)(-2)(-2)(-2)$ 16. 6

17. 7 cubed   18. $4 \cdot 4 \cdot 4 \cdot 6 \cdot 6$ 19. $2 \cdot 3 \cdot 5 \cdot 2 \cdot 3 \cdot 3$

20. $a \cdot a \cdot a \cdot a \cdot b \cdot b$ 21. $3 \cdot x \cdot x \cdot y \cdot y$ 22. $(-5)(m)(m)(m)(n)$

Homework Help

For Exercises	See Examples
14–22	1–4
23–30	5–7
31–41, 43	9, 10

Extra Practice
See page 707.

Write each power as a multiplication expression.

23. 3^5 24. $(-2)^2$ 25. $2^4 \cdot 3^2$ 26. $2 \cdot 3^5$

27. y^3 28. $x^2 y^2$ 29. $6ab^4$ 30. $-2y^4$

Evaluate each expression if $x = -2$, $y = 3$, $z = -1$, and $w = 0.5$.

31. x^5 32. $4y^2$ 33. $-2x^6$ 34. $x^2 - y^2$

35. $3(y^2 + z)$ 36. $-2(x^3 + 1)$ 37. $2w^2$ 38. wx^3y

39. Find the value of $x^2 + 2x + 1$ if $x = -3$.

40. Which is greater, 2^5 or 5^2?

Applications and Problem Solving

41. **Number Theory** The prime factorization of a number is $2 \cdot 3^5$. Find the number.

42. **Geometry** Use exponents to write an expression that represents the total number of unit cubes in the large cube. Then evaluate the expression.

Exercise 42

43. **Landscaping** Landscape architects use the formula $A = \pi r^2$ to find the area of circular flower beds. In the formula, $\pi \approx 3.14$, and r is the radius of the circle.

   a. Estimate the area of a circular flower bed with a radius of 8 feet.

   b. About how many bags of mulch will the landscape architect need to cover the bed if each bag covers about 10 square feet?

44. **Critical Thinking** Suppose you raise a negative integer to a positive power. When is the result negative? When is the result positive?

Mixed Review

45. Write an equation of the line that is parallel to the graph of $y = -2x + 3$ and passes through $(1, 4)$. *(Lesson 7–7)*

Graph each equation using the slope and y-intercept. *(Lesson 7–6)*

46. $y = 3x - 2$ 47. $y = x + 2$ 48. $-x + 2y = 8$

Solve each problem. *(Lesson 5–4)*

49. Find 26% of 120.   50. 17 is 40% of what number?

51. 9 is what percent of 18?   52. 98% of 40 is what number?

Standardized Test Practice
Ⓐ Ⓑ Ⓒ Ⓓ

53. **Multiple Choice** A mechanic charges an initial fee of $40 plus $30 for each hour she works. Which equation represents the cost c of a repair job that lasts h hours? *(Lesson 4–5)*

   Ⓐ $c = 30 + 40h$

   Ⓑ $c = 40 + 30h$

   Ⓒ $c = 40 - 30h$

   Ⓓ $c = 30 - 40h$

www.algconcepts.com/self_check_quiz

8-2 Multiplying and Dividing Powers

What You'll Learn
You'll learn to multiply and divide powers.

Why It's Important
Movie Industry
The intensity of sound is measured using *decibels*, a unit that is based on powers of ten. *See page 346.*

Powers can be multiplied and divided. In the example below, we will use powers of 2 to form a rule about multiplying exponents. The table shows several powers of 2 and their values.

Power of 2	2^1	2^2	2^3	2^4	2^5	2^6
Value	2	4	8	16	32	64

You can use the table to substitute exponents for the factors of multiplication equations. What do you notice about the exponents in the following products?

Numerical Products	$4 \cdot 2 = 8$	$4 \cdot 8 = 32$	$8 \cdot 8 = 64$
Products of Powers	$2^2 \cdot 2^1 = 2^3$	$2^2 \cdot 2^3 = 2^5$	$2^3 \cdot 2^3 = 2^6$

These examples suggest that you can multiply powers with the same base by adding the exponents. Think about $a^2 \cdot a^3$.

$a^2 \cdot a^3 = (a \cdot a)(a \cdot a \cdot a)$ *a^2 has two factors; a^3 has three factors.*

$ = a \cdot a \cdot a \cdot a \cdot a$ *Substitution Property*

$ = a^5$ *The product has 2 + 3 or 5 factors.*

	Words:	You can multiply powers with the same base by adding the exponents.
Product of Powers	**Numbers:**	$3^3 \cdot 3^2 = 3^{3 + 2}$ or 3^5
	Symbols:	$a^m \cdot a^n = a^{m + n}$

Examples

Reading Algebra

An expression is *simplified* when each base appears only once and all of the fractions are in simplest form.

Simplify each expression.

1 $4^3 \cdot 4^5$

$4^3 \cdot 4^5 = 4^{3 + 5}$ *To multiply powers that have the same base, write*
$ = 4^8$ *the common base, then add the exponents.*

2 $x^3 \cdot x^4$

$x^3 \cdot x^4 = x^{3 + 4}$ *To multiply powers that have the same base, write*
$ = x^7$ *the common base, then add the exponents.*

3 $(4y^2)(3y)$

$$(4y^2)(3y) = (4 \cdot 3)(y^2 \cdot y) \quad \textit{Use the Commutative and Associative Properties.}$$
$$= 12y^{2\,+\,1} \qquad\qquad y = y^1$$
$$= 12y^3 \qquad\qquad\quad \textit{Simplify.}$$

4 $(a^3b^2)(a^2b^4)$

$$(a^3b^2)(a^2b^4) = (a^3 \cdot a^2)(b^2 \cdot b^4) \quad \textit{Use the Commutative and Associative}$$
$$= a^{3\,+\,2} \cdot b^{2\,+\,4} \qquad\quad \textit{Properties.}$$
$$= a^5b^6 \qquad\qquad\qquad \textit{Simplify.}$$

Your Turn

a. $10^4 \cdot 10^2$ **b.** $y^4 \cdot y^2$ **c.** $(-3x^2)(5x)$ **d.** $(x^5y^2)(x^4y^6)$

You can use powers of 2 to help find a rule for dividing powers. Study each quotient in the table. What do you notice about the exponents?

Numerical Quotients	$16 \div 8 = 2$	$32 \div 4 = 8$	$64 \div 2 = 32$
Quotient of Powers	$2^4 \div 2^3 = 2^1$	$2^5 \div 2^2 = 2^3$	$2^6 \div 2^1 = 2^5$

These examples suggest that you can divide powers with the same base by subtracting the exponents. Think about $a^5 \div a^2$. Remember that you can write a division expression as a fraction.

$$\frac{a^5}{a^2} = \frac{a \cdot a \cdot a \cdot a \cdot a}{a \cdot a} \quad \textit{a^5 has five factors; a^2 has two factors.}$$

$$= \frac{\overset{1}{\cancel{a}} \cdot \overset{1}{\cancel{a}} \cdot a \cdot a \cdot a}{\underset{1}{\cancel{a}} \cdot \underset{1}{\cancel{a}}} \quad \textit{Notice that $\frac{a \cdot a}{a \cdot a} = 1$.}$$

$$= a \cdot a \cdot a \qquad\quad \textit{The quotient has $5 - 2$ or 3 factors.}$$

$$= a^3 \qquad\qquad\qquad \textit{Write using an exponent.}$$

Quotient of Powers	**Words:**	You can divide powers with the same base by subtracting the exponents.
	Numbers:	$\dfrac{5^7}{5^4} = 5^{7\,-\,4}$ or 5^3
	Symbols:	$\dfrac{a^m}{a^n} = a^{m\,-\,n}$ *The value of a cannot be zero.*

Simplify each expression.

5 $\dfrac{4^3}{4^2}$

$\dfrac{4^3}{4^2} = 4^{3\,-\,2}$ *To divide powers that have the same base, write the common base. Then subtract the exponents.*

$= 4^1$ or 4 *Simplify.*

6 $\dfrac{x^6}{x^4}$

$\dfrac{x^6}{x^4} = x^{6\,-\,4}$ *Write the common base. Then subtract the exponents.*

$= x^2$ *Simplify.*

7 $\dfrac{8m^4n^5}{2m^3n^2}$

$\dfrac{8m^4n^5}{2m^3n^2} = \left(\dfrac{8}{2}\right)\left(\dfrac{m^4}{m^3}\right)\left(\dfrac{n^5}{n^2}\right)$ *Group the powers that have the same base.*

$= 4m^{4\,-\,3}n^{5\,-\,2}$ *Divide.*

$= 4m^1n^3$ *Subtract the exponents.*

$= 4mn^3$ *Simplify.*

Your Turn

e. $\dfrac{10^5}{10^2}$ **f.** $\dfrac{y^5}{y^4}$ **g.** $\dfrac{a^4b^3}{ab^2}$ **h.** $\dfrac{-30m^5n^2}{10m^3n}$

 Personal Tutor at algconcepts.com

A special case results when you divide a power by itself. Consider the following two ways to simplify $\dfrac{b^4}{b^4}$, where $b \neq 0$.

Method 1   Definition of Power

$$\dfrac{b^4}{b^4} = \dfrac{\overset{1}{\cancel{b}}\cdot\overset{1}{\cancel{b}}\cdot\overset{1}{\cancel{b}}\cdot\overset{1}{\cancel{b}}}{\underset{1}{\cancel{b}}\cdot\underset{1}{\cancel{b}}\cdot\underset{1}{\cancel{b}}\cdot\underset{1}{\cancel{b}}}$$

$$= 1$$

Method 2   Quotient of Powers

$$\dfrac{b^4}{b^4} = b^{4\,-\,4}$$

$$= b^0$$

COncepts in MOtion
BrainPOP®
algconcepts.com

Since $\dfrac{b^4}{b^4}$ cannot have two different values, you can conclude that $b^0 = 1$. Therefore, any nonzero number raised to the zero power is equal to 1.

8 **Simplify** $\dfrac{x^4 y^2}{x y^2}$.

$$\dfrac{x^4 y^2}{x y^2} = \left(\dfrac{x^4}{x^1}\right)\left(\dfrac{y^2}{y^2}\right) \quad \textit{Group the powers that have the same base.}$$

$$= x^{4-1} y^{2-2} \quad \textit{Divide.}$$

$$= x^3 y^0 \quad y^0 = 1$$

$$= x^3 \cdot 1 \text{ or } x^3 \quad \textit{Simplify.}$$

Your Turn

i. $\dfrac{a^3 b^4}{a^3 b}$ 　　　　**j.** $\dfrac{10 x^4 y^3}{5 x^4 y^2}$ 　　　　**k.** $\dfrac{m^{10} n^5}{m^{10} n^5}$

Check for Understanding

Communicating Mathematics

1. **Explain** why $a^4 \cdot a^7$ can be simplified but $a^4 \cdot b^7$ cannot.

2. **You Decide?** Tonia says $10^3 \times 10^2 = 100^5$, but Emilio says $10^3 \times 10^2 = 10^5$. Who is correct? Explain your reasoning.

Guided Practice

Simplify each expression.

Examples 1 & 2

3. $y^2 \cdot y^5$ 　　　**4.** $m^5(m)$ 　　　**5.** $(t^2)(t^2)(t)$

Examples 3 & 4

6. $(x^3 y)(x y^3)$ 　　**7.** $(3a^2)(4a^3)$ 　　**8.** $(-5x^3)(4x^4)$

Examples 5 & 6

9. $\dfrac{n^8}{n^5}$ 　　　**10.** $\dfrac{b^6 c^5}{b^3 c^2}$ 　　　**11.** $\dfrac{x y^7}{y^4}$

Examples 7 & 8

12. $\dfrac{12 x^5}{4 x^4}$ 　　**13.** $\dfrac{a b^5 c}{a c}$ 　　**14.** $\dfrac{22 a^4 b^5 c^7}{-11 a b c^2}$

Example 1

15. Measurement There are 10^1 millimeters in 1 centimeter and 10^2 centimeters in 1 meter. How many millimeters are in 1 meter? Write your answer as a power and then evaluate the expression.

Exercises

Practice

Simplify each expression.

16. $2^6 \cdot 2^8$ 　**17.** $5^3 \cdot 5$ 　**18.** $y^7 \cdot y^7$ 　**19.** $d \cdot d^5$

Homework Help	
For Exercises	See Examples
16–31, 45, 46, 48	1–4
32–43, 47	5–7
45	8
Extra Practice	
See page 708.	

20. $(b^4)(b^2)$ 　**21.** $(a^2 b)(a b^4)$ 　**22.** $(m^3 n)(m n^2)$ 　**23.** $(r^3 t^4)(r^4 t^4)$

24. $(2xy)(3x^2 y^2)$ 　**25.** $(-5a)(-3a)$ 　**26.** $(-10 x^3 y)(2 x^2)$ 　**27.** $(4 x^2 y^3)(2 x y^2)$

28. $(-8 x^3 y)(2 x^4)$ 　**29.** $m^4(m^3 b^2)$ 　**30.** $(ab)(ac)(bc)$ 　**31.** $(m^2 n)(am)(a n^2)$

32. $\dfrac{10^9}{10^3}$ 　**33.** $\dfrac{9^6}{9^5}$ 　**34.** $\dfrac{w^7}{w^3}$ 　**35.** $\dfrac{k^9}{k^4}$

36. $\dfrac{a^4b^8}{ab^2}$ **37.** $\dfrac{24a^3b^6}{-2a^2b^2}$ **38.** $\dfrac{24x^2y^7z^3}{-6x^2y^3z}$ **39.** $\dfrac{-40mn^2}{-10mn^2}$

40. $\dfrac{3}{4}a(12b^2)$ **41.** $\dfrac{5}{6}c(18a^3)$ **42.** $\left(\dfrac{1}{2}a^2\right)\left(6ab^2\right)$ **43.** $x^0(2x^3)$

44. Evaluate 5^0.

45. Find the product of $2x$ and $-8x$.

Applications and Problem Solving

46. Geometry The measure of the length of a rectangle is $5x$ and the measure of the width is $3x$. Find the measure of the area.

47. Earth Science At a distance of 10^7 meters from Earth, a satellite can see almost all of our planet. At a distance of 10^{13} meters, a satellite can see all of our solar system. How many times as great is a distance of 10^{13} meters as 10^7 meters?

48. Manufacturing The Pizza Parlor uses square boxes to package their pizzas. The drawing shows that a pizza with radius r just fits inside the box. Write an expression for the area of the bottom of the box.

49. Critical Thinking Study the following pattern.

$$(5^2)^3 = (5^2)(5^2)(5^2) \text{ or } 5^6$$

Simplify each expression.

a. $(10^3)^4$ **b.** $(4^5)^3$ **c.** $(x^2)^4$

d. Write a rule for finding the *power of a power*.

Mixed Review

Evaluate each expression if $x = -1$, $y = 2$, and $z = -3$. *(Lesson 8–1)*

50. z^3 **51.** $3x^4$ **52.** $5xy^3z$ **53.** $3(x^2 + y^2)$

54. Write an equation of the line that is perpendicular to the graph of $y = 3x + 5$ and passes through $(0, 0)$. *(Lesson 7–7)*

Solve each proportion. *(Lesson 5–1)*

55. $\dfrac{96}{6} = \dfrac{152}{x}$ **56.** $\dfrac{9}{m} = \dfrac{15}{10}$ **57.** $\dfrac{8.6}{25.8} = \dfrac{1}{n}$ **58.** $\dfrac{3}{7} = \dfrac{2.1}{d}$

Standardized Test Practice
Ⓐ Ⓑ Ⓒ Ⓓ

59. Multiple Choice Choose the expression that has a value of 28. *(Lesson 1–2)*

 Ⓐ $4 + 3 \cdot 4$ Ⓑ $75 \div 3 - 2$ Ⓒ $8 \cdot 4 - 8$ Ⓓ $(5 + 3) \cdot 7 \div 2$

Broadcast Technician

Do you dream of being in the movies? If you don't make it onto the big screen, you may find a career behind the scenes as a sound mixer. Sound mixers are broadcast technicians who develop movie sound tracks. Using a process called *dubbing*, they sit at sound consoles and fade in and fade out each sound by regulating its volume. All of the sounds for each scene are blended on a master sound track.

Sound intensity is measured in *decibels*. The decibel scale is based on powers of ten. The softest audible sound is represented by 10^0. The chart lists several common sounds and their intensity as compared to the softest audible sound.

Sound	Decibels	Intensity
jet airplane	140	10^{14}
rock band	120	10^{12}
motorcycle	110	10^{11}
circular saw	100	10^{10}
busy traffic	80	10^8
vacuum cleaner	70	10^7
noisy office	60	10^6
talking	40	10^4
whispering	20	10^2
breathing	10	10^1
softest sound	0	10^0

Find how many times as intense the first sound is as the second.

1. vacuum cleaner, noisy office
2. motorcycle, busy traffic
3. rock band, talking
4. jet airplane, whispering

5. Find two sounds, one of which is 10^2 times as intense as the other.

FAST FACTS About Broadcast Technicians

Working Conditions
- usually work indoors in pleasant conditions
- work a 40-hour week, but some overtime required to meet deadlines
- evening, weekend, and holiday work

Education
- high school math, physics, and electronics
- postsecondary training in engineering or electronics at a technical school or community college

Earnings

Median Salary by Station Type

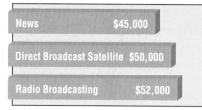

Station Type	Salary
News	$45,000
Direct Broadcast Satellite	$50,000
Radio Broadcasting	$52,000

Salary

Source: *PayScale, Inc.*

---

*inter*NET CONNECTION
Career Data For up-to-date information about a career as a broadcast technician,
visit: www.algconcepts.com

You'll learn to simplify expressions containing negative exponents.

Why It's Important

Electronics Electric current is measured using the prefixes *micro*, which is 10^{-6}, and *milli*, which is 10^{-3}. *See Exercise 14.*

Not all exponents are positive integers. Some exponents are negative integers. Study the pattern at the right to find the value of 10^{-1} and 10^{-2}. Extending the pattern suggests that $10^{-1} = \frac{1}{10}$ and $10^{-2} = \frac{1}{10^2}$ or $\frac{1}{100}$.

$$
\begin{aligned}
10^3 &= 1000 \\
10^2 &= 100 \\
10^1 &= 10 \\
10^0 &= 1 \\
10^{-1} &= ? \\
10^{-2} &= ?
\end{aligned}
\qquad
\begin{aligned}
&\div 10 \\
&\div 10 \\
&\div 10 \\
&\div 10 \\
&\div 10
\end{aligned}
$$

When you multiply by the base, the exponent in the result increases by one. For example, $10^3 \times 10 = 10^4$. When you divide by the base, the exponent in the result decreases by one. For example, $10^{-2} \div 10 = 10^{-3}$.

You can use the Quotient of Powers rule and the definition of power to simplify the expression $\frac{x^3}{x^5}$ and write a definition of negative exponents.

Method 1
Quotient of Powers

$$\frac{x^3}{x^5} = x^{3-5}$$

$$= x^{-2}$$

Method 2
Definition of Power

$$\frac{x^3}{x^5} = \frac{\overset{1}{\cancel{x}} \cdot \overset{1}{\cancel{x}} \cdot \overset{1}{\cancel{x}}}{\underset{1}{\cancel{x}} \cdot \underset{1}{\cancel{x}} \cdot \underset{1}{\cancel{x}} \cdot x \cdot x}$$

$$= \frac{1}{x \cdot x} \text{ or } \frac{1}{x^2}$$

You can conclude that x^{-2} and $\frac{1}{x^2}$ are equal because $\frac{x^3}{x^5}$ cannot have two different values. This and other examples suggest the following definition.

Negative Exponents	**Numbers:** $5^{-2} = \frac{1}{5^2}$; $\frac{1}{4^{-3}} = 4^3$
	Symbols: $a^{-n} = \frac{1}{a^n}$; $\frac{1}{a^{-n}} = a^n$

The value of a cannot be zero.

1 Write 10^{-3} using positive exponents. Then evaluate the expression.

$10^{-3} = \dfrac{1}{10^3}$ *Definition of negative exponent*

$\phantom{10^{-3}} = \dfrac{1}{1000}$ or 0.001   *$10 \cdot 10 \cdot 10 = 1000$*

Your Turn

a. 2^{-4} **b.** 10^{-2} **c.** 5^{-1}

To simplify an expression with a negative exponent, write an equivalent expression that has positive exponents. Each base should appear only once and all fractions should be in simplest form.

Examples

Simplify each expression.

2 xy^{-2}

$xy^{-2} = x \cdot y^{-2}$ *Group the powers by base.*

$\phantom{xy^{-2}} = x \cdot \dfrac{1}{y^2}$ *Definition of negative exponent*

$\phantom{xy^{-2}} = \dfrac{x}{y^2}$ *Simplify.*

3 $\dfrac{a^5 b}{a^3 b^4}$

$\dfrac{a^5 b}{a^3 b^4} = \dfrac{a^5}{a^3} \cdot \dfrac{b^1}{b^4}$ *Group the powers by base.*

$\phantom{\dfrac{a^5 b}{a^3 b^4}} = a^{5-3} \cdot b^{1-4}$ *Quotient of powers*

$\phantom{\dfrac{a^5 b}{a^3 b^4}} = a^2 \cdot b^{-3}$ *Simplify.*

$\phantom{\dfrac{a^5 b}{a^3 b^4}} = a^2 \cdot \dfrac{1}{b^3}$ *Definition of negative exponent*

$\phantom{\dfrac{a^5 b}{a^3 b^4}} = \dfrac{a^2}{b^3}$ *Simplify.*

Your Turn

d. mn^{-3} **e.** $\dfrac{x^4 y}{x^2 y^5}$

4 Simplify $\dfrac{-6r^3 s^5}{18 r^{-7} s^5 t^{-2}}$.

$\dfrac{-6r^3 s^5}{18 r^{-7} s^5 t^{-2}} = \left(\dfrac{-6}{18}\right)\left(\dfrac{r^3}{r^{-7}}\right)\left(\dfrac{s^5}{s^5}\right)\left(\dfrac{1}{t^{-2}}\right)$ *Group the powers by base.*

$\phantom{\dfrac{-6r^3 s^5}{18 r^{-7} s^5 t^{-2}}} = \left(\dfrac{-1}{3}\right)\left(\dfrac{r^3}{r^{-7}}\right)\left(\dfrac{s^5}{s^5}\right)\left(\dfrac{t^0}{t^{-2}}\right)$ *$\dfrac{-6}{18} = -\dfrac{1}{3}, 1 = t^0$*

$\phantom{\dfrac{-6r^3 s^5}{18 r^{-7} s^5 t^{-2}}} = -\dfrac{1}{3} r^{3-(-7)} s^{5-5} t^{0-(-2)}$ *Quotient of powers*

$\phantom{\dfrac{-6r^3 s^5}{18 r^{-7} s^5 t^{-2}}} = -\dfrac{1}{3} r^{10} s^0 t^2$ *$3 - (-7) = 10, 0 - (-2) = 2$*

$\phantom{\dfrac{-6r^3 s^5}{18 r^{-7} s^5 t^{-2}}} = -\dfrac{r^{10} t^2}{3}$ *$s^0 = 1$*

Simplify each expression.

f. $\dfrac{5a^4b^6}{-25a^{-2}b^6c^{-3}}$

g. $\dfrac{8x^3y^5}{10x^{-3}y^5z}$

Biology Link

Real World

5 The *E. coli* bacteria has a width of 10^{-3} millimeter. The head of a pin has a diameter of 1 millimeter. How many *E. coli* bacteria can fit across the head of a pin?

Photo Graphic

To find the number of bacteria, divide 1 by 10^{-3}.

$\dfrac{1}{10^{-3}} = \dfrac{10^0}{10^{-3}}$ *$1 = 10^0$*

$= 10^{0 - (-3)}$ *Quotient of powers*

$= 10^3$ *Simplify.*

Since $10^3 = 1000$, about 1000 bacteria could fit across the head of a pin.

Your Turn

h. The figure below shows the electromagnetic wave spectrum. An ultraviolet wave has a length of 10^{-5} centimeter. An FM radio wave has a length of 10^2 centimeters. How many times as long is the FM wave as the ultraviolet wave?

Wavelength (centimeters)

Online Personal Tutor at algconcepts.com

Check for Understanding

Communicating Mathematics

1. **Express** 5^{-2} using positive integers. Then evaluate the expression.

2. **Evaluate** $6t^{-2}$ if $t = 3$.

3. **You Decide!** Booker and Antonio both correctly simplified $a^{-2} \cdot a^3$.

Booker's Method

$a^{-2} \cdot a^3 = a^{-2+3}$

$= a^1$ or a

Antonio's Method

$a^{-2} \cdot a^3 = \dfrac{1}{a^2} \cdot a^3$

$= \dfrac{a^3}{a^2}$ or a

a. Which student used the Product of Powers rule?

b. Which student used the definition of negative exponents?

c. Whose method do you prefer? Explain your reasoning.

Guided Practice

Example 1

Write each expression using positive exponents. Then evaluate the expression.

4. 10^{-4}

5. 3^{-3}

Simplify each expression.

Example 2

6. z^{-1}　　**7.** n^{-3}　　**8.** $s^{-2}t^3$　　**9.** $p^{-1}q^{-2}r^2$

Examples 3–5

10. $\dfrac{x^2}{x^3}$　　**11.** $\dfrac{a^3}{a^{-4}}$　　**12.** $\dfrac{m^4 n^{-2}}{m^6 n^2}$　　**13.** $\dfrac{3a^3 bc^5}{27a^4 bc^2}$

Example 1

14. Electronics Electric current can be measured in amperes, milliamperes, or microamperes. The prefixes *milli* and *micro* mean 10^{-3} and 10^{-6}, respectively. Express 10^{-3} and 10^{-6} using positive exponents.

Exercises　• • • •　• • • • • • • • • • • • • • • •

Practice

Write each expression using positive exponents. Then evaluate the expression.

15. 2^{-5}　　**16.** 10^{-5}　　**17.** 4^{-1}　　**18.** 6^{-2}

Homework Help

For Exercises	See Examples
15–18, 45–47	1
19–26	2
27–42	3

Extra Practice

See page 708.

Simplify each expression.

19. r^{-10}　　**20.** p^{-6}　　**21.** $a^4(a^{-2})$　　**22.** $x^{-7}(x^4)$

23. $s^{-2}t^4$　　**24.** $a^0 b^{-1} c^{-2}$　　**25.** $15rs^{-2}$　　**26.** $10x^{-4}y^{-5}z$

27. $\dfrac{m^2}{m^{-4}}$　　**28.** $\dfrac{x^2}{x^3}$　　**29.** $\dfrac{k^{-2}}{k^6}$　　**30.** $\dfrac{1}{r^{-3}}$

31. $\dfrac{an^3}{n^5}$　　**32.** $\dfrac{bm^2}{m^6}$　　**33.** $\dfrac{x^3 y^{-3}}{x^3 y^6}$　　**34.** $\dfrac{a^5 b^{-3}}{a^7 b^3}$

35. $\dfrac{12b^5}{4b^{-4}}$　　**36.** $\dfrac{24c^6}{4c^{-2}}$　　**37.** $\dfrac{7x^4}{28x}$　　**38.** $\dfrac{20y^5}{40y^{-2}}$

39. $\dfrac{4x^3}{28x}$　　**40.** $\dfrac{-15r^5 s^8}{5r^5 s^2}$　　**41.** $\dfrac{5ac}{8ab^5 c^2}$　　**42.** $\dfrac{12c^3 d^4 f^6}{60cd^6 f^3}$

43. Evaluate $4x^{-3}y^2$ if $x = 2$ and $y = 6$.

44. Find the value of $(2b)^{-3}$ if $b = -2$.

45. Which is greater, 2^{-4} or 2^{-6}?

Applications and Problem Solving

46. Physical Science Visible light waves have wavelengths between 10^{-5} centimeter and 10^{-4} centimeter. Express 10^{-5} and 10^{-4} using positive exponents. Then evaluate each expression.

47. Physical Science Refer to the figure on page 349 in Your Turn, part h. Some infrared waves have lengths of 10^{-3} centimeter. Which kind of wave has a length that is 1000 times as long as an infrared wave?

48. Critical Thinking Which point on the number line could be the graph of n^{-2} if n is a positive integer?

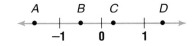

Mixed Review

Simplify each expression. *(Lesson 8–2)*

49. $(5a^3)(-2a^4)$ **50.** $x^4 \cdot x$ **51.** $\dfrac{n^{10}}{n^4}$ **52.** $\dfrac{15a^3b^6c^2}{-5ab^2c^2}$

Write each expression using exponents. *(Lesson 8–1)*

53. $z \cdot z \cdot z \cdot z$ **54.** $(3)(3)(-2)(-2)(-2)$ **55.** 9

Standardized Test Practice
Ⓐ Ⓑ Ⓒ Ⓓ

56. Grid In Determine the slope of the line passing through $(1, -2)$ and $(6, 2)$. *(Lesson 7–1)*

57. Multiple Choice The Jaguars softball team played 8 games and scored a total of 96 runs. What was the mean number of runs scored per game? *(Lesson 3–3)*

Ⓐ 8   Ⓑ 12   Ⓒ 88   Ⓓ 104

Quiz 1   Lessons 8–1 through 8–3

▶ **Write each expression using exponents.** *(Lesson 8–1)*

1. $6 \cdot 6 \cdot 6$ **2.** $x \cdot x \cdot x \cdot x$ **3.** 10 **4.** $(-3)(-3)(-3)$

5. Number Theory The prime factorization of 96 is $2 \cdot 2 \cdot 2 \cdot 2 \cdot 2 \cdot 3$. Write the prime factorization using exponents. *(Lesson 8–1)*

Simplify each expression. *(Lessons 8–2 & 8–3)*

6. $(m^4)(m^6)$ **7.** $\dfrac{x^7}{x^2}$ **8.** $(-3x^2y^3)(-2x^4y^{-1})$ **9.** $\dfrac{3a^2b}{-9a^2b^4}$

10. Biology The length of a *Euglena* protist is about 10^{-2} centimeter. Express 10^{-2} using positive exponents. Then evaluate the expression.

What You'll Learn
You'll learn to express numbers in scientific notation.

Why It's Important
Fiber Optics
Laser technicians use scientific notation when they deal with the speed of light, 3×10^8 meters per second.
See Example 7.

The prefixes *mega*, *giga*, and *kilo* are metric prefixes. They are used with very large measures. Other prefixes are used with very small measures. The chart shows some metric prefixes.

Metric Prefixes					
Prefix	Power of 10	Meaning	Prefix	Power of 10	Meaning
tera	10^{12}	1,000,000,000,000	pico	10^{-12}	0.000000000001
giga	10^9	1,000,000,000	nano	10^{-9}	0.000000001
mega	10^6	1,000,000	micro	10^{-6}	0.000001
kilo	10^3	1000	milli	10^{-3}	0.001

Scientific fields and industry use metric units because calculations are easier with powers of ten. This is also why metric units, based on powers of ten, are widely used in science.

When you multiply a number by a power of ten, the nonzero digits in the original number and the product are the same. Only the position of the decimal point is different.

$$5 \times 10^2 = 5 \times 100$$
$$= 500 \quad \text{2 places right}$$

Prerequisite Skills Review
Operations with Decimals, p. 684

$$8.23 \times 10^4 = 8.23 \times 10,000$$
$$= 82,300 \quad \text{4 places right}$$

$$4 \times 10^{-1} = 4 \times 0.1$$
$$= 0.4 \quad \text{1 place left}$$

$$1.23 \times 10^{-3} = 1.23 \times 0.001$$
$$= 0.00123 \quad \text{3 places left}$$

These and similar examples suggest the following rules for multiplying a number by a power of ten.

Multiplying by Powers of 10	• If the exponent is *positive*, move the decimal point to the *right*. • If the exponent is *negative*, move the decimal point to the *left*.

Express each measurement in standard form.

1 **2 megabytes**

2 megabytes = 2×10^6 bytes   *The prefix mega- means 10^6.*
            = 2,000,000 bytes   *Move the decimal point 6 places right.*

Reading Algebra

The absolute value of the exponent tells the number of places to move the decimal point.

2 **3.6 nanoseconds**

3.6 nanoseconds = 3.6×10^{-9} seconds   *The prefix nano- means 10^{-9}.*
                = 0.0000000036 seconds
                                   Move the decimal point 9 places left.

Your Turn

a. 2 gigabytes   **b.** 3.4 milliseconds

When you use very large numbers like 5,800,000 or very small numbers like 0.000076, it is difficult to keep track of the place value. Numbers such as these can be written in **scientific notation**.

Scientific Notation	A number is expressed in scientific notation when it is in the form $a \times 10^n$, where $1 \le a < 10$ and n is an integer.

Follow these steps to write a number in scientific notation.
- First, move the decimal point after the first nonzero digit.
- Then, find the power of ten by counting the decimal places.
- When the number is greater than one, the exponent of 10 is *positive*.
- When the number is between zero and one, the exponent of 10 is *negative*.

Examples

Express each number in scientific notation.

3 **5,800,000**

5,800,000 = $5.8 \times 10^?$ *The decimal point moves 6 places.*
          = 5.8×10^6 *Since 5,800,000 is greater than one, the exponent is positive.*

4 **0.000076 in scientific notation.**

0.000076 = $7.6 \times 10^?$ *The decimal point moves 5 places.*
         = 7.6×10^{-5} *Since 0.000076 is between zero and one, the exponent is negative.*

Your Turn

c. 3,900,000,000   **d.** 0.0000035

You can use scientific notation to simplify computation.

Examples

Evaluate each expression.

5 400 × 2,000,000,000

First express each number in scientific notation. Then use the Associative and Commutative Properties to regroup terms.

$$400 \times 2,000,000,000 = (4 \times 10^2)(2 \times 10^9) \quad \textit{Write in scientific notation.}$$
$$= (4 \times 2)(10^2 \times 10^9) \quad \textit{Associative and Commutative Properties}$$
$$= 8 \times 10^{11} \quad \textit{Multiply.}$$
$$= 800,000,000,000 \quad \textit{Write in standard form.}$$

Technology Tip

On a graphing calculator, 8×10^{11} is shown as 8E11.

6 $\dfrac{4.8 \times 10^3}{1.6 \times 10^1}$

$$\frac{4.8 \times 10^3}{1.6 \times 10^1} = \left(\frac{4.8}{1.6}\right)\left(\frac{10^3}{10^1}\right) \qquad \frac{4.8}{1.6} = 3$$
$$= 3 \times 10^2 \text{ or } 300 \quad \textit{Simplify.}$$

Your Turn

e. 2000 × 3,000,000,000

f. $\dfrac{7.5 \times 10^7}{1.5 \times 10^4}$

Physics Link

Real World

7 The light from a laser beam travels at a speed of 300,000,000 meters per second. How far does the light travel in 2 nanoseconds? Use the formula $d = rt$, where d is the distance in meters, r is the speed of light, and t is the time in seconds.

Express 300,000,000 in scientific notation.
Express 2 nanoseconds in seconds.

$300,000,000 = 3 \times 10^8$ and 2 nanoseconds $= 2 \times 10^{-9}$ seconds

Method 1   Paper and Pencil
$d = rt$ *Formula for distance*
$d = (3 \times 10^8)(2 \times 10^{-9})$ $r = 3 \times 10^8$ and $t = 2 \times 10^{-9}$
$d = (3 \times 2)(10^8 \times 10^{-9})$ *Associative Property*
$d = 6 \times 10^{-1}$ *Multiply.*

Method 2   Calculator

3 [2nd] [EE] 8 [×] 2 [2nd] [EE] [(−)] 9 [ENTER] *0.6*

The light travels 6×10^{-1} meter, or 0.6 meter, in 2 nanoseconds.

🌐nline **Personal Tutor at** algconcepts.com

Check for Understanding

1. **Tell** whether 23.5×10^3 is expressed in scientific notation. Explain your answer.

2. **Explain** an advantage of expressing very large or very small numbers in scientific notation.

3. **Writing Math** Find two very large numbers and two very small numbers in a newspaper. Write each number in standard form and in scientific notation.

Vocabulary
scientific notation

Guided Practice

⏰ **Getting Ready** Find each product.

Sample 1: 2×10^3
Solution: 2000 *three places right*

Sample 2: 3.4×10^{-1}
Solution: 0.34 *one place left*

4. 2.45×10^2
5. 6.8×10^4
6. 2×10^6
7. 6.4×10^{-1}
8. 9.23×10^{-2}
9. 3×10^{-4}

Examples 1 & 2

Express each measure in standard form.

10. 5 megaohms
11. 6.5 milliamperes

Examples 3 & 4

Express each number in scientific notation.

12. 9500
13. 56.9
14. 0.0087
15. 0.000023

Examples 5–7

Evaluate each expression. Express each result in scientific notation and standard form.

16. $(2 \times 10^5)(3 \times 10^{-8})$
17. $\dfrac{5.2 \times 10^5}{2 \times 10^2}$

Example 4

18. **Health** The length of the virus that causes AIDS is 0.00011 millimeter. Express 0.00011 in scientific notation.

Exercises • • • • • • • • • • • • • • • • • •

Practice

Homework Help	
For Exercises	See Examples
19–24, 45–47	1, 2
25–36, 48	3, 4
37–40	5
41–43	6
44, 49	7
Extra Practice	
See page 708.	

Express each measure in standard form.

19. 5.8 billion dollars
20. 4 megahertz
21. 3.9 nanoseconds
22. 82 kilobytes
23. 9 milliamperes
24. 2.3 micrograms

Express each number in scientific notation.

25. 5280
26. 240,000
27. 268.3
28. 25,000,000
29. 0.00032
30. 0.08
31. 0.004296
32. 15.9
33. 0.012
34. 1,000,000
35. 0.000000022
36. 0.0000946

Evaluate each expression. Express each result in scientific notation and standard form.

37. $(3 \times 10^2)(2 \times 10^4)$
38. $(4 \times 10^2)(1.5 \times 10^6)$
39. $(3 \times 10^{-2})(2.5 \times 10^4)$
40. $(7.8 \times 10^{-6})(1 \times 10^{-2})$

Evaluate each expression. Express each result in scientific notation and standard form.

41. $\dfrac{6.4 \times 10^5}{3.2 \times 10^2}$

42. $\dfrac{8 \times 10^2}{2 \times 10^{-3}}$

43. $\dfrac{13.2 \times 10^{-6}}{2.4 \times 10^{-1}}$

44. Find the product of (1.2×10^5) and (5×10^{-4}) mentally.

45. Express 5×10^6 in standard form.

Applications and Problem Solving

Orchid •·················

46. **Biology**   The mass of an orchid seed is 3.5×10^{-6} grams. Express 3.5×10^{-6} in standard form.

47. **Astronomy**   The diameter of Venus is 1.218×10^4 km, the diameter of Earth is 1.276×10^4 km, and the diameter of Mars is 6.76×10^3 km. List the planets in order from greatest to least diameter.

48. **Electronics**   Engineering notation is similar to scientific notation. However, in engineering notation, the powers of ten are always multiples of 3, such as 10^3, 10^6, 10^{-9}, and 10^{-12}. For example, 240,000 is expressed as 240×10^3. Express 15,000 ohms in scientific and engineering notation.

49. **Biology**   Laboratory technicians look at bacteria through microscopes. A microscope set on 1000× makes an organism appear to be 1000 times larger than its actual size. Most bacteria are between 3×10^{-4} and 2×10^{-3} millimeter in diameter. How large would the bacteria appear under a microscope set on 1000×?

50. **Critical Thinking**   Express each number in scientific notation.
    a. 32×10^5 b. 284×10^3 c. 0.76×10^{-2} d. 0.09×10^{-3}

Mixed Review

Simplify each expression.   *(Lessons 8–2, 8–3)*

51. $y^5(y^{-2})$

52. $15a^{-1}b^3c^{-2}$

53. $\dfrac{30c^4}{-5c^{-2}}$

54. $\dfrac{5r^2s^7}{25r^2s^{10}}$

55. $x^2 \cdot x^3 \cdot x$

56. $\dfrac{2}{3}r(15s^2)$

57. $\dfrac{a^4b^5}{ab^2}$

58. $(-15y^2z)(-2y^2z^0)$

59. $(3x^5y^2)(-2x^{-3}y^4)$

Standardized Test Practice
Ⓐ Ⓑ Ⓒ Ⓓ

60. **Short Response**   Julia's wages vary directly as the number of hours she works. If her wages for 5 hours are $34.75, how much will they be for 30 hours in dollars?   *(Lesson 6–5)*

61. **Multiple Choice**   Choose the graph that represents a function. *(Lesson 6–4)*

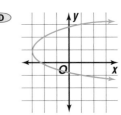

www.algconcepts.com/self_check_quiz

8-5 Square Roots

What You'll Learn
You'll learn to simplify radicals by using the Product and Quotient Properties of Square Roots.

Why It's Important
Aviation Pilots can use the formula $d = 1.5\sqrt{h}$ to determine the distance to the horizon. The formula contains a square root symbol. *See Example 6.*

Can you help this character from *Shoe* take his math test?

SHOE

You will find the square root of 225 in Example 3.

In Lesson 8–1, you learned that *squaring* a number means using that number as a factor twice. The opposite of squaring is finding a **square root**. To find a square root of 36, you must find *two equal factors* whose product is 36.

$6 \times 6 = 36 \quad \rightarrow \quad$ The square root of 36 is 6.

Square Root	**Words:** A square root of a number is one of its two equal factors.
	Symbols: $\sqrt{a} = b$, where $a = b \cdot b$.

The symbol $\sqrt{}$, called a **radical sign**, is used to indicate the square root.

Reading Algebra
Read $\sqrt{a}$ as *the square root of a.*

$\sqrt{36} = 6 \quad$ *$\sqrt{36}$ indicates the positive square root of 36.*

$-\sqrt{36} = -6 \quad$ *$-\sqrt{36}$ indicates the negative square root of 36.*

Exponents can also be used to indicate the square root. $9^{\frac{1}{2}}$ means the same thing as $\sqrt{9}$. $9^{\frac{1}{2}}$ is read *nine to the one half power.* $9^{\frac{1}{2}} = 3$.

Examples

Simplify each expression.

1 $\sqrt{49}$
Since $7^2 = 49$, $\sqrt{49} = 7$.

2 $-\sqrt{64}$
Since $8^2 = 64$, $-\sqrt{64} = -8$.

Your Turn

a. $\sqrt{25}$ b. $\sqrt{121}$ c. $-\sqrt{25}$ d. $-\sqrt{9}$

A **radical expression** is an expression that contains a square root. You can simplify a radical expression like $\sqrt{225}$ by using prime numbers.

A **prime number** is a whole number that has exactly two factors, the number itself and 1. A **composite number** is a whole number that has more than two factors. Every composite number can be written as the product of prime numbers. The tree diagram shows one way to find the prime factors of 225.

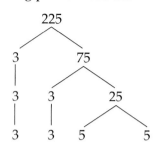

When a number is expressed as a product of prime factors, the expression is called the **prime factorization** of the number. Since 3 and 5 are prime numbers, the prime factorization of 225 is $3 \times 3 \times 5 \times 5$.

To simplify $\sqrt{225}$, use the following property.

Product Property of Square Roots	**Words:**	The square root of a product is equal to the product of each square root.
	Numbers:	$\sqrt{4 \cdot 9} = \sqrt{4} \cdot \sqrt{9}$
	Symbols:	$\sqrt{ab} = \sqrt{a} \cdot \sqrt{b} \quad a \geq 0, b \geq 0$

Examples

Simplify each expression.

❸ $\sqrt{225}$

$$\sqrt{225} = \sqrt{3 \cdot 3 \cdot 5 \cdot 5} \quad \textit{Find the prime factorization of 225.}$$
$$= \sqrt{9 \cdot 25} \quad \textit{3} \times \textit{3} = \textit{9, 5} \times \textit{5} = \textit{25}$$
$$= \sqrt{9} \cdot \sqrt{25} \quad \textit{Use the Product Property of Square Roots.}$$
$$= 3 \cdot 5 \text{ or } 15 \quad \textit{Simplify each radical.}$$

❹ $\sqrt{576}$

$$\sqrt{576} = \sqrt{2 \cdot 2 \cdot 2 \cdot 2 \cdot 2 \cdot 2 \cdot 3 \cdot 3} \quad \textit{Find the prime factorization of 576.}$$
$$= \sqrt{64 \cdot 9} \quad \textit{2} \cdot \textit{2} \cdot \textit{2} \cdot \textit{2} \cdot \textit{2} \cdot \textit{2} = \textit{64, 3} \cdot \textit{3} = \textit{9}$$
$$= \sqrt{64} \cdot \sqrt{9} \quad \textit{Use the Product Property of Square Roots.}$$
$$= 8 \cdot 3 \text{ or } 24 \quad \textit{Simplify each radical.}$$

e. $\sqrt{144}$

f. $\sqrt{324}$

A similar property for quotients can be used to simplify radicals.

Quotient Property of Square Roots		
	Words:	The square root of a quotient is equal to the quotient of each square root.
	Numbers:	$\sqrt{\dfrac{4}{9}} = \dfrac{\sqrt{4}}{\sqrt{9}}$
	Symbols:	$\sqrt{\dfrac{a}{b}} = \dfrac{\sqrt{a}}{\sqrt{b}}$ $a \geq 0, b > 0$

Example

5 Simplify $\sqrt{\dfrac{81}{121}}$.

$\sqrt{\dfrac{81}{121}} = \dfrac{\sqrt{81}}{\sqrt{121}}$ *Use the Quotient Property of Square Roots.*

$= \dfrac{9}{11}$ *Simplify.*

Your Turn

g. $\sqrt{\dfrac{64}{81}}$

h. $\sqrt{\dfrac{36}{4}}$

 Personal Tutor at algconcepts.com

You can use a graphing calculator to evaluate square roots. Press 2nd [√] and then the number to find its positive square root.

Example

Aviation Link

6 Pilots use the formula $d = 1.5\sqrt{h}$ to determine the distance in miles that an observer can see under ideal conditions. In the formula, d is the distance in miles and h is the height in feet of the plane. If an observer is in a plane that is flying at a height of 3600 feet, how far can he or she see?

$d = 1.5\sqrt{h}$ *Original formula*

$d = 1.5 \times \sqrt{3600}$ *Replace h with 3600.*

1.5 2nd [√] 3600 ENTER   *90*

The observer can see a distance of 90 miles.

On a graphing calculator, it is not necessary to use the ✕ key when you multiply square roots.

Check for Understanding

Communicating Mathematics

1. **Find** the first ten perfect squares.

2. **Write** the symbol for the negative square root of 9.

3. Writing Math Explain why finding a square root and squaring are inverse operations.

Vocabulary

square root
radical sign
radical expression
prime number
composite number
prime factorization

Guided Practice

⊙ **Getting Ready** Find the prime factorization of each number.

Sample: 81   **Solution:** $81 = 3 \cdot 3 \cdot 3 \cdot 3$ or 3^4

4. 100   5. 121   6. 169   7. 196   8. 256

Simplify.

Examples 1 & 2
   9. $\sqrt{4}$ 10. $-\sqrt{49}$ 11. $-\sqrt{121}$

Examples 2 & 3
   12. $\sqrt{256}$ 13. $\sqrt{\dfrac{1}{4}}$ 14. $-\sqrt{\dfrac{49}{121}}$

Example 5
   15. **Buildings** A famous 1933 movie about a gorilla helped make the Empire State Building in New York City a popular tourist attraction. If the gorilla's eyes were at a height of 1225 feet, how far could he see in the distance on a clear day? Use the formula $d = 1.5\sqrt{h}$, where d is the visible distance in miles and h is the height in feet.

Exercises

Practice

Simplify.

16. $-\sqrt{81}$ 17. $\sqrt{100}$ 18. $\sqrt{144}$ 19. $-\sqrt{196}$

20. $-\sqrt{169}$ 21. $\sqrt{529}$ 22. $\sqrt{25}$ 23. $-\sqrt{676}$

24. $\sqrt{441}$ 25. $-\sqrt{484}$ 26. $-\sqrt{1024}$ 27. $\sqrt{289}$

28. $\sqrt{\dfrac{81}{64}}$ 29. $-\sqrt{\dfrac{9}{100}}$ 30. $\sqrt{\dfrac{36}{196}}$ 31. $-\sqrt{\dfrac{25}{400}}$

32. $\sqrt{\dfrac{225}{25}}$ 33. $\sqrt{\dfrac{144}{196}}$ 34. $\sqrt{\dfrac{196}{289}}$ 35. $\sqrt{\dfrac{0.09}{0.16}}$

36. $\sqrt{0.16}$ 37. $-\sqrt{0.0025}$ 38. $\sqrt{0.0036}$ 39. $\sqrt{0.0009}$

40. Find the negative square root of 49.

41. If $x = \sqrt{36}$, what is the value of x?

Homework Help

For Exercises	See Examples
16–27, 36–41	1, 2
28–35	5

Extra Practice
See page 709.

Applications and Problem Solving

42. **Geometry** The area of a square is 25 square inches. Find the length of one of its sides.

25 in²

43. **Geometry** The formula for the perimeter P of a square is $P = 4s$, where s is the length of a side. The area of a square is 169 square meters. Find its perimeter.

44. **Firefighting** The velocity of water sprayed from a nozzle is given by the formula $V = 12.14\sqrt{P}$, where V is the velocity in feet per second and P is the pressure at the nozzle in pounds per square inch. Find the velocity of water if the nozzle pressure is 64 pounds per square inch.

45. **Critical Thinking** *True* or *false*: $\sqrt{-36} = -6$. Explain.

Mixed Review

Express each number in scientific notation. *(Lesson 8–4)*

46. 350   47. 63,000   48. 0.023   49. 0.00076

Write each expression using positive exponents. *(Lesson 8–3)*

50. 7^{-2} 51. x^{-4} 52. $ab^{-2}c$ 53. $x^{-2}y^{-3}$

Math Data
Online Update
For the latest information about high school enrollment, visit: algconcepts.com

54. **Business** Among high school students ages 15–18, 41% say they are employed. The graph shows the number of hours per week these students work. If you survey 50 high school students who have jobs, predict how many work between 11 and 20 hours per week. *(Lesson 5–3)*

High School Workers

Hours per week

10 or less 29%
11–20 40%
21–35 22%
36 or more 9%

Source: Michaels Opinion Research

Standardized Test Practice
Ⓐ Ⓑ Ⓒ Ⓓ

55. **Short Response** Write an equation with variables on both sides in which the solution is -5. *(Lesson 4–6)*

Quiz 2   Lessons 8–4 and 8–5

Express each number in scientific notation. *(Lesson 8–4)*

1. 700,000   2. 0.000053

Simplify each expression. *(Lesson 8–5)*

3. $\sqrt{441}$ 4. $-\sqrt{\dfrac{9}{36}}$

5. **Architecture** A square house is the most energy-efficient because it has the least outside wall space for its area. What is the length and width of the most energy-efficient house you could build with an area of 900 square feet? *(Lesson 8–5)*

What You'll Learn
You'll learn to estimate square roots.

Why It's Important
Law Enforcement
Police officers use the formula $s = \sqrt{30df}$ when they investigate accidents.
See Exercise 38.

Numbers like 25 and 81 are perfect squares because their square roots are whole numbers ($\sqrt{25} = 5$ and $\sqrt{81} = 9$). But there are many other numbers that are *not* perfect squares.

Notice what happens when you find $\sqrt{2}$ and $\sqrt{15}$ with a calculator.

[2nd] [$\sqrt{}$] 2 [ENTER] *1.414213562...*

[2nd] [$\sqrt{}$] 15 [ENTER] *3.872983346...*

Numbers like $\sqrt{2}$ and $\sqrt{15}$ are not integers or rational numbers because their decimal values do not terminate or repeat. They are **irrational numbers**. You can estimate irrational square roots by using perfect squares.

Hands-On Algebra

Materials: base-ten tiles

You can use base-ten tiles to estimate the square root of 60.

Step 1 Arrange 60 tiles into the largest square possible. The square has 49 tiles, with 11 left over.

Step 2 Add tiles until you have the next larger square. You need to add 4 tiles. This square has 64 tiles.

Step 1   Step 2

Step 3 Now use these models to estimate $\sqrt{60}$.

- 60 is between 49 and 64.
- $\sqrt{60}$ is between 7 and 8.
- Since 60 is closer to 64 than to 49, $\sqrt{60}$ is closer to 8 than to 7.
- To the nearest whole number, $\sqrt{60} \approx 8$.

Try These

For each number, arrange base-ten tiles into the largest square possible. Then add tiles until you have the next larger square. To the nearest whole number, estimate the square root of each number.

1. 20   **2.** 76   **3.** 150   **4.** 3

Estimate each square root to the nearest whole number.

1 $\sqrt{22}$

List some perfect squares to find the two perfect squares closest to 22.

$$1, 4, 9, 16, 25, 36, \ldots$$

22 is between 16 and 25.

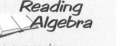

Reading Algebra

The expression $16 < 22 < 25$ means that *16 is less than 22 and 22 is less than 25.* It also means that 22 is *between* 16 and 25.

$$16 < 22 < 25$$
$$\sqrt{16} < \sqrt{22} < \sqrt{25}$$
$$4 < \sqrt{22} < 5$$

Since 22 is closer to 25 than to 16, the best whole number estimate for $\sqrt{22}$ is 5.

2 $\sqrt{130}$

$$121 < 130 < 144$$
$$\sqrt{121} < \sqrt{130} < \sqrt{144}$$
$$11 < \sqrt{130} < 12$$

$\ldots 64, 81, 100, 121, 144, \ldots$

Since 130 is closer to 121 than to 144, the best whole number estimate for $\sqrt{130}$ is 11.

Your Turn

a. $\sqrt{45}$ **b.** $\sqrt{190}$

Gardening Link

3 A box of fertilizer covers 250 square feet of garden. Find the length in whole feet of the largest square garden that can be fertilized with one box of fertilizer.

area ≤ 250 ft^2

Find the largest perfect square that is less than 250. Then find its square root.

$225 < 250$ $\begin{aligned} 15^2 &= 225 \\ 16^2 &= 256 \end{aligned}$

The square root of 225 is 15. Therefore, the largest square garden that can be fertilized with one box of fertilizer has a length of 15 feet.

 Online Personal Tutor at algconcepts.com

Check for Understanding

Communicating Mathematics

1. **Explain** why $\sqrt{10}$ is an irrational number.

2. **Graph** $\sqrt{75}$ on a number line.

3. **Write a problem** that can be solved by estimating $\sqrt{200}$.

> *Vocabulary*
> irrational number

Guided Practice

 Getting Ready Find two consecutive perfect squares between which each number lies.

| **Sample:** 60 | **Solution:** 60 is between 49 and 64. |

4. 56   **5.** 85   **6.** 175   **7.** 500

Examples 1 & 2

Estimate each square root to the nearest whole number.

8. $\sqrt{85}$ **9.** $\sqrt{71}$ **10.** $\sqrt{149}$ **11.** $\sqrt{255}$

Example 3

12. **Geometry** The area of a square is 200 square inches. Find the length of each side. Round to the nearest whole number.

Exercises

Practice

Estimate each square root to the nearest whole number.

13. $\sqrt{3}$ **14.** $\sqrt{7}$ **15.** $\sqrt{13}$ **16.** $\sqrt{19}$ **17.** $\sqrt{33}$

18. $\sqrt{56}$ **19.** $\sqrt{113}$ **20.** $\sqrt{175}$ **21.** $\sqrt{410}$ **22.** $\sqrt{500}$

23. $\sqrt{575}$ **24.** $\sqrt{1000}$ **25.** $\sqrt{60.3}$ **26.** $\sqrt{94.5}$ **27.** $\sqrt{131.4}$

28. $\sqrt{2.314}$ **29.** $\sqrt{152.75}$ **30.** $\sqrt{189.2}$ **31.** $\sqrt{0.08}$ **32.** $\sqrt{0.76}$

Homework Help	
For Exercises	See Examples
13–34	1, 2
35–38	3
Extra Practice	
See page 709.	

33. Tell whether 6 is closer to $\sqrt{34}$ or $\sqrt{44}$.

34. Which is closer to $\sqrt{43}$, 6 or 7?

Applications and Problem Solving

35. **Weather** Meteorologists use the formula $t = \sqrt{\dfrac{D^3}{216}}$ to describe violent storms such as hurricanes. In the formula, D is the diameter of the storm in miles, and t is the number of hours it will last. A typical hurricane has a diameter of about 40 miles. Estimate how long a typical hurricane lasts.

40 mi

Photo Graphic

36. **Aviation**   The British Airways Concorde flew at a height of 60,000 feet. To the nearest mile, about how far could the pilot see when he or she looked out the window on a clear day? Use the formula $d = 1.5\sqrt{h}$, where d is the visible distance in miles and h in the height in feet.

37. **Geometry**   You can use the formula $A = \sqrt{s(s-a)(s-b)(s-c)}$ to find the area of a triangle given the measures of its sides. In the formula, a, b, and c, are the measures of the sides, A is the area, and s is one-half the perimeter. The figure shows a triangular plot of land that contains a large lake. Estimate the area of the triangle to the nearest square mile.

38. **Law Enforcement**   When police officers investigate traffic accidents, they need to determine the speed of the vehicles involved in the accident. They can measure the skid marks and then use the formula $s = \sqrt{30df}$. In the formula, s is the speed in miles per hour, d is the length of the skid marks in feet, and f is the friction factor that depends on road conditions. The table gives some values for f.

Friction Factor (*f*)		
	concrete	asphalt
wet	0.4	0.5
dry	0.8	1.0

Use the formula to determine the speed in miles per hour for each skid length and road condition. Round to the nearest tenth.
   a. 40-foot skid, wet concrete   **b.** 150-foot skid, dry asphalt
   c. 350-foot skid, wet asphalt   **d.** 45-foot skid, dry concrete

39. **Critical Thinking**   Find three numbers that have square roots between 3 and 4.

Mixed Review

Find each square root.   *(Lesson 8–5)*
40. $\sqrt{0.36}$ 41. $\sqrt{256}$ 42. $\sqrt{729}$ 43. $\sqrt{\dfrac{169}{121}}$

Express each measure in standard form.   *(Lesson 8–4)*
44. 2 gigabytes   45. 4 nanoseconds   46. 6.5 megahertz

Write an equation in slope-intercept form of the line having the given slope that passes through the given point.   *(Lesson 7–3)*
47. 5; (3, −2)   48. $\dfrac{1}{4}$; (0, 8)   49. −5; (5, 4)

Standardized Test Practice
Ⓐ Ⓑ Ⓒ Ⓓ

50. **Grid In**   Tim drove 4 hours at an average speed of 60 miles per hour. How many hours would it take Tim to drive the same distance at an average speed of 50 miles per hour?   *(Lesson 6–6)*

51. **Multiple Choice**   Which statement is *not* correct?
    (Lesson 3–1)
    Ⓐ $\dfrac{1}{2} > \dfrac{3}{8}$ Ⓑ $\dfrac{4}{5} < \dfrac{5}{6}$ Ⓒ $\dfrac{8}{11} < \dfrac{7}{13}$ Ⓓ $\dfrac{1}{3} < \dfrac{10}{13}$

What You'll Learn
You'll learn to use the Pythagorean Theorem to solve problems.

Why It's Important
Carpentry
Carpenters use the Pythagorean Theorem to determine whether the corners of a deck are right angles.
See Example 4.

The sides of the right triangle below have lengths of 3, 4, and 5 units. The relationship among these lengths forms the basis for one of the most famous theorems in mathematics.

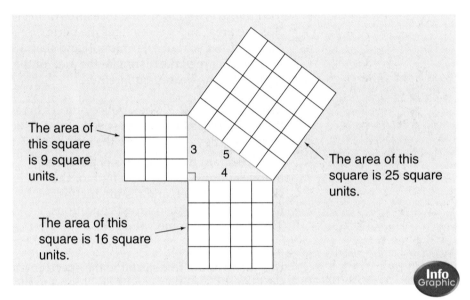

The area of this square is 9 square units.

3   5

4

The area of this square is 25 square units.

The area of this square is 16 square units.

Info Graphic

The two sides that form the right angle are called the **legs**. In the triangle above, the lengths of the legs are 3 units and 4 units. The side opposite the right angle is called the **hypotenuse**. The hypotenuse of this triangle has a length of 5 units.

The squares drawn along each side of the triangle illustrate the Pythagorean Theorem geometrically. Study the areas of the squares. Do you notice a relationship between them? The area of the larger square is equal to the total area of the two smaller squares.

$$25 = 9 + 16$$

$$5^2 = 3^2 + 4^2$$

This relationship is true for *any* right triangle and is called the **Pythagorean Theorem**.

Pythagorean Theorem	**Words:** In a right triangle, the square of the length of the hypotenuse, c, is equal to the sum of the squares of the lengths of the legs, a and b.
	Model: *(right triangle with legs a and b, hypotenuse c)* **Symbols:** $c^2 = a^2 + b^2$

Example ➊ **Find the length of the hypotenuse of the right triangle.**

$c^2 = a^2 + b^2$ *Pythagorean Theorem*

$c^2 = 15^2 + 8^2$ *Replace a with 15 and b with 8.*

$c^2 = 225 + 64$ *Multiply.*

$c^2 = 289$ *Add.*

$c = \sqrt{289}$ *Find the square root of each side.*

$c = 17$ *Simplify.*

The length of the hypotenuse is 17 feet.

Your Turn

a. Find the length of the hypotenuse of a right triangle if the lengths of the legs are 6 meters and 8 meters.

You can also use the Pythagorean Theorem to find the length of a leg of a right triangle.

Example ➋ **Find the length of one leg of a right triangle if the length of the hypotenuse is 14 meters and the length of the other leg is 6 meters. Round to the nearest tenth.**

 Reading Algebra

The hypotenuse is *always* the longest side of a right triangle.

$c^2 = a^2 + b^2$ *Pythagorean Theorem*

$14^2 = 6^2 + b^2$ *Replace c with 14 and a with 6.*

$196 = 36 + b^2$ *Multiply.*

$196 - 36 = 36 - 36 + b^2$ *Subtract 36 from each side.*

$160 = b^2$ *Estimate: Since $10^2 = 100$ and*

$\sqrt{160} = b$ *$15^2 = 225$, $\sqrt{160}$ is between 10 and 15.*

160 [2nd] [√] [ENTER] *12.64911064*

To the nearest tenth, the length of the leg is 12.6 meters.

Your Turn

Find each missing measure. Round to the nearest tenth.

b.

c.

A **converse** of a theorem is the reverse, or opposite, of the theorem. You can use the converse of the Pythagorean Theorem to test whether a triangle is a right triangle.

Converse of the Pythagorean Theorem	If c is the measure of the longest side of a triangle and $c^2 = a^2 + b^2$, then the triangle is a right triangle.

Examples ③ **The measures of the three sides of a triangle are 5, 7, and 9. Determine whether this triangle is a right triangle.**

$c^2 = a^2 + b^2$ *Pythagorean Theorem*

$9^2 \stackrel{?}{=} 5^2 + 7^2$ *Replace c with 9, a with 5, and b with 7.*

$81 \stackrel{?}{=} 25 + 49$ *Multiply.*

$81 \neq 74$ *Add.*

Since $c^2 \neq a^2 + b^2$, the triangle is *not* a right triangle.

Your Turn

The measures of three sides of a triangle are given. Determine whether each triangle is a right triangle.

d. 20, 21, 28   **e.** 37, 12, 34

Carpentry Link

Real World

④ **A carpenter checks whether the corners of a deck are square by using a 3-4-5 system. He or she measures along one side in 3-foot units and along the adjacent side in the same number of 4-foot units. If the measure of the hypotenuse is the same number of 5-foot units, the corner of the deck is square.** *If a corner is square, it is formed by a right angle. So, the triangle is a right triangle.*

Suppose a carpenter measures along one side of a deck, a distance of 9 feet, and along the adjacent side, a distance of 12 feet. The measure of the hypotenuse is 15 feet. Is the corner of the deck square?

Explore   You know that the measures of the sides are 9, 12, and 15. You want to know if the sides form a right triangle.

Plan   The measures of the legs are 9 and 12, and the measure of the hypotenuse is 15. Use these numbers in the Pythagorean Theorem.

Solve	$c^2 = a^2 + b^2$	*Pythagorean Theorem*
	$15^2 \stackrel{?}{=} 9^2 + 12^2$	*Replace c with 15, a with 9, and b with 12.*
	$225 \stackrel{?}{=} 81 + 144$	*Multiply.*
	$225 = 225$	*Add.*

Since $c^2 = a^2 + b^2$, the triangle is a right triangle and the corner of the deck is square.

Examine   The measures of the sides are 9, 12, and 15. They are multiples of a 3-4-5 triangle. The answer is reasonable.

 Personal Tutor at algconcepts.com

Check for Understanding

Communicating Mathematics

1. **Draw** a right triangle and label the right angle, the hypotenuse, and the legs.

2. **Explain** how you know whether a triangle is a right triangle if you know the lengths of the three sides.

┌─ *Vocabulary* ─┐
hypotenuse
leg
Pythagorean Theorem
converse

Guided Practice

 Getting Ready   **Determine whether each sentence is *true* or *false*.**

Sample: $4^2 + 5^2 = 6^2$ **Solution:** $16 + 25 \neq 36$; false

3. $9^2 + 10^2 = 11^2$ **4.** $7^2 + 24^2 = 25^2$ **5.** $3^2 + 12^2 = 20^2$

If *c* is the measure of the hypotenuse and *a* and *b* are the measures of the legs, find each missing measure. Round to the nearest tenth if necessary.

Example 1

6.

6 ft   *c* ft   8 ft

7.
a cm   15 cm   5 cm

Example 2

8. $a = 30, c = 34, b = ?$ **9.** $a = 7, b = 4, c = ?$

Example 3

The lengths of three sides of a triangle are given. Determine whether each triangle is a right triangle.

10. 9 ft, 16 ft, 20 ft   **11.** 9 mm, 40 mm, 41 mm

Example 4

12. Construction   A builder is laying out the foundation for a house. The measure along one side is 12 feet, the adjacent side is 16 feet, and the hypotenuse is 21 feet. Determine whether the corner is a right angle.

Practice

If *c* is the measure of the hypotenuse and *a* and *b* are the measures of the legs, find each missing measure. Round to the nearest tenth if necessary.

13.
5 cm   *c* cm
12 cm

14.
16 ft   20 ft
b ft

15.
2 km   *c* km
8 km

16.
25 in.
24 in.   *a* in.

17.
3 yd
c yd   3 yd

18.
5 m
10 m   *c* m

19. $a = 6, b = 3, c = ?$
20. $b = 10, c = 11, a = ?$
21. $c = 29, a = 20, b = ?$
22. $a = 5, c = 30, b = ?$
23. $a = 7, b = 9, c = ?$
24. $a = 11, c = 20, b = ?$

The lengths of three sides of a triangle are given. Determine whether each triangle is a right triangle.

25. 11 in., 12 in., 16 in.
26. 11 cm, 60 cm, 61 cm
27. 6 ft, 8 ft, 9 ft
28. 6 mi, 7 mi, 12 mi
29. 45 m, 60 m, 75 m
30. 1 mm, 1 mm, 2 mm

31. Is a triangle with measures 30, 40, and 50 a right triangle?

32. Find the length of the hypotenuse of a right triangle if the lengths of the legs are 6 miles and 11 miles. Round to the nearest tenth.

Applications and Problem Solving

33. **Geometry**   Find the length of the diagonal of a rectangle whose length is 8 meters and whose width is 5 meters. Round to the nearest tenth.

34. **Carpentry**   The rise, the run, and the rafter of a pitched roof form a right triangle. Find the length of the rafter that is needed if the rise of the roof is 6 feet and the run is 12 feet. Round your answer to the nearest tenth.

Rafter Length
6-ft rise
12-ft run

35. Architecture Architects often use arches when they design windows and doors. The Early English, or pointed, arch is based on an *equilateral triangle*. An equilateral triangle has three sides of equal measure.

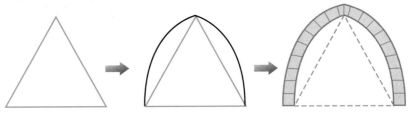

In the figure at the right, the height h separates the equilateral triangle into two right triangles of equal size. Find the height of the arch. Round to the nearest tenth.

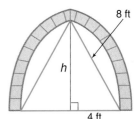

8 ft

h

4 ft

36. Critical Thinking The figure shows three squares. The two smaller squares each have an area of 1 square unit.

a. Find the area of the third square.

b. Draw a square on rectangular dot paper with an area of 5 square units.

Mixed Review

Estimate each square root to the nearest whole number. *(Lesson 8–6)*

37. 65   **38.** 95   **39.** 200   **40.** 500

Simplify. *(Lesson 8–5)*

41. $\sqrt{225}$ **42.** $-\sqrt{49}$ **43.** $\sqrt{100}$ **44.** $-\sqrt{\dfrac{9}{25}}$

45. Finance To save for a new bicycle, Katie begins a savings plan. Her plan can be described by the equation $s = 5w + 100$, where s represents the total savings in dollars and w represents the number of weeks since the start of the savings plan. *(Lesson 7–3)*

a. How much money had Katie already saved when the plan started?

b. How much does Katie save each week?

Standardized Test Practice
Ⓐ Ⓑ Ⓒ Ⓓ

46. Short Response Write the point-slope form of an equation for the line that passes through the point at $(2, -4)$ and has a slope of -1. *(Lesson 7–2)*

47. Multiple Choice Which set is the domain of the relation shown on the graph? *(Lesson 6–1)*

Ⓐ $\{-3, 0, 2\}$

Ⓑ $\{-4, -2, 3, 4\}$

Ⓒ neither A nor B

Ⓓ both A and B

Chapter 8

Investigation

Must Be TV

Using the Pythagorean Theorem

Materials

 uncooked spaghetti
 ruler
 grid paper
 scissors
 calculator

Look at an advertisement for an electronics store and you will find television sets that have 10-inch, 32-inch, and 36-inch screens. Did you ever wonder what these numbers represent? They represent the hypotenuse of a right triangle.

A television set is shaped like a rectangle with length ℓ and width w. The diagonal is a line segment from one corner to the opposite corner, which separates the screen into two right triangles of equal size. The diagonal is the hypotenuse of the right triangles. So, a 10-inch television set has a screen in which the diagonal is 10 inches.

Investigate

1. Use a piece of uncooked spaghetti and grid paper to find several rectangles that have a 10-inch diagonal.

   a. Break the spaghetti so that one piece measures 10 inches.

   b. Place the spaghetti diagonally on the grid paper as shown below.

   c. Using the ends of the spaghetti as endpoints of the diagonal, draw horizontal and vertical lines from the endpoints to complete the rectangle.

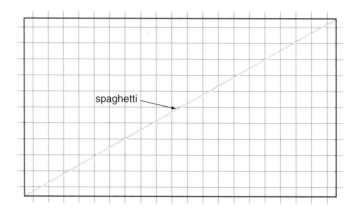
spaghetti

d. Repeat Steps 1b and 1c four more times using the same piece of spaghetti and different pieces of grid paper. For each rectangle, change the angle at which you place the spaghetti on the grid paper.

e. Cut out the rectangles. Choose the rectangle that looks most like a television screen. Describe how this rectangle is different from the others.

2. Television manufacturers have set a standard ratio for the size of a television screen. The ratio of the height to the width is 3:4. For which of your rectangles is the ratio of the height to the width closest to 3:4?

Extending the Investigation

In this extension, you will predict the dimensions of a 19-inch television screen.

- Use a calculator and the Pythagorean Theorem to find the measures of five different rectangles that have a 19-inch diagonal.

- Find the areas of each rectangle and the ratios of each height to width.

- Analyze the data and make a prediction about the dimensions of a 19-inch television screen.

Presenting Your Investigation

Here are some ideas to help you present your conclusions to the class.

- Make a poster that shows scale drawings of your rectangles.

- Write a report in which you explain how you made your prediction. Be sure to include the terms *ratio* and *Pythagorean Theorem* in your report. Then compare your prediction to the dimensions of an actual 19-inch television.

*inter*NET
CONNECTION **Investigation**   For more information on ratios, visit: www.algconcepts.com

Understanding and Using the Vocabulary

After completing this chapter, you should be able to define each term, property, or phrase and give an example or two of each.

*inter*NET
CONNECTION **Review Activities**
For more review activities, visit:
www.algconcepts.com

Algebra

base *(p. 336)*
composite number *(p. 358)*
exponent *(p. 336)*
irrational numbers *(p. 362)*
negative exponent *(p. 348)*
perfect square *(p. 336)*
power *(p. 336)*

prime factorization *(p. 358)*
prime number *(p. 358)*
radical expression *(p. 358)*
radical sign *(p. 357)*
scientific notation *(p. 353)*
square root *(p. 357)*

Geometry

hypotenuse *(p. 366)*
leg *(p. 366)*
Pythagorean Theorem *(p. 366)*

Logic

converse *(p. 268)*

Complete each sentence using a term from the vocabulary list.

1. When a number is expressed as a product of factors that are all prime, the expression is called the ___?___ of the number.
2. The symbol $\sqrt{}$ is called a ___?___ .
3. The square roots of numbers such as 2, 3, and 7 are ___?___ .
4. The ___?___ tells the relationship among the measures of the legs of a right triangle and the hypotenuse.
5. In the expression 2^3, the 3 is called the ___?___ . It tells how many times 2 is used as a factor.
6. The ___?___ of a number is one of its two equal factors.
7. The longest side of a right triangle is called the ___?___ .
8. When you express 5,000,000 as 5×10^6, you are using ___?___ .
9. Numbers like 25, 36, and 100 are called ___?___ .
10. Numbers that are expressed using exponents are called ___?___ .

Skills and Concepts

Objectives and Examples	Review Exercises

• **Lesson 8–1**   Use powers in expressions.

Write $n \cdot n \cdot n \cdot n$ using exponents.
$n \cdot n \cdot n \cdot n = n^4$

Write $2^3 \cdot 3^2$ as a multiplication expression.
$(2^3)(3^2) = (2)(2)(2)(3)(3)$

Write each expression using exponents.
11. $9 \cdot 9 \cdot 9 \cdot 9 \cdot 9$ 12. 5 squared
13. $2 \cdot 2 \cdot 3 \cdot 3 \cdot 3$ 14. $(-2)(x)(x)(y)(y)$

Write each power as a multiplication expression.
15. 12^3 16. $(-2)^5$ 17. y^2
18. a^3b^2 19. $5x^2y^3z^2$ 20. $-5m^2$

 www.algconcepts.com/vocabulary_review

Objectives and Examples

Review Exercises

• **Lesson 8–2** Multiply and divide powers.

$$(2y^4)(5y^5) = (2 \cdot 5)(y^4 \cdot y^5)$$
$$= (10)(y^{4+5})$$
$$= 10y^9$$

$$\frac{x^7y^4}{x^3y^3} = \left(\frac{x^7}{x^3}\right)\left(\frac{y^4}{y^3}\right)$$
$$= (x^{7-3})(y^{4-3})$$
$$= x^4y$$

Simplify each expression.

21. $b^2 \cdot b^5$ **22.** $y^3 \cdot y^3 \cdot y$

23. $(a^2b)(a^2b^2)$ **24.** $(3ab)(-4a^2b^3)$

25. $\dfrac{y^{10}}{y^6}$ **26.** $\dfrac{a^2b^4}{a^3b^2}$

27. $\dfrac{42x^7}{14x^4}$ **28.** $\dfrac{-15x^3y^5z^4}{5x^2y^2z}$

• **Lesson 8–3** Simplify expressions containing negative exponents.

$$a^2b^{-3} = a^2 \cdot \frac{1}{b^3}$$
$$= \frac{a^2}{b^3}$$

Write each expression using positive exponents. Then evaluate the expression.

29. 2^{-3} **30.** 10^{-2}

Simplify each expression.

31. x^{-4} **32.** $m^{-2}n^5$

33. $\dfrac{y^3}{y^{-4}}$ **34.** $\dfrac{rs^2}{s^3}$

35. $\dfrac{-25a^5b^6}{5a^5b^3}$ **36.** $\dfrac{27b^{-2}}{14b^{-3}}$

• **Lesson 8–4** Express numbers in scientific notation.

$$3600 = 3.6 \times 10^3$$

$$0.023 = 2.3 \times 10^{-2}$$

Express each measure in standard form.

37. 1.5 nanoseconds **38.** 5 kilobytes

Express each number in scientific notation.

39. 240,000 **40.** 0.000314

41. 4,880,000,000 **42.** 0.000015

Evaluate each expression.

43. $(2 \times 10^5)(3 \times 10^6)$ **44.** $\dfrac{8 \times 10^{-3}}{2 \times 10^4}$

• **Lesson 8–5** Simplify radicals.

$$\sqrt{400} = \sqrt{2 \cdot 2 \cdot 2 \cdot 2 \cdot 5 \cdot 5}$$
$$= \sqrt{16 \cdot 25}$$
$$= \sqrt{16} \cdot \sqrt{25}$$
$$= 4 \cdot 5 \text{ or } 20$$

Simplify.

45. $\sqrt{121}$ **46.** $-\sqrt{324}$

47. $\sqrt{\dfrac{4}{81}}$ **48.** $-\sqrt{\dfrac{100}{225}}$

Mixed Problem Solving
See pages 724–731.

Objectives and Examples

Review Exercises

• **Lesson 8–6** Estimate square roots.

Estimate $\sqrt{150}$.

$144 < 150 < 169$ ... *100, 121, 144,*
169, . . .

$\sqrt{144} < \sqrt{150} < \sqrt{169}$

$12 < \sqrt{150} < 13$ *150 is closer to 144*
than to 169.

The best whole number estimate for $\sqrt{150}$
is 12.

Estimate each square root to the nearest
whole number.

49. $\sqrt{19}$

50. $\sqrt{108}$

51. $\sqrt{200}$

52. $\sqrt{125.52}$

53. Which is closer to $\sqrt{50}$, 7 or 8?

• **Lesson 8–7** Use the Pythagorean Theorem
to solve problems.

Find the length of the missing side.

$$c^2 = a^2 + b^2$$
$$25^2 = 15^2 + b^2$$
$$625 = 225 + b^2$$
$$625 - 225 = 225 - 225 + b^2$$
$$400 = b^2$$
$$20 = b$$

15 ft
25 ft
b ft

If *c* is the measure of the hypotenuse and
a and *b* are the measures of the legs, find
each missing measure. Round to the
nearest tenth if necessary.

54.

9 cm
c cm
12 cm

55.

15 m
10 m
b m

56. $a = 6$ mi, $b = 10$ mi, $c = ?$

57. $b = 6$, $c = 12$, $a = ?$

Applications and Problem Solving

58. Biology *E. coli* is a fast-growing bacteria
that splits into two identical cells every
15 minutes. If you start with one *E. coli*
bacterium, how many bacteria will there
be at the end of 2 hours? *(Lesson 8–1)*

59. Geometry The area of a square is
90 square meters. Estimate the length
of its side to the nearest whole
number. *(Lesson 8–6)*

60. Construction Park managers want to
construct a road from the park entrance
directly to a campsite. Use the figure below
to find the length of the proposed road.
(Lesson 8–7)

Park Entrance
Proposed Road
8 mi
Campsite
15 mi

1. **Write** a multiplication problem whose product is $12x^5$.
2. **Explain** why 3 is the square root of 9.
3. **Draw** a figure that can be used to estimate $\sqrt{96}$.

Evaluate each expression if $x = 2$, $y = 3$, $z = -1$, and $w = -2$.

4. $3x^5$
5. $5w^3y$
6. $12y^{-2}$
7. z^0

Simplify each expression.

8. $(a^3b^2)(a^5b)$
9. $\dfrac{xy^3}{y^5}$
10. m^{-10}
11. $\dfrac{-3r^2s^8}{12r^2s^5}$

Express each number in scientific notation.

12. 0.00000125
13. $36,000,000,000$

Evaluate each expression. Express each result in scientific notation and standard form.

14. $(8.2 \times 10^{-5})(1 \times 10^{-3})$
15. $\dfrac{6 \times 10^3}{2 \times 10^{-2}}$

Estimate each square root to the nearest whole number.

16. $\sqrt{5}$
17. $\sqrt{20}$
18. $\sqrt{63}$
19. $\sqrt{395}$

Find each missing measure. Round to the nearest tenth if necessary.

20.

21.

22.
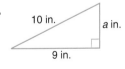

23. **Chemistry** The diameter of an atom is about 10^{-8} centimeter and the diameter of its nucleus is about 10^{-13} centimeter. How many times as large is the diameter of the atom as the diameter of its nucleus?

24. **Sports** Gymnastic routines are performed on a 40-foot by 40-foot square mat. Gymnasts usually use the diagonal of the mat because it gives them more distance to complete their routine. To the nearest whole number, how much longer is the diagonal of the mat than one of its sides?

25. **Sequences** The first term of the sequence $1, \dfrac{1}{2}, \dfrac{1}{4}, \dfrac{1}{8}, \ldots$ can be written as 2^0. Write the next three terms using negative exponents.

Pythagorean Theorem Problems

All standardized tests contain several problems that you can solve using the Pythagorean Theorem. The **Pythagorean Theorem** states that in a right triangle, the sum of the squares of the measures of the legs equals the square of the measure of the hypotenuse.

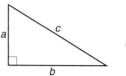

$$a^2 + b^2 = c^2$$

Example 1

Use the information in the figure below. Find BC to the nearest centimeter.

  Ⓐ 31 cm   Ⓑ 41 cm

  Ⓒ 51 cm   Ⓓ 80 cm

Hint Use the Pythagorean Theorem twice.

Solution The figure is made up of two right triangles, $\triangle ABD$ and $\triangle BDC$. $\overline{BD}$ is the hypotenuse of $\triangle ABD$. $\overline{BD}$ is also a leg of $\triangle BDC$.

First, find BD. Then use BD to find BC.

$(BD)^2 = 9^2 + 40^2$	$(BC)^2 = 30^2 + 41^2$
$(BD)^2 = 81 + 1600$	$(BC)^2 = 900 + 1681$
$(BD)^2 = 1681$	$(BC)^2 = 2581$
$BD = 41$	$BC \approx 50.80$

The answer is C.

Example 2

A 25-foot ladder is placed against a vertical wall of a building with the bottom of the ladder standing on concrete 7 feet from the base of the building. If the top of the ladder slips down 4 feet, then the bottom of the ladder will slide out how many feet?

  Ⓐ 5 ft   Ⓑ 6 ft

  Ⓒ 7 ft   Ⓓ 8 ft

Hint Start by drawing a diagram.

Solution The ladder placed against the wall forms a 7-24-25 right triangle. After the ladder slips down 4 feet, the new right triangle has sides that are multiples of a 3-4-5 right triangle, 15-20-25.

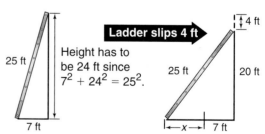

The bottom of the ladder will be 15 feet from the wall. This means the bottom of the ladder will slide out $15 - 7$ or 8 feet.

The answer is D.

After you work each problem, record your answer on the answer sheet provided or on a sheet of paper.

Multiple Choice

1. Diego bought 3 fish. Every month, the number of fish doubles. The formula $f = 3 \cdot 2^m$ represents the number of fish he will have after m months. How many fish will he have after 4 months if he keeps all of them and none of them dies? *(Lesson 8–1)*
   - Ⓐ 12
   - Ⓑ 24
   - Ⓒ 48
   - Ⓓ 1296

2. Which number is greater than 2.8? *(Lesson 8–6)*
   - Ⓐ 175%
   - Ⓑ $\sqrt{12}$
   - Ⓒ −3.5
   - Ⓓ 5.6×10^{-1}

3. A wire from the top of a 25-foot flagpole is attached to a point 20 feet from the base of the flagpole. To the nearest whole number, find the length of the wire. *(Lesson 8–7)*

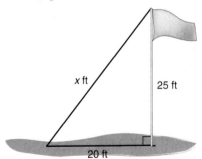
x ft   25 ft   20 ft

   - Ⓐ 45 ft
   - Ⓑ 32 ft
   - Ⓒ 15 ft
   - Ⓓ 35 ft

4. Which of the following are measures of three sides of a right triangle? *(Lesson 8–7)*
   - Ⓐ 4, 7, 8
   - Ⓑ 10, 15, 20
   - Ⓒ 3, 7, 9
   - Ⓓ 9, 12, 15

5. Which equation does *not* have 6 as a solution? *(Lesson 4–5)*
   - Ⓐ $2(x - 6) = 0$
   - Ⓑ $4x + 3 = 27$
   - Ⓒ $2x + 1 = 14$
   - Ⓓ $2x + x = 18$

6. $\triangle ABC$ and $\triangle ACD$ are right triangles. If $AB = 20$, $BC = 15$, and $AD = 7$, then $CD = $ — *(Lesson 8–7)*

   - Ⓐ 22.
   - Ⓑ 23.
   - Ⓒ 24.
   - Ⓓ 25.

7. A farmer wants to put a fence around a square garden that has a total area of 1600 square feet. If fencing costs $1.50 per foot, how much will it cost to fence the garden? To solve this problem, begin by— *(Lesson 1–5)*
   - Ⓐ dividing 1600 by $1.50.
   - Ⓑ multiplying 1600 by $1.50 and taking the square root of the product.
   - Ⓒ dividing 1600 by 4 and multiplying by $1.50.
   - Ⓓ finding the square root of 1600 and multiplying by 4.

8. If $x = 3$, $y = -2$, $z = -1$, and $w = 5$, then the value of $9xy^2z^{10}w^0$ is— *(Lesson 8–1)*
   - Ⓐ −590.
   - Ⓑ −108.
   - Ⓒ 1.
   - Ⓓ 108.

Grid In

9. On the first test of the semester, Jennifer scored a 60. On the last test of the semester, she scored a 75. By what percent did Jennifer's score improve? *(Lesson 5–5)*

Short Response

10. The table shows the squares of six whole numbers. Describe one way that the digits in the squares are related to the digits in the original number. *(Lesson 1–5)*

$15^2 = 225$	$55^2 = 3025$	$135^2 = 18{,}225$
$25^2 = 625$	$95^2 = 9025$	$175^2 = 30{,}625$

Polynomials

What You'll Learn

Key Ideas

- Identify and classify polynomials. *(Lesson 9–1)*
- Add and subtract polynomials. *(Lesson 9–2)*
- Multiply polynomials. *(Lessons 9–3 and 9–4)*
- Develop and use the patterns for $(a + b)^2$, $(a - b)^2$, and $(a + b)(a - b)$. *(Lesson 9–5)*

Key Vocabulary

binomial *(p. 383)*

FOIL method *(p. 401)*

monomial *(p. 382)*

polynomial *(p. 383)*

Why It's Important

Insects With colors ranging from pale green to bright orange, many moths are as beautiful as butterflies. But the lifespan of an adult moth is short. They live only a few days, just long enough to produce eggs for the next generation.

Polynomials can help you solve nonlinear equations. You will use a polynomial to model the number of eggs a certain type of moth can produce in Lesson 9–1.

Check Your Readiness

Simplify each expression.

1. $5a + 2a$ **2.** $6x + 8x$ **3.** $12g + 4g$

4. $8n - 4n$ **5.** $15t - 7t$ **6.** $20k - 5k + 1$

7. $3(m + 2)$ **8.** $6(d - 1)$ **9.** $3(p + q)$

10. $8(m + 2n)$ **11.** $a(b + c)$ **12.** $7x(y - 4)$

13. $5a + 2b + a - b$ **14.** $14c - 1 + 2c - 3$ **15.** $3(y - 1) + 6$

Simplify each expression.

16. $x \cdot x^2$ **17.** $n^2 \cdot n^4$ **18.** $(w^4)(w^3)$

19. $(5d)(3d)$ **20.** $(-3r)(r)$ **21.** $(2a)(6a^2)$

22. $(7x)(10x^3)$ **23.** $k(12jk)$ **24.** $(ab)(ab^2)$

25. $(12n)(4m^2)$ **26.** $p^3(p^2q^2)$ **27.** $(2g)(3g^2h)(4gh^3)$

FOLDABLES™
Study Organizer

Make this Foldable to help you organize your Chapter 9 notes. Begin with a sheet of notebook paper.

❶ Fold lengthwise to the holes.

❷ Cut along the top line and then cut three tabs.

❸ Label the tabs with the lesson topics as shown.

Reading and Writing Store the foldable in a 3-ring binder. As you read and study the chapter, write notes and examples under the tabs.

What You'll Learn

You'll learn to identify and classify polynomials and find their degree.

Why It's Important

Medicine Doctors can use polynomials to study the heart. *See Exercise 57.*

A **monomial** is a number, a variable, or a product of numbers and variables that have only positive exponents. A monomial cannot have a variable as an exponent. The tables below show examples of expressions that are and are not monomials.

Monomials	
-4	a number
y	a variable
a^2	the product of variables
$\frac{1}{2}x^2y$	the product of numbers and variables

Not Monomials	
2^x	has a variable as an exponent
$x^2 + 3$	a sum
$5a^{-2}$	includes a negative exponent
$\frac{3}{x}$	a quotient

Examples

Determine whether each expression is a monomial. Explain why or why not.

 1 $-6ab$

$-6ab$ is a monomial because it is the product of a number and variables.

 2 $m^2 - 4$

$m^2 - 4$ is *not* a monomial because it includes subtraction.

Your Turn

a. 10   **b.** $5z^{-3}$ **c.** $\frac{6}{x}$ **d.** x^2

 nline **Personal Tutor at** algconcepts.com

A monomial or the sum of one or more monomials is called a **polynomial**. For example, $x^3 + x^2 + 3x + 2$ is a polynomial. Each monomial is a term. The terms of the polynomial are x^3, x^2, $3x$, and 2. *Recall that to subtract, you add the opposite.*

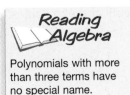
Special names are given to polynomials with two or three terms. A polynomial with two terms is a **binomial**. A polynomial with three terms is a **trinomial**. Here are some examples.

Binomial	Trinomial
$x + 2$	$a + b + c$
$5c - 4$	$x^2 + 5x - 7$
$4w^2 - w$	$3a^2 + 5ab + 2b^2$

Examples

State whether each expression is a polynomial. If it is a polynomial, identify it as a *monomial*, *binomial*, or *trinomial*.

3 $2m - 7$

The expression $2m - 7$ can be written as $2m + (-7)$. So, it is a polynomial. Since it can be written as the sum of two monomials, $2m$ and -7, it is a binomial.

4 $x^2 + 3x - 4 - 5$

The expression $x^2 + 3x - 4 - 5$ can be written as $x^2 + 3x + (-9)$. So, it is a polynomial. Since it can be written as the sum of three monomials, it is a trinomial.

5 $\dfrac{5}{2x} - 3$

The expression $\dfrac{5}{2x} - 3$ is not a polynomial since $\dfrac{5}{2x}$ is not a monomial.

Your Turn

e. $5a - 9 + 3$ **f.** $4m^{-2} + 2$ **g.** $3y^2 - 6 + 7y$

The terms of a polynomial are usually arranged so that the powers of one variable are in descending or ascending order.

Polynomial	Descending Order	Ascending Order	
$2x + x^2 + 1$	$x^2 + 2x + 1$	$1 + 2x + x^2$	
$3y^2 + 5y^3 + y$	$5y^3 + 3y^2 + y$	$y + 3y^2 + 5y^3$	
$x^2 + y^2 + 3xy$	$x^2 + 3xy + y^2$	$y^2 + 3xy + x^2$	*in terms of x*
$2xy + y^2 + x^2$	$y^2 + 2xy + x^2$	$x^2 + 2xy + y^2$	*in terms of y*

The **degree** of a monomial is the sum of the exponents of the variables.

Monomial	Degree
$-3x^2$	2
$5pq^2$	$1 + 2 = 3$
2	0

$p = p^1$
$2 = 2x^0$

Look Back

Zero Power:
Lesson 8–2

To find the degree of a polynomial, find the degree of each term. The greatest of the degrees of its terms is the degree of the polynomial.

Polynomial	Terms	Degree of the Terms	Degree of the Polynomial
$2n + 7$	$2n$, 7	1, 0	1
$3x^2 + 5x$	$3x^2$, $5x$	2, 1	2
$a^6 + 2a^3 + 1$	a^6, $2a^3$, 1	6, 3, 0	6
$5x^4 - 4a^2b^6 + 3x$	$5x^4$, $-4a^2b^6$, $3x$	4, 8, 1	8

Examples

Find the degree of each polynomial.

6 $5a^2 + 3$

Term	Degree
$5a^2$	2
3	0

$3 = 3x^0$

The degree of $5a^2 + 3$ is 2.

7 $6x^2 - 4x^2y - 3xy$

Term	Degree
$6x^2$	2
$-4x^2y$	2 + 1 or 3
$-3xy$	1 + 1 or 2

$y = y^1$
$x = x^1, y = y^1$

The degree of $6x^2 - 4x^2y - 3xy$ is 3.

 Your Turn

h. $3x^2 - 7x$

i. $8m^3 - 2m^2n^2 + 5$

COncepts in MOtion

BrainPOP®

algconcepts.com

Polynomials can be used to represent many real-world situations.

Example

Science Link

8 **The expression $14x^3 - 17x^2 - 16x + 34$ can be used to estimate the number of eggs that a certain type of female moth can produce. In the expression, x represents the width of the abdomen in millimeters. About how many eggs would you expect this type of moth to produce if her abdomen measures 3 millimeters?**

Explore	You know the width of the abdomen. You need to determine the number of eggs that the moth can produce.
Plan	In the expression $14x^3 - 17x^2 - 16x + 34$, replace x with 3 to determine the number of eggs that the moth can produce.

Female Cecropia Moth

Solve
$$14x^3 - 17x^2 - 16x + 34 = 14(3)^3 - 17(3)^2 - 16(3) + 34$$
$$= 14(27) - 17(9) - 16(3) + 34$$
$$= 378 - 153 - 48 + 34$$
$$= 211$$

Examine Check your computations by adding in a different order.
$$378 - 153 - 48 + 34 \stackrel{?}{=} 378 + 34 + (-153) + (-48)$$
$$\stackrel{?}{=} 412 + (-201)$$
$$= 211 \quad \checkmark$$

The moth can produce about 211 eggs.

Check for Understanding

Communicating Mathematics

1. **Write** an expression that is *not* a monomial. Explain why it is *not* a monomial.

2. **Explain** how to find the degree of a polynomial.

3. **Arrange** the terms of the polynomial $3x^2 + 5xy + 2y^2$ so that the powers of x are in ascending order.

4. Writing Math Write a definition for *monomial, polynomial, binomial,* and *trinomial.* Give three examples of each.

> **Vocabulary**
> monomial
> polynomial
> binomial
> trinomial
> degree

Guided Practice

Examples 1 & 2

Determine whether each expression is a monomial. Explain why or why not.

5. $7xy$

6. $a^2 + 5$

Examples 3–5

State whether each expression is a polynomial. If it is a polynomial, identify it as a *monomial, binomial,* or *trinomial.*

7. $2mn + 3$

8. 0

9. $\frac{4}{d} - d^2$

10. $h^2 - 3h + 8 + 2$

Find the degree of each polynomial.

11. $25x$

12. 5

13. $15m^2 + 4n$

14. $-6y^3z - 4y^2z$

Example 8

15. Geometry To find the number of diagonals d in an n-sided figure, you can use the formula $d = 0.5n^2 - 1.5n$.

    a. Find the number of diagonals in an octagon.

    b. How many diagonals are in a hexagon?

Exercises

Practice

Homework Help	
For Exercises	See Examples
16–21	1, 2
22–33, 54, 55	3, 4, 5
33–45, 57	6, 7
56	8
Extra Practice	
See page 710.	

Determine whether each expression is a monomial. Explain why or why not.

16. $5z$

17. $7a + 2$

18. $\dfrac{5}{x}$

19. $8x^2y$

20. y^{-3}

21. $-10a^2b^2c$

State whether each expression is a polynomial. If it is a polynomial, identify it as a *monomial*, *binomial*, or *trinomial*.

22. y^3

23. $2y^2 + 5y - 7$

24. $a^3 - 3a$

25. $2r + 3s^{-3} - t$

26. $17 + 12x^3 - 3x^4$

27. $2m^2 + 5 - 1$

28. $-4j + 2k^2 - 3$

29. $x^2 - \dfrac{1}{2}x$

30. $\dfrac{x}{y}$

31. $-12a^2b^3cd^5e^3$

32. $2.5w^5 - v^3w^2$

33. $3x + 4x$

Find the degree of each polynomial.

34. 1

35. $8x^2$

36. $-14x^3$

37. $5p^2 + 3p + 5$

38. $4s^2 - 6t^8$

39. $p - 4q + 5r$

40. $25a + a^2$

41. $a^3 + a^2 + a$

42. $15g^3h^4 - 10gh^5$

43. $10v^4w^2 + vw^5$

44. $3rs^4 - 2r^2 + 7$

45. $5cd^4 - 2bcd$

Arrange the terms of each polynomial so that the powers of *x* are in descending order.

46. $2x^2 + x^4 - 6$

47. $x^2 - x + 25 - x^5$

48. $3x - 4x^2y + 5 - 3x^5$

49. $6w^3x + 3wx^2 - 10x^6 + 5x^7$

Arrange the terms of each polynomial so that the powers of *x* are in ascending order.

50. $2 + x^4 + x^5 + x^2$

51. $5x^3 - x^2 + 7 + 2x$

52. $a^3x - 6a^3x^3 + 0.5x^5 + 4x^2$

53. $3x^3y + 5xy^4 - x^2y^3 + y^6$

54. *True* or *false*: Every monomial is a polynomial.

55. *True* or *false*: Every polynomial is a monomial.

56. Geometry The surface area of a right cylinder is given by the polynomial $2\pi rh + 2\pi r^2$, where r is the radius of the base of the cylinder and h is the height of the cylinder.

  a. Classify the polynomial.

  b. Find the surface area of a right circular cylinder with a height of 5 feet and a radius of 3 feet. Round to the nearest tenth.

  c. The volume of a cylinder is given by the expression $\pi r^2 h$. Using the same height and radius as in part b, find the volume of the cylinder. Round to the nearest tenth.

57. Medicine Doctors study the heart of a potential heart attack patient by injecting a dye in a vein near the heart. In a healthy heart, the amount of dye in the bloodstream after t seconds is given by the expression $-0.006t^4 + 1.79t - 0.53t^2 + 0.14t^3$.

  a. Arrange the terms of the polynomial so that the powers of t are in descending order.

  b. Find the degree of the polynomial.

58. Critical Thinking A number in base 10 can be written in polynomial form. For example, $3247 = 3(10)^3 + 2(10)^2 + 4(10) + 7(10)^0$.

  a. Write the year of your birth in polynomial form.

  b. Suppose 2356 is a number in base a. Write 2356 in polynomial form.

  c. What is the degree of the polynomial in part b?

Mixed Review

The lengths of three sides of a triangle are given. Determine whether each triangle is a right triangle. *(Lesson 8–7)*

59. 5 cm, 12 cm, 13 cm   **60.** 2 in., 3 in., 4 in.   **61.** 8 ft, 15 ft, 17 ft

Estimate each square root to the nearest whole number. *(Lesson 8–6)*

62. $\sqrt{20}$ **63.** $\sqrt{50}$ **64.** $\sqrt{75}$ **65.** $\sqrt{120}$

66. Population In 2050, Earth's population is expected to be 8,900,000,000. Write this number in scientific notation. *(Lesson 8–4)*

Simplify each expression. *(Lesson 8–3)*

67. xy^{-3} **68.** $\dfrac{k^{-3}}{k^4}$

69. Multiple Choice The graph of $y = 3x + 6$ is shown at the right. Which is the slope of a line parallel to the graph of $y = 3x + 6$? *(Lesson 7–7)*

  Ⓐ -3 Ⓑ 3

  Ⓒ $-\dfrac{1}{3}$ Ⓓ $\dfrac{1}{6}$

What You'll Learn
You'll learn to add and subtract polynomials.

Why It's Important
Framing Addition of polynomials can be used to find the size of a picture.
See Exercise 43.

You can use algebra tiles like the ones below to model polynomials.

A model of the polynomial $2x^2 + 3x + 1$ is shown below.

Red tiles are used to represent -1, $-x$, and $-x^2$. The model of the polynomial $-x^2 - 2x - 3$ is shown below.

Once you know how to model polynomials using algebra tiles, you can use algebra tiles to add and subtract polynomials.

Hands-On Algebra
Algebra Tiles

Materials: algebra tiles

Add $x^2 + 2x - 3$ and $x^2 - x + 4$.

Step 1 Model each polynomial.

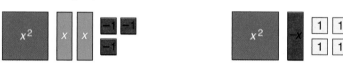

$$x^2 + 2x - 3 \qquad\qquad x^2 - x + 4$$

Look Back

Zero Pair:
Lesson 2–3

Step 2 Combine like shapes and remove all zero pairs. Recall that a zero pair is formed by pairing one tile with its opposite.

Therefore, $(x^2 + 2x - 3) + (x^2 - x + 4) = 2x^2 + x + 1$.

Try These

Use algebra tiles to find each sum.

1. $(3x + 5) + (2x - 3)$
2. $(-2x + 3) + (4x - 3)$
3. $(2x^2 + 2x - 4) + (x^2 + 3x + 7)$
4. $(3x^2 + x - 4) + (x^2 - 2x - 1)$
5. $(x^2 - 1) + (2x + 3)$
6. $(2x^2 + 3) + (x^2 - 2x - 1)$

7. **Write** a rule for adding polynomials without models.

You can add polynomials by grouping the like terms together and then finding their sum.

Examples

Find each sum.

1 $(4x - 3) + (2x + 5)$

Look Back

Like Terms: Lesson 1–4

Method 1

Group the like terms together.

$(4x - 3) + (2x + 5) = (4x + 2x) + (-3 + 5)$ *Group the like terms.*
$= (4 + 2)x + (-3 + 5)$ *Distributive Property*
$= 6x + 2$ *Simplify.*

Method 2

Add in column form.

$$\begin{array}{r} 4x - 3 \\ (+)\ 2x + 5 \\ \hline 6x + 2 \end{array}$$ *Align the like terms.*

2 $(x^2 + 2x - 5) + (3x^2 - x + 4)$

$(x^2 + 2x - 5) + (3x^2 - x + 4) = (x^2 + 3x^2) + (2x - x) + (-5 + 4)$
$= (1 + 3)x^2 + (2 - 1)x + (-5 + 4)$
$= 4x^2 + 1x - 1 \text{ or } 4x^2 + x - 1$

Check your answer by adding in column form.

3 $(2x^2 + 5xy + 3y^2) + (8x^2 - 7y^2)$

$$\begin{array}{r} 2x^2 + 5xy + 3y^2 \\ (+)\ \ \ 8x^2 \qquad\ - 7y^2 \\ \hline 10x^2 + 5xy - 4y^2 \end{array}$$ *Align the like terms.*

Your Turn

a. $(3x + 9) + (5x + 3)$

b. $(-2x^2 + x + 5) + (x^2 - 3x + 2)$

c. $$\begin{array}{r} a^2 - 2ab + 4b^2 \\ (+)\ 7a^2 \qquad\quad - 2b^2 \end{array}$$

d. $(7m^2 - 6) + (5m - 2)$

Look Back

Subtracting
Integers:
Lesson 2–4

Recall that you can subtract an integer by adding its additive inverse or opposite.

$$2 - 3 = 2 + (-3) \quad \textit{The additive inverse of 3 is } -3.$$

$$5 - (-4) = 5 + 4 \quad \textit{The additive inverse of } -4 \textit{ is 4.}$$

Similarly, you can subtract a polynomial by adding its additive inverse. To find the additive inverse of a polynomial, replace each term with its additive inverse.

Polynomial	Additive Inverse
$a + 2$	$-a - 2$
$x^2 + 3x - 1$	$-x^2 - 3x + 1$
$2x^2 - 5xy + y^2$	$-2x^2 + 5xy - y^2$

$-(a + 2) = -a - 2$

$-(x^2 + 3x - 1) = -x^2 - 3x + 1$

$-(2x^2 - 5xy + y^2) = -2x^2 + 5xy - y^2$

Examples

Find each difference.

4 $(6x + 5) - (3x + 1)$

Method 1
Find the additive inverse of $3x + 1$. Then group the like terms together and add.

The additive inverse of $3x + 1$ is $-(3x + 1)$ or $-3x - 1$.

$$
\begin{aligned}
(6x + 5) - (3x + 1) &= (6x + 5) + (-3x - 1) && \textit{Add the additive inverse.}\\
&= (6x - 3x) + (5 - 1) && \textit{Group the like terms.}\\
&= (6 - 3)x + (5 - 1) && \textit{Distributive Property}\\
&= 3x + 4
\end{aligned}
$$

Method 2
Arrange like terms in column form.

$$
\begin{array}{r}
6x + 5 \\
(-)\ 3x + 1 \\
\hline
\end{array}
\qquad \textit{Add the additive inverse.} \qquad
\begin{array}{r}
6x + 5 \\
(+)\ -3x - 1 \\
\hline
3x + 4
\end{array}
$$

5 $(2y^2 - 3y + 5) - (y^2 + 2y + 8)$

The additive inverse of $y^2 + 2y + 8$ is $-(y^2 + 2y + 8)$ or $-y^2 - 2y - 8$.

$$
\begin{aligned}
(2y^2 - 3y + 5) - (y^2 + 2y + 8) &= (2y^2 - 3y + 5) + (-y^2 - 2y - 8)\\
&= (2y^2 - 1y^2) + (-3y - 2y) + (5 - 8)\\
&= (2 - 1)y^2 + (-3 - 2)y + (5 - 8)\\
&= 1y^2 + (-5y) + (-3) \text{ or } y^2 - 5y - 3
\end{aligned}
$$

To check the answer, add $y^2 - 5y - 3$ and $y^2 + 2y + 8$. The sum should be $2y^2 - 3y + 5$.

6 $(3x^2 + 5) - (-4x + 2x^2 + 3)$

Reorder the terms of the second polynomial so that the powers of x are in descending order.
$-4x + 2x^2 + 3 = 2x^2 - 4x + 3$

Then arrange like terms in column form.

$$
\begin{array}{r}
3x^2 \qquad + 5 \\
(-)\ 2x^2 - 4x + 3 \\
\hline
\end{array}
$$
Add the additive inverse.
$$
\begin{array}{r}
3x^2 \qquad + 5 \\
(+)\ -2x^2 + 4x - 3 \\
\hline
x^2 + 4x + 2
\end{array}
$$

Your Turn

e. $(3x - 2) - (5x - 4)$

f. $(10x^2 + 8x - 6) - (3x^2 + 2x - 9)$

g.
$$
\begin{array}{r}
6m^2 \qquad + 7 \\
(-)\ -2m^2 + 2m - 3 \\
\hline
\end{array}
$$

h. $(5x^2 - 4x) - (2 - 3x)$

Polynomials can be used to represent measures of geometric figures.

Example

Geometry Link

7 **The perimeter of triangle ABC is $7x + 2y$. Find the measure of the third side of the triangle.**

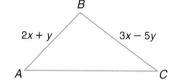

Explore You know the perimeter of the triangle and the measures of two sides. You need to find AC, the measure of the third side.

Plan The perimeter of a triangle is the sum of the measures of the three sides. To find AC, subtract the two given measures from the perimeter.

Solve $AC = Perimeter - AB - BC$
$$
\begin{aligned}
AC &= (7x + 2y) - (2x + y) - (3x - 5y) \\
&= (7x + 2y) + (-2x - y) + (-3x + 5y) \\
&= [7x + (-2x) + (-3x)] + [2y + (-y) + 5y] \\
&= [7 + (-2) + (-3)]x + [2 + (-1) + 5]y \\
&= 2x + 6y
\end{aligned}
$$

The measure of the third side of the triangle, AC, is $2x + 6y$.

Examine Check by adding the measures of the three sides.

$$
\begin{array}{r}
2x +\ \ y \\
3x - 5y \\
(+)\ 2x + 6y \\
\hline
7x + 2y
\end{array}
$$
The perimeter is $7x + 2y$. The answer checks.

 Personal Tutor at algconcepts.com

Check for Understanding

Communicating Mathematics

1. **Describe** the first step you take when you add or subtract polynomials in column form.
2. **Explain** how addition and subtraction of polynomials are related.

Guided Practice

⏱ **Getting Ready**   **Find the additive inverse of each polynomial.**

Sample 1: $6y - 3z$ **Sample 2:** $-a^2 + 2ab - 3b^2$

Solution: $-6y + 3z$ **Solution:** $a^2 - 2ab + 3b^2$

3. $2a + 9b$
4. $-4m + 6n$
5. $x^2 + 8x + 5$
6. $-3h^2 - 2h - 3$
7. $4xy^2 + 6x^2y - y^3$
8. $-2c^3 + c^2 - 3c$

Examples 1–3

Find each sum.

9. $6y - 5$
    $\underline{(+)\; 2y + 7}$

10. $x^2 - 6x + 5$
    $\underline{(+)\; x^2 + 4x - 7}$

11. $(2x^2 + 4x + 5) + (x^2 - 5x - 3)$
12. $(2x^2 - 5x + 4) + (3x^2 - 1)$

Examples 4–6

Find each difference.

13. $3x + 4$
    $\underline{(-)\; x + 2}$

14. $6m^2 - 5m + 3$
    $\underline{(-)\; 5m^2 + 2m - 7}$

15. $(5x^2 + 4x - 1) - (4x^2 + x + 2)$
16. $(5x^2 - 4x) - (-3x + 2)$

Example 7

17. **Geometry**   The perimeter of triangle DEF is $12x^2 - 7x + 9$. Find the measure of the third side of the triangle.

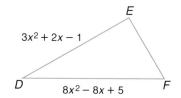

Exercises

• • • • •   • • • • • • • • • • • • • • • •

Practice

Find each sum.

18. $2x + 3$
    $\underline{(+)\; x - 1}$

19. $2x^2 - 5x + 4$
    $\underline{(+)\; 2x^2 + 8x - 1}$

20. $2x^2 + 3xy - 4y^2$
    $\underline{(+)\; 4x^2 \qquad + 4y^2}$

21. $(12x + 8y) + (2x - 7y)$
22. $(-7y + 3x) + (4x + 3y)$
23. $(n^2 + 5n + 3) + (2n^2 + 8n + 8)$
24. $(5x^2 - 7x + 9) + (3x^2 + 4x - 6)$
25. $(5n^2 - 4) + (2n^2 - 8n + 9)$
26. $(-2x^2 + 3xy - 3y^2) + (2x^2 - 5xy)$

Homework Help	
For Exercises	See Examples
18–26, 37, 38, 40	1, 2, 3
27–36, 39, 41, 44	4, 5, 6
42	7
Extra Practice	
See page 710.	

Find each difference.

27. $9x + 5$
    $\underline{(-)\; 4x - 3}$

28. $5a^2 + 7a + 9$
    $\underline{(-)\; 3a^2 + 4a + 1}$

29. $3x^2 + 4x - 1$
    $\underline{(-)\; 4x^2 + \; x + 2}$

30. $(3x - 4) - (5x + 2)$

31. $(9x - 4y) - (12x - 5y)$

32. $(2x^2 + 5x + 3) - (x^2 - 2x + 3)$

33. $(3a^2 + 2a - 5) - (2a^2 - a - 4)$

34. $(3x^2 + 5x + 4) - (-1 + x^2)$

35. $(5x^2 - 4xy) - (2y^2 - 3xy)$

Find each sum or difference.

36. $(6x^3 - 10) - (2x^3 + 5x + 7)$

37. $(2pq + 3qr - 4pr) + (5pr - 3qr)$

38. $(3 + 2a - a^2) + (a^2 + 8a + 5)$

39. $(-x^2y - 4x^2y^2) - (x^2y + 3x^2y^2)$

40. Find the sum of $(x^2 + x + 5)$, $(3x^2 - 4x - 2)$, and $(2x^2 + 2x - 1)$.

41. What is $x^2 + 6x$ minus $3x^2 + 7$?

Applications and Problem Solving

42. Geometry   The measure of each side of a triangle is $3x + 1$. Find its perimeter.

43. Framing   The standard measurement for a custom-made picture frame is the *united inch*. It refers to the sum of the length and the width of a picture to be framed. Suppose a picture's length is 4 inches longer than its width. Then the width is w and the length would be $w + 4$.

a. Write a polynomial that represents the size of the picture in united inches.

b. If the width is 15 inches, use the polynomial from part a to find the size of the picture in united inches.

44. Critical Thinking   One polynomial is subtracted from a second polynomial and the difference is $3m^2 + m - 5$. What is the difference when the second polynomial is subtracted from the first?

Mixed Review

Find the degree of each polynomial.   *(Lesson 9–1)*

45. $y^2 + 3y + 2$

46. $n^6 + 3n^3 + 2$

47. $3x^2 + 2x^2y$

48. Sailing   A rope from the top of a mast on a sailboat extends to a point 6 feet from the base of the mast. Suppose the rope is 24 feet long. To the nearest foot, how high is the mast?   *(Lesson 8–7)*

Simplify each expression.   *(Lesson 8–2)*

49. $(x^2y)(x^3y^2)$

50. $(4a^2)(2a^4)$

51. $(-2y)(-5xy)$

Standardized Test Practice
Ⓐ Ⓑ Ⓒ Ⓓ

52. Multiple Choice   Which relation is a function?   *(Lesson 6–4)*

Ⓐ $\{(3, 1), (4, 2), (2, 3), (1, 2)\}$

Ⓑ $\{(1, 1), (-1, 5), (3, 3), (-1, 2)\}$

Ⓒ $\{(-1, 3), (1, 4), (3, 2), (1, 2)\}$

Ⓓ $\{(-2, 1), (4, 2), (3, 3), (3, 2)\}$

What You'll Learn

You'll learn to multiply a polynomial by a monomial.

Why It's Important

Recreation You can use monomials and polynomials to solve problems involving recreation.

See Exercise 65.

Suppose you have a square whose length and width are x units. If you increase the length by 3 units, what is the area of the new figure?

You can model this problem by using algebra tiles. The figures show how to make a rectangle whose length is $x + 3$ units and whose width is x units.

The area of any rectangle is the product of its length and its width. The area can also be found by adding the areas of the tiles.

Formula

$A = \ell w$
$= (x + 3)x$ or $x(x + 3)$
$= x^2 + 3x$

Algebra Tiles

$A = x^2 + x + x + x$
$= x^2 + 3x$

Since the areas are equal, $x(x + 3) = x^2 + 3x$. Another expression for the same area is $x^2 + 3x$ square units.

The example above shows how the Distributive Property can be used to multiply a polynomial by a monomial.

Multiplying a Polynomial by a Monomial	**Words:**	To multiply a polynomial by a monomial, use the Distributive Property.
	Symbols:	$a(b + c) = ab + ac$
	Model:	

Examples

Find each product.

❶ $y(y + 5)$

$y(y + 5) = y(y) + y(5)$
$= y^2 + 5y$

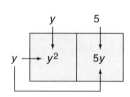

Look Back

Multiplying Powers:
Lesson 8–2

❷ $b(2b^2 + 3)$

$b(2b^2 + 3) = b(2b^2) + b(3)$
$= 2b^3 + 3b$

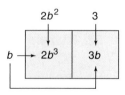

③ $-2n(7 - 5n^2)$

$-2n(7 - 5n^2) = -2n(7) + (-2n)(-5n^2)$

$\qquad\qquad\quad\ = -14n + 10n^3$

	7	$-5n^2$
$-2n$	$-14n$	$10n^3$

④ $3x^3(2x^2 - 5x + 8)$

$3x^3(2x^2 - 5x + 8) = 3x^3(2x^2) + 3x^3(-5x) + 3x^3(8)$

$\qquad\qquad\qquad\quad\ = 6x^5 - 15x^4 + 24x^3$

Your Turn

a. $7(2x + 5)$ **b.** $4x(3x^2 - 7)$

c. $-5a(6 - 3a^2)$ **d.** $2m^2(5m^2 - 7m + 8)$

Many equations contain polynomials that must be multiplied.

Examples

Solve each equation.

⑤ $11(y - 3) + 5 = 2(y + 22)$

$11(y - 3) + 5 = 2(y + 22)$	*Original equation*
$11y - 33 + 5 = 2y + 44$	*Distributive Property*
$11y - 28 = 2y + 44$	*Combine like terms.*
$11y - 28 - 2y = 2y + 44 - 2y$	*Subtract 2y from each side.*
$9y - 28 = 44$	*Simplify.*
$9y - 28 + 28 = 44 + 28$	*Add 28 to each side.*
$9y = 72$	*Simplify.*
$\dfrac{9y}{9} = \dfrac{72}{9}$	*Divide each side by 9.*
$y = 8$	The solution is 8.

Look Back

Solving Equations
with Grouping
Symbols:
Lesson 4–7

⑥ $w(w + 12) = w(w + 14) + 12$

$w(w + 12) = w(w + 14) + 12$	*Original equation*
$w^2 + 12w = w^2 + 14w + 12$	*Distributive Property*
$w^2 + 12w - w^2 = w^2 + 14w + 12 - w^2$	*Subtract w^2 from each side.*
$12w = 14w + 12$	*Simplify.*
$12w - 14w = 14w + 12 - 14w$	*Subtract 14w from each side.*
$-2w = 12$	*Simplify.*
$\dfrac{-2w}{-2} = \dfrac{12}{-2}$	*Divide each side by -2.*
$w = -6$	The solution is -6.

Your Turn

e. $2(5x - 12) = 6(-2x + 3) + 2$ **f.** $a(a + 2) + 3a = a(a - 3) + 8$

Lesson 9–3 Multiplying a Polynomial by a Monomial

You can apply multiplication of a polynomial by a monomial to problems involving area.

Example **7**

Geometry Link

Find the area of the shaded region in simplest form.

Subtract the area of the smaller rectangle from the area of the larger rectangle.

area of larger rectangle: $3x(x + 2)$ $A = \ell w$
area of smaller rectangle: $2x(x)$
area of shaded region: $3x(x + 2) - 2x(x)$

$$A = 3x(x + 2) - 2x(x) \quad \textit{Area of shaded region}$$
$$= 3x(x) + 3x(2) - 2x(x) \quad \textit{Distributive Property}$$
$$= 3x^2 + 6x - 2x^2 \quad \textit{Multiply.}$$
$$= 1x^2 + 6x \text{ or } x^2 + 6x \quad \textit{Combine like terms.}$$

The area of the shaded region is $x^2 + 6x$.

🌐nline **Personal Tutor at** algconcepts.com

Check for Understanding

Communicating Mathematics

1. **Name** the property used to express $3n(2n - 6)$ as $6n^2 - 18n$.

2. **Write** an expression for the area of the rectangle at the right as:

   a. a product of a monomial and a polynomial

   b. a sum of monomials

3. Consuelo says that $2x(3x + 4) = 6x^2 + 8x$ is a true statement. Shawn says that $2x(3x + 4) = 6x^2 + 4$ is a true statement. Who is correct? Explain.

Guided Practice
Examples 1–4

Find each product.

4. $2(y + 5)$ 5. $x(x + 2)$ 6. $3a(a - 1)$

7. $-4(x - 2)$ 8. $-3n(5 - 2n)$ 9. $4z(z^2 - 2z)$

10. $2(2d^2 + 3d + 8)$ 11. $3(8y^2 + 3y - 5)$ 12. $-2a(4a^2 - 3a - 1)$

Examples 5 & 6

Solve each equation.

13. $7(x + 2) = 42$ 14. $4(y - 8) + 10 = 2y + 12$

15. $-3(2a - 4) + 9 = 3(a + 1)$ 16. $x(x + 3) + 5x = x(x + 5) + 9$

Example 7

17. **Geometry** Find the area of the shaded region in simplest form.

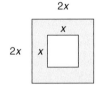

Exercises ● ● ● ● ● ● ● ● ● ● ● ● ● ● ● ● ●

Practice

Homework Help	
For Exercises	See Examples
18–26, 62, 65–66	1, 2
27–44, 57–61	3, 4
45–58	5, 6
63, 64	7
Extra Practice	
See page 710.	

Find each product.

18. $2(x + 6)$

19. $-3(y + 3)$

20. $7(2a + 3)$

21. $x(x - 5)$

22. $n(n + 4)$

23. $z(3z - 2)$

24. $2m(m + 4)$

25. $4x(2x - 3)$

26. $5y(y + 1)$

27. $-2a(a - 2)$

28. $-3x(x - 5)$

29. $-5y(6 - 2y)$

30. $3s(4s^2 - 7)$

31. $5d(d^2 + 3)$

32. $-7p(-3p - 6)$

33. $-2a(5a^2 - 7a + 2)$

34. $4x(-2x + 7x^2)$

35. $5n(8n^3 + 7n^2 - 3n)$

36. $-3y(6 - 9y + 4y^2)$

37. $7(-2a^2 + 5a - 11)$

38. $-5x^2(3x^2 - 8x - 12)$

39. $1.2(c^2 - 10)$

40. $0.1(4x^2 - 7x)$

41. $0.25x(4x - 8)$

42. $\frac{1}{2}(2x^2 - 6x)$

43. $\frac{2}{3}(6y^2 - 9y + 3)$

44. $\frac{1}{4}x(12x^2 + 8x)$

Solve each equation.

45. $8(x - 3) = 16$

46. $4(y - 3) = -8$

47. $40 = 5(a + 10)$

48. $2(x + 4) + 9 = 5x - 1$

49. $6(a + 2) - 5 = 2a + 3$

50. $8x + 14 = 3(x - 2) + 10$

51. $2(5a - 13) = 6(-2a + 3)$

52. $3(2y - 4) = -2(2y - 9)$

53. $5(n + 7) = 3(-2n + 1) - 1$

54. $b(b + 12) = b(b + 14) + 12$

55. $c(c + 8) - c(c + 3) - 23 = 3c + 11$

56. $m(m - 8) + 3m = -2 + m(m - 9)$

57. What is the product of $4x$ and $x^2 - 2x + 1$?

58. Multiply $-2n$ and $3 - 5n$.

Simplify.

59. $2a(a^2 - 5a + 4) - 6(a^3 + 3a - 11)$

60. $5t^2(t + 2) - 5t(4t^2 - 3t + 6) + 3(t^2 - 6)$

61. $3n^2(n - 4) + 6n(3n^2 + n - 7) - 4(n - 7)$

Applications and Problem Solving

62. Sports The length of a pool table is twice as long as its width. Suppose the width of a pool table is $4x - 1$. What is the length of the table?

Geometry Find the area of the shaded region for each figure.

63.

64.

65. Recreation On the Caribbean island of Trinidad, children play a form of hopscotch called Jumby. The pattern for this game is shown at the right. Suppose the length of each rectangle is $y + 5$ units long and y units wide.

a. Write an expression in simplest form for the area of the pattern.

b. If y represents 10 inches, find the area of the pattern.

66. Critical Thinking Write three different multiplication expressions whose product is $6a^2 + 8a$.

Mixed Review

Find each sum or difference. *(Lesson 9–2)*

67. $(2x + 1) + (5x - 3)$

68. $(3a - 2) - (a + 4)$

69. $(x^2 + 2x - 3) + (2x^2 - 4x)$

70. $(y^2 + 8y) - (2y^2 + 5)$

71. Ecology The deer population of the Kaibab Plateau in Arizona from 1905 to 1930 can be estimated by the polynomial $-0.13x^5 + 3.13x^4 + 4000$, where x is the number of years after 1900. Find the degree of the polynomial. *(Lesson 9–1)*

Simplify. *(Lesson 8–5)*

72. $\sqrt{81}$

73. $-\sqrt{121}$

74. $\sqrt{256}$

75. $\sqrt{\dfrac{9}{100}}$

Find the odds of each outcome if a die is rolled. *(Lesson 5–6)*

76. a number less than 5

77. an odd number

Standardized Test Practice
Ⓐ Ⓑ Ⓒ Ⓓ

78. Multiple Choice Which equation is equivalent to $12 + (x - 7) = 21$? *(Lesson 4–7)*

Ⓐ $12x - 7 = 21$

Ⓑ $12x - 84 = 21$

Ⓒ $x + 5 = 21$

Ⓓ $19 + x = 21$

Quiz 1   Lessons 9–1 through 9–3

1. Find the degree of $3y^4 + 2y^3 - y^2 + 5y - 3$. *(Lesson 9–1)*

2. Arrange the terms of $2xy + x^2 - y^2$ so that the powers of x are in descending order. *(Lesson 9–1)*

Find each sum or difference. *(Lesson 9–2)*

3. $(4x^2 + 3x) - (6x^2 - 5x + 2)$

4. $(2w^2 - 4w - 12) + (15 - 3w^2 + 2w)$

5. Sports In the National Football League, the length of the playing field is 40 feet longer than twice its width. *(Lesson 9–3)*

a. Express the width and length of the playing field as polynomials.

b. Find the polynomial that represents the area of the playing field.

---

www.algconcepts.com/self_check_quiz

What You'll Learn

You'll learn to multiply two binomials.

Why It's Important

Packaging The FOIL method can be used to find the dimensions of a cereal box. *See Exercise 55.*

Katie has a square herb garden in which she grows parsley. The measure of each side is x feet. She wants to increase the length by 2 feet and the width by 1 foot so she can grow sage, rosemary, and thyme. A plan for the garden is shown at the right. Find two different expressions for the area of the new garden.

One way to find the area of the garden is to use the formula for area. Find the product of the length and width of the new garden.

Formula

$$A = \ell w$$
$$= (x + 2)(x + 1)$$

You can also find the area by adding the areas of the smaller regions.

Sum of Regions

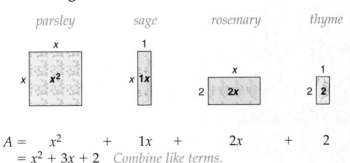

$$A = \quad x^2 \quad + \quad 1x \quad + \quad 2x \quad + \quad 2$$
$$= x^2 + 3x + 2 \quad \textit{Combine like terms.}$$

Since the areas are equal, both expressions are equal. Therefore, $(x + 2)(x + 1) = x^2 + 3x + 2$.

The multiplication expression can also be shown in the model below. Notice that the Distributive Property is used twice.

$$x(x + 1) = x(x) + x(1) \quad \textit{Area of top row}$$
$$= x^2 + 1x$$

$$2(x + 1) = 2(x) + 2(1) \quad \textit{Area of bottom row}$$
$$= 2x + 2$$

So, $(x + 2)(x + 1) = x^2 + 1x + 2x + 2$ or $x^2 + 3x + 2$.

Materials: ✏ straightedge

Use a model to find the product of $(x + 2)$ and $(x - 1)$.

Step 1 Put the $x + 2$ and $x - 1$ outside the box as shown. *Note that $x - 1 = x + (-1)$.*

	x	-1
x	x^2	$-1x$
2		

Step 2 Use the Distributive Property to multiply x by $x - 1$ and place the products inside the boxes.

Step 3 Use the Distributive Property to multiply 2 by $x - 1$ and place the products inside the boxes.

	x	-1
x	x^2	$-1x$
2	$2x$	-2

Step 4 Find the sum of the terms inside the boxes: $x^2 - 1x + 2x - 2 = x^2 + 1x - 2$ or $x^2 + x - 2$.

Therefore, $(x + 2)(x - 1) = x^2 + x - 2$.

Try These Use a model to find each product.

1. $(x + 2)(x + 3)$ **2.** $(x + 3)(x + 4)$ **3.** $(2x + 1)(x + 1)$

4. $(x - 2)(x + 1)$ **5.** $(x - 3)(x - 2)$ **6.** $(x - 1)(x + 1)$

You can also use the Distributive Property to multiply binomials.

Examples

Find each product.

1 $(x + 3)(x - 4)$

$$
\begin{aligned}
(x + 3)(x - 4) &= x(x - 4) + 3(x - 4) & \text{\textit{Distributive Property}}\\
&= x(x) + x(-4) + 3(x) + 3(-4) & \text{\textit{Distributive Property}}\\
&= x^2 - 4x + 3x - 12 & \text{\textit{Simplify.}}\\
&= x^2 - 1x - 12 \text{ or } x^2 - x - 12 & \text{\textit{Combine like terms.}}
\end{aligned}
$$

2 $(2y - 1)(y - 3)$

$(2y - 1)(y - 3)$

$$
\begin{aligned}
&= 2y(y - 3) + (-1)(y - 3) & \text{\textit{Distributive Property}}\\
&= 2y(y) + 2y(-3) + (-1)(y) + (-1)(-3) & \text{\textit{Distributive Property}}\\
&= 2y^2 - 6y - 1y + 3 & \text{\textit{Simplify.}}\\
&= 2y^2 - 7y + 3 & \text{\textit{Combine like terms.}}
\end{aligned}
$$

Your Turn

a. $(y + 4)(y - 2)$ **b.** $(m - 3)(m + 1)$

c. $(x - 5)(x - 2)$ **d.** $(2a - 3)(a - 2)$

Two binomials can always be multiplied using the Distributive Property. However, the following shortcut can also be used. It is called the **FOIL method**.

$$(3x + 1)(x + 2) = (3x)(x) \quad + \quad (3x)(2) \quad + \quad (1)(1x) \quad + \quad (1)(2)$$

F	O	I	L
product of FIRST terms	product of OUTER terms	product of INNER terms	product of LAST terms

$$= 3x^2 + 6x + 1x + 2$$
$$= 3x^2 + 7x + 2$$

COncepts in MOtion
Animation
algconcepts.com

FOIL Method for Multiplying Two Binomials	To multiply two binomials, find the sum of the products of **F** the First terms, **O** the Outer terms, **I** the Inner terms, and **L** the Last terms.

Examples

Find each product.

3 $(y + 4)(y + 6)$

$$(y + 4)(y + 6) = \overset{F}{(2x)(2x)} + \overset{O}{(2x)(2)} + \overset{I}{(-3)(2x)} + \overset{L}{(-3)(2)}$$
$$= y^2 + 6y + 4y + 24 \quad \textit{Multiply.}$$
$$= y^2 + 10y + 24 \quad \textit{Combine like terms.}$$

4 $(2x - 3)(2x + 2)$

$$(2x - 3)(2x + 2) = \overset{F}{(2x)(2x)} + \overset{O}{(2x)(2)} + \overset{I}{(-3)(2x)} + \overset{L}{(-3)(2)}$$
$$= 4x^2 + 4x - 6x - 6 \quad \textit{Multiply.}$$
$$= 4x^2 - 2x - 6 \quad \textit{Combine like terms.}$$

5 $(3a - b)(2a + 4b)$

$$(3a - b)(2a + 4b) = (3a)(2a) + (3a)(4b) + (-b)(2a) + (-b)(4b)$$
$$= 6a^2 + 12ab - 2ab - 4b^2$$
$$= 6a^2 + 10ab - 4b^2$$

Your Turn

e. $(n + 3)(n + 5)$ **f.** $(x - 4)(2x + 3)$ **g.** $(2x + y)(3x - 2y)$

Example ──⑥ Find the product of $x^2 - 4$ and $x + 3$.

$$(x^2 - 4)(x + 3) = (x^2)(x) + (x^2)(3) + (-4)(x) + (-4)(3)$$
$$= x^3 + 3x^2 - 4x - 12 \quad \textit{There are no like terms.}$$

Your Turn

h. Find the product of $x - 2$ and $x^2 + 3$.

You can use the FOIL method to solve problems involving volume.

Example ──⑦
Geometry Link

The volume V of a rectangular prism is equal to the area of the base B times the height h. Express the volume of the prism as a polynomial. Use $V = Bh$.

$x - 1$ units

x units

$x + 6$ units

First, find the area of the base. The base is a rectangle.

$B = \ell w$
$\quad = (x + 6)(x) \quad \textit{Replace } \ell \textit{ with } x + 6 \textit{ and } w \textit{ with } x.$
$\quad = x^2 + 6x \quad \textit{Use the Distributive Property.}$

To find the volume, multiply the area of the base by the height.

$V = Bh$
$\quad = (x^2 + 6x)(x - 1) \quad \textit{Replace } B \textit{ with } x^2 + 6x \textit{ and } h \textit{ with } x - 1.$
$\quad = (x^2)(x) + (x^2)(-1) + (6x)(x) + (6x)(-1) \quad \textit{Use the FOIL method.}$
$\quad = x^3 - x^2 + 6x^2 - 6x \quad \textit{Multiply.}$
$\quad = x^3 + 5x^2 - 6x \quad \textit{Combine like terms.}$

The volume of the prism is $x^3 + 5x^2 - 6x$ cubic units.

🌐nline **Personal Tutor at** algconcepts.com

Check for Understanding

Communicating Mathematics

1. **Draw** a model that represents the product of $x - 2$ and $x + 3$. Then find the product.

Vocabulary

FOIL method

2. **Write** the multiplication expression represented by each model.

a.

	x	-4
x	x^2	$-4x$
-3	$-3x$	$+12$

b.

	x	-4
$2x$	$2x^2$	$-8x$
5	$5x$	-20

3. **Writing Math** Compare and contrast the procedure for multiplying two binomials and the procedure for multiplying a binomial and a monomial.

Guided Practice

Getting Ready Find the sum of the products of the inner terms and the outer terms.

Sample: $(x - 6)(x + 2)$ **Solution:** $(x - 6)(x + 2)$

$$-6x + 2x = -4x$$

4. $(a - 5)(a - 3)$ 5. $(x + 1)(x + 5)$ 6. $(g + 1)(g - 9)$
7. $(2m + 3)(3m + 2)$ 8. $(2j + 1)(j - 3)$ 9. $(y + 5)(2y + 4)$

Examples 1–5

Find each product. Use the Distributive Property or the FOIL method.
10. $(x + 2)(x + 4)$ 11. $(w - 7)(w + 5)$
12. $(a - 3)(a - 6)$ 13. $(2y + 3)(y - 1)$
14. $(3x - 4)(2x - 3)$ 15. $(2y + 7)(3y - 5)$
16. $(x + 2y)(x + 3y)$ 17. $(2a - 5b)(a + 2b)$
18. $(y^2 + 3)(y - 2)$ 19. $(m^3 - 2m)(m + 3)$

Example 6

20. **Geometry** Express the volume of the rectangular prism as a polynomial. Use the formula $V = Bh$, where B is the area of the base and h is the height of the prism.

$x - 3$ in.

$2x$ in.

$x + 4$ in.

Exercises

Find each product. Use the Distributive Property or the FOIL method.
21. $(x + 4)(x + 8)$ 22. $(r + 7)(r + 2)$ 23. $(a - 3)(a + 7)$

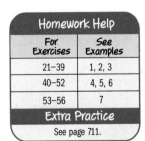

Homework Help	
For Exercises	**See Examples**
21–39	1, 2, 3
40–52	4, 5, 6
53–56	7
Extra Practice	
See page 711.	

24. $(x - 3)(x - 7)$ 25. $(n - 11)(n - 5)$ 26. $(y - 4)(y + 15)$
27. $(z + 6)(z - 4)$ 28. $(s - 11)(s + 5)$ 29. $(x - 4)(3x + 2)$
30. $(2a + 5)(a - 7)$ 31. $(z - 4)(3z - 5)$ 32. $(y - 3)(2y - 5)$
33. $(5n + 2)(n - 3)$ 34. $(x + 7)(2x - 3)$ 35. $(3a + 1)(3a + 1)$
36. $(2x + 5)(5x + 3)$ 37. $(4h - 3)(3h + 2)$ 38. $(7a - 1)(2a - 3)$
39. $(2x + 7)(x - 3)$ 40. $(8m + 2n)(6m + 5n)$ 41. $(2y - 4z)(3y - 6z)$
42. $(3a - b)(2a + b)$ 43. $(2x + 5y)(3x - y)$ 44. $(5n - 2p)(5n + 2p)$
45. $(x^2 + 1)(x - 2)$ 46. $(y + 3)(y^2 - 4)$ 47. $(x^2 + 3)(3x^2 - 1)$
48. $(2y^2 + 1)(y + 1)$ 49. $(x^2 + 2x)(x - 3)$ 50. $(x + a)(x + b)$

51. Find the product of $(x + 3)$ and $(x - 3)$.
52. What is the product of x, $(2x + 1)$, and $(x - 3)$?

53. **Geometry** Find the area of each shaded region in simplest form.

a.

$x + 2$

$x + 1$ | x | x

b.

$2x + 1$

$2x$ | $x - 1$ | $x - 1$

54. **Number Theory** Use the FOIL method to find the product of 16 and 38. (*Hint*: Write 16 as $10 + 6$ and 38 as $30 + 8$.)

55. **Packaging** A cereal box has a length of $2x$ inches, a width of $x - 2$ inches, and a height of $2x + 5$ inches.
    a. Express the volume of the package as a polynomial.
    b. Find the volume of the box if $x = 5$.

56. **Critical Thinking** Use the Distributive Property to find the product of $x + 2$ and $x^2 - 3x + 2$.

Mixed Review

57. **Clocks** Before mechanical clocks were invented, candles were used to keep track of time. One formula that was used was $c = 2(5 - t)$, where c is the height of the candle in inches and t is the time in hours that the candle burns. *(Lesson 9–3)*
    a. Use the Distributive Property to multiply 2 and $5 - t$.
    b. Find the height of the candle when $t = 2$.

58. Find the sum of $x^2 + 5x$ and $-3x - 7$. *(Lesson 9–2)*

Write each expression using exponents. *(Lesson 8–1)*

59. $2 \cdot 2 \cdot 2 \cdot x \cdot x \cdot x$ 60. 10 61. $(3)(3)(-2)(-2)(-2)$

Graph each equation. *(Lesson 7–5)*

62. $y = 2x + 3$ 63. $y = -x + 2$ 64. $4x + 5y = 20$

Math Data
nline Update
For the latest information on breakfast foods, visit:
algconcepts.com

65. **Food** The graph shows the favorite breakfast food for students ages 9–12. Suppose you interview 200 students who are in this age range. How many would you expect to like cereal for breakfast? *(Lesson 5–4)*

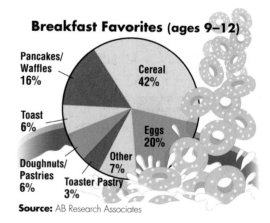

Breakfast Favorites (ages 9–12)

Pancakes/ Waffles 16%

Cereal 42%

Toast 6%

Eggs 20%

Doughnuts/ Pastries 6%

Other 7%

Toaster Pastry 3%

Source: AB Research Associates

Standardized Test Practice
Ⓐ Ⓑ Ⓒ Ⓓ

66. **Extended Response** Draw and label a rectangle whose perimeter is 20 centimeters. *(Lesson 1–5)*

Some products of polynomials appear frequently in real-life problems. Expressions like $(a + b)^2$, $(a - b)^2$, and $(a + b)(a - b)$ occur so often that it is helpful to develop patterns for their products.

To develop these patterns, we will model a simpler expression, $(x + 1)^2$, geometrically.

The area of a square with a side length of $x + 1$ is $(x + 1)(x + 1)$ or $(x + 1)^2$. The total area can also be found by adding the areas of the inner regions together. The right side of the equation below represents the area of the square as the sum of the areas of the four small squares.

$$(x + 1)^2 = x^2 + 1x + 1x + 1$$
$$= x^2 + 2x + 1$$

→ *twice the product of 1 and x*

You can use a similar model to find $(x - 1)^2$.   *x − 1 = x + (−1)*

$$(x - 1)^2 = x^2 - 1x - 1x + 1$$
$$= x^2 - 2x + 1$$

→ *twice the product of −1 and x*

The square of a sum and the square of a difference can be found by using the following rules.

	Symbols:	$(a + b)^2 = a^2 + 2ab + b^2$
Square of a Sum and Square of a Difference		$(a - b)^2 = a^2 - 2ab + b^2$
	Models:	

	a	b
a	a^2	ab
b	ab	b^2

	a	−b
a	a^2	$-ab$
−b	$-ab$	b^2

Find each product.

1 $(x + 4)^2$

$(a + b)^2 = a^2 + 2ab + b^2$ *Square of a Sum*
$(x + 4)^2 = x^2 + 2(x)(4) + 4^2$ *Replace a with x and b with 4.*
$\qquad\quad = x^2 + 8x + 16$ *Simplify.*

2 $(4m + n)^2$

$(a + b)^2 = a^2 + 2ab + b^2$ *Square of a Sum*
$(4m + n)^2 = (4m)^2 + 2(4m)(n) + n^2$ *Replace a with 4m and b with n.*
$\qquad\qquad = 16m^2 + 8mn + n^2$ *Simplify.*

3 $(w - 5)^2$

$(a - b)^2 = a^2 - 2ab + b^2$ *Square of a Difference*
$(w - 5)^2 = w^2 - 2(w)(5) + 5^2$ *Replace a with w and b with 5.*
$\qquad\quad = w^2 - 10w + 25$ *Simplify.*

4 $(3p - 2q)^2$

$(a - b)^2 = a^2 - 2ab + b^2$ *Square of a Difference*
$(3p - 2q)^2 = (3p)^2 - 2(3p)(2q) + (2q)^2$ *Replace a with 3p and b with 2q.*
$\qquad\qquad = 9p^2 - 12pq + 4q^2$ *Simplify.*

Your Turn

a. $(y + 3)^2$ 　　　　　　　　　**b.** $(3g + h)^2$
c. $(d - 2)^2$ 　　　　　　　　　**d.** $(4x - 3y)^2$

Biologists use a method that is similar to squaring a sum to find the characteristics of offspring based on genetic information.

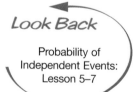

Example

Biology Link

5 **In a certain population, a parent has a 10% chance of passing the gene for brown eyes to its offspring. If an offspring receives one eye-color gene from its mother and one from its father, what is the probability that an offspring will receive at least one gene for brown eyes?**

There is a 10% chance of passing the gene for brown eyes. Therefore, there is a 90% chance of *not* passing the gene.

Look Back

Probability of Independent Events: Lesson 5–7

Use the model at the right to show all possible combinations. In the model, B represents the gene for brown eyes and b represents the gene for *not* brown eyes. *Note that the percents are written as decimals.*

	Father	
	B = 0.1	b = 0.9
Mother B = 0.1	BB (0.1)(0.1)= 0.01	Bb (0.1)(0.9)= 0.09
b = 0.9	bB (0.9)(0.1)= 0.09	bb (0.9)(0.9)= 0.81

www.algconcepts.com/extra_examples

Three of the four small squares in the model contain a B. Add their probabilities.

$0.01 + 0.09 + 0.09 = 0.19$

So, the probability of an offspring receiving at least one gene for brown eyes is 0.19 or 19%.

 Personal Tutor at algconcepts.com

You can use the FOIL method to find product of the sum and difference of the same two terms. Consider $(x + 3)(x - 3)$.

	x	-3
x	x^2	$-3x$
3	$3x$	-9

$$(x + 3)(x - 3) = x^2 - 3x + 3x - 9$$
$$= x^2 - 9$$

→ *square of the second term*
→ *square of the first term*

Note that the product is the difference of the squares of the terms. The product of a sum and a difference can be found by using the following rule.

Product of a Sum and a Difference

Symbols: $(a + b)(a - b) = a^2 - b^2$

Model:

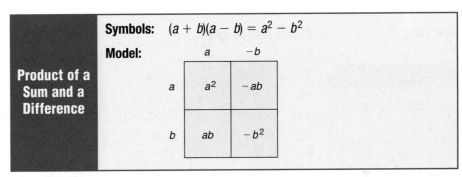

	a	$-b$
a	a^2	$-ab$
b	ab	$-b^2$

Examples

Find each product.

6 $(y + 2)(y - 2)$

$(a + b)(a - b) = a^2 - b^2$ *Product of a Sum and a Difference*
$(y + 2)(y - 2) = y^2 - (2)^2$ *Replace a with y and b with 2.*
$= y^2 - 4$ *Simplify.*

7 $(2r + s)(2r - s)$

$(a + b)(a - b) = a^2 - b^2$ *Product of a Sum and a Difference*
$(2r + s)(2r - s) = (2r)^2 - s^2$ *Replace a with 2r and b with s.*
$= 4r^2 - s^2$ *Simplify.*

Your Turn

e. $(x + y)(x - y)$ **f.** $(5m - 6n)(5m + 6n)$

Check for Understanding

Communicating Mathematics

1. **Match** each description with an expression.
   a. square of a sum
   b. square of a difference
   c. product of a sum and difference

   i. $(r + t)(r - t)$
   ii. $(z + 1)^2$
   iii. $(6 - a)^2$

2. **You Decide** Jessica says that the product of two binomials is always a trinomial. Hector disagrees. Who is correct? Explain your reasoning.

Guided Practice

Examples 1–4, 6, & 7

Find each product.

3. $(y + 2)^2$

4. $(2a + 4b)^2$

5. $(m - 9)^2$

6. $(j - 7k)^2$

7. $(x + 7)(x - 7)$

8. $(5r + 9s)(5r - 9s)$

Example 5

9. **Biology** In a certain population, a parent has a 5% chance of passing the gene for cystic fibrosis to its offspring. Use a model to find the probability that an offspring will *not* receive the gene for cystic fibrosis from either of its parents.

Exercises

Practice

Find each product.

10. $(r + 2)^2$

11. $(x + 5)^2$

12. $(a + 2b)^2$

13. $(w - 8)^2$

14. $(2x - y)^2$

15. $(m - 3n)^2$

16. $(p + 2)(p - 2)$

17. $(x + 4y)(x - 4y)$

18. $(2x - 3)(2x + 3)$

19. $(5 + k)^2$

20. $(3x - 2y)^2$

21. $(y + 3z)(y - 3z)$

22. $(6 - 2m)^2$

23. $(3x + 5)(3x - 5)$

24. $(5 + 2p)(5 + 2p)$

25. $(4 + 2x)^2$

26. $(5a - 2b)(5a + 2b)$

27. $(a - 3b)(a - 3b)$

28. $y(y + 1)(y - 1)$

29. $x(x + 1)^2$

30. $(x + 2)(x - 2)(x - 3)$

31. Find the product of $x - 2y$ and $x + 2y$.

32. What is the product of n, $n + 1$, and $n + 1$?

For Exercises	See Examples
10–12, 19, 24, 25, 29, 32, 34	1, 2
13–15, 20, 22, 27	3, 4
16–18, 21, 23, 26, 28, 30, 31, 33, 35	6, 7

Homework Help

Extra Practice
See page 711.

Applications and Problem Solving

33. **Number Theory** Explain how to find the product of 39 and 41 mentally. (*Hint*: Write 39 as $40 - 1$ and 41 as $40 + 1$.)

34. **Photography** Kareem cut off a 1-inch strip from each side of a square photograph so that it would fit into a square picture frame. He removed a total of 20 square inches.
   a. Draw and label a diagram that represents this situation. Let x represent the side length of the original photograph.
   b. Write an equation that could be used to find the dimensions of the original photograph.
   c. Find the original dimensions of the photograph.

35. Geometry The area of a triangle is given by the expression $\frac{1}{2}bh$, where b represents the length of the base and h represents the height. Suppose a right triangle has a base that measures $x - 3$ units and a height of $x + 3$ units.

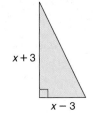

a. Express the area of the triangle as a sum of two monomials.

b. Find the area of the triangle if $x = 5$.

c. What is the length of the hypotenuse if $x = 6$?

36. Critical Thinking Use a model to find $(a + b + c)^2$.

Mixed Review

Geometry **Find the area of each shaded region in simplest form.** *(Lessons 9–3 & 9–4)*

37.

38.

Write each equation in slope-intercept form. *(Lesson 7–3)*

39. $2y = 6x + 8$ **40.** $x + y = -4$ **41.** $2x - 5y = 15$

Standardized Test Practice
Ⓐ Ⓑ Ⓒ Ⓓ

42. Grid In Enrique has 3 dimes, 2 quarters, and 5 nickels in his pocket. He takes one coin from his pocket at random. What is the probability that the coin is either a dime or a nickel? *(Lesson 5–7)*

43. Multiple Choice For what value(s) of b is the equation $3 + 2\lvert b \rvert = 7$ true? *(Lesson 3–7)*

Ⓐ -5 and 5 Ⓑ -2 and 2 Ⓒ -2 only   Ⓓ 2 only

Quiz 2   Lessons 9–4 and 9–5

Find each product. *(Lessons 9–4 & 9–5)*

1. $(c + 2)(c + 8)$ **2.** $(5y - 3)(y + 2)$
3. $(c - 1)^2$ **4.** $(3x - 8)(3x + 8)$

5. Geometry Find the area of the shaded region if the length of a side of the large square is $x + 3$ centimeters and the length of a side of the smaller square is $x - 3$ centimeters. *(Lesson 9–5)*

It's Greek to Me!

Areas and Ratio

Materials

 ruler

 calculator

 grid paper

Early Greek mathematicians enjoyed solving problems that required a lot of thought and investigation. Some of these problems still fascinate people today. For example, the Greeks wanted to construct a cube whose volume was twice the volume of a given cube. First, let's look at similar problems. What happens to the area of a square when you double the length of the sides of that square? Can you construct a square whose area is twice as great as a given square?

Temple of Poseidon

Investigate

Draw squares with different lengths on grid paper and find the ratios of their areas. Use a table or spreadsheet to record the lengths, areas, and ratios.

1. Draw a 3-centimeter by 3-centimeter square. Label it A. Find the area.

2. Draw a second square that has a side length twice that of square A. Label it B. Find the area.

3. Record your information in a table like the one below.

Square A		Square B		Ratios	
Length of Side	Area	Length of Side	Area	Length B / Length A	Area B / Area A
3	9	6	36	2:1	4:1

4. Draw three more pairs of squares and label them A and B. Find the areas and ratios for each pair of squares.

5. Make a conjecture about the ratio of areas of squares if the ratio of the lengths of their sides is 3:1.

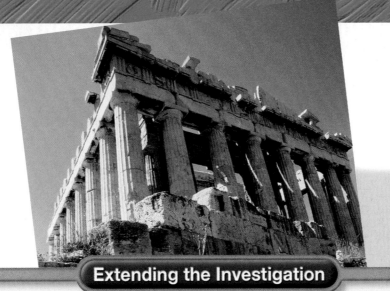

Extending the Investigation

In this extension, you will investigate the ratios of the areas of other squares. You will also try to find the length of the sides of two squares so that the area of one square is twice the other.

- Draw pairs of squares so that the ratio of the side length of one square to the side length of the second square is 3:1. **Make a conjecture** about the ratio of their areas.

- Draw pairs of squares so that the ratio of the side length of one square to the side length of the second square is *not* 2:1 or 3:1. Find the ratio of the lengths of their sides. **Make a conjecture** about the ratio of their areas.

- Draw pairs of squares so that the lengths of their sides differ by 1. For example, one square might have a side length of 4 units and the second a side length of 5 units. Find the ratio of the lengths of their sides. **Make a conjecture** about the ratios of their areas.

- Draw pairs of squares so that the lengths of their sides differ by a number greater than 1. Find the ratio of the lengths of their sides. **Make a conjecture** about the ratios of their areas.

- Draw a 2-centimeter by 2-centimeter square. Then sketch a square whose area is twice as great as the area of the first square. Find the side length of the second square.

Presenting Your Conclusions

Here are some ideas to help you present your conclusions to the class.

- Make a bulletin board showing the results of this investigation.

- Write a report about your conjectures. Include diagrams or computations that help to explain your findings.

- Make a video showing the ratios that you discovered. Present your conjectures in a creative way.

 Investigation For more information on three classical problems in Greek mathematics, visit: www.algconcepts.com

Understanding and Using the Vocabulary

*inter***NET**
CONNECTION **Review Activities**
For more review activities, visit:
www.algconcepts.com

After completing this chapter, you should be able to define each term, property, or phrase and give an example or two of each.

binomial *(p. 383)* FOIL method *(p. 401)* polynomial *(p. 382)*
degree *(p. 384)* monomial *(p. 382)* trinomial *(p. 383)*

Choose the letter of the term that best matches each statement or phrase. Each letter is used once.

1. $(x - y)^2$
2. a polynomial with two terms
3. a polynomial with three terms
4. a monomial or a sum of monomials
5. the sum of the exponents of the variables
6. a number, a variable, or a product of numbers and variables
7. Subtract polynomials by adding this.
8. Add polynomials by grouping these together.
9. Use this to multiply any two binomials.
10. Use this to multiply a polynomial by a monomial.

a. additive inverse
b. binomial
c. degree of a polynomial
d. Distributive Property
e. FOIL method
f. like terms
g. monomial
h. polynomial
i. square of a difference
j. trinomial

Skills and Concepts

Objectives and Examples	Review Exercises

Objectives and Examples

• **Lesson 9–1** Identify and classify polynomials and find their degree.

Identify $3x^3 - 2x^2$ as a *monomial, binomial,* or *trinomial.*

$3x^3 - 2x^2$ can be written as a sum of two monomials, $3x^3$ and $-2x^2$. It is a binomial.

Review Exercises

State whether each expression is a polynomial. If it is a polynomial, identify it as either a *monomial, binomial,* or *trinomial.*

11. $7m$
12. $5g^{-2}$
13. $2x^2 + 3x - 4$
14. $3y + 5y + 8$

Find the degree of each polynomial.

15. 7
16. $-4m^5$
17. $10x^2 - x^3y$
18. $2abc + 9a^5b - 4bc^3$

19. Arrange $3d^3 - d^2 + 7cd$ so that the powers of d are in ascending order.

 www.algconcepts.com/vocabulary_review

Objectives and Examples	Review Exercises

• **Lesson 9-2** Add and subtract polynomials.

$(2x + 4) + (5x - 7)$
$= (2x + 5x) + (4 - 7)$
$= (2 + 5)x + (4 - 7)$
$= 7x - 3$

$(7s^2 + 3s - 4) - (3s^2 - 2s + 5)$

The additive inverse of $(3s^2 - 2s + 5)$
is $(-3s^2 + 2s - 5)$.

$(7s^2 + 3s - 4) - (3s^2 - 2s + 5)$
$= (7s^2 + 3s - 4) + (-3s^2 + 2s - 5)$
$= (7s^2 - 3s^2) + (3s + 2s) + (-4 - 5)$
$= (7 - 3)s^2 + (3 + 2)s + (-4 - 5)$
$= 4s^2 + 5s - 9$

Find each sum or difference.

20. $\begin{array}{r} 8x - 7 \\ (+)\ 3x + 5 \\ \hline \end{array}$

21. $\begin{array}{r} 6x^2 \qquad + 4 \\ (+)\ 3x^2 - 3x + 2 \\ \hline \end{array}$

22. $\begin{array}{r} 9m^2 + 6m - 6 \\ (-)\ 7m^2 - 3m + 3 \\ \hline \end{array}$

23. $\begin{array}{r} 15s^2 \qquad - 6 \\ (-)\ 8s^2 + 6s - 4 \\ \hline \end{array}$

24. $(17x - 3y) + (-2x + 5y)$
25. $(10g^2 + 5g) - (6g^2 - 3g)$
26. $(14x^2 + 3x - 6) - (8 - 2x)$
27. $(-3s^2t^2 - 4st^2) + (-7s^2t^2 + 6st^2)$
28. What is $5n^2 + 3$ minus $2n^2 - 4$?

• **Lesson 9-3** Multiply a polynomial by a monomial.

$2x^2(3x^3 + 2x^2 - x)$
$= 2x^2(3x^3) + 2x^2(2x^2) + 2x^2(-x)$
$= 6x^5 + 4x^4 - 2x^3$

Find each product.

29. $3(x - 5)$
30. $n(n + 3)$
31. $2m(m^2 + 4m)$
32. $x^4(3x^2 - 2x)$
33. $5(2x^2 + 3x - 2)$
34. $-3m^2(m^2 - 2m + 1)$

Solve each equation.

35. $x(x + 8) = x(x + 11) + 9$
36. $9(w - 4) + 10 = 2w + 16$

• **Lesson 9-4** Multiply two binomials.

Find the product of $x + 4$ and $x - 3$.

$(x + 4)(x - 3)$

$= (x)(x) + (x)(-3) + (4)(x) + (4)(-3)$
$= x^2 - 3x + 4x - 12$
$= x^2 + x - 12$

Find each product.

37. $(y - 2)(y + 6)$
38. $(m + 5)(m + 7)$
39. $(x - 1)(x - 3)$
40. $(a + 11)(a - 4)$
41. $(2m - 1)(m - 4)$
42. $(x + 4)(3x - 2)$
43. $(4y + 2)(3y + 1)$
44. $(6x - 3)(2x + 5)$
45. Find the product of $5x + 4$ and $2x - 3$.

Objectives and Examples	**Review Exercises**

• **Lesson 9–5** Develop and use the patterns for $(a + b)^2$, $(a - b)^2$, and $(a + b)(a - b)$.

$$(x + 3)^2 = x^2 + 2(x)(3) + 3^2 \quad \textit{Square of}$$
$$= x^2 + 6x + 9 \qquad\qquad \textit{a sum}$$

$$(y - 5)^2 = y^2 - 2(y)(5) + 5^2 \quad \textit{Square of}$$
$$= y^2 - 10y + 25 \qquad\qquad \textit{a difference}$$

$$(s + 4)(s - 4) = s^2 - 4^2 \quad \textit{Product of a sum}$$
$$= s^2 - 16 \qquad \textit{and difference}$$

Find each product.

46. $(w + 6)^2$ **47.** $(2x + 3)^2$

48. $(8 + g)^2$ **49.** $(x - 2)^2$

50. $(3y - 1)^2$ **51.** $(5m - 3n)^2$

52. $(y + 4)(y - 4)$ **53.** $(2a + 3)(2a - 3)$

54. $(2p - 3q)(2p + 3q)$ **55.** $(5a + 1)(5a + 1)$

Applications and Problem Solving

56. Recreation Bocce is a game similar to lawn bowling, but it is played on a rectangular dirt court. If the length of the rectangle is $12x$ and the width is $2x + 2$, find the area of the court in terms of x. Write your answer as a polynomial in simplest form. *(Lesson 9–3)*

57. Geometry The area of a triangle is given by $\frac{1}{2}bh$, where b is the length of the base of the triangle and h is the measure of the height of the triangle. In triangle ABC, $b = 3x + 8$ and $h = 7x - 4$. Write the polynomial representing the measure of the area of triangle ABC. *(Lesson 9–4)*

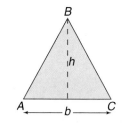

58. Gardening A rectangular garden is 10 feet longer than it is wide. It is surrounded by a brick walkway 3 feet wide, as shown at the right. Suppose the total area of the walkway is 396 square feet. *(Lesson 9–4)*

    a. Find the area of the garden.

    b. What are the dimensions of the garden?

59. Quilting Kirsten is making a quilt to enter in the arts festival. The diagram gives the dimensions of each block of the quilt. *(Lesson 9–5)*

    a. Find the area of each block.

    b. The completed quilt is a square of 36 blocks. Write a product of binomials that could be used to determine the area of the entire quilt.

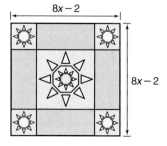

1. **Explain** how to use the FOIL method to multiply two binomials.

2. **Explain** how to subtract polynomials.

3. **You Decide?** Mark says that $-4wx^{-2}y^3$ is a monomial of degree 2. Linda disagrees. Who is correct? Explain your reasoning.

State whether each expression is a polynomial. If it is a polynomial, identify it as either a monomial, binomial, or trinomial, and state its degree.

4. $4x^3 + 3x^2$
5. $12r^{-2} - 3r + 6$
6. m^0
7. $5w^3yz^2$

Find each sum or difference.

8. $\begin{array}{r} 3x^2 - 5x + 4 \\ (+)\ 5x^2 + 7x + 8 \\ \hline \end{array}$

9. $\begin{array}{r} 8z^2 - 5z \\ (-)\ 6z^2 - 3z + 4 \\ \hline \end{array}$

10. $(y^3 + 6y^2 + 4y) - (2y^3 - 7y)$

11. $(-9h^2 - 5g) + (2g + 7h^2)$

Find each product.

12. $3x(2x^2 + 4x - 5)$
13. $5t^2(-2t + 3t^3 + 4)$

14. $(x + 2)(x + 3)$
15. $(y - 4)(y - 5)$

16. $(x - 2)(x + 5)$
17. $(3n + 4)(2n + 3)$

18. $(x + 5)^2$
19. $(2w - 3)^2$

20. $(7r - 4)(7r + 4)$
21. $(m - 3n)(m + 3n)$

Solve.

22. $2(y - 3) + 9 = 5y - 6$

23. $x(x + 4) + 5x = x(x - 1) + 20$

24. **Geometry** Find the area of the shaded region.

Exercise 24

25. **Family Life** Mrs. Douglas wants her five children, Aaron, Briana, Casey, Danielle, and Eddie, to spend a total of 35 hours on chores this week. Aaron, the oldest, works twice as many hours as the others. Danielle has earned an hour off from chores by getting an A on her algebra test. If x is the number of hours Danielle will work, then Briana, Casey, and Eddie will each work $x + 1$ hours, and Aaron will work twice as long, $2x + 2$ hours.

   a. Write an equation to represent the number of hours the children will work.

   b. Use the equation to determine the number of hours each child will work on chores this week.

Percent Problems

Standardized tests contain several types of percent problems. They may be numeric problems, word problems, or data analysis problems.

Test-Taking Tip

A percent is a fraction whose denominator is 100.

Familiarize yourself with common fractions, decimals, and percents. Below are some examples of common equivalents.

$$0.01 = \frac{1}{100} = 1\%$$ $$0.1 = \frac{1}{10} = 10\%$$ $$0.333... = \frac{1}{3} \approx 33.3\%$$

$$0.25 = \frac{1}{4} = 25\%$$ $$0.5 = \frac{1}{2} = 50\%$$ $$0.75 = \frac{3}{4} = 75\%$$

Example 1

The table shows millions of U.S. travelers to visit foreign countries. What percent of the travelers in 2001 visited Canada? Round to the nearest percent.

U.S. Travelers to Foreign Countries (millions)			
Year	Area		
	Canada	Mexico	Overseas
1999	15.3	17.7	24.6
2000	15.2	18.8	26.9
2001	15.6	17.2	25.2

Source: U.S. Dept. of Commerce

Hint Write a ratio, and then write the fraction as a percent.

Solution First, find the total travelers in 2001.

$$15.6 + 17.2 + 25.2 = 58.0$$

Next, write the ratio of travelers to Canada to total travelers.

$$\frac{15.6}{58.0}$$

Then write this fraction as a decimal and as a percent. Use your calculator.

$$\frac{15.6}{58.0} \approx 0.2690 \text{ or } 27\% \text{ to the nearest percent}$$

Travelers to Canada were about 27% of foreign travel in 2001.

Example 2

A bus company charges $5 each way to shuttle passengers between the hotel and a shopping mall. On a given day, the bus company has a total capacity of 250 people on the way to the mall and back. If the bus company runs at 90% of capacity, how much money would it collect that day?

(A) $1147.50   (B) $1250   (C) $2250

(D) $2500   (E) $2625

Hint Avoid partial answers. Be sure you answer the question that is asked.

Solution First determine how much money the bus company makes when it runs at total capacity. The 250 passengers would pay $10 each, because the charge is $5 each way.

$$250(\$10) = \$2500$$

Notice that this total amount is answer choice D, but it is a wrong answer. The question asks for the amount when the bus runs at 90% of capacity.

The word *of* is a clue to multiply. Find 90% of $2500.

$$(90\%)(\$2500) = (0.90)(2500) \text{ or } 2250$$

The bus company would collect $2250. Therefore, the answer is C.

After you work each problem, record your answer on the answer sheet provided or on a sheet of paper.

Multiple Choice

1. The areas of two rooms are 150 square feet and 135 square feet. If the total area of the home is 2000 square feet, what percent of the total area is the area of the two rooms? *(Lesson 5–3)*

   Ⓐ $6\frac{3}{4}\%$ Ⓑ $7\frac{1}{2}\%$ Ⓒ $14\frac{1}{4}\%$ Ⓓ $85\frac{3}{4}\%$

2. A $9.95 calendar is marked down 40%. Before tax, how much is saved on the purchase of one calendar? *(Lesson 5–4)*

   Ⓐ $1.99   Ⓑ $3.98
   Ⓒ $4.00   Ⓓ $5.97

3. On a 16-question quiz, Tom answered 2 questions incorrectly. If each question is worth the same number of points, what percent of the total points is his point total? *(Lesson 5–4)*

   Ⓐ 12.5%   Ⓑ 16%   Ⓒ 85%
   Ⓓ 87.5%   Ⓔ 94%

4. An increase in prices has made the cost of remodeling an office 12% more than the original cost. The original cost was $7145. What is the best estimate of the new cost? *(Lesson 5–5)*

   Ⓐ $700   Ⓑ $7700
   Ⓒ $10,000   Ⓓ $70,000

5. The graph shows the attendance at a park. Predict the attendance in the year 2013. *(Lesson 6–1)*

   Ⓐ 38 million   Ⓑ 48 million
   Ⓒ 60 million   Ⓓ 100 billion

6. A pair of shoes that regularly sold for $44 is now on sale for $33. What is the discount rate? *(Lesson 5–5)*

   Ⓐ 11%   Ⓑ 25%   Ⓒ 75%
   Ⓓ $33\frac{1}{3}\%$ Ⓔ $66\frac{2}{3}\%$

7. There are 60 students in the band. How many play a percussion instrument? *(Lesson 5–4)*

   Instruments in Bay City Band

   Ⓐ 15   Ⓑ 21   Ⓒ 24   Ⓓ 25

8. Which expression can be used to find the value of y? *(Lesson 7–3)*

   Ⓐ $2x + 1$ Ⓑ $1 - 3x$
   Ⓒ $3x - 1$ Ⓓ $3x + 1$

x	y
1	2
2	5
3	8
4	11

Grid In

9. A CD player is on sale for $250. If there is a 6% sales tax, what is the total cost? *(Lesson 5–4)*

Extended Response

10. The total land area of a state is 23,159,000 acres. Of that, 13,513,000 acres are cropland, 1,866,000 acres are pastureland, and 3,626,000 acres are forest. *(Lesson 5–4)*

    Part A   To the nearest percent, what percent of the state's area is cropland?

    Part B   What percent of the state's area is *not* cropland, pastureland, or forest?

www.algconcepts.com/standardized_test

Chapter 9 Preparing for Standardized Tests   **417**

Factoring

What You'll Learn

Key Ideas

- Find greatest common factor (GCF). *(Lesson 10–1)*

- Use the GCF and the Distributive Property to factor polynomials. *(Lesson 10–2)*

- Factor trinomials of the forms $x^2 + bx + c$ and $ax^2 + bx + c$. *(Lessons 10–3 and 10–4)*

- Recognize and factor the differences of squares and perfect square trinomials. *(Lesson 10–5)*

Key Vocabulary

difference of squares *(p. 447)*

factoring *(p. 428)*

greatest common factor (GCF) *(p. 422)*

perfect square trinomial *(p. 430)*

Why It's Important

Landscaping Landscape architects design parks, homes, and businesses to be both functional and beautiful. They plan the locations of buildings, roads, and walkways and the arrangement of plants. They are also often involved with protecting the environment and historical sites.

Factoring polynomials allows you to solve real-world problems involving polynomial equations. You will use factoring to plan a walkway as a landscape architect would in Lesson 10–2.

✓ Check Your Readiness

Lesson 8–1,
pp. 336–340

Write each expression using exponents.

1. $3 \cdot 3 \cdot 3 \cdot 3$

2. $9 \cdot 9 \cdot 9$

3. $10 \cdot 10 \cdot 10 \cdot 10 \cdot 10$

4. $2 \cdot 2 \cdot 2 \cdot 3 \cdot 3$

5. $5 \cdot 2 \cdot 2 \cdot 5 \cdot 5 \cdot 5$

6. $n \cdot n \cdot n \cdot n \cdot n \cdot n$

7. $p \cdot p \cdot p \cdot q \cdot q \cdot q$

8. $12 \cdot a \cdot a \cdot a \cdot a \cdot b$

Lesson 8–2,
pp. 341–345

Simplify each expression.

9. $\dfrac{5^3}{5}$

10. $\dfrac{x^6}{x^2}$

11. $\dfrac{k^{12}}{k^3}$

12. $\dfrac{a^3 b^6}{ab}$

13. $\dfrac{12m^4 n}{2m^2}$

14. $\dfrac{45cd^5}{9cd^2}$

Lesson 1–4,
pp. 19–23

Simplify each expression.

15. $2(x + y)$

16. $3(4d + 1)$

17. $8(6b + 2)$

18. $15a - 3a$

19. $6f + 9f - 8$

20. $(9m + 3n) + (2m - 6n)$

Make this Foldable to help you organize your Chapter 10 notes. Begin with a sheet of plain $8\frac{1}{2}$" by 11" paper.

❶ **Fold** in half lengthwise.

❷ **Fold** again in thirds.

❸ **Open.** Cut along the second fold to make three tabs.

❹ **Label** each tab as shown.

Reading and Writing As you read and study the chapter, unfold each tab and fill the journal with notes and examples.

What You'll Learn
You'll learn to find the greatest common factor of a set of numbers or monomials.

Why It's Important
Crafts Quilters use greatest common factors when they cut fabric.
See Exercise 57.

Recall that when two or more numbers are multiplied, each number is a *factor* of the product. For example, 12 can be expressed as the product of different pairs of whole numbers. Factors can be shown geometrically.

1 [============ 12 ============]
$1 \times 12 = 12$

2 [6]
$2 \times 6 = 12$

3 [4]
$3 \times 4 = 12$

The whole numbers 1, 12, 2, 6, 3, and 4 are the factors of 12.

Some whole numbers have exactly two factors, the number itself and 1. Recall that these numbers are called *prime numbers*. Whole numbers that have more than two factors, such as 12, are called *composite numbers*.

Prime Numbers Less Than 20	2, 3, 5, 7, 11, 13, 17, 19
Composite Numbers Less Than 20	4, 6, 8, 9, 10, 12, 14, 15, 16, 18
Neither Prime nor Composite	0, 1

Examples

Find the factors of each number. Then classify each number as *prime* or *composite*.

❶ **72**

To find the factors of 72, list all pairs of whole numbers whose product is 72.

1×72 2×36 3×24 4×18 6×12 8×9

The factors of 72 are 1, 2, 3, 4, 6, 8, 9, 12, 18, 24, 36, and 72. Since 72 has more than two factors, it is a composite number.

❷ **37**

There is only one pair of whole numbers whose product is 37.

1×37

The factors of 37 are 1 and 37. Therefore, 37 is a prime number.

**Math Data
nline Update**
For the latest information on prime numbers, visit:
algconcepts.com

Your Turn

a. 25   **b.** 23   **c.** 79   **d.** 51

You can use a graphing calculator to investigate factor patterns.

Graphing
Calculator Tutorial
See pp. 750–753.

Graphing Calculator Exploration

The table below shows the numbers 2 through 12 and their factors arranged by the number of factors.

2 Factors	3 Factors	4 Factors	5 Factors	6 Factors
2: 1, 2 **3:** 1, 3 **5:** 1, 5 **7:** 1, 7 **11:** 1, 11	**4:** 1, 2, 4 **9:** 1, 3, 9	**6:** 1, 2, 3, 6 **8:** 1, 2, 4, 8 **10:** 1, 2, 5, 10	none	**12:** 1, 2, 3, 4, 6, 12

Step 1: Copy the table above.

Step 2: Use the graphing calculator program below to find the factors of the numbers 13 through 20.

```
PROGRAM:FACTOR
:Input "ENTER NUMBER", N
:For (D, 1, N)
:If iPart (N/D) = (N/D)
:Disp D
:END
```

Try These

1. Place the numbers 13 through 20 in the correct column of the table.

2. Predict a number from 21 through 100 for each column. Check your prediction by using the calculator program.

3. Explain the pattern in each column.

Since $4 \cdot 3 = 12$, 4 is a factor of 12. However, it is not a prime factor of 12 because 4 is not a prime number. Recall that when a number is expressed as a product of prime factors, the expression is called the *prime factorization* of the number.

You can use a *factor tree* to find the prime factorization of a number. Two different factor trees are shown for the prime factorization of 12.

All of the factors in the last row are prime numbers. The factors are in a different order, but the result is the same. Except for the order of the factors, there is only one prime factorization of a number. Thus, the prime factorization of 12 is $2 \cdot 2 \cdot 3$ or $2^2 \cdot 3$.

You can use prime factorization to factor monomials. A monomial is in *factored form* when it is expressed as the product of prime numbers and variables and no variable has an exponent greater than 1.

Look Back

Monomials:
Lesson 9–1

Examples

Factor each monomial.

3 $12a^2b$

$12a^2b = 2 \cdot 2 \cdot 3 \cdot a \cdot a \cdot b$ *$12 = 2 \cdot 2 \cdot 3, a^2 = a \cdot a$*

4 $100mn^3$

$100mn^3 = 2 \cdot 2 \cdot 5 \cdot 5 \cdot m \cdot n \cdot n \cdot n$ *$100 = 2 \cdot 2 \cdot 5 \cdot 5, n^3 = n \cdot n \cdot n$*

5 $-25x^2$

To factor a negative integer, first express it as the product of a whole number and -1. Then find the prime factorization.
$$-25x^2 = -1 \cdot 25x^2 \qquad -25 = -1 \cdot 25$$
$$= -1 \cdot 5 \cdot 5 \cdot x \cdot x \quad 25 = 5 \cdot 5$$

Your Turn

e. $15ab^2$ f. $84yz^2$ g. $-36b^3$

Two or more numbers may have some common prime factors. Consider the prime factorization of 36 and 42.

$$36 = \boxed{2} \cdot 2 \cdot \boxed{3} \cdot 3 \qquad \textit{Line up the common factors.}$$
$$42 = \boxed{2} \cdot \quad \boxed{3} \cdot \quad 7$$

The integers 36 and 42 have 2 and 3 as common prime factors. The product of these prime factors, $2 \cdot 3$ or 6, is called the **greatest common factor (GCF)** of 36 and 42. The GCF is the greatest number that is a factor of both original numbers.

Greatest Common Factor	The greatest common factor of two or more integers is the product of the prime factors common to the integers.

The GCF of two or more monomials is the product of their common factors when each monomial is expressed in factored form.

Find the GCF of each set of numbers or monomials.

6 **24, 60, and 72**

$24 = 2 \cdot 2 \cdot 2 \cdot 3$ *Find the prime factorization of each number.*
$60 = 2 \cdot 2 \cdot\ \ \ \ \ 3 \cdot 5$ *Line up as many factors as possible.*
$72 = 2 \cdot 2 \cdot 2 \cdot 3 \cdot 3$ *Circle the common factors.*

The GCF of 24, 60, and 72 is $2 \cdot 2 \cdot 3$ or 12.

7 **15 and 8**

$15 = 3 \cdot 5$ *Find the prime factorization of each number.*
$\ 8 = \ \ \ \ \ 2 \cdot 2 \cdot 2$ *Look for common factors.*

There are no common prime factors. The only common factor is 1. So, the GCF of 15 and 8 is 1.

8 **$15a^2b$ and $18ab$**

$15a^2b = \ \ \ \ \ \ \ 3 \cdot 5 \cdot a \cdot a \cdot b$ *Find the prime factorization of each monomial.*
$18ab = 2 \cdot 3 \cdot 3 \cdot\ \ \ a \cdot\ \ \ b$ *Circle the common factors.*

The GCF of $15a^2b$ and $18ab$ is $3 \cdot a \cdot b$ or $3ab$.

Your Turn

h. 75, 100, and 150   **i.** $5a$ and $8b$ **j.** $24ab^2c$ and $60a^2bc$

Knowing the factors of a number can help you with geometry.

9 **The area of a rectangle is 18 square inches. Find the length and width so that the rectangle has the least perimeter. Assume that the length and width are both whole numbers.**

Explore   You know that the area of the rectangle is 18 square inches. You want to find the length and width so that the rectangle has the least perimeter.

Plan   Find the factors of 18 and draw rectangles with each length and width. Then find each perimeter.

Solve

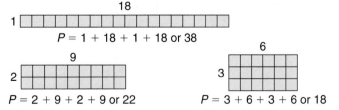

$P = 1 + 18 + 1 + 18$ or 38

$P = 2 + 9 + 2 + 9$ or 22 $P = 3 + 6 + 3 + 6$ or 18

The least perimeter is 18 inches. The rectangle has a length of 6 inches and a width of 3 inches. *Examine this solution.*

 Personal Tutor at algconcepts.com

Check for Understanding

Communicating Mathematics

1. **List** the prime numbers between 20 and 50.

2. **Name** two numbers whose GCF is 4.

3. **Explain** how to find the GCF of $8x^2$ and $16x$.

4. Jennifer believes that $2 \cdot 3 \cdot 4 \cdot 5$ is the prime factorization of 120, but Arturo disagrees. Who is correct? Explain.

Vocabulary

greatest common factor (GCF)

Guided Practice

⟳ **Getting Ready**   Find the prime factorization of each number.

Sample 1: 28   **Sample 2:** 60
Solution: $28 = 2 \cdot 2 \cdot 7$ **Solution:** $60 = 2 \cdot 2 \cdot 3 \cdot 5$

5. 21   6. 72   7. 51
8. 150   9. 108   10. 110

Examples 1 & 2

Find the factors of each number. Then classify each number as *prime* or *composite*.

11. 42   12. 47

Examples 3–5

Factor each monomial.

13. $24x^2y$ 14. $-16ab^2c$

Examples 6–8

Find the GCF of each set of numbers or monomials.

15. 15, 70   16. 16, 24, 28   17. 20, 21
18. $2x, 5y$ 19. $7y^2, 14y^3$ 20. $-12ab, 4a^2b^3$

Example 9

21. **Geometry**   The area of a rectangle is 72 square centimeters. Find the length and width so that the rectangle has the greatest perimeter. Assume that the length and width are both whole numbers.

Exercises

Practice

Find the factors of each number. Then classify each number as *prime* or *composite*.

22. 19   23. 20   24. 61
25. 45   26. 49   27. 91

Factor each monomial.

28. $20x^2$ 29. $-15a^2b$ 30. $-24c^3$
31. $50m^2n^2$ 32. $44r^2s$ 33. $90yz^2$

Find the GCF of each set of numbers or monomials.

34. 24, 40
35. 12, 8
36. 17, 21
37. 18, 36
38. 20, 30
39. 45, 72
40. $3x^2, 3x$
41. $18y^2, 3y$
42. $-5ab, 6b^2$
43. $-18, 45mn$
44. $24a^2, 60ab$
45. $9x^2y, 10m^2n$
46. 6, 8, 12
47. 20, 21, 25
48. 18, 30, 54
49. $5m^2, 15n^2, 25mn$
50. $6ax^2, 18ay^2, 9az^3$
51. $15r^2, 35s^2, 70rs$

52. What is the greatest prime number less than 90?

53. Find the greatest common factor of $5x^2, 5y^2$, and $10xy$.

54. *Twin primes* are prime numbers that differ by 2, such as 5 and 7. Find two other sets of twin primes that are between 25 and 45.

Homework Help

For Exercises	See Examples
22–27, 52, 54, 57	1, 2
28–33	3, 4, 5
34–51, 53	6, 7, 8
56	1, 2, 6, 7, 8

Extra Practice
See page 711.

Applications and Problem Solving

55. **Crafts**   Ashley wants to make a quilt from two different kinds of fabric. One is 60 inches wide, and the other is 48 inches wide. What are the dimensions of the largest squares she can cut from both fabrics so that no fabric is wasted?

56. **Math History**   In 1880, English mathematician John Venn (1834–1923) developed a way to show how sets of numbers are related. The *Venn diagram* shows the prime factors of 12 and 28. The common factors are in the overlapping circles, and the GCF of 12 and 28 is $2 \cdot 2$ or 4.

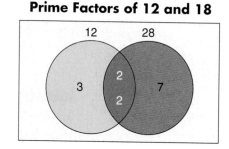

Prime Factors of 12 and 18

a. Draw a Venn diagram showing the prime factors of 36 and 45.
b. Find the GCF of 36 and 45.

57. **Critical Thinking**   Explain why 2 is the only even prime number.

Mixed Review

Find each product.   *(Lessons 9–3, 9–4, and 9–5)*

58. $(x + 3)(x - 3)$
59. $(2y + 1)(2y + 1)$
60. $(3a - 2)^2$
61. $(z + 4)(z + 3)$
62. $(x - 5)(x - 4)$
63. $(2n + 1)(n + 4)$
64. $3(x - 5)$
65. $2a(3 - a^2)$
66. $4x^2y(3x - 2y)$

Standardized Test Practice
(A) (B) (C) (D)

67. **Short Response**   In 2025, there are expected to be 817,000,000 cars in use worldwide. Write 817,000,000 in scientific notation.   *(Lesson 8–4)*

68. **Multiple Choice**   Which graph below is *not* the graph of a function? *(Lesson 6–4)*

A Puzzling Perimeter Problem

Materials

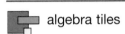
algebra tiles

Perimeter and Area

You know that the perimeter of a rectangle is the distance around the outside of the figure. It is measured in units like inches, centimeters, or feet. The area of a rectangle is the number of square units needed to cover the surface. It is measured in units like square inches or square centimeters. Are there any rectangles in which the measure of the perimeter is equal to the measure of the area? Let's investigate.

Investigate

1. Use the 1-tiles from a set of algebra tiles.

   a. Make a table like the one below with 21 rows.

Area of Rectangle	Possible Dimensions of Rectangle	Perimeter of Rectangle
1		
2		

   A 1-by-2 rectangle and a 2-by-1 rectangle have the same perimeter. So, they are listed only once in the table.

   b. Select one tile. The measure of its area is 1. The dimension of this rectangle is 1 by 1. Find the measure of the perimeter of the rectangle and enter it in column 3.

   c. Select two tiles. You can form a 1-by-2 rectangle with the tiles. This rectangle has an area of 2. Write 1 by 2 in column 2. What is the perimeter of this rectangle? Enter it in column 3.

   d. Repeat this process using three tiles.

   e. When you select four tiles, you have two options: a 1-by-4 rectangle or a 2-by-2 rectangle.

   Write 1 by 4 and 2 by 2 in column 2 as the dimensions of the rectangles. Find the perimeters of the two rectangles and enter the results in column 3.

Your table should look like this.

Area of Rectangle	Possible Dimensions of Rectangle	Perimeters of Rectangle
1	1 by 1	4
2	1 by 2	6
3	1 by 3	8
4	1 by 4, 2 by 2	10, 8

2. Use tiles to form all possible rectangles with areas from 5 through 20. Write the dimensions in column 2 and the perimeters in column 3.

3. Analyze the data. Which rectangle(s) has the same numerical values for perimeter and area?

4. Make a conjecture about whether there are other rectangles that have the same numerical values for perimeter and area.

Extending the Investigation

In this extension, you will continue to investigate the area and perimeter of rectangles. Extend your table through at least 30 squares.

- Study the perimeters for only the rectangles whose dimensions are 1 by *n*. Make an ordered list of the perimeters. Describe the pattern of the perimeters.

- Study the perimeters for only the rectangles whose dimensions are 2 by *n*, but not 2 by 2. Make an ordered list of the perimeters. Describe the pattern of the perimeters.

- Study the perimeters for only the rectangles whose dimensions are 3 by *n*, but not 3 by 3. Make an ordered list of the perimeters. Describe the pattern of the perimeters.

- Study the perimeters for rectangles that are also squares. Make an ordered list of these perimeters. Describe the pattern of the perimeters.

Presenting Your Conclusions

Here are some ideas to help you present your conclusions to the class.

- Make a brochure describing your findings. Include figures and tables to illustrate your results.

- Make a video showing the patterns you discovered. You may want to have your classmates take on the roles of rectangles with various dimensions.

 Investigation For more information on perimeter and area, visit: www.algconcepts.com

What You'll Learn

You'll learn to use the GCF and the Distributive Property to factor polynomials.

Why It's Important

Marine Biology You can find the height a dolphin jumps out of the water by evaluating an expression that is written in factored form.
See Exercise 46.

Sometimes, you know the product and are asked to find the factors. This process is called **factoring**.

For example, suppose you want to paint a rectangle on a wall and you only have enough paint to cover 20 square feet. If the length of each side must be an integer, what are the dimensions of all the possible rectangles you could paint?

Recall that the formula for the area of a rectangle is $A = \ell w$. If $A = 20$ square feet, then the measures of the length and width of the painted rectangle must be factor pairs of 20. The factor pairs of 20 are 1 and 20, 2 and 10, and 4 and 5. The figures below show rectangles with these factors as measures of length and width.

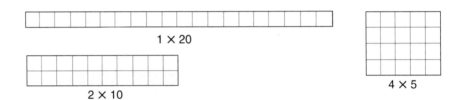

1 × 20

2 × 10

4 × 5

Hands-On Algebra
Algebra Tiles

Materials: ▭ algebra tiles ▢ product mat

Use algebra tiles to factor $2x + 8$.

Step 1 Model the polynomial $2x + 8$.

Step 2 Arrange the tiles into a rectangle. The total area of the tiles represents the product. Its length and width represent the factors. The rectangle has a width of 2 and a length of $x + 4$. So, $2x + 8 = 2(x + 4)$.

Try These

Use algebra tiles to factor each binomial.

1. $3x + 9$ **2.** $4x + 10$ **3.** $x^2 + 5x$ **4.** $3x^2 + 4x$

In Chapter 9, you used the Distributive Property to multiply a polynomial by a monomial.

Look Back

Distributive Property: Lesson 1–4

$$2y(4y + 5) = 2y(4y) + 2y(5)$$
$$= 8y^2 + 10y$$

You can reverse this process to express a polynomial in *factored form*. A polynomial is in factored form when it is expressed as the product of polynomials. For example, to factor $8y^2 + 10y$, find the greatest common factor of $8y^2$ and $10y$.

$$8y^2 = 2 \cdot 2 \cdot 2 \cdot y \cdot y$$

$$10y = 2 \cdot 5 \cdot y$$

The GCF of $8y^2$ and $10y$ is $2y$. Write each term as a product of the GCF and its remaining factors. Then use the Distributive Property.

$$8y^2 + 10y = 2y(4y) + 2y(5)$$
$$= 2y(4y + 5) \qquad \textit{Distributive Property}$$

$8y^2 + 10y$ written in factored form is $2y(4y + 5)$.

Examples

Factor each polynomial.

1 $30x^2 + 12x$

First, find the GCF of $30x^2$ and $12x$.

$$30x^2 = 2 \cdot 3 \cdot 5 \cdot x \cdot x$$
$$12x = 2 \cdot 2 \cdot 3 \cdot x$$

The GCF of $30x^2$ and $12x$ is $6x$. Write each term as a product of the GCF and its remaining factors.

$$30x^2 + 12x = 6x(5x) + 6x(2)$$
$$= 6x(5x + 2) \qquad \textit{Distributive Property}$$

2 $15ab^2 - 25abc$

$$15ab^2 = 3 \cdot 5 \cdot a \cdot b \cdot b$$
$$25abc = 5 \cdot 5 \cdot a \cdot b \cdot c$$

The GCF is $5ab$.

$$15ab^2 - 25abc = 5ab(3b) - 5ab(5c)$$
$$= 5ab(3b - 5c) \qquad \textit{Distributive Property}$$

Factor each polynomial.

③ $18x^2y + 12xy^2 + 6xy$

$18x^2y = 2 \cdot 3 \cdot 3 \cdot x \cdot x \cdot y$
$12xy^2 = 2 \cdot 2 \cdot 3 \cdot x \cdot y \cdot y$
$6xy = 2 \cdot 3 \cdot x \cdot y$

	$3x$	$2y$	1
$6xy$	$18x^2y$	$12xy^2$	$6xy$

The GCF is $6xy$. When $6xy$ is factored from $6xy$ the remaining factor is 1.

$18x^2y + 12xy^2 + 6xy = 6xy(3x) + 6xy(2y) + 6xy(1)$
$= 6xy(3x + 2y + 1)$ *Distributive Property*

④ $7x^2 + 9yz$

$7x^2 = 7 \cdot x \cdot x$
$9yz = 3 \cdot 3 \cdot y \cdot z$

There are no common factors of $7x^2$ and $9yz$ other than 1. Therefore, $7x^2 + 9yz$ cannot be factored using the GCF. It is a prime polynomial.

Reading Algebra

A polynomial that cannot be written as a product of two polynomials with integral coefficients is called a *prime polynomial*.

Your Turn

a. $12n^2 - 8n$

b. $16a^2b + 10ab^2$

c. $20rs^2 - 15r^2s + 5rs$

d. $21x + 5y + 16z$

If you know a product and one of its factors, you can use division to find the other factor. To divide a polynomial by a monomial, divide each term of the polynomial by the monomial.

Example

⑤ **Divide $15x^3 + 12x^2$ by $3x$.**

$(15x^3 + 12x^2) \div 3x = \dfrac{15x^3}{3x} + \dfrac{12x^2}{3x}$ *Divide each term by $3x$.*

$= \dfrac{\overset{5}{15} \cdot \overset{1}{x} \cdot x \cdot x}{\underset{1}{3} \cdot \underset{1}{x}} + \dfrac{\overset{4}{12} \cdot \overset{1}{x} \cdot x}{\underset{1}{3} \cdot \underset{1}{x}}$ *Simplify.*

$= 5x^2 + 4x$

Therefore, $(15x^3 + 12x^2) \div 3x = 5x^2 + 4x$.

Look Back

Dividing Powers: Lesson 8–2

Your Turn

Find each quotient.

e. $(9b^2 - 15) \div 3$

f. $(10x^2y^2 + 5xy) \div 5xy$

Factoring a polynomial can help simplify computations.

Example 6

Landscaping Link

A stone walkway is to be built around a square planter that contains a shade tree.

A. If the walkway is 2 meters wide, write an expression in factored form that represents the area of the walkway.

Let x represent the length and width of the planter. You can find the area of the walkway by finding the sum of the areas of the 8 rectangular sections shown in the figure.

The resulting expression can be simplified by using the Distributive Property to combine like terms and then factoring.

Regions

$$\underbrace{}_{1} \quad \underbrace{}_{2} \quad \underbrace{}_{3} \quad \underbrace{}_{4} \quad \underbrace{}_{5} \quad \underbrace{}_{6} \quad \underbrace{}_{7} \quad \underbrace{}_{8}$$

$$A = 2 \cdot 2 + 2 \cdot x + 2 \cdot 2 + 2 \cdot x + 2 \cdot 2 + 2 \cdot x + 2 \cdot 2 + 2 \cdot x$$
$$= 4 + 2x + 4 + 2x + 4 + 2x + 4 + 2x$$
$$= 16 + 8x \quad \textit{4 + 4 + 4 + 4 = 16 and 2x + 2x + 2x + 2x = 8x}$$
$$= 8(2) + 8(x) \quad \textit{The GCF of 16 and 8x is 8.}$$
$$= 8(2 + x) \quad \textit{Distributive Property}$$

B. If the dimensions of the square planter are 1.5 meters by 1.5 meters, find the area of the walkway.

$$A = 8(2 + x) \quad \textit{Area of a rectangle}$$
$$= 8(2 + 1.5) \quad \textit{Replace x with 1.5.}$$
$$= 8(3.5) \text{ or } 28 \quad \textit{Simplify.}$$

The area of the walkway is 28 square meters.

Check for Understanding

Communicating Mathematics

1. **Illustrate** with algebra tiles or a drawing how to factor $x^2 + 2x$.

Vocabulary

factoring

2. **Explain** what it means to factor a polynomial.

3. Writing Math Write a few sentences explaining how the Distributive Property is used to factor polynomials. Include at least two examples.

Getting Ready Find the GCF of the terms in each expression.

Sample: $9a^2 + 3a$ **Solution:** $9a^2 = 3 \cdot 3 \cdot a \cdot a$
$3a = \quad 3 \cdot a$ The GCF is $3a$.

4. $8xy + 12mn$ **5.** $x^2 + 2x$ **6.** $5xy + y^2$
7. $18y^2 + 30y$ **8.** $3ab - 2a^2b$ **9.** $15m^2n - 20m^2n$

Examples 1–4

Factor each polynomial. If the polynomial cannot be factored, write *prime*.

10. $3x + 6$ **11.** $2x^2 + 4x$ **12.** $12a^2b + 6a$
13. $7mn - 13yz$ **14.** $3x^2y + 6xy + 9y^2$ **15.** $2a^3b^2 + 8ab + 16a^2b^3$

Example 5

Find each quotient.

16. $(12m^2 - 15m) \div 3m$ **17.** $(3c^2d + 9cd) \div 3cd$

18. Landscaping Kiyoshi is planning to build a walkway around her square koi pond. The walkway is 6 feet wide.

Example 6

  a. If x represents the measure of one side of the pond, write an expression in factored form that represents the area of the walkway.

Example 7

  b. If the dimensions of the pond are 8 feet by 8 feet, find the area of the walkway.

Exercises • • • • • • • • • • • • • • • • •

Practice

Factor each polynomial. If the polynomial cannot be factored, write *prime*.

Homework Help	
For Exercises	See Examples
19–30, 46	1, 2
31–36	3, 4
37–43, 45	5
47	6

Extra Practice
See page 712.

19. $9x + 15$ **20.** $6x + 3x^2$ **21.** $8x + 2x^2y$
22. $7a^2b^2 + 3ab^3$ **23.** $3c^2d - 6c^2d^2$ **24.** $7x - 3y$
25. $36mn - 11mn^2$ **26.** $18xy^2 + 24x^2y$ **27.** $19ab + 21xy$
28. $14mn^2 - 2mn$ **29.** $12xy^3 + y^4$ **30.** $3a^2b - 6a^2b^2$
31. $24xy + 18xy^2 - 3y$ **32.** $3x^3y + 9xy + 36xy^2$
33. $x + x^2y^3 + x^3y^2$ **34.** $6x^2 + 9xy + 24x^2y^2$
35. $12axy - 14ay + 20ax$ **36.** $42xyz - 12x^2y^2 + 3x^3y^3$

Find each quotient.

37. $(27x^2 - 21y^2) \div 3$ **38.** $(5abc + c) \div c$
39. $(14ab + 28b) \div 14b$ **40.** $(16x + 24xy) \div 8x$
41. $(4x^2y^2z + 6xz^2) \div 2xz$ **42.** $(3x^2y + 12xyz^2) \div 3xy$

43. Divide $6x^2 + 9$ by 3.

44. What is the GCF of $14abc^2$ and $18c$?

Applications and Problem Solving

45. Geometry The area of a rectangle is $(16x + 4y)$ square feet. If the width is 4 feet, find the length.

46. Marine Biology   In a pool at a water park, a dolphin jumps out of the water traveling at 24 feet per second. Its height h, in feet, above the water after t seconds is given by the formula $h = 24t - 16t^2$.

   a. Factor the expression $24t - 16t^2$.

   b. Find the height of a dolphin when $t = 0.75$ second.

47. Geometry   Write an expression in factored form that represents the area of the shaded region.

   a.   **b.**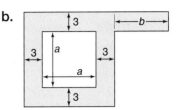

48. Critical Thinking   The length and width of a rectangle are represented by $2x$ and $9 - 4x$. If x must be an integer, what are the possible measures for the area of this rectangle?

Mixed Review

Classify each number as *prime* or *composite*.   *(Lesson 10–1)*

49. 2   **50.** 21   **51.** 49   **52.** 53   **53.** 90

54. Geometry   The length of a side of a square is $3x + 5$ units. What is the area of the square?   *(Lesson 9–5)*

Add or subtract.   *(Lesson 9–2)*

55. $(x^2 + 4x - 3) + (2x^2 - 6x - 9)$ **56.** $(2y^2 - 5y + 3) - (5y^2 - 4)$

Standardized Test Practice
Ⓐ Ⓑ Ⓒ Ⓓ

57. Short Response   Write a second degree polynomial. *(Lesson 9–1)*

Quiz 1   Lessons 10–1 and 10–2

1. Find the prime factorization of 24.   *(Lesson 10–1)*

Find the GCF of the terms in each expression. Then factor the expression.   *(Lessons 10–1, 10–2)*

2. $20s + 40s^2$ **3.** $ax^3 + 7bx^3 + 11cx^3$ **4.** $6x^3 + 12x^2 + 6x$

5. Landscaping   A 2-foot wide stone path is to be built along each side of a rectangular flower garden. The length of the garden is twice the width. If the flower garden is bordered on one side by a house, write an expression in factored form to represent the area of the path.   *(Lesson 10–2)*

What You'll Learn

You'll learn to factor trinomials of the form $x^2 + bx + c$.

Why It's Important

Biology Geneticists use Punnett squares, which are similar to the models for factoring trinomials.

See Exercise 51.

In biology, *Punnett squares* are used to show possible ways that traits can be passed from parents to their offspring.

Each parent has two genes for each trait. The letters representing the parent's genes are placed on the outside of the Punnett square. The letters inside the boxes show the possible gene combinations for their offspring.

	G	g
G	GG	Gg
g	Gg	gg

The Punnett square at the right shows the gene combinations for fur color in rabbits.

- *G* represents the dominant gene for gray fur.
- *g* represents the recessive gene for white fur.

Notice that the Punnett square is similar to the model for multiplying binomials. The model below shows the product of $(x + 1)$ and $(x + 3)$.

$$(x + 1)(x + 3) = x^2 + 3x + 1x + 3$$
$$= x^2 + 4x + 3$$

	x	3
x	x^2	$3x$
1	$1x$	3

In this lesson, you will factor a trinomial into the product of two binomials.

Hands-On Algebra

Materials: straightedge

Use a model to factor $x^2 + 5x + 4$.

Step 1 Draw a square with four sections. Put the first and last terms into the boxes as shown.

x^2	
	4

Step 2 Factor x^2 as $x \cdot x$ and place the factors outside the box.

Now, think of factors of 4 to place outside the box.

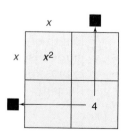

Step 3 The number 4 has two different factor pairs, 2 and 2, and 4 and 1. Try the factor pairs until you find the one that results in a middle term of $5x$.

Try 2 and 2.

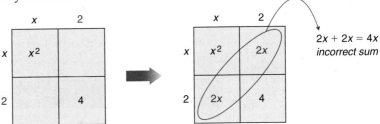

$2x + 2x = 4x$
incorrect sum

Try 4 and 1.

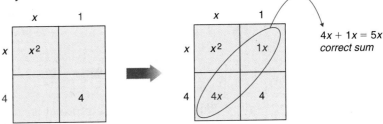

$4x + 1x = 5x$
correct sum

Step 4 The integers 4 and 1 result in the correct middle term, $5x$. Therefore, $x^2 + 5x + 4 = (x + 4)(x + 1)$.

Try These

Use a model to factor each trinomial.

1. $x^2 + 6x + 5$
2. $x^2 + 7x + 6$
3. $x^2 + 8x + 12$
4. $x^2 - 3x + 2$
5. $x^2 - 6x + 8$
6. $x^2 - 6x + 9$

Look Back

FOIL Method:
Lesson 9–4

The FOIL method will help you factor trinomials without models. Use the following method to factor $x^2 + 6x + 8$.

Step 1 x^2 is the product of the First terms, and 8 is the product of the Last terms.

$$x^2 + 6x + 8 = (x + \blacksquare)(x + \blacksquare)$$

Step 2 Try several factor pairs of 8 until the sum of the products of the Outer and Inner terms is $6x$. Check by using FOIL.

Try 1 and 8. $(x + 1)(x + 8) = x^2 + 8x + 1x + 8$
$= x^2 + 9x + 8$ *9x is not the correct term.*

Try 2 and 4. $(x + 2)(x + 4) = x^2 + 4x + 2x + 8$
$= x^2 + 6x + 8$ ✓

Therefore, $x^2 + 6x + 8 = (x + 2)(x + 4)$.

Factor each trinomial.

1 $x^2 - 7x + 10$

$$x^2 - 7x + 10 = (x + \blacksquare)(x + \blacksquare)$$

Find integers whose product is 10 and whose sum is -7. *Recall that the product of two negative integers is positive.*

Product	Integers	Sum
10	$-1, -10$	$-1 + (-10) = -11$
10	$-2, -5$	$-2 + (-5) = -7$ ✓

Therefore, $x^2 - 7x + 10 = (x - 2)(x - 5)$.

2 $x^2 + 5x - 6$

$$x^2 + 5x - 6 = (x + \blacksquare)(x + \blacksquare)$$

Find integers whose product is -6 and whose sum is 5. *Recall that the product of a positive integer and a negative integer is negative.*

Product	Integers	Sum
-6	$-2, 3$	$-2 + 3 = 1$
-6	$2, -3$	$2 + (-3) = -1$
-6	$-1, 6$	$-1 + 6 = 5$ ✓

You can stop listing factors when you find a pair that works.

Therefore, $x^2 + 5x - 6 = (x - 1)(x + 6)$.

3 $x^2 - 7 - 3x$

First, write the trinomial as $x^2 - 3x - 7$.

$$x^2 - 3x - 7 = (x + \blacksquare)(x + \blacksquare)$$

Find two integers whose product is -7 and whose sum is -3.

Product	Integers	Sum
-7	$-1, 7$	$-1 + 7 = 6$
-7	$1, -7$	$1 + (-7) = -6$

There are no factors of -7 whose sum is -3. Therefore, $x^2 - 3x - 7$ is a prime polynomial.

Your Turn

a. $x^2 + 3x + 2$ **b.** $a^2 + 4a + 3$ **c.** $b^2 + 4b + 4$

d. $y^2 - 7y + 12$ **e.** $n^2 - 5n - 14$ **f.** $m^2 - m + 1$

In the previous lesson, you learned that the terms of a polynomial might have a GCF that can be factored using the Distributive Property. When you factor trinomials, always check for a GCF first.

Example **4** **Factor $2x^2 - 20x - 22$.**

First, check for a GCF.
$2x^2 - 20x - 22 = 2(x^2 - 10x - 11)$ *The GCF is 2.*

Now, factor $x^2 - 10x - 11$.   *Find two integers whose product is -11 and whose sum is -10.*

Product	Integers	Sum
-11	$-1, 11$	$-1 + 11 = 10$
-11	$1, -11$	$1 + (-11) = -10$ ✓

So, $x^2 - 10x - 11 = (x + 1)(x - 11)$.
Therefore, $2x^2 - 20x - 22 = 2(x + 1)(x - 11)$.   *Check by using FOIL.*

Your Turn

Factor each polynomial.
g. $3y^2 - 9y - 54$ **h.** $5m^2 + 45m + 100$

The area of a figure can often be expressed as a trinomial.

Example **5**

Gardening Link

Tammy is planning a rectangular garden in which the width will be 4 feet less than its length. She has decided to put a birdbath within the garden, occupying a space 3 feet by 4 feet. How many square feet are now left for planting? Express the answer in factored form.

Let ℓ = the length of the original rectangle.
Let $\ell - 4$ = the width of the original rectangle.

Find the area of the original rectangle.
$A = \ell w$ *Area = length × width*
$A = \ell(\ell - 4)$ *Replace w with $\ell - 4$.*
$A = \ell^2 - 4\ell$ *Distributive Property*

Find the area of the small rectangle.
$A = 4(3)$ or 12

Remaining area = area of original rectangle − area of small rectangle
$=$ $\ell^2 - 4\ell$ $-$ 12

The remaining area is $\ell^2 - 4\ell - 12$ or $(\ell - 6)(\ell + 2)$.

Check for Understanding

Communicating Mathematics

1. **Illustrate** how to factor $x^2 + 7x + 6$ using a model.
2. **Explain** why the trinomial $x^2 + x + 5$ cannot be factored.
3. **Complete** the following sentence.
   When you factor $m^2 - 3m - 10$, you want to find two integers whose product is ___?___ and whose sum is ___?___.

Guided Practice

 Find two integers whose product is the first number and whose sum is the second number.

Sample: 10, 7   **Solution:** $2 \times 5 = 10,\ 2 + 5 = 7$

4. 30, 11   5. 12, −7   6. −10, 3   7. −6, −5   8. −30, −7

Examples 1–4

Factor each trinomial. If the trinomial cannot be factored, write *prime*.

9. $x^2 + 5x + 6$ 10. $y^2 + 9y + 20$ 11. $a^2 - 5a + 4$
12. $z^2 - 8z + 16$ 13. $x^2 + 3x - 10$ 14. $m^2 - 4m - 21$
15. $w^2 + w + 2$ 16. $3a^2 + 15a + 12$ 17. $2c^2 - 12c - 14$

Example 5

18. **Geometry** Find the area of the shaded region. Express the area in factored form.

Exercises

Practice

Factor each trinomial. If the trinomial cannot be factored, write *prime*.

Homework Help	
For Exercises	See Examples
19–39, 46, 48, 50, 52	1–3
40–45	4
49	5

Extra Practice
See page 712.

19. $b^2 + 5b + 4$ 20. $x^2 + 10x + 25$ 21. $a^2 + 7a + 12$
22. $a^2 + 3a + 5$ 23. $y^2 + 12y + 27$ 24. $z^2 + 13z + 40$
25. $x^2 - 8x + 15$ 26. $a^2 - 4a + 4$ 27. $c^2 - 13c + 36$
28. $d^2 - 11d + 28$ 29. $m^2 - 5m + 1$ 30. $y^2 - 12y + 32$
31. $c^2 + 2c - 3$ 32. $x^2 - 5x - 24$ 33. $r^2 - 3r - 18$
34. $m^2 - 2m - 24$ 35. $n^2 + 13n - 30$ 36. $m^2 + 11m - 12$
37. $x^2 - 17x + 72$ 38. $a^2 - a - 90$ 39. $r^2 + 22r - 48$
40. $4x^2 + 28x + 40$ 41. $3y^2 - 21y + 36$ 42. $z^3 + z^2 - 12z$
43. $m^3 + 3m^2 + 2m$ 44. $3y^3 - 24y^2 + 36y$ 45. $2a^3 + 14a^2 - 16a$

46. Express $x^2 + 24x + 95$ as the product of two binomials.
47. Write a trinomial that cannot be factored.
48. Complete the trinomial $x^2 + 6x + $ ___?___ with a positive integer so that the resulting trinomial can be factored.

Applications and Problem Solving

49. **Geometry** Refer to the figure at the right.
    a. Express the area of the shaded region as a polynomial.
    b. Express the area in factored form.

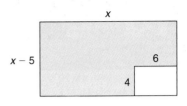

50. **Geometry** The volume of a rectangular prism is $x^3 + 4x^2 + 3x$. Find the length, width, and height of the prism if each dimension can be written as a monomial or binomial with integral coefficients. (*Hint*: Use the formula $V = \ell wh$.)

51. **Genetics** In guinea pigs, a black coat is a dominant trait over a white coat. Let C represent a black coat and c represent a white coat in the Punnett squares below. Find the missing genes or gene pair.

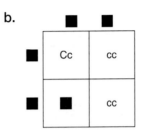

52. **Critical Thinking** Find all values of k so that the trinomial $x^2 + kx + 10$ can be factored.

Mixed Review

Find each quotient. (*Lesson 10–2*)

53. $(10x^2 + 25y^2) \div 5$

54. $(2y^2 + 4y) \div y$

55. $(6a^2 + 8ab - 6b^2) \div 2$

56. $(3x^2y^2 + 9x^3y^2z) \div 3x^2y^2$

Find the GCF of each set of numbers or monomials. (*Lesson 10–1*)

57. $12a, 16b$

58. $6a^2b, 9ab^2$

59. $15x, 7y$

60. $15, 60, 75$

Standardized Test Practice
Ⓐ Ⓑ Ⓒ Ⓓ

61. **Extended Response** The graph shows the value of riding lawn mower shipments in 1997 and 2000. (*Lesson 7–4*)
    a. Write an equation of the line in slope-intercept form.
    b. What does the slope represent?
    c. Use the equation to predict the value of riding lawn mower shipments in 2005.

Riding Lawn Mower Shipments (billion dollars)

Source: Freedonia Group

62. **Multiple Choice** Evaluate $8x + 3y$ if $x = 9$ and $y = -2$. (*Lesson 2–5*)
    Ⓐ 66   Ⓑ 57   Ⓒ 78   Ⓓ 11

In this lesson, you will learn to factor trinomials in which the coefficient of x^2 is a number other than 1.

Hands-On Algebra

Materials: straightedge

Use a model to factor $2x^2 + 7x + 6$.

Step 1 Draw a square with four sections. Put the first and last terms into the boxes as shown.

Step 2 Factor $2x^2$ as $2x \cdot x$ and place the factors outside the box.

Think of factors of 6 to place outside the box.

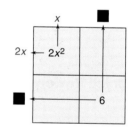

Step 3 The number 6 has two different factor pairs, 2 and 3, and 1 and 6. Try the factor pairs until you find the one that results in a middle term of $7x$. First, try 2 and 3. Note that there are two different ways of placing the 2 and 3 outside of the box.

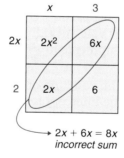

$2x + 6x = 8x$
incorrect sum

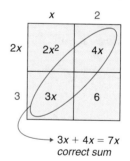

$3x + 4x = 7x$
correct sum

Step 4 The second model results in the correct middle term, $7x$. Therefore, $2x^2 + 7x + 6 = (2x + 3)(x + 2)$.

Try These Use a model to factor each trinomial.
1. $2x^2 + 7x + 3$
2. $2x^2 + 5x + 3$
3. $3x^2 + 7x + 2$
4. $3x^2 + 8x + 5$
5. $4x^2 + 8x + 3$
6. $4x^2 + 13x + 3$

The FOIL method will help you factor trinomials without models.

Examples

Factor each trinomial.

1 $2x^2 - 7x + 3$

$2x^2$ is the product of the First terms, and 3 is the product of the Last terms.

$$2x^2 - 7x + 3 = (2x + \blacksquare)(x + \blacksquare)$$

The last term, 3, is positive. The sum of the inside and outside terms, -7, is negative. So, both factors of 3 must be negative. Try factor pairs of 3 until the sum of the products of the **O**uter and **I**nner terms is $-7x$.

Try -3 and -1. $(2x - 3)(x - 1) = 2x^2 - 2x - 3x + 3$
$\qquad\qquad\qquad\qquad\qquad = 2x^2 - 5x + 3$ *$-5x$ is not the correct middle term.*
$\qquad\qquad\quad (2x - 1)(x - 3) = 2x^2 - 6x - 1x + 3$
$\qquad\qquad\qquad\qquad\qquad = 2x^2 - 7x + 3$ ✓

Therefore, $2x^2 - 7x + 3 = (2x - 1)(x - 3)$.

2 $3y^2 + 2y - 5$

$3y^2$ is the product of the First terms, and -5 is the product of the Last terms.

$$3y^2 + 2y - 5 = (3y + \blacksquare)(y + \blacksquare)$$

Find integers whose product is -5. Try factor pairs of -5 until the sum of the products of the **O**uter and **I**nner terms is $2y$.

Try -5 and 1. $(3y - 5)(y + 1) = 3y^2 + 3y - 5y - 5$
$\qquad\qquad\qquad\qquad\qquad = 3y^2 - 2y - 5$ *$-2y$ is not the correct middle term.*
$\qquad\qquad\quad (3y + 1)(y - 5) = 3y^2 - 15y + 1y - 5$
$\qquad\qquad\qquad\qquad\qquad = 3y^2 - 14y - 5$ *$-14y$ is not the correct middle term.*
Try 5 and -1. $(3y + 5)(y - 1) = 3y^2 - 3y + 5y - 5$
$\qquad\qquad\qquad\qquad\qquad = 3y^2 + 2y - 5$ ✓

Therefore, $3y^2 + 2y - 5 = (3y + 5)(y - 1)$.

Your Turn

a. $2x^2 + 3x + 1$ **b.** $5y^2 + 2y - 3$ **c.** $3z^2 - 8z + 4$

Sometimes the coefficient of x^2 can be factored into more than one pair of integers.

Example ③ Factor $4x^2 + 12x + 5$.

Number	Factor Pairs
4	4 and 1, 2 and 2
5	5 and 1

Reading Algebra

When factoring this kind of trinomial, it is important to keep an organized list of the factors.

Try 4 and 1. $(4x + 5)(1x + 1) = 4x^2 + 4x + 5x + 5$
$= 4x^2 + 9x + 5$ *9x is not the correct middle term.*

$(4x + 1)(1x + 5) = 4x^2 + 20x + 1x + 5$
$= 4x^2 + 21x + 5$ *21x is not the correct middle term.*

Try 2 and 2. $(2x + 5)(2x + 1) = 4x^2 + 2x + 10x + 5$
$= 4x^2 + 12x + 5$ ✓

Therefore, $4x^2 + 12x + 5 = (2x + 5)(2x + 1)$.

Your Turn

d. $6x^2 + 17x + 5$ **e.** $4x^2 - 8x - 5$

Recall that the first step in factoring any polynomial is to factor out any GCF other than 1.

Real World

Example ④
Manufacturing Link

The volume of a rectangular shipping crate is $6x^3 - 15x^2 - 36x$. Find possible dimensions for the crate.

The formula for the volume of a rectangular prism is $V = \ell wh$. Find three factors of $6x^3 - 15x^2 - 36x$. First, look for a GCF.

$6x^3 - 15x^2 - 36x = 3x(2x^2 - 5x - 12)$ *The GCF is 3x.*

$3x$ is one factor of $6x^3 - 15x^2 - 36x$. Factor $2x^2 - 5x - 12$ to find the other two factors.

$2x^2 - 5x - 12 = (2x + \blacksquare)(x + \blacksquare)$

The factors of -12 are -3 and 4, 3 and -4, -2 and 6, 2 and -6, -1 and 12, and 1 and -12. Check several combinations; the correct factors are 3 and -4.

$2x^2 - 5x - 12 = (2x + 3)(x - 4)$

So, $6x^3 - 15x^2 - 36x = 3x(2x + 3)(x - 4)$. Therefore, the dimensions can be $3x$, $2x + 3$, and $x - 4$.

Communicating Mathematics

1. **Write** the trinomial and its binomial factors shown by each model.

a.

	2x	3
x	$2x^2$	3x
−4	−8x	−12

b.

	3x	−1
2x	$6x^2$	−2x
−3	−9x	3

c.

	x	5
5x	$5x^2$	25x
3	3x	15

2. **You Decide** Jamal factored the trinomial $18k^2 - 24k + 8$ as $(3k - 2)(6k - 4)$. Jacqui disagrees with Jamal's answer. She says that he did not factor the trinomial completely. Who is correct? Explain your reasoning.

Guided Practice

Examples 1–4

Factor each trinomial. If the trinomial cannot be factored, write *prime*.

3. $2a^2 + 5a + 3$
4. $3y^2 + 7y + 2$
5. $5x^2 + 13x + 6$
6. $2x^2 + x - 3$
7. $2x^2 + x - 21$
8. $2n^2 - 11n + 7$
9. $10a^2 - 9a + 2$
10. $6y^2 - 11y + 4$
11. $6x^2 + 16x + 10$

Example 4

12. **Geometry** The measure of the volume of a rectangular prism is $2x^3 + x^2 - 15x$. Find possible dimensions for the prism.

Exercises

Practice

Factor each trinomial. If the trinomial cannot be factored, write *prime*.

Homework Help	
For Exercises	See Examples
13–21, 23–25, 36, 39, 44	3, 4
22, 26–35, 37–38, 40–41	1, 2
42	4
Extra Practice	
See page 712.	

13. $2y^2 + 7y + 3$
14. $2x^2 + 11x + 5$
15. $4a^2 + 8a + 3$
16. $2x^2 - 9x - 5$
17. $2q^2 - 9q - 18$
18. $5x^2 - 13x - 6$
19. $7a^2 + 22a + 3$
20. $3y^2 + 7y + 15$
21. $3x^2 + 14x + 8$
22. $2z^2 - 11z + 15$
23. $3x^2 + 14x + 15$
24. $3m^2 + 10m + 8$
25. $3x^2 + 5x + 1$
26. $4x^2 - 8x + 3$
27. $14x^2 + 33x - 5$
28. $6y^2 - 11y + 4$
29. $8m^2 - 10m + 3$
30. $6r^2 + 9r - 42$
31. $6x^2 + 3x - 30$
32. $4x^2 + 10x - 6$
33. $2x^3 + 5x^2 - 12x$
34. $7x - 5 + 6x^2$
35. $11y + 6y^2 - 2$
36. $15x^3 - 11x^2 - 12x$
37. $2a^2 + 5ab - 3b^2$
38. $15x^2 - 13xy + 2y^2$
39. $9k^2 + 30km + 25m^2$

40. Factor $2x^2 + 5x - 25$.
41. What are the factors of the trinomial $6x^3 + 15x^2 - 9x$?

Applications and Problem Solving

42. **Measurement** The volume of a rectangular prism is 60 cubic feet. If the measure of the length, width, and height are consecutive integers, find the dimensions.

43. **Manufacturing** The dimensions of a rectangular piece of metal are shown at the right.

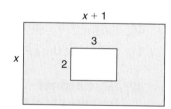

2y in.

y − 7 in.

    a. If a 1-inch by 1-inch square is removed from each corner, write an expression that represents the area of the remaining piece of metal. Express the area in factored form.

    b. If the metal is folded along the dashed lines, an open box is formed. Write an expression that represents the volume of the box.

    c. If $y = 10$ inches, find the area of the metal and the volume of the box.

44. **Critical Thinking** Find all values of k so that the trinomial $4y^2 + ky + 5$ can be factored.

Mixed Review

Factor each polynomial. *(Lessons 10–2 & 10–3)*

45. $x^2 + 14x - 32$

46. $y^2 - 7x + 12$

47. $3a^3 - 15a^2 + 6a$

48. $2n^2 + 2n - 24$

49. **Geometry** Find the length of the diagonal of a rectangle whose length is 24 feet and whose height is 7 feet. *(Lesson 8–7)*

Solve. Assume that y varies directly as x. *(Lesson 6–5)*

50. If $y = 28$ when $x = 7$, find x when $y = 52$.

51. Find x when $y = 45$, if $y = 27$ when $x = 6$.

Standardized Test Practice

Ⓐ Ⓑ Ⓒ Ⓓ

52. **Multiple Choice** What is the solution of $10 - 3(x + 4) = 16$? *(Lesson 4–7)*

    Ⓐ -6 Ⓑ $-\dfrac{12}{7}$ Ⓒ 2 Ⓓ 6

Quiz 2 Lessons 10–3 and 10–4

Factor each trinomial.

1. $x^2 + 3x - 10$

2. $x^2 - 5x - 24$ *(Lesson 10–3)*

3. $2x^2 + 9x + 7$

4. $8x^2 - 16x - 10$ *(Lesson 10–4)*

5. **Geometry** Find the area of the shaded region. Express the area in factored form. *(Lesson 10–3)*

x + 1

3

x

2

www.algconcepts.com/self_check_quiz

10-5 Special Factors

What You'll Learn

You'll learn to recognize and factor the differences of squares and perfect square trinomials.

Why It's Important

Manufacturing You can find the area of a washer by using the difference of squares. *See Exercise 52.*

In this lesson, you will learn to recognize and factor polynomials that are **perfect square trinomials**. They have two equal binomial factors.

The model below shows the product $(x + 3)^2$.

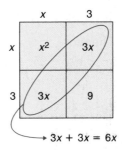

$$3x + 3x = 6x$$

You can also use the FOIL method to find the product.

$$(x + 3)(x + 3) = x^2 + 3x + 3x + 9$$

perfect squares

$$= x^2 + 6x + 9$$

twice the product of x and 3

The square of $(x + 3)$ is the sum of

- the square of the first term of the binomial,
- the square of the last term of the binomial, and
- twice the product of the terms of the binomial.

Look Back

Square of a Sum: Lesson 9–5

These observations will help you recognize when a trinomial is a perfect square trinomial. They can be factored as shown.

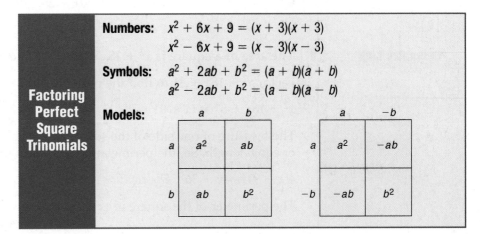

Factoring Perfect Square Trinomials

Numbers: $x^2 + 6x + 9 = (x + 3)(x + 3)$
$x^2 - 6x + 9 = (x - 3)(x - 3)$

Symbols: $a^2 + 2ab + b^2 = (a + b)(a + b)$
$a^2 - 2ab + b^2 = (a - b)(a - b)$

Models:

Determine whether each trinomial is a perfect square trinomial. If so, factor it.

❶ $x^2 + 10x + 25$

To determine whether $x^2 + 10x + 25$ is a perfect square trinomial, answer each question.

- Is the first term a perfect square? Yes, x^2 is the square of x.
- Is the last term a perfect square? Yes, 25 is the square of 5.
- Is the middle term twice the product of x and 5? Yes, $10x = 2(5x)$.

Therefore, $x^2 + 10x + 25$ is a perfect square trinomial.
$x^2 + 10x + 25 = (x + 5)^2$

❷ $4n^2 - 4n + 1$

- Is the first term a perfect square? Yes, $4n^2$ is the square of $2n$.
- Is the last term a perfect square? Yes, 1 is the square of 1 and -1.
- Is the middle term twice the product of $2n$ and -1? Yes, $2(-2n) = -4n$.

Therefore, $4n^2 - 4n + 1$ is a perfect square trinomial.
$4n^2 - 4n + 1 = (2n - 1)^2$

❸ $4p^2 - 12p + 36$

- Is the first term a perfect square? Yes, $4p^2$ is the square of $2p$.
- Is the last term a perfect square? Yes, 36 is the square of 6 and -6.
- Is the middle term twice the product of $2p$ and -6? No, $2(-12p) \neq -12p$.

Therefore, $4p^2 - 12p + 36$ is *not* a perfect square trinomial.

Your Turn

a. $a^2 + 2a + 1$ **b.** $16x^2 + 20x + 25$ **c.** $49x^2 - 14x + 1$

Geometry Link

❹ **The area of a square is $x^2 + 18x + 81$. Find the perimeter.**

Factor $x^2 + 18x + 81$ to find the measure of one side of the square.

$x^2 + 18x + 81 = (x + 9)^2$

The measure of one side of the square is $x + 9$. A square has four sides of equal length. So, the perimeter is four times the length of a side.

$4(x + 9) = 4x + 36$ *Distributive Property*

The perimeter of the square is $4x + 36$.

www.algconcepts.com/extra_examples

A polynomial like $x^2 - 9$ is called the **difference of squares**. Although this is not a trinomial, it *can* be factored into two binomials. The model shows how to factor $x^2 - 9$.

Look Back

Product of a Sum and a Difference:
Lesson 9–5

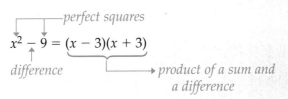

perfect squares

$$x^2 - 9 = (x - 3)(x + 3)$$

difference   product of a sum and a difference

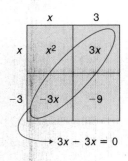

$3x - 3x = 0$

A difference of squares can be factored as shown.

COncepts in MOtion

Animation

algconcepts.com

Factoring a Difference of Squares	**Numbers:** $x^2 - 9 = (x - 3)(x + 3)$
	Symbols: $a^2 - b^2 = (a - b)(a + b)$
	Model: 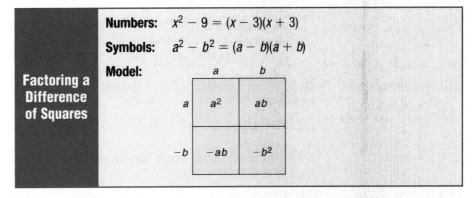

Examples

Determine whether each binomial is the difference of squares. If so, factor it.

5 $a^2 - 25$

a^2 and 25 are both perfect squares, and $a^2 - 25$ is a difference.

$a^2 - 25 = (a)^2 - (5)^2$ *$a \cdot a = a^2, 5 \cdot 5 = 25$*

$\qquad\quad = (a - 5)(a + 5)$ *Difference of Squares*

6 $y^2 + 100$

y^2 and 100 are both perfect squares. But $y^2 + 100$ is a sum, not a difference. Therefore $y^2 + 100$ is *not* a difference of squares. It is a prime polynomial.

7 $3n^2 - 48$

First, look for a GCF. Then, determine whether the remaining factor is a difference of squares.

$3n^2 - 48 = 3(n^2 - 16)$ *The GCF of $3n^2$ and 48 is 3.*

$\qquad\quad = 3[(n)^2 - (4)^2]$ *$n \cdot n = n^2, 4 \cdot 4 = 16$*

$\qquad\quad = 3(n - 4)(n + 4)$ *Difference of Squares*

Your Turn

d. $121 - p^2$ **e.** $25x^3 - 100x$ **f.** $4a^2 + 49$

The following chart summarizes factoring methods.

	Factoring Method	Number of Terms		
		Two	Three	Four or more
Concept Summary	greatest common factor	✓	✓	✓
	difference of squares	✓		
	perfect square trinomials		✓	
	trinomial with two binomial factors		✓	

Check for Understanding

Communicating Mathematics

1. **State** whether $4c^2 - 7$ can be factored as a difference of squares. Explain.

2. **Writing Math** Copy the chart shown above into your math journal. Then write a polynomial that can be factored by each method.

Vocabulary

Vocabulary
perfect square trinomials
difference of squares

Guided Practice

Examples 1–3

Determine whether each trinomial is a perfect square trinomial. If so, factor it.

3. $y^2 + 14y + 49$

4. $x^2 - 10x + 100$

5. $a^2 - 10a + 25$

Examples 5–7

Determine whether each binomial is the difference of squares. If so, factor it.

6. $16x^2 - 25$

7. $8x^2 - 50y^2$

8. $49m^2 + 16$

Factor each polynomial. If the polynomial cannot be factored, write *prime*.

9. $3x^2 + 15$

10. $y^2 + 6y - 9$

11. $3y^2 + 21y - 24$

Example 4

12. **Geometry** The area of a square is $4x^2 + 20xy + 25y^2$. Find the perimeter.

Exercises

Practice

Determine whether each trinomial is a perfect square trinomial. If so, factor it.

13. $r^2 + 8r + 16$

14. $x^2 - 16x + 64$

15. $a^2 + 2a + 1$

16. $4a^2 + 4a + 1$

17. $4z^2 - 20z + 25$

18. $9m^2 + 15m + 25$

19. $9a^2 + 24a + 16$

20. $d^2 - 22d + 121$

21. $49 + 14z + z^2$

Homework Help	
For Exercises	See Examples
13–21, 32	1, 2, 3
22–31	5, 6, 7
33–53	3, 4, 5
Extra Practice	
See page 713.	

Determine whether each binomial is the difference of squares. If so, factor it.

22. $x^2 - 16$

23. $a^2 - 36$

24. $y^2 - 20$

25. $1 - 9m^2$

26. $16m^2 - 25n^2$

27. $y^2 + z^2$

28. $8a^2 - 18$

29. $2z^2 - 98$

30. $49 - a^2b^2$

31. Write a polynomial that is the difference of two squares. Then factor it.

32. Is $x^2 + x - 1$ a perfect square trinomial? Explain.

Factor each polynomial. If the polynomial cannot be factored, write prime.

33. $5x^2 + 25$

34. $a^2 - 16b^2$

35. $y^2 - 5y + 6$

36. $m^2 + 8m + 16$

37. $2x^2 - 72$

38. $3a^2b + 6ab + 9ab^2$

39. $8xy^2 - 13x^2y$

40. $2r^2 + 3r + 1$

41. $x^2 - 6x - 9$

42. $z^3 + 6z^2 + 9z$

43. $20n^2 + 34n + 6$

44. $b^2 + 6 - 7b$

45. $8w^2 + 14w - 15$

46. $a^3 - 17a^2 + 72a$

47. $5x^2 + 15x + 10$

48. $7a^2 - 21a$

49. $2x^3 - 32x$

50. $2x^2 - 11x - 21$

Applications and Problem Solving

51. Number Theory The difference of two numbers is 2. The difference of their squares is 12. Find the numbers.

52. Manufacturing A metal washer is manufactured by stamping out a circular hole from a metal disk. In the figure, r represents the radius of the metal disk. The radius of the hole is 1 centimeter.

a. Write an expression in factored form for the area of the washer. (*Hint*: Use $A = \pi r^2$.)

b. If $r = 2$ centimeters, find the area of the washer to the nearest hundredth.

53. Critical Thinking The area of a square is $81 - 90x + 25x^2$. If x is a positive integer, what is the least possible measure for the square's perimeter?

Mixed Review

Factor each polynomial. *(Lessons 10–3 & 10–4)*

54. $4y^2 + 16y + 15$

55. $4x^2 + 11x - 3$

56. $a^3 - 7a^2 + 12a$

57. $m^2 - 5m - 14$

Simplify each expression. *(Lesson 8–3)*

58. n^{-2}

59. $a^5(a^{-3})$

60. $\dfrac{1}{r^{-3}}$

61. $\dfrac{3c^2d^3f^4}{9c^4d^2f^4}$

Standardized Test Practice

Ⓐ Ⓑ Ⓒ Ⓓ

62. Extended Response Describe the difference between the graphs of $y = 4x$ and $y = 4x - 5$. *(Lesson 7–6)*

63. Multiple Choice Which graph is the best example of data that exhibit a linear relationship between the variables x and y? *(Lesson 6–3)*

Understanding and Using the Vocabulary

After completing this chapter, you should be able to define each term, property, or phrase and give an example or two of each.

*inter*NET
CONNECTION **Review Activities**
For more review activities, visit:
www.algconcepts.com

difference of squares *(p. 447)*
factoring *(p. 428)*
greatest common factor (GCF) *(p. 422)*
perfect square trinomials *(p. 445)*
prime polynomial *(p. 430)*

State whether each sentence is *true* or *false*. If false, replace the underlined word or number to make a true sentence.

1. The prime factorization of 12 is $\underline{3 \cdot 4}$.

2. $\underline{3x}$ is the greatest common factor of $6x^2$ and $9x$.

3. When two or more numbers are multiplied, each number is a $\underline{\text{factor}}$ of the product.

4. $2y$ and $(y + 3)$ are factors of $\underline{2y^2 + 3}$.

5. When you factor trinomials, always check for a $\underline{\text{GCF}}$ first.

6. The number $\underline{51}$ is an example of a prime number.

7. $(x - 3)(x + 3)$ is the factored form of $\underline{x^2 + 9}$.

8. Whole numbers that have more than two factors are called $\underline{\text{composite numbers}}$.

9. A polynomial is in $\underline{\text{factored form}}$ when it is expressed as the product of polynomials.

10. $4a^2 - b^2$ is an example of a $\underline{\text{perfect square trinomial}}$.

Skills and Concepts

Objectives and Examples	Review Exercises

• **Lesson 10–1**   Find the greatest common factor of a set of numbers or monomials.

Find the GCF of $12x^2y$ and $30xy^2$.
$12x^2y = 2 \cdot \boxed{2} \cdot \boxed{3} \cdot \boxed{x} \cdot x \cdot \boxed{y}$
$30xy^2 = \boxed{2} \cdot \boxed{3} \cdot 5 \cdot \boxed{x} \cdot \boxed{y} \cdot y$

The GCF of $12x^2y$ and $30xy^2$ is
$2 \cdot 3 \cdot x \cdot y$ or $6xy$.

Find the GCF of each set of numbers or monomials.

11. 20, 25

12. 12, 18, 42

13. 20, 25, 28

14. $5xy, 10x$

15. $9x^2, 9x$

16. $6a^2b, 18a^2b^2, 9ab^2$

Objectives and Examples

- **Lesson 10–2** Use the GCF and the Distributive Property to factor polynomials.

  Factor $12x^2 - 8xy$.

  The GCF of $12x^2$ and $8xy$ is $4x$. Write each term as a product of the GCF and its remaining factors.

  $12x^2 - 8xy = 4x(3x) - 4x(2y)$
  $\qquad\qquad\quad = 4x(3x - 2y)$ *Distributive Property*

Review Exercises

Factor each polynomial. If the polynomial cannot be factored, write *prime*.

17. $5x + 30y$

18. $16a^2 + 32b^2$

19. $12ab - 18a^2$

20. $5mn^2 + 10mn$

21. $3xy + 12x^2y^2$

Find each quotient.

22. $(20x^3 + 15x^2) \div 5x$

23. $(40a^2b^2 - 8ab) \div 8ab$

- **Lesson 10–3** Factor trinomials of the form $x^2 + bx + c$.

  Factor $x^2 + 3x - 10$.

  Find integers whose product is -10 and whose sum is 3.

Product	Integers	Sum
-10	$2, -5$	$2 + (-5) = -3$
-10	$-2, 5$	$-2 + 5 = 3$ ✓

  Therefore, $x^2 + 3x - 10 = (x - 2)(x + 5)$.

Factor each trinomial. If the trinomial cannot be factored, write *prime*.

24. $y^2 + 9y + 14$

25. $x^2 - 8x + 15$

26. $a^2 + 5a - 7$

27. $x^2 - 2x - 8$

28. $y^2 + 7y + 12$

29. $x^2 + 2x - 35$

30. $a^2 - a - 1$

31. $2n^2 - 8n - 24$

- **Lesson 10–4** Factor trinomials of the form $ax^2 + bx + c$.

  Factor $2x^2 + 5x + 3$.

  $2x^2 + 5x + 3 = (2x + \blacksquare)(x + \blacksquare)$

  $(2x + 3)(x + 1) = 2x^2 + 2x + 3x + 3$
  $\qquad\qquad\qquad\;\; = 2x^2 + 5x + 3$

  Therefore, $2x^2 + 5x + 3 = (2x + 3)(x + 1)$.

Factor each trinomial. If the trinomial cannot be factored, write *prime*.

32. $2z^2 + 7z + 5$

33. $3x^2 + 8x + 5$

34. $3a^2 + 8a + 4$

35. $6a^2 - a - 2$

36. $3x^2 - 7x - 6$

37. $2y^2 - 9y - 18$

38. $2x^2 + 5x + 6$

39. $15a^2 - 20a + 5$

Mixed Problem Solving
See pages 724–731.

Objectives and Examples

- **Lesson 10–5** Recognize and factor the differences of squares and perfect square trinomials.

Factor $a^2 + 6a + 9$.
$$a^2 + 6a + 9 = a^2 + 2(3a) + 3^2$$
$$= (a + 3)^2$$

Factor $x^2 - 25$.
$$x^2 - 25 = x^2 - 5^2$$
$$= (x + 5)(x - 5)$$

Review Exercises

Factor each polynomial. If the polynomial cannot be factored, write *prime*.

40. $y^2 + 8y + 16$

41. $a^2 - 12a + 36$

42. $n^2 + 2n - 1$

43. $25x^2 + 20x + 4$

44. $y^2 - 81$

45. $4x^2 - 9$

46. $a^2 + 49$

47. $12c^2 - 12$

Applications and Problem Solving

48. Physics If a flare is launched into the air, its height h feet above the ground after t seconds is given by the formula $h = vt - 16t^2$. In the formula, v represents the initial velocity in feet per second. *(Lesson 10–2)*

    a. Factor the expression $vt - 16t^2$.

    b. If the flare is launched with an initial velocity of 144 feet per second, find the height after 2 seconds.

49. Genetics The Punnett square below represents the possible gene combinations for hair length in dogs. H represents long hair, and h represents short hair. Find the missing genes for the parents. *(Lesson 10–3)*

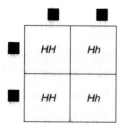

50. Geometry The area of a square is $(25x^2 + 30x + 9)$ square units. Find the perimeter. *(Lesson 10–5)*

1. **Explain** what it means to *factor* a polynomial.

2. **Write** two monomials whose GCF is 1.

3. **Write** the trinomial and its binomial factors shown by the model.

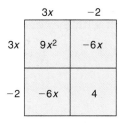

Exercise 3

4. **List** two different methods of factoring polynomials.

5. **Classify** the number 15 as *prime* or *composite*. Explain your reasoning.

Factor each monomial.

6. $25x^2y^2$

7. $-15b^3$

8. $24a^2b$

Find the GCF of each set of numbers or monomials.

9. $24, 60$

10. $16a^2, 30a^3$

11. $20a^2b, 25a^2b^2$

Factor each polynomial. If the polynomial cannot be factored, write *prime*.

12. $12x^2 + 18x$

13. $3x^2y - 12xy^2$

14. $6a^3 + 8a^2 + 2a$

15. $x^2 + 9x + 8$

16. $m^2 - 10m + 24$

17. $y^2 - 3y - 18$

18. $3x^2 + x - 14$

19. $3m^2 + 17m + 10$

20. $2x^2 - 18$

21. $n^2 - 8n - 16$

22. $y^2 + 10y + 25$

23. $25m^2 - 16$

24. $3r^2 + r + 1$

25. $6x^3 + 15x^2 - 9x$

26. **Geometry** Find the area of the shaded region. Express the area in factored form.

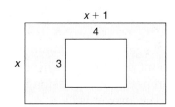

Function and Graph Problems

All standardized tests include problems with functions and graphs.

Familiarize yourself with the concepts below.

function

table of values

graph of a line

equation of a line

y-intercept, x-intercept

slope

Example 1

Martin is paid by the hour for babysitting. His hourly wage is a fixed amount plus an additional amount for each child. The graph shows his hourly wage for up to 5 children. If x represents the number of children, which expression can be used to find Martin's hourly wage, y?

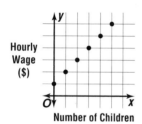

Number of Children

Ⓐ x

Ⓑ $x - 1$

Ⓒ $2x + 1$

Ⓓ $x + 1$

Hint Study the graph. The points lie on a line. Find the y-intercept and the slope.

Solution The fixed amount is represented by the y-intercept. It is the point that represents 0 children. The y-intercept is 1.

The slope of the line shows the amount for each child. Moving left to right, each point is one unit higher than the previous point. So the slope is 1.

Martin's hourly wage is the fixed amount, $1, plus the amount per child, $1, times the number of children, x. The expression is $1x + 1$ or $x + 1$. The answer is D.

Example 2

What is the equation of a line that is parallel to the line whose equation is $y = \frac{2}{3}x + 5$ and passes through the point at $(-6, 2)$?

Ⓐ $y = \frac{2}{3}x + 5z$

Ⓑ $y = \frac{2}{3}x - 2$

Ⓒ $y = \frac{2}{3}x + 6$

Ⓓ $y = \frac{2}{3}x - \frac{22}{3}$

Ⓔ $y = -\frac{3}{2}x - 7$

Hint Memorize the slope-intercept and point-slope forms of linear equations.

Solution The equation of the given line is in slope-intercept form. The slope is $\frac{2}{3}$. Parallel lines have the same slope. So, the slope of the parallel line must also be $\frac{2}{3}$. This eliminates answer choice E.

The parallel line must pass through $(-6, 2)$. Write the equation of the line in point-slope form.

$y - 2 = \frac{2}{3}[x - (-6)]$ *Point-slope form*

$y - 2 = \frac{2}{3}(x + 6)$ *Simplify.*

$y - 2 = \frac{2}{3}x + \frac{2}{3}(6)$ *Distributive Property*

$y - 2 = \frac{2}{3}x + 4$ *Multiply.*

$y = \frac{2}{3}x + 6$ *Add 2 to each side.*

The answer is C.

After you work each problem, record your answer on the answer sheet provided or on a sheet of paper.

Multiple Choice

1. Use the function table to find the value of y when $x = 5$. *(Lesson 6–1)*

x	y
0	3
1	17
2	31
3	45

Ⓐ 59 Ⓑ 60
Ⓒ 73 Ⓓ 75

2. What is the y-intercept of the line determined by the equation $5x + 2 = 7y - 3$? *(Lesson 7–5)*

Ⓐ -1 Ⓑ $-\frac{1}{7}$ Ⓒ $\frac{1}{7}$ Ⓓ $\frac{5}{7}$ Ⓔ 5

3. Which expression can be used to find the value of y in the graph? *(Lesson 7–3)*

Ⓐ $1 - 2x$ Ⓑ $1 - 3x$
Ⓒ $2x + 1$ Ⓓ $3x - 1$

4. The charge to enter a nature reserve is a fixed amount per vehicle plus a fee for each person in it. The table shows some charges. What would the charge be for a vehicle with 8 people? *(Lesson 6–1)*

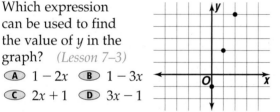

People	Charge
1	$1.50
2	$2.00
3	$2.50
4	$3.00

Ⓐ $3.50 Ⓑ $4.00 Ⓒ $5.00 Ⓓ $6.00

5. At what point does the line MN cross the y-axis? *(Lesson 7–5)*

Ⓐ $(-4, 0)$
Ⓑ $(0, -4)$
Ⓒ $(-2, 0)$
Ⓓ $(0, -2)$

6. The average of two numbers x and y is A. Which of the following is an expression for y? *(Lesson 3–3)*

Ⓐ $\frac{A + x}{2}$ Ⓑ $\frac{A}{2} - x$ Ⓒ $2A - x$
Ⓓ $A - x$ Ⓔ $x - A$

7. Write 4^{-4} without using an exponent. *(Lesson 8–3)*

Ⓐ 0.00039 Ⓑ 0.0039
Ⓒ 0.016 Ⓓ 256

8. Which expression should come next in the pattern $2x, 4x^2, 8x^3, 16x^4, \ldots$? *(Lesson 8–1)*

Ⓐ $24x^5$ Ⓑ $32x^5$
Ⓒ $24x^6$ Ⓓ $32x^6$

Grid In

9. The graph of $y = 4x - 2$ is shown. What is the x-intercept? *(Lesson 7–5)*

Extended Response

10. The graph shows the distance traveled by an African elephant. *(Lesson 7–1)*

Distance Traveled

Part A What is the slope of the line?
Part B Explain what the slope represents.

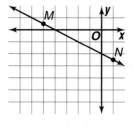

Quadratic and Exponential Functions

What You'll Learn

Key Ideas

- Graph quadratic functions. *(Lesson 11–1)*

- Investigate families of parabolas. *(Lesson 11–2)*

- Solve quadratic equations by graphing, by factoring, by completing the square, and by using the Quadratic Formula. *(Lessons 11–3 to 11–6)*

- Graph exponential functions. *(Lesson 11–7)*

Key Vocabulary

completing the square *(p. 478)*

exponential function *(p. 489)*

parabola *(p. 458)*

quadratic equation *(p. 468)*

Quadratic Formula *(p. 484)*

roots *(p. 468)*

zeros *(p. 468)*

Why It's Important

Space From the Apollo Moon landings to robotic surveys of the Sun, space exploration has changed our view of the universe. It has also led to inventions for everyday life. For example, better athletic shoes, X-ray machines, and scratch-resistant sunglasses were developed from discoveries in the space program.

Quadratic functions are used to solve real-world problems involving gravity. You will use a quadratic function to analyze the effect of gravity on an astronaut in Lesson 11–6.

Check Your Readiness

Graph each equation.

1. $y = x + 1$

2. $y = 3x$

3. $y = -2x$

4. $y = 2x + 4$

5. $y = 6 - x$

6. $y = \frac{1}{2}x - 1$

Evaluate each expression.

7. $2x^2$ if $x = 3$

8. $x^2 + 2x$ if $x = 2$

9. $-x^2 - x + 4$ if $x = -1$

10. $\frac{1}{2}x^2 + x - 1$ if $x = -4$

11. $-\dfrac{b}{2a}$ if $a = 1$ and $b = -4$

12. $-\dfrac{b}{2a}$ if $a = -2$ and $b = 6$

Factor each trinomial. If the trinomial cannot be factored, write prime.

13. $x^2 + 10x + 16$

14. $n^2 + 4n - 12$

15. $b^2 - 2b - 15$

16. $p^2 - 11p + 30$

17. $2c^2 + 5c - 3$

18. $3m^2 - 5m - 8$

 Study Organizer

Make this Foldable to help you organize your Chapter 11 notes. Begin with four sheets of plain $8\frac{1}{2}$" by 11" paper.

❶ **Fold** each sheet in half along the width.

❷ **Unfold** each sheet and tape to form one long sheet.

❸ **Label** each page with a lesson number as shown.

❹ **Refold** to form a booklet. Label the front cover "Quadratic and Exponential Functions."

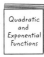

Reading and Writing As you read and study the chapter, write notes and examples for each lesson on each page of the journal.

What You'll Learn

You'll learn to graph quadratic functions.

Why It's Important

Architecture
Architects graph quadratic functions when designing arches.
See Exercise 15.

The shape of a parabola is like an arch. **Parabolas** are modeled by **quadratic functions** of the form $y = ax^2 + bx + c$. The first term cannot equal zero because it will make the function linear, or a straight line. Therefore, the coefficient a cannot be zero.

	Words:	A quadratic function is a function that can be described by an equation of the form $y = ax^2 + bx + c$, where $a \neq 0$.
Quadratic Function	**Models:**	

Graphs of all quadratic functions have the shape of a parabola.

Examples

Graph each quadratic function by making a table of values.

1 $y = x^2 + 1$

First, choose integer values for x. Evaluate the function for each x-value. Graph the points and connect them with a smooth curve.

Look Back

Graphing Relations:
Lesson 6–3

x	$x^2 + 1$	y	(x, y)
-2	$(-2)^2 + 1$	5	$(-2, 5)$
-1	$(-1)^2 + 1$	2	$(-1, 2)$
0	$0^2 + 1$	1	$(0, 1)$
1	$1^2 + 1$	2	$(1, 2)$
2	$2^2 + 1$	5	$(2, 5)$

2 $y = -x^2$

x	$-x^2$	y	(x, y)
-2	$-(-2)^2$	-4	$(-2, -4)$
-1	$-(-1)^2$	-1	$(-1, -1)$
0	$-(0)^2$	0	$(0, 0)$
1	$-(1)^2$	-1	$(1, -1)$
2	$-(2)^2$	-4	$(2, -4)$

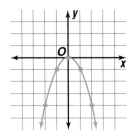

Your Turn

a. $y = x^2 + 6$

b. $y = -2x^2 - 1$

In Example 1, the *lowest point*, or **minimum**, of the graph of $y = x^2 + 1$ is at (0, 1). Since the coefficient of x^2 is positive, the graph opens *upward*. In Example 2, the *highest point*, or **maximum**, of the graph of $y = -x^2$ is at (0, 0). Since the coefficient of x^2 is negative, the graph opens *downward*.

As with linear functions, in a quadratic function x is the independent variable and y is the dependent variable. Since the graph of a quadratic function extends forever to the left and to the right, the domain (x values) of a quadratic function is the set of all real numbers. For a quadratic function with a graph that opens upward, the range (y values) is all real numbers greater than or equal to the minimum value. For a quadratic function with a graph that opens downward, the range is all real numbers less than or equal to the maximum value.

The maximum or minimum point of a parabola is called the **vertex**. The vertical line containing the vertex of a parabola is called the **axis of symmetry**. If you fold the graph of $y = x^2 + 1$ or $y = -x^2$ along the axis of symmetry, the two halves of each graph will coincide. In both examples, the axis of symmetry is the line $x = 0$.

You can use the rule below to find the equation of the axis of symmetry.

Equation of the Axis of Symmetry	**Words:** The equation of the axis of symmetry for the graph of $y = ax^2 + bx + c$, where $a \neq 0$, is $x = -\dfrac{b}{2a}$. **Model:**

Example ③ Use characteristics of quadratic functions to graph $y = x^2 - 4x - 1$.
  A. Find the equation of the axis of symmetry.
  B. Find the coordinates of the vertex of the parabola.
  C. Graph the function.

  A. First identify a, b, and c.
  $$y = ax^2 + bx + c$$

  $$y = 1x^2 - 4x - 1 \quad x^2 = 1x^2$$
  So, $a = 1$, $b = -4$, and $c = -1$.

  Now, find the equation of the axis of symmetry.

  $x = -\dfrac{b}{2a}$ *Equation of axis of symmetry*

  $x = -\dfrac{-4}{2(1)}$ *$a = 1$, $b = -4$*

  $x = 2$ *Simplify.*

  (continued on the next page)

B. Next, find the vertex. Since the equation of the axis of symmetry is $x = 2$, the x-coordinate of the vertex must be 2. Substitute 2 for x in the equation $y = x^2 - 4x - 1$ to solve for y.

$$y = x^2 - 4x - 1$$
$$= (2)^2 - 4(2) - 1$$
$$= 4 - 8 - 1 \text{ or } -5$$

The point at $(2, -5)$ is the vertex. *This point is a minimum.*

C. Construct a table. Choose some values for x that are less than 2 and some that are greater than 2. This ensures that points on each side of the axis of symmetry are graphed.

x	$x^2 - 4x - 1$	y	(x, y)
0	$0^2 - 4(0) - 1$	-1	$(0, -1)$
1	$1^2 - 4(1) - 1$	-4	$(1, -4)$
2	$2^2 - 4(2) - 1$	-5	$(2, -5)$
3	$3^2 - 4(3) - 1$	-4	$(3, -4)$
4	$4^2 - 4(4) - 1$	-1	$(4, -1)$

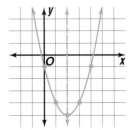

Because the graph continues indefinitely, the domain of the function is all real numbers. Because the vertex is a minimum, the range of the function is all real numbers greater than or equal to 5.

Your Turn Find the coordinates of the vertex and the equation of the axis of symmetry for the graph of each equation. Then graph the function and state the domain and range.

c. $y = x^2 + x$

d. $y = -x^2 + 2x - 3$

Example
4

Architecture Link

The Exchange House in London, England, is supported by an arch shaped like a parabola. It can be modeled by $y = -0.025x^2 + 2x$, where y represents the height of the arch and x represents the horizontal distance from one end of the base in meters. What is the highest point of the parabolic arch? State the domain and range of the function.

The highest point of the arch is the y-coordinate of the vertex. Find the equation of the axis of symmetry for $h(x) = -0.025x^2 + 2x$. State the domain and range of the function.

$$x = -\frac{b}{2a} \qquad \textit{Equation of axis of symmetry}$$
$$= -\frac{2}{2(-0.025)} \qquad a = -0.025, b = 2$$
$$= -\frac{2}{-0.05} \text{ or } 40 \quad \textit{Simplify.}$$

Next, find the vertex. Since the equation of the axis of symmetry is $x = 40$, the x-coordinate of the vertex must be 40. Substitute 40 for x in the function $h(x) = -0.025x^2 + 2x$. Then solve for h.

$$h(x) = -0.025x^2 + 2x \qquad \textit{Original function}$$
$$h(40) = -0.025(40)^2 + 2(40) \qquad \textit{Replace x with 40.}$$
$$= -40 + 80 \text{ or } 40 \qquad \textit{Simplify.}$$

The point (40, 40) is the vertex. So, the maximum height is 40 meters. The domain of the function is all real numbers greater than or equal to 0 and less than or equal to 80. The range of the function is all real numbers greater than or equal to 0 and less than or equal to 40.

Check for Understanding

Communicating Mathematics

1. **Compare and contrast** quadratic and linear functions.

2. **Explain** what effect a negative coefficient of x^2 has on the orientation of a parabola.

3. **Name** the point on the graph of a quadratic function that has a unique y-coordinate.

Vocabulary

parabola
quadratic function
minimum
maximum
vertex
axis of symmetry

Guided Practice

Getting Ready Identify the values for *a*, *b*, and *c* for each quadratic function in the form $y = ax^2 + bx + c$.

Sample 1: $y = x^2 - 9$

Solution: $a = 1, b = 0, c = -9$

Sample 2: $y = -4x^2 - 9x + 13$

Solution: $a = -4, b = -9, c = 13$

4. $y = 6x^2 - 3x + 1$

5. $y = x^2 + 4$

6. $y = -x^2 - x - 12$

7. $y = 3x^2 + 4x$

Examples 1 & 2

Graph each quadratic function by making a table of values.

8. $y = -2x^2$

9. $y = x^2 + 3x$

10. $y = -x^2 - 2x + 5$

Example 3

Write the equation of the axis of symmetry and the coordinates of the vertex of the graph of each quadratic function. Then graph the function and state its domain and range.

11. $y = x^2 + 2$

12. $y = x^2 + 8x + 12$

13. $y = -x^2 + 4x + 3$

14. $y = -x^2 + 10x$

Example 4

15. **Architecture** Mr. Kwan is drafting the windows for a building. Their shape is a parabola modeled by the equation $h = -w^2 + 9$, where h is the height of the window and w is the width in feet.

a. Graph the function.

b. Find the maximum height of each window.

c. Find the width of each window at its base.

Practice

Graph each quadratic function by making a table of values.

16. $y = 2x^2$ **17.** $y = 3x^2$ **18.** $y = -5x^2$

19. $y = x^2 + 4$ **20.** $y = 2x^2 - 1$ **21.** $y = -x^2 + 8x$

22. $y = -x^2 + 4x + 1$ **23.** $y = 3x^2 - 6x + 1$ **24.** $y = \frac{1}{2}x^2 - 6x + 5$

Write the equation of the axis of symmetry and the coordinates of the vertex of the graph of each quadratic function. Then graph the function and state the domain and range.

25. $y = x^2$ **26.** $y = -2x^2$ **27.** $y = 4x^2$

28. $y = 2x^2 + 3$ **29.** $y = -x^2 + 1$ **30.** $y = x^2 - 6x$

31. $y = x^2 - 4x + 2$ **32.** $y = x^2 - 4x + 10$ **33.** $y = -3x^2 - 6x + 4$

34. $y = -\frac{1}{4}x^2 + x - 2$ **35.** $y = x^2 + 5x$ **36.** $y = 3x^2 - 3x - 2$

Match each function with its graph.

37. $y = x^2 + 2x + 1$ **38.** $y = x^2 - 2x - 1$ **39.** $y = -x^2 + 4x - 3$

A   B   C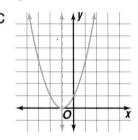

40. Suppose the equation of the axis of symmetry for a quadratic function is $x = -3$ and one of the x-intercepts is -8. What is the other x-intercept?

41. Suppose the points at $(-8, 5)$ and $(6, 5)$ are on the graph of a parabola. What is the equation of the axis of symmetry?

Applications and Problem Solving

42. Sports Mark wanted to know the angle at which he should kick a football for maximum distance. He used a device that kicked a football at a constant velocity at varying angles. He recorded his results in the table.

Angle	Distance (ft)
30°	140
35°	152
40°	160
45°	162
50°	160
55°	152

  a. Graph the data in the table.

  b. Which angle gives the maximum distance?

  c. Predict how far the football will go if it is kicked at a 60° angle.

43. **Business**   The profit function of a small business can be expressed as $P(x) = -x^2 + 300x$, where x represents the number of employees.
    a. How many employees will yield the maximum profit?
    b. What is the maximum profit?
    c. What are the domain and range of $P(x)$, and what do they represent?

44. **Critical Thinking**   Graph $y = x^2$. Then graph $y = x^2 - 4$ on the same axes. Describe the difference between the graphs.

Mixed Review

Factor each polynomial.   *(Lesson 10–5)*

45. $x^2 - 49$
46. $a^2 + 10a + 25$
47. $2g^2 - 16g + 32$

48. **Geometry**   The volume of a rectangular prism is $3x^3 - 6x^2 - 24x$. Find dimensions for the prism in terms of x.   *(Lesson 10–4)*

49. Arrange the terms of the polynomial $-4x^2 + 5 - x^3 - 8x$ so that the powers of x are in descending order. Then state the degree of the polynomial.   *(Lesson 9–1)*

Use the map for Exercises 50–51.

50. You can use the letters and numbers on the map to form ordered pairs and name square areas. State the locations of Fancyburg and Northam Parks as ordered pairs.   *(Lesson 2–2)*

51. Find the straight-line distance in miles between Fancyburg and Northam Parks.   *(Lesson 5–2)*

52. **Multiple Choice**   The line graph shows the amount of money spent on magazine advertising of soft drinks over a five-year period. During which of the following periods was the greatest change in spending?   *(Lesson 1–7)*

    Ⓐ 1998–1999
    Ⓑ 1999–2000
    Ⓒ 2000–2001
    Ⓓ 2001–2002

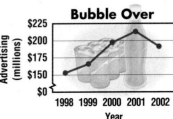

Bubble Over

$225
$200
$175
$150
$0

1998 1999 2000 2001 2002
Year

Source: Publishers Information Bureau, Inc.

11-2 Families of Quadratic Functions

What You'll Learn

You'll learn the characteristics of families of parabolas.

Why It's Important

Animation Digital artists use families of parabolas to create the illusion of motion.
See Example 5.

In families of linear graphs, lines either have the same slope or the same y-intercept. However, in families of parabolas, graphs either share a vertex or an axis of symmetry, or both. Also, a family can consist of parabolas of the same shape.

Families of Parabolas

Share the same vertex | Share the same axis of symmetry | Have the same shape

Examples

Graph each group of equations on the same screen. Compare and contrast the graphs. What conclusions can be drawn?

1 $y = x^2$, $y = 0.2x^2$, $y = 3x^2$

Technology Tip

Use the Y= screen to enter functions into the graphing calculator.

Each graph opens upward and has its vertex at the origin. Therefore, these equations are a family of parabolas. The graph of $y = 0.2x^2$ is wider than the graph of $y = x^2$. The graph of $y = 3x^2$ is more narrow than the graph of $y = x^2$.

The parent graph for each family is $y = x^2$.

The shape of the parabola narrows as the coefficient of x^2 becomes greater. The shape widens as the coefficient of x^2 becomes smaller.

2 $y = x^2$, $y = x^2 - 6$, $y = x^2 + 3$

Each graph opens upward and has the same shape as $y = x^2$ so they form a family. Yet, each parabola has a different vertex located along the y-axis.

A constant greater than 0 shifts the graph upward, and a constant less than 0 shifts the graph downward along the axis of symmetry.

3 $y = x^2$, $y = (x + 2)^2$, $y = (x - 4)^2$

$y = (x + 2)^2$

$y = x^2$

$y = (x - 4)^2$

Each graph opens upward and has the same shape as $y = x^2$. However, each parabola has a different vertex located along the x-axis.

Find the number for x that results in 0 inside the parentheses. The graph shifts this number of units to the left or right.

4 $y = x^2$, $y = (x - 7)^2 + 2$

$y = x^2$

$y = (x - 7)^2 + 2$

The graph of $y = (x - 7)^2 + 2$ has the same shape as the graph of $y = x^2$. However, it shifts to the right 7 units because a positive 7 will result in zero inside the parentheses. It also shifts upward 2 units because of the constant 2 outside the parentheses.

Your Turn

COncepts in MOtion
Animation
algconcepts.com

a. $y = x^2$, $2x^2$, $4x^2$

b. $y = x^2$, $x^2 - 1$, $x^2 - 8$

c. $y = x^2$, $-x^2$

d. $y = -x^2$, $y = -(x + 2)^2$, $y = -(x + 4)^2$

Sometimes computers are used to generate families of graphs.

Example
Computer Animation Link

5 In a computer game, a player dodges space shuttles that are shaped like parabolas. Suppose the vertex of one shuttle is at the origin. The shuttle's initial shape and position are given by the equation $y = 0.5x^2$. It leaves the screen with its vertex at (6, 5). Find an equation to model the final shape and position of the shuttle.

The shape of the shuttle remains the same. However, the vertex shifts from (0, 0) to the right 6 units and up 5 units.

Begin with the original equation.

$y = 0.5x^2$

$y = 0.5(x - 6)^2$

If $x = 6$, then $x - 6 = 0$. Shift the vertex to the right 6 units.

$y = 0.5(x - 6)^2 + 5$

The 5 outside the parentheses shifts the entire parabola up 5 units.

Math Online Data Update For the latest information on computer animation, visit: algconcepts.com

So, the final shape and position of the shuttle can be described by the equation $y = 0.5(x - 6)^2 + 5$.

Communicating Mathematics

1. **Describe** the parabola whose equation is $y = 100x^2$.

2. Vanessa says that the graphs of $y = (x - 2)^2$ and $y = x^2 - 2$ are the same. Vickie says that they are different. Who is correct? Explain your answer and sketch the graphs.

3. **Match** each equation with its corresponding graph.

$y = -3x^2$ $\qquad\qquad$ $y = x^2 + 3$ $\qquad\qquad$ $y = (x - 3)^2$

A $\qquad\qquad\qquad$ **B** $\qquad\qquad\qquad$ **C**

 $\qquad$ $\qquad$

Guided Practice

Example 1

4. Graph $y = x^2$, $y = 0.5x^2$, and $y = 0.1x^2$ on the same axes. Compare and contrast the graphs.

Examples 1–4

Describe how each graph changes from the parent graph of $y = x^2$. Then name the vertex of each graph.

5. $y = 0.7x^2$ $\qquad\qquad\qquad\qquad$ 6. $y = x^2 + 10$

7. $y = (x + 4)^2$ $\qquad\qquad\qquad\qquad$ 8. $y = (x + 2)^2 - 9$

Example 5

9. **Computer Animation** Refer to Example 5. Suppose the shuttle is programmed to move to point (4, 3) before leaving the screen. Write the equation that describes its location.

Exercises • • • • • • • • • • • • • • • • • • •

Practice

Graph each group of equations on the same screen. Compare and contrast the graphs.

10. $y = -x^2$ $\qquad\quad$ 11. $y = (x + 1)^2$ $\qquad\quad$ 12. $y = -x^2 - 1$

$\quad$ $y = -4x^2$ $\qquad\qquad$ $y = (x + 2)^2$ $\qquad\qquad$ $y = -x^2 - 3$

$\quad$ $y = -6x^2$ $\qquad\qquad$ $y = (x + 3)^2$ $\qquad\qquad$ $y = -x^2 - 5$

Homework Help	
For Exercises	See Examples
10, 13, 17, 18	1
11, 15, 16, 25, 27	3
12, 14, 26	2
23, 28, 29	4, 5
Extra Practice	
See page 713.	

Describe how each graph changes from the parent graph of $y = x^2$. Then name the vertex of each graph.

13. $y = 5x^2$ $\qquad\quad$ 14. $y = x^2 - 8$ $\qquad\quad$ 15. $y = (x - 7)^2$

16. $y = (x - 3)^2$ $\qquad$ 17. $y = -2x^2$ $\qquad\quad$ 18. $y = -0.9x^2$

19. $y = 2x^2 + 1$ $\qquad$ 20. $y = -(x + 4)^2$ $\qquad$ 21. $y = 0.4x^2 - 8$

22. $y = -x^2 + 6$ $\qquad$ 23. $y = (x + 1)^2 - 5$ $\qquad$ 24. $y = [x - (-2)]^2 + 3$

25. What is the equation of the parabola that moves the parent graph $y = x^2$ 8 units to the left ?

26. Suppose the parent graph is $y = x^2 + 3$. Write the equation of the parabola that would move it down 5 units.

Applications and Problem Solving

Fireworks display •············

··27. Fireworks   In Cincinnati, fireworks are launched from a large barge on the Ohio River for a Labor Day celebration. Their flight can be modeled by the function $h(t) = -4.9(t - 4)^2 + 80$, where h is the height in meters and t is the time in seconds. Suppose that buildings obstruct the fireworks. So, the barge is relocated 30 meters to the east. Write a function to model the path of the fireworks in the new location.

28. Sports   A baseball player hits a pop-up. The height of the ball can be modeled by the function $h(t) = -16(t - 3.5)^2 + 200$, where h is the height in feet and t is the time in seconds. Another pop-up is hit with the same velocity, but from a half-foot higher.

a. Write equations to model the height of the ball for each hit.

b. Graph both equations and find each vertex.

c. Does the higher height affect the maximum height? Does the higher height affect the time the ball takes to reach its maximum height?

29. Critical Thinking   Graphs of two quadratic functions are shown. The graph of $y = x^2$ is the parent graph. Write the equation for the other graph.

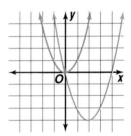

Mixed Review

Find the coordinates of the vertex for each quadratic function.
(Lesson 11–1)

30. $y = -4x^2 + 8x + 13$ **31.** $y = 2x^2 + 6x + 3$

32. Finance   Tracy invested $500, earning an annual interest rate r. The value of her investment at the end of two years can be represented by the polynomial $500r^2 + 1000r + 500$.   *(Lesson 10–5)*

a. Write an expression in factored form for the value of her investment at the end of two years.

b. If $r = 6\%$, find the value of her investment at the end of two years.

Find each product.   *(Lesson 9–4)*

33. $(a + 7)(a - 4)$ **34.** $(3n + 6)(n - 4)$

Standardized Test Practice

Ⓐ Ⓑ Ⓒ Ⓓ

35. Multiple Choice   Nine tickets, numbered 3 through 11, are placed in an empty hat. If one ticket is drawn at random from the hat, what is the probability that a prime number will be on the ticket?
(Lesson 5–6)

Ⓐ $\frac{1}{9}$ Ⓑ $\frac{4}{9}$ Ⓒ $\frac{5}{9}$ Ⓓ $\frac{1}{3}$

In a **quadratic equation,** the value of the related quadratic function is 0. For example, if you substitute 0 for y in the quadratic function, the result is the quadratic equation $0 = ax^2 + by + c$. The solutions of a quadratic equation are called the **roots** of the equation. The roots of a quadratic equation can be found by finding the x-intercepts or **zeros** of the related quadratic function.

Examples

Landmark Link

1 The Buckingham Fountain in Chicago has 133 jets through which water flows. The path of water streaming from a jet is in the shape of a parabola. Find the distance from the jet where the water hits the ground by graphing. Use the function $h(d) = -2d^2 + 4d + 6$, where $h(d)$ represents the height of a stream of water at any distance d from its jet in feet.

Explore Graph the parabola to determine where the stream of water will hit the ground.

Plan Find the solution of the equation by looking at the values of d where $h(d)$ is 0.

Solve Make a table of values to graph the related function $h(d) = -2d^2 + 4d + 6$.

d	$-2d^2 + 4d + 6$	$h(d)$
-1	$-2(-1)^2 + 4(-1) + 6$	0
0	$-2(0)^2 + 4(0) + 6$	6
1	$-2(1)^2 + 4(1) + 6$	8
2	$-2(2)^2 + 4(2) + 6$	6
3	$-2(3)^2 + 4(3) + 6$	0

Buckingham Fountain, Chicago

The roots of $0 = -2d^2 + 4d + 6$ are -1 and 3.

Examine Since d represents distance, it cannot be negative. Therefore, $d = -1$ is not a solution, and the only reasonable solution is 3. The water will hit the ground 3 feet from the jet.

Press 2nd [TABLE]
2 to find the roots of an
equation graphed on the
calculator screen.

2 **Find the roots of $x^2 - 10x + 16 = 0$ by graphing the related function.**

Graph the related function $f(x) = x^2 - 10x + 16$. Before making a table of values, find the equation of the axis of symmetry. This will make selecting x-values for your table easier.

$x = -\dfrac{b}{2a}$ *Equation of the axis of symmetry*

$x = -\dfrac{-10}{2(1)}$ or 5   *$a = 1$ and $b = -10$*

The equation of the axis of symmetry is $x = 5$. Now, make a table using x-values around 5. Graph each point on a coordinate plane.

x	$x^2 - 10x + 16$	$f(x)$
2	$2^2 - 10(2) + 16$	0
4	$4^2 - 10(4) + 16$	-8
5	$5^2 - 10(5) + 16$	-9
6	$6^2 - 10(6) + 16$	-8
8	$8^2 - 10(8) + 16$	0

The zeros of the function appear to be 2 and 8. So, the roots are 2 and 8.

Check: Substitute 2 and 8 for x in the equation $x^2 - 10x + 16 = 0$.

$x^2 - 10x + 16 = 0$ $x^2 - 10x + 16 = 0$

$2^2 - 10(2) + 16 \overset{?}{=} 0$ $8^2 - 10(8) + 16 \overset{?}{=} 0$

$4 - 20 + 16 \overset{?}{=} 0$ $64 - 80 + 16 \overset{?}{=} 0$

$0 = 0$ ✓ $0 = 0$ ✓

Your Turn

a. Find the roots of $0 = x^2 - 5x + 4$ by graphing the related function.

Sometimes exact roots cannot be found by graphing. In this case, estimate solutions by stating the consecutive integers between which the roots are located.

Example   **3** **Estimate the roots of $-x^2 + 2x + 1 = 0$.**

Find the equation of the axis of symmetry.

$x = -\dfrac{b}{2a}$ *Equation of the axis of symmetry*

$x = -\dfrac{2}{2(-1)}$ or 1   *$a = -1$ and $b = 2$*

The equation of the axis of symmetry is $x = 1$. Now, make a table using x-values around 1. Graph each point on a coordinate plane.

(continued on the next page)

x	$-x^2 + 2x + 1 = 0$	$f(x)$
−1	$-(-1)^2 + 2(-1) + 1$	−2
0	$-0^2 + 2(0) + 1$	1
1	$-(1)^2 + 2(1) + 1$	2
2	$-(2)^2 + 2(2) + 1$	1
3	$-(3)^2 + 2(3) + 1$	−2

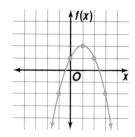

The *x*-intercepts of the graph are between −1 and 0 and between 2 and 3. So, one root of the equation is between −1 and 0, and the other root is between 2 and 3.

COncepts in MOtion

Animation

algconcepts.com

Your Turn

b. Estimate the roots of $y = x^2 - 2x - 9$.

Example

4

Number Theory Link

Find two numbers whose sum is 4 and whose product is 5.

Explore Let x = one of the numbers.
Then $4 - x$ = the other number.

Plan Since the product of the two numbers is 5, you know that $5 = x(4 - x)$.

$5 = x(4 - x)$

$5 = 4x - x^2$ *Distributive Property*

$0 = -x^2 + 4x - 5$ *Subtract 5 from each side.*

Solve You can solve $0 = -x^2 + 4x - 5$ by graphing the related function $f(x) = -x^2 + 4x - 5$. The equation of the axis of symmetry is $x = -\dfrac{4}{2(-1)}$ or 2. So, choose *x*-values around 2 for your table.

x	$-x^2 + 4x - 5$	$f(x)$
0	$-0^2 + 4(0) - 5$	−5
1	$-(1)^2 + 4(1) - 5$	−2
2	$-(2)^2 + 4(2) - 5$	−1
3	$-(3)^2 + 4(3) - 5$	−2
4	$-(4)^2 + 4(4) - 5$	−5

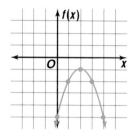

The graph has no *x*-intercepts since it does not cross the *x*-axis. This means the equation $x^2 - 4x + 5 = 0$ has no real roots. Thus, it is *not* possible for two numbers to have a sum of 4 and a product of 5. *Examine this solution by testing several pairs of numbers.*

The graphing calculator is a useful tool when making tables in order to solve functions.

Graphing Calculator Tutorial
See pp. 750–753.

Solve $x^2 + x - 2 = 0$ by making a table.

Enter the equation in the $\boxed{Y=}$ screen. Then press $\boxed{2nd}$ [TABLE]. A table with two columns labeled X and Y1 will appear on your screen.

Begin entering values for x. For each x-value entered, a corresponding y-value is calculated using the equation $y = x^2 + x - 2$.

Since $y = 0$ when $x = -2$ and $x = 1$, the roots of the equation are -2 and 1.

Try These
Make a table to find the roots of each equation.

1. $x^2 - 18x + 81 = 0$　　　　　　**2.** $x^2 - 2x - 15 = 0$

Check for Understanding

Communicating Mathematics

1. Explain why finding the x-intercepts of the graph of a quadratic function can be used to solve the related quadratic equation.

2. Write the related function for $4 = x^2 - 3x$.

3. Sketch a parabola that has one root.

Vocabulary

quadratic equation
roots
zeros

Guided Practice

 Getting Ready **State the roots of each quadratic equation.**

Sample:

Solution: The x-intercepts of the graph are -2 and 3. So, the roots are -2 and 3.

4.

5.

6.

Solve each equation by graphing the related function. If exact roots cannot be found, state the consecutive integers between which the roots are located.

7. $x^2 - 5x + 6 = 0$ **8.** $x^2 + 2x - 15 = 0$ **9.** $2x^2 - 3x - 7 = 0$

Example 4
10. Number Theory Use a quadratic equation to find two numbers whose sum is 4 and whose product is −12.

Exercises •

Practice

Homework Help	
For Exercises	See Examples
11–19	1, 2
20, 21, 24, 25	3, 4
27	2, 3
Extra Practice	
See page 714.	

Solve each equation by graphing the related function. If exact roots cannot be found, state the consecutive integers between which the roots are located.

11. $x^2 + 2x + 1 = 0$ **12.** $x^2 - 4x + 3 = 0$ **13.** $x^2 - 5x - 14 = 0$

14. $-x^2 + 6x + 7 = 0$ **15.** $x^2 - 10x + 25 = 0$ **16.** $x^2 + 3x - 2 = 0$

17. $x^2 + 4x - 2 = 0$ **18.** $x^2 - 2x + 2 = 0$ **19.** $-2x^2 + 3x + 4 = 0$

Use a quadratic equation to determine the two numbers that satisfy each situation.

20. Their sum is 18 and their product is 81.

21. Their difference is 4 and their product is 32.

Use the given roots and vertex of a quadratic equation to graph the related quadratic function.

22. roots: −2, 4
vertex: (1, 9)

23. roots: −7, −3
vertex: (−5, −4)

Applications and Problem Solving

24. Business Mr. Jamison owns a manufacturing company that produces key rings. Last year, he collected data about the number of key rings produced per day and the corresponding profit. He then modeled the data using the function $P(k) = -2k^2 + 12k - 10$, where P is the profit in thousands of dollars and k is the number of key rings in thousands.

a. Graph the profit function.

b. How many key rings must be produced per day so that there is no profit and no loss?

c. How many key rings must be produced for the maximum profit?

d. What is the maximum profit?

25. Skydiving In a recent year, Adrian Nicholas broke two world records by flying 10 miles in 4 minutes 55 seconds without a plane. He jumped from an airplane at 35,000 feet and did not activate his parachute until he was 500 feet above the ground. If air resistance is ignored, how long did Nicholas free-fall? Use the formula $h(t) = -16t^2 + h_0$, where t is the time in seconds and h_0 is the initial height in feet.

26. **Geometry** Mrs. Parker wants to enclose a rectangular running yard at her dog kennel with 200 feet of fencing.

    a. Make a drawing of the enclosed yard. Then write a formula for the perimeter and solve for the width.

    b. Write a quadratic equation for the area A of the yard. Use the expression for the width from part a.

    c. Show graphically how the area of the yard changes when the length changes. Can the yard have a length of 100 feet? Explain.

27. **Critical Thinking** Suppose the value of a quadratic function is negative when $x = 1$ and positive when $x = 2$. Explain why it is reasonable to assume that the related equation has a root between 1 and 2.

Mixed Review

Describe how each graph changes from its parent graph of $y = x^2$. Then name the vertex. *(Lesson 11–2)*

28. $y = -x^2$
         29. $y = x^2 + 4$
         30. $y = (x + 6)^2$

Standardized Test Practice
Ⓐ Ⓑ Ⓒ Ⓓ

31. **Grid In** If the equation of the axis of symmetry of a quadratic function is $x = 0$, and one of the x-intercepts is -4, what is the other x-intercept? *(Lesson 11–1)*

32. **Multiple Choice** Evaluate $4^x + (x^2)^3$ if $x = 2$. *(Lesson 8–1)*

    Ⓐ 6
        Ⓑ 20
        Ⓒ 32
        Ⓓ 80

Quiz 1   Lessons 11–1 through 11–3

Write the equation of the axis of symmetry and the coordinates of the vertex for each quadratic function. Then graph the function. *(Lesson 11–1)*

  1. $y = x^2 + 6x - 5$
         2. $y = -2x^2 - 9$
         3. $y = -3x^2 - 6x + 4$

Describe how each graph changes from its parent graph of $y = x^2$. Then name the vertex of each graph. *(Lesson 11–2)*

  4. $y = x^2 + 7$
         5. $y = (x - 6)^2$
         6. $y = -0.8x^2$

Solve each equation by graphing the related function. If exact roots cannot be found, state the consecutive integers between which the roots are located. *(Lesson 11–3)*

  7. $x^2 - 5x + 6 = 0$
         8. $x^2 + 6x + 10 = 0$
         9. $x^2 - 2x - 1 = 0$

10. **Sports** Anna is doing a back dive from a 10-meter platform. Her path of descent is given by the graph of $h(d) = -d^2 + 2d + 10$, where the height h and distance from the platform d are both in meters. *(Lessons 11–1 & 11–3)*

    a. Graph the function.

    b. At her maximum height, how far away from the platform is she?

    c. About how far away from the platform will she enter the water?

Solving Quadratic Equations by Factoring

In the previous lesson you learned to solve quadratic equations by graphing. Quadratic equations can also be solved by factoring. For example, the equation $8x^2 - 4x = 0$ can be solved by finding factors.

$$8x^2 - 4x = 0$$

$$4x(2x - 1) = 0 \qquad \textit{Factor } 8x^2 - 4x.$$

To solve this equation, find values of x that make the product $4x(2x - 1)$ equal to 0. Since the product of 0 and any number is 0, *at least one* of the factors in the expression must be zero.

$$4x = 0 \qquad \text{or} \qquad 2x - 1 = 0$$

$$x = 0 \qquad\qquad\qquad x = \frac{1}{2}$$

The solutions of $8x^2 - 4x = 0$ are 0 and $\frac{1}{2}$.

This method of solving quadratic equations uses the **Zero Product Property**.

Zero Product Property	For all numbers a and b, if $ab = 0$, then $a = 0$, $b = 0$ or both a and b equal 0.

We can use this property to solve any equation that is written in the form $ab = 0$.

Example

1 Solve $3x(x - 1) = 0$. Check your solution.

If $3x(x - 1) = 0$, then $3x = 0$ or $x - 1 = 0$. *Zero Product Property*
$3x = 0$ or $x - 1 = 0$ *Solve each equation.*
$x = 0$ $\qquad\quad x = 1$

Check: Substitute 0 and 1 for x in the original equation.

$$3x(x - 1) = 0 \qquad\qquad \text{or} \qquad\qquad 3x(x - 1) = 0$$
$$3(0)(0 - 1) \stackrel{?}{=} 0 \qquad\qquad\qquad\qquad 3(1)(1 - 1) \stackrel{?}{=} 0$$
$$0(-1) \stackrel{?}{=} 0 \qquad\qquad\qquad\qquad\qquad 3(0) \stackrel{?}{=} 0$$
$$0 = 0 \; \checkmark \qquad\qquad\qquad\qquad\qquad 0 = 0 \; \checkmark$$

The solutions are 0 and 1.

Your Turn Solve each equation. Check your solution.

a. $z(z - 8) = 0$

b. $(a - 4)(4a + 3) = 0$

Example — 2

Recreation Link

— 26 ft

h

At an adventure park, you can jump off a 26-foot cliff into a pool of water below. The equation $h = -16t^2 + 4t + 26$ describes your height h in feet t seconds after your jump. If you jump up and off the cliff, what time will you pass the height of 26 feet again?

To find the time at which this occurs, let $h = 26$ and solve for t.

$26 = -16t^2 + 4t + 26$
$0 = -16t^2 + 4t$ *Subtract 26 from each side.*
$0 = 4t(-4t + 1)$ *Factor* $-16t^2 + 4t$.

If $4t(-4t + 1) = 0$, then $4t = 0$ or $-4t + 1 = 0$.   *Zero Product Property*
$4t = 0$ or $-4t + 1 = 0$ *Solve each equation.*
$t = 0$ $-4t = -1$
                        $t = \dfrac{1}{4}$

Check: Graph the equation and find the x-intercepts. We will graph the equation $h = 4t(-4t + 1)$ on a graphing calculator.

Enter the equation in the Y= screen. Press GRAPH. To find the zeros, press 2nd [CALC]

▼ 2. Use the arrow keys to move the cursor to the left of one of the x-intercepts. Press ENTER. Move the cursor to the right of the same x-intercept. Press ENTER twice. Coordinates of the root will appear. Repeat for the other root.

Set viewing window for x: [−1, 1] by 0.25 and y: [−1, 1] by 0.25.

The solutions are 0 and $\dfrac{1}{4}$. The solution 0 represents the beginning of the jump. So, you would return to your original location after a fourth of a second.

You will often need to factor an equation before using the Zero Product Property to solve the equation.

Example — 3

Look Back

Factoring Trinomials: Lessons 10–3 & 10–4

Solve $x^2 + 4x - 12 = 0$. Check your solution.

$x^2 + 4x - 12 = 0$
$(x - 2)(x + 6) = 0$ *Factor.*
$x - 2 = 0$ or $x + 6 = 0$ *Zero Product Property*
    $x = 2$ $x = -6$ *Check this solution.*

The solutions are -6 and 2.

 Your Turn

c. Solve $x^2 - 2x = 3$. Check your solution.

4 The length ℓ of a rectangle is 2 feet more than twice its width w. The area is 144 square feet. Find the measures of the sides.

Explore   The formula for the area of a rectangle is $A = \ell w$ and $\ell = 2w + 2$.

Plan

Area	equals	length	times	width.
144	=	$2w + 2$	·	w

Solve $144 = 2w^2 + 2w$ *Distributive Property*

$0 = 2w^2 + 2w - 144$ *Subtract 144 from each side.*

$0 = 2(w^2 + w - 72)$ *Factor out the GCF, 2.*

$0 = w^2 + w - 72$ *Divide each side by 2.*

$0 = (w + 9)(w - 8)$ *Factor.*

$w + 9 = 0$ or $w - 8 = 0$ *Zero Product Property*

$w = -9$ $w = 8$ *Solve each equation.*

Examine   Since width cannot be negative, the width must be equal to 8 feet. Substituting 8 for w, the length of the rectangle is $2(8) + 2$ or 18 feet.

🌐nline **Personal Tutor at** algconcepts.com

Check for Understanding

Communicating Mathematics

1. **Define** the Zero Product Property in your own words.

2. **Explain** why you should check your answers by substituting values into the original equation rather than into a simplified version.

3. Lashonda said that she should solve $2x^2 - 3x - 35 = 0$ by graphing. Jerome disagreed. He said that she should solve the equation by factoring. Who is correct, and why?

Guided Practice

Examples 1 & 3

Solve each equation. Check your solution.

4. $(x + 1)(x - 7) = 0$ 5. $3b(b - 5) = 0$ 6. $(2m - 1)(m + 4) = 0$

7. $x^2 - 6x + 9 = 0$ 8. $x^2 = x$ 9. $7x^2 - 14x = 56$

Example 2

10. **Geometry**   The height of a triangle measures 5 centimeters more than its base. The area of the triangle is 18 square centimeters. Find the measures of the base and the height of the triangle.

Exercises

Practice

Solve each equation. Check your solution.

11. $3m(m + 2) = 0$ 12. $5z(z + 8) = 0$ 13. $(p + 7)(p - 6) = 0$

14. $(s - 2)(s + 11) = 0$ 15. $(r + 1)(2r - 8) = 0$ 16. $(x - 2)(3x + 4) = 0$

17. $x^2 + 10x + 16 = 0$ 18. $z^2 + 5z - 6 = 0$ 19. $p^2 - 11p + 24 = 0$

20. $n^2 + 9n + 18 = 0$ 21. $m^2 - m - 12 = 0$ 22. $r^2 + 14r = 0$

Homework Help

For Exercises	See Examples
11–16	1, 2
17–34	3, 4

Extra Practice
See page 714.

23. $3w^2 - 9w = 0$ **24.** $15 = x^2 - 2x$ **25.** $2x^2 - 8x + 8 = 0$
26. $2a^2 - 70 = 4a$ **27.** $3b^2 - 12b - 15 = 0$ **28.** $2v^2 - 17v = 9$

For each problem, define a variable. Then use an equation to solve the problem.

29. The length of Luis' house is ten feet longer than it is wide. The area in square feet is 875. Find the dimensions of the house.

30. Find two consecutive even integers whose product is 120.

31. Find two integers whose difference is 3 and whose product is 88.

Applications and Problem Solving

32. Physics   A flare is launched from a life raft with an initial upward velocity of 192 feet per second. How many seconds will it take for the flare to return to the sea? Use the formula $h = 192t - 16t^2$, where h is the height of the flare in feet and t is the time in seconds.

33. Geometry   Four corners are cut from a rectangular piece of cardboard that measures 3 feet by 5 feet. The cuts are x feet from the corners as shown in the figure at the right. After the cuts are made, the sides are folded up to form an open box. The area of the bottom of the box is 3 square feet. Find the dimensions of the box.

34. Photography   A rectangular photograph is 4 inches wide and 6 inches long. The photograph is enlarged by increasing the length and width by an equal amount. If the area of the new photograph is twice as large as the original area, what are the dimensions of the new photograph?

35. Critical Thinking   The graph of a quadratic function is shown.
  a. What are the zeros of the function?
  b. Write an equation for the graph.

Mixed Review

Solve each equation by graphing the related function. *(Lesson 11–3)*
36. $x^2 - 10x + 21 = 0$ **37.** $x^2 - 2x + 5 = 0$ **38.** $x^2 + 4x = 12$

39. Graph $y = 0.25x^2$, $y = x^2$, and $y = 4x^2$ on the same set of axes. Compare and contrast the graphs. *(Lesson 11–2)*

Factor. *(Lesson 10–3)*
40. $y^2 + 7y + 6$ **41.** $b^2 + b - 42$

Standardized Test Practice
Ⓐ Ⓑ Ⓒ Ⓓ

42. Multiple Choice   Which of the following expressions is equivalent to $2^3 \times 4$? *(Lesson 8–2)*
  Ⓐ -8^4 Ⓑ 8^3 Ⓒ 6^4 Ⓓ 2^6 Ⓔ 2^5

What You'll Learn
You'll learn to solve quadratic equations by completing the square.

Why It's Important
Construction You can use quadratic equations to determine the dimensions of a room.
See Exercise 13.

Carmen planted daylilies on a square piece of land with an area of 49 square feet. She wants to plant petunias around the daylilies to form a border whose area is 72 square feet. What length should she make the sides of the outer square?

Petunias

Daylilies

Let x represent the length of a side of the outer square. Find a relationship between the variable and given numerical information to write and solve an equation.

$$\underbrace{\textit{Total area}}\ \ \underbrace{\textit{minus}}\ \ \underbrace{\textit{the daylily area}}\ \ \underbrace{\textit{is}}\ \ \underbrace{\textit{the petunia area.}}$$

$$x^2 \qquad\quad - \qquad\quad 49 \qquad\quad = \qquad\quad 72$$

$$x^2 = 121 \qquad \textit{Add 49 to each side.}$$
$$\sqrt{x^2} = \pm\sqrt{121} \qquad \textit{Find the square root of each side.}$$
$$x = \pm 11$$

Length cannot be negative, so the sides of the outer square measure 11 feet.

We were able to solve the problem easily because 121 is a perfect square. However, few quadratic expressions are perfect squares. To make *any* quadratic expression a perfect square, a method called **completing the square** can be used.

Look Back

Perfect Square Trinomials:
Lesson 10–5

When given an equation, you can complete the square by writing one side of the equation as a perfect square. This is modeled below.

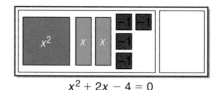

Hands-On Algebra
Algebra Tiles

Materials: algebra tiles   equation mat

Use algebra tiles to complete the square for $x^2 + 2x - 4 = 0$.

Step 1 Model
$x^2 + 2x - 4 = 0$
on the mat.

$x^2 + 2x - 4 = 0$

Step 2 Add 4 to each side of the mat. Remove the zero pairs.

$x^2 + 2x - 4 + 4 = 0 + 4$

Step 3 Begin to arrange the x^2 and x-tiles into a square.

$x^2 + 2x = 4$

Step 4 To complete the square, add 1 yellow tile to each side of the mat. So, the equation is $x^2 + 2x + 1 = 5$ or $(x + 1)^2 = 5$.

$x^2 + 2x + 1 = 5$

Try These

Use algebra tiles to complete the square of each equation.

1. $x^2 + 4x + 1 = 0$ **2.** $x^2 - 6x = -5$

To complete the square for any quadratic expression of the form $x^2 + bx$, you can follow the steps below.

Step 1 Take $\frac{1}{2}$ of b, the coefficient of x.

Step 2 Square the result of Step 1.

Step 3 Add the result of Step 2, $(\frac{b}{2})^2$, to $x^2 + bx$.

Example ❶ **Find the value of c that makes $x^2 + 14x + c$ a perfect square.**

Step 1 Find one half of 14. $\frac{14}{2} = 7$

Step 2 Square the result of Step 1. $7^2 = 49$

Step 3 Add the result of Step 2 to $x^2 + 14x$. $x^2 + 14x + 49$

Thus, $c = 49$. Notice that $x^2 + 14x + 49 = (x + 7)^2$.

Your Turn

a. Find the value of c that makes $x^2 - 6x + c$ a perfect square.

Once a perfect square is found, we can solve the equation by taking the square root of each side.

Example ❷ **Solve $x^2 + 12x - 13 = 0$ by completing the square.**

Reading Algebra

Read $\pm\sqrt{49}$ as *plus or minus the square root of 49.*

$x^2 + 12x - 13 = 0$ *$x^2 + 12x - 13$ is not a perfect square.*

$x^2 + 12x = 13$ *Add 13 to each side.*

$x^2 + 12x + 36 = 13 + 36$ *Since $(\frac{12}{2})^2 = 36$, add 36 to each side.*

$(x + 6)^2 = 49$ *Factor $x^2 + 12x + 36$.*

$\sqrt{(x + 6)^2} = \pm\sqrt{49}$ *Take the square root of each side.*

(continued on the next page)

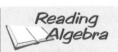

$$x + 6 = \pm 7 \qquad \text{Simplify.}$$
$$x + 6 - 6 = \pm 7 - 6 \qquad \text{Subtract 6 from each side.}$$
$$x = 7 - 6 \quad \text{or} \quad x = -7 - 6 \qquad \text{Simplify each equation.}$$
$$x = 1 \qquad\qquad x = -13 \qquad \text{Simplify.}$$

The solutions are -13 and 1. *Check the solution by using the graphing calculator or by substituting -13 and 1 for x in the original equation.*

Your Turn

b. Solve $x^2 + 6x - 16 = 0$ by completing the square.

Example **3**

Construction Link

You can complete the square only if the coefficient of the first term is 1. If the coefficient is *not* 1, first divide each term by the coefficient.

A school wants to redesign its nurse's station. Because of building codes, the maximum length of the rectangular room is 20 meters less than twice its width. Find the dimensions, to the nearest tenth of a meter, for the widest possible nurse's station if its area is to be 60 square meters.

Explore   Draw a picture of the room.
Let w = the maximum width.
Then $2w - 20$ = the maximum length.

Plan   Area equals length times width.
$$60 = 2w - 20 \cdot w$$

Solve
$$60 = w(2w - 20) \qquad \text{Original equation}$$
$$60 = 2w^2 - 20w \qquad \text{Distributive Property}$$
$$30 = w^2 - 10w \qquad \text{Divide each side by 2.}$$
$$30 + 25 = w^2 - 10w + 25 \qquad \left(\frac{-10}{2}\right)^2 = 25. \text{ Add 25 to each side.}$$
$$55 = (w - 5)^2 \qquad \text{Factor } w^2 - 10w + 25.$$
$$\sqrt{55} = \sqrt{(w-5)^2} \qquad \text{Take the square root of each side.}$$
$$\pm\sqrt{55} = w - 5 \qquad \text{Simplify.}$$
$$5 \pm \sqrt{55} = w - 5 + 5 \qquad \text{Add 5 to each side.}$$
$$5 \pm \sqrt{55} = w \qquad \text{Simplify.}$$

The roots are $5 + \sqrt{55}$ and $5 - \sqrt{55}$. You can use a calculator to find decimal approximations for these numbers.
$$5 + \sqrt{55} \approx 12.4 \qquad \text{and} \qquad 5 - \sqrt{55} \approx -2.4$$

The solution of -2.4 is not reasonable since there cannot be negative width. The dimensions of the nurse's station should be about 12.4 meters by $2(12.4) - 20$ or 4.8 meters.

Examine   Since $12.4(4.8) = 59.52$, the answer seems reasonable.

 Personal Tutor at algconcepts.com

Check for Understanding

Communicating Mathematics

1. **Explain** why completing the square is a good strategy to use in solving the equation $x^2 - 5x - 7 = 0$.

Vocabulary

completing the square

2. **Writing Math** List the steps necessary to complete the square of the expression $x^2 - 10x$.

Guided Practice

> **Getting Ready** State whether each trinomial is a perfect square.
>
> **Sample 1:** $x^2 - 4x - 4$ **Sample 2:** $x^2 + 6x + 9$
>
> **Solution:** No, it is not factorable.   **Solution:** Yes, $x^2 + 6x + 9 = (x + 3)^2$.

3. $x^2 - 2x + 1$ 4. $x^2 - 18x - 9$
5. $x^2 + 7x - 14$ 6. $x^2 + 8x + 16$

Example 1

Find the value of c that makes each trinomial a perfect square.

7. $x^2 + 2x + c$ 8. $z^2 - 18z + c$ 9. $m^2 + 3m + c$

Example 2

Solve each equation by completing the square. *(Example 2)*

10. $p^2 - 6p = 0$ 11. $r^2 + 7r + 12 = 0$ 12. $x^2 - 4x - 9 = 0$

Example 3

13. **Construction** The Thompsons' rectangular bathroom measures 6 feet by 9 feet. They want to double the area of the bathroom by increasing the length and width by the same amount. Find the dimensions, to the nearest foot, of the new bathroom. *(Example 3)*

Exercises • • • • • • • • • • • • • • • •

Practice

Find the value of c that makes each trinomial a perfect square.

14. $z^2 - 10z + c$ 15. $f^2 + 12f + c$ 16. $h^2 - 20h + c$
17. $s^2 - 2s + c$ 18. $q^2 + 14q + c$ 19. $x^2 - 24x + c$
20. $z^2 + 16z + c$ 21. $k^2 + 5k + c$ 22. $r^2 + 9r + c$

Homework Help	
For Exercises	See Examples
14–22	1
23–34, 35	2
36	3

Extra Practice
See page 714.

Solve each equation by completing the square.

23. $x^2 + 2x - 3 = 0$ 24. $x^2 - 8x + 7 = 0$ 25. $p^2 + 6p = -5$
26. $r^2 - 16r = 0$ 27. $s^2 - 14s = 0$ 28. $x(x + 10) - 1 = 0$
29. $q^2 - 2q = 30$ 30. $4x^2 + 16x + 24 = 0$ 31. $3z^2 - 18z = 30$
32. $m^2 - 2m = 6$ 33. $d^2 + 3d - 10 = 0$ 34. $a^2 - \frac{7}{2}a + \frac{3}{2} = 0$

Applications and Problem Solving

35. **Physical Science** Omar set off a rocket in a field. The height h of the rocket in feet can be roughly estimated using the formula $h = -16t^2 + 128t$, where t is the time in seconds. How long will it take the rocket to reach a height of 240 feet?

36. Photography Sheila places a 54 square inch photo behind a 12-inch-by-12-inch piece of matting. The photograph is positioned so that the matting is twice as wide at the top and bottom as the sides.

12 in.

a. Write an equation for the area of the photo in terms of x.

b. Find the dimensions of the photo.

37. Critical Thinking Find the values of b that make $x^2 + bx + 81$ a perfect square.

Mixed Review

Solve each quadratic equation by factoring. *(Lesson 11–4)*

38. $a^2 + 3a - 28 = 0$

39. $x^2 - 7x + 10 = 0$

40. $z^2 + 5z = -4$

41. $2b^2 + 12b + 18 = 0$

42. Rockets Jon is launching rockets in an open field. The path of the rocket can be modeled by the quadratic function $h(t) = -16t^2 + 96t$, where $h(t)$ is the height in feet any time t in seconds. *(Lesson 11–3)*

a. Graph the quadratic function.

b. After how many seconds will the rocket reach a height of 80 feet for the *second* time?

c. After how many seconds will the rocket hit the ground?

d. What is the rocket's maximum height?

Standardized Test Practice

Ⓐ Ⓑ Ⓒ Ⓓ

43. Extended Response The area of a rectangle is represented by $(2ab + 8b^2)$.

a. Factor the expression to find the dimensions of the rectangle. *(Lesson 10–2)*

b. Using the dimensions from part a, write an expression for the perimeter of the rectangle in simplest form. *(Lessons 9–2 & 9–3)*

c. Find the dimensions, area, and perimeter of the rectangle if $a = 5$ inches and $b = 1$ inch. *(Lesson 1–2)*

44. Multiple Choice The graph shows how each Ohio tax dollar was divided among programs and services in a recent year. If Maria paid $32 in state taxes on her last paycheck, what amount of her money went towards education? *(Lessons 5–3 & 5–4)*

Ⓐ $5.12

Ⓑ $10.24

Ⓒ $15.36

Ⓓ $24.96

Where the Tax Dollar Goes

Property Tax Relief 7%
Other 1%
Law Enforcement 11%
Elementary/Secondary Schools 32%
Human Services 29%
Colleges and Universities 16%
General Government 4%

Source: Ohio Dept. of Taxation

 www.algconcepts.com/self_check_quiz

What You'll Learn

You'll learn to solve quadratic equations by using the Quadratic Formula.

Why It's Important

Science You can use the Quadratic Formula to solve problems dealing with physics.
See Exercise 10.

The table below summarizes the ways you have learned to solve quadratic equations. Although these methods can be used to solve all quadratic equations, each method is most useful in particular situations.

Method	When Is the Method Useful?
Graphing	Use only to estimate solutions.
Factoring	Use when the quadratic expression is easy to factor.
Completing the Square	Use when the coefficient of x^2 is 1 and all other coefficients are fairly small.

An alternative method is to develop a general formula for solving *any* quadratic equation. Begin with the general form of a quadratic equation, $ax^2 + bx + c = 0$, where $a \neq 0$, and complete the square.

$$ax^2 + bx + c = 0 \qquad \text{\textit{General quadratic equation}}$$

$$x^2 + \frac{b}{a}x + \frac{c}{a} = 0 \qquad \text{\textit{Divide by a so the coefficient of }} x^2 \text{\textit{ is 1.}}$$

$$x^2 + \frac{b}{a}x = -\frac{c}{a} \qquad \text{\textit{Subtract }} \frac{c}{a} \text{\textit{ from each side.}}$$

Now complete the square.

$$x^2 + \frac{b}{a}x + \left(\frac{b}{2a}\right)^2 = -\frac{c}{a} + \left(\frac{b}{2a}\right)^2 \qquad \frac{1}{2} \cdot \frac{b}{a} = \frac{b}{2a}$$

$$\left(x + \frac{b}{2a}\right)^2 = -\frac{c}{a} + \frac{b^2}{4a^2} \qquad \text{\textit{Factor the left side of the equation.}}$$

$$\left(x + \frac{b}{2a}\right)^2 = \frac{4a}{4a}\left(-\frac{c}{a}\right) + \frac{b^2}{4a^2} \qquad \text{\textit{The common denominator is }} 4a^2.$$

$$\left(x + \frac{b}{2a}\right)^2 = \frac{b^2 - 4ac}{4a^2} \qquad \text{\textit{Simplify the right side of the equation.}}$$

Finally, take the square root of each side and solve for x.

$$\sqrt{\left(x + \frac{b}{2a}\right)^2} = \pm\sqrt{\frac{b^2 - 4ac}{4a^2}} \qquad \text{\textit{Take the square root of each side.}}$$

$$x + \frac{b}{2a} = \frac{\pm\sqrt{b^2 - 4ac}}{2a} \qquad \text{\textit{Simplify: }} \sqrt{4a^2} = 2a.$$

$$x = \frac{\pm\sqrt{b^2 - 4ac}}{2a} - \frac{b}{2a} \qquad \text{\textit{Subtract }} \frac{b}{2a} \text{\textit{ from each side.}}$$

$$x = \frac{-b \pm \sqrt{b^2 - 4ac}}{2a} \qquad \text{\textit{Combine the terms.}}$$

The result is called the **Quadratic Formula** and can be used to solve *any* quadratic equation.

The Quadratic Formula	$x = \dfrac{-b \pm \sqrt{b^2 - 4ac}}{2a}, a \neq 0$

Examples

Use the Quadratic Formula to solve each equation.

1 $x^2 - 4x + 3 = 0$

$$x = \frac{-b \pm \sqrt{b^2 - 4ac}}{2a} \qquad \textit{Quadratic Formula}$$

$$= \frac{-(-4) \pm \sqrt{(-4)^2 - 4(1)(3)}}{2(1)} \qquad a = 1, b = -4, \text{ and } c = 3$$

$$= \frac{4 \pm \sqrt{16 - 12}}{2} \qquad (-4)^2 = 16 \text{ and } 4(1)(3) = 12$$

$$= \frac{4 \pm \sqrt{4}}{2} \qquad \textit{Subtract.}$$

$$= \frac{4 \pm 2}{2} \qquad \textit{Simplify.}$$

$$= \frac{4 + 2}{2} \quad \text{or} \quad x = \frac{4 - 2}{2} \qquad \textit{Find two solutions.}$$

$$= \frac{6}{2} \text{ or } 3 \qquad x = \frac{2}{2} \text{ or } 1 \qquad \textit{Simplify.}$$

Check: Substitute values into the original equation.

$$x^2 - 4x + 3 = 0 \quad \text{or} \quad x^2 - 4x + 3 = 0$$
$$3^2 - 4(3) + 3 \stackrel{?}{=} 0 \qquad 1^2 - 4(1) + 3 \stackrel{?}{=} 0$$
$$9 - 12 + 3 \stackrel{?}{=} 0 \qquad 1 - 4 + 3 \stackrel{?}{=} 0$$
$$0 = 0 \; \checkmark \qquad\qquad 0 = 0 \; \checkmark$$

The roots are 3 and 1.

2 $-2x^2 + 3x - 5 = 0$

$$x = \frac{-b \pm \sqrt{b^2 - 4ac}}{2a} \qquad \textit{Quadratic Formula}$$

$$= \frac{-3 \pm \sqrt{3^2 - 4(-2)(-5)}}{2(-2)} \qquad a = -2, b = 3, \text{ and } c = -5$$

$$= \frac{-3 \pm \sqrt{9 - 40}}{-4} \qquad 3^2 = 9 \text{ and } 4(-2)(-5) = 40$$

$$= \frac{-3 \pm \sqrt{-31}}{-4} \qquad \textit{Subtract.}$$

The square root of a negative number, such as $\sqrt{-31}$, is not a real number. So, there are no real solutions for x.

www.algconcepts.com/extra_examples

Check: Use a graphing calculator. It is clear that the graph of $f(x) = -2x^2 + 3x - 5$ never crosses the x-axis. Therefore, there are no real roots for the equation $-2x^2 + 3x - 5 = 0$.

Your Turn

a. $-3x^2 + 6x + 9 = 0$

b. $x^2 + 4x + 2 = 0$

 Personal Tutor at algconcepts.com

When the solutions are irrational numbers, use a calculator to estimate.

Example

Sports Link

3 "Hang time" is the total amount of time a ball stays in the air. Manuel can kick a football with an upward velocity of 64 ft/s. His foot meets the ball 2 feet off the ground. What is Manuel's hang time if the ball is not caught? Use the formula $h(t) = -16t^2 + 64t + 2$, where $h(t)$ is the ball's height in feet for any time t, in seconds, after the ball is kicked.

$-16t^2 + 64t + 2 = 0$ *Final height is 0.*

$t = \dfrac{-b \pm \sqrt{b^2 - 4ac}}{2a}$ *Use the Quadratic Formula.*

$= \dfrac{-64 \pm \sqrt{64^2 - 4(-16)(2)}}{2(-16)}$ *$a = -16$, $b = 64$, and $c = 2$*

$= \dfrac{-64 \pm \sqrt{4096 + 128}}{-32}$ *$64^2 = 4096$ and $4(-16)(2) = -128$*

$= \dfrac{-64 \pm \sqrt{4224}}{-32}$ *$4096 - (-128) = 4096 + 128$*

$t = \dfrac{-64 + \sqrt{4224}}{-32}$ or $t = \dfrac{-64 - \sqrt{4224}}{-32}$

≈ -0.031 ≈ 4.031

Since time cannot be negative, the only approximate solution is 4.031. The football has a hang time of about 4 seconds.

Check for Understanding

Communicating Mathematics

1. **Choose** the correct term and justify your answer. Quadratic equations (*sometimes*, *always*, *never*) have at least one real solution.

> **Vocabulary**
>
> Quadratic Formula

2. **Writing Math** List the steps you take when you use the Quadratic Formula to solve a quadratic equation.

Getting Ready Find the value of $b^2 - 4ac$ for each equation.

Sample 1: $-3z^2 + 8z + 5 = 0$

Solution: $8^2 - 4(-3)(5)$
$= 64 + 60$
$= 124$

Sample 2: $6a^2 = 3$

Solution: Rewrite $6a^2 = 3$ as
$6a^2 - 3 = 0$.
$0^2 - 4(6)(-3) = 72$

3. $x^2 + 10x + 13 = 0$

4. $3x^2 - 9x + 2 = 0$

5. $4x^2 = 12$

6. $-2x^2 - 7x = -5$

Examples 1 & 2   Use the Quadratic Formula to solve each equation.

7. $z^2 - 25 = 0$ 8. $r^2 - 5r + 7 = 0$ 9. $d^2 + 4d = -3$

Example 3   10. **Physical Science** Josefina tosses a ball upward with an initial velocity of 76 feet per second. She tosses it from an initial height of 2 feet. Find the approximate time that the ball hits the ground. Use the function $h(t) = -16t^2 + 76t + 2$.

Exercises •

Practice

Use the Quadratic Formula to solve each equation.

11. $x^2 + 7x + 6 = 0$

12. $r^2 + 10r + 9 = 0$

13. $-a^2 + 5a - 6 = 0$

14. $-b^2 + 6b + 7 = 0$

15. $9d^2 - 4d + 3 = 0$

16. $3v^2 + 5v - 8 = 0$

17. $x^2 - 8x = 0$

18. $-p^2 + 6p - 5 = 0$

19. $w^2 + w - 7 = 0$

20. $z^2 + 9z - 2 = 0$

21. $4g^2 + 8g - 3 = 0$

22. $-2t^2 + 4t = 3$

23. $2c^2 + 14c = 10$

24. $5z^2 + 12 = -6z$

25. $8k^2 - 5k = 7$

Homework Help

For Exercises	See Examples
11–25	1, 2
26–27	3

Extra Practice
See page 715.

Applications and Problem Solving

26. **Skiing** Brandi, an amateur skier, begins on the "bunny hill," which has a vertical drop of 200 feet. Her speed at the bottom of the hill is given by the equation $v^2 = 64h$, where the velocity v is in feet per second and the height h of the hill is in feet. Assuming there is no friction, approximate Brandi's velocity at the bottom of the hill.

27. **Space** Suppose an astronaut can jump vertically with an initial velocity of 5 m/s. The time that it takes him to touch the ground is given by the equation $0 = 5t - 0.5at^2$. The time t is in seconds and the acceleration due to gravity a is in m/s^2.

   a. If the astronaut jumps on the moon where $a = 1.6$ m/s^2, how long will it take him to reach the ground?

   b. How long will it take him to reach the ground if he jumps on Earth where $a = 9.8$ m/s^2?

28. Critical Thinking   The expression $b^2 - 4ac$ is called the **discriminant** of a quadratic equation. It determines the number of real solutions you can expect when solving a quadratic equation.

a. Copy and complete the table below.

Equation	$x^2 - 4x + 1 = 0$	$x^2 + 6x + 11 = 0$	$x^2 - 4x + 4 = 0$
Value of the Discriminant			
Number of x-intercepts			
Number of Real Solutions			

b. Use the table above to describe the discriminant of a quadratic equation for each type of real solution.

   i. two solutions   **ii.** one solution   **iii.** no solutions

Mixed Review

Find the value of c to make a perfect square.   *(Lesson 11–5)*

29. $a^2 + 4a + c$ **30.** $x^2 + 8x + c$ **31.** $v^2 - 18v + c$

32. Geometry   The area of a circle varies directly as the square of the radius r. If the radius of a circle is doubled, how many times as large is the area of the circle?   *(Lesson 6–5)*

Standardized Test Practice
Ⓐ Ⓑ Ⓒ Ⓓ

33. Multiple Choice   If four times a number is equal to that number decreased by 5, what is the number?   *(Lesson 4–6)*

   Ⓐ -5 Ⓑ $-\dfrac{5}{3}$ Ⓒ $-\dfrac{1}{4}$ Ⓓ $\dfrac{1}{3}$ Ⓔ 4

Quiz 2   Lessons 11–4 through 11–6

Solve each quadratic equation by factoring.   *(Lesson 11–4)*

   1. $x^2 + 4x - 12 = 0$ **2.** $x^2 - 8x + 16 = 0$ **3.** $x^2 - 10x = -21$

Solve each quadratic equation by completing the square.   *(Lesson 11–5)*

   4. $x^2 + 4x + 3 = 0$ **5.** $x^2 - 6x + 7 = 0$ **6.** $2x^2 - 8x = 4$

Solve each equation by using the Quadratic Formula.   *(Lesson 11–6)*

   7. $x^2 + 3x + 3 = 0$ **8.** $x^2 + 8x + 15 = 0$ **9.** $x^2 + 5x + 3 = 0$

10. Construction   A company is designing a parabolic arch bridge to span a creek. The height of the bridge is given by the equation $h = -0.02d^2 + 0.8d$, where the height h and the distance d from one end of the base are in feet. Find the width of the bridge.   *(Lesson 11–6)*

Civil Engineer

Civil engineers work in the oldest branch of engineering. They design and supervise the construction of buildings, bridges, roads, and sewer systems. One of the most innovative structures in bridges, the parabolic arch, was first used extensively by the Romans. Civil engineers continue to use the parabolic arch today for its simple beauty.

A parabolic arch supports the Hulme Arch Bridge in England. The Hulme Arch can be modeled by the quadratic function $h(x) = -0.037x^2 + 25$, where $h(x)$ represents the height of the arch at any point x to the left or right of the center. All measures are in meters.

1. Use the Quadratic Formula to approximate the zeros of the quadratic function $h(x) = -0.037x^2 + 25$ to the nearest whole number.
2. What do the zeros represent on the Hulme Arch?
3. Find the width of the Hulme Arch to the nearest meter.
4. What is the height of the Hulme Arch?
5. Graph the Hulme Arch on a coordinate plane.

FAST FACTS About Civil Engineers

Working Conditions
- usually work in office buildings, industrial plants, or outside on construction sites
- 40-hour work weeks, except around deadlines when more hours are expected

Education
- college degree in engineering, physical science, or mathematics
- creative and detail-oriented personalities are well-suited for the profession

Earnings

Starting Salaries, 2003

Bachelor's Degree — $41,669
Master's Degree — $47,245

0   20,000   40,000   60,000

Source: Bureau of Labor Statistics, 2004

*inter*NET
CONNECTION
www.algconcepts.com

Career Data For the latest information on civil engineers, visit:

Often, patterns can be used to describe and classify types of functions.

Hands-On Algebra

Materials: ☐ large sheet of paper

Step 1 Fold a large sheet of paper in half. Unfold it and record how many sections are formed by the creases. Refold the paper.

Step 2 Fold the paper in half again. Record how many sections are formed by the creases. Refold the paper.

Step 3 Continue folding in half and recording the number of sections until you can no longer fold the paper.

Try These

1. How many folds could you make?
2. How many sections were formed?

The data collected in the Hands-On Algebra activity can be represented in a table to see a pattern more easily. Each fold increases the number of sections by a factor of 2.

Folds	Sections	Pattern
1	2	$2 = 2^1$
2	4	$2(2) = 2^2$
3	8	$2(2)(2) = 2^3$
4	16	$2(2)(2)(2) = 2^4$
5	32	$2(2)(2)(2)(2) = 2^5$

Let x equal the number of folds and let y equal the number of sections. Then the function $y = 2^x$ represents the number of sections for any number of folds. This is an example of an **exponential function**. An exponential function is of the form $y = a^x$, where $a > 0$ and $a \neq 1$. Some people also refer to equations of the form $y = ab^x + c$ as exponential functions. For this form of the exponential function, the value a is called the *coefficient*.

You can make a table to help graph exponential functions.

Example **Graph $y = 2^x$.**

x	2^x	y
-1	2^{-1}	0.5
0	2^0	1
1	2^1	2
2	2^2	4
3	2^3	8

The graph of an exponential function changes little for small x values. However, as the values of x increase, the y values increase quickly.

The **initial value** of an exponential function is the value of the function when $x = 0$. This is the same as the y-intercept. Exponential functions of the form $y = a^x$ have an initial value of 1. Exponential functions of the form $y = ab^x + c$ have an initial value of c.

Examples

Graph each exponential function. Then state the y-intercept.

2 $y = 5^x$

x	5^x	y
-2	5^{-2}	0.04
-1	5^{-1}	0.2
0	5^0	1
1	5^1	5
2	5^2	25

To find the y-intercept, let $x = 0$ and solve for y.

$$y = 5^0 \text{ or } 1$$

In this case, the y-intercept is 1.

3 $y = 2^x + 3$

Reading Algebra

Read $2^x + 3$ as *two to the x plus three.*

x	$2^x + 3$	y
-2	$2^{-2} + 3$	3.25
-1	$2^{-1} + 3$	3.5
0	$2^0 + 3$	4
1	$2^1 + 3$	5
2	$2^2 + 3$	7
3	$2^3 + 3$	11

The y-intercept is 4. *The constant is 3: $1 + 3 = 4$.*

Your Turn **Graph each function. Then state the y-intercept.**

a. $y = 2.5^x$ **b.** $y = 4^x + 3$

Quantities that increase rapidly have *exponential growth*. For instance, money in the bank may grow at an exponential rate.

Real World

Example

Finance Link

4 **When Taina was 10 years old, she received a certificate of deposit (CD) for $2000 with an annual interest rate of 5%. After eight years, how much money will she have in the account?**

After the first year, the CD is worth the initial deposit plus interest.

$2000 + 2000(0.05) = 2000(1 + 0.05)$ *Distributive Property*

$= 2000(1.05)$

 www.algconcepts.com/extra_examples

Taina can make a table to look for patterns in the growth of her CD.

Year	Balance	Pattern
0	2000	$2000(1.05)^0$
1	2000(**1.05**) = 2100	$2000(1.05)^1$
2	(2000 · **1.05**)**1.05** = 2205.00	$2000(1.05)^2$
3	(2000 · **1.05** · **1.05**)**1.05** = 2315.25	$2000(1.05)^3$

Let x represent the number of years. Then the function that represents the balance in Taina's CD is $B(x) = 2000(1.05)^x$. After eight years, it will have a balance of $2954.91.

 Personal Tutor at algconcepts.com

For exponential data, you can create a best-fit curve that passes through most of the data points. You can then use this curve to make predictions.

Graphing Calculator Tutorial
See pp. 750–753.

Graphing Calculator Exploration

A city Web site tracks the number of "hits". The table shows the average number of hits per week since the launch.

Years After Launch	"Hits" per Week
4	10,000
5	20,000
6	37,500
7	62,500
8	190,000

Step 1 Use the Edit option in the STAT menu to enter the year data in L1 and the hits data in L2.

Step 2 Find an equation for a best-fit curve.

Enter: STAT ▶ 0 2nd [L1] ,
2nd [L2] , VARS ▶ ENTER ENTER ENTER

The screen displays the equation $y = a \cdot b^x$ and gives values for a and b.

Make sure that the STATPLOT is turned on.

Step 3 The exponential equation has been stored in the Y= menu. To see a graph of the function in the proper window, press ZOOM 9.

Try This

Use the graph of the best-fit curve to predict the number of weekly hits during the 15th year after the launch.

Check for Understanding

Communicating Mathematics

1. **Describe** the general shape of an exponential function.

2. **Explain** why the y-intercept of an exponential function in the form $y = a^x$ is always 1.

Vocabulary
exponential function
initial value

Getting Ready Use a calculator to find each value to the nearest hundredth.

Sample 1: 3^{-4}

Solution:

3 [∧] [(−)] 4 [ENTER] *0.01*

Sample 2: 1.25^5

Solution:

1.25 [∧] 5 [ENTER] *3.05*

3. 2^8 **4.** 4^{-1} **5.** 1.08^3

Examples 1–3

Graph each exponential function. Then state the *y*-intercept.

6. $y = 2^x + 2$ **7.** $y = 2^x - 6$ **8.** $y = 3^x + 5$

Example 4

9. Finance When Lindsay was born, her parents opened a savings account for her with $500. The account pays an annual interest of 2%.

     a. Make a table showing the account's growth for 3 years.

     b. Write a function that represents the amount of money in Lindsay's account after *x* number of years.

     c. How much money will be in Lindsay's account after 18 years?

     d. If the initial deposit had been $1000, how much would Lindsay's account be worth after 18 years?

Exercises • • • • • • • • • • • • • • •

Practice

Graph each exponential function. Then state the *y*-intercept.

10. $y = 3^x$ **11.** $y = 4^x$ **12.** $y = 2^x - 5$

13. $y = 2^x + 1$ **14.** $y = 3^x + 2$ **15.** $y = 4^x - 3$

16. $y = 2^{2x}$ **17.** $y = 3^{2x} - 1$ **18.** $y = 2^{0.5x}$

Homework Help

For Exercises	See Examples
10, 11, 16, 18, 25, 26	1, 2
12–15, 17	3
19–21, 24	4

Extra Practice
See page 715.

Find the amount of money in a bank account given the following conditions.

19. initial deposit = $5000, annual rate = 3%, time = 2 years

20. initial deposit = $1500, annual rate = 10%, time = 10 years

21. initial deposit = $3000, annual rate = 5.5%, time = 5 years

Applications and Problem Solving

22. Taxes In a recent year, the average tax refund had risen to about $1700. The table shows the nest egg you would make for yourself by investing that refund at a 10% interest rate. Find your savings after 30 years.

Compounding Tax Refunds

Year	Savings
0	1700
1	1700(1.1)
2	1700(1.1)(1.1)

Source: National Tax Services

23. Finance Nuno bought a new car for $18,000. Once the car is driven off the lot, the car depreciates in value 15% each year. How much will the car be worth in 5 years?

Downtown Atlanta ·················

24. Population The Atlanta, Georgia, metropolitan area experienced population growth throughout the 1990s. If the population in 1990 was 2,960,000 and the annual rate of increase was 3%, predict the population in the year 2002. Use the function $P(t) = 2,960,000(1.03)^t$, where t is the number of years since 1990.

25. Money Suppose on the first day of the year, your parents offered you $500 for the month of January. If you did not take the $500, they would give you a penny on the first day and then continue to double the amount each following day. Which is the better deal?

  a. Copy and complete the table.

  b. Write a function that represents the amount of money you will receive after x days.

  c. Use the equation in part b to find the day when you receive a payment of more than $500.

Day	Cents	Pattern
1	1	
2	$2 \cdot 1$	
3	$2 \cdot 2 \cdot 1$	
4		
5		

26. Critical Thinking Make a graph of each exponential function listed below. Then describe the shape of an exponential function when a number less than 1 is raised to the x power compared to a number greater than 1.

  a. $y = 2^x$ **b.** $y = 1.5^x$ **c.** $y = 0.5^x$ **d.** $y = 0.25^x$

Mixed Review

Use the Quadratic Formula to solve each equation. *(Lesson 11–6)*

27. $3p^2 - 12 = 0$ **28.** $x^2 + 4x = 10$

29. Flooring There are 30 square yards in a roll of carpet. If the carpet is unrolled, it forms a rectangle. The length is four yards more than twice its width. *(Lesson 11–5)*

  a. Write a quadratic equation to represent the carpet's area.

  b. Find the dimensions.

Write the point-slope form of an equation for each line passing through the given point and having the given slope. *(Lesson 7–2)*

30. $(4, -3), 2$ **31.** $(-2, 6), 0$

Standardized Test Practice

Ⓐ Ⓑ Ⓒ Ⓓ

32. Extended Response During a road-trip, Javonte recorded the number of miles he had driven by the end of each hour. He graphed the number of miles driven on the number line shown. *(Lesson 7–1)*

  a. Find the slope of the line between hours 0 and 2.

  b. Find the slope of the line between hours 2 and 3.

  c. Find the slope of the line between hours 3 and 6.

  d. What does the slope represent? (*Hint:* Look at the units.)

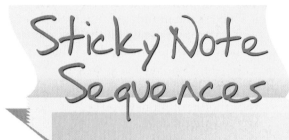

Sticky Note Sequences

Materials:

five 3-in. by 3-in. self-adhesive notes

scissors

Geometric Sequences

In the Chapter 3 Investigation, you examined arithmetic sequences. In this investigation, you will look at a different type of sequence.

Investigate

1. Use a self-adhesive note to represent 1 whole unit. Label it "1." Cut a second self-adhesive note in half to represent one half of a unit. Label one of the resulting pieces "$\frac{1}{2}$." Cut the other piece in half again to represent one-fourth of a unit. Label one of the resulting pieces "$\frac{1}{4}$." Continue this process. Arrange the self-adhesive notes as shown below.

1    $\frac{1}{2}$ $\frac{1}{4}$ $\frac{1}{8}$ $\frac{1}{16}$ $\frac{1}{32}$

a. Make a table like the one shown. Write the numbers 1–10 in column 1. Record the fractions from each note in column 2.

b. Find the ratio of the new portion and the previous portion to complete column 3.

Term Number	Portion of Self-Adhesive Note	New Portion / Previous Portion
1	1	—
2	$\frac{1}{2}$	$\frac{1}{2} \div 1 = ?$
3	$\frac{1}{4}$	$\frac{1}{4} \div \frac{1}{2} = ?$

c. List the numbers in column 2 from greatest to least.

d. The sequence in part c is called a **geometric sequence** because the quotient between any two consecutive terms, called the **common ratio**, is always the same. What is the common ratio?

2. A geometric sequence can be written as a formula in several ways.
   a. Copy and complete a table like the one below for terms 1–10.

Term	Portion of Self-Adhesive Note	Form 1	Form 2	Form 3
1	1	1	1	$1 \cdot \left(\frac{1}{2}\right)^0$
2	$\frac{1}{2}$	$1 \cdot \frac{1}{2}$	$1 \cdot \frac{1}{2}$	$1 \cdot \left(\frac{1}{2}\right)^1$
3	$\frac{1}{4}$	$\frac{1}{2} \cdot \frac{1}{2}$	$1 \cdot \frac{1}{2} \cdot \frac{1}{2}$	$1 \cdot \left(\frac{1}{2}\right)^2$

   b. You can describe Form 1 by using the formula *portion = previous portion times one half*. Another way to write this formula is $f_{n+1} = f_n \times \frac{1}{2}$. f_{n+1} is the value you are trying to find. f_n is the value of the previous term. How could you describe Forms 2 and 3 using a formula like $f_{n+1} = f_n \times \frac{1}{2}$?

3. Some bacteria reproduce exponentially by *fission*. One cell splits to become two cells. There are now two bacteria instead of one.

Term Number	Number of Bacteria	New Number of Bacteria / Previous Number of Bacteria
1	3	—
2	6	$6 \div 3 = ?$

   a. Copy and complete a table like the one at the left for terms 1–10. Use 3 self-adhesive notes to represent 3 bacteria. Cut each in half to produce 6 bacteria. Continue this process.

   b. Write a sequence that models the bacteria's growth. Then write a recursive formula representing the sequence.

Extending the Investigation

In this extension, you will examine other sequences and their formulas.

1. Graph the sequence in Exercise 1. Label the x-axis "term number" and the y-axis "portion of self-adhesive note." What type of function would describe the graph?

2. Use the sequence in Exercise 1 to find the sum of the first 10 terms. Suppose the sequence continued for 100 terms. What would you expect the sum to be?

3. Repeat 1 and 2 for Exercise 3.

Presenting Your Investigation

Here are some ideas to help you present your conclusions to the class.

• Make a brochure showing the sequences you wrote in this investigation.

• Describe a situation that results in a sequence similar to those in this investigation.

*inter*NET
CONNECTION   **Investigation**   For more information on geometric sequences, visit: www.algconcepts.com

Understanding and Using the Vocabulary

After completing this chapter, you should be able to define each term, property, or phrase and give an example or two of each.

*inter*NET
CONNECTION **Review Activities**
For more review activities, visit:
www.algconcepts.com

axis of symmetry *(p. 459)*
completing the square *(p. 478)*
discriminant *(p. 487)*
exponential function *(p. 489)*
geometric sequence *(p. 494)*

initial value *(p. 490)*
maximum *(p. 459)*
minimum *(p. 459)*
parabola *(p. 458)*
quadratic equation *(p. 468)*

Quadratic Formula *(p. 484)*
quadratic function *(p. 458)*
roots *(p. 468)*
vertex *(p. 459)*
zeros *(p. 468)*

Choose the letter of the term that best matches each statement or phrase. Each letter is used once.

1. the maximum or minimum point of a parabola
2. Use this to find the exact roots of any quadratic equation.
3. the method of making a quadratic expression into a perfect square
4. the solutions of a quadratic equation
5. the shape of the graph of any quadratic function
6. a vertical line containing the vertex of a parabola
7. $f(x) = a^x$, $a > 0$ and $a \neq 1$
8. $f(x) = ax^2 + bx + c$, $a \neq 0$
9. the x-intercepts of a quadratic function
10. the highest point on the graph of a quadratic function

a. exponential function
b. Quadratic Formula
c. completing the square
d. zeros
e. maximum
f. vertex
g. roots
h. axis of symmetry
i. quadratic function
j. parabola

Skills and Concepts

Objectives and Examples

• **Lesson 11–1** Graph quadratic functions.

Write the equation of the axis of symmetry and the coordinates of the vertex of $f(x) = x^2 - 8x + 12$.

$$x = -\frac{-8}{2(1)} \text{ or } 4$$

Since $4^2 - 8(4) + 12 = -4$, the graph has a vertex at $(4, -4)$.

Review Exercises

Write the equation of the axis of symmetry and the coordinates of the vertex of each quadratic function.

11. $y = x^2 - 3$
12. $y = -x^2 + 4x + 5$
13. $y = 3x^2 + 6x - 17$
14. $y = x^2 - 3x - 4$

 www.algconcepts.com/vocabulary_review

Objectives and Examples

Review Exercises

• **Lesson 11-1** Graph quadratic functions.

Graph
$f(x) = x^2 - 8x + 12$.

Use the results from Exercises 11-14 and a table to graph each quadratic function.

15. $y = x^2 - 3$

16. $y = -x^2 + 4x + 5$

17. $y = 3x^2 + 6x - 17$

18. $y = x^2 - 3x - 4$

• **Lesson 11-2** Recognize characteristics of families of parabolas.

Describe how the graph of $y = (x + 12)^2 - 5$ changes from the graph of $y = x^2$.

If $x = -12$, then $x + 12 = 0$. The graph shifts 12 units to the left. The -5 outside the parentheses shifts the graph 5 units down. The vertex is at $(-12, -5)$.

Describe how each graph changes from its parent graph of $y = x^2$. Then name the vertex.

19. $y = (x - 8)^2$

20. $y = x^2 - 10$

21. $y = 0.4x^2$

22. $y = (x + 3)^2$

23. $y = (x - 1)^2 + 7$

24. $y = -1.5x^2$

• **Lesson 11-3** Locate the roots of quadratic equations by graphing.

Find the roots of $0 = x^2 - 8x + 12$.

Based on the graph of the related function shown above, the roots are 2 and 6.

Solve each equation by graphing the related function.

25. $x^2 - x - 12 = 0$

26. $x^2 + 10x + 25 = 0$

27. $2x^2 - 5x + 4 = 0$

28. $x^2 + x - 4 = 0$

29. $6x^2 - 13x = 15$

• **Lesson 11-4** Solve quadratic equations by factoring.

Solve $x^2 - 8x - 20 = 0$.

$x^2 - 8x - 20 = 0$
$(x - 10)(x + 2) = 0$
$x - 10 = 0$ or $x + 2 = 0$
$x = 10$ $x = -2$

Solve each equation. Check your solution.

30. $y(y + 11) = 0$

31. $(t + 4)(2t - 10) = 0$

32. $z^2 - 36 = 0$

33. $x^2 + 16x + 64 = 0$

34. $b^2 - 5b - 14 = 0$

35. $4g^2 - 4 = 0$

36. $2x^2 + 10x + 12 = 0$

Mixed Problem Solving
See pages 724–731.

Objectives and Examples

Review Exercises

• **Lesson 11–5** Solve quadratic equations by completing the square.

$$x^2 + 6x + 2 = 0$$
$$x^2 + 6x = -2 \qquad \textit{Subtract 2.}$$
$$x^2 + 6x + 9 = -2 + 9 \qquad \textit{Complete the square.}$$
$$(x + 3)^2 = 7$$
$$x + 3 = \pm\sqrt{7} \qquad \textit{Take the square root.}$$
$$x = -3 \pm \sqrt{7} \qquad \textit{Subtract 3.}$$

Find the value of c that makes each trinomial a perfect square.

37. $r^2 + 4r + c$ **38.** $t^2 - 12t + c$

Solve each quadratic equation by completing the square.

39. $t^2 - 4t - 21 = 0$ **40.** $x^2 - 16x + 23 = 0$

• **Lesson 11–6** Solve equations by using the Quadratic Formula.

Solve $-x^2 + 3x + 40 = 0$.

$$x = \frac{-b \pm \sqrt{b^2 - 4ac}}{2a}$$

$$= \frac{-3 \pm \sqrt{3^2 - 4(-1)(40)}}{2(-1)}$$

$$= \frac{-3 \pm \sqrt{169}}{-2}$$

$$= \frac{-3 + 13}{-2} \quad \text{or} \quad x = \frac{-3 - 13}{-2}$$

$$= -5 \qquad\qquad\qquad x = 8$$

Solve each equation by using the Quadratic Formula. Check your solution.

41. $r^2 + 10r + 9 = 0$

42. $-2d^2 + 8d = 0$

43. $3n^2 - 2n - 1 = 0$

44. $s^2 + 2s + 2 = 0$

45. $m^2 + 5m = -3$

46. $-4x^2 + 8x + 3 = 0$

• **Lesson 11–7** Graph exponential functions.

Graph $y = 2^x - 1$.

$y = 2^x - 1$

Graph each exponential function. Then state the y-intercept.

47. $y = 2^x$

48. $y = 2.7^x$

49. $y = 3^x + 3$

50. $y = 4^x - 4$

Applications and Problem Solving

51. Finance Game room tickets cost $5. The function $I(x) = -50x^2 + 100x + 6000$ represents the weekly income when x is the number of 50¢ price increases. Graph the function to find the ticket price that results in the maximum income. *(Lesson 11–1)*

52. Tuition A college board voted to increase tuition prices a maximum of 4% annually. If tuition is $6678, in how many years will it exceed $10,000? Use the function $P(t) = 6678(1.04)^t$, where $P(t)$ is tuition and t is the number of years. *(Lesson 11–7)*

For each situation, state whether it is *always* true, *sometimes* true, or *never* true.

1. The graph of a quadratic function opens upward in a "u-shape."
2. You can use the Quadratic Formula to solve quadratic equations.
3. There are two real solutions to any quadratic equation.

Write the equation of axis of symmetry and the coordinates of the vertex of the graph of each quadratic function. Then graph the function.

4. $y = x^2 + 6x + 8$

5. $y = -x^2 + 3x$

Describe how each graph changes from its parent graph of $y = x^2$. Then name the vertex of each graph.

6. $y = x^2 - 10$

7. $y = (x - 9)^2$

8. $y = (x + 7)^2 + 3$

Solve each equation by graphing the related function. If exact roots cannot be found, state the consecutive integers between which the roots are located.

9. $y = x^2 - 8x + 16$

10. $y = x^2 - 5x - 1$

Solve each equation by factoring. Check your solution.

11. $(r - 3)(r - 8) = 0$

12. $s^2 - 5s + 4 = 0$

13. $x^2 + 7x + 10 = 0$

Find the value of c that makes each trinomial a perfect square.

14. $b^2 + 10b + c$

15. $x^2 + 8x + c$

16. $g^2 - 2g + c$

Solve each equation by completing the square.

17. $a^2 + 14a = -45$

18. $v^2 - 6v + 3 = 0$

Solve each equation by using the Quadratic Formula.

19. $2n^2 - n - 3 = 0$

20. $3x^2 + x + 5 = 0$

Graph each exponential function. Then state the y-intercept.

21. $y = 1.5^x$

22. $y = 2^x - 3$

23. **Physics** A plane flying 400 feet over Antarctica drops equipment needed by research scientists below. How long will it take the equipment to touch ground? Use the formula $h = -16t^2 + 400$, where the height h is in feet and the time t is in seconds.

24. **Geometry** The length of a rectangle is 4 inches more than its width. The area of the rectangle is 60 square inches. Find the dimensions of the rectangle.

25. **Finance** A school district has a $64 million budget. The school board has voted to increase the budget by only 1% each year. After how many years will the budget be over $70 million? Use the function $B(t) = 64(1.01)^t$, where $B(t)$ is the budget and t is the time in years.

Polynomial and Factoring Problems

State proficiency tests may include a few polynomial problems. In addition, many questions on the SAT and ACT ask you to factor or evaluate polynomials.

Memorize these polynomial relationships.

1. difference of squares $\qquad a^2 - b^2 = (a + b)(a - b)$
2. perfect square trinomials $\qquad a^2 + 2ab + b^2 = (a + b)^2$;
$\qquad\qquad\qquad\qquad\qquad\qquad a^2 - 2ab + b^2 = (a - b)^2$

Example 1

Evaluate $2x^2y - 4y + x^4$ if $x = 2$ and $y = -1$.

(A) 12   (B) 16   (C) 20   (D) 28

Hint Apply the rules for operations carefully when negative numbers are involved.

Solution Substitute 2 for x and -1 for y. Then simplify the expression.

$$
\begin{aligned}
2x^2y - 4y + x^4 &= 2(2)^2(-1) - 4(-1) + (2)^4 \\
&= 2(4)(-1) - 4(-1) + 16 \\
&= 8(-1) + 4 + 16 \\
&= -8 + 20 \\
&= 12
\end{aligned}
$$

The answer is A.

Example 2

If $y < 0$, which of the polynomials must be less than $y^2 + 150y + 75^2$?

(A) $y^2 + 160y + 75^2$

(B) $(y + 75)^2$

(C) $y^2 - 10y + 75^2$

(D) $(y - 75)^2$

Hint Consider the sign of y and the coefficients of each polynomial.

Solution First expand the binomials in choices B and D. Since $(y + 75)^2 = y^2 + 150y + 75^2$, the polynomial in choice B is equal to the original polynomial, not less than it. This eliminates answer choice B. The binomial in choice D expands to $y^2 - 150y + 75^2$.

The original polynomial and each of the three remaining choices, A, C, and D, have terms y^2 and 75^2. The only differences are in the y terms. The y terms for the original polynomial and the remaining answer choices are $150y$, $160y$, $-10y$, and $-150y$. Since y is negative, the least term is $160y$.

Therefore, the polynomial that must be less than $y^2 + 150y + 75^2$ is $y^2 + 160y + 75^2$.

The answer is A.

After you work each problem, record your answer on the answer sheet provided or on a sheet of paper.

Multiple Choice

1. Evaluate $h^2 - 5h$ if $h = -3$. *(Lesson 8–2)*

   (A) -24 (B) -6

   (C) 6 (D) 24

2. A swimming pool is shaped like a rectangular prism. Its volume is 1200 cubic feet. The dimensions of a smaller pool are half the size of the larger pool's dimensions. What is the volume, in cubic feet, of the smaller pool? *(Lesson 8–1)*

   (A) 150 (B) 300 (C) 600 (D) 1200

3. For all $x \neq -3$, which of the following is equivalent to the expression $\dfrac{x^2 - x - 12}{x + 3}$? *(Lesson 10–3)*

   (A) $x - 4$

   (B) $x - 2$

   (C) $x + 2$

   (D) $x + 4$

   (E) $x + 6$

4. If $x = -2$, then find $3x^2 - 5x - 6$. *(Lesson 8–1)*

   (A) -8 (B) -4 (C) 10 (D) 16

5. The figures below represent a sequence of numbers. What is an expression for the nth term of this sequence? *(Lesson 8–1)*

   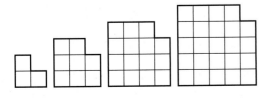

   (A) $n^2 + 2$ (B) $n^2 - 1$

   (C) $n(n + 2)$ (D) $2n^3 - 1$

6. Which quadratic equation has roots $\dfrac{1}{2}$ and $\dfrac{1}{3}$? *(Lesson 11–4)*

   (A) $5x^2 - 5x - 2 = 0$

   (B) $5x^2 - 5x + 1 = 0$

   (C) $6x^2 + 5x + 1 = 0$

   (D) $6x^2 - 6x + 1 = 0$

   (E) $6x^2 - 5x + 1 = 0$

7. Use the function table to find the value of y when $x = 8$. *(Lesson 7–3)*

   (A) 9 (B) 10.5

   (C) 12 (D) 13.5

x	y
1	1.5
2	3.0
3	4.5
4	6.0
5	7.5

8. If $a > 0$ and $a \neq b$, which statement must be true? *(Lesson 10–5)*

   (A) $\dfrac{a^2 - b^2}{a - b} - b$ is positive.

   (B) $\dfrac{a^2 - b^2}{a - b} + b$ is positive.

   (C) $\dfrac{a^2 - b^2}{a - b}$ is positive.

   (D) $\dfrac{a^2 - b^2}{a - b}$ is negative.

Grid In

9. Solve for x if $\dfrac{x^2 + 7x + 12}{x + 4} = 5$. *(Lesson 10–3)*

Extended Response

10. Consider $2x + 3y = 15$. *(Lesson 6–3)*

    Part A Make a table of at least six pairs of values that satisfy the equation above.

    Part B Use your table from Part A to graph the equation.

Inequalities

What You'll Learn

Key Ideas

- Graph inequalities on a number line. *(Lesson 12–1)*
- Solve inequalities involving one or two operations. *(Lessons 12–2 to 12–4)*
- Solve compound inequalities. *(Lesson 12–5)*
- Solve inequalities involving absolute value. *(Lesson 12–6)*
- Graph inequalities on the coordinate plane. *(Lesson 12–7)*

Key Vocabulary

boundary *(p. 535)*

compound inequality *(p. 524)*

quadratic inequality *(p. 540)*

Why It's Important

Cycling The Tour de France is the world's most famous and difficult bicycle race. Approximately 20 professional teams of nine riders each participate in the 3-week-long race every July. They race approximately 3600 kilometers (about 2235 miles) traveling through France, Belgium, Italy, Germany, and Spain.

Inequalities represent real-world situations in which a quantity falls within a range of values. You will use an inequality to represent the size of parts for a bicycle in Lesson 12–6.

Study these lessons
to improve your skills.

 **Lesson 3–1,
pp. 94–99**

Replace each • with <, >, or = to make a true sentence.

1. $2 • 2.5$ **2.** $-7 • 5$ **3.** $0.56 • 0.05$

4. $\dfrac{4}{12} • \dfrac{1}{3}$ **5.** $-6 • 0(6)$ **6.** $14 - 7 • (-4)(3)$

✔ **Lesson 2–1,
pp. 52–57**

Graph each set of numbers on a number line.

7. $\{1, 3, 6\}$ **8.** $\{-2, 3, 4\}$

9. $\{4, 0, -1\}$ **10.** $\{-5, -3, 2\}$

Solve each equation. Check your solution.

✔ **Lesson 3–6,
pp. 122–127**

11. $n + 7 = 12$ **12.** $d - 12 = 7$ **13.** $-2 + r = 8$

14. $m + 14 = 20$ **15.** $p - 18 = 21$ **16.** $-8 + w = 11$

✔ **Lesson 3–7,
pp. 128–131**

17. $|x| = 6$ **18.** $5 + |h| = 9$ **19.** $14 = |j - 2|$

20. $-|y| = -9$ **21.** $|g| - 2 = 8$ **22.** $|k + 4| = 10$

✔ **Lesson 4–4,
pp. 160–164**

23. $4b = 36$ **24.** $-16 = 4t$ **25.** $\dfrac{w}{5} = 3$

26. $96 = 12d$ **27.** $-5v = -35$ **28.** $-12 = \dfrac{b}{3}$

✔ **Lesson 4–5,
pp. 165–170**

29. $3z + 11 = 23$ **30.** $1 + 6c = -5$ **31.** $\dfrac{1}{3}r - 5 = 2$

 FOLDABLES™
Study Organizer

Make this Foldable to help you organize your Chapter 12 notes. Begin with four
sheets of grid paper.

❶ **Fold** each sheet
in half from top to
bottom.

❷ **Cut** each sheet along
the fold. Staple the eight
half-sheets together to
form a booklet.

❸ **Label** each page
with a lesson number
and title.

Inequalities

Reading and Writing As you read and study the chapter, write notes and examples for each
lesson in the journal.

What You'll Learn
You'll learn to graph inequalities on a number line.

Why It's Important
Postal Service
Inequalities can be used to describe postal regulations. See Exercise 12.

You already know that equations are mathematical statements that describe two expressions with equal values. When the values of the two expressions are *not* equal, their relationship can be described in an inequality.

Verbal phrases like *greater than* or *less than* describe inequalities. For example, 6 is *greater than* 2. This is the same as saying 2 is *less than* 6.

The chart below lists other phrases that indicate inequalities and their corresponding symbols.

Look Back
Inequalities:
Lesson 3–1

Inequalities			
$<$	$\leq$	$>$	$\geq$
• less than • fewer than	• less than or equal to • at most • no more than • a maximum of	• greater than • more than	• greater than or equal to • at least • no less than • a minimum of

Example

1 **Suppose the minimum driving age in your state is 16. Write an inequality to describe people who are *not* of legal driving age in your state.**

Let *d* represent the ages of people who are *not* of legal driving age.

The ages of all drivers   are greater than or equal to   16 years.
  d $\geq$ 16

$d \geq 16$ is the same as $16 \leq d$.

Then *d* is *less than* 16, or $d < 16$.

Your Turn

a. Lisa can carry no more than 50 newspapers on her paper route. Express the number of papers that Lisa can carry as an inequality.

Not only can inequalities be expressed through words and symbols, but they can also be graphed.

Consider the inequality $d \geq 16$.

Step 1 Graph a bullet at 16 to show that d can *equal* 16.

Step 2 Graph all numbers *greater than* 16 by drawing a line and an arrow to the right.

$$d \geq 16$$

The graph shows all values that are *greater than or equal to* 16.

Now suppose we want to graph all numbers that are *less than* 16. We want to include all numbers up to 16, but not including 16.

Step 1 Graph a circle at 16 to show that d *does not equal* 16.

Step 2 Graph all numbers *less than* 16 by drawing a line and an arrow to the left.

$$d < 16$$

The graph shows all values that are *less than* 16.

Examples

Graph each inequality on a number line.

2 $x > -4$

Since x cannot equal -4, graph a circle at -4 and shade to the right.

3 $n \geq 1.5$

Since n can equal 1.5, graph a bullet at 1.5 and shade to the right.

Your Turn

b. $y < 3$

c. $p \leq -\dfrac{1}{2}$

You can also write inequalities given their graphs.

Examples

Write an inequality for each graph.

4

Locate where the graph begins. This graph begins at 1, but 1 is not included. Also note that the arrow is to the left. The graph describes values that are less than 1.

So, $x < 1$.

⑤

$$0 \quad \frac{1}{4} \quad \frac{2}{4} \quad \frac{3}{4} \quad 1$$

Locate where the graph begins. This graph begins at $\frac{1}{4}$ and includes $\frac{1}{4}$. Note that the arrow is to the right. The graph describes values greater than or equal to $\frac{1}{4}$.

So, $x \geq \frac{1}{4}$.

Your Turn

d.

$$-4 \quad -3 \quad -2 \quad -1 \quad 0$$

e.

$$0 \quad \frac{1}{3} \quad \frac{2}{3} \quad 1 \quad 1\frac{1}{3}$$

Inequalities are commonly used in the real world.

Example

Sports Link

⑥ **To play junior league soccer, you must be at least 14 years of age.**

A. Write an inequality to represent this situation.

Let a represent the ages of people who can play junior league soccer. Then write an inequality using $\geq$ since the soccer players have to be greater than or equal to 14 years of age.

$a \geq 14$

B. Graph the inequality on a number line.

To graph the inequality, first graph a bullet at 14. Then include all ages *greater than* 14 by drawing a line and an arrow to the right.

$$12 \quad 13 \quad 14 \quad 15 \quad 16$$

C. The Valdez children are 10, 13, 14, and 16 years old. Which of the children can play junior league soccer?

The set of the children's ages, {10, 13, 14, and 16}, can be called a *replacement set*. It includes possible values of the variable a. In this case, only 14 and 16 satisfy the inequality $a \geq 14$ and are members of the *solution set*. So, the two Valdez children that are 14 and 16 years old can play junior league soccer.

Check for Understanding

Communicating Mathematics

1. Match each inequality symbol with its description.

   i.   less than or equal to   **a.** $\geq$

   ii. at least   **b.** $\leq$

   iii. fewer than   **c.** $>$

   iv. greater than   **d.** $<$

2. **Describe** the graph of $x \geq 12$.

3. **Explain** the difference between the graphs of $x \leq 4$ and $x < 4$.

4. Soto says that $x \geq 7$ is the same as $7 \leq x$. Darrell says that it is not. Who is correct? Explain.

Guided Practice

Example 1

Examples 2 & 3

Examples 4 & 5

Example 6

Write an inequality to describe each number.

5. a number less than 14

6. a number no less than 0

Graph each inequality on a number line.

7. $x > 6$

8. $x \leq 2$

9. $5 > x$

Write an inequality for each graph.

10.

11.

12. **Mail** All letters mailed with one first-class stamp can weigh no more than 1 ounce. Write an inequality to represent this situation. Then graph the inequality.

Exercises

· · · · ● ● ● ● ● ● ● ● ● ● ● ● ● ● ●

Practice

Homework Help	
For Exercises	See Examples
13–18	1
19–30	2, 3
31–38	4, 5
39–42	6
Extra Practice	
See page 715.	

Write an inequality to describe each number.

13. a number greater than 4

14. a number less than or equal to 13

15. a number less than 9

16. a number no more than 7

17. a number that is at least -8

18. a number more than -6

Graph each inequality on a number line.

19. $x < 7$

20. $y > 3$

21. $b \leq 0$

22. $a \leq 5$

23. $x > -4$

24. $1 \leq z$

25. $x \geq 2.5$

26. $d < 1.9$

27. $x < -3.4$

28. $g > \frac{3}{4}$

29. $1\frac{1}{2} \geq x$

30. $h \leq -\frac{1}{3}$

Write an inequality for each graph.

31.

32.

33.

34.

35.

36.

37.

38.

**For Exercises 39–41, write an inequality to represent each situation.
Then graph the inequality on a number line.**

39. **Conservation** At a pond, you must return any fish that you catch if it
weighs less than 3 pounds.

40. **Entertainment** A roller coaster requires a person to be at least
42 inches tall to ride it.

41. **Safety** The elevators in an office building have been approved to
hold a maximum of 3600 pounds.

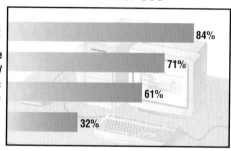

Internet Use

Surfing Internet — 84%

Going online more
than once a day — 71%

Spending 1–4 hours
a day online — 61%

Regularly visiting
shopping sites — 32%

Source: *USA TODAY*

42. **Critical Thinking** The graph
shows survey results on the
Internet use of 100 college
students. Write an inequality
to represent the number of
students that may spend at
least 5 hours online each day.

43. **Finance** Mr. Johnson invested $10,000 at an annual interest rate of
6%. Graph the function $B(t) = 10,000(1.06)^t$, where t is the time in
years. How many years will it take before the balance is over $15,000?
(Lesson 11–7)

Solve each equation by using the Quadratic Formula. *(Lesson 11–6)*

44. $t^2 - 3t - 40 = 0$ 45. $3a^2 + a + 1 = 0$ 46. $x^2 - 6x + 7 = 0$

47. Find the prime factorization of 24. How many distinct prime factors
are there? *(Lesson 10–1)*

Simplify each expression.

48. $-\sqrt{16}$ *(Lesson 8–5)*   49. $\dfrac{-18x^3y}{3x}$ *(Lesson 8–2)*

50. **Multiple Choice** If two lines are perpendicular to each other, which
of the following statements could be true? *(Lesson 7–7)*

I.   The slopes of both lines are equal.

II. One slope is positive, and the other slope is negative.

III. The slope of one line is the reciprocal of the slope of the
other line.

Ⓐ I only   Ⓑ II only   Ⓒ III only

Ⓓ I and II   Ⓔ II and III

12-2 Solving Addition and Subtraction Inequalities

What You'll Learn
You'll learn to solve inequalities involving addition and subtraction.

Why It's Important
Sports You can use inequalities to set goals in competitions. *See Example 2.*

The Iditarod, nicknamed "The Last Great Race on Earth," is a dogsled race in Alaska. The 12- to 16-dog teams cover over 1150 miles in subzero weather. The record time for finishing the race is 218 hours (9 days, 2 hours).

A dogsled race

Suppose a team takes 73 hours to reach the first checkpoint of the Iditarod and 98 hours to reach the second. How much time can be spent on the last leg of the race to beat the record time?

Let t represent the time for the last leg of the race. Write an inequality.

The sum of the times	*must be less than*	*the record time.*
$73 + 98 + t$	$<$	218

$$171 + t < 218$$

If this were an equation, we would subtract 171 from each side to solve for t. The same procedure can be used with inequalities, as explained by the properties below. *This problem will be solved in Example 2.*

These properties are also true for $\leq$ and $\geq$.

	Words:	For any inequality, if the same quantity is added or subtracted to each side, the resulting inequality is true.
Addition and Subtraction Properties for Inequalities	**Symbols:**	For all numbers a, b, and c, 1. if $a > b$, then $a + c > b + c$ and $a - c > b - c$. 2. if $a < b$, then $a + c < b + c$ and $a - c < b - c$.
	Numbers:	$-5 < 1$ $\qquad\qquad$ $2 > -4$ $-5 + 2 < 1 + 2$ $\qquad$ $2 - 3 > -4 - 3$ $-3 < 3$ $\qquad\qquad$ $-1 > -7$

Example

1 Solve $x + 14 \geq 5$. Check your solution.

$x + 14 \geq 5$	*Original inequality*
$x + 14 - 14 \geq 5 - 14$	*Subtract 14 from each side.*
$x \geq -9$	*Simplify.*

(continued on the next page)

Check: Substitute a number less than −9, the number −9, and a number greater than −9 into the inequality.

Let $x = -10$.	Let $x = -9$.	Let $x = 0$.
$x + 4 \geq 5$	$x + 14 \geq 5$	$x + 14 \geq 5$
$-10 + 14 \overset{?}{\geq} 5$	$-9 + 14 \overset{?}{\geq} 5$	$0 + 14 \overset{?}{\geq} 5$
$4 \geq 5$ false	$5 \geq 5$ true	$14 \geq 5$ true

The solution is {all numbers greater than or equal to −9}.

Your Turn   Solve each inequality. Check your solution.

a. $x + 2 < 7$ **b.** $x - 6 \geq 12$

 Personal Tutor at algconcepts.com

A more concise way to express the solution to an inequality is to use **set-builder notation**. The solution in Example 1 in set-builder notation is $\{x \mid x \geq -9\}$.

$\{x$ $\mid$ $x \geq -9\}$

The set of all numbers x   *such that*   *x is greater than or equal to −9.*

In Lesson 12–1, you learned that you can show the solution to an inequality on a line graph. The solution, $\{x \mid x \geq -9\}$, is shown below.

$$-11 \quad -10 \quad -9 \quad -8 \quad -7$$

Real World

Examples

Sports Link

❷ Refer to the application at the beginning of the lesson. Solve $171 + t < 218$ to find the time needed to finish the last leg of the Iditarod and beat the record.

$171 + t < 218$	*Original inequality*
$171 + t - 171 < 218 - 171$	*Subtract 171 from each side.*
$t < 47$	*This means all numbers less than 47.*

The solution can be written as $\{t \mid t < 47\}$. So any time less than 47 hours will beat the record.

❸ Solve $7y + 4 > 8y - 12$. Graph the solution.

$7y + 4 > 8y - 12$	*Original inequality*
$7y + 4 - 7y > 8y - 12 - 7y$	*Subtract 7y from each side.*
$4 > y - 12$	*Simplify.*
$4 + 12 > y - 12 + 12$	*Add 12 to each side.*
$16 > y$	*Simplify.*

Since $16 > y$ is the same as $y < 16$, the solution is $\{y \mid y < 16\}$.

The graph of the solution has a circle at 16, since 16 is not included. The arrow points to the left.

$$14 \quad 15 \quad 16 \quad 17 \quad 18$$

 www.algconcepts.com/extra_examples

Your Turn Solve each inequality. Graph the solution.

c. $5y - 3 > 6y - 9$

d. $3r + 7 \le 2r + 4$

You can also use inequalities to describe some geometry concepts.

Hands-On Algebra

Materials: straws ✂ scissors ≈ pipe cleaners 📏 ruler

Step 1 Cut one straw so that it is 3 inches long. Cut a second straw 4 inches long. Cut a third straw 5 inches long.

Step 2 Insert a pipe cleaner into each straw. Then form a triangle. Label each side like the one shown at the right.

Try These

1. Repeat Steps 1 and 2 for each of the side lengths listed below. In each case, can a triangle be formed? Explain.

   a. $x = 4$ in., $y = 8$ in., $z = 6$ in.

   b. $x = 3$ in., $y = 5$ in., $z = 2$ in.

   c. $x = 1$ in., $y = 6$ in., $z = 3$ in.

2. Study the triangles in Exercise 1. Use an inequality to describe what must be true about the side lengths to form a triangle.

3. Can a triangle be formed with side lengths 9 cm, 12 cm, and 15 cm? Explain.

Check for Understanding

Communicating Mathematics

1. **Compare and contrast** solving inequalities by using addition and subtraction with solving equations by using addition and subtraction.

> **Vocabulary**
> set-builder notation

2. **Write** two inequalities that each have $\{y \mid y > 4\}$ as their solution.

3. **Writing Math** List three situations in which solving an inequality may be helpful.

Guided Practice

⏱ **Getting Ready** Write an inequality for each statement.

> **Sample:** Five more than a number is greater than sixteen.
> **Solution:** $n + 5 > 16$

4. A number minus three is greater than or equal to ten.

5. The sum of 8 and a number is at most 12.

Solve each inequality. Check your solution.

6. $x + 9 < 12$

7. $v - 12 \geq 5$

8. $a - 6 \leq -14$

9. $h - 7 \leq 14$

10. $4.6 + x > 2.1$

11. $y - \dfrac{2}{3} < 1\dfrac{1}{6}$

Example 3

Solve each inequality. Graph the solution.

12. $z - 12 \geq 2z + 4$

13. $4x - 1 > 5x$

Math Data
Online Update
For the latest prices on computer hardware, visit: algconcepts.com

14. Budgeting Antonio can spend no more than $1000 on a new computer system. The hard drive he wants costs $220. The monitor costs $300. How much money does Antonio have to spend on other components? **Example 2**

Exercises • • • • • • • • • • • • • • • • • •

Practice

Homework Help

For Exercises	See Examples
15–32, 39,	1
33–38, 40,	3
41–44	2

Extra Practice

See page 716.

Solve each inequality. Check your solution.

15. $n + 6 > 9$

16. $x - 7 > -3$

17. $-3 + b < -8$

18. $g + 12 \leq 5$

19. $r - 8 < 11$

20. $x + 6 \geq 14$

21. $w - 9 > 13$

22. $4 + p \leq 1$

23. $t - 5 \leq -5$

24. $x + 3 \geq 19$

25. $-2 < c + 2$

26. $11 \leq -4 + m$

27. $d + 1.4 < 6.8$

28. $-3 + x > 11.9$

29. $-0.2 \geq 0.3 + z$

30. $s - (-2) > \dfrac{3}{4}$

31. $3\dfrac{3}{8} + v \leq 5\dfrac{7}{8}$

32. $\dfrac{1}{2} > x - \dfrac{2}{3}$

Solve each inequality. Graph the solution.

33. $7 > 8g - 7g - 6$

34. $5n < 4n + 10$

35. $3x + 12 \geq 2x - 4$

36. $6a \leq 5(a - 1)$

37. $-(-t + 9) \geq 0$

38. $3(v - 5) > 2(v - 2)$

Write an inequality for each statement. Then solve.

39. Eight less than a number is not greater than 12.

40. The sum of 5, 10, and a number is more than 25.

Applications and Problem Solving

41. Fund-raising Westfield High School is having a raffle to raise money for Habitat for Humanity. Any homeroom that sells at least 150 tickets will get to help build a home. Ms. Martinez' homeroom is keeping a table of the number of tickets sold each day. How many more tickets do they need to sell to help build a home?

Day	Tickets
1	32
2	19
3	?
4	?
5	?

42. Sports Carlos weighs 175 pounds. At least how much weight must he lose to wrestle in the 171-pound weight class?

43. Academics Alissa must earn 475 out of 550 points to receive a grade of B. So far, she has earned 244 test points, 82 quiz points, and 50 homework points. How many points must she score on her final exam to earn at least a B in the class?

44. Volunteerism Each summer, 70 men bicycle in the Journey of Hope to raise money for people with disabilities. They begin their journey in San Francisco, California, and end in Washington, D.C. The map shows the miles they cycle in California. It takes them at most 285 miles to cycle to Nevada. How many miles is it from Jackson, California, to their first stop in Nevada?

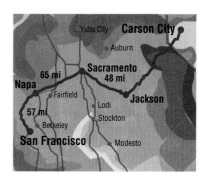

45. Critical Thinking Choose the correct term and justify your answer. The solution set of an inequality is (*sometimes*, *always*, *never*) the empty set (∅).

Mixed Review

Write an inequality to describe each number. *(Lesson 12–1)*

46. a number no more than 1   **47.** a number less than −8

48. a number greater than 3   **49.** a minimum number of 5

50. The graph of which of the following equations is shown at the right: $y = 2^x$, $y = 2^x + 1$, $y = 2^x - 1$, or $y = -2^x$? *(Lesson 11–7)*

51. Factor the trinomial $3x^2 - 13x - 10$. *(Lesson 10–4)*

52. Write $3x + y = -8$ in slope-intercept form. *(Lesson 7–3)*

53. Grid In If y varies inversely as x and $y = 8$ when $x = 24$, find y when $x = 6$. *(Lesson 6–6)*

54. Multiple Choice Suppose **x** = $4x - 2$. If **x** = 10, then what is the value of x? *(Lesson 4–4)*

Ⓐ 2   Ⓑ 3   Ⓒ 38   Ⓓ 40

Solving Multiplication and Division Inequalities

You'll learn to solve inequalities involving multiplication and division.

Why It's Important

Savings You can use inequalities when you are trying to budget your money.
See Exercise 37.

In the previous lesson, you learned that addition and subtraction inequalities are solved with the same procedures as addition and subtraction equations. Can you use the procedures for solving multiplication and division equations to solve inequalities as well?

For example, to solve $4x = 36$, we would divide each side by 4. Will this work when solving inequalities? Consider the inequality $4 < 36$.

$$4 < 36$$
$$\frac{4}{4} \overset{?}{<} \frac{36}{4} \qquad \textit{Divide each side by 4.}$$
$$1 < 9 \quad \text{true}$$

So, it is possible to solve an inequality when dividing by a positive number. What happens when you divide by a negative number? Consider the inequality $-4 < 36$.

$$-4 < 36$$
$$\frac{-4}{-4} \overset{?}{<} \frac{36}{-4} \qquad \textit{Divide each side by } -4.$$
$$1 < -9 \quad \text{false}$$

When dividing by a negative number, you must reverse the symbol for the inequality to remain true. This example leads us to the Division Property for Inequalities.

This property is also true for $\leq$ and $\geq$.

Division Property for Inequalities	**Words:**	If you divide each side of an inequality by a positive number, the inequality remains true. If you divide each side of an inequality by a negative number, the inequality symbol must be reversed for the inequality to remain true.
	Symbols:	For all numbers a, b, and c, 1. if c is positive and $a < b$, then $\frac{a}{c} < \frac{b}{c}$, and if c is positive and $a > b$, then $\frac{a}{c} > \frac{b}{c}$. 2. if c is negative and $a < b$, then $\frac{a}{c} > \frac{b}{c}$, and if c is negative and $a > b$, then $\frac{a}{c} < \frac{b}{c}$.
	Numbers:	If $9 > 6$, then $\frac{9}{3} > \frac{6}{3}$ or $3 > 2$. If $9 > 6$, then $\frac{9}{-3} < \frac{6}{-3}$ or $-3 < -2$.

Examples

Sports Link

1 Carmen runs at least 15 miles in the park every day. If she runs 5 miles per hour, how long does she run every day? Recall that rate times time equals distance, or $rt = d$.

$5t \geq 15$ *In $rt = d$, $r = 5$ and $d = 15$.*

$\dfrac{5t}{5} \geq \dfrac{15}{5}$ *Divide each side by 5.*

$t \geq 3$ *Keep the symbol facing the same direction.*

Carmen runs at least 3 hours every day.

Prerequisite Skills Review

Operations with Decimals, p. 684

2 Solve $-10x \leq 25.6$. Check your solution.

$-10x \leq 25.6$ *Original inequality*

$\dfrac{-10x}{-10} \geq \dfrac{25.6}{-10}$ *Divide each side by -10 and reverse the symbol.*

$x \geq -2.56$ *Simplify.*

Check: Substitute -2.56 and a number greater than -2.56, such as 0, into the inequality.

Let $x = -2.56$.	Let $x = 0$.
$-10x \leq 25.6$	$-10x \leq 25.6$
$-10(-2.56) \stackrel{?}{\leq} 25.6$	$-10(0) \stackrel{?}{\leq} 25.6$
$25.6 \leq 25.6$ true	$0 \leq 25.6$ true

The solution set is $\{x \mid x \geq -2.56\}$.

Your Turn Solve each inequality. Check your solution.

a. $8x > 40$

b. $-x \leq 4.7$

 Personal Tutor at algconcepts.com

We can also solve inequalities by multiplying. Use the Multiplication Property for Inequalities.

This property is also true for $\leq$ and $\geq$.

Multiplication Property for Inequalities	**Words:**	If you multiply each side of an inequality by a positive number, the inequality remains true. If you multiply each side of an inequality by a negative number, the inequality symbol must be reversed for the inequality to remain true.
	Symbols:	For all numbers a, b, and c, **1.** if c is positive and $a < b$, then $ac < bc$, and if c is positive and $a > b$, then $ac > bc$. **2.** if c is negative and $a < b$, then $ac > bc$, and if c is negative and $a > b$, then $ac < bc$.
	Numbers:	If $3 > -7$, then $3(2) > -7(2)$ or $6 > -14$. If $3 > -7$, then $3(-2) < -7(-2)$ or $-6 < 14$.

③ Solve $-\frac{x}{3} < -6$. **Check your solution.**

$$-\frac{x}{3} < -6 \qquad \textit{Original inequality}$$

$$-3\left(-\frac{x}{3}\right) > -3(-6) \qquad \textit{Multiply each side by –3 and reverse the symbol.}$$

$$x > 18 \qquad \textit{Simplify.}$$

Check: Substitute 18 and a number greater than 18, such as 21, into the inequality.

Let $x = 18$.   Let $x = 21$.

$$-\frac{x}{3} < -6 \qquad\qquad -\frac{x}{3} < -6$$

$$-\frac{18}{3} \overset{?}{<} -6 \qquad\qquad -\frac{21}{3} \overset{?}{<} -6$$

$$-6 < -6 \quad \text{false} \qquad\qquad -7 < -6 \quad \text{true}$$

The solution is $\{x \mid x > 18\}$.

Your Turn

Solve each inequality. Check your solution.

c. $\frac{x}{-4} \leq 8$ **d.** $\frac{2}{3}x > 22$

Check for Understanding

**Communicating
Mathematics**

1. **Compare and contrast** the Division Property for Inequalities and the Multiplication Property for Inequalities.

2. **Explain** why 5 is *not* a solution of the inequality $4x > 20$.

3. **Graph** the solution of $-2x \geq -8$ on a number line.

Guided Practice

Solve each inequality. Check your solution.

Examples 1 & 2

4. $2z > 12$ **5.** $-3x \leq 27$ **6.** $-7b < 14$

Example 3

7. $\frac{x}{4} \geq 5$ **8.** $-\frac{r}{6} < 6$ **9.** $\frac{2}{5}a \leq -12$

Example 1

10. Time Cherise drives to the restaurant where she works by different routes, but the trip is at most 10 miles. Considering the stops at traffic lights, she thinks her average driving speed is about 40 miles per hour. How much time does it take Cherise to get to work?

Practice

Homework Help	
For Exercises	See Examples
11–13, 17–19, 23–25, 29–31, 35–38	1, 2
14–16, 20–22, 26–29, 32–34	3
Extra Practice	
See page 716.	

Solve each inequality. Check your solution.

11. $-6h \leq 12$

12. $-9n < 18$

13. $8d \leq -24$

14. $\frac{a}{6} > 5$

15. $\frac{g}{3} > -12$

16. $-\frac{b}{8} \geq 5$

17. $7x \geq 49$

18. $-4y \leq -40$

19. $3z > -9$

20. $-\frac{p}{2} > -1$

21. $\frac{c}{9} > -6$

22. $-\frac{a}{8} \leq -4$

23. $3k > 5$

24. $-2t \leq 11$

25. $-3 > -6w$

26. $\frac{3}{4}x < 3$

27. $5 \leq \frac{5}{6}y$

28. $-\frac{2}{5}v \leq 20$

29. $5.3v \geq 10.6$

30. $4.1x < -6.15$

31. $-28 \leq -0.1s$

32. $-\frac{h}{3.8} \geq 2$

33. $\frac{n}{10.5} < 10$

34. $-\frac{x}{0.5} < -7$

Applications and Problem Solving

35. **Geometry** An *acute angle* has a measure less than $90°$. If the measure of an acute angle is $2x$, what is the value of x?

$2x°$

36. **Production** The ink cartridge that Bill just bought for his printer can print up to 900 pages of text. Bill is printing handbooks that are 32 pages each. How many complete handbooks can he print with this cartridge?

37. **Budgeting** Jenny mows lawns to earn money. She wants to earn at least $200 to buy a new stereo system. If she charges $12 a lawn, at least how many lawns does she need to mow?

38. **Critical Thinking** Use a counterexample to show that if $x < y$, then $x^2 < y^2$ is not always true.

Mixed Review

Solve each inequality. Check your solution. *(Lesson 12–2)*

39. $z + 1 \leq 5$

40. $-3 \geq b + 11$

41. **Contests** At a beach museum in San Pedro, California, more than 600 people built a life-size sand sculpture of a whale. Use an inequality to represent the number of people who built the sculpture. *(Lesson 12–1)*

Whale sculpture in San Pedro, California

42. Solve $x^2 + 5x + 4 = 0$ by factoring. *(Lesson 11–3)*

43. Find the product of $2v - 1$ and $2v + 1$. *(Lesson 9–5)*

44. Earthquakes In a recent year, the state of Illinois experienced an earthquake tremor. It measured 3.5 on the Richter scale, releasing about 1.6×10^7 Joules of energy. The largest earthquake ever recorded in Illinois measured 5.5 on the Richter scale, releasing about 5.7×10^{11} Joules of energy. How many times as strong was the largest earthquake as the tremor? *(Lesson 8–4)*

Standardized Test Practice
Ⓐ Ⓑ Ⓒ Ⓓ

45. Short Response There are 20 students in a class. Each student's name is written on a separate slip of paper and placed in a box. A name is randomly drawn to determine who will read the daily announcements. The slip of paper is then returned to the box. What is the probability that the same name is drawn two days in a row? *(Lesson 5–7)*

46. Multiple Choice If the figure at the right is a square, then what is the value of x? *(Lesson 4–4)*

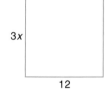
$3x$

12

Ⓐ 3
Ⓑ 4
Ⓒ 12
Ⓓ 36

Quiz 1   Lessons 12–1 through 12–3

1. Write an inequality for the graph. *(Lesson 12–1)*

$-3 \quad -2 \quad -1 \quad 0 \quad 1$

Solve each inequality. Graph the solution. *(Lesson 12–2)*

2. $a + 2 < 10$

3. $-4 + x > 1$

4. $-5 \leq s - (-3)$

5. $-3 + 13z > 14z$

Solve each inequality. Check your solution. *(Lesson 12–3)*

6. $3b \geq 30$

7. $-5x < 25$

8. $-\dfrac{k}{2} > -8$

9. $\dfrac{4}{5}v \leq 6$

10. Weather The record high temperature in Jackson, Mississippi, is 100° F. Suppose a recent temperature in Jackson is 84° F. How many degrees must the temperature rise to break the record? *(Lesson 12–2)*

www.algconcepts.com/self_check_quiz

What You'll Learn

You'll learn to solve inequalities involving more than one operation.

Why It's Important

School You can determine what score is needed to receive a certain class grade. See Example 5.

Some inequalities involve more than one operation. The best strategy to solve multi-step inequalities is to undo the operations in reverse order. In other words, work backward just as you did to solve multi-step equations. For example, $2x + 5 > 11$ is a multi-step inequality. You can solve this inequality by following these steps.

Step 1: Undo addition. *Subtract 5 from each side.*

$$2x + 5 - 5 > 11 - 5$$
$$2x > 6$$

Step 2: Undo multiplication. *Divide each side by 2.*

$$\frac{2x}{2} > \frac{6}{2}$$
$$x > 3$$

Example 1

Look Back

Solving Multi-Step Equations: Lesson 4–5

Solve $9 + 3x < 27$. Check your solution.

$$9 + 3x < 27 \quad \textit{Original inequality}$$
$$9 + 3x - 9 < 27 - 9 \quad \textit{Subtract 9 from each side.}$$
$$3x < 18 \quad \textit{Simplify.}$$
$$\frac{3x}{3} < \frac{18}{3} \quad \textit{Divide each side by 3.}$$
$$x < 6 \quad \textit{Simplify.}$$

Check: Substitute 0 and 6 into the inequality.

Let $x = 0$.
$$9 + 3x < 27 \quad \textit{Original inequality}$$
$$9 + 3(0) \overset{?}{<} 27 \quad \textit{Substitute.}$$
$$9 + 0 \overset{?}{<} 27 \quad \textit{Simplify.}$$
$$9 < 27 \quad \text{true}$$

Let $x = 6$.
$$9 + 3x < 27 \quad \textit{Original inequality}$$
$$9 + 3(6) \overset{?}{<} 27 \quad \textit{Substitute.}$$
$$9 + 18 \overset{?}{<} 27 \quad \textit{Simplify.}$$
$$27 < 27 \quad \text{false}$$

The solution is $\{x \mid x < 6\}$.

Your Turn

Solve each inequality. Check your solution.

a. $4 + 2x \leq 12$ **b.** $8x - 5 \geq 11$

Example ❷

Weather Link

During Rafael's trip to Mexico, the temperature was always warmer than 30º Celsius. Use the formula $\frac{5}{9}(F - 32) = C$, where F is degrees Fahrenheit and C is degrees Celsius, to write and solve an inequality for the temperature in degrees Farenheit.

$\frac{5}{9}(F - 32) > 30$ *The temperature is greater than 30.*

$\frac{9}{5} \cdot \frac{5}{9}(F - 32) > \frac{9}{5} \cdot 30$ *Multiply each side by $\frac{9}{5}$, the reciprocal of $\frac{5}{9}$.*

$F - 32 > 54$ *Simplify.*

$F - 32 + 32 > 54 + 32$ *Add 32 to each side.*

$F > 86$ *Check your solution.*

Therefore, Rafael can expect it to be warmer than 86°F in Mexico.

Remember to *reverse the inequality symbol* if you multiply or divide by a negative number.

Example ❸

Solve $-4x + 3 \geq 23 + 6x$. Check your solution.

$-4x + 3 \geq 23 + 6x$ *Original inequality*

$-4x + 3 - 6x \geq 23 + 6x - 6x$ *Subtract 6x from each side.*

$-10x + 3 \geq 23$ *Simplify.*

$-10x + 3 - 3 \geq 23 - 3$ *Subtract 3 from each side.*

$-10x \geq 20$ *Simplify.*

$\frac{-10x}{-10} \leq \frac{20}{-10}$ *Divide each side by –10 and reverse the symbol.*

$x \leq -2$ *Simplify.*

The solution is $\{x \mid x \leq -2\}$.   *Check your solution.*

Your Turn   Solve each inequality. Check your solution.

c. $10 - 5x < 25$ **d.** $-3x + 1 > -17$

To solve inequalities that contain grouping symbols, you may use the Distributive Property first.

Example ❹

Solve $8 \leq -2(x - 5)$. Check your solution.

$8 \leq -2(x - 5)$ *Original inequality*

$8 \leq -2x + 10$ *Distributive Property*

$8 - 10 \leq -2x + 10 - 10$ *Subtract 10 from each side.*

$-2 \leq -2x$ *Simplify.*

$\frac{-2}{-2} \geq \frac{-2x}{-2}$ *Divide each side by -2 and reverse the symbol.*

$1 \geq x$ *Simplify.*

The solution is $\{x \mid x \leq 1\}$.   *Check your solution.*

      www.algconcepts.com/extra_examples

e. $2 > -(x + 7)$ f. $3(x - 4) \leq x - 5$

You can use a graphing calculator to solve multi-step inequalities.

Graphing Calculator Tutorial
See pp. 750–753.

Graphing Calculator Exploration

Solve $-5 - 8x \geq 43$ by using a graphing calculator.

Step 1 Clear the $\boxed{Y=}$ list to enter the inequality $-5 - 8x \geq 43$. (The $\geq$ symbol is item 4 on the TEST menu.)

Step 2 Press $\boxed{\text{ZOOM}}$ 6. Use the $\boxed{\text{TRACE}}$ and arrow keys to to move the cursor along the graph. You should see a line above the x-axis for values of x that are less than or equal to -6.

This represents the solution $\{x \mid x \leq -6\}$.

Try These

Solve each inequality. Check your solution with a graphing calculator.

1. $-9x + 2 \leq 20$ 2. $-5(x + 4) \geq 3(x - 4)$

Real World

Example ⑤

School Link

Hannah's scores on the first three of four 100-point tests were 85, 92, and 90. What score must she receive on the fourth test to have a mean score of more than 92 for all tests?

Explore   Let s = Hannah's score on the fourth test.

Plan   The sum of Hannah's four test scores, divided by 4, will give the mean score. The mean must be *more than* 92.

Solve $\dfrac{85 + 90 + 92 + s}{4} > 92$ *The mean is greater than 92.*

$4\left(\dfrac{85 + 90 + 92 + s}{4}\right) > 4(92)$ *Multiply each side by 4.*

$85 + 90 + 92 + s > 368$ *Simplify.*

$267 + s > 368$ *Add.*

$267 + s - 267 > 368 - 267$ *Subtract 267 from each side.*

$s > 101$ *Simplify.*

(continued on the next page)

Examine Substitute a number greater than 101, such as 102, into the original problem. Hannah's average would be $\frac{85 + 90 + 92 + 102}{4}$ or 92.25. Since $92.25 > 92$ is a true statement, the solution is correct. Hannah must score more than 101 points out of a 100-point test. Without extra credit, this is not possible. So, Hannah cannot have a mean score over 92.

 Personal Tutor at algconcepts.com

Check for Understanding

Communicating Mathematics

1. **Name** the operations used to solve $5 - 2x < -9$.

2. **Write** an inequality requiring more than one operation to solve.

Guided Practice

 State which operation you would perform first to solve the inequality.

| **Sample:** $12x + 4 > 20$ | **Solution:** Subtract 4 from each side. |

3. $-15z + 7 \geq -10$

4. $24 < 8b - 3$

5. $4.5a - (-3.1) > 8.2$

6. $\frac{n + 3}{9} \leq -11$

Examples 1–4 **Solve each inequality. Check your solution.**

7. $2y + 4 > 12$

8. $8 - 3h \leq 20$

9. $5 - 2x > 7$

10. $7z - 4 \geq 10$

11. $10 - 5w < w + 22$

12. $3(n - 4) > 2(n + 6)$

Example 5 13. **School** Kira wants her average math grade to be at least 90. Her test scores are 88, 93, and 87. What score does she need on her fourth test to earn an average score of at least 90?

Exercises

Practice

Solve each inequality. Check your solution.

14. $4t - 8 > 16$

15. $3b + 9 < 45$

16. $2 - 3n \leq 17$

17. $-20 \geq 8 - 7x$

18. $5 \leq 7c - 2$

19. $-6g - 1 < -13$

20. $-12 + 11r \leq 54$

21. $8 - 4v \geq 6$

22. $7 + 0.1a > 9$

23. $0.3m - 2.1 \leq -3.0$

24. $\frac{s}{4} - 6 < -11$

25. $\frac{11 - 6d}{5} > 1$

26. $3h - 5 \geq 2h + 4$

27. $5x + 3 \leq 2x + 9$

28. $6j - 9 > j + 6$

29. $2(7 - 2y) > 10$

30. $-\frac{1}{6}(z + 2) \leq -1$

31. $\frac{2}{3}(b + 1) < \frac{1}{2}(b + 5)$

Homework Help	
For Exercises	See Examples
14–25, 29, 30, 32–36	1, 2, 4
26–28, 31, 37	3
Extra Practice	
See page 716.	

Write and solve an inequality for each situation.

32. The sum of twice a number and 17 is no greater than 41.

33. Five times the sum of a number and 6 is less than 35.

34. Two thirds of a number decreased by 7 is at least 9.

Applications and Problem Solving

35. Employment Jeremy is a receptionist at a hair salon. He earns $7.00 an hour, plus 10% of any hair products he sells. Suppose he works 22 hours a week. How much money in hair products must he sell to earn at least $180?

36. Recreation The admission fee to a state fair is $5.00. Each ride costs an additional $1.50.

  a. Suppose Pilar does not want to spend more than $20. How many rides can she go on?

  b. The fair has a special admission price for $14, which includes unlimited rides. For how many rides is this a better deal than paying for each ride separately?

37. Finance At a bank, an advertisement reads, "In one year, your earnings will be greater than your original investment plus 6% of the investment." Suppose Diego invests $1400. How much money can he expect to have at the end of the year?

38. Critical Thinking Would the solution of $x^2 > 4$ be $x > 2$? Justify your answer.

Mixed Review

Solve each inequality. Check your solution. *(Lesson 12–3)*

39. $5p > 35$

40. $-24 \leq -8v$

41. $-\dfrac{x}{4} \geq 10$

42. $\dfrac{2}{5}z < -6$

43. Budgeting Haley earned $36 babysitting. She plans to buy two books that cost $8.25 each. With the rest of the money, she plans to go to dinner and see a movie with her friends. At most, how much money can she spend on a movie and dinner? *(Lesson 12–2)*

Standardized Test Practice
Ⓐ Ⓑ Ⓒ Ⓓ

44. Grid In Find the value of c that makes $x^2 - 6x + c$ a perfect trinomial square. *(Lesson 11–5)*

45. Multiple Choice If $x \neq 5$, then $\dfrac{x^2 - 3x - 10}{x - 5}$ is equivalent to which of the following? *(Lesson 10–3)*

  Ⓐ $x + 2$
  Ⓑ $x - 3$
  Ⓒ $x - 5$
  Ⓓ $x + 10$

What You'll Learn
You'll learn to solve compound inequalities.

Why It's Important
Nutrition
Pharmacists use inequalities to write prescriptions.
See Example 2.

Lamar is buying vitamins for his dog. The daily dose for the vitamins is based on the dog's weight. Lamar's dog weighs 32 pounds. Since 32 is greater than 25, but less than or equal to 50, he will give his dog 2 tablets.

Another way to write this information is to use an inequality. Let w represent the weight that requires 2 tablets.

Daily Dose	Dog's Weight (pounds)
1 tablet	25 or less
2 tablets	greater than 25 and less than or equal to 50
3 tablets	more than 50

Weight is greater than 25.

$$w > 25$$

and

Weight is less than or equal to 50.

$$w \leq 50$$

These two inequalities form a **compound inequality**. The compound inequality $w > 25$ and $w \leq 50$ can be written without using the word *and*.

Method 1 $25 < w \leq 50$

This can be read as 25 is less than w, which is less than or equal to 50.

Method 2 $50 \geq w > 25$

This can be read as 50 is greater than or equal to w, which is greater than 25.

Note that in each, both inequality symbols are facing the same direction.

Example **Write $x \geq 2$ and $x < 7$ as a compound inequality without using *and*.**

$x \geq 2$ and $x < 7$ can be written as $2 \leq x < 7$ or as $7 > x \geq 2$.

Your Turn

a. $x < 10$ and $x \geq -4$ **b.** $x \leq 6$ and $x \geq 2$

A compound inequality using *and* is true if and only if *both* inequalities are true. Thus, the graph of a compound inequality using *and* is the **intersection** of the graphs of the two inequalities.

Consider the inequality $-2 < x < 3$. It can be written using *and*: $x > -2$ and $x < 3$. To graph, follow the steps below.

Step 1   Graph $x > -2$.

Step 2   Graph $x < 3$.

Step 3   Find the intersection of the graphs.

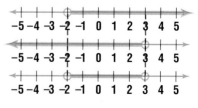

The solution, shown by the graph of the intersection, is $\{x \mid -2 < x < 3\}$.

Example

Pet Care Link

2 Refer to the application at the beginning of the lesson. Graph the solution of $25 < w \le 50$.

Rewrite the compound inequality using *and*.
$25 < w \le 50$ is the same as $w > 25$ and $w \le 50$.

Step 1 Graph $w > 25$.

Step 2 Graph $w \le 50$.

Step 3 Find their intersection.

The solution is $\{w \mid 25 < w \le 50\}$.

Reading Algebra

Most of the time, $<$ or $\le$ symbols are used with compound inequalities.

Your Turn

c. Graph the solution of $3 \le x \le 5$.

 Personal Tutor at algconcepts.com

Often, you must solve a compound inequality before graphing it.

Example

3 Solve $4 < x + 3 \le 12$. Graph the solution.

Step 1 Rewrite the compound inequality using *and*.
$$4 < x + 3 \le 12$$
$$x + 3 > 4 \qquad \text{and} \qquad x + 3 \le 12$$

Step 2 Solve each inequality.
$$x + 3 > 4 \qquad \text{and} \qquad x + 3 \le 12$$
$$x + 3 - 3 > 4 - 3 \qquad\qquad x + 3 - 3 \le 12 - 3$$
$$x > 1 \qquad\qquad\qquad x \le 9$$

Step 3 Rewrite the inequality as $1 < x \le 9$.

The solution is $\{x \mid 1 < x \le 9\}$.
The graph of the solution is
shown at the right.

Your Turn

d. Solve $-2 < x - 4 < 2$. Graph the solution.

Another type of compound inequality uses the word *or*. This type of inequality is true if one or more of the inequalities is true. The graph of a compound inequality using *or* is the **union** of the graphs of the two inequalities.

Graph the solution of $x > 0$ or $x \leq -1$.

Step 1 Graph $x > 0$.

```
←—+—+—+—+—+—⊕—+—+—+—+—+—→
 -5 -4 -3 -2 -1  0  1  2  3  4  5
```

Step 2 Graph $x \leq -1$.

```
←—+—+—+—+—●—+—+—+—+—+—+—→
 -5 -4 -3 -2 -1  0  1  2  3  4  5
```

Step 3 Find the union of the graphs.

```
←—+—+—+—+—●—+—⊕—+—+—+—+—→
 -5 -4 -3 -2 -1  0  1  2  3  4  5
```

Your Turn

d. Graph the solution of $x > -3$ or $x < -5$.

Sometimes you must solve compound inequalities containing the word *or* before you are able to graph the solution.

Solve $3x \geq 15$ or $-2x < 4$. Graph the solution.

$$3x \geq 15 \qquad\qquad \text{or} \qquad\qquad -2x < 4$$
$$\frac{3x}{3} \geq \frac{15}{3} \qquad\qquad\qquad\qquad \frac{-2x}{-2} > \frac{4}{-2}$$
$$x \geq 5 \qquad\qquad\qquad\qquad\qquad x > -2$$

Now graph the solution.

Step 1 Graph $x \geq 5$.

Step 2 Graph $x > -2$

Step 3 Find the union of the graphs.

```
←—+—⊕—+—+—+—+—+—+—+—+—+—→
 -3 -2 -1  0  1  2  3  4  5  6  7
```

The last graph shows the solution $\{x \mid x > -2\}$.

COncepts in MOtion

Animation

algconcepts.com

Your Turn

e. Solve $-6x > 18$ or $x - 2 < 1$. Graph the solution.

Check for Understanding

Communicating Mathematics

1. **Define** *compound inequality* in your own words.

2. **Write** a compound inequality for *x is greater than 3 and less than or equal to 5.*

Vocabulary

compound inequality
intersection
union

Guided Practice

⊕ **Getting Ready**   State whether to find the intersection or union of the two inequalities to graph the solution.

Sample 1: $-8 \leq x + 4 < 12$ **Solution:** Find the intersection since the inequalities can be joined by the word *and*.

Sample 2: $x > -9$ or $x < 5$ **Solution:** Find the union since the inequalities are joined by the word *or*.

3. $x > -10$ and $x < 3$ **4.** $x < 7$ or $x < 4$
5. $x - 5 < 2$ or $x \leq 15$ **6.** $-13 < 9x \leq -11$

Example 1

Write each compound inequality without using *and*.

7. $\ell > 5$ and $\ell < 10$ **8.** $b < 3$ and $b \geq -2$

Examples 2 & 4

Graph the solution of each compound inequality.

9. $y < 5$ and $y \geq 0$ **10.** $x > 3$ or $x < -7$

Examples 3 & 5

Solve each compound inequality. Graph the solution.

11. $10 > c + 2 > 5$ **12.** $2 > s + 4 \geq -3$
13. $0 \leq 2v \leq 8$ **14.** $-5x > 10$ or $7x < 28$
15. $j + 6 > 6$ or $-4j \geq 4$ **16.** $-1 + r < -4$ or $-1 + r > 3$

Example 2

17. Construction   Odyssey of the Mind competitions encourage students to use creativity in solving difficult problems. One year, students had to construct a balsa-wood structure. The structure needed to be at least 9 inches tall and no more than 11.5 inches tall. Write a compound inequality describing the height of the structure. Graph the solution.

Exercises

Practice

Write each compound inequality without using *and*.

18. $b > 0$ and $b < 5$ **19.** $h > -8$ and $h < 8$
20. $y \leq 4$ and $y \geq -1$ **21.** $g \geq 2$ and $g \leq 5$
22. $-2 < r$ and $1 > r$ **23.** $6 < x$ and $x \leq 8$

Homework Help	
For Exercises	See Examples
18–23	1
24–29	2, 4
30–35	3
36–55	3, 4, 5
56–59	2, 3
Extra Practice	
See page 717.	

Graph the solution of each compound inequality.

24. $x > 0$ or $x \leq -4$ **25.** $z \leq 2$ and $z > -2$
26. $k < 7$ and $k > 5$ **27.** $y \geq 16$ and $y \leq 21$
28. $b > 4$ or $b < 0$ **29.** $m \leq 10$ or $m < 6$

Solve each compound inequality. Graph the solution.

30. $2 \leq a + 3 < 7$ **31.** $9 \geq x + 1 \geq 5$
32. $-16 < 8s < 16$ **33.** $9 \geq 3w > 0$
34. $-6 > r - 2 > -10$ **35.** $2 \leq h + 5 \leq 8$

Solve each compound inequality. Graph the solution.

36. $-5 < y - 1 < -4$

37. $16 < -4c < 20$

38. $6 \geq x - (-5) \geq 0$

39. $z + 3 > 7$ or $z - 5 \leq -12$

40. $v - 8 > 12$ or $v + 4 < 20$

41. $r \leq -1$ or $-8r < 0$

42. $3j > 18$ or $j - 3 \geq 5$

43. $p - 5 > -3$ or $-2p \leq 2$

44. $c - 2.4 \geq 7.6$ or $c - 8.8 \leq 0$

45. $d + 2.1 > 3.6$ or $d - 4 > 0.5$

46. $8x \geq 4$ or $x + \dfrac{3}{4} < 1$

47. $\dfrac{w}{3} \leq -2$ or $-w < 2$

Write a compound inequality for each solution shown below.

48.

```
<++|++|+O++|++|++●|++|++>
 -5 -4 -3 -2 -1  0  1  2  3  4  5
```

49.

```
<+●++O++|++|++|++|++O++|++>
 -5 -4 -3 -2 -1  0  1  2  3  4  5
```

50.

```
<++|++|++|++●|++|++●|++|++>
 -8 -7 -6 -5 -4 -3 -2 -1  0  1  2
```

51.

```
<+|+●++|++|++|++|++|++|++O+|+>
 -4 -3 -2 -1  0  1  2  3  4  5  6
```

Solve each compound inequality.

52. $2 > 3y + 2 > -13$

53. $4x - 7 \leq 5$ or $-(x - 8) < -1$

54. $2x - 1 > -5$ or $-3(x + 1) > 6$

55. $10 \leq 2(k - 6) \leq 14$

Applications and Problem Solving

56. **Taxes** Matthew Brooks is single and has a part-time job while attending college. Last year, he paid \$649 in federal income tax. Write an inequality for his taxable income. Use the table that describes the different tax brackets.

If Form 1040A, line 24, is —		And you are —			
At least	But less than	Single	Married filing jointly	Married filing sepa- rately	Head of a house- hold
			Your tax is —		
4,200	4,250	634	634	634	634
4,250	4,300	641	641	641	641
4,300	4,350	649	649	649	649
4,350	4,400	656	656	656	656
4,400	4,450	664	664	664	664
4,450	4,500	671	671	671	671
4,500	4,550	679	679	679	679
4,550	4,600	686	686	686	686

Source: Ohio Department of Taxation

57. **Cooking** A box of macaroni and cheese lists two sets of directions for cooking. It says to heat the macaroni for 11 to 13 minutes on the stove or 12 to 14 minutes in the microwave.

   a. Write an inequality that represents possible heating times.

   b. Graph the solution.

58. **Geometry** To construct any triangle, the sum of the lengths of two sides must be greater than the length of the third. Suppose that two sides of a triangle have lengths of 4 inches and 12 inches. What are the possible values for the length of the third side? Express your answer as a compound inequality.

59. Chemistry Soil pH is measured on a scale from 0 to 14. The pH level describes the soil as shown below.

pH SCALE

←─── INCREASING ACIDITY ──── NEUTRAL ──── DECREASING ACIDITY ───→
0   1   2   3   4   5   6   7   8   9   10   11   12   13   14

Most plants grow best if the soil pH is between 6.5 and 7.2. Mr. Cohen took samples of soil from his garden and found pH values of 6.3, 6.4, and 6.7. What range of values must the fourth sample have if the soil pH is the best for growing plants? (*Hint:* Find the mean soil pH.)

60. Critical Thinking Graph each compound inequality. Then state the solution.

a. $x > 4$ or $x \leq 4$
b. $2 > 3z + 2 > 14$

Mixed Review

Solve each inequality. Check your solution. *(Lessons 12–3 & 12–4)*

61. $2y + 4 < 4$
62. $-3n - 8 > 22$
63. $-2 \leq 0.6x - 5$
64. $\frac{2}{3}p - 3 \leq 7$
65. $-10b < -60$
66. $3r \geq -12$

67. Welding Maxwell is welding two pieces of iron together. During this process, the iron melts and begins to boil. Small droplets erupt and follow the paths of parabolic arcs. The paths can be modeled by the quadratic function $h(d) = -d^2 + 4d + 30$, where $h(d)$ represents the height of the arc above the ground at any horizontal distance d from the two pieces of iron. All measures are in inches. *(Lesson 11–1)*

a. Graph the function.

b. How high above the pieces of iron do the iron droplets jump?

68. Find the GCF of $8x^2$, $2x$, and $4xy$. *(Lesson 10–1)*

69. Simplify $3(b - 6)$. *(Lesson 9–3)*

70. Short Response Simplify $(c^2 + 5) - (c^2 - 8c + 1)$. *(Lesson 9–2)*

71. Multiple Choice Emily drove 8 miles in 12 minutes. At this rate, how many miles will she drive in 1 hour? *(Lesson 5–1)*

Ⓐ 4 mi
Ⓑ 20 mi
Ⓒ 40 mi
Ⓓ 56 mi
Ⓔ 96 mi

There are three types of open sentences that can involve absolute value. They are listed below. Note that in each case, n is nonnegative since the absolute value of a number can only equal 0 or a positive number.

$$|x| = n \qquad\qquad |x| < n \qquad\qquad |x| > n$$

You have already studied equations involving absolute value. Inequalities involving absolute value are similar. Consider the graphs and solutions of the three open sentences below.

Look Back

Absolute Value:
Lesson 3–7

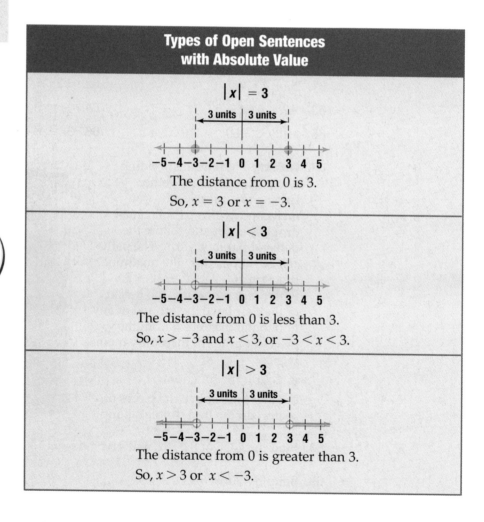

Types of Open Sentences with Absolute Value

$|x| = 3$

3 units | 3 units

−5 −4 −3 −2 −1 0 1 2 3 4 5

The distance from 0 is 3.
So, $x = 3$ or $x = -3$.

$|x| < 3$

3 units | 3 units

−5 −4 −3 −2 −1 0 1 2 3 4 5

The distance from 0 is less than 3.
So, $x > -3$ and $x < 3$, or $-3 < x < 3$.

$|x| > 3$

3 units | 3 units

−5 −4 −3 −2 −1 0 1 2 3 4 5

The distance from 0 is greater than 3.
So, $x > 3$ or $x < -3$.

For both equations and inequalities involving absolute value, there are two cases to consider.

Case 1 The value within the absolute value symbols is positive.

Case 2 The value within the absolute value symbols is negative.

Example ━━① **Solve** $|x - 5| \leq 2$. **Graph the solution.**

Case 1 $x - 5$ is positive.
$$x - 5 \leq 2$$
$$x - 5 + 5 \leq 2 + 5 \quad \textit{Add 5.}$$
$$x \leq 7$$

Case 2 $x - 5$ is negative.
$$-(x - 5) \leq 2$$
$$-(x - 5)(-1) \geq 2(-1) \qquad \textit{Multiply by } -1$$
$$\textit{and reverse}$$
$$x - 5 \geq -2 \qquad \textit{the symbol.}$$
$$5 + 5 \geq -2 + 5 \quad \textit{Add 5.}$$
$$x \geq 3$$

So, the solution is $\{x \mid 3 \leq x \leq 7\}$.

The solution makes sense since 3 and 7 are at most 2 units from 5.

Your Turn

a. Solve $|x - 7| < 4$. Graph the solution.

 Personal Tutor at algconcepts.com _____

    As in Example 1, when solving an inequality involving absolute value and the symbols $<$ or $\leq$, the solution can be written as an inequality using *and*. However, when solving an inequality involving absolute value and the symbols $>$ or $\geq$, the solution can be written as an inequality using *or*.

Example ━━② **Solve** $|6x| > 18$. **Graph the solution.**

Case 1 $6x$ is positive.
$$6x > 18$$
$$\frac{6x}{6} > \frac{18}{6} \quad \textit{Divide by 6.}$$
$$x > 3$$

Case 2 $6x$ is negative.
$$-6x > 18$$
$$\frac{-6x}{-6} < \frac{18}{-6} \qquad \textit{Divide by } -6 \textit{ and}$$
$$\textit{reverse the symbol.}$$
$$x < -3$$

So, the solution is $\{x \mid x < -3 \text{ or } x > 3\}$.

              -5 -4 -3 -2 -1 0 1 2 3 4 5

Your Turn

b. Solve $|x - 2| \geq 3$. Graph the solution.

    Inequalities involving absolute value are often used to indicate *tolerance*. Tolerance is the amount of error or uncertainty that is allowed when taking measurements.

Example 3
Measurement Link

When producing $\frac{1}{2}$-inch bolts for bicycle parts, the tolerance is 0.005 inch. What is the range of acceptable bolt measures?

Explore   The difference in the actual size of the bolt and its expected size has to be less than or equal to 0.005 inch.

Plan   Let m = the actual measure of the bolt.
Then, $\left| m - \frac{1}{2} \right| \leq 0.005$.

Solve $|m - 0.5| \leq 0.005$ *Write $\frac{1}{2}$ as 0.5.*

Case 1 $m - 0.5$ is positive.	**Case 2** $m - 0.5$ is negative.
$m - 0.5 \leq 0.005$	$-(m - 0.5) \leq 0.005$
$m - 0.5 + 0.5 \leq 0.005 + 0.5$	$-(m - 0.5)(-1) \geq 0.005(-1)$
$m \leq 0.505$	$m - 0.5 \geq -0.005$
	$m - 0.5 + 0.5 \geq -0.005 + 0.5$
	$m \geq 0.495$

The solution is $\{m \,|\, 0.495 \leq m \leq 0.505\}$.

Examine   An acceptable bolt must measure from 0.495 inch to 0.505 inch, inclusive. To check the solution, choose a value for m within this range and one outside of this range. Substitute them into the original problem. Which value results in a true inequality?

Check for Understanding

Communicating Mathematics

1. **Compare and contrast** the graphs of the solutions for $|x| < 7$ and $|x| > 7$.

2. **Graph** the solutions for $|x - 2| > 1$ and $|x - 2| < 1$. Which inequality is the *intersection* of the graphs of two inequalities? Which inequality is the *union* of the graphs of two inequalities?

3.   Madison says that the solution for $|x| \leq 0$ is the same as the solution for $|x| \geq 0$. Mia says it is not. Who is correct? Explain.

Guided Practice

⏱ Getting Ready   **Write two inequalities to describe the solution.**

Sample 1: $|x| \leq 5$	**Sample 2:** $|x| > 1$
Solution: $x \leq 5$ and $x \geq -5$	**Solution:** $x > 1$ or $x < -1$

4. $|x| < 10$ 5. $|x| \leq 3$
6. $|x| \geq 2$ 7. $|x| > 8$

Solve each inequality. Graph the solution.

8. $|n - 4| < 5$

9. $|x - 2| < 6$

10. $|3j| \leq 12$

11. $|t - 5| \geq 3$

12. $|2y| > 2$

13. $|s + 4| \geq 3$

Example 3

14. **Measurement** Refer to Example 3. What are the possible measures for the bolt if the tolerance is 0.05 inch? Does a lesser or greater tolerance ensure more accurate measurements? Explain.

Exercises •

Practice

Solve each inequality. Graph the solution.

15. $|m + 1| < 5$

16. $|3v| < 15$

17. $|z + 7| \leq 2$

18. $|x + 3| \leq 8$

19. $|p - 1| < 2$

20. $|r + 4| < 4$

21. $|7t| \leq 14$

22. $|a - 3| \leq 4$

23. $|k + 2| < 3.5$

24. $|5n| \geq 30$

25. $|y + 3| \geq 6$

26. $|z - 1| \geq 1$

27. $|9x| > 18$

28. $|w + 2| > 5$

29. $|r - 4| \geq 1$

30. $|a - 8| \geq 3$

31. $|h - 3| > 9$

32. $|d + 9| > 0.2$

Homework Help	
For Exercises	See Examples
15, 17, 18–20 22, 23, 25, 26, 28–35, 38–43	1, 3
18, 21, 24, 27	2
Extra Practice	
See page 717.	

Write an inequality involving absolute value for each statement. Do not solve.

33. Quincy's golf score s was within 4 strokes of his average score of 90.

34. The measure m of a board used to build a cabinet must be within $\frac{1}{4}$ inch of 46 inches, inclusive, to fit properly.

35. The cruise control of a car set at 65 mph should keep the speed s within 3 mph, inclusive, of 65 mph.

For each graph, write an inequality involving absolute value.

36.
```
  +—○—+—+—+—+—+—+—+—○—+→
 −5−4−3−2−1  0  1  2  3  4  5
```

37.
```
←+—+—+—+—+—+—+—+—+—+—+→
 −5−4−3−2−1  0  1  2  3  4  5
```

Solve each inequality. Graph the solution.

38. $|2x - 11| \geq 7$

39. $|3x - 12| < 12$

40. $4|x + 3| \leq 8$

41. $10|x + 1| > 90$

Applications and Problem Solving

For Exercises 42–43, write and solve an absolute value inequality.

42. **Finance** Ms. Gibson is a bank teller. She must balance her drawer and be within $2 of the expected balance. If Ms. Gibson's expected balance is $8758.20, what are acceptable balances for her drawer?

43. **Chemistry** For a chemistry project, Marvin must pour 3.25 milliliters of solution into a beaker. If he does not pour within 0.05, inclusive, of 3.25 milliters, the results will be inaccurate. How many milliliters of solution can Marvin use?

44. Critical Thinking The *percent of error* is the ratio of the greatest possible error (tolerance) to a measurement. You can find the percent of error using this formula.

$$\text{percent of error} = \left| \frac{\text{greatest possible error}}{\text{measurement}} \right| \cdot 100$$

One rating system for in-line skate bearings is based on tolerances. The table shows tolerances for the outside diameter of bearings measuring 22 millimeters.

Rating	Tolerance (mm)
1	0.010
3	0.008
5	0.005

a. Find the percent of error for each rating to the nearest hundredth.

b. What can you conclude about the rating system?

Mixed Review

Graph the solution of each compound inequality. *(Lesson 12–5)*

45. $m < -7$ or $m \geq 0$ **46.** $x \geq -2$ and $x \leq 5$

47. $y > -5$ and $y < 0$ **48.** $r > 2$ or $r < -2$

49. Solve $1 \geq 4y + 5$. *(Lesson 12–4)*

50. Sales Grant bought a sweater for $44.52. This cost included 6% sales tax. What was the cost of the sweater before tax? *(Lesson 5–5)*

51. Solve $|x - 2| = 4$. *(Lesson 3–7)*

Standardized Test Practice
Ⓐ Ⓑ Ⓒ Ⓓ

52. Short Response Write three fractions whose sum is $1\frac{5}{8}$. *(Lesson 3–2)*

Quiz 2   Lessons 12–4 through 12–6

Solve each inequality. *(Lesson 12–4)*

1. $3x + 8 < 11$ **2.** $5 - 6n > -19$ **3.** $9d - 4 \geq 8 - d$

Solve each compound inequality. Graph the solution. *(Lesson 12–5)*

4. $1 + x < -4$ or $1 + x < 4$ **5.** $-2 \leq n + 3 < 4$ **6.** $-6 < -2f < 10$

Solve each inequality. Graph the solution. *(Lesson 12–6)*

7. $|y - 7| < 2$ **8.** $|a + 8| \geq 1$ **9.** $|5x| < 20$

10. Travel Before the meter begins ticking, the charge for a taxi is $1.60. For each mile driven, there is an additional charge of 80 cents. What is the greatest distance you can travel in this taxi if you do not want to pay more than $10.00? *(Lesson 12–4)*

www.algconcepts.com/self_check_quiz

You'll learn to graph inequalities on the coordinate plane.

Why It's Important

Budgeting By graphing inequalities, you can solve problems where there are many solutions.
See Example 3.

Mr. Wheat is planning to take the Jazz Club to a music festival. Lawn tickets cost $20, and pavilion tickets cost $30. If he plans to spend at most $300, how many of each ticket can Mr. Wheat purchase?

Let x represent the number of lawn tickets. Let y represent the number of pavilion tickets. Then the inequality below represents the solution.

$$\underbrace{The\ cost\ of}_{20x}\ \ \underbrace{plus}_{+}\ \ \underbrace{the\ cost\ of}_{30y}\ \ \underbrace{is\ at\ most}_{\leq}\ \ \underbrace{\$300.}_{300}$$

The inequality is written in two variables. It is similar to an equation written in two variables. An easy way to show the solution of an inequality is to graph it in the coordinate plane. *This problem will be solved in Example 3.*

The solution set of an inequality in two variables contains many ordered pairs. The graph of these ordered pairs fills an area on the coordinate plane called a **half-plane**. The graph of an equation defines the **boundary** or edge for each half-plane. Use these steps to graph $y > 3$.

The related equation for $y > 3$ is $y = 3$.

Step 1 Determine the boundary by graphing the related equation, $y = 3$.

Step 2 Draw a *dashed* line since the boundary is *not* part of the graph.

Step 3 Determine which half-plane is the solution. To do this, substitute a point from each half-plane into the inequality. Find which point results in a true statement.

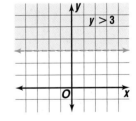

Test the point at (5, 8).
$y > 3$
$8 > 3$ *Replace y with 8.*
true

Test the point at (−3, 1).
$y > 3$
$1 > 3$ *Replace y with 1.*
false

The half-plane that contains (5, 8) is the solution. Shade that half-plane. Any point in the shaded region is a solution of the inequality $y > 3$.

Graph $y > -x + 3$.

Step 1   Determine the boundary by graphing the related equation, $y = -x + 3$.

x	$-x + 3$	y
-2	$-(-2) + 3$	5
-1	$-(-1) + 3$	4
0	$-(0) + 3$	3
1	$-1 + 3$	2
2	$-2 + 3$	1

Step 2   Draw a *dashed* line since the boundary is not included.

Step 3   Test any point to find which half-plane is the solution. Use $(0, 0)$ since it is the easiest point to use in calculations.

$y > -x + 3$ *   Original inequality*
$0 > -(0) + 3$ * $x = 0, y = 0$*
$0 > 3$ *   false*

Since $(0, 0)$ does *not* result in a true inequality, the half-plane containing $(0, 0)$ is *not* the solution. Thus, shade the other half-plane.

Your Turn

a. Graph $y < x - 7$.

When graphing inequalities, the boundary line is not always dashed. Consider the graph of $y \geq 3$. Since the inequality means $y > 3$ or $y = 3$, the boundary is part of the solution. This is indicated by graphing a solid line.

Example ── **2**   **Graph $4x + y \leq 12$.**

To make a table or graph for the boundary line, solve the inequality for y in terms of x.

$4x + y \leq 12$ *   Original inequality*
$4x + y - 4x \leq 12 - 4x$ *   Subtract 4x from each side.*
$y \leq -4x + 12$ *   Rewrite 12 − 4x as −4x + 12.*

www.algconcepts.com/extra_examples

Step 1 Determine the boundary by graphing $y = -4x + 12$.

Step 2 Draw a *solid* line since the boundary is included.

Step 3 Test $(0, 0)$ to find which half-plane contains the solution.

$$4x + y \le 12 \quad \textit{Original inequality}$$
$$4(0) + 0 \le 12 \quad \textit{x = 0, y = 0}$$
$$0 \le 12 \quad \text{true}$$

The half-plane that contains $(0, 0)$ should be in the solution, which is indicated by the shaded region.

Your Turn

b. Graph $-2x + y \le 10$.

When solving real-life inequalities, the domain and range of the inequality are often restricted to nonnegative numbers or whole numbers.

Example **3**

Budgeting Link

Refer to the application at the beginning of the lesson. How many lawn and pavilion tickets can Mr. Wheat purchase?

First, solve for y in terms of x.

$$20x + 30y \le 300 \qquad \textit{Original inequality}$$
$$20x + 30y - 20x \le 300 - 20x \quad \textit{Subtract 20x from each side.}$$
$$30y \le -20x + 300 \quad \textit{Simplify.}$$
$$\frac{30y}{30} \le \frac{-20x + 300}{30} \quad \textit{Divide each side by 30.}$$
$$y \le \frac{-20x}{30} + \frac{300}{30} \quad \textit{Simplify.}$$
$$y \le -\frac{2}{3}x + 10 \quad \textit{Simplify.}$$

Step 1 Determine the boundary by graphing $y = -\frac{2}{3}x + 10$.

Step 2 Draw a *solid* line since the boundary is included.

Step 3 Test $(0, 0)$ to find which half-plane contains the solution.

COncepts in MOtion

Interactive Lab

algconcepts.com

(continued on the next page)

Mr. Wheat cannot buy a negative number of tickets, nor can he buy portions of tickets. The solution is positive ordered pairs that are whole numbers beneath or on the graph of the line $y = -\frac{2}{3}x + 10$.

One solution is (12, 2). This represents 12 lawn tickets and 2 pavilion tickets costing $300.

🌐nline **Personal Tutor** at algconcepts.com

Check for Understanding

Communicating Mathematics

1. **Explain** how to determine whether the boundary is a solid line or dashed line when graphing inequalities in two variables.

2. Writing Math Describe how you could check whether a point is part of the solution of an inequality.

Vocabulary

half-plane
boundary

Guided Practice

⏱ **Getting Ready** **Test (0, 0) to find which half-plane is the solution of each inequality.**

Sample: $3x + y > 4$

Solution: $3x + y > 4$

$3(0) + 0 > 4$

$0 > 4$ false

Shade the half-plane *not* containing the point at (0, 0).

3. $x < -2$

4. $y \leq 1$

5. $5x - y > 5$

Examples 1 & 2

Graph each inequality.

6. $y \leq -1$

7. $y \geq x - 7$

8. $y < 3x + 1$

9. $2x + y < 0$

10. $-4x + y > -8$

11. $x + 3y \geq 9$

Example 3

12. **Sales** Tickets for the winter dance are $5 for singles and $8 for couples. To cover the deejay, photographer, and decoration expenses, a minimum of $1000 must be made from ticket sales. Write and graph an inequality that describes the number of singles' and couples' tickets that must be sold. Name at least one solution.

Exercises •

Practice

Homework Help	
For Exercises	See Examples
13–21, 27, 28, 32	1
22–26, 29–31, 33–34	2, 3
Extra Practice	
See page 717.	

Graph each inequality.

13. $y \leq 7$ **14.** $x \geq 5$ **15.** $y < x - 6$

16. $y < -x + 3$ **17.** $x + y > 8$ **18.** $y > x$

19. $-y \leq x$ **20.** $y \leq 2x$ **21.** $y < -3x + 4$

22. $-x + y < -5$ **23.** $x + 2y < 10$ **24.** $2x + y \leq 6$

25. $-3x + y \geq -1$ **26.** $3x - 2y > -12$ **27.** $y \leq 2(2x + 1)$

28. $3y > 3(4x - 3)$ **29.** $2(x + y) \leq 14$ **30.** $-3(5x - y) > 0$

For Exercises 31–32, write an inequality and graph the solution.

31. The sum of two numbers is greater than four.

32. Twice a number is less than or equal to another number.

Applications and Problem Solving

33. **Animals** Amara and Toshi have set a goal to find homes for more than twelve pets through the Humane Society.

   a. Write and graph an inequality to determine how many homes each girl must find to reach the goal.

   b. List three of the solutions.

   c. Describe the limitations on the solution set.

34. **Airlines** For each mile that Mr. Burnett flies, he earns one point toward a free flight. His credit card company is associated with the airline, and he earns one-half point for each dollar charged. Suppose he needs at least 20,000 points for a free flight. Show with a graph the miles that Mr. Burnett must fly and the money he must charge to get a free flight.

35. **Critical Thinking** Graph the intersection of the solutions of $4x + 2y \geq 8$ and $y < x$.

Mixed Review

Solve each inequality. Graph the solution. *(Lessons 12–5 & 12–6)*

36. $|t + 4| \geq 3$ **37.** $|h - 7| < 2$

38. $-1 < p + 1 \leq 6$ **39.** $-5b > 10$ or $b + 4 > 5$

40. **Travel** Ben and Pam left the park at the same time. Ben traveled north at 45 miles per hour. Pam traveled east at 60 miles per hour. After 1 hour, how far apart are Ben and Pam? *(Lesson 8–7)*

Standardized Test Practice
Ⓐ Ⓑ Ⓒ Ⓓ

41. **Short Response** 848.3 in scientific notation. *(Lesson 8–4)*

42. **Multiple Choice** Suppose you toss 2 coins at the same time. What is the probability that both land heads up? *(Lesson 5–6)*

   Ⓐ 0   Ⓑ 0.25   Ⓒ 0.50   Ⓓ 0.75   Ⓔ 1

Parabolas and Pavilions

Quadratic Inequalities

Materials

 grid paper

 ruler

yellow and blue colored pencils

In this investigation, you will learn how to solve and graph **quadratic inequalities**.

Investigate

1. Graph the quadratic equation $y = x^2 + 2x - 3$ on a piece of graph paper.

   a. With a yellow colored pencil, shade the region inside the parabola.

   b. Make a table like the one below. Fill in column 1 with five points that appear in the yellow region, such as (0, 0). Compare the value of the y-coordinate with the value of $y = x^2 + 2x - 3$ when it is evaluated at the x-coordinate. When the quadratic equation is evaluated at 0, the result is -3. The y-coordinate, 0, is greater than -3. Place the correct inequality symbol in column 3 for each of the other four points that fall in the yellow region.

Point	y-coordinate	< or >	$y = x^2 + 2x - 3$
(0, 0)	0	>	$y = (0)^2 + 2(0) - 3$ or -3

   c. Write a quadratic inequality that compares the points of the yellow region with the quadratic equation $y = x^2 + 2x - 3$.

   d. With a blue colored pencil, shade the region outside the parabola. Repeat parts b–c for the blue region.

2. Graph the quadratic equation $y = -x^2 - 2x + 3$ on a separate piece of graph paper.

   a. Shade the region inside the parabola yellow. Then shade the region outside the parabola blue. Make tables similar to those in Step 1. Write inequalities describing the yellow and blue regions.

   b. Suppose you wanted to include the values on the boundary line of a quadratic inequality. Explain how you would write the inequality to include the boundary line.

   c. Now suppose you did *not* want to include the boundary line of a quadratic inequality. Explain how you would draw the graph to show that the boundary line was *not* included in the solution.

3. Spring Town is building a community center, the Spring Town Pavilion. The building is to be 20 feet longer than it is wide and to have an area greater than or equal to 1500 square feet.

a. Let w represent the width of the building. Then $w + 20$ represents the length. The area of the building, $w(w + 20)$, must be greater than or equal to 1500 square feet. That is, $w^2 + 20w - 1500 \geq 0$. Graph $f(w) = w^2 + 20w - 1500$.

b. Find the values for w that satisfy the inequality in part a. To do this, choose values for w both inside and outside of your parabola.

c. What restrictions are placed on your solution for w since it represents width?

d. Use your graph to find two possible dimensions for the Spring Town Pavilion.

Extending the Investigation

In this extension, you will continue to investigate quadratic inequalities.

Graph each quadratic inequality.

1. $y < x^2 + 4x - 8$
2. $y \geq -x^2 + 2x + 15$
3. $y \leq -3x^2 + 3$

4. The Spring Town Pavilion will have a deck for small outdoor gatherings. The length of the deck is to be 6 feet more than the width. The area of the deck is at least one-fifth of the area of the community center building. Write and graph a quadratic inequality that describes this situation. Then give three possible dimensions for the new deck.

Presenting Your Investigation

Here are some ideas to help you present your conclusions to the class.

• Make a poster that describes how to solve quadratic inequalities. Include at least two different inequalities. Show how the graphs are shaded to represent the solutions.

• Write and solve a problem similar to the one in Step 3. Make a brochure that describes the problem, the graph of its solution, and a list of possible solutions.

*inter*NET **Investigation** For more information on graphing quadratic
CONNECTION inequalities, visit: www.algconcepts.com

Understanding and Using the Vocabulary

After completing this chapter, you should be able to define each term, property, or phrase and give an example or two of each.

*inter*NET
CONNECTION **Review Activities**
For more review activities, visit:
www.algconcepts.com

boundary *(p. 535)*
compound inequality *(p. 524)*
half-plane *(p. 535)*

intersection *(p. 524)*
quadratic inequalities *(p. 540)*

set-builder notation *(p. 510)*
union *(p. 525)*

Complete each sentence using a term from the vocabulary list.

1. The solution of a compound inequality using *or* can be found by the ___?___ of the graphs of the two inequalities.

2. Graph the related equation of an inequality to find the ___?___ of the half-plane.

3. Use a test point from each ___?___ to find the solution of an inequality in two variables.

4. An inequality of the form $x < y < z$ is called a(n) ___?___ .

5. The ___?___ for the graph of an inequality in two variables will either be a dashed or solid line.

6. The ___?___ of two graphs is the area where they overlap.

7. A(n) ___?___ is an area on the coordinate plane representing the solution for an inequality in two variables.

8. $2x + 3 < 5$ or $x \geq -1$ is an example of a(n) ___?___ .

9. A solution written in the form $\{x \mid x \leq 3\}$ is written in ___?___ .

10. The solution of a compound inequality using *and* can be found by the ___?___ of the graphs of the two inequalities.

Skills and Concepts

Objectives and Examples

- **Lesson 12–1** Graph inequalities on a number line.

  Graph $x < 5$ on a number line.

  The graph begins at 5, but 5 is not included. The arrow is to the left. The graph describes values that are less than 5.

  ```
  ◄──┼──┼──┼──◇──┼──┼──►
     2  3  4  5  6  7
  ```

Review Exercises

Graph each inequality on a number line.

11. $x > -3$

12. $z \leq 2$

13. $3.5 > x$

14. $a \geq -1\frac{1}{2}$

Write an inequality for each graph.

15.
```
◄──┼──┼──◆──┼──┼──►
  -7 -6 -5 -4 -3
```

16.
```
◄──┼──┼──┼──◇──┼──┼──►
   5  6  7  8  9
```

 www.algconcepts.com/vocabulary_review

Objectives and Examples

- **Lesson 12–2** Solve inequalities involving addition and subtraction.

  Solve $x - 6 > 2$.
  $$x - 6 > 2 \qquad \text{\textit{Original inequality}}$$
  $$x - 6 + 6 > 2 + 6 \qquad \text{\textit{Add 6 to each side.}}$$
  $$x > 8 \qquad \text{\textit{Simplify.}}$$
  The solution is $\{x \mid x > 8\}$.

- **Lesson 12–3** Solve inequalities involving multiplication and division.

  Solve $-4x < -8$.
  $$-4x < -8 \qquad \text{\textit{Original inequality}}$$
  $$\frac{-4x}{-4} > \frac{-8}{-4} \qquad \text{\textit{Divide each side by -4 and}}$$
  $$\qquad\qquad\qquad \text{\textit{reverse the symbol.}}$$
  $$x > 2 \qquad \text{\textit{Simplify.}}$$
  The solution is $\{x \mid x > 2\}$.

- **Lesson 12–4** Solve inequalities involving more than one operation.

  Solve $2(x + 1) < 4 - 3x$.
  $$2(x + 1) < 4 - 3x \qquad \text{\textit{Original inequality}}$$
  $$2x + 2 < 4 - 3x \qquad \text{\textit{Distributive Property}}$$
  $$5x + 2 < 4 \qquad \text{\textit{Add 3x to each side.}}$$
  $$5x < 2 \qquad \text{\textit{Subtract 2 from each side.}}$$
  $$x < \frac{2}{5} \qquad \text{\textit{Divide each side by 5.}}$$
  The solution is $\left\{x \mid x < \frac{2}{5}\right\}$.

- **Lesson 12–5** Solve compound inequalities and graph the solution.

  Solve $3 < x + 2 \leq 8$.
  $$x + 2 > 3 \qquad \text{and} \qquad x + 2 \leq 8$$
  $$x + 2 - 2 > 3 - 2 \qquad\qquad x + 2 - 2 \leq 8 - 2$$
  $$x > 1 \qquad\qquad\qquad x \leq 6$$
  The solution is $\{x \mid 1 < x \leq 6\}$.

Review Exercises

Solve each inequality. Check your solution.

17. $x + 3 > 7$

18. $a - 4 \geq 2$

19. $\frac{2}{3} \leq y - \frac{1}{2}$

20. $12x - 11x + 5 < 3$

21. $3(x + 1) \geq 4x$

Solve each inequality. Check your solution.

22. $3y \geq 12$

23. $-6n > 30$

24. $0.2w \leq -1.8$

25. $\frac{x}{5} > 3$ **26.** $-\frac{t}{2} \geq -14$

27. $\frac{3}{4}w \leq 6$ **28.** $\frac{h}{2.3} < -7$

Solve each inequality. Check your solution.

29. $2x + 6 \geq 14$ **30.** $9 < -0.2y - 1$

31. $\frac{2}{3}t - 4 \leq 2$ **32.** $3(4 + n) < 21$

Write and solve an inequality.

33. Four times a number decreased by 3 is greater than 25.

34. Seven minus two times a number is no less than nine.

Solve each compound inequality. Graph the solution.

35. $-4 \leq y + 3 < 2$

36. $2 \leq 3 - t$ or $3 - t > 5$

37. $3a \geq 6$ or $-5 - a \geq -6$

38. $9 < 2x + 1 < 13$

Mixed Problem Solving
See pages 724–731.

Objectives and Examples

- **Lesson 12–6** Solve inequalities involving absolute value and graph the solution.

Solve $|x + 1| < 1$.

Case 1 $x + 1$ is positive.
$x + 1 < 1$
$\quad x < 0$ *Subtract 1 from each side.*

Case 2 $x + 1$ is negative.
$\quad -(x + 1) < 1$
$-(x + 1)(-1) > 1(-1)$ *Multiply by -1 and*
$\quad\quad x + 1 > -1$ *reverse the symbol.*
$\quad\quad\quad x > -2$ *Subtract 1 from each side.*

The solution is $\{x \mid -2 < x < 0\}$.

Review Exercises

Solve each inequality. Graph the solution.

39. $|t - 1| \leq 5$

40. $|a + 7| < -2$

41. $|x + 3| < 1$

42. $|s + 4| \geq 3$

43. $|y - 2| > 0$

Write an inequality involving absolute value for each statement. Do not solve.

44. Bianca's guess g was within $6 of the actual value of $25.

45. The difference between Greg's score s on his final exam and his mean grade of 80 is more than 5 points.

- **Lesson 12–7** Graph inequalities in the coordinate plane.

Graph $2x + y < 5$.

First, solve for y.
$\quad\quad 2x + y < 5$
$2x + y - 2x < 5 - 2x$
$\quad\quad\quad y < -2x + 5$

Graph the related function $y = -2x + 5$ as a dashed line and shade.

Graph each inequality.

46. $y \geq x - 1$

47. $y < -2x + 4$

48. $\dfrac{y}{2} > x - 2$

49. $x - y \geq 3$

Applications and Problem Solving

50. Shipping An empty book crate weighs 30 pounds. The weight of a book is 1.5 pounds. For shipping, the crate can weigh no more than 60 pounds. What is the acceptable number of books that can be packed in a crate? *(Lesson 12–4)*

51. Geometry An obtuse angle measures more than 90° but less than 180°. If the measure of an obtuse angle is $3x$, what are the possible values for x? *(Lesson 12–5)*

$3x°$

1. **List** at least three verbal phrases that are used to describe inequalities.

2. If you multiply or divide each side of an inequality by a negative number, what must happen to the symbol for the inequality to remain true?

3. Before solving an inequality involving absolute value, which two cases must you consider?

Write an inequality for each graph.

4.
$$1 \quad 2 \quad 3 \quad 4 \quad 5$$

5.
$$-4 \quad -3 \quad -2 \quad -1 \quad 0$$

Solve each inequality. Check your solution.

6. $2 + x \geq 12$

7. $5t + 6 \leq 4t - 3$

8. $8 < -4t$

9. $-0.2x > -6$

10. $\frac{t}{4} > 1$

11. $-\frac{2}{5}m \leq 10$

12. $-3r - 1 \geq -16$

13. $7x - 12 < 30$

14. $2(h - 3) > 6$

15. $8(1 - 2z) \leq 25 + z$

Solve each inequality. Graph the solution.

16. $x + 1 > -2$ and $x + 1 < 6$

17. $4 < 3j - 2 \leq 7$

18. $2n + 5 \geq 15$ or $2n + 5 \leq 3$

19. $-6c > -24$ or $c + 0.25 < 1.3$

20. $|x + 3| \geq 4$

21. $|4b| \leq 16$

Graph each inequality.

22. $y \geq 5x - 6$

23. $4x - 2y > -6$

24. **Car Rental** Justine is renting a car that costs $32 a day with free unlimited mileage. Since she is under the age of 25, it costs her an additional $10 per day. Justine does not want to pay any more than $200 on car rental costs. For what number of days can she rent a car?

25. **Manufacturing** Ball bearings are used to connect moving parts and minimize friction. Ball bearings for an automobile will work properly only if their diameter is within 0.01 inch, inclusive, of 5 inches. Write and solve an inequality to represent the range of acceptable diameters for these ball bearings.

Angle, Triangle, and Quadrilateral Problems

You are already aware of some geometry concepts that you need to know for standardized tests. Be sure that you also know the meanings and definitions of the following terms.

right	obtuse	acute	triangle	quadrilateral
equilateral	isosceles	similar ($\approx$)	congruent ($\cong$)	collinear
reflection	translation	rotation	dilation	

Example 1

The triangles below are similar. Find the length of side $\overline{KL}$.

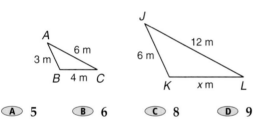

Ⓐ 5	Ⓑ 6	Ⓒ 8	Ⓓ 9

Hint   The measures of corresponding sides of similar polygons are proportional.

Solution   Find the corresponding sides of the two triangles. Side $\overline{AB}$ of $\triangle ABC$ corresponds to side $\overline{JK}$ of $\triangle JKL$. Side $\overline{BC}$ of $\triangle ABC$ corresponds to side $\overline{KL}$ of $\triangle JKL$. Using these two pairs of corresponding sides, write a proportion.

$\dfrac{AB}{JK} = \dfrac{BC}{KL}$ *Sides are proportional.*

$\dfrac{3}{6} = \dfrac{4}{x}$ *Substitute side measures.*

$3x = 4(6)$ *Cross multiply.*

$3x = 24$ *Multiply.*

$x = 8$ *Divide each side by 3.*

Side $\overline{KL}$ measures 8 meters. The answer is C.

Example 2

In the figure below, $\overline{ON}$ is congruent to $\overline{LN}$, $m\angle LON = 30$, and $m\angle LMN = 40$. What is $m\angle NLM$?

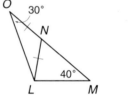

Ⓐ 40	Ⓑ 80	Ⓒ 90	Ⓓ 120

Hint   Examine *all* of the triangles in the figure.

Solution   Label the unknown angles as 1 and 2. Find $m\angle 2$.

Since $\overline{ON} \cong \overline{LN}$, $\triangle ONL$ is isosceles and the base angles are equal. So, $m\angle 1 = 30$. Since the sum of the angle measures in any triangle is 180, find the sum of the angle measures in $\triangle OML$.

$180 = 30 + 40 + (30 + m\angle 2)$

$180 = 100 + m\angle 2$

$80 = m\angle 2$ *Subtract 100 from each side.*

The answer is B.

After you work each problem, record your answer on the answer sheet provided or on a sheet of paper.

Multiple Choice

1. The rectangles are similar. Find the value of x.
(Lesson 5–1)

21 in.   7 in. x in.   1 in.

- **A** 3
- **B** 7
- **C** 14
- **D** 15

2. $\triangle ABC$ is isosceles. What is $m\angle C$?
(Lesson 4–5)

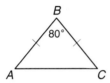

- **A** 45
- **B** 50
- **C** 80
- **D** 100

3. What is the sum of a, b, and c?   *(Basic Skill)*

- **A** 180
- **B** 240
- **C** 270
- **D** 360
- **E** It cannot be determined from the information given.

4. The formula for the area of a trapezoid is $A = \frac{1}{2}h(b_1 + b_2)$, where b_1 and b_2 are the lengths of the bases and h is the height. If the area of trapezoid $QRST$ is 72 square centimeters, then what is the height?
(Lesson 1–5)

R 6 cm S
h
Q 12 cm T

- **A** 1 cm
- **B** 2 cm
- **C** 4 cm
- **D** 8 cm

5. Find the x-intercept of $y = -\frac{2}{3}x + 4$.
(Lesson 7–5)

- **A** 6
- **B** −6
- **C** 4
- **D** 0

6. Simplify the expression $\dfrac{(-5)(4)\,|-6|}{-3}$.
(Lesson 2–6)

- **A** −120
- **B** −40
- **C** 40
- **D** 120

7. The square of a number is 255 greater than twice the number. What is the number?
(Lesson 11–4)

- **A** 15 or −17
- **B** 17 or −15
- **C** 31 or −33
- **D** 33 or −31

8. What is the value of $b + c$?   *(Basic Skill)*

110° $a°$
$c°$
111°
$b°$ $d°$

- **A** 179
- **B** 180
- **C** 181
- **D** 221

Grid In

9. In the figure, what is the sum of a, b, and c?   *(Basic Skill)*

$a°$ $b°$
$c°$

Extended Response

10. Draw a triangle that fits each description. If a drawing is not possible, explain why.
(Basic Skill)

Part A
i. an obtuse equilateral triangle
ii. an acute equilateral triangle

Part B
i. an obtuse isosceles triangle
ii. an acute isosceles triangle

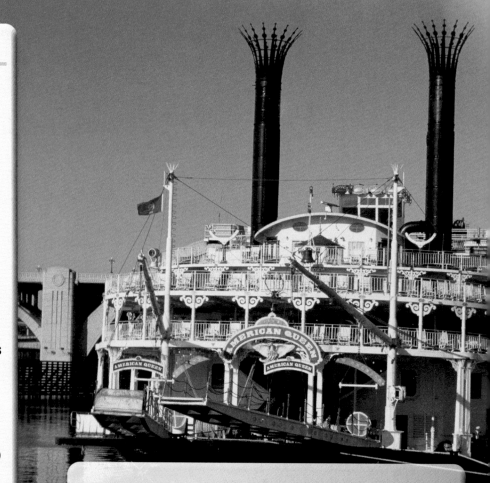

CHAPTER

13

Systems of Equations and Inequalities

What You'll Learn

Key Ideas

- Solve systems of equations by graphing. *(Lesson 13–1)*
- Determine how many solutions a system of equations has by graphing. *(Lesson 13–2)*
- Solve systems of equations by substitution and by elimination. *(Lessons 13–3 to 13–5)*
- Solve systems of quadratic and linear equations. *(Lesson 13–6)*
- Solve systems of inequalities by graphing. *(Lesson 13–7)*

Key Vocabulary

elimination *(p. 566)*

quadratic-linear system *(p. 580)*

substitution *(p. 560)*

system of equations *(p. 550)*

system of inequalities *(p. 586)*

Why It's Important

Travel After the Louisiana Purchase in 1803, a flood of settlers that included riverboat captains, loggers, trappers, and millers came to live on the Mississippi River. Today the river still provides transportation for industry.

Systems of equations and inequalities are used to represent real-world situations in which there are many conditions that must be met. You will use a system of equations to find the rate of the current of the Mississippi River in Lesson 13–5.

✓ Check Your Readiness

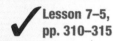
Graph each equation.

1. $y = x + 2$ **2.** $y = 2x - 1$ **3.** $y = \frac{1}{4}x + 2$

4. $x + y = 4$ **5.** $x - 2y = 6$ **6.** $2x + y = 4$

Solve each equation. Check your solution.

7. $3n + 2 = 11$ **8.** $2c - 6 = -8$ **9.** $2 + 5r = 22$

10. $6x - 8 = 22$ **11.** $5d - 1 = 9$ **12.** $-14 = 4k + 2$

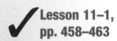
Graph each quadratic function by making a table of values.

13. $y = x^2$ **14.** $y = x^2 + 2$ **15.** $y = 3x^2$

16. $y = -x^2 + 3$ **17.** $y = x^2 - 3x$ **18.** $y = \frac{1}{2}x^2$

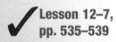
Graph each inequality.

19. $y \le x + 2$ **20.** $y > 4x + 5$ **21.** $x \ge 3$

22. $y < \frac{1}{3}x + 2$ **23.** $x + y \le 3$ **24.** $3x - y > 5$

FOLDABLES™
Study Organizer

Make this Foldable to help you organize your Chapter 13 notes. Begin with four sheets of grid paper.

❶ **Stack** sheets of paper with edges 4 grids apart to create tabs.

❷ **Fold** up the bottom edges. All tabs should be the same size.

❸ **Staple** along the fold.

❹ **Label** the tabs using lesson numbers and titles.

Reading and Writing As you read and study the chapter, use each page to write notes and examples.

What You'll Learn

You'll learn to solve systems of equations by graphing.

Why It's Important

Sales Systems of equations can be used to determine how many items need to be sold in order to make a profit.
See Example 3.

A **system of equations** is a set of two or more equations with the same variables. Below are some examples of systems.

$$x + y = 5$$
$$3x - 4y = -13$$

$$\frac{a}{b} = 2$$
$$a + 2b = 8$$

The solution of a system of equations in two variables is an ordered pair that satisfies both equations.

System of Equations	**Words:** A system of equations is a set of two or more equations with the same variables. The solution is the ordered pair that satisfies all of the equations.
	Model:

One method for solving a system of equations is to graph the equations on the same coordinate plane. The coordinates of the point of intersection are the solution.

Examples

Solve each system of equations by graphing.

1 $y = 2x$

$y = -x + 3$

The graphs appear to intersect at $(1, 2)$. Check this estimate by substituting the coordinates into each equation.

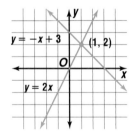

Look Back

Graphing Linear Equations: Lesson 7–5

Check:

$y = 2x$	*Original equation*	$y = -x + 3$	*Original equation*
$2 \stackrel{?}{=} 2(1)$	*Replace x with 1 and y with 2.*	$2 \stackrel{?}{=} -1 + 3$	*Replace x with 1 and y with 2.*
$2 = 2$ ✓		$2 = 2$ ✓	

The solution of the system of equations is $(1, 2)$.

2 $x - y = -2$
$x + y = 4$

The graphs appear to intersect at (1, 3). *Check this estimate.*

The solution of the system of equations is (1, 3).

$x + y = 4$

$(1, 3)$

$x - y = -2$

Your Turn

a. $y = 2x$
$y = -x + 6$

b. $x + y = 1$
$2x + y = 4$

A graphing calculator can be used to solve systems of equations or to check solutions.

Graphing Calculator Tutorial

See pp. 750–753.

 Graphing Calculator Exploration

You can use a graphing calculator to solve systems of equations.

$3x + y = 1$
$-x + 2y = 16$

Step 1 Solve each equation for y.

$$3x + y = 1 \rightarrow y = -3x + 1$$
$$-x + 2y = 16 \rightarrow y = \frac{1}{2}x + 8$$

Step 2 Make a table of values for the two equations.

Press [Y=] to enter $-3x + 1$ as Y_1. Scroll down to enter $\frac{1}{2}x + 8$ as Y_2. Then press [2nd] [TBLSET] and enter -5 as TblStart and 1 as ΔTbl.

X	Y₁	Y₂
-5	16	5.5
-4	13	6
-3	10	6.5
-2	7	7
-1	4	7.5
0	1	8
1	-2	8.5

X= -5

Press [2nd] [TABLE] to display the table.

Step 3 Look in the table for an x value that has the same y value in both equations. The solution is $(-2, 7)$. *Check by graphing on the calculator and finding the intersection point.*

Try These

Use a graphing calculator to solve each system of equations.

1. $y = x + 7$
$y = -x + 9$

2. $y = -3x$
$4x + y = 2$

3. $2x - y = 5$
$x + y = 16$

The Math Club is selling T-shirts for a profit of $4 each and caps for a profit of $5 each. The club wants to sell 50 items and make a profit of $230. How many of each item should the club try to sell?

Let x = the number of T-shirts and y = the number of caps. You can write two equations to represent this situation.

$x + y = 50$ ← *the number of items*
$4x + 5y = 230$ ← *the total profit*

Graph $x + y = 50$ and $4x + 5y = 230$. The graphs appear to intersect at $(20, 30)$. Check this estimate.

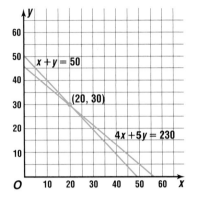

Check:

$x + y = 50$ *Original equation*
$20 + 30 \stackrel{?}{=} 50$ $(x, y) = (20, 30)$
$50 = 50$ ✓

$4x + 5y = 230$ *Original equation*
$4(20) + 5(30) \stackrel{?}{=} 230$ $(x, y) = (20, 30)$
$80 + 150 \stackrel{?}{=} 230$
$230 = 230$ ✓

They should try to sell 20 T-shirts and 30 caps.

Online Personal Tutor at algconcepts.com

Check for Understanding

Communicating Mathematics

1. **Describe** the solution of a system of equations.

2. **Sketch** a system of linear equations that has $(2, 3)$ as its solution.

3. **Determine** the solution of the system of equations represented by each pair of lines.
   a. ℓ and m b. m and n c. n and ℓ

Guided Practice

Examples 1–3

Solve each system of equations by graphing.

4. $y = x - 3$
   $y = -x + 3$

5. $x = -4$
   $y = \frac{1}{2}x + 3$

6. $x + y = 6$
   $y = 2$

7. $x + y = -2$
   $x - y = 4$

Example 3

8. **Shopping** Randall wants to buy 10 bouquets. The standard bouquet costs $7, and the deluxe one costs $12. He can afford to spend $100.

   a. Write a system of equations for the number of standard bouquets x and the number of deluxe bouquets y that he can buy.

   b. Use a graphing calculator to find the number of each type of bouquet he can buy.

Exercises • • • • • • • • • • • • • • • •

Practice

Homework Help

For Exercises	See Examples
9–14	1
15–24	2, 3

Extra Practice
See page 718.

Solve each system of equations by graphing.

9. $x = 5$
  $y = -4$

10. $x = 4$
  $y = x - 5$

11. $y = x + 2$
  $y = -x + 2$

12. $y = -2x + 3$
  $y = -\frac{1}{2}x$

13. $y = 6$
  $y = \frac{4}{3}x + 2$

14. $y = 2x - 7$
  $y = \frac{3}{2}x - 6$

15. $y = 2$
  $x + y = 7$

16. $x + y = 3$
  $x - y = 1$

17. $x - y = 1$
  $y = -2x - 7$

18. $2x - y = 4$
  $x + y = -4$

19. $x + y = -3$
  $\frac{1}{2}x + y = 0$

20. $5x + 4y = 10$
  $2x + y = 1$

21. Find the solution of the system $y = x + 4$ and $3x + 2y = 18$.

22. What is the solution of the system $2x + 10y = 0$ and $x + y = 4$?

Applications and Problem Solving

23. Submarines Two submarines began dives in the same vertical position to meet at a designated point. If one submarine was on a course approximated by the equation $x + 4y = -14$ and the other was on a course approximated by the equation $x + 3y = -8$, at what location would they meet? Write the coordinates of the point.

24. Geometry The graphs of the equations $y = x + 2$, $3x + y = 6$, and $y = 5x + 6$ contain the sides of a triangle.
  a. Graph the equations.
  b. Find the coordinates of the vertices of the triangle.

25. Critical Thinking Use your knowledge of slope and y-intercepts to write a system of equations with a solution of $(0, 4)$.

Mixed Review

26. Which ordered pair, $(-2, 0)$, $(-1, 2)$, $(0, -1)$, or $(2, -2)$, is a solution of $3x + 4y \geq 2$? *(Lesson 12–7)*

Solve each inequality. Then graph the solution set. *(Lesson 12–6)*

27. $|x - 3| < 2$

28. $|a + 5| \geq 3$

29. $|2m + 2| > 6$

30. School Everyone in Mr. McClain's algebra class is at least 15 years old. Write an inequality to represent this information and then make a graph of the inequality. *(Lesson 12–1)*

Standardized Test Practice
Ⓐ Ⓑ Ⓒ Ⓓ

31. Short Response Solve $2x^2 - 7x + 6 = 0$ by factoring. *(Lesson 11–4)*

32. Multiple Choice Which graph represents the function $y = x^2 + 2x - 3$? *(Lesson 11–1)*

Ⓐ

Ⓑ

Ⓒ

Graphs of systems of linear equations may be intersecting lines, parallel lines, or the same line. Systems of equations can be described by the number of solutions they have.

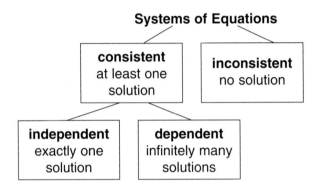

The different possibilities for the graphs of two linear equations are summarized in the following table.

Graph	Description of Graph	Slopes and Intercepts of Lines	Number of Solutions	Type of System
$y = -x + 3$ $\;(1, 2)$ $\;y = x + 1$	intersecting lines	different slopes	1	consistent and independent
$x + 2y = 2$ $\;2x + 4y = 4$	same line	same slope, same intercepts	infinitely many	consistent and dependent
$y = -2x + 4$ $\;y = -2x$	parallel lines	same slope, different intercepts	0	inconsistent

State whether each system is *consistent and independent,* **consistent and dependent,** **or** *inconsistent.*

1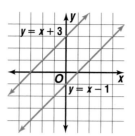

The graphs appear to be parallel lines. Since they do not intersect, there is no solution. This system is inconsistent.

2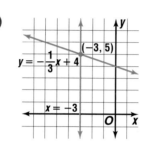

The graphs appear to intersect at the point at $(-3, 5)$. Because there is one solution, this system of equations is consistent and independent.

3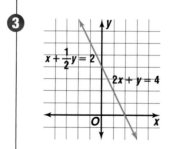

Both equations have the same graph. Because any ordered pair on the graph will satisfy both equations, there are infinitely many solutions. The system is consistent and dependent.

Your Turn

a.

b.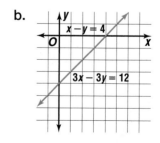

You can determine the number of solutions to a system of equations by graphing.

Look Back

Graphing Linear
Equations:
Lesson 7–5

Determine whether each system of equations has *one* solution, *no* solution, or *infinitely many* solutions by graphing. If the system has one solution, name it.

④ $y = x + 2$
$y = -3x - 6$

The graphs appear to intersect at $(-2, 0)$. Therefore, this system of equations has one solution, $(-2, 0)$. Check that $(-2, 0)$ is a solution to each equation.

Check:

$y = x + 2$	*Original equation*	$y = -3x - 6$	*Original equation*
$0 \stackrel{?}{=} -2 + 2$	*Replace x with -2*	$0 \stackrel{?}{=} -3(-2) - 6$	*Replace x with -2*
$0 = 0$ ✓	*and y with 0.*	$0 = 0$ ✓	*and y with 0.*

The solution of the system of equations is $(-2, 0)$.

⑤ $2x + y = 4$
$2x + y = 6$

Write each equation in slope-intercept form.

$2x + y = 4 \quad \rightarrow \quad y = -2x + 4$
$2x + y = 6 \quad \rightarrow \quad y = -2x + 6$

The graphs have the same slope and different y-intercepts. The system of equations has no solution.

c. $y = x + 3$
$\quad y = -2x + 3$

d. $2x + y = 6$
$\quad 4x + 2y = 12$

e. $3x - y = 3$
$\quad 3x - y = 0$

f. $y - 2x = 0$
$\quad y + x = 6$

Example 6

Transportation Link

The system of equations below represents the tracks of two trains. Do the tracks intersect, run parallel, or are the trains running on the same track? Explain.

$x + 2y = 4$
$3x + 6y = 12$

$x + 2y = 4$ $3x + 6y = 12$
$3(x + 2y) = 3(4)$
$x + 2y = 4$ *Divide each side by 3.*

One equation is a multiple of the other. Each equation has the same graph and there are infinitely many solutions. Therefore, the trains are running on the same track.

Check for Understanding

Communicating Mathematics

1. **Describe** the possible graphs of a system of two linear equations.

2. **State** the number of solutions for each system of equations described below.

   a. One equation is a multiple of the other.

   b. The equations have the same *y*-intercept and different slopes.

   c. The equations have the same slope and different *y*-intercepts.

3. Writing Math   Create memory devices or other ways to remember the definitions of the terms in the Vocabulary box. Describe them in your journal.

Vocabulary

consistent
independent
dependent
inconsistent

Guided Practice

Examples 1–3

State whether each system is *consistent and independent, consistent and dependent,* or *inconsistent.*

4.

5.

Examples 4 & 5

Determine whether each system of equations has *one* solution, *no* solution, or *infinitely many* solutions by graphing. If the system has one solution, name it.

6. $y = x$
$y = x + 5$

7. $y = -x$
$y = 3x - 4$

8. $2x + y = -3$
$6x + 3y = -9$

Example 6

9. Animals A dog sees a cat 60 feet away and starts running after it at 50 feet per second. At the same time, the cat runs away at 30 feet per second. This situation can be represented by the following system of equations and the graph at the right.

$y = 50x \leftarrow dog \qquad y = 30x + 60 \leftarrow cat$

a. Is this system of equations *consistent and independent, consistent and dependent,* or *inconsistent*? Explain.

b. Explain what the point at (3, 150) represents.

Exercises • • • • • • • • • • • • • • • • • •

Practice

State whether each system is *consistent and independent, consistent and dependent,* or *inconsistent*.

Homework Help	
For Exercises	See Examples
10–15	1, 2, 3
16–26, 28	4, 5
27, 29	6
Extra Practice	
See page 718.	

10.

11.

12.

13.

14.

15.

Determine whether each system of equations has *one* solution, *no* solution, or *infinitely many* solutions by graphing. If the system has one solution, name it.

16. $y = 4x - 6$
$y = 4x - 1$

17. $y = 3x - 2$
$4y = 12x - 8$

18. $y = \frac{1}{2}x$
$y = -2x + 5$

19. $y = -x + 3$
$y = \frac{1}{5}x - 3$

20. $x = 3$
$2x - 3y = 0$

21. $3x - 2y = -6$
$3x - 2y = 6$

22. $x + y = 5$
$2x + 2y = 8$

23. $6x + y = -3$
$-x + y = 4$

24. $x - 4y = -4$
$\frac{1}{2}x - 2y = -2$

25. Does the system $x - y = 4$ and $x - 3y = 2$ have *one* solution, *no* solution, or *infinitely many* solutions? If the system has one solution, name it.

26. Without graphing, determine whether the system $x - 3y = 11$ and $2x - 6y = -5$ has *one* solution, *no* solution, or *infinitely many* solutions. Explain how you know.

Applications and Problem Solving

27. Animals Refer to Exercise 9. Suppose the dog is chasing another dog whose distance y can be represented by the equation $y = 50x + 20$. What is the solution? Explain what it represents in terms of the dogs.

28. Ballooning A hot air balloon is 10 meters above the ground and rising at a rate of 15 meters per minute. Another balloon is 150 meters above the ground and descending at a rate of 20 meters per minute.
  a. Write a system of equations to represent the balloons.
  b. What is the solution of the system of equations?
  c. Explain what the solution means.

29. Critical Thinking Write an equation of a line in slope-intercept form that, together with the equation $x + 3y = 9$, forms a system that is inconsistent.

Mixed Review

30. What is the solution of the system $y = x + 4$ and $y = -3x - 4$? *(Lesson 13–1)*

31. Graph $y \geq 2x + 1$. *(Lesson 12–7)*

32. Number Theory Find two numbers if one number is 6 less than the other and whose product is 7. *(Lesson 11–5)*

Factor each polynomial. If the polynomial cannot be factored, write prime. *(Lesson 10–2)*

33. $4x - 8$

34. $13x + 2m$

35. $3a^2b^2 + 6ab - 9a$

Standardized Test Practice
Ⓐ Ⓑ Ⓒ Ⓓ

36. Short Response Write a square root whose best whole number estimate is 12. *(Lesson 8–6)*

What You'll Learn

You'll learn to solve systems of equations by the substitution method.

Why It's Important

Metallurgy Systems of equations can be used to make metal alloys.
See Example 6.

Look Back

Solving Equations with Algebra Tiles: Lesson 3–5

In Lesson 13–2, you learned to solve systems of equations by graphing. But sometimes the exact coordinates of the point where lines intersect cannot be easily determined from a graph. The solution of a system can also be found by using an algebraic method called **substitution**.

Hands-On Algebra
Algebra Tiles

Materials: algebra tiles equation mat

Use a model to solve the system $y = x + 1$ and $3x + y = 9$.

Step 1 Let a green tile represent x. Then, since $y = x + 1$, 1 green tile and 1 yellow tile represent y.

$$x = \boxed{x} \qquad y = \boxed{x}\ \boxed{1}$$

Step 2 Represent $3x + y = 9$ on the equation mat. On one side of the mat, place three green tiles for $3x$ and 1 green tile and 1 yellow tile for y. On the other side, place 9 yellow tiles.

$$3x + y \quad = \quad 9$$

Try These

1. Use what you know about equation mats and zero pairs to solve the equation. What is the value of x?
2. Use the value of x from Exercise 1 and the equation $y = x + 1$ to find the value of y.
3. What is the solution of the system of equations?

Use substitution to solve each system of equations.

1 $y = 2x$
$3x + y = 5$

The first equation tells you that y is equal to $2x$. So, substitute $2x$ for y in the second equation. Then solve for x.

$3x + y = 5$ *Original equation*
$3x + 2x = 5$ *Replace y with $2x$.*
$5x = 5$ *Simplify.*
$\dfrac{5x}{5} = \dfrac{5}{5}$ *Divide each side by 5.*
$x = 1$ *Simplify.*

Now substitute 1 for x in either equation and solve for y. *Choose the equation that is easier to solve.*

$y = 2x$ *Original equation*
$y = 2(1)$ or 2   *Replace x with 1.*

The solution of this system of equations is (1, 2). You can see from the graph that the solution is correct. *You can also check by substituting (1, 2) into each of the original equations.*

2 $x + y = 1$
$x = y + 6$

Substitute $y + 6$ for x in the first equation. Then solve for y.

$x + y = 1$ *Original equation*
$y + 6 + y = 1$ *Replace x with $y + 6$.*
$2y + 6 = 1$ *Simplify.*
$2y + 6 - 6 = 1 - 6$ *Subtract 6 from each side.*
$2y = -5$ *Simplify.*
$\dfrac{2y}{2} = \dfrac{-5}{2}$ *Divide each side by 2.*
$y = -\dfrac{5}{2}$ *Simplify.*

Now substitute $-\dfrac{5}{2}$ for y in either equation and solve for x.

$x = y + 6$ *Original equation*
$x = -\dfrac{5}{2} + 6$ *Replace y with $-\dfrac{5}{2}$.*
$x = \dfrac{7}{2}$ *Simplify.*

The solution of this system of equations is $\left(\dfrac{7}{2}, -\dfrac{5}{2}\right)$. *Check by substituting $\left(\dfrac{7}{2}, -\dfrac{5}{2}\right)$ into each of the original equations.*

 www.algconcepts.com/extra_examples

$x - 3y = 3$
$2x - y = 11$

Solve the first equation for x since the coefficient of x is 1.
$$x - 3y = 3 \quad \rightarrow \quad x = 3 + 3y$$

Next, find the value of y by substituting $3 + 3y$ for x in the second equation.
$$2x - y = 11$$
$$2(3 + 3y) - y = 11$$
$$6 + 6y - y = 11$$
$$6 + 5y = 11$$
$$6 + 5y - 6 = 11 - 6$$
$$5y = 5$$
$$\frac{5y}{5} = \frac{5}{5}$$
$$y = 1$$

Now substitute 1 for y in either equation and solve for x.
$$x - 3y = 3$$
$$x - 3(1) = 3 \quad \textit{Replace y with 1.}$$
$$x - 3 = 3$$
$$x - 3 + 3 = 3 + 3$$
$$x = 6$$

The solution is (6, 1).

Your Turn

a. $x = 1 + 6y$
$\quad x + 2y = 9$

b. $x + y = 3$
$\quad -2x - 7y = 4$

In Lesson 13–2, you learned how to tell whether a system has one solution, no solution, or infinitely many solutions by looking at the graph. You can also determine this information algebraically.

Use substitution to solve each system of equations.

❹ $y = 4x + 1$
$4x - y = 7$

Find the value of x by substituting $4x + 1$ for y in the second equation.
$$4x - y = 7 \quad \textit{Original equation}$$
$$4x - (4x + 1) = 7 \quad \textit{Replace y with 4x + 1.}$$
$$4x - 4x - 1 = 7 \quad \textit{Distributive Property}$$
$$-1 = 7 \quad \textit{Simplify.}$$

The statement $-1 = 7$ is false. This means that there are no ordered pairs that are solutions to both equations. Compare the slope-intercept forms of the equations, $y = 4x + 1$ and $y = 4x - 7$. Notice that the graphs of these equations have the same slope but different y-intercepts. Thus, the lines are parallel, and the system has no solution.

⑤ $x = 3 - 2y$
$2x + 4y = 6$

$$\begin{aligned}
2x + 4y &= 6 & \textit{Original equation} \\
2(3 - 2y) + 4y &= 6 & \textit{Replace x with } 3 - 2y. \\
6 - 4y + 4y &= 6 & \textit{Distributive Property} \\
6 &= 6 & \textit{Simplify.}
\end{aligned}$$

The statement $6 = 6$ is true. This means that an ordered pair for any point on either line is a solution to both equations. The system has infinitely many solutions.

Your Turn

Use substitution to solve each system of inequalities.

c. $y = 2x + 1$
$4x - 2y = -9$

d. $x - 6y = 5$
$2x = 12y + 10$

Systems of equations can be used to solve mixture problems.

Example ⑥

Metals Link

A certain metal alloy is 25% copper. Another metal alloy is 50% copper. How much of each alloy should be used to make 1000 grams of a metal alloy that is 45% copper?

Explore   Let a = the number of grams of the 25% copper alloy.
Let b = the number of grams of the 50% copper alloy.

	25% Copper	50% Copper	45% Copper
Total Grams	a	b	1000
Grams of Copper	$0.25a$	$0.50b$	$0.45(1000)$

Prerequisite Skills Review
Decimals and Percents, p. 689

Plan   Write two equations to represent the information.

$a + b = 1000$ ← *total grams of alloy*
$0.25a + 0.50b = 0.45(1000)$ ← *grams of copper*

Solve   Use substitution to solve this system. Since $a + b = 1000$, $a = 1000 - b$.

$$\begin{aligned}
0.25a + 0.50b &= 0.45(1000) & \textit{Original equation} \\
0.25(1000 - b) + 0.50b &= 0.45(1000) & \textit{Replace a with } 1000 - b. \\
250 - 0.25b + 0.50b &= 450 & \textit{Distributive Property} \\
250 + 0.25b &= 450 & \textit{Simplify.} \\
250 + 0.25b - 250 &= 450 - 250 & \textit{Subtract 250.} \\
0.25b &= 200 & \textit{Simplify.} \\
\frac{0.25b}{0.25} &= \frac{200}{0.25} & \textit{Divide each side by 0.25.} \\
b &= 800 & \textit{Simplify.}
\end{aligned}$$

(continued on the next page)

Now substitute 800 for b in either equation and solve for a.

$$a + b = 1000 \qquad \textit{Original equation}$$
$$a + 800 = 1000 \qquad \textit{Replace b with 800.}$$
$$a + 800 - 800 = 1000 - 800 \qquad \textit{Subtract 800 from each side.}$$
$$a = 200 \qquad \textit{Simplify.}$$

So, 200 grams of the 25% copper alloy and 800 grams of the 50% copper alloy should be used.

Examine   There should be more of the 50% alloy than the 25% alloy to make a 45% alloy. So the solution (200, 800) is reasonable.

🌐nline **Personal Tutor at** algconcepts.com

Check for Understanding

Communicating Mathematics

1. **Explain** when you might use substitution instead of graphing to solve a system of equations.

┌─── *Vocabulary* ───┐
│   substitution   │
└──────────────────┘

2. **State** what you would conclude if the solution of a system of equations yields the equation $4 = 4$.

3.   Faith and Todd are using substitution to solve the system $x + 3y = 8$ and $4x - y = 9$. Faith says that the first step is to solve for x in $x + 3y = 8$. Todd disagrees. He says to solve for y in $4x - y = 9$. Who is right and why?

Guided Practice

Examples 1–5

Use substitution to solve each system of equations.

4. $y = x - 4$
   $3x + 2y = 2$

5. $y = 3x$
   $7x - y = 16$

6. $x = 2y$
   $4x + 2y = 15$

7. $3x - 7y = 12$
   $x - 2y = 4$

8. $x = 2y + 5$
   $3x - 6y = 15$

9. $4y = -3x + 8$
   $3x + 4y = 6$

Example 6

10. **Mixtures**   MX Labs needs to make 500 gallons of a 34% acid solution. The only solutions available are 25% acid and 50% acid. How many gallons of each should be mixed to make the 34% solution?

Exercises   • • • • • • • • • • • • • • • • • •

Practice

Use substitution to solve each system of equations.

┌─────────────────────────┐
│   **Homework Help**   │
├───────────┬─────────────┤
│ For   │ See   │
│ Exercises │ Examples   │
├───────────┼─────────────┤
│ 11, 13, 15–17, 31 │ 1, 4 │
├───────────┼─────────────┤
│ 12, 14, 18 │ 2, 5   │
├───────────┼─────────────┤
│ 19–30   │ 3   │
├───────────┴─────────────┤
│   **Extra Practice**   │
│   See page 718.   │
└─────────────────────────┘

11. $y = x$
    $3x + y = 4$

12. $x = 2y$
    $x + y = 3$

13. $y = 3x$
    $x + 2y = -21$

14. $2y = x$
    $x - y = 10$

15. $y = 3x - 8$
    $3x - y = 12$

16. $y = x + 7$
    $x + y = 1$

17. $y = \frac{1}{2}x + 3$
    $y - 5 = x$

18. $x = 7 - y$
    $2x - y = 8$

19. $x - 3y = -9$
    $5x - 2y = 7$

20. $x + y = 0$
    $4x + 4y = 0$

21. $x - 2y = -2$
    $5x - 4y = 2$

22. $2x - y = 6$
    $3x - 5y = 9$

23. $x - y = 12$
$3x - y = 16$

24. $4x - 3y = -6$
$x + 5y = 10$

25. $x - 3y = 3$
$2x + 9y = 11$

26. $4x - y = -3$
$y + 2 = x$

27. $x - 6y = 5$
$2x - 12y = 10$

28. $x - 5y = 11$
$3x + y = 7$

29. Use substitution to solve $2x - y = -4$ and $-3x + y = -9$.

30. What is the solution of the system $x - 2y = 5$ and $3x - 5y = 8$?

Applications and Problem Solving

31. Driving Anita and Tionna were both driving the same route from college to their hometown. Anita left an hour before Tionna. Anita drove at an average speed of 55 miles per hour, and Tionna averaged 65 miles per hour.

a. After how many hours did Tionna catch up with Anita?

b. How many miles did she drive?

32. Exercise Laura and Ji-Yong were jogging on a 10-mile path. Laura had a 2-mile head start on Ji-Yong. If Laura ran at an average rate of 5 miles per hour and Ji-Yong ran at an average rate of 8 miles per hour, how long would it take for Ji-Yong to catch up with Laura?

33. Critical Thinking Suppose you have three equations A, B, and C. Suppose a system containing any two of the equations is consistent and independent. Must the three equations together be consistent? Give an example or a counterexample to support your answer.

Mixed Review

34. Determine whether the system $y = -2x$ and $2y + 4x = 0$ has *one* solution, *no* solution, or *infinitely many* solutions by graphing. If the system has one solution, name it. *(Lesson 13–2)*

35. Solve $x - y = -5$ and $x + y = 3$ by graphing. *(Lesson 13–1)*

Write a compound inequality for each solution set shown. *(Lesson 12–5)*

36.   **37.**

$$-5\ -4\ -3\ -2\ -1\ \ 0\ \ 1\ \ 2\ \ 3\ \ 4\ \ 5 \qquad\qquad -5\ -4\ -3\ -2\ -1\ \ 0\ \ 1\ \ 2\ \ 3\ \ 4\ \ 5$$

Standardized Test Practice
(A) (B) (C) (D)

38. Short Response Solve $-5 + x > 12.8$. *(Lesson 12–2)*

39. Grid In Mr. Drew is a salesperson. The formula for his daily income is $t(s) = 0.15s + 25$, where s is his total sales for the day. Suppose Mr. Drew earns $94 on Monday. What were his total sales in dollars for that day? *(Lesson 6–4)*

Quiz 1   Lessons 13–1 through 13–3

Determine whether each system of equations has *one* solution, *no* solution, or *infinitely many* solutions by graphing. If the system has one solution, name it.
(Lessons 13–1 & 13–2)

1. $y = x - 5$
$y = -\dfrac{1}{4}x$

2. $y = 2x + 5$
$y = x + 3$

3. $x + y = 3$
$x + y = 1$

4. $y = -2x + 3$
$2y = -4x + 6$

5. Mixtures Ms. Williams mixed nuts that cost $3.90 per pound with nuts that cost $4.30 per pound to make a 50-pound mixture of nuts worth $4.20 per pound. How many pounds of each type of nut did she use? *(Lesson 13–3)*

13-4 Elimination Using Addition and Subtraction

What You'll Learn

You'll learn to solve systems of equations by the elimination method using addition and subtraction.

Why It's Important

Entertainment
Solving systems of equations using elimination could be used to determine entertainment costs. *See Exercise 35.*

Another algebraic method for solving systems of equations is called **elimination**. You can eliminate one of the variables by adding or subtracting the equations.

Use elimination when the coefficients of one of the variables are equal or additive inverses. For example, consider the system below.

$$5x + 11y = 12 \qquad 2x + 11y = 36$$

Notice that the coefficient of y in both equations is the same. You can solve this system of equations in three steps.

Step 1: Subtract the two equations, so that y is eliminated. The result is an equation only in x: $3x = -24$.

Step 2: The next step is to solve for x: $x = -8$.

Step 3: Substitute the value of x in any one of the two original equations and solve for y: $y = \dfrac{52}{11}$.

Example ① **Use elimination to solve the system of equations.**

$2x + y = 3$
$x + y = 1$

$$
\begin{array}{ll}
2x + y = 3 & \textit{Write the equations in column form.} \\
\underline{(-)\ x + y = 1} & \textit{Subtract the equations to eliminate the y terms.} \\
x + 0 = 2 & \\
x = 2 & \textit{The value of x is 2.}
\end{array}
$$

Now substitute in either equation to find the value of y. *Choose the equation that is easier for you to solve.*

$$
\begin{array}{ll}
x + y = 1 & \textit{Original equation} \\
2 + y = 1 & \textit{Replace x with 2.} \\
2 + y - 2 = 1 - 2 & \textit{Subtract 2 from each side.} \\
y = -1 & \textit{The value of y is } -1.
\end{array}
$$

The solution of the system of equations is $(2, -1)$.

$$
\begin{array}{lll}
\textbf{Check:} & 2x + y = 3 & \textit{Original equation} \\
& 2(2) + (-1) \stackrel{?}{=} 3 & \textit{Replace } (x, y) \textit{ with } (2, -1). \\
& 4 + (-1) \stackrel{?}{=} 3 & \textit{Multiply.} \\
& 3 = 3 \ \checkmark & \\
& & \\
& x + y = 1 & \textit{Original equation} \\
& 2 + (-1) \stackrel{?}{=} 1 & \textit{Replace } (x, y) \textit{ with } (2, -1). \\
& 1 = 1 \ \checkmark &
\end{array}
$$

Your Turn Use elimination to solve each system of equations.

a. $x + y = 5$
$2x + y = 4$

b. $3x + 5y = -2$
$3x - 2y = -16$

Example **2**

Entertainment Link

A group of 3 adults and 10 students paid $102 for a cavern tour. Another group of 3 adults and 7 students paid $84 for the tour. Find the admission price for an adult and for a student.

Let a = the price for an adult and s = the price for a student.

Write two equations to represent this situation.

$3a + 10s = 102$ ← *total cost for the first van*
$3a + 7s = 84$ ← *total cost for the second van*

$\;\; 3a + 10s = 102$ *Write the equations in column form.*
$\underline{(-)\, 3a + 7s = 84}$ *Subtract the equations to eliminate the a terms.*
$\;\; 0 + 3s = 18$
$ 3s = 18$ *Simplify.*
$ \dfrac{3s}{3} = \dfrac{18}{3}$ *Divide each side by 3.*
$ s = 6$ The value of s is 6.

Now substitute in either equation to find the value of a.

$3a + 7s = 84$ *Original equation*
$3a + 7(6) = 84$ *Replace s with 6.*
$3a + 42 = 84$ *Multiply.*
$3a + 42 - 42 = 84 - 42$ *Subtract 42 from each side.*
$3a = 42$ *Simplify.*
$\dfrac{3a}{3} = \dfrac{42}{3}$ *Divide each side by 3.*
$a = 14$ The value of a is 14.

The solution of the system of equations is (14, 6). This means that the cost for adults was $14 and the cost for students was $6.

Check: $3a + 10s = 102$ *Original equation*
$3(14) + 10(6) \stackrel{?}{=} 102$ *Replace (a, s) with (14, 6).*
$42 + 60 \stackrel{?}{=} 102$ *Multiply.*
$102 = 102$ $\checkmark$

$3a + 7s = 84$ *Original equation*
$3(14) + 7(6) \stackrel{?}{=} 84$ *Replace (a, s) with (14, 6).*
$42 + 42 \stackrel{?}{=} 84$ *Multiply.*
$84 = 84$ $\checkmark$

In some systems of equations, the coefficients of one of the variables are additive inverses. For these systems, apply the elimination method by adding the equations.

Example **3** **Use elimination to solve the system of equations.**

$$4x - 6y = 10$$
$$3x + 6y = 4$$

$$
\begin{array}{ll}
4x - 6y = 10 & \textit{$-6y$ and $+6y$ are additive inverses.} \\
\underline{(+)\ 3x + 6y = \ \ 4} & \textit{Add the equations to eliminate the y terms.} \\
7x + 0\ = 14 &
\end{array}
$$

$$
\begin{aligned}
7x &= 14 && \textit{Simplify.} \\
\frac{7x}{7} &= \frac{14}{7} && \textit{Divide each side by 7.} \\
x &= 2 && \text{The value of x is 2.}
\end{aligned}
$$

Now substitute x into either equation to find the value of y.

$$
\begin{aligned}
4x - 6y &= 10 && \textit{Original equation} \\
4(2) - 6y &= 10 && \textit{Replace x with 2.} \\
8 - 6y &= 10 && \textit{Multiply.} \\
8 - 6y - 8 &= 10 - 8 && \textit{Subtract 8 from each side.} \\
-6y &= 2 && \textit{Simplify.} \\
\frac{-6y}{-6} &= \frac{2}{-6} && \textit{Divide each side by -6.} \\
y &= -\frac{2}{6} && \textit{Simplify.} \\
y &= -\frac{1}{3} && \text{The value of y is $-\frac{1}{3}$.}
\end{aligned}
$$

The solution of the system is $\left(2, -\frac{1}{3}\right)$. *Check this result.*

Prerequisite Skills Review
Simplifying Fractions, p. 685

Your Turn

Use elimination to solve each system of equations.

c. $x + y = 8$
$x - y = -2$

d. $-4x + y = 15$
$3y = 5 - 4x$

Systems of equations can be used to solve **digit problems**. Digit problems explore the relationships between digits of a number.

Example **4**
Number Theory Link

The sum of the digits of a two-digit number is 8. If the tens digit is 4 more than the units digit, what is the number?

Let t represent the tens digit and let u represent the units digit.

$t + u = 8$ ← *the sum of the digits*
$t = u + 4$ ← *the relationship between the digits*

Rewrite the second equation so that the t and u are on the same side of the equation.

$$t = u + 4 \rightarrow t - u = 4$$

Then use elimination to solve.

$t + u = 8$ *Write the equations in column form.*
$\underline{(+)\ t - u = 4}$ *Add the equations to eliminate the u terms.*
$2t + 0 = 12$
$2t = 12$ *Simplify.*
$\dfrac{2t}{2} = \dfrac{12}{2}$ *Divide each side by 2.*
$t = 6$ The tens digit is 6.

Now substitute to find the units digit.

$t + u = 8$ *Original equation*
$6 + u = 8$ *Replace t with 6.*
$6 + u - 6 = 8 - 6$ *Subtract 6 from each side.*
$u = 2$ The units digit is 2.

Since t is 6 and u is 2, the number is 62. *Check this solution.*

 Personal Tutor at algconcepts.com

Check for Understanding

Communicating Mathematics

1. **Explain** when you would use elimination to solve a system of equations.

2. **Describe** the result when you add each pair of equations. What does the result tell you about the system of equations?

   a. $2x - y = 12$
   $-2x + y = 14$

   b. $x + 5y = 3$
   $-x - 5y = -3$

Guided Practice

⏱ Getting Ready   State whether *addition*, *subtraction*, *either*, or *neither* should be used to solve each system.

Sample: $3x + y = 6$
$4x + y = 7$

Solution: $3x + y = 6$
$\underline{(-)\ 4x + y = 7}$ Subtraction should be used because it
$-x = -1$ eliminates the *y* terms.

3. $x - 2y = 3$
$3x + 2y = 8$

4. $x + y = 5$
$x - y = 9$

5. $4x - 5y = 18$
$2x - 5y = 1$

6. $x + 7y = 3$
$2x + y = 4$

Use elimination to solve each system of equations.

7. $x + y = 7$
   $x - y = 9$

8. $x + y = 5$
   $2x + y = 4$

9. $x + 2y = 6$
   $3x - 2y = 2$

10. $-7x + 4y = 6$
    $7x + y = 19$

11. $3x - 2y = 10$
    $3x - 5y = 1$

12. $4x = 5y - 9$
    $4x + 3y = -1$

Example 4

13. **Number Theory** The sum of two numbers is 42. The greater number is three more than twice the lesser number.

    a. Write a system of equations to represent the problem.

    b. What are the numbers?

Exercises •

Practice

Use elimination to solve each system of equations.

14. $x + y = 3$
    $x - y = 3$

15. $x - y = 9$
    $x + y = 19$

16. $x + 4y = 6$
    $x + 3y = 5$

17. $x + y = 10$
    $3x + y = 0$

18. $3x + y = 13$
    $2x - y = 2$

19. $9x + 2y = 12$
    $7x - 2y = -12$

20. $-3x + 4y = 5$
    $3x - 2y = -7$

21. $2x - 3y = 19$
    $2x + 3y = 13$

22. $11x - 3y = 10$
    $-2x + 3y = 8$

23. $2x - y = -8$
    $x - y = 7$

24. $2x - 5y = 9$
    $2x - 3y = 11$

25. $3x - 2y = 10$
    $2x - 2y = 5$

26. $x + 2y = 8$
    $3x = 6 - 2y$

27. $x + y = 7$
    $21 - 3x = 3y$

28. $y = 3x - 5$
    $x - y = 13$

29. $x = y - 7$
    $2x - 5y = -2$

30. $5x - y = -3$
    $2y = 10x - 7$

31. $x = 10 - y$
    $2x - y = -4$

Homework Help

For Exercises	See Examples
14, 15, 18–25, 32	3
16, 17, 34, 35	1, 2
26–31, 33, 36, 37	4

Extra Practice
See page 719.

32. What is the solution of the system $3x + 5y = -16$ and $3x - 2y = -2$?

33. Find the solution of the system $5s + 4t = 12$ and $3s = 4 + 4t$.

Applications and Problem Solving

34. **Spas** The Feel Better Spa has two specials for new members. They can receive 3 facials and 5 manicures for $114 or 3 facials and 2 manicures for $78. What are the prices for facials and manicures?

35. **Entertainment** The cost of admission to Water World was $137.50 for 13 children and 2 adults in one party. The admission was $103.50 for 9 children and 2 adults in another party.

    a. Write a system of equations to represent this problem.

    b. How much is admission for children and adults?

36. **Number Theory** The difference between two numbers is 15. The greater number is two less than twice the lesser number.
 a. Write a system of equations to represent this situation.
 b. Find the numbers.

37. **Number Theory** The sum of a number and twice a second number is 29. The second number is ten less than three times the first number.
 a. Write a system of equations to represent this situation.
 b. Find both numbers.

38. **Critical Thinking** The solution of a system of equations is $(-2, 5)$, and the first equation is $3x + 4y = 14$.
 a. Write a second equation for this system.
 b. Is this the only equation that could be in the system? Explain.

Mixed Review

39. Use substitution to solve the system of equations. *(Lesson 13–3)*
$$y = -3x - 5$$
$$4x + y = 6$$

State whether each system is *consistent and independent*, *consistent and dependent*, or *inconsistent*. *(Lesson 13–2)*

40.

41.

42.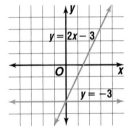

Solve each inequality. *(Lesson 12–4)*
43. $3y < 12$
44. $6 - 4m \leq 22$
45. $9 - 3x > 15$

Find each product. *(Lesson 9–4)*
46. $(m + 4)(m + 7)$
47. $(a - 3)(a - 3)$
48. $(2x - 2)(3x + 6)$

49. **Short Response** Write the point-slope form of an equation for a line passing through points $(6, 1)$ and $(7, -4)$. *(Lesson 7–2)*

50. **Multiple Choice**
In a scale model of a classic car, the front bumper measures 1.5 inches. If the actual bumper measures 3 feet, what is the scale of the model?
(Lesson 5–2)

Ⓐ 1 in. = 16 in. Ⓑ 1 in. = 18 in.
Ⓒ 1 in. = 24 in. Ⓓ 1 in. = 43 in.

13-5 Elimination Using Multiplication

What You'll Learn

You'll learn to solve systems of equations by the elimination method using multiplication and addition.

Why It's Important

Transportation
Systems of equations can be used to determine the rate of a river's current. See Example 5.

You have learned when and how to solve systems of equations by graphing, substitution, and elimination using addition or subtraction. The best times to use these methods are summarized in the table below.

Method	The Best Time to Use	Example
graphing	if the equations are easy to graph, or if you want an estimate because graphing usually does not give an exact solution	$y = x + 4$ $y = -x + 1$
substitution	if one of the variables in either equation has a coefficient of 1 or -1	$y = 2x + 3$ $3x - 5y = 8$
elimination using addition	if the coefficients of one variable are additive inverses	$4x - y = 9$ $-4x + 3y = 6$
elimination using subtraction	if the coefficients of one variable are the same	$2x + y = 4$ $3x + y = 1$

Sometimes neither of the variables in a system of equations can be eliminated by simply adding or subtracting the equations. In this case, another method is to multiply one or both of the equations by some number so that adding or subtracting eliminates one of the variables.

Examples

1 **Use elimination to solve the system of equations.**

$$2x + 3y = 9$$
$$8x - 5y = 19$$

Multiply the first equation by -4 so that the x terms are additive inverses.

$$2x + 3y = 9$$
$$8x - 5y = 19$$

 Multiply by –4.

$$-8x - 12y = -36$$
$$\underline{(+)\ 8x - \ 5y = \ \ \ \ 19}$$
$$0\ - 17y = -17$$
$$\frac{-17y}{-17} = \frac{-17}{-17}$$
$$y = 1$$

Now find the value of x by replacing y with 1 in either equation.

$2x + 3y = 9$	*Original equation*
$2x + 3(1) = 9$	*Replace y with 1.*
$2x + 3 = 9$	*Multiply.*
$2x + 3 - 3 = 9 - 3$	*Subtract 3 from each side.*
$2x = 6$	*Simplify.*
$\frac{2x}{2} = \frac{6}{2}$	*Divide each side by 2.*
$x = 3$	*Simplify.*

The solution of this system of equations is (3, 1). *Check this solution.*

2 Use elimination to solve the system of equations.

$5x - 6y = 25$

$4x + 2y = 3$

Multiply the second equation by 3 so that the y terms are additive inverses.

$5x - 6y = 25$
$4x + 2y = 3$ → **Multiply by 3.** →

$$5x - 6y = 25$$
$$(+)\ 12x + 6y = \ \ 9$$
$$\overline{17x + 0\ \ = 34}$$
$$\frac{17x}{17} = \frac{34}{17}$$
$$x = 2$$

Now find the value of y by replacing x with 2 in either equation.

$4x + 2y = 3$	*Original equation*
$4(2) + 2y = 3$	*Replace x with 2.*
$8 + 2y = 3$	*Multiply.*
$8 + 2y - 8 = 3 - 8$	*Subtract 8 from each side.*
$2y = -5$	*Simplify.*
$\dfrac{2y}{2} = \dfrac{-5}{2}$	*Divide each side by 2.*
$y = -\dfrac{5}{2}$	*Simplify.*

The solution of the system of equations is $\left(2, -\dfrac{5}{2}\right)$. *Check this solution.*

Your Turn

a. $2x - y = 16$
$\ \ 5x + 3y = -4$

b. $4x - 3y = -2$
$\ \ 2x + 7y = 16$

c. $2x - 6y = -8$
$\ \ 4x - 3y = 11$

Real World

Example

Transportation Link

3 Chris and Alana both take the Metro train to work. In May, Chris took the train 15 times during rush hour and 29 times during non-rush hour for $64.80. Alana took the train 30 times during rush hour and 14 times during non-rush hour for $76.80. What are the rush hour and non-rush hour fares?

r = the rush hour fare $\qquad\qquad$ n = the non-rush hour fare

$15r + 29n = 64.80$ ← *Chris' expenses*

$30r + 14n = 76.80$ ← *Alana's expenses*

Multiply the first equation by -2 to eliminate the r terms.

Prerequisite Skills Review

Operations with Decimals, p. 684

$15r + 29n = 64.80$
$30r + 14n = 76.80$ → **Multiply by –2.** →

$$-30r - 58n = -129.60$$
$$(+)\ 30r + 14n = \ \ \ \ 76.80$$
$$\overline{0 - 44n = \ -52.80}$$
$$\frac{-44n}{-44} = \frac{-52.80}{-44}$$
$$n = 1.20$$

(continued on the next page)

Now find the value of *r* by replacing *n* with 1.20 in either equation.

$$15r + 29n = 64.80 \qquad \textit{Original equation}$$
$$15r + 29(1.20) = 64.80 \qquad \textit{Replace n with 1.20.}$$
$$15r + 34.80 = 64.80 \qquad \textit{Multiply.}$$
$$15r + 34.80 - 34.80 = 64.80 - 34.80 \qquad \textit{Subtract 34.80 from each side.}$$
$$15r = 30 \qquad \textit{Simplify.}$$
$$\frac{15r}{15} = \frac{30}{15} \qquad \textit{Divide each side by 15.}$$
$$r = 2 \qquad \textit{Simplify.}$$

The solution is (2, 1.20). This means that the rush hour fare is $2.00 and the non-rush hour fare is $1.20. Do these fares seem reasonable? *Check this solution by substituting (2, 120) for (r, n) in the original equations.*

🌐nline Personal Tutor at algconcepts.com

Sometimes it is necessary to multiply each equation by a different number and then add in order to eliminate one of the variables.

Example ④ **Use elimination to solve the system of equations.**
$$3x + 4y = -25$$
$$2x - 3y = 6$$

Multiply the first equation by 2 and the second equation by -3 so that the *x* terms are additive inverses.

$$3x + 4y = -25 \quad \boxed{\text{Multiply by 2.}}\blacktriangleright \quad 6x + 8y = -50$$

$$2x - 3y = 6 \quad \boxed{\text{Multiply by –3.}}\blacktriangleright \quad \underline{(+)-6x + 9y = -18}$$
$$0 + 17y = -68$$
$$\frac{17y}{17} = \frac{-68}{17}$$
$$y = -4$$

Now find the value of *x* by replacing *y* with -4 in either equation.

$$3x + 4y = -25 \qquad \textit{Original equation}$$
$$3x + 4(-4) = -25 \qquad \textit{Replace y with } -4.$$
$$3x - 16 = -25 \qquad \textit{Multiply.}$$
$$3x - 16 + 16 = -25 + 16 \qquad \textit{Add 16 to each side.}$$
$$3x = -9 \qquad \textit{Simplify.}$$
$$\frac{3x}{3} = \frac{-9}{3} \qquad \textit{Divide each side by 3.}$$
$$x = -3 \qquad \textit{Simplify.}$$

The solution of the system of equations is $(-3, -4)$. You can also solve this system of equations by multiplying the first equation by 3 and the second equation by 4. *Why?*

Use elimination to solve each system of equations.

d. $5x + 3y = 12$
  $4x - 5y = 17$

e. $4x + 3y = 19$
  $3x - 4y = 8$

Systems of equations can be used to solve rate problems.

Real World

Example ➎

Transportation Link

A barge on the Mississippi River travels 36 miles upstream in 6 hours. The return trip takes the barge only 4 hours. Find the rate of the current.

Explore Let r represent the rate of the barge in still water.
Let c represent the rate of the current.

The rate of the barge traveling downstream *with* the current is $r + c$, and the rate of the barge traveling upstream *against* the current is $r - c$.

Plan Use the formula distance = rate × time, or $d = rt$, to write a system of equations. Then solve the system to find c.

	d	r	t	d = rt	
Downstream	36	$r + c$	4	$36 = 4r + 4c$	$4(r + c) = 4r + 4c$
Upstream	36	$r - c$	6	$36 = 6r - 6c$	$6(r - c) = 6r - 6c$

Solve $4r + 4c = 36$ **Multiply by 3.** ➤

$6r - 6c = 36$ **Multiply by -2.** ➤

$$12r + 12c = 108$$
$$(+)\; -12r + 12c = -72$$
$$0 + 24c = 36$$
$$24c = 36$$
$$\frac{24c}{24} = \frac{36}{24}$$
$$c = 1.5$$

The rate of the current is 1.5 miles per hour.

Examine Find the value of r for this system and then check the solution.

Barge on the Mississippi

Check for Understanding

Communicating Mathematics

1. **Explain** when you would use elimination with multiplication to solve a system of equations.

2. **Write** a system of equations in which you can eliminate the variable x by multiplying one equation by 3 and then adding the equations.

3. **Match** each system of equations with the method listed below that would best solve the system.

a. $y = 3x$
   $6x - 2y = 9$

b. $2x + 4y = 5$
   $3x - 2y = -3$

c. $10x - 2y = 12$
   $-10x + 3y = -13$

d. $x + 7y = 5$
   $3x + 7y = 1$

substitution
elimination (+)
elimination (−)
elimination (×)

Guided Practice

⏱ **Getting Ready** Explain the steps you would take to eliminate the variable *y* in each system of equations.

Sample: $4x + 3y = -7$ **Solution:** Multiply second equation by −3.
        $3x + y = 1$ Then add.

4. $2x + 6y = 10$
   $5x + 3y = 1$

5. $3x - 2y = 3$
   $8x + y = 27$

6. $2x + y = 4$
   $7x - 4y = 29$

Examples 1–4

Use elimination to solve each system of equations.

7. $x + 2y = 6$
   $-4x + 5y = 2$

8. $x - 5y = 0$
   $2x - 3y = 7$

9. $5x + 8y = 1$
   $2x - 4y = -14$

10. $x + 2y = 5$
    $3x + y = 7$

11. $3x - y = -5$
    $6x - 2y = 8$

12. $2x - 5y = -10$
    $7x - 3y = -6$

Example 5

13. **Recreation** A fishing boat traveled 48 miles upstream in 4 hours. Returning at the same rate, it took 3 hours.

a. Find the rate of the current.

b. Find the rate of the boat in still water.

Exercises

• • • • •   • • • • • • • • • • • • • •

Practice

Use elimination to solve each system of equations.

14. $x + 8y = 3$
    $2x + 3y = -7$

15. $4x + y = 8$
    $x - 7y = 2$

16. $2x + y = 6$
    $3x - 7y = 9$

17. $2x + 5y = 13$
    $4x - 3y = -13$

18. $-3x + 2y = 10$
    $-2x - y = -5$

19. $3x - 2y = 0$
    $x + 6y = 5$

20. $-5x + 8y = 21$
    $10x + 3y = 15$

21. $6x - 4y = 11$
    $2x + 2y = 7$

22. $6x - 5y = -6$
    $3x + 10y = 72$

23. $2x - 7y = 9$
    $-3x + 4y = 6$

24. $5x + 3y = 4$
    $-4x + 5y = -18$

25. $6x - 3y = 7$
    $18x - 9y = 21$

26. $7x - 4y = 16$
    $2x + 3y = 17$

27. $-5x - 2y = 12$
    $11x - 5y = -17$

28. $7x - 3y = -9$
    $5x - 4y = -25$

Homework Help	
For Exercises	See Examples
14, 15, 17, 20, 29	1
16, 18, 30	2
23, 24, 26–28	4
34	5
35	3
Extra Practice	
See page 719.	

29. What is the solution of the system $9x + 8y = 7$ and $18x - 15y = 14$?

30. Use elimination to find the solution of the system $\frac{1}{3}x - y = -1$ and $\frac{1}{5}x - \frac{2}{5}y = -1$.

Determine the best method to solve each system of equations. Then solve.

31. $x + 2y = 6$
$2x - 4y = -20$

32. $y = 3x - 2$
$x - 5y = -4$

33. $y = \frac{1}{2}x + 6$
$2x - 4y = 8$

Applications and Problem Solving

34. Travel   A riverboat on the Mississippi River travels 30 miles upstream in 2 hours and 30 minutes. The return trip downstream takes only 2 hours.
   a. Find the rate of the current.
   b. Find the rate of the riverboat in still water.

35. Entertainment   The science club purchased tickets for a magic show. They paid $108 for 6 tickets in section A and 10 tickets in section B. The following week, they paid $104 for 4 tickets in section A and 12 tickets in section B.
   a. Write a system of equations to represent the problem.
   b. What are the prices of the tickets in section A and B?

36. Critical Thinking   The solution of the system $5x + 6y = -9$ and $10x + 8y = c$ is $(9, b)$. Find values for c and b.

Mixed Review

Use elimination to solve each system of equations.   *(Lesson 13–4)*

37. $x - y = 8$
$x + y = 6$

38. $x + y = -4$
$4x + y = 2$

39. $2x + 4y = 7$
$2x - y = 2$

40. Use substitution to solve $x = 5 + 2y$ and $3x - 4y = 3$.   *(Lesson 13–3)*

Solve each inequality.   *(Lesson 12–3)*

41. $-4m > 16$

42. $\frac{x}{2} < 6$

43. $\frac{3a}{4} \geq -3$

44. Use the Quadratic Formula to solve $n^2 - 5n + 12 = 0$.   *(Lesson 11–6)*

45. Grid In   Suppose a baseball player hits a fly ball into right field above first base. The player hits the ball with his bat at a height of 2 feet above the ground and sends the ball upward at a velocity of 25 meters per second. The height of the ball t seconds after the hit can be approximated by the formula $h = -5t^2 + 25t + 2$. If the ball is not caught, how many seconds will it take to hit the ground? Round to the nearest second.   *(Lesson 11–3)*

46. Multiple Choice   The graph shows the top turkey producing states. What percent of turkeys does the state of North Carolina produce? Round to the nearest percent.
(Lesson 5–3)

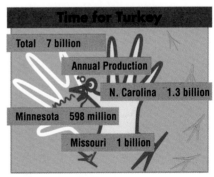

Time for Turkey

Total   7 billion
Annual Production
N. Carolina   1.3 billion
Minnesota   598 million
Missouri   1 billion

Source: U.S. Census Bureau

   Ⓐ 5%
   Ⓑ 19%
   Ⓒ 33%
   Ⓓ 45%

Chapter 13 Investigation

Crazy Computers

Matrices

In this investigation, you will use matrices to solve systems of equations. Consider the following problem.

Look Back

Matrices:
Pages 80–81

Kristin and Scott are sales associates at Crazy Computers. On Monday, Kristin sold 4 Model A computers and 1 Model B computer for a total of $6395. Scott sold 2 Model A computers and 3 Model B computers for a total of $7195. What was the price of each computer?

Investigate

You can write a system of equations to solve this problem. You may want to use matrices to solve this system of equations.

1. Write each equation in standard form. Then write the coefficients and constants of the system in a special matrix called an **augmented matrix**. Write the coefficients in the matrix to the left of the dashed line. Write the constants to the right of the dashed line.

$$3x + 5y = 7$$
$$6x - 1y = -8$$
$$\Rightarrow \begin{bmatrix} 3 & 5 & | & 7 \\ 6 & -1 & | & -8 \end{bmatrix}$$

   a. Write $2x - 3y = 5$ and $x + 4y = -7$ as an augmented matrix.
   b. Write $x + 6y = -1$ and $y = 5$ as an augmented matrix.

2. You can use **row operations** to simplify an augmented matrix.

   • Switch any two rows.
   • Replace any row with a nonzero multiple of that row.
   • Replace any row with the sum or difference of that row and a multiple of another row.

   a. Switch rows 1 and 2 of $\begin{bmatrix} 0 & -5 & | & 2 \\ 1 & 3 & | & 4 \end{bmatrix}$.

   b. Multiply row 2 of $\begin{bmatrix} 1 & -3 & | & 5 \\ 0 & 2 & | & -8 \end{bmatrix}$ by 0.5.

   c. Replace row 2 of $\begin{bmatrix} 1 & -2 & | & -5 \\ -1 & 3 & | & 4 \end{bmatrix}$ with the sum of rows 1 and 2.

   d. In $\begin{bmatrix} 1 & -3 & | & 2 \\ 0 & 3 & | & -5 \end{bmatrix}$, what row operation would you use to get a 0 in the second column of row 1?

3. The goal of using row operations to solve a system of equations is to get to the **identity matrix**, which is $\begin{bmatrix} 1 & 0 \\ 0 & 1 \end{bmatrix}$. Suppose the solution of a system of equations is $\begin{bmatrix} 1 & 0 & -4 \\ 0 & 1 & 2 \end{bmatrix}$. What does this matrix represent?

4. Solve the problem. Let x = the cost of the Model A computer and let y = the cost of the Model B computer.

$$4x + 1y = 6395 \quad \leftarrow Kristin's\ sales$$
$$2x + 3y = 7195 \quad \leftarrow Scott's\ sales$$

Step 1 Write the system as an augmented matrix.

$$\begin{bmatrix} 4 & 1 & 6395 \\ 2 & 3 & 7195 \end{bmatrix}$$

Step 2 Multiply row 2 by 2.
$2 \times 2 = 4; 2 \times 3 = 6; 2 \times 7195 = 14{,}390$

$$\begin{bmatrix} 4 & 1 & 6395 \\ 4 & 6 & 14{,}390 \end{bmatrix}$$

Step 3 Replace row 2 with the difference of row 1 and row 2. $4 - 4 = 0; 1 - 6 = -5;$ $6395 - 14{,}390 = -7995$

$$\begin{bmatrix} 4 & 1 & 6395 \\ 0 & -5 & -7995 \end{bmatrix}$$

Step 4 Divide row 2 by -5. $0 \div (-5) = 0;$ $-5 \div (-5) = 1; -7995 \div (-5) = 1599$

$$\begin{bmatrix} 4 & 1 & 6395 \\ 0 & 1 & 1599 \end{bmatrix}$$

Step 5 Replace row 1 with the difference of row 1 and row 2. $4 - 0 = 4; 1 - 1 = 0;$ $6395 - 1599 = 4796$

$$\begin{bmatrix} 4 & 0 & 4796 \\ 0 & 1 & 1599 \end{bmatrix}$$

Step 6 Divide row 1 by 4.
$4 \div 4 = 1; 0 \div 4 = 0; 4796 \div 4 = 1199$

$$\begin{bmatrix} 1 & 0 & 1199 \\ 0 & 1 & 1599 \end{bmatrix}$$

This means that Model A computer is $1199 and Model B computer is $1599.

Extending the Investigation

In this extension, you will use matrices as a problem-solving tool.

1. Suppose Scott sells 5 Model TX laser printers and 6 Model DM ink jet printers for a total of $4185. Kristin sells 6 Model TX laser printers and 5 Model DM ink jet printers for a total of $4549. What is the cost of each printer?

2. At Crazy Computers, there are two prices for computer software, Price A and Price B. Suppose Aislyn purchases 3 Price A software and 1 Price B software for a total of $250 and Devin purchases 5 Price A software and 2 Price B software for a total of $450. What are the two prices for the software?

Presenting Your Investigation

Write a real-world problem that can be solved by using matrices and a system of equations. Present your problem and solution on a poster board.

*inter*NET
CONNECTION **Investigation** For more information on matrices, visit: www.algconcepts.com

In late November, the jet stream moving across North America could be described by the quadratic equation $y = \frac{1}{4}x^2 - 12$, where Chicago was at the origin. Suppose a plane's route is described by the linear equation $y = -\frac{1}{2}x$. What are the coordinates of the point at which turbulence will occur? *This problem will be solved in Example 7.*

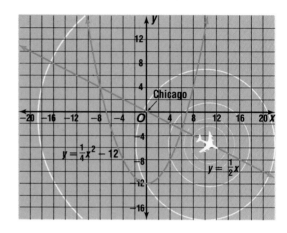

Like a linear system of equations, the solution of a **quadratic-linear** system of equations is the ordered pair that satisfies both equations. A quadratic-linear system can have 0, 1, or 2 solutions, as shown below.

Quadratic–Linear Systems

no solution
graphs do not intersect

one solution
graphs intersect at one point

two solutions
graphs intersect at two points

You can solve quadratic-linear systems of equations by using some of the methods you used for solving systems of linear equations. One method is graphing.

Examples

Look Back

Graphing Quadratic Equations: Lesson 11–1

Determine whether each system of equations has *one* solution, *two* solutions, or *no* solution by graphing. If the system has one solution or two solutions, name them.

1 $y = x^2$
$y = x + 2$

The graphs appear to intersect at $(-1, 1)$ and $(2, 4)$. Check this estimate by substituting the coordinates into each equation.

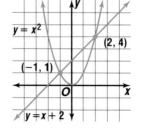

Check: $y = x^2$ $\qquad\qquad\qquad\qquad\qquad$ $y = x + 2$

$1 \overset{?}{=} (-1)^2$ $\quad (x, y) = (-1, 1)$ $\qquad$ $1 \overset{?}{=} -1 + 2$ $\quad (x, y) = (-1, 1)$

$1 = 1$ $\checkmark$ $\qquad\qquad\qquad\qquad\qquad$ $1 = 1$ $\checkmark$

Check that the ordered pair (2, 4) satisfies both equations.

The solutions of the system of equations are $(-1, 1)$ and $(2, 4)$.

2 $y = 2x^2 + 5$

$y = -x + 3$

Because the graphs do not intersect, there is no solution to this system of equations.

3 $y = -x^2 + 3$

$y = 2x + 4$

The graphs appear to intersect at $(-1, 2)$.

Check: $y = -x^2 + 3$ $\qquad\qquad\qquad$ $y = 2x + 4$

$2 \overset{?}{=} -(-1)^2 + 3$ $\;(x, y) = (-1, 2)$ $\quad$ $2 \overset{?}{=} 2(-1) + 4$ $\;(x, y) = (-1, 2)$

$2 \overset{?}{=} -(1) + 3$ $\quad$ *Simplify.* $\qquad$ $2 \overset{?}{=} -2 + 4$ $\qquad$ *Simplify.*

$2 = 2$ $\checkmark$ $\qquad\qquad\qquad\qquad\qquad$ $2 = 2$ $\checkmark$

The solution of the system of equations is $(-1, 2)$.

Your Turn

a. $y = x^2$ $\qquad\qquad$ **b.** $y = -2x^2 + 1$ $\qquad\qquad$ **c.** $y = -x^2 + 3$

$\quad\;\; y = x - 2$ $\qquad\qquad\qquad\; y = 1$ $\qquad\qquad\qquad\qquad\;\; y = -2x$

You can also solve quadratic-linear systems of equations by using the substitution method.

Examples

4 **Use substitution to solve each system of equations.**

$y = -4$

$y = x^2 - 4$

Substitute -4 for y in the second equation. Then solve for x.

$\qquad\qquad y = x^2 - 4$ $\qquad\qquad$ *Original equation*

$\qquad\qquad -4 = x^2 - 4$ $\qquad\qquad$ *Replace y with -4.*

$\quad -4 + 4 = x^2 - 4 + 4$ $\quad$ *Add 4 to each side.*

$\qquad\qquad\quad 0 = x^2$ $\qquad\qquad\quad$ *Simplify.*

$\qquad\qquad\quad 0 = x$ $\qquad\qquad\quad$ *Take the square root of each side.*

The solution of the system of equations is $(0, -4)$.

(continued on the next page)

Check:

Sketch the graphs of the equations. The parabola and line appear to intersect at $(0, -4)$. The solution is correct.

⑤ $y = x^2$
$y = -3$

Substitute -3 for y in the first equation.

$y = x^2$ *Original equation*
$-3 = x^2$ *Replace y with −3.*
 $\sqrt{-3} = x$ *Take the square root of each side.*

There is no real solution because the square root of a negative number is not a real number. *Check by graphing.*

⑥ $y = x^2 - 4x + 6$
$y = -x + 4$

Substitute $-x + 4$ for y in the first equation.

$y = x^2 - 4x + 6$ *Original equation*
$-x + 4 = x^2 - 4x + 6$ *Replace y with −x + 4.*
$-x + 4 - 4 = x^2 - 4x + 6 - 4$ *Subtract 4 from each side.*
$-x = x^2 - 4x + 2$ *Simplify.*
$-x + x = x^2 - 4x + 2 + x$ *Add x to each side.*
$0 = x^2 - 3x + 2$ *Simplify.*
$0 = (x - 2)(x - 1)$ *Factor.*

$0 = x - 2$ or $0 = x - 1$ *Zero Product Property*
$x = 2$ $x = 1$ *Simplify.*

Substitute the values of x in either equation to find the corresponding values of y. *Choose the equation that is easier for you to solve.*

$y = -x + 4$ *Original equation*
$y = -2 + 4$ *Replace x with 2.*
$y = 2$ *Add.*

$y = -x + 4$ *Original equation*
$y = -1 + 4$ *Replace x with 1.*
$y = 3$ *Add.*

The solutions of the system of equations are $(2, 2)$ and $(1, 3)$. The graph shows that the solutions are probably correct. *You can also check by substituting the ordered pairs into the original equations.*

Your Turn

Use substitution to solve each system of equations.

d. $y = x^2 - 12$
$x = 3$

e. $y = -x^2$
$y = x + 2$

f. $y = x^2 - 3x$
$y = 2x$

Example ⑦
Meteorology Link

Refer to the application at the beginning of the lesson. What are the coordinates of the point at which turbulence will occur?

Use substitution to solve the system $y = \frac{1}{4}x^2 - 12$ and $y = -\frac{1}{2}x$.

Substitute $-\frac{1}{2}x$ for y in the first equation.

Math Online Data Update
For the latest information about the jet stream, visit:
algconcepts.com

$$y = \frac{1}{4}x^2 - 12 \qquad \textit{Original equation}$$

$$-\frac{1}{2}x = \frac{1}{4}x^2 - 12 \qquad \textit{Replace } y \textit{ with } -\frac{1}{2}x.$$

$$4\left(-\frac{1}{2}x\right) = 4\left(\frac{1}{4}x^2 - 12\right) \qquad \textit{Multiply each side by 4.}$$

$$-2x = x^2 - 48 \qquad \textit{Simplify.}$$

$$-2x + 2x = x^2 - 48 + 2x \qquad \textit{Add } 2x \textit{ to each side.}$$

$$0 = x^2 + 2x - 48 \qquad \textit{Simplify.}$$

$$0 = (x + 8)(x - 6) \qquad \textit{Factor.}$$

$$0 = x + 8 \quad \text{or} \quad 0 = x - 6 \qquad \textit{Zero Product Property}$$

$$x = -8 \qquad\qquad x = 6 \qquad \textit{Simplify.}$$

Substitute the values of x to find the corresponding values of y.

$$y = -\frac{1}{2}x \qquad \textit{Original equation}$$

$$y = -\frac{1}{2}(-8) \text{ or } 4 \quad \textit{Replace } x \textit{ with } -8.$$

$$y = -\frac{1}{2}x \qquad \textit{Original equation}$$

$$y = -\frac{1}{2}(6) \text{ or } -3 \quad \textit{Replace } x \textit{ with } 6.$$

The solutions of the system are $(-8, 4)$ and $(6, -3)$. This means that turbulence will occur as the plane passes through points having these coordinates. *Check by substituting the coordinates into each equation and by looking at the graph at the beginning of the lesson.*

Check for Understanding

Communicating Mathematics

1. **State** the solution of the system of equations shown at the right.

2. **Sketch** a system of quadratic-linear equations that has solutions $(-2, 1)$ and $(2, 1)$.

3. **Writing Math** List the different methods of solving linear and quadratic-linear systems of equations. Describe the situation in which each method is most useful.

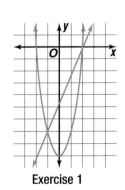

Exercise 1

Solve each system of equations by graphing.

4. $y = x^2$
   $y = -2x - 1$

5. $y = x^2$
   $y = 2x$

Use substitution to solve each system of equations.

6. $y = x^2 - 6$
   $y = -5x$

7. $y = x^2 + 5$
   $y = -3$

8. **Business** Students in an Algebra I class at Banneker High School are simulating the start-up of a company. The income y can be described by the equation $y = \frac{1}{8}x^2$, where x represents the time in months. The expenses have been growing at a constant rate and can be defined by the equation $y = x$. When will income equal expenses?

Exercises • • • • • • • • • • • • • • • • • • •

Practice

Homework Help	
For Exercises	See Examples
9–14	1, 2, 3
15–22, 24	4, 5, 6
23–25	7
Extra Practice	
See page 719.	

Solve each system of equations by graphing.

9. $y = x^2 - 4$
   $y = 2x - 4$

10. $x = 2$
    $y = x^2 + 1$

11. $y = 2x + 1$
    $y = x^2 + 4$

12. $y = x^2 - 6$
    $y = x - 4$

13. $y = -x^2 + 5$
    $y = -x + 3$

14. $y = -x^2 + 4$
    $y = \frac{1}{2}x + 5$

Use substitution to solve each system of equations.

15. $y = \frac{1}{8}x^2 + 5$
    $x = -4$

16. $y = -3x^2$
    $y = 2$

17. $y = \frac{1}{2}x^2$
    $y = 3x + 8$

18. $y = \frac{1}{2}x^2 - 4$
    $y = 3x + 4$

19. $y = x^2 + 3$
    $y = -\frac{1}{2}x - 5$

20. $y = x^2 - x - 3$
    $y = -1$

21. What is the solution of the system $y = x^2 - 9$ and $y = 3x - 9$?

22. Find the solution of the system $y = -x^2 + 2x - 3$ and $y = -2x + 1$.

Applications and Problem Solving

23. **Geometry** Four corners are cut from a rectangular piece of cardboard that is 10 feet by 4 feet. The cuts are x feet from the corners, as shown in the figure at the right. After the cuts are made, the sides of the rectangle are folded to form an open box. The area of the bottom of the box is 16 square feet.

   a. Write two equations to represent the area A of the bottom of the box.

   b. What are the dimensions of the box?

   c. What is the volume of the box?

24. **Meteorology**   Refer to the application at the beginning of the lesson. If a plane's route is described by the linear equation $y = x - 14$, will it experience turbulence by the jet stream that is described by the quadratic equation $y = \frac{1}{4}x^2 - 12$? Explain.

25. **Critical Thinking**   Suppose the perimeter and area of a square have the same measure. Find the length of the sides of the square. Explain how you can use a system of quadratic-linear equations to find the answer.

Mixed Review

26. **Sports**   Diego kayaks 16 miles downstream in 2 hours. It takes him 8 hours to make the return trip. What is the rate of the current?   *(Lesson 13–5)*

27. Solve the system $4x + 3y = 5$ and $-3x - 2y = -4$ by using elimination.   *(Lesson 13–4)*

28. Describe how the graph of $y = -(x - 3)^2$ changes from its parent graph of $y = x^2$.   *(Lesson 11–2)*

Solve each equation.   *(Lesson 3–7)*

29. $2 + |x| = 9$

30. $|y| + 4 = 3$

31. $12 = 6 + |w + 2|$

Standardized Test Practice
Ⓐ Ⓑ Ⓒ Ⓓ

32. **Multiple Choice**   Which expression shows *3 less than the product of 5 and d*?   *(Lesson 1–1)*

   Ⓐ $3 - 5d$ Ⓑ $5d + 3$ Ⓒ $5d - 3$ Ⓓ $5 + d - 3$

Quiz 2   Lessons 13–4 through 13–6

Use elimination to solve each system of equations.   *(Lessons 13–4 & 13–5)*

1. $x + y = 6$
   $2x - y = 6$

2. $2x - 3y = -9$
   $x - 2y = -5$

3. $3x - 5y = 8$
   $4x - 7y = 10$

4. Use substitution to find the solution of the system $y = x^2 + 5$ and $y = 6x - 3$.   *(Lesson 13–6)*

5. **Number Theory**   The difference between two numbers is 38. The greater number is three times the lesser number minus two.   *(Lesson 13–4)*
   a. Write a system of equations to represent the situation.
   b. What are the numbers?

What You'll Learn

You'll learn to solve systems of inequalities by graphing.

Why It's Important

Money Systems of inequalities can be useful in helping you get the most for your money.
See Example 3.

Thomas Edison was a newsboy during the Civil War. One day, he persuaded the editor to give him 300 copies of the paper instead of the usual 100. He went to the train station where people were eager for news of the war. He was able to sell the papers for 10¢ and 25¢ instead of the usual 5¢. If he wanted to earn at least $52, how many papers could he have sold for 10¢ and 25¢?

x = the number of 10¢ papers y = the number of 25¢ papers

The following **system of inequalities** can be used to represent the conditions of this problem.

$x + y \leq 300$ *He can sell as many as 300 papers.*
$0.10x + 0.25y \geq 52$ *He wants to earn at least $52.*

Since both x and y represent the number of papers, neither can be a negative number. Thus, $x \geq 0$ and $y \geq 0$. The solution of the system is the set of all ordered pairs that satisfy both inequalities and lie in the first quadrant. The solution can be determined by graphing each inequality in the same coordinate plane as shown below.

Look Back

Graphing Linear Inequalities in Two Variables: Lesson 12–7

Recall that the graph of each inequality is called a *half-plane*.

- Any point in the yellow region satisfies $x + y \leq 300$.
- Any point in the blue region satisfies $0.10x + 0.25y \geq 52$.
- Any point in the green region satisfies both inequalities. The intersection of the two half-planes represents the solution to the system of inequalities.

- The graphs of $x + y = 300$ and $0.10x + 0.25y = 52$ are the boundaries of the region and are included in the graph of the system.

This solution is a region that contains the graphs of an infinite number of ordered pairs. An example is (50, 225). This means that Edison could have sold 50 papers at 10¢ and 225 papers at 25¢ to earn at least $52.

Check:

$x + y \leq 300$ *Original inequality*
$50 + 225 \overset{?}{\leq} 300$ *(x, y) = (50, 225)*
$275 \leq 300$ ✓

$0.10x + 0.25y \geq 52$ *Original inequality*
$0.10(50) + 0.25(225) \overset{?}{\geq} 52$ *(x, y) = (50, 225)*
$61.25 \geq 52$ ✓

COncepts in MOtion

Animation
algconcepts.com

Remember that the boundary lines of inequalities are only included in the solution if the inequality symbol is greater than or equal to, $\geq$, or less than or equal to, $\leq$.

 Examples

Solve each system of inequalities by graphing. If the system does not have a solution, write *no solution*.

1 $y \leq -1$
$y > x - 3$

The solution is the ordered pairs in the intersection of the graphs of $y \leq -1$ and $y > x - 3$. The region is shaded in green at the right. The graphs of $y = -1$ and $y = x - 3$ are the boundaries of this region. The graph of $y = x - 3$ is a dashed line and is not included in the solution of the system. *Choose a point and check the solution.*

2 $2y < x - 2$
$-3x + 6y \geq 12$

The graphs of $2y = x - 2$ and $-3x + 6y = 12$ are parallel lines. *Check this by graphing or by comparing the slopes.*

Because the regions in the solution of $2y < x - 2$ and $-3x + 6y \geq 12$ have no points in common, the system of inequalities has no solution.

Your Turn

a. $x \geq -1$
$y < x - 2$

b. $y \geq 4x + 5$
$x + y \geq 3$

Online **Personal Tutor at** algconcepts.com

Real World

Example

Spending Link

3 Luisa has $96 to spend on gifts for the holidays. She must buy at least 9 gifts. She plans to buy puzzles that cost $8 or $12. How many of each puzzle can she buy?

x = the number of $8 puzzles y = the number of $12 puzzles

The following system of inequalities can be used to represent the conditions of this problem.

$x + y \geq 9$ *She wants to buy at least 9 puzzles.*
$8x + 12y \leq 96$ *The total cost must be no more than $96.*

Because the number of puzzles she can buy cannot be negative, both $x \geq 0$ and $y \geq 0$.

(continued on the next page)

The solutions are all of the ordered pairs in the intersection of the graphs of these inequalities. Only the first quadrant is used because $x \geq 0$ and $y \geq 0$.

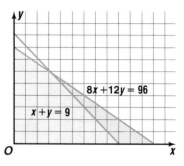

Any point with whole-number coordinates in the region is a possible solution. For example, since the point at (9, 1) is in the region, Luisa could buy 9 puzzles for $8 and 1 for $12. The greatest number of gifts that she could buy would be 12 puzzles for $8 and no puzzles for $12.

COncepts in MOtion

Interactive Lab

algconcepts.com

A graphing calculator can be helpful in solving systems of inequalities or in checking solutions.

Graphing Calculator Tutorial

See pp. 750–753.

 Graphing Calculator Exploration

Use a graphing calculator to solve the system of inequalities.

$y \geq 4x - 3$
$y \leq -2x + 1$

The graphing calculator graphs equations and shades above the first equation entered and below the second equation entered. (Note that inequalities that have $>$ or $\geq$ are lower boundaries and inequalities that have $<$ or $\leq$ are upper boundaries.)

Step 1 Press ⟨2nd⟩ [DRAW] 7 to choose the SHADE feature.

Step 2 Enter the equation that is the lower boundary of the region to be shaded, 4 ⟨X,T,θ,n⟩ ⟨−⟩ 3.

Step 3 Press ⟨,⟩ and enter the equation that is the upper boundary of the region to be shaded, ⟨(−)⟩ 2 ⟨X,T,θ,n⟩ ⟨+⟩ 1 ⟨)⟩ ⟨ENTER⟩.

Try These

Use a graphing calculator to graph each system of inequalities. State one possible solution.

1. $y \geq x - 1$
    $y < x + 1$

2. $y \geq x - 2$
    $y \leq 3$

3. $y > 2x$
    $y \leq -x + 1$

Check for Understanding

Communicating Mathematics

1. **Compare and contrast** the solution of a system of equations and the solution of a system of inequalities.

2. **Sketch** a system of linear equations that has no solution.

3. Kyle says that the solution of $y \geq 2x + 2$ and $y \leq -x - 1$ is all of the points in region B. Tarika says that the solution is all of the points in region D. Who is correct? Explain.

Vocabulary

system of inequalities

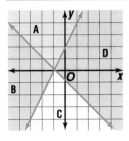

Exercise 3

Guided Practice

🕐 **Getting Ready** State whether each ordered pair is a solution of the system of inequalities $x \geq -2$ and $y \leq 5$.

Sample: $(-3, 5)$ **Solution:** $x \geq -2$
$-3 \overset{?}{\geq} -2$ *Replace x with −3.*
$(-3, 5)$ is not a solution.

4. $(6, -1)$ 5. $(8, 7)$ 6. $(0, 0)$

Examples 1 & 2

Solve each system of inequalities by graphing. If the system does not have a solution, write *no solution*.

7. $x \geq -1$
   $y \geq -4$

8. $y \leq 5$
   $y \geq -x + 1$

9. $x + y > 2$
   $y < x + 6$

10. $y \geq \frac{1}{2}x - 2$
    $x - y < 3$

Example 3

11. **Shopping** Nathaniel must buy two types of cookies for a banquet dinner. He only has $25 to spend and needs at least 6 dozen cookies. The grocery store has small sugar cookies for $3 a dozen and chocolate chip cookies for $4 a dozen. How many cookies of each type can he buy? List three possible solutions.

Exercises

Practice

Solve each system of inequalities by graphing. If the system does not have a solution, write *no solution*.

Homework Help	
For Exercises	See Examples
12–25	1, 2
26	3
28	2
Extra Practice	
See page 720.	

12. $x < 2$
    $y > -1$

13. $y > 0$
    $x \leq 0$

14. $y > 2$
    $y > -x + 2$

15. $y \geq x$
    $y \leq -x$

16. $y \leq 2x$
    $y \geq -3x$

17. $y \geq -3$
    $y < -3x + 1$

18. $x > 2$
    $y > -2x + 3$

19. $y \leq 2x - 1$
    $y \geq 2x + 2$

20. $x + y \geq 2$
    $x + y \leq 6$

21. $y \leq 2x + 2$
    $2x + y \leq 4$

22. $y + 4 < x$
    $2y + 4 > -3x$

23. $2y + x < 4$
    $3x - y > 1$

24. Use the graphs of $y \geq x - 3$ and $y \geq -x - 1$ to determine the solution of the system.

25. Find the solution of the system $3x + 2y \leq -6$ and $y > x + 2$ by graphing.

26. Sports   Nykia must score at least 20 points in the last basketball game of the season to tie the school record for total points in a season. Usually she has no more than 40 opportunities to shoot the ball (both field goals and free throws combined) and makes half of her shots. What combination of shots can Nykia make to score the 20 points? Field goals are worth 2 points and free throws are worth 1 point.

27. Communication   A long-distance carrier offers three fee plans to their customers.

- 15 cents a minute with no monthly service charge
- 10 cents a minute and a monthly service charge of $5.25
- 5 cents a minute and a monthly service charge of $7.50

Depending on how much people use the phone each month, which plan should they select?

28. Critical Thinking   Write an inequality involving absolute value that has no solution.

Mixed Review

Determine whether each system of equations has *one* solution, *two* solutions, or *no* solution by graphing. If the system has one or two solutions, name them.   *(Lesson 13–6)*

29. $y = x^2$
$y = 4$

30. $y = x^2 + 3$
$y = 2x - 1$

31. $y = -x^2 + 2$
$y = -x + 5$

32. Use elimination to find the solution of the system $5x + 4y = -2$ and $3x - 2y = 1$.   *(Lesson 13–5)*

33. Mixtures   The Coffee Hut mixes coffee beans that cost $5.25 per pound with coffee beans that cost $6.50 per pound. They need a mixture of 5 pounds of coffee beans that costs $5.50 per pound. How many pounds of each type of coffee bean should they use? *(Lesson 13–3)*

Solve each equation.   *(Lesson 4–4)*

34. $-4w = 32$

35. $13 = 2.6a$

36. $\frac{2}{5}y = -8$

37. Multiple Choice   Find $-15 + (-3) + (-7) + 6$.   *(Lesson 2–3)*

Ⓐ -19

Ⓑ 1

Ⓒ 11

Ⓓ 16

www.algconcepts.com/self_check_quiz

Car Dealer

Do you work well with people? Do cars interest you? If so, you may want to consider a career as a car dealer. Car dealers must keep track of the number of cars they need to sell each week to ensure maximum profit.

Suppose a car dealer receives a profit of $500 for each mid-sized car m sold and $750 for each sport-utility vehicle s sold. The dealer must sell at least two mid-sized cars for each sport-utility vehicle and must earn at least $3500 per week. The table and graph shown represent this situation.

Situation	Inequality
At least 2 mid-sized cars must be sold for each sport-utility vehicle.	$m \geq 2s$
Earn a profit of at least $3500.	$500m + 750s \geq 3500$

Weekly Sales

$m \geq 2s$

Mid-Sized Cars Sold

$500m + 750s \geq 3500$

Sport-Utility Vehicles Sold

1. Suppose a car dealer sells 2 sport-utility vehicles. How many mid-sized cars must be sold to earn at least $3500?

2. If a car dealer sells only one sport-utility vehicle, how many mid-sized cars must be sold to meet the goal?

FAST FACTS About Car Dealers

Working Conditions

- usually work in a comfortable environment
- evening, weekend, and holiday work
- work a 40-hour week, with some overtime

Education

- good communication skills and computer skills are helpful
- high school math and business classes
- college degree in business is helpful

Job Outlook

Expected Growth in Jobs

| 2002 | 31.0 |
| 2012 | 37.4 |

Employment (thousands)

Source: Bureau of Labor Statistics

interNET
CONNECTION
Career Data For the latest information on car dealers, visit:
www.algconcepts.com

Understanding and Using the Vocabulary

After completing this chapter, you should be able to define each term, property, or phrase and give an example or two of each.

augmented matrix *(p. 578)*
consistent *(p. 554)*
dependent *(p. 554)*
digit problems *(p. 568)*
elimination *(p. 566)*

identity matrix *(p. 579)*
inconsistent *(p. 554)*
independent *(p. 554)*
quadratic-linear *(p. 580)*

row operations *(p. 578)*
substitution *(p. 560)*
system of equations *(p. 550)*
system of inequalities *(p. 586)*

State whether each sentence is *true* or *false*. If false, replace the underlined word(s) to make a true statement.

1. Inconsistent is the description used for a system of equations that has <u>no solution</u>.

2. A system of inequalities can be solved by <u>graphing</u> the inequalities.

3. The solution to a system of equations is the ordered pair that satisfies <u>two</u> of the equations in the system.

4. A system of linear equations that has <u>at least</u> one solution is called independent.

5. The exact coordinates of the point where two or more lines intersect can be determined by using <u>substitution</u>.

6. A set of two or more equations is called a <u>system of equations</u>.

7. <u>Dependent</u> systems of equations have infinitely many solutions.

8. The description of a system of linear equations depends on the number of <u>solutions</u>.

9. To solve a system of equations by <u>substitution</u>, one of the variables is eliminated by either adding or subtracting the equations.

10. A system of equations that has at least one solution can be described as <u>inconsistent</u>.

Skills and Concepts

Objectives and Examples	**Review Exercises**

• **Lesson 13–1**   Solve systems of equations by graphing.

$y = x + 1$
$x + y = 3$

The solution is (1, 2).

Solve each system of equations by graphing.

11. $y = 4x$
    $y = -x + 5$

12. $x = y$
    $y = 2x - 3$

13. $y = \frac{1}{2}x + 3$
    $y = -x$

14. $y = 5$
    $2y = x + 3$

15. $x - 3y = -6$
    $y = -\frac{1}{2}x - 3$

16. $y = 3x + 1$
    $y = -3x + 1$

    www.algconcepts.com/vocabulary_review

Objectives and Examples	Review Exercises

• **Lesson 13–2** Determine whether a system of equations has one solution, no solution, or infinitely many solutions by graphing.

$y = x + 5$
$y = -4x$

This system has one solution, $(-1, 4)$.

Determine whether each system of equations has *one* solution, *no* solution, or *infinitely many* solutions by graphing. If the system has one solution, name it.

17. $y = -x + 1$
    $y = x - 5$

18. $y = 2x + 3$
    $y = 2x - 2$

19. $x + y = -2$
    $4x + 4y = -8$

20. $y = \frac{1}{2}x - 1$
    $y = x$

• **Lesson 13–3** Solve systems of equations by the substitution method.

To solve $y = 3x$ and $2x + y = 10$, substitute $3x$ for y in the second equation.

$\begin{aligned} 2x + 3x &= 10 \\ 5x &= 10 \\ x &= 2 \end{aligned}$ $\quad \bigg| \quad \begin{aligned} y &= 3x \\ y &= 3(2) \text{ or } 6 \\ &\text{The solution is } (2, 6). \end{aligned}$

Use substitution to solve each system of equations.

21. $y = 2x$
    $4x + y = 3$

22. $3y = x$
    $4x + 2y = 14$

23. $y = 3x + 1$
    $3x - y = 4$

24. $y = x + 1$
    $3x - 2y = -12$

25. $2x - 3y = 4$
    $4x - 6y = 8$

26. $y - 3 = x$
    $2x + y = 4$

• **Lesson 13–4** Solve systems of equations by the elimination method using addition and subtraction.

$\begin{aligned} 3x - 4y &= 5 \\ (-)\ 3x + 2y &= -7 \\ \hline -6y &= 12 \\ y &= -2 \end{aligned}$ $\quad \bigg| \quad \begin{aligned} 3x - 4(-2) &= 5 \\ 3x + 8 &= 5 \\ 3x &= -3 \\ x &= -1 \end{aligned}$

The solution is $(-1, -2)$.

Use elimination to solve each system of equations.

27. $2x + y = 2$
    $x + y = -2$

28. $x - 2y = 4$
    $2x + 2y = 2$

29. $x - y = 5$
    $-x + y = 3$

30. $5x + 3y = 11$
    $2x + 3y = 2$

31. $4x + 3y = 3$
    $-4x - 3y = -3$

32. $y = 3 - 4x$
    $x + y = 2$

• **Lesson 13–5** Solve systems of equations by the elimination method using multiplication and addition.

$\begin{aligned} 3x + 2y &= 5 \\ x + y &= 1 \end{aligned}$ **Multiply by −2.** $\quad \begin{aligned} 3x + 2y &= 5 \\ (+)\ -2x - 2y &= -2 \\ \hline x &= 3 \end{aligned}$

Substitute to find the value of y. The solution is $(3, -2)$.

Use elimination to solve each system of equations.

33. $2x + 3y = 4$
    $x + 2y = 2$

34. $4x + 3y = 3$
    $-2x + 4y = 4$

35. $3x + 4y = 2$
    $2x - y = 1$

36. $x - 3y = 2$
    $-3x + 5y = 2$

37. $3x - 2y = 4$
    $4x - 3y = -5$

38. $5x + 3y = 4$
    $6x + 4y = 3$

Mixed Problem Solving
See pages 724–731.

Objectives and Examples

- **Lesson 13–6** Solve systems of quadratic and linear equations.

$$y = 2x + 3$$
$$y = x^2 - 5$$

Substitute $x^2 - 5$ for y in the first equation.

$$y = 2x + 3$$
$$x^2 - 5 = 2x + 3$$
$$x^2 - 5 - 2x - 3 = 2x + 3 - 2x - 3$$
$$x^2 - 2x - 8 = 0$$
$$(x + 2)(x - 4) = 0$$

$$x + 2 = 0 \quad \text{or} \quad x - 4 = 0$$
$$x = -2 \qquad\qquad x = 4$$

Solve by finding the corresponding values of y. The solutions are $(-2, -1)$ and $(4, 11)$.

Review Exercises

Determine whether each system of equations has *one* solution, *two* solutions, or *no* solution by graphing. If the system has one or two solutions, name them.

39. $y = -x^2$
  $y = x - 2$

40. $y = x^2 + 2$
  $y = 6$

41. $y = 2x^2$
  $y = -x - 4$

42. $y = x^2 - 1$
  $y = -4x - 5$

Use substitution to solve each system of equations.

43. $y = x^2 + 2$
  $y = 3x + 6$

44. $y = -x^2 + 4$
  $y = 4x + 10$

45. Solve the system $y = x^2 - 2x$ and $y = 4x - 9$ by using substitution.

- **Lesson 13–7** Solve systems of inequalities by graphing.

Solve the system of inequalities $x \geq -2$ and $y < x - 3$ by graphing.

Solve each system of inequalities by graphing.

46. $x < 4$
  $y \geq -2$

47. $x \geq -3$
  $y \leq x + 2$

48. $x + y > -4$
  $x + y \leq 2$

49. $y \geq 2x + 1$
  $y > -x - 1$

50. Use the graphs of $-6x + 3y > 9$ and $y < 2x - 4$ to determine the solution of the system. If the system does not have a solution, write *no solution*.

Applications and Problem Solving

51. Mixtures Mr. Collins mixed almonds that cost $4.25 per pound with cashews that cost $6.50 per pound. He now has a mixture of 20 pounds of nuts that costs $5.60 per pound. How many pounds of each type of nut did he use? *(Lesson 13–3)*

52. Number Theory The difference between the tens digit and the units digit of a two-digit number is 3. Suppose the tens digit is one less than twice the units digit. What is the number? *(Lesson 13–4)*

53. Geometry The difference between the length and width of a rectangle is 7 feet. Find the dimensions of the rectangle if its perimeter is 50 feet. *(Lesson 13–5)*

1. **Graph** a system of equations that has infinitely many solutions.

2. **Explain** when you would use elimination with subtraction to solve a system of equations.

Solve each system of equations by graphing.

3. $y = 3$
$y = x + 4$

4. $x + y = -2$
$2x - y = -4$

5. $y = -x^2 - 1$
$y = -5$

State whether each system is *consistent and independent, consistent and dependent*, or *inconsistent*.

6.

7.

8. Use graphing to determine whether the system $y = -2x$ and $2x + y = 4$ has *one* solution, *no* solution, or *infinitely many* solutions. If the solution has one solution, name it.

Use substitution to solve each system of equations.

9. $y = 3x$
$x + y = 4$

10. $x + y = -2$
$x = y + 10$

11. $y = 5x - 3$
$10x - 2y = -2$

12. $y = x^2 - 15$
$y = 2x$

13. $y = 5x + 4$
$y = x^2 + 5x$

14. $y = 3x + 2$
$y = x^2 + 6$

Use elimination to solve each system of equations.

15. $x + y = 5$
$x - y = -9$

16. $4x - 5y = 7$
$x + 5y = 8$

17. $2x - y = 32$
$y = 60 - 2x$

18. $x + 3y = -1$
$2x + 4y = -2$

19. $5x - 2y = 3$
$15x - 6y = 9$

20. $-5x + 8y = 21$
$10x + 3y = 15$

Solve each system of inequalities by graphing.

21. $y \le -3$
$y > -x - 2$

22. $y < x + 4$
$y > x - 2$

23. $x \le 2y$
$2x + 3y \le 6$

24. Find two numbers whose sum is 64 and whose difference is 42.

25. **Transportation** Two trains travel toward each other on parallel tracks at the same time from towns 450 miles apart. Suppose one train travels 6 miles per hour faster than the other train. What is the rate of each train if they meet in 5 hours?

Perimeter, Area, and Volume Problems

Standardized tests often include questions on perimeter, area, circumference, and volume. You'll need to know and apply formulas for each of these measurements. Be sure you know these terms.

area   circumference   height   perimeter   width
base   diameter   length   radius

Example 1

Richard plans to increase the floor area of his health club's weight room. The figure below shows the existing floor area with a solid line and the additional floor area with a dotted line. Find the length, in feet, of the new weight room.

Hint Familiarize yourself with the formulas for the area and perimeter of quadrilaterals.

Solution The existing room has an area of 360 square feet and a width of 18 feet. The addition has an area of 180 square feet and a width of 18 feet.

To find the length of the new weight room, first find the area of the new room.

$$A = 360 + 180 \text{ or } 540 \text{ square feet}$$

Next, use the formula for the area of a rectangle to find the length of the room.

$A = \ell w$ *Area formula*
$540 = \ell \cdot 18$ *Replace A with 540 and w with 18.*
$\dfrac{540}{18} = \dfrac{18\ell}{18}$ *Divide each side by 18.*
$30 = \ell$ *Simplify.*

The length of the new room is 30 feet.

Example 2

What is the diameter of a circle with a circumference of 5 inches?

   A $\dfrac{5}{\pi}$ in.   **B** $\dfrac{10}{\pi}$ in.   **C** 5 in.
   D 5π in.   **E** 10π in.

Hint If a geometry problem has no figure, sketch one.

Solution

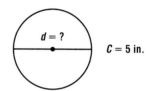

You know the circumference of the circle. You need to find the diameter of the circle.

Use the formula $C = \pi d$, where C is the circumference and d is the diameter to find the circumference of the circle.

First, replace C with 5. Then solve the equation for d.

$C = \pi d$ *Circumference formula*
$5 = \pi d$ *Replace C with 5.*
$\dfrac{5}{\pi} = \dfrac{\pi d}{\pi}$ *Divide each side by π.*
$\dfrac{5}{\pi} = d$ *Simplify.*

The answer is A.

After you work each problem, record your
answer on the answer sheet provided or on
a sheet of paper.

Multiple Choice

1. Quinn's neighborhood is a rectangle 2 miles
by 1.5 miles. How many miles does Quinn
jog if he jogs around the boundary of
his neighborhood? *(Basic Skill)*

(A) 3 (B) 3.5 (C) 6 (D) 7

2. If $AC = 4$, what is the area of $\triangle ABC$?
(Lesson 1–5)

(A) $\frac{1}{2}$ (B) 2
(C) 4 (D) 8

3. Micela is making a poster that has a length
of 36 inches. If the maximum perimeter is
96 inches, which inequality can be used
to determine the width w of the poster?
(Lesson 12–4)

(A) $96 \geq 2(36) + 2w$ (B) $96 \leq 2(36) + 2w$
(C) $96 \geq 36 + 2w$ (D) $96 \geq 36 + w$

4. If the area of a circle is 16 square meters,
what is its radius in meters? *(Lesson 8–1)*

(A) $\frac{8}{\pi}$ (B) $\frac{16}{\pi}$ (C) $\frac{4\sqrt{\pi}}{\pi}$
(D) 12π (E) $144\pi^2$

5. If you double the length and the width of a
rectangle, how does its perimeter change?
(Lesson 8–1)

(A) It increases by $1\frac{1}{2}$.
(B) It doubles.
(C) It quadruples.
(D) It does not change.

6. What is the area of the figure? *(Lesson 1–5)*

(A) 192 ft²
(B) 360 ft²
(C) 456 ft²
(D) 720 ft²

7. The table shows the speed of a car and the
distance needed to safely stop the car.
Which equation represents the data?
(Lesson 11–2)

(A) $y = x + 25$
(B) $y = x^2 + 20$
(C) $y = x^2 \times 20$
(D) $y = x^2 \div 20$
(E) $y = x^2 \div 25$

Speed (x)	Distance (y)
20	20
30	45
40	80
50	125

8. Which number could be the diameter of
the circle? *(Lesson 1–5)*

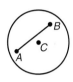

$AB = 8$

(A) 12 (B) 8 (C) 6 (D) 4

Grid In

9. Allie sketched her walking route. The
curved part is a semicircle with a radius of
10.5 meters. If Allie walks this route 10
times, how many meters does she walk?
Round to the nearest meter. *(Lesson 1–5)*

Extended Response

10. Suppose you have 120 feet of fence.
Part A What is the greatest possible
rectangular area you can enclose?
Part B What are the dimensions of the
rectangle that has this area? *(Lesson 1–5)*

What You'll Learn

Key Ideas

- Describe the relationships among sets of numbers. *(Lesson 14–1)*

- Find the distance between two points in the coordinate plane. *(Lesson 14–2)*

- Simplify, add, and subtract radical expressions. *(Lessons 14–3 and 14–4)*

- Solve simple radical equations. *(Lesson 14–5)*

Key Vocabulary

Distance Formula *(p. 616)*

radical equations *(p. 624)*

real numbers *(p. 600)*

rationalizing the denominator *(p. 615)*

Why It's Important

Meteorology At any moment about 2000 thunderstorms are in progress on Earth. Some storms are a few miles in diameter, and some span hundreds of miles. Scientists use sophisticated equipment and mathematical models to monitor and predict these storms.

Radical expressions are used to represent some real-world situations that cannot be modeled with linear equations. You will use a radical equation to find how long a thunderstorm is expected to last in Lesson 14–1.

✓ Check Your Readiness

Lesson 8–5, pp. 357–361

Simplify.

1. $\sqrt{\dfrac{1}{4}}$

2. $\sqrt{\dfrac{4}{9}}$

3. $\sqrt{\dfrac{16}{4}}$

4. $\sqrt{\dfrac{36}{49}}$

5. $-\sqrt{\dfrac{1}{121}}$

6. $-\sqrt{\dfrac{25}{144}}$

Lesson 8–6, pp. 362–365

Estimate each square root to the nearest whole number.

7. $\sqrt{2}$

8. $\sqrt{6}$

9. $\sqrt{14}$

10. $\sqrt{26}$

11. $\sqrt{52}$

12. $\sqrt{104}$

Lesson 1–4, pp. 19–23

Simplify each expression.

13. $6p + 7p$

14. $14c - 3c$

15. $5v - v$

16. $7d - 3d + 5d$

17. $12r - 6r + 3$

18. $5m - 4n + m + 6n$

Lesson 9–5, pp. 405–409

Find each product.

19. $(y - 3)(y + 3)$

20. $(q + 4)(q - 4)$

21. $(z - 5)(z + 5)$

22. $(r - t)(r + t)$

23. $(3c - 5)(3c + 5)$

24. $(2a + 7b)(2a - 7b)$

FOLDABLES™ Study Organizer

Make this Foldable to help you organize your Chapter 14 notes. Begin with a sheet of 11" by 17" paper.

❶ **Fold** the short sides to meet in the middle.

❷ **Fold** the top to the bottom.

❸ **Open.** Cut along the second fold to make four tabs.

❹ **Label** each tab as shown.

Reading and Writing As you read and study the chapter, write notes and examples under each tab.

What You'll Learn

You'll learn to describe the relationships among sets of numbers.

Why It's Important

Meteorology
Meteorologists use real numbers when determining the duration of a thunderstorm.
See Exercise 16.

Look Back

Venn Diagram:
Lesson 2–1

In Lesson 3–1, you learned about *rational numbers*. Natural numbers, whole numbers, and integers are all rational numbers. These sets are listed below.

Natural Numbers: $\{1, 2, 3, 4, \ldots\}$
Whole Numbers: $\{0, 1, 2, 3, \ldots\}$
Integers: $\{\ldots, -2, -1, 0, 1, 2, \ldots\}$
Rational Numbers: {all numbers that can be expressed in the form $\frac{a}{b}$, where a and b are integers and $b \neq 0$}

Recall that repeating or terminating decimals are also rational numbers because they can be expressed as $\frac{a}{b}$, where a and b are integers and $b \neq 0$. The square roots of perfect squares are also rational numbers. For example, $\sqrt{0.16}$ is a rational number since $\sqrt{0.16} = 0.4$. However, $\sqrt{21}$ is irrational because 21 is not a perfect square.

The Venn diagram shows the relationship among the different types of rational numbers. For example, the set of whole numbers is a subset of the integers. This means that all whole numbers are integers. Similarly, all rational numbers are real numbers.

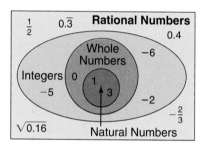

In Lesson 8–6, you learned about *irrational numbers*. A few examples of irrational numbers are shown below.

$0.1010010001\ldots$ π $0.153768\ldots$ $-\sqrt{23}$

The set of rational numbers and the set of irrational numbers together form the set of **real numbers**. *Numbers such as $\sqrt{-1}$ and $4 + \sqrt{-9}$ are called **complex numbers**. The set of complex numbers includes all of the real numbers as well as numbers involving square roots of negative numbers. We will not deal with complex numbers in this text.*

Real Numbers

		Rational Numbers	Irrational Numbers	
$\frac{1}{2}$	$0.\overline{3}$		π 2π	$\frac{\pi}{4}$
		Whole Numbers 0.4		
		−6		
Integers 0			$\sqrt{5}$ $2\sqrt{5}$	$\frac{1}{\sqrt{3}}$
−5		1 3		
		−2		
		$-\frac{2}{3}$	$\sqrt{2}+\sqrt{3}$ $4+\sqrt{3}$	$\pi + 2$
$\sqrt{0.16}$		Natural Numbers		

Name the set or sets of numbers to which each real number belongs.

1   −5   This number is an integer and a rational number.

2 $\dfrac{12}{3}$ Since $\dfrac{12}{3} = 4$, this number is a natural number, a whole number, an integer, and a rational number.

3 $\sqrt{95}$ $\sqrt{95} = 9.746794345\ldots$ It is not the square root of a perfect square. So, it is irrational.

4 $0.\overline{7}$ This repeating decimal is a rational number since it is equivalent to $\dfrac{7}{9}$. $7 \div 9 = 0.7777777\ldots$

5 $-\sqrt{4}$ Since $-\sqrt{4} = -2$, this number is an integer and a rational number.

Your Turn   Name the set or sets of numbers to which each real number belongs. Let N = natural numbers, W = whole numbers, Z = integers, Q = rational numbers, and I = irrational numbers.

a. $3.141592\ldots$ **b.** $-\sqrt{9}$ **c.** $0.1666666\ldots$ **d.** $\dfrac{20}{4}$ **e.** 0

If you graph all of the rational numbers, you will still have some "holes" in the number line. The irrational numbers "fill in" the number line. The graph of all real numbers is the entire number line without any "holes."

This property of real numbers is called the Completeness Property.

Completeness Property for Points on the Number Line	Each real number corresponds to exactly one point on the number line. Each point on the number line corresponds to exactly one real number.

You have learned how to graph rational numbers. Irrational numbers can also be graphed. Therefore, every real number can be graphed. Use a calculator or a table of squares and square roots to find approximate values of square roots that are irrational. These values can be used to approximate the graphs of square roots.

Find an approximation, to the nearest tenth, for each square root. Then graph the square root on a number line.

6 $\sqrt{3}$

Enter: [2nd] [$\sqrt{}$] 3 [ENTER] *1.732050808*

An approximate value for $\sqrt{3}$ is 1.7.

7 $-\sqrt{12}$

Enter: [(—)] [2nd] [$\sqrt{}$] 12 [ENTER] *−3.464101615*

An approximate value for $-\sqrt{12}$ is −3.5.

Your Turn

f. $\sqrt{6}$ g. $-\sqrt{23}$

Determine whether each number is rational or irrational. If it is irrational, find two consecutive integers between which its graph lies on the number line.

8 $\sqrt{47}$

$\sqrt{36} < \sqrt{47} < \sqrt{49}$ *$\sqrt{47}$ is between $\sqrt{36}$ and $\sqrt{49}$.*
$\quad 6 < \sqrt{47} < 7$ *$\sqrt{36} = 6$ and $\sqrt{49} = 7$.*

47 is not a perfect square. So, its square root is irrational. The graph of $\sqrt{47}$ lies between 6 and 7.

9 $-\sqrt{26}$

$-\sqrt{36} < -\sqrt{26} < -\sqrt{25}$ *$-\sqrt{26}$ is between $-\sqrt{36}$ and $-\sqrt{25}$.*
$\quad -6 < -\sqrt{26} < -5$ *$-\sqrt{36} = -6$ and $-\sqrt{25} = -5$.*

26 is not a perfect square. So, its square root is irrational. The graph of $-\sqrt{26}$ lies between −6 and −5.

Your Turn

h. $\sqrt{64}$ i. $-\sqrt{28}$

You have solved equations with rational number solutions. Some equations have solutions that are irrational numbers.

Example 10
Science Link

The time t in seconds it takes for a pendulum to complete one full swing (back and forth) is given by the equation $t = 2\pi\sqrt{\dfrac{\ell}{9.8}}$, where ℓ is the length of the pendulum in meters. Suppose a pendulum has a length of 4.9 meters. How long does it take the pendulum to complete one full swing?

$t = 2\pi\sqrt{\dfrac{\ell}{9.8}}$ *Original formula*

$\quad = 2\pi\sqrt{\dfrac{4.9}{9.8}}$ *Replace ℓ with 4.9.*

$\quad = 2\pi\sqrt{0.5}$ *Simplify.*

Enter: 2 [2nd] [π] [2nd] [√] 0.5 [ENTER] 4.442882938

The pendulum will complete one full swing in about 4.4 seconds.

Check for Understanding

Communicating Mathematics

1. **Give an example** of a number that is an integer and a rational number.

2. **Give** two counterexamples for the statement *all square roots are irrational numbers.*

3. **Write** a square root that is an irrational number.

Guided Practice

Examples 1–5

Name the set or sets of numbers to which each real number belongs. Let N = natural numbers, W = whole numbers, Z = integers, Q = rational numbers, and I = irrational numbers.

4. $-\sqrt{36}$ 5. $0.3131131113\ldots$ 6. 4 7. $0.\overline{2}$

Examples 6 & 7

Find an approximation, to the nearest tenth, for each square root. Then graph the square root on a number line.

8. $\sqrt{7}$ 9. $-\sqrt{20}$ 10. $\sqrt{32}$ 11. $\sqrt{54}$

Examples 8 & 9

Determine whether each number is *rational* or *irrational*. If it is irrational, find two consecutive integers between which its graph lies on the number line.

12. $\sqrt{63}$ 13. $-\sqrt{25}$ 14. $-\sqrt{147}$ 15. $\sqrt{49}$

Example 10

16. **Meteorology**   To estimate the amount of time t in hours that a thunderstorm will last, meteorologists use the formula $t = \sqrt{\dfrac{d^3}{216}}$, where d is the diameter of the storm in miles.

   a. Estimate how long a thunderstorm will last if its diameter is 8.4 miles.

   b. Find the diameter of a storm that will last for 1 hour.

Exercises • • • • • • • • • • • • • • • • •

Practice

Name the set or sets of numbers to which each real number belongs. Let N = natural numbers, W = whole numbers, Z = integers, Q = rational numbers, and I = irrational numbers.

17. $-\dfrac{1}{2}$ 18. $\sqrt{17}$ 19. 0 20. $\dfrac{15}{3}$

21. -5 22. $1.202002\ldots$ 23. $\sqrt{81}$ 24. 0.125

25. $-\sqrt{121}$ 26. $\dfrac{6}{18}$ 27. $0.834834\ldots$ 28. $-\dfrac{3}{1}$

Homework Help

For Exercises	See Examples
17–28	1–5
29–40, 53, 54, 57, 58	6, 7
41–52, 56	8, 9
55	10

Extra Practice
See page 720.

Find an approximation, to the nearest tenth, for each square root. Then graph the square root on a number line.

29. $\sqrt{3}$ 30. $\sqrt{8}$ 31. $-\sqrt{10}$ 32. $\sqrt{17}$

33. $-\sqrt{40}$ 34. $\sqrt{37}$ 35. $\sqrt{52}$ 36. $-\sqrt{99}$

37. $\sqrt{108}$ 38. $-\sqrt{112}$ 39. $\sqrt{250}$ 40. $\sqrt{300}$

Determine whether each number is *rational* or *irrational*. If it is irrational, find two consecutive integers between which its graph lies on the number line.

41. $\sqrt{16}$ 42. $-\sqrt{12}$ 43. $\sqrt{41}$ 44. $\sqrt{24}$

45. $-\sqrt{81}$ 46. $\sqrt{66}$ 47. $-\sqrt{7}$ 48. $\sqrt{98}$

49. $\sqrt{125}$ 50. $\sqrt{144}$ 51. $-\sqrt{169}$ 52. $\sqrt{220}$

53. Graph $\sqrt{2}$, π, and $-\sqrt{5}$ on a number line.

54. Write $2.\overline{4}$, 2.41, and $\sqrt{6}$ in order from least to greatest.

Applications and Problem Solving

55. **Science**   Refer to Example 10. Suppose a pendulum has a length of 18 meters. How long does it take the pendulum to complete one full swing?

56. **Geometry**   The radius of a circle is given by $r = \sqrt{\dfrac{A}{\pi}}$, where A is the area of the circle. Suppose a circle has an area of 25 square centimeters. What is the measure of the radius?

57. Electricity The voltage V in a circuit is given by $V = \sqrt{PR}$, where P is the power in watts and R is the resistance in ohms. A circuit is designed with two resistance settings, 4.6 ohms and 5.2 ohms, and two power settings, 1200 watts and 1500 watts. Which settings are required so the voltage is between 75 and 85 volts?

58. Geometry Find the area of each rectangle. Round answers to the nearest tenth. (*Hint:* Use the Pythagorean Theorem.)

a.

18 in. 9 in.

b.

15 cm 22 cm

59. Critical Thinking Roots can also be indicated with *rational exponents*. For any real number b, $\sqrt[n]{b} = b^{\frac{1}{n}}$. So, a square root is given by $\sqrt{b}$ or $b^{\frac{1}{2}}$. A *cube root* is given by $\sqrt[3]{b}$ or $b^{\frac{1}{3}}$. For example, $\sqrt[3]{216} = 216^{\frac{1}{3}} = 6$ because $6 \times 6 \times 6 = 216$. To what power do you raise

a. 9 to get 3? **b.** 64 to get 4? **c.** x^4 to get x?

Mixed Review

60. Find the solution of the system of inequalities $y \leq x + 3$ and $y > -x - 4$. *(Lesson 13–7)*

61. What is the solution of the system of equations $y = x^2 + 2$ and $y = 4x - 1$? *(Lesson 13–6)*

62. Graph the solution of $|4n| > 16$ on a number line. *(Lesson 12–6)*

Solve each inequality. *(Lesson 12–2)*

63. $x + 3 < -4$ **64.** $w - 14 \geq -8$ **65.** $3.4 + m \leq 1.6$

66. Write an inequality for the graph shown at the right. *(Lesson 12–1)*

−3 −2 −1 0 1 2 3 4 5 6 7

67. Graph $y = 2^x - 1$. Then state the y-intercept. *(Lesson 11–7)*

Factor each polynomial. If the polynomial cannot be factored, write *prime*. *(Lesson 10–5)*

68. $x^2 - 25y^2$ **69.** $4a^2 - 36$ **70.** $m^2 + 2m + 4$

71. Find the degree of $5x^3 - 3x^3y - 6xy$. *(Lesson 9–4)*

72. Short Response Cleavon has $35 to spend on CDs. Suppose CDs cost $13.85 each, including tax. Write an equation that can be used to determine how many CDs Cleavon can buy. *(Lesson 4–4)*

73. Multiple Choice What is the value of a if $6 - (-8) = a$? *(Lesson 2–3)*

Ⓐ 14 Ⓑ 6 Ⓒ −14 Ⓓ −6

What You'll Learn

You'll learn to find the distance between two points in the coordinate plane.

Why It's Important

Engineering
Engineers can use the distance formula to determine the amount of cable needed to install a cable system. *See Exercise 9.*

A coordinate system is superimposed over a map of Washington, D.C. Jessica and Omar walk from the Metro Center to 15th Street and then up 15th Street to McPherson Square. How far is the Metro center from McPherson Square? *This problem will be solved in Example 2.*

Recall that subtraction is used to find the distance between two points that lie on a vertical or horizontal line.

In the following activity, you will find the distance between two points that do not lie on a horizontal or vertical line.

Hands-On Algebra

Look Back

Graphing Ordered Pairs: Lesson 2–2

Materials: grid paper   straightedge

Step 1   Graph $M(-4, 3)$ and $N(2, -5)$ on a coordinate plane.

Step 2   Draw a vertical segment from M and a horizontal segment from N. Label the point of intersection P.

Step 3   Find the coordinates of P.

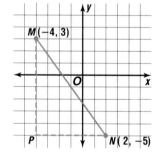

Try These

1. Find the distance between M and P.
2. Find the distance between N and P.
3. What kind of triangle is $\triangle MNP$?
4. What theorem can be used to find MN if MP and NP are known?
5. Find the distance between M and N.

In the activity, you discovered that $(MN)^2 = (MP)^2 + (NP)^2$. You can solve this equation for MN to find the distance between points M and N.

$$(MN)^2 = (MP)^2 + (NP)^2$$
$$\sqrt{(MN)^2} = \sqrt{(MP)^2 + (NP)^2} \quad \textit{Take the positive square root of each side.}$$
$$MN = \sqrt{(MP)^2 + (NP)^2}$$

The **Distance Formula** can be used to find the distance between any two points (x_1, y_1) and (x_2, y_2).

The Distance Formula	**Words:** The distance d between any two points with coordinates (x_1, y_1) and (x_2, y_2) is given by $$d = \sqrt{(x_2 - x_1)^2 + (y_2 - y_1)^2}.$$ **Model:**

Example ① Find the distance between points $A(1, 2)$ and $B(-3, 7)$.

$$d = \sqrt{(x_2 - x_1)^2 + (y_2 - y_1)^2} \quad \textit{Distance Formula}$$
$$= \sqrt{(-3 - 1)^2 + (7 - 2)^2} \quad (x_1, y_1) = (1, 2), \textit{ and } (x_2, y_2) = (-3, 7)$$
$$= \sqrt{(-4)^2 + 5^2} \quad \textit{Subtract.}$$
$$= \sqrt{16 + 25} \quad (-4)^2 = 16 \textit{ and } 5^2 = 25.$$
$$= \sqrt{41} \text{ or about 6.4 units}$$

Your Turn

a. Find the distance between points $C(3, 5)$ and $D(6, 4)$.

Example ②

Map Link

Refer to the application at the beginning of the lesson. The Metro Center is located 4 blocks east of the intersection of 15th Street and G Street. McPherson Square is located 2 blocks north of the intersection. How far is the Metro Center from McPherson Square?

(continued on the next page)

McPherson Square,
Washington, D.C.

Let the Metro Center be represented by (4, 0) and McPherson Square by (0, 2). So, $x_1 = 4$, $y_1 = 0$, $x_2 = 0$, and $y_2 = 2$.

$d = \sqrt{(x_2 - x_1)^2 + (y_2 - y_1)^2}$ *Distance Formula*

$ = \sqrt{(0 - 4)^2 + (2 - 0)^2}$ *$(x_1, y_1) = (4, 0)$, and $(x_2, y_2) = (0, 2)$*

$ = \sqrt{(-4)^2 + (2)^2}$ *Subtract.*

$ = \sqrt{16 + 4}$ *$(-4)^2 = 16$ and $2^2 = 4$.*

$ = \sqrt{20}$ or about 4.5 blocks

The Metro Center is about 4.5 blocks from McPherson Square.

Suppose you know the coordinates of a point, one coordinate of another point, and the distance between the two points. You can use the Distance Formula to find the missing coordinate.

Example ❸ **Find the value of a if $G(4, 7)$ and $H(a, 3)$ are 5 units apart.**

$d = \sqrt{(x_2 - x_1)^2 + (y_2 - y_1)^2}$ *Distance Formula*

$5 = \sqrt{(a - 4)^2 + (3 - 7)^2}$ *$(x_1, y_1) = (4, 7)$, and $(x_2, y_2) = (a, 3)$*

$5 = \sqrt{(a - 4)^2 + (-4)^2}$ *Subtract.*

$5 = \sqrt{a^2 - 8a + 16 + 16}$ *$(a - 4)^2 = a^2 - 8a + 16$ and $(-4)^2 = 16$*

$5 = \sqrt{a^2 - 8a + 32}$ *Simplify.*

$5^2 = (\sqrt{a^2 - 8a + 32})^2$ *Square each side.*

$25 = a^2 - 8a + 32$ *Simplify.*

$0 = a^2 - 8a + 7$ *Subtract 25 from each side.*

$0 = (a - 7)(a - 1)$ *Factor.*

$a - 7 = 0$ or $a - 1 = 0$ *Zero Product Property*

$\phantom{a - 7 = 0 \text{ or }} a = 7 a = 1$ *Solve.*

The value of a is 7 or 1.

Look Back

Zero Product Property: Lesson 11–4

Your Turn

b. Find the value of a if $J(a, 5)$ and $K(-7, 3)$ are $\sqrt{29}$ units apart.

Check for Understanding

Communicating Mathematics

1. **Name** the theorem that is used to derive the Distance Formula.

2. Tell how to find the distance between the points $A(10, 3)$ and $B(2, 3)$ without using the Distance Formula.

3. **You Decide?** Anna says that to find the distance between $A(5, -8)$ and $B(7, -6)$, you should take the square root of $(7 - 5)^2 + [-6 - (-8)]^2$. Nick disagrees. He says to take the square root of $(5 - 7)^2 + [-8 - (-6)]^2$. Who is correct? Explain your reasoning.

Guided Practice

Find the distance between each pair of points. Round to the nearest tenth, if necessary.

Example 1

4. $C(5, -1)$, $D(2, 2)$ **5.** $M(6, 8)$, $N(3, 4)$ **6.** $R(-3, 8)$, $S(5, 4)$

Example 3

Find the value of *a* if the points are the indicated distance apart.

7. $A(3, -1)$, $B(a, 7)$; $d = 10$ **8.** $J(5, a)$, $K(6, 1)$; $d = \sqrt{10}$

Example 2

9. Engineering The MAXTEC Company is installing a fiber optic cable system between two of their office buildings. One of the buildings, MAXTEC West, is located 5 miles west and 2 miles north of the corporate office. The other building, MAXTEC East, is located 4 miles east and 5 miles north of the corporate office. How much cable will be needed to connect MAXTEC West and MAXTEC East? Round to the nearest tenth.

Exercises

Practice

Find the distance between each pair of points. Round to the nearest tenth, if necessary.

10. $X(5, 0)$, $Y(12, 0)$ **11.** $A(-6, -4)$, $B(-6, 8)$

12. $M(-2, -5)$, $N(3, 7)$ **13.** $P(-4, 0)$, $Q(3, -3)$

14. $V(-3, -4)$, $W(-1, -2)$ **15.** $C(7, 2)$, $D(-4, 10)$

16. $E(3, -6)$, $F(9, -2)$ **17.** $G(-4, -6)$, $H(-7, -3)$

Homework Help	
For Exercises	**See Examples**
10–17, 24, 25	1, 2
18–23, 26, 27	3
Extra Practice	
See page 720.	

Find the value of *a* if the points are the indicated distance apart.

18. $A(a, -5)$, $B(-3, -2)$; $d = 5$ **19.** $D(-3, a)$, $E(5, 2)$; $d = 17$

20. $Q(7, 2)$, $R(-1, a)$; $d = 10$ **21.** $G(7, -3)$, $H(5, a)$; $d = \sqrt{85}$

22. $T(6, -3)$, $U(-3, a)$; $d = \sqrt{130}$ **23.** $U(1, -6)$, $V(10, a)$; $d = \sqrt{145}$

24. Find the distance between $J(-9, 5)$ and $K(-4, -2)$.

25. What is the distance between $C(-8, 1)$ and $D(5, 6)$?

26. What is the value of c if $W(1, c)$ and $V(-4, 9)$ are 13 units apart?

27. Suppose $M(b, 9)$ and $N(20, -5)$ are $\sqrt{340}$ units apart. What is the value of b?

Applications and Problem Solving

28. Geometry Triangle MNP has vertices $M(-6, 14)$, $N(2, -1)$, and $P(-2, 2)$. Find the perimeter of $\triangle MNP$. Round to the nearest tenth.

29. Geometry An isosceles triangle is a triangle with at least two congruent sides. Determine whether $\triangle CDE$ with vertices $C(1, 6)$, $D(2, 1)$, and $E(4, 1)$ is isosceles. Explain your reasoning.

30. School At Dannis State University, Webber Hall is located 35 meters west and 55 meters south of the student union. Packard Hall is located 42 meters east and 30 meters south of the student union.

a. Draw a diagram on a coordinate grid to represent this situation.

b. How far is Webber Hall from Packard Hall? Round to the nearest tenth.

31. Engineering Refer to Exercise 9. Suppose the MAXTEC Company builds another office building, MAXTEC North, 4 miles east and 2 miles north of the corporate office. How much cable will be needed to connect each of the buildings to the corporate office? Round to the nearest tenth.

32. Critical Thinking If the diagonals of a trapezoid have the same length, then the trapezoid is isosceles. The vertices of trapezoid $ABCD$ are $A(-2, 2)$, $B(10, 6)$, $C(9, 8)$, and $D(0, 5)$. Is the trapezoid isosceles? Explain.

Find an approximation, to the nearest tenth, for each square root. Then graph the square root on a number line. *(Lesson 14–1)*

33. $\sqrt{12}$ **34.** $-\sqrt{27}$ **35.** $\sqrt{135}$

36. Solve $y \geq -4$ and $y < -2x + 1$ by graphing. *(Lesson 13–7)*

Solve each inequality. Check your solution. *(Lesson 12–3)*

37. $\frac{k}{3} \geq 2$ **38.** $\frac{-n}{4} > 3$ **39.** $\frac{3}{4}b \leq -6$

40. Factor $12m^2 + 7m - 10$. *(Lesson 10–4)*

Standardized Test Practice
Ⓐ Ⓑ Ⓒ Ⓓ

41. Short Response Find the solution of $|3 + w| + 4 = 2$. *(Lesson 3–7)*

42. Multiple Choice The table shows the number of pets that have lived at the White House. Which two pets account for exactly half of the pets that have lived at the White House? *(Lesson 1–7)*

White House Pets	
dog	23
bird	16
horse	11
cat	10
cow	4
goat	4

Source: *USA TODAY*

Ⓐ dog and cat   Ⓑ horse and bird
Ⓒ bird and dog   Ⓓ dog and horse

Quiz 1   Lessons 14–1 and 14–2

Name the set or sets of numbers to which each real number belongs. Let N = natural numbers, W = whole numbers, Z = integers, Q = rational numbers, and I = irrational numbers. *(Lesson 14–1)*

1. 0   **2.** $\sqrt{36}$ **3.** $0.121231234\ldots$

4. Find an approximation, to the nearest tenth, for $-\sqrt{35}$. *(Lesson 14–1)*

5. Write $\sqrt{2}$, 1.22, and $1.\overline{2}$ in order from least to greatest. *(Lesson 14–1)*

Find the distance between each pair of points. Round to the nearest tenth, if necessary. *(Lesson 14–2)*

6. $A(2, 0)$, $B(-1, 3)$ **7.** $C(4, 5)$, $D(-3, -2)$ **8.** $E(0, -4)$, $F(8, 7)$

9. What is the value of m if $A(m, 8)$ and $B(3, 4)$ are 5 units apart? *(Lesson 14–2)*

10. Geometry A scalene triangle is a triangle with no congruent sides. Determine whether $\triangle XYZ$ with vertices $X(5, 4)$, $Y(1, 5)$, and $Z(-1, 1)$ is scalene. Explain. *(Lesson 14–2)*

A Cut-Out Caper

Materials

grid paper

ruler

scissors

Midpoints

In Lesson 14–2, you learned how to find the distance between two points in a coordinate plane. In this Investigation, you will learn how to find the **midpoint** of a segment. The midpoint of a line segment is the point on the segment that separates it into two segments of equal length.

Investigate

1. Graph $A(-2, -1)$, $B(6, -1)$, $C(10, 3)$, and $D(2, 3)$ on a coordinate plane.

   a. Connect the points. What type of figure is formed?

   b. Make a table and record the coordinates as shown.

Coordinates of First Point	Coordinates of Second Point	Coordinates of Midpoint
$A(-2, -1)$	$B(6, -1)$	$E(\quad,\quad)$
$B(6, -1)$	$C(10, 3)$	$F(\quad,\quad)$
$C(10, 3)$	$D(2, 3)$	$G(\quad,\quad)$
$D(2, 3)$	$A(-2, -1)$	$H(\quad,\quad)$

   c. Fold the paper so that points A and B coincide. Crease the paper to mark the midpoint of $\overline{AB}$. Label this point E. Find the coordinates of point E. Why is point E the midpoint of $\overline{AB}$?

   d. Fold the paper so that points B and C coincide. Crease the paper to mark the midpoint of $\overline{BC}$. Label this point F. Find the coordinates of point F. Continue this process with points C and D and points D and A. Label the midpoints G and H, respectively. Record the coordinates of these midpoints in your table.

   e. Draw $\overline{AC}$ and $\overline{BD}$. These are the diagonals of the figure. Find the midpoints of $\overline{AC}$ and $\overline{DB}$. Label the midpoints M and N, respectively. What are the coordinates of M and N? What appears to be true about the midpoints of the diagonals?

   f. Draw $\overline{EG}$ and $\overline{FH}$. What are the coordinates of their midpoints?

   g. Notice the triangles formed on the inside of the figure. Explain any relationships that exist between the triangles. You may cut out the triangles to help see the relationships.

2. Graph $X(-3, -3)$, $Y(3, -3)$, and $Z(3, 5)$ on a coordinate plane.

   a. Connect the points. What type of figure is formed?

   b. Make a table and record the coordinates as shown.

Coordinates of First Point	Coordinates of Second Point	Coordinates of Midpoint
$X(-3, -3)$	$Y(3, -3)$	$U(\ ,\ )$
$Y(3, -3)$	$Z(3, 5)$	$V(\ ,\ )$
$Z(3, 5)$	$X(-3, -3)$	$W(\ ,\ )$

   c. Use paper folding to find the midpoints of the segments. Label the midpoints U, V, and W as shown in the table. What are the coordinates of the midpoints? Record the results.

   d. Draw $\overline{YW}$. Compare the measures of $\overline{YW}$, $\overline{XW}$, and $\overline{ZW}$. What appears to be true? Use the Distance Formula to verify your results.

   e. Draw $\overline{WU}$ and $\overline{WV}$. Four triangles are formed. Explain any relationships that exist between the triangles. You may cut out the triangles to help see the relationships.

3. Refer to the tables. Study each pair of first point coordinates, second point coordinates, and the corresponding midpoint coordinates.

   a. Explain how the concept of *mean*, or average, can be used to find the midpoint of a segment.

   b. Write a formula to find the midpoint of two points $X(a, b)$ and $Y(c, d)$.

Extending the Investigation

In this extension, you will continue to investigate midpoints.

- Graph points $A(3, 2)$, $B(-1, -6)$, and $C(-5, -4)$ on a coordinate plane. Connect the points to form a figure. Using the Midpoint Formula you discovered in Exercise 3b, determine the coordinates of the midpoints of each side of the figure. Label the points D, E, and F. Use the Distance Formula to verify that the midpoints separate each segment into two equal lengths.

- Use the Midpoint Formula to determine the midpoint of each pair of points.
  - a. $M(4, 3)$, $N(2, 5)$ b. $R(1, 0)$, $S(-3, 2)$ c. $C(3, -6)$, $D(1, -4)$

Presenting Your Investigation

Here are some ideas to help you present your conclusions to the class.

- Make a poster showing the results of your research. Include the tables, graphs, and formula for finding the midpoint of a segment. List any relationships you discovered.

- Discuss any similarities in the triangles formed inside of the figures.

*inter*NET
CONNECTION   **Investigation** For more information on midpoints, visit: www.algconcepts.com

What You'll Learn

You'll learn to simplify radical expressions.

Why It's Important

Science Scientists can use radical expressions to find the speed of a river. *See Exercise 41.*

Why does the character from the comic yell *"the square root of sixteen"* when hitting the golf ball?

FoxTrot

FOXTROT©1998 Bill Amend. Reprinted with permission of UNIVERSAL PRESS SYNDICATE. All rights reserved.

The expression $\sqrt{16}$ is a **radical expression**. Since the radicand, 16, is a perfect square, $\sqrt{16} = 4$. In Lesson 8–5, you learned to simplify radical expressions using the Product Property of Square Roots and prime factorization. You can simplify radical expressions in which the radicand is not a perfect square in a similar manner. *Recall that the radicand is the number or expression under the square root symbol.*

Examples

1 **Simplify $\sqrt{75}$.**

$$\sqrt{75} = \sqrt{3 \cdot 5 \cdot 5} \quad \text{\textit{Prime factorization}}$$
$$= \sqrt{3 \cdot 25} \quad \text{\textit{5 × 5 = 25}}$$
$$= \sqrt{3} \cdot \sqrt{25} \quad \text{\textit{Product Property of Square Roots}}$$
$$= \sqrt{3} \cdot 5 \text{ or } 5\sqrt{3} \quad \text{\textit{Simplify } } \sqrt{25}.$$

Your Turn Simplify each square root. Leave in radical form.

a. $\sqrt{68}$ **b.** $\sqrt{375}$

Sports Link

2 **On a softball field, the distance from second base to home plate is $\sqrt{800}$ meters. Express $\sqrt{800}$ in simplest radical form.**

$$\sqrt{800} = \sqrt{2 \cdot 2 \cdot 2 \cdot 2 \cdot 2 \cdot 5 \cdot 5} \quad \text{\textit{Prime factorization}}$$
$$= \sqrt{16 \cdot 2 \cdot 25} \quad \text{\textit{4 × 4 = 16, 5 × 5 = 25}}$$
$$= \sqrt{16} \cdot \sqrt{2} \cdot \sqrt{25} \quad \text{\textit{Product Property of Square Roots}}$$
$$= 4 \cdot \sqrt{2} \cdot 5 \quad \text{\textit{Simplify } } \sqrt{16} \text{ and } \sqrt{25}.$$
$$= 20\sqrt{2} \quad \text{\textit{Multiply.}}$$

The distance is $20\sqrt{2}$ meters.

*inter*NET
C O N N E C T I O N

Data Update For the latest information on softball, visit:
www.algconcepts.com

The Product Property can also be used to multiply square roots.

Example ❸ **Simplify** $\sqrt{5} \cdot \sqrt{35}$.

$$\sqrt{5} \cdot \sqrt{35} = \sqrt{5} \cdot \sqrt{5 \cdot 7} \qquad \textit{Prime factorization}$$
$$= \sqrt{5} \cdot \sqrt{5} \cdot \sqrt{7} \qquad \textit{Product Property of Square Roots}$$
$$= \sqrt{5^2} \cdot \sqrt{7} \qquad \sqrt{5} \times \sqrt{5} = \sqrt{5^2}$$
$$= 5 \cdot \sqrt{7} \text{ or } 5\sqrt{7} \qquad \textit{Simplify.}$$

Your Turn **Simplify each expression. Leave in radical form.**

c. $\sqrt{3} \cdot \sqrt{15}$ **d.** $\sqrt{10} \cdot \sqrt{30}$

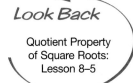

Look Back

Quotient Property
of Square Roots:
Lesson 8–5

To divide square roots and simplify radical expressions that involve division, use the Quotient Property of Square Roots. A fraction containing radicals is in simplest form if no radicals are left in the denominator.

Example ❹ **Simplify** $\dfrac{\sqrt{32}}{\sqrt{4}}$.

$$\frac{\sqrt{32}}{\sqrt{4}} = \sqrt{\frac{32}{4}} \qquad \textit{Quotient Property of Square Roots}$$
$$= \sqrt{8} \qquad 32 \div 4 = 8$$
$$= \sqrt{2 \cdot 2 \cdot 2} \qquad \textit{Prime factorization}$$
$$= \sqrt{2^3} \qquad 2 \cdot 2 \cdot 2 = 2^3$$
$$= \sqrt{2^2} \cdot \sqrt{2} \qquad \textit{Product Property of Square Roots}$$
$$= 2\sqrt{2} \qquad \textit{Simplify.}$$

Your Turn **Simplify each expression. Leave in radical form.**

e. $\dfrac{\sqrt{72}}{\sqrt{6}}$ **f.** $\dfrac{\sqrt{160}}{\sqrt{2}}$

To eliminate radicals from the denominator of a fraction, you can use a method for simplifying radical expressions called **rationalizing the denominator**.

Example ❺ **Simplify** $\dfrac{\sqrt{5}}{\sqrt{10}}$.

$$\frac{\sqrt{5}}{\sqrt{10}} = \frac{\sqrt{5}}{\sqrt{10}} \cdot \frac{\sqrt{10}}{\sqrt{10}} \qquad \frac{\sqrt{10}}{\sqrt{10}} = 1$$
$$= \frac{\sqrt{5 \cdot 10}}{\sqrt{10 \cdot 10}} \qquad \textit{Product Property of Square Roots}$$

(continued on the next page)

$$= \frac{\sqrt{50}}{\sqrt{100}} \qquad \textit{Simplify.}$$

$$= \frac{\sqrt{25 \cdot 2}}{10} \qquad \textit{50 = 25 · 2}$$

$$= \frac{\sqrt{25} \cdot \sqrt{2}}{10} \qquad \textit{Product Property of Square Roots}$$

$$= \frac{5 \cdot \sqrt{2}}{10} \text{ or } \frac{\sqrt{2}}{2} \qquad \textit{Simplify.}$$

Your Turn **Simplify each expression. Leave in radical form.**

g. $\dfrac{\sqrt{6}}{\sqrt{8}}$

h. $\dfrac{\sqrt{32}}{\sqrt{3}}$

Binomials of the form $a\sqrt{b} + c\sqrt{d}$ and $a\sqrt{b} - c\sqrt{d}$ are **conjugates** of each other because their product is a rational number.

Look Back

Product of a Sum and a Difference: Lesson 9–5

$$(6 + \sqrt{3})(6 - \sqrt{3}) = 6^2 - (\sqrt{3})^2 \qquad \textit{Use the pattern } (a + b)(a - b) = a^2 - b^2$$
$$\textit{to simplify the product.}$$
$$= 36 - 3$$
$$= 33$$

Conjugates are useful for simplifying radical expressions because their product is always a rational number.

Example ⑥ **Simplify $\dfrac{6}{3 - \sqrt{2}}$.**

To rationalize the denominator, multiply both the numerator and denominator by $3 + \sqrt{2}$, which is the conjugate of $3 - \sqrt{2}$.

$$\frac{6}{3 - \sqrt{2}} = \frac{6}{3 - \sqrt{2}} \cdot \frac{3 + \sqrt{2}}{3 + \sqrt{2}} \qquad \textit{Notice that } \frac{3 + \sqrt{2}}{3 + \sqrt{2}} = 1.$$

$$= \frac{6(3) + 6\sqrt{2}}{3^2 - (\sqrt{2})^2} \qquad \textit{Distributive Property}$$
$$\textit{(a − b)(a + b) = a}^2 - b^2$$

$$= \frac{18 + 6\sqrt{2}}{9 - 2} \qquad \textit{Simplify.}$$

$$= \frac{18 + 6\sqrt{2}}{7} \qquad \textit{Subtract.}$$

Your Turn **Simplify each expression. Leave in radical form.**

i. $\dfrac{3}{3 - \sqrt{5}}$

j. $\dfrac{4}{5 + \sqrt{6}}$

🌐nline **Personal Tutor at** algconcepts.com

Radical expressions are in simplest form if the following conditions are met.

Simplified Form for Radicals	A radical expression is in simplest form when the following three conditions have been met. **1.** No radicands have perfect square factors other than 1. **2.** No radicands contain fractions. **3.** No radicals appear in the denominator of a fraction.

Consider the expression $\sqrt{x^2}$. It appears that $\sqrt{x^2} = x$. However, if $x = -3$, then $\sqrt{(-3)^2}$ is 3, not -3. For radical expressions like $\sqrt{x^2}$, use absolute value to ensure nonnegative results. The results of simplifying a few radical expressions are listed below.

$$\sqrt{x^2} = |x| \quad \sqrt{x^3} = x\sqrt{x} \quad \sqrt{x^4} = x^2 \quad \sqrt{x^5} = x^2\sqrt{x} \quad \sqrt{x^6} = |x^3|$$

For $\sqrt{x^3}$, absolute value is not necessary. If x were negative, then x^3 would be negative, and $\sqrt{x^3}$ is not a real number. *Why is absolute value not used for $\sqrt{x^4}$?*

Example ⑦ **Simplify $\sqrt{98ab^2c^4}$. Use absolute value symbols if necessary.**

$\sqrt{98ab^2c^4}$

$= \sqrt{2 \cdot 7 \cdot 7 \cdot a \cdot b^2 \cdot c^4}$ *Prime factorization*

$= \sqrt{2 \cdot 49 \cdot a \cdot b^2 \cdot c^4}$ *$7 \times 7 = 49$*

$= \sqrt{2} \cdot \sqrt{49} \cdot \sqrt{a} \cdot \sqrt{b^2} \cdot \sqrt{c^4}$ *Product Property of Square Roots*

$= \sqrt{2} \cdot 7 \cdot \sqrt{a} \cdot |b| \cdot c^2$ *Simplify.*

$= 7|b|c^2\sqrt{2a}$ *The absolute value of b ensures a nonnegative result.*

Your Turn

Simplify each expression. Use absolute value symbols if necessary.

k. $\sqrt{63ab^2}$ **l.** $\sqrt{200x^2y^3}$

Check for Understanding

Communicating Mathematics

1. **Explain** why absolute values are sometimes needed when simplifying radical expressions containing variables.

2. **Explain** how you can show whether $8 + \sqrt{2}$ and $8 - \sqrt{2}$ are conjugates.

Vocabulary

radical expression
rationalizing the denominator
conjugates

3. LaToya says that, in simplest form, the expression $\dfrac{2}{3 + \sqrt{5}}$ is written as $\dfrac{6 - 2\sqrt{5}}{4}$. Greg disagrees. He says it should be written as $\dfrac{3 - \sqrt{5}}{2}$. Who is correct? Explain your reasoning.

Guided Practice

> ⏱ **Getting Ready** **State the conjugate of each expression. Then multiply the expression by its conjugate.**
>
> **Sample:** $1 + \sqrt{7}$ **Solution:** The conjugate is $1 - \sqrt{7}$.
> $$(1 + \sqrt{7})(1 - \sqrt{7}) = 1^2 - (\sqrt{7})^2$$
> $$= 1 - 7 \text{ or } -6$$

4. $5 - \sqrt{6}$ **5.** $2 + \sqrt{8}$ **6.** $2\sqrt{7} - 3\sqrt{2}$

Examples 1–6   **Simplify each expression. Leave in radical form.**

7. $\sqrt{75}$ **8.** $\sqrt{96}$ **9.** $\sqrt{6} \cdot \sqrt{15}$

10. $\dfrac{\sqrt{36}}{\sqrt{3}}$ **11.** $\dfrac{\sqrt{3}}{\sqrt{7}}$ **12.** $\dfrac{1}{6 - \sqrt{3}}$

Example 7   **Simplify each expression. Use absolute value symbols if necessary.**

13. $\sqrt{36x^2y}$ **14.** $\sqrt{50m^4n^5}$

Example 2   **15. Geometry** The radius of the circle is $\sqrt{32}$ units. Express the radius in simplest form.

Exercises • • • •   • • • • • • • • • • • • • •

Practice

Simplify each expression. Leave in radical form.

Homework Help	
For Exercises	See Examples
16–20, 40, 41	1, 2
21–24	3
25–30	4, 5
31–33	6
34–39	7
Extra Practice	
See page 721.	

16. $\sqrt{45}$ **17.** $\sqrt{98}$ **18.** $\sqrt{280}$

19. $\sqrt{500}$ **20.** $\sqrt{1000}$ **21.** $\sqrt{2} \cdot \sqrt{8}$

22. $\sqrt{6} \cdot \sqrt{18}$ **23.** $3\sqrt{5} \cdot \sqrt{5}$ **24.** $\sqrt{8} \cdot \sqrt{12}$

25. $\dfrac{\sqrt{48}}{\sqrt{3}}$ **26.** $\dfrac{\sqrt{52}}{\sqrt{4}}$ **27.** $\dfrac{\sqrt{80}}{\sqrt{2}}$

28. $\dfrac{\sqrt{3}}{\sqrt{8}}$ **29.** $\dfrac{\sqrt{4}}{\sqrt{5}}$ **30.** $\dfrac{\sqrt{5}}{\sqrt{12}}$

31. $\dfrac{2}{4 - \sqrt{3}}$ **32.** $\dfrac{5}{3 + \sqrt{2}}$ **33.** $\dfrac{4}{6 - \sqrt{7}}$

Simplify each expression. Use absolute value symbols if necessary.

34. $\sqrt{16gh^4}$

35. $\sqrt{40m^2}$

36. $\sqrt{47c^6d}$

37. $\sqrt{54a^2b^3}$

38. $\sqrt{125rst}$

39. $\sqrt{36x^3y^4z^5}$

Applications and Problem Solving

40. Quilting The quilt pattern *Sky Rocket* is shown at the right. Suppose the legs of the indicated triangle each measure 8 centimeters. Use the Pythagorean Theorem to find the measure of the hypotenuse. Express the answer as a radical in simplest form.

Sky Rocket

41. Science An L-shaped tube like the one shown can be used to measure the speed V in miles per hour of water in a river. By using the formula $V = \sqrt{2.5h}$, where h is the height in inches of the column of water above the surface, the speed of the water can be found.

a. Suppose the tube is placed in a river and the height of the column of water is 4.8 inches. What is the speed of the water?

b. What would h be if the speed is exactly 5 miles per hour?

42. Critical Thinking Is the sentence $\sqrt{a \cdot b} = \sqrt{a} \cdot \sqrt{b}$ true for negative values of a and b? Explain your reasoning.

Mixed Review

43. Point J is located at $(-1, 5)$, and point K is located at $(m, 2)$. What is the value of m if the points are $\sqrt{18}$ units apart? *(Lesson 14–2)*

Name the set or sets of numbers to which each real number belongs. Let N = natural numbers, W = whole numbers, Z = integers, Q = rational numbers and I = irrational numbers. *(Lesson 14–1)*

44. 0.75

45. $-\sqrt{144}$

46. $\dfrac{20}{4}$

47. Use substitution to solve $y = 2x + 3$ and $y - 2x = -5$. *(Lesson 13–3)*

48. Solve $15 > a - 3 > 10$. Graph the solution. *(Lesson 12–5)*

Standardized Test Practice
Ⓐ Ⓑ Ⓒ Ⓓ

49. Short Response Tell whether 9 is closer to $\sqrt{79}$ or $\sqrt{89}$. *(Lesson 8–6)*

50. Multiple Choice Two dice are rolled. Find the probability that a number less than 5 is rolled on one die and an odd number is rolled on the other die. *(Lesson 5–7)*

Ⓐ $\dfrac{1}{2}$ Ⓑ $\dfrac{2}{3}$ Ⓒ $\dfrac{1}{3}$ Ⓓ $\dfrac{3}{4}$

14-4 Adding and Subtracting Radical Expressions

What You'll Learn
You'll learn to add and subtract radical expressions.

Why It's Important
Hobbies Knowing how to add radical expressions can help you find the perimeter of a sail on a sailboat. *See Exercise 37.*

To find the exact perimeter of quadrilateral *ABCD*, you will need to add radical expressions.

Radical expressions with the same radicands can be added or subtracted in the same way that monomials are added or subtracted.

Monomials	**Radical Expressions**

$$5x + 3x = (5 + 3)x$$
$$\qquad = 8x$$

$$5\sqrt{2} + 3\sqrt{2} = (5 + 3)\sqrt{2}$$
$$\qquad = 8\sqrt{2}$$

Look Back

Adding and Subtracting Monomials: Lesson 9–2

$$8y - 2y = (8 - 2)y$$
$$\qquad = 6y$$

$$8\sqrt{3} - 2\sqrt{3} = (8 - 2)\sqrt{3}$$
$$\qquad = 6\sqrt{3}$$

Notice that the Distributive Property was used to simplify each radical expression.

Examples

Simplify each expression.

1 $6\sqrt{7} - 2\sqrt{7}$

$6\sqrt{7} - 2\sqrt{7} = (6 - 2)\sqrt{7}$ *Distributive Property*

$\qquad = 4\sqrt{7}$ *Simplify.*

2 $5\sqrt{5} + 3\sqrt{5} - 18\sqrt{5}$

$5\sqrt{5} + 3\sqrt{5} - 18\sqrt{5} = (5 + 3 - 18)\sqrt{5}$ *Distributive Property*

$\qquad = -10\sqrt{5}$ *Simplify.*

Your Turn

a. $8\sqrt{6} + 3\sqrt{6}$ b. $5\sqrt{2} - 12\sqrt{2}$

c. $4\sqrt{3} + 7\sqrt{3} - 2\sqrt{3}$ d. $3\sqrt{13} - 2\sqrt{13} - 6\sqrt{13}$

3 Refer to the beginning of the lesson. Find the exact perimeter of quadrilateral *ABCD*.

$$P = 6\sqrt{3} + 8\sqrt{2} + 8\sqrt{3} + 5\sqrt{2}$$ *Like terms: $6\sqrt{3}$ and $8\sqrt{3}$; $8\sqrt{2}$ and $5\sqrt{2}$*

$$= 6\sqrt{3} + 8\sqrt{3} + 8\sqrt{2} + 5\sqrt{2}$$ *Commutative Property*

$$= (6 + 8)\sqrt{3} + (8 + 5)\sqrt{2}$$ *Distributive Property*

$$= 14\sqrt{3} + 13\sqrt{2}$$ *Simplify.*

The exact perimeter of quadrilateral *ABCD* is $14\sqrt{3} + 13\sqrt{2}$ centimeters.

In Example 3, the expression $14\sqrt{3} + 13\sqrt{2}$ cannot be simplified further for the following reasons.

- The radicands are different.
- There are no common factors.
- Each radicand is in simplest form.

If the radicals in a radical expression are not in simplest form, simplify them first. Then use the Distributive Property wherever possible to further simplify the expression.

Example

4 Simplify $2\sqrt{20} - \sqrt{45}$.

$$2\sqrt{20} - \sqrt{45} = 2\sqrt{2^2 \cdot 5} - \sqrt{3^2 \cdot 5}$$ *Prime factorization*

$$= 2(2\sqrt{5}) - (3\sqrt{5})$$ *Simplify.*

$$= 4\sqrt{5} - 3\sqrt{5}$$ *Multiply.*

$$= (4 - 3)\sqrt{5}$$ *Distributive Property*

$$= 1\sqrt{5} \text{ or } \sqrt{5}$$ *Simplify.*

Your Turn Simplify each expression.

e. $4\sqrt{27} - 5\sqrt{12}$ f. $7\sqrt{18} + 3\sqrt{50}$

Check for Understanding

1. **Describe** in your own words how to add radical expressions.

2. **Explain** why you should simplify each radical in a radical expression before adding or subtracting.

3. Writing Math Explain how you use the Distributive Property to simplify the sum or difference of like radicals.

Getting Ready Express each radical in simplest form. Then determine the like radicals.

Sample: $5\sqrt{6}, 3\sqrt{5}, 3\sqrt{20}$ **Solution:** $5\sqrt{6} = 5\sqrt{6}$
$3\sqrt{5} = 3\sqrt{5}$ ✓
$3\sqrt{20} = 6\sqrt{5}$ ✓

4. $4\sqrt{2}, -3\sqrt{18}, 2\sqrt{6}$ 5. $-3\sqrt{10}, 4\sqrt{5}, -5\sqrt{3}$

6. $-2\sqrt{7}, 6\sqrt{14}, 5\sqrt{28}$ 7. $4\sqrt{12}, -7\sqrt{6}, -6\sqrt{27}$

Examples 1, 2, & 4   **Simplify each expression.**

8. $7\sqrt{6} + 4\sqrt{6}$ 9. $4\sqrt{5} - 2\sqrt{5}$

10. $8\sqrt{3} - 3\sqrt{3} + 7\sqrt{3}$ 11. $3\sqrt{7} - 6\sqrt{7} - 2\sqrt{7}$

12. $5\sqrt{27} + 2\sqrt{48}$ 13. $2\sqrt{50} - 4\sqrt{32}$

Example 3   14. **Geometry** Find the exact perimeter of triangle ABC.

Exercises

• • • • • • • • • • • • • • • •

Practice

Simplify each expression.

15. $4\sqrt{3} + 2\sqrt{3}$ 16. $12\sqrt{2} + 3\sqrt{2}$

17. $15\sqrt{5} - \sqrt{5}$ 18. $5\sqrt{2} - 8\sqrt{3}$

Homework Help	
For Exercises	See Examples
15–17, 19–22 24–26, 37	1, 2
18, 23, 27–34	4
35	3
Extra Practice	
See page 721.	

19. $-6\sqrt{7} + 10\sqrt{7}$ 20. $-7\sqrt{3} - 2\sqrt{3}$

21. $5\sqrt{5} + 3\sqrt{5} - 10\sqrt{5}$ 22. $6\sqrt{7} - 5\sqrt{7} + 8\sqrt{7}$

23. $7\sqrt{2} - 16\sqrt{2} + 8\sqrt{5}$ 24. $5\sqrt{11} + 2\sqrt{11} - 4\sqrt{11}$

25. $-\sqrt{6} - 4\sqrt{6} + 12\sqrt{6}$ 26. $-9\sqrt{14} + 2\sqrt{14} - 6\sqrt{14}$

27. $4\sqrt{3} + 5\sqrt{12}$ 28. $-2\sqrt{28} + 3\sqrt{7}$

29. $2\sqrt{18} - 5\sqrt{8}$ 30. $3\sqrt{63} - \sqrt{112}$

31. $\sqrt{50} + \sqrt{108} + \sqrt{18}$ 32. $-4\sqrt{27} - 5\sqrt{32} + 2\sqrt{75}$

33. Simplify $2\sqrt{20} - \sqrt{180} - 3\sqrt{24}$.

34. Write $-2\sqrt{50} + 5\sqrt{48} + 7\sqrt{98}$ in simplest form.

Applications and Problem Solving

35. **Geometry** The perimeter of quadrilateral $DEFG$ is $14\sqrt{7}$ feet. What is the missing measure?

Close to Home

COUNT OFF BY THE SQUARE ROOT OF 7!

UHH....

36. **Media**   Refer to the comic. Suppose the first team member says "$\sqrt{7}$," the second says "$2\sqrt{7}$," and so on.
   a. Write an expression that could be used to determine the response of the fourth student in line.
   b. Determine the response of the fourth person in line.
   c. Suppose the coach asks the students to count off by $\sqrt{5}$. Find the response of the last person in line.

37. **Hobbies**   Ling is making a model of a sailboat. The dimensions of one of the sails are shown in the diagram.
   a. Find the missing measure. (*Hint:* Use the Pythagorean Theorem.)
   b. Determine the exact perimeter of the sail.

6 in.

x in.

$2\sqrt{3}$ in.

38. **Critical Thinking**   Is the set of irrational numbers closed under addition? Explain your reasoning.

Mixed Review

39. Simplify $\sqrt{45mn^3p^2}$.   (*Lesson 14–3*)

Find the distance between each pair of points. Round to the nearest tenth, if necessary.   (*Lesson 14–2*)

40. $C(-3, 5)$, $D(0, 8)$

41. $J(4, -2)$, $K(-6, -1)$

42. **Geometry**   The base of a triangle measures 3 feet more than its height. The area of the triangle is 20 square feet. Write and solve a quadratic equation to find the height and base of the triangle. (*Lesson 11–4*)

Factor each monomial.   (*Lesson 10–1*)

43. $-26a^3b^2$

44. $36xy^4z$

Standardized Test Practice
Ⓐ Ⓑ Ⓒ Ⓓ

45. **Short Response**   Find the product of $a + 2c$ and $a - 2c$.   (*Lesson 9–5*)

46. **Multiple Choice**   The scatter plot shows the relationship between the cost of a stereo system and the rating it received by a consumer group. Which of the following best describes the relationship? (*Lesson 7–4*)

   Ⓐ positive
   Ⓑ dependent
   Ⓒ no pattern
   Ⓓ negative

Stereo Systems

Cost (dollars)

850
800
750
700
650
600
550
500

0 1 2 3 4 5 6 7 8 9 10
Rating

Lesson 14–4 Adding and Subtracting Radical Expressions

The speed of a roller coaster as it travels through a loop depends on the height of the hill from which the coaster has just descended. The equation $s = 8\sqrt{h - 2r}$ gives the speed s in feet per second where h is the height of the hill and r is the radius of the loop.

Suppose the owner of an amusement park wants to design a roller coaster that will travel at a speed of 40 feet per second as it goes through a loop with a radius of 30 feet. How high should the hill be? *This problem will be solved in Example 5.*

Equations like $s = 8\sqrt{h - 2r}$ that contain radicals with variables in the radicand are called **radical equations**. To solve these equations, first isolate the radical on one side of the equation. Then square each side of the equation to eliminate the radical.

Examples

Solve each equation. Check your solution.

① $\sqrt{x} + 5 = 8$

$$\sqrt{x} + 5 = 8 \qquad \text{\textit{Original equation}}$$
$$\sqrt{x} + 5 - 5 = 8 - 5 \qquad \text{\textit{Subtract 5 from each side.}}$$
$$\sqrt{x} = 3 \qquad \text{\textit{Simplify.}}$$
$$(\sqrt{x})^2 = 3^2 \qquad \text{\textit{Square each side.}}$$
$$x = 9 \qquad \text{\textit{Simplify.}}$$

Check: $\sqrt{x} + 5 = 8 \qquad$ *Original equation*
$$\sqrt{9} + 5 \overset{?}{=} 8 \qquad \text{\textit{Replace x with 9.}}$$
$$3 + 5 \overset{?}{=} 8 \qquad \sqrt{9} = 3$$
$$8 = 8 \quad \checkmark$$

The solution is 9.

② $\sqrt{m + 3} + 2 = 6$

$$\sqrt{m + 3} + 2 = 6 \qquad \text{\textit{Original equation}}$$
$$\sqrt{m + 3} + 2 - 2 = 6 - 2 \qquad \text{\textit{Subtract 2 from each side.}}$$
$$\sqrt{m + 3} = 4 \qquad \text{\textit{Simplify.}}$$
$$(\sqrt{m + 3})^2 = 4^2 \qquad \text{\textit{Square each side.}}$$
$$m + 3 = 16 \qquad \text{\textit{Simplify.}}$$
$$m + 3 - 3 = 16 - 3 \qquad \text{\textit{Subtract 3 from each side.}}$$
$$m = 13 \qquad \text{\textit{Check this result.}}$$

The solution is 13.

Your Turn Solve each equation. Check your solution.

a. $\sqrt{y} - 6 = 4$

b. $\sqrt{a - 1} + 5 = 7$

You can use a graphing calculator to solve radical equations.

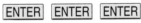

Graphing Calculator Tutorial
See pp. 750–753.

Graphing Calculator Exploration

Find the solution of $\sqrt{x} + 5 = 8$.

Step 1 Set the viewing window for x: [0, 15] by 1 and y: [0, 15] by 1.

Step 2 Press [Y=]. Enter the left side of the equation, $\sqrt{x} + 5$, as Y_1. Enter the right side of the equation, 8, as Y_2.

Step 3 Press [GRAPH] to graph the equations.

Step 4 The solution of the equation is the intersection point of the two lines. Use the INTERSECT feature to find the intersection point.

Enter: [2nd] [CALC] 5

[ENTER] [ENTER] [ENTER]

The graph intersects at the point (9, 8). The solution is the x-coordinate, 9.

Try These

1. Solve the equation in Example 2 by using a graphing calculator.
2. Use a graphing calculator to find the solution of Your Turn Exercise b.
3. What changes must be made to the graphing calculator before finding the solution of Your Turn Exercise a?

Squaring each side of an equation may produce results that do not satisfy the *original* equation. So, you must check all solutions when you solve radical equations.

Examples

Solve each equation. Check your solution.

3 $\sqrt{n + 2} = n - 4$

$$\sqrt{n + 2} = n - 4 \qquad \text{\textit{Original equation}}$$
$$(\sqrt{n + 2})^2 = (n - 4)^2 \qquad \text{\textit{Square each side.}}$$
$$n + 2 = n^2 - 8n + 16 \qquad \text{\textit{Simplify.}}$$
$$n - n + 2 - 2 = n^2 - 8n + 16 - n - 2 \qquad \text{\textit{Subtract n and 2 from each side.}}$$
$$0 = n^2 - 9n + 14 \qquad \text{\textit{Simplify.}}$$
$$0 = (n - 7)(n - 2) \qquad \text{\textit{Factor.}}$$

(continued on the next page)

$$n - 7 = 0 \quad \text{or} \quad n - 2 = 0 \quad \textit{Use the Zero Product Property.}$$
$$n = 7 \qquad\qquad n = 2 \quad \textit{Solve.}$$

Check: $\sqrt{n + 2} = n - 4$ $\qquad\qquad$ $\sqrt{n + 2} = n - 4$

$\qquad\quad \sqrt{7 + 2} \stackrel{?}{=} 7 - 4$ $\qquad\qquad$ $\sqrt{2 + 2} \stackrel{?}{=} 2 - 4$

$\qquad\qquad\quad \sqrt{9} \stackrel{?}{=} 3$ $\qquad\qquad\qquad$ $\sqrt{4} \stackrel{?}{=} -2$

$\qquad\qquad\qquad 3 = 3 \;\; \checkmark$ $\qquad\qquad\qquad$ $2 \neq -2$

Since 2 does not satisfy the original equation, 7 is the only solution.

④ $\sqrt{3h - 5} + 5 = h$

$\sqrt{3h - 5} + 5 = h$	*Original equation*
$\sqrt{3h - 5} + 5 - 5 = h - 5$	*Subtract 5 from each side.*
$\sqrt{3h - 5} = h - 5$	*Simplify.*
$(\sqrt{3h - 5})^2 = (h - 5)^2$	*Square each side.*
$3h - 5 = h^2 - 10h + 25$	*Simplify.*
$3h - 3h - 5 + 5 = h^2 - 10h + 25 - 3h + 5$	*Add $-3h$ and 5 to each side.*
$0 = h^2 - 13h + 30$	*Simplify.*
$0 = (h - 10)(h - 3)$	*Factor.*

$$h - 10 = 0 \quad \text{or} \quad h - 3 = 0 \qquad \textit{Use the Zero Product}$$
$$h = 10 \qquad\qquad h = 3 \qquad\qquad \textit{Property.}$$

Check: $\sqrt{3h - 5} + 5 = h$ $\qquad\qquad$ $\sqrt{3h - 5} + 5 = h$

$\qquad\quad \sqrt{3(10) - 5} + 5 \stackrel{?}{=} 10$ $\qquad\qquad$ $\sqrt{3(3) - 5} + 5 \stackrel{?}{=} 3$

$\qquad\qquad \sqrt{30 - 5} + 5 \stackrel{?}{=} 10$ $\qquad\qquad$ $\sqrt{9 - 5} + 5 \stackrel{?}{=} 3$

$\qquad\qquad\quad \sqrt{25} + 5 \stackrel{?}{=} 10$ $\qquad\qquad\qquad$ $\sqrt{4} + 5 \stackrel{?}{=} 3$

$\qquad\qquad\qquad 5 + 5 \stackrel{?}{=} 10$ $\qquad\qquad\qquad\quad$ $2 + 5 \stackrel{?}{=} 3$

$\qquad\qquad\qquad\quad 10 = 10 \;\; \checkmark$ $\qquad\qquad\qquad\qquad$ $7 \neq 3$

Since 3 does not satisfy the original equation, 10 is the only solution.

Your Turn

c. $\sqrt{d + 1} = d - 1$ $\qquad\qquad\qquad$ **d.** $\sqrt{3x - 14} + x = 6$

Radical equations are used in many real-life situations.

Example

Engineering Link

⑤ **Refer to the application at the beginning of the lesson. How high should the hill be?**

Explore You know the speed of the coaster and the radius of the loop. You need to know the height of the hill.

Data Update For the latest information on roller coasters, visit: www.algconcepts.com

Plan   Use the equation $s = 8\sqrt{h - 2r}$ to find the height of the hill.

Solve

$s = 8\sqrt{h - 2r}$	*Original equation*
$40 = 8\sqrt{h - 2(30)}$	*Replace s with 40 and r with 30.*
$40 = 8\sqrt{h - 60}$	*Multiply.*
$\dfrac{40}{8} = \dfrac{8\sqrt{h - 60}}{8}$	*Divide each side by 8.*
$5 = \sqrt{h - 60}$	*Simplify.*
$5^2 = (\sqrt{h - 60})^2$	*Square each side.*
$25 = h - 60$	*Simplify.*
$25 + 60 = h - 60 + 60$	*Add 60 to each side.*
$85 = h$	*Simplify.*

The height of the hill is 85 feet.

Examine   Check the result.

$s = 8\sqrt{h - 2r}$	*Original equation*
$40 \overset{?}{=} 8\sqrt{85 - 2(30)}$	*Replace s with 40, h with 85, and r with 30.*
$40 \overset{?}{=} 8\sqrt{85 - 60}$	*Multiply.*
$40 \overset{?}{=} 8\sqrt{25}$	*Subtract.*
$40 \overset{?}{=} 8 \cdot 5$	$\sqrt{25} = 5$
$40 = 40$ ✓	

The answer is correct.

Check for Understanding

Communicating Mathematics

1. **Tell** the first step you should take when solving a radical equation.

2. **Explain** why it is important to check radical equations after finding possible solutions.

3. **Writing Math** Write an example of a radical equation. Then write the steps for solving the equation.

Vocabulary
radical equation

Guided Practice

Getting Ready **Square each side of the following equations.**

Sample: $\sqrt{y - 1} = 4$

Solution: $\sqrt{y - 1} = 4$
$(\sqrt{y - 1})^2 = 4^2$
$y - 1 = 16$

4. $\sqrt{x} = 7$ 5. $\sqrt{a + 5} = 2$ 6. $9 = \sqrt{2c - 3}$

Solve each equation. Check your solution.

7. $\sqrt{x} - 4 = 7$

8. $\sqrt{a-2} - 5 = 3$

9. $\sqrt{m+5} + 1 = m$

10. $\sqrt{1+2x} = x + 1$

Example 5

11. **Engineering** Refer to the application at the beginning of the lesson. Suppose the owner of an amusement park wants to design a roller coaster that will travel through a loop at a speed of 56 feet per second after descending a 120-foot hill. What should the radius of the loop be?

Exercises

Practice

Solve each equation. Check your solution.

12. $\sqrt{x} = 3$

13. $\sqrt{a} = -4$

14. $7 = \sqrt{7m}$

15. $\sqrt{-3d} = 6$

16. $\sqrt{y} + 5 = 0$

17. $0 = \sqrt{2c} - 2$

18. $\sqrt{m-4} = 6$

19. $\sqrt{w+6} = 9$

20. $3 = \sqrt{4n+1}$

21. $11 = \sqrt{2z-5}$

22. $\sqrt{8h+1} - 5 = 0$

23. $2 = \sqrt{3b-5} + 6$

24. $\sqrt{x+6} = x$

25. $p = \sqrt{5p-6}$

26. $\sqrt{8-b} = b - 2$

27. $\sqrt{k-2} + 4 = k$

28. $t - 1 = \sqrt{2t+6}$

29. $5 + \sqrt{3j-5} = j$

Homework Help	
For Exercises	See Examples
12–23, 30–33	1, 2
24–29	3, 4
Extra Practice	
See page 721.	

30. Solve $\sqrt{\dfrac{a}{4}} = 6$.

31. Find the solution of $\sqrt{\dfrac{5x}{7}} - 8 = 2$.

Applications and Problem Solving

32. **Science** The speed of sound S, in meters per second, near Earth's surface can be determined using the formula $S = 20\sqrt{t + 273}$, where t is the surface temperature in degrees Celsius. Suppose a racing team has designed a car that can travel 340 meters per second, in hopes of breaking the sound barrier. At what temperature will the speed of sound be 340 meters per second?

33. **Science** The formula $t = \sqrt{\dfrac{2s}{g}}$ can be used to determine the time t, in seconds, it takes an object initially at rest to fall s meters. In this formula, g is the acceleration due to gravity in meters per second squared.

a. Suppose a rock falls 7.2 meters in 3 seconds on the moon. What is the acceleration due to gravity on the moon?

b. Suppose a rock falls 78.4 meters in 4 seconds on Earth. What is the acceleration due to gravity on Earth?

34. Critical Thinking Find two numbers such that the square root of their sum is 5 and the square root of their product is 12.

Simplify each expression. *(Lesson 14–4)*

35. $-7\sqrt{5} + 2\sqrt{5}$ **36.** $4\sqrt{7} - 3\sqrt{28}$ **37.** $-\sqrt{54} - \sqrt{18} + \sqrt{24}$

Simplify each expression. Leave in radical form. *(Lesson 14–3)*

38. $\sqrt{5} \cdot \sqrt{12}$ **39.** $\dfrac{\sqrt{15}}{\sqrt{3}}$ **40.** $\dfrac{5}{4 - \sqrt{7}}$

41. Find the solution of $y = \frac{1}{2}x - 4$ and $y = -3$ by graphing. *(Lesson 13–1)*

Write an inequality for each graph. *(Lesson 12–1)*

42.
```
<---+--●--+--+--+--+--+--->
   -4 -3 -2 -1  0  1  2
```
43.
```
<---+--+--+--+--+--◇--+--+--->
   -3 -2 -1  0  1  2  3
```

44. Short Response There are two field mice living in a barn. Suppose the number of mice triples every 4 months. How many mice will there be after 3 years if none of them die? *(Lesson 11–7)*

45. Grid In Suppose y varies inversely as x and $y = 18$ when $x = 15$. Find y when $x = 12$. *(Lesson 6–6)*

Quiz 2   Lessons 14–3 through 14–5

Simplify each expression. Leave in radical form. *(Lesson 14–3)*

1. $\sqrt{90}$ **2.** $\sqrt{5} \cdot \sqrt{25}$ **3.** $\dfrac{1}{7 + \sqrt{5}}$

4. Write $\sqrt{16x^2y^3}$ in simplest form. Use absolute value symbols if necessary. *(Lesson 14–3)*

Simplify each expression. *(Lesson 14–4)*

5. $3\sqrt{5} + 9\sqrt{5}$

6. $2\sqrt{40} - 7\sqrt{2} + 6\sqrt{10}$

7. Geometry The perimeter of $\triangle ABC$ is $16\sqrt{2}$ inches. Find the value of x. *(Lesson 14–4)*

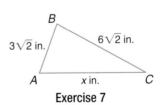

Exercise 7

Solve each equation. Check your solution. *(Lesson 14–5)*

8. $\sqrt{x} = 4$ **9.** $\sqrt{a} + 8 = 0$ **10.** $\sqrt{m + 3} = 5$

Understanding and Using the Vocabulary

*inter***NET**
CONNECTION **Review Activities**
For more review activities, visit:
www.algconcepts.com

After completing this chapter, you should be able to define each term, property, or phrase and give an example or two of each.

conjugates *(p. 616)*
Distance Formula *(p. 607)*

radical equations *(p. 624)*
real numbers *(p. 600)*

rationalizing the
denominator *(p. 615)*

Choose the correct term to complete each sentence.

1. To find the distance d between any two points (x_1, y_1) and (x_2, y_2), use the formula $(\sqrt{(x_2 - x_1)^2 - (y_2 - y_1)^2}, \sqrt{(x_2 - x_1)^2 + (y_2 - y_1)^2})$.

2. The product of conjugates is (sometimes, always) a rational number.

3. Rationalizing the denominator is a way of eliminating (radicals, perfect squares) from the denominator of a fraction.

4. Natural numbers, whole numbers, and integers are all (irrational, rational) numbers.

5. The expression $(\sqrt{7ab}, \sqrt{8mn})$ is in simplest form.

6. Radical equations (sometimes, always) have more than one solution.

7. The number 0 is a (whole, natural) number.

8. Binomials of the form $a\sqrt{b} + c\sqrt{d}$ and $a\sqrt{b} - c\sqrt{d}$ are (conjugates, radical equations).

9. The equation $\sqrt{x} + 4 = 0$ has (one, no) real solution.

10. The set of (integers, real numbers) contains the sets of rational and irrational numbers.

Skills and Concepts

Objectives	Examples
• **Lesson 14–1** Describe the relationships among sets of numbers.	Name the set or sets of numbers to which each real number belongs. Let **N** = natural numbers, **W** = whole numbers, **Z** = integers, **Q** = rational numbers, and **I** = irrational numbers.
$\sqrt{16} = 4$; so $\sqrt{16}$ is a natural number, a whole number, an integer, and a rational number.	11. $-\frac{1}{4}$ 12. π
The number -3 is an integer and a rational number.	13. $-\sqrt{25}$ 14. $\frac{24}{6}$
2.151151115 . . . is not the square root of a perfect square. So, it is an irrational number.	Find an approximation, to the nearest tenth, for each square root. Then graph the square root on a number line.
	15. $\sqrt{5}$ 16. $-\sqrt{15}$
	17. $\sqrt{111}$ 18. $-\sqrt{260}$

 www.algconcepts.com/vocabulary_review

Objectives and Examples

Review Exercises

• **Lesson 14-2** Find the distance between two points in the coordinate plane.

Find the distance between points $A(3, -2)$ and $B(5, 8)$.

$$d = \sqrt{(x_2 - x_1)^2 + (y_2 - y_1)^2}$$
$$= \sqrt{(5 - 3)^2 + (8 - (-2))^2}$$
$$= \sqrt{(2)^2 + (10)^2}$$
$$= \sqrt{4 + 100}$$
$$= \sqrt{104}$$
$$\approx 10.2$$

Find the distance between each pair of points. Round to the nearest tenth, if necessary.

19. $P(1, 2), Q(4, 6)$
20. $J(-6, -3), K(1, 0)$
21. $X(4, -1), Y(-2, 5)$
22. $R(-9, -5), S(-2, 19)$

Find the value of a if the points are the indicated distance apart.

23. $G(a, 2), H(-3, 5); d = \sqrt{13}$
24. $C(4, -7), D(-6, a); d = \sqrt{116}$

• **Lesson 14-3** Simplify radical expressions.

$$\sqrt{288} = \sqrt{12^2 \cdot 2}$$
$$= 12\sqrt{2}$$

$$\sqrt{32x^3y^2} = \sqrt{32x^3y^2}$$
$$= \sqrt{2 \cdot 2 \cdot 2 \cdot 2 \cdot 2 \cdot x^3 \cdot y^2}$$
$$= \sqrt{2 \cdot 16 \cdot x^3 \cdot y^2}$$
$$= \sqrt{2} \cdot \sqrt{16} \cdot \sqrt{x^3} \cdot \sqrt{y^2}$$
$$= \sqrt{2} \cdot 4 \cdot x \cdot \sqrt{x} \cdot |y|$$
$$= 4x|y|\sqrt{2x}$$

Simplify each expression. Leave in radical form.

25. $\sqrt{50}$ **26.** $\sqrt{3} \cdot \sqrt{6}$

27. $\dfrac{\sqrt{54}}{\sqrt{2}}$ **28.** $\dfrac{\sqrt{7}}{\sqrt{13}}$

29. $\dfrac{1}{4 - \sqrt{5}}$ **30.** $\dfrac{3}{13 - \sqrt{2}}$

Simplify each expression. Use absolute value symbols if necessary.

31. $\sqrt{300a^2bc^4}$ **32.** $\sqrt{121m^5n^4p^3}$

• **Lesson 14-4** Add and subtract radical expressions.

$$8\sqrt{7} - 6\sqrt{7} = (8 - 6)\sqrt{7}$$
$$= 2\sqrt{7}$$

$$13\sqrt{2} + 4\sqrt{2} - 7\sqrt{2} = (13 + 4 - 7)\sqrt{2}$$
$$= 10\sqrt{2}$$

$$\sqrt{54} + \sqrt{96} = \sqrt{3^2 \cdot 6} + \sqrt{4^2 \cdot 6}$$
$$= 3\sqrt{6} + 4\sqrt{6}$$
$$= (3 + 4)\sqrt{6} \text{ or } 7\sqrt{6}$$

Simplify each expression.

33. $2\sqrt{5} + 5\sqrt{5}$
34. $3\sqrt{8} - 6\sqrt{8}$
35. $-5\sqrt{6} + 14\sqrt{6} - 9\sqrt{6}$
36. $12\sqrt{3} - 14\sqrt{3} + 6\sqrt{3}$
37. $7\sqrt{2} + 5\sqrt{18}$
38. $-5\sqrt{40} + 10\sqrt{10}$

39. Simplify $2\sqrt{45} - 5\sqrt{80}$.
40. Write $2\sqrt{45} + 3\sqrt{75} - \sqrt{50}$ in simplest form.

Mixed Problem Solving
See pages 724–731.

Objectives and Examples

- **Lesson 14–5** Solve simple radical equations in which only one radical contains a variable.

Solve $\sqrt{x + 4} - 8 = -5$.

$$\sqrt{x + 4} - 8 = -5$$
$$\sqrt{x + 4} - 8 + 8 = -5 + 8$$
$$\sqrt{x + 4} = 3$$
$$(\sqrt{x + 4})^2 = 3^2$$
$$x + 4 = 9$$
$$x + 4 - 4 = 9 - 4$$
$$x = 5$$

The solution is 5.

Review Exercises

Solve each equation. Check your solution.

41. $\sqrt{y} = 5$ **42.** $\sqrt{m} = -3$

43. $\sqrt{9h} = 9$ **44.** $\sqrt{-4a} = 8$

45. $7 = \sqrt{a + 10}$ **46.** $\sqrt{n - 12} = 15$

47. $t - 1 = \sqrt{3t + 7}$ **48.** $\sqrt{4x - 3} = x$

49. Solve $\sqrt{\dfrac{g}{6}} = 5$.

50. Find the solution of $\sqrt{\dfrac{3x}{4}} + 2 = 5$.

Applications and Problem Solving

51. Weather The graph shows record low temperatures. Name the set or sets of numbers to which the temperatures belong. *(Lesson 14–1)*

Record Lows

Lowest in USA · Lowest in Contiguous 48 States · Lowest East of Miss. River

Prospect Creek, Alaska — Jan. 23, 1971
Rogers Pass, Montana — Jan. 20, 1954
Couderay, Wisconsin — Feb. 4, 1996

Source: National Oceanic and Atmospheric Administration

52. Geometry Quadrilateral *EFGH* is a square with vertices $E(2, 5)$, $F(6, 1)$, $G(2, -3)$, and $H(-2, 1)$. *(Lesson 14–2)*
  a. Draw quadrilateral *EFGH* on a coordinate plane.
  b. Determine the perimeter of the figure. Round to the nearest tenth.

53. Gems The weight of the largest pearl ever found in a giant clam is about $\sqrt{200}$ pounds. Express the weight in simplest radical form. *(Lesson 14–3)*

54. Geometry Triangle *XYZ* has a perimeter of $28\sqrt{5}$ inches. What is the measure of side *YZ*? *(Lesson 14–4)*

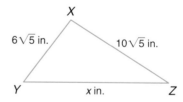

55. Nature The equation $S = 3\sqrt{d}$ gives the speed *S* of a tidal wave in meters per second if *d* is the depth of the water in meters. *(Lesson 14–5)*
  a. Suppose a tidal wave is traveling 186 meters per second. What is the depth of the water?
  b. What is the depth of the water if a tidal wave is traveling 153 meters per second?

1. **Name** two examples of each of the following.
   a. integer **b.** irrational number **c.** rational number

2. **Write** a pair of conjugates.

Name the set or sets of numbers to which each real number belongs. Let N = natural numbers, W = whole numbers, Z = integers, Q = rational numbers, and I = irrational numbers.

3. $\dfrac{5}{3}$ 　　4. 0 　　5. $-0.45275563\ldots$ 　　6. $-\sqrt{81}$

7. Graph $\sqrt{5}$, $-\sqrt{3}$, and 4.5 on a number line.

8. Write 3.22, -3.22, $\sqrt{11}$, and $3.\overline{2}$ in order from least to greatest.

Find the distance between each pair of points. Round to the nearest tenth, when necessary.

9. $G(7, 4)$, $H(-2, 4)$ 　　10. $X(3, -3)$, $Y(-9, 2)$ 　　11. $M(1, -1)$, $N(-5, 1)$

12. Suppose $A(5, m)$ and $B(8, 1)$ are 5 units apart. What is the value of m?

Simplify each expression. Leave in radical form.

13. $\sqrt{80}$ 　　14. $\sqrt{3} \cdot \sqrt{27}$ 　　15. $4\sqrt{3} \cdot \sqrt{6}$

16. $\dfrac{\sqrt{96}}{\sqrt{8}}$ 　　17. $\dfrac{\sqrt{7}}{\sqrt{11}}$ 　　18. $\dfrac{7}{7 + \sqrt{5}}$

19. $-4\sqrt{7} + 6\sqrt{7} - 12\sqrt{7}$ 　　20. $5\sqrt{18} - 2\sqrt{50}$

Simplify each expression. Use absolute value symbols if necessary.

21. $\sqrt{25a^3b^2}$ 　　22. $\sqrt{56m^4np}$

Solve each equation. Check your solution.

23. $\sqrt{b} + 8 = 5$ 　　24. $\sqrt{4x - 3} = 6 - x$

25. **Science** The distance d in miles a person can see on any planet is given by the formula $d = \sqrt{\dfrac{rh}{2640}}$, where r is the radius of the planet in miles and h is the height of the person in feet. Suppose a person 6 feet tall is standing on Mars. If the radius of the planet is 2109 miles, how far can the person see? Round to the nearest tenth.

Systems of Equations Problems

On standardized tests, you will often need to translate problems into systems of equations and inequalities. You will also need to solve systems of equations using substitution or addition methods.

Test-Taking Tip
To solve systems of equations on standardized tests, try adding them first. This technique may work because the systems often are of the form $x + y = 6$ and $x - y = 4$.

Example 1

Hector has 20 coins. They are all quarters and nickels. The total value is $2.20. Which system of equations can be used to determine the number of quarters q and the number of nickels n?

- **A** $q + n = 20$
  $0.30qn = 2.20$
- **B** $q + n = 20$
  $0.25q + 0.05n = 2.20$
- **C** $q + n = 20$
  $0.05q + 0.25n = 2.20$
- **D** $q + n = 20$
  $0.25q + 0.05n = 20$

Hint   First write the equations in words and then translate them into symbols.

Solution   The number of quarters plus the number of nickels is 20.

$$q + n = 20$$

The value of the quarters plus the value of the nickels equals the total value. The value of the quarters is 25 cents times the number of quarters or $0.25q$. The value of the nickels is 5 cents times the number of nickels or $0.05n$.

$$0.25q + 0.05n = 2.20$$

The answer is B.

Example 2

If $2x + 3y = 20$ and $3x + 2y = 40$, what is the value of $x + y$?

Hint   Look for ways to solve a problem without lengthy calculations.

Solution   Read carefully. You must find the value of $x + y$, not the individual values of x and y.

Add the two equations.

$$\begin{aligned} 2x + 3y &= 20 \\ +\ 3x + 2y &= 40 \\ \hline 5x + 5y &= 60 \end{aligned}$$

Notice that the coefficients of x and y are both 5. Divide each side of the equation by 5. Simplify each term.

$$\frac{5x + 5y}{5} = \frac{60}{5}$$

$$\frac{\overset{1}{5}(x + y)}{\underset{1}{5}} = \frac{\overset{12}{60}}{\underset{1}{5}}$$

$$x + y = 12$$

The value of $x + y$ is 12. The answer is 12.

After you work each problem, record your answer on the answer sheet provided or on a sheet of paper.

Multiple Choice

1. Jared's total SAT score was 1340. His math score m was 400 points less than twice his verbal score v. Which system of equations can determine his scores? *(Lesson 13–1)*

- **A** $m + v = 1340$
  $m = 2v - 400$
- **B** $m + v = 1340$
  $m = 400 - 2v$
- **C** $m + v = 1340$
  $400m = 2v$
- **D** $m - v = 1340$
  $m = 2v - 400$

2. The flag for Monroe High School has an area of 120 square feet. A new flag will have a width $1\frac{1}{2}$ times the width of the old flag. The length is the same. What is the area, in square feet, of the new flag? *(Lesson 13–2)*

- **A** 160
- **B** 180
- **C** 240
- **D** 360

3. Where do the lines with equations $y = 2x - 2$ and $7x - 3y = 11$ intersect? *(Lesson 13–3)*

- **A** $(5, 8)$
- **B** $(8, 5)$
- **C** $\left(\frac{5}{8}, -1\right)$
- **D** $\left(\frac{5}{8}, 1\right)$
- **E** $\left(\frac{25}{16}, \frac{9}{8}\right)$

4. $\triangle LMN \sim \triangle PQR$. What is PQ? *(Lesson 5–1)*

- **A** 4.4 in.
- **B** 7 in.
- **C** 10 in.
- **D** 11.2 in.

5. Which equation represents the line graphed? *(Lesson 7–5)*

- **A** $y = -\frac{1}{3}x - 2$
- **B** $y = -\frac{1}{3}x + 2$
- **C** $y = \frac{1}{3}x - 2$
- **D** $y = \frac{1}{3}x + 2$

6. Gina has x marbles and Dawn has y marbles. Together they have q marbles. If Gina gives Dawn 3 of her marbles, then they will have an equal number. How many marbles does Dawn have? *(Lesson 1–1)*

- **A** $\frac{q - 6}{2}$
- **B** $\frac{q - 3}{2}$
- **C** $\frac{q - 6}{2}$
- **D** $q - 3$
- **E** $q + 3$

7. What is the slope of a line parallel to the line graphed? *(Lesson 7–7)*

- **A** -3
- **B** $-\frac{1}{3}$
- **C** $\frac{1}{3}$
- **D** 3

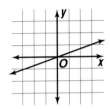

8. What value of n makes the equation true? *(Lesson 8–1)*

$$9^n - 8^n = 1^n$$

- **A** -1
- **B** 0
- **C** 1
- **D** 2

Grid In

9. $\angle ABC$ is a right angle, $\overrightarrow{BC}$ bisects $\angle EBD$. If $m\angle ABD$ is 130, then what is x? *(Lesson 3–6)*

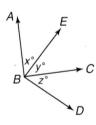

Extended Response

10. Talent show tickets are $3 for adults and $2 for students. Ticket sales of at least $600 are needed to cover costs. At least 5 times as many students as adults attend. *(Lesson 13–7)*

Part A Write a system of inequalities to find the number of adult and student tickets that need to be sold. Let x represent the number of adult tickets and let y represent the number of student tickets.

Part B Graph the system of inequalities. Give one example of adult and student ticket sales that would solve the system.

Rational Expressions and Equations

What You'll Learn

Key Ideas

- Simplify, multiply, and divide rational expressions.
  (Lessons 15–1 and 15–2)
- Divide polynomials by binomials. *(Lesson 15–3)*
- Add and subtract rational expressions. *(Lessons 15–4 and 15–5)*
- Solve rational equations. *(Lessons 15–6)*

Key Vocabulary

rational equation *(p. 668)*

rational expression *(p. 638)*

rational function *(p. 638)*

Why It's Important

Transportation As the United States spread from east to west in the 1800s, systems were built to connect the nation. Rivers and roads and then canals and railroads moved people and goods. By 1930, more than half of American families owned a car. Now airplanes provide even faster transportation.

Rational expressions are fractions in which the numerator and denominator are polynomials. You will use rational expressions to find travel times in Lesson 15–3.

Study these lessons to improve your skills.

✓ Check Your Readiness

Lesson 11–4, pp. 474–477

Solve each equation. Check your solution.

1. $4n(n - 1) = 0$ **2.** $2b(b + 4) = 0$ **3.** $(h - 1)(h + 2) = 0$

4. $(2x - 1)(x + 6) = 0$ **5.** $a^2 - 8a + 12 = 0$ **6.** $x^2 + 4x - 45 = 0$

Lesson 8–2, 341–345

Simplify each expression.

7. $k \cdot k^6$ **8.** $b \cdot 6b^3$ **9.** $3d^2 \cdot 4d^3$

10. $(p^3q)(p^2q^2)$ **11.** $5m^2 \cdot 6n^2$ **12.** $6a^2 \cdot 10ab$

Lesson 9–3, 394–398

Find each product.

13. $w(w + 4)$ **14.** $3k(k + 2)$ **15.** $5(3c - 4)$

16. $5m(n + 2)$ **17.** $r(r^2 + 3r - 1)$ **18.** $5t(4t^3 + t - 6)$

Lesson 9–4, pp. 399–404

19. $(y - 1)(y + 3)$ **20.** $(w - 2)(w - 4)$ **21.** $(p - 6)(p + 5)$

22. $(2r + 5)(r - 6)$ **23.** $(s - 2)(2s + 5)$ **24.** $(2g^2 + 1)(g - 1)$

FOLDABLES™ Study Organizer

Make this Foldable to help you organize your Chapter 15 notes. Begin with a sheet of notebook paper.

❶ Fold lengthwise to the holes.

❷ Cut along the top line and then cut ten tabs.

❸ Label the tabs using vocabulary words from the chapter as shown.

Reading and Writing Store the Foldable in a 3-ring binder. As you read and study the chapter, write notes and examples under each tab.

What You'll Learn

You'll learn to simplify rational expressions.

Why It's Important

Photography
Photographers use a rational expression to find the amount of light that falls on an object.
See Exercise 63.

A fraction denotes a quotient. In algebra, the fraction $\dfrac{2x}{x+3}$ is called a **rational expression.** In a rational expression, both the numerator and denominator are polynomials.

Rational Expression	A rational expression is an algebraic fraction whose numerator and denominator are polynomials.

Every polynomial is a rational expression because it can be written as a quotient with 1 in the denominator.

Zero cannot be the denominator of a fraction because division by zero is undefined. In the expression $\dfrac{2x}{x+3}$, if $x = -3$, the denominator equals zero. So, any value assigned to a variable that results in a denominator of zero must be excluded from the domain of the variable. These values are called **excluded values.**

A function that contains a rational expression is called a **rational function.** You can use the graph of the rational function of a rational expression to investigate excluded values of the variable.

Graphing Calculator Tutorial
See pp. 750–753.

 Graphing Calculator Exploration

The screen shows the graph of the rational function $y = \dfrac{1}{x}$. The graph is made up of two branches. One branch is in the first quadrant, and the other branch is in the third quadrant.

Try These
1. Explain why the graph has two branches.
2. Describe what happens to the graph as the value of x approaches 0.
3. Explain why the two branches of the graph do not touch each other. How does this relate to the excluded value of x?

4. Graph each function. State the excluded value(s) of x.

   a. $y = \dfrac{1}{x - 1}$

   b. $y = \dfrac{1}{x - 2}$

   c. $y = \dfrac{1}{x + 2}$

   d. $y = \dfrac{1}{x(x - 1)}$

   e. $y = \dfrac{1}{(x - 1)(x + 2)}$

   f. $y = \dfrac{1}{(x + 3)(x + 4)}$

5. **Make a conjecture** about how to find the excluded values of a rational expression without using a calculator.

Examples

Find the excluded value(s) for each rational expression.

❶ $\dfrac{5m}{2 + m}$

Exclude the values for which $2 + m = 0$.

$$2 + m = 0$$
$$m = -2$$

So, m cannot equal -2.

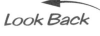
Look Back

Zero Product
Property:
Lesson 11–4

❷ $\dfrac{6}{a(a - 4)}$

Exclude the values for which $a(a - 4) = 0$.

$a(a - 4) = 0$

$a = 0$ or $a - 4 = 0$ *Zero Product Property*

So, a cannot equal 0 or 4.

❸ $\dfrac{5n}{n^2 - 25}$

Exclude the values for which $n^2 - 25 = 0$.

$$n^2 - 25 = 0$$
$(n + 5)(n - 5) = 0$ *Factor $n^2 - 25$.*
$n = -5$ or $n = 5$ *Zero Product Property*

So, n cannot equal -5 or 5.

   a. $\dfrac{2s}{s - 3}$

   b. $\dfrac{3x}{x(x + 2)}$

   c. $\dfrac{2a - 3}{a^2 - a - 6}$

Recall that you can simplify a fraction by using the following steps.

- First, factor the numerator and denominator.

- Then, divide the numerator and denominator by the greatest common factor.

Example

History Link

4 In the 1960 presidential election, more than 60% of the registered voters cast ballots. No presidential election since 1960 has had a greater voter turnout. Express 60% as a fraction in simplest form.

$$60\% = \frac{60}{100}$$

$$= \frac{2 \cdot 2 \cdot 3 \cdot 5}{2 \cdot 2 \cdot 5 \cdot 5} \qquad \textit{Factor 60 and 100.}$$

$$= \frac{\overset{1}{2} \cdot \overset{1}{2} \cdot 3 \cdot \overset{1}{5}}{\underset{1}{2} \cdot \underset{1}{2} \cdot 5 \cdot \underset{1}{5}} \text{ or } \frac{3}{5} \qquad \textit{The GCF of 60 and 100 is } 2 \cdot 2 \cdot 5.$$

Your Turn Simplify each fraction.

d. $\frac{9}{24}$ **e.** $\frac{15}{20}$ **f.** $\frac{12}{30}$

You can use the same procedure to simplify rational expressions that have polynomials in the numerator and denominator. To *simplify* means that the numerator and denominator have no factors in common, except 1.

Examples

From this point on, you can assume that all values that give a denominator of zero are excluded.

5 $\frac{8a^2b}{12ab^3}$

$$\frac{8a^2b}{12ab^3} = \frac{2 \cdot 2 \cdot 2 \cdot a \cdot a \cdot b}{2 \cdot 2 \cdot 3 \cdot a \cdot b \cdot b \cdot b} \qquad \textit{Note that } a \neq 0 \textit{ and } b \neq 0.$$

$$= \frac{\overset{1}{2} \cdot \overset{1}{2} \cdot 2 \cdot \overset{1}{a} \cdot a \cdot \overset{1}{b}}{\underset{1}{2} \cdot \underset{1}{2} \cdot 3 \cdot \underset{1}{a} \cdot \underset{1}{b} \cdot b \cdot b} \text{ or } \frac{2a}{3b^2} \quad \textit{The GCF is 4ab.}$$

6 $\frac{2x - 2}{5x - 5}$

$$\frac{2x - 2}{5x - 5} = \frac{2(x - 1)}{5(x - 1)} \qquad \textit{Factor } 2x - 2 \textit{ and } 5x - 5.$$

$$= \frac{2(x \overset{1}{\cancel{- 1}})}{5(x \underset{1}{\cancel{- 1}})} \text{ or } \frac{2}{5} \quad \textit{The GCF is } (x - 1).$$

Look Back

Factoring Trinomials: Lesson 10–3

7 $\frac{a(a + 1)}{a^2 + 3a + 2}$

$$\frac{a(a + 1)}{a^2 + 3a + 2} = \frac{a(a + 1)}{(a + 2)(a + 1)} \qquad \textit{Factor } a^2 + 3a + 2.$$

$$= \frac{a(a \overset{1}{\cancel{+ 1}})}{(a + 2)(a \underset{1}{\cancel{+ 1}})} \text{ or } \frac{a}{a + 2} \quad \textit{The GCF is } (a + 1).$$

8 $\dfrac{x^2 - 9}{x^2 - x - 6}$

$\dfrac{x^2 - 9}{x^2 - x - 6} = \dfrac{(x - 3)(x + 3)}{(x + 2)(x - 3)}$ *Factor $x^2 - 9$ and $x^2 - x - 6$.*

$= \dfrac{\overset{1}{\cancel{(x - 3)}}(x + 3)}{(x + 2)\underset{1}{\cancel{(x - 3)}}}$ or $\dfrac{x + 3}{x + 2}$ *The GCF is $(x - 3)$.*

9 $\dfrac{4 - 2x}{x^2 - 4x + 4}$

$\dfrac{4 - 2x}{x^2 - 4x + 4} = \dfrac{2(2 - x)}{(x - 2)(x - 2)}$ *Factor $4 - 2x$ and $x^2 - 4x + 4$.*

$= \dfrac{2(-1)(x - 2)}{(x - 2)(x - 2)}$ *Factor -1 from $(2 - x)$.*

$= \dfrac{2(-1)\overset{1}{\cancel{(x - 2)}}}{(x - 2)\underset{1}{\cancel{(x - 2)}}}$ or $\dfrac{-2}{x - 2}$ *The GCF is $(x - 2)$.*

Your Turn   Simplify each rational expression.

g. $\dfrac{14y^3 z}{8y^2 z^4}$ **h.** $\dfrac{-12a^2 b^2}{18a^5}$ **i.** $\dfrac{3y - 9}{4y - 12}$

j. $\dfrac{x(x + 3)}{x^2 + 6x + 9}$ **k.** $\dfrac{a^2 - 2a - 8}{a^2 + 3a + 2}$ **l.** $\dfrac{x^2 - 4x}{16 - x^2}$

Check for Understanding

Communicating Mathematics

1. **Explain** why 2 is an excluded value for $\dfrac{x}{x - 2}$.

2. **List** the steps you would use to simplify a rational expression.

3. **You Decide?** Sam and Darnell are trying to simplify $\dfrac{x + 3}{x + 4}$. Sam says the answer is $\dfrac{3}{4}$. Darnell says it is already in simplest form. Who is correct? Explain your reasoning.

Vocabulary

rational expression
excluded values
rational function

Guided Practice

Getting Ready   Factor each expression.

Sample 1: $5y + 15$	**Sample 2:** $x^2 + 5x + 6$
Solution: $5(y + 3)$	**Solution:** $(x + 3)(x + 2)$

4. $4x - 20$ **5.** $16x^2 - 20x$ **6.** $y^2 + 6y + 8$

7. $a^2 - a - 20$ **8.** $n^2 - 3n - 18$ **9.** $x^2 - 10x + 24$

Examples 1–3   Find the excluded value(s) for each rational expression.

10. $\dfrac{7}{x - 2}$ **11.** $\dfrac{4a}{a(a + 6)}$ **12.** $\dfrac{2x + 3}{x^2 - 2x - 15}$

Simplify each rational expression.

13. $\dfrac{8}{20}$

14. $\dfrac{5x}{25x^2}$

15. $\dfrac{12y^2z}{18yz^4}$

16. $\dfrac{x(x+3)}{7(x+3)}$

17. $\dfrac{3y+9}{4y+12}$

18. $\dfrac{x^2-3x}{x^2+x-12}$

19. $\dfrac{x^2+2x-8}{x^2+5x+4}$

20. $\dfrac{x^2+6x+5}{x^2+3x-10}$

21. $\dfrac{25-x^2}{x^2+x-30}$

Example 4

22. Sports The National Hockey League team that wins the first game of the best-of-seven series for the Stanley Cup has about an 80% chance of winning the series. Express 80% as a fraction in simplest form. **Source:** NHL

Exercises • • • • • • • • • • • • • • • • •

Practice

Find the excluded value(s) for each rational expression.

23. $\dfrac{x}{x+5}$

24. $\dfrac{3a}{2a+4}$

25. $\dfrac{n+6}{n-10}$

26. $\dfrac{5}{a(a+8)}$

27. $\dfrac{x+2}{(x+2)(x-3)}$

28. $\dfrac{5m}{(m-2)(m-3)}$

29. $\dfrac{9x}{x^2+5x}$

30. $\dfrac{x^2+6x+5}{x^2+3x-10}$

31. $\dfrac{y+3}{y^2-16}$

Homework Help	
For Exercises	See Examples
23–31, 60, 64	1–3
32–44, 54	5, 6
45–58	7–9
61–63	4
Extra Practice	
See page 722.	

Simplify each rational expression.

32. $\dfrac{10}{16}$

33. $\dfrac{12}{36}$

34. $\dfrac{15}{20}$

35. $\dfrac{4c}{6d}$

36. $\dfrac{8xy}{24x^2}$

37. $\dfrac{-36abc^2}{9ac}$

38. $\dfrac{2(x+1)}{8(x+1)}$

39. $\dfrac{n-3}{5(n-3)}$

40. $\dfrac{x(x+5)}{y(x+5)}$

41. $\dfrac{(x+3)(x-2)}{(x-2)(x+1)}$

42. $\dfrac{(y+6)(y-6)}{(y+6)(y+6)}$

43. $\dfrac{x^2-3x}{2(x-3)}$

44. $\dfrac{a^2-a}{a-1}$

45. $\dfrac{r^2-r-6}{3r-9}$

46. $\dfrac{x^2+3x+2}{x^2+2x+1}$

47. $\dfrac{y^2+7y+12}{y^2-16}$

48. $\dfrac{x-3}{x^2+x-12}$

49. $\dfrac{y^2+4y+4}{y^2+y-2}$

50. $\dfrac{r^2-4r-5}{r^2-2r-15}$

51. $\dfrac{z^2-z-20}{z^2+7z+12}$

52. $\dfrac{6x^2+24x}{x^2+8x+16}$

53. $\dfrac{8m^2-16m}{m^2-4m+4}$

54. $\dfrac{r^2+6r+5}{2r^2-2}$

55. $\dfrac{c^2-c-20}{c^3+10c^2+24c}$

56. $\dfrac{9-3x}{x^2-6x+9}$

57. $\dfrac{x^2-4x}{16-x^2}$

58. $\dfrac{12-4y}{y^2+y-12}$

59. Write $\dfrac{4a}{3a+a^2}$ in simplest form.

60. What are the excluded values of x for $\dfrac{x^2-9}{x^2+5x+6}$?

Applications and Problem Solving

61. **Investing** The equation $y = \frac{72}{x}$ can be used to estimate how long it will take to double an investment. In the equation, y is the number of years, and x is the annual interest rate expressed as a percent. How long would it take to double your money if it is invested at an annual rate of 8%?

Double Your Money

62. **Entertainment** The graph shows the number of films that adults see in a theater in a typical month.
    a. What fraction of adults see one movie?
    b. What fraction of adults see three or more movies?

At the Movies
(number seen by adults per month)

Source: TELENATION/Market Facts Inc

63. **Photography** The intensity of light that falls on an object is given by the formula $I = \frac{P}{d^2}$. In the formula, d is the distance from the light source, P is the power in lumens, and I is the intensity of light in lumens per square meter. An object being photographed is 3 meters from a 72-lumen light source. Find the intensity.

64. **Critical Thinking** Write rational expressions that have the following as excluded values.
    a. 0   b. −3   c. −2 and 7

Mixed Review

Solve each equation. *(Lesson 14–6)*

65. $\sqrt{x} = 4$

66. $\sqrt{y} = 2\sqrt{5}$

67. $\sqrt{n - 3} = 4$

68. $\sqrt{a} + 4 = 29$

Simplify each expression. *(Lesson 14–5)*

69. $2\sqrt{3} + \sqrt{3} + 4\sqrt{3}$

70. $6\sqrt{5} - 3\sqrt{5} + 10\sqrt{5}$

71. $4\sqrt{3} + \sqrt{12}$

72. $2\sqrt{50} + 3\sqrt{5}$

73. Find the distance between $J(2, 1)$ and $K(5, 5)$. *(Lesson 14–2)*

Standardized Test Practice

Ⓐ Ⓑ Ⓒ Ⓓ

74. **Short Response** The sum of two numbers is 9. Their difference is 3. Find the numbers. *(Lesson 13–4)*

75. **Multiple Choice** Which property or properties of inequalities allow(s) you to prove that if $2y + 8 > 18$, then $y > 5$? *(Lesson 12–4)*
    Ⓐ Distributive Property
    Ⓑ Addition Property
    Ⓒ Division Property
    Ⓓ Addition and Division Properties

Two rational expressions can be multiplied or divided just like two rational numbers.

You can use two methods to multiply rational numbers.

- **Method 1**   Multiply numerators and multiply denominators. Then divide each numerator and denominator by the greatest common factor.
- **Method 2**   Divide numerators and denominators by any common factors. Then multiply numerators and denominators.

Method 1	**Method 2**
Multiply, then simplify.	Simplify, then multiply.

Method 1

$$\frac{2}{5} \cdot \frac{1}{2} = \frac{2}{10}$$

$$= \frac{\overset{1}{\cancel{2}}}{\underset{5}{\cancel{10}}} \text{ or } \frac{1}{5} \quad \textit{The GCF is 2.}$$

Method 2

$$\frac{2}{5} \cdot \frac{1}{2} = \frac{2}{5} \cdot \frac{1}{\underset{1}{\cancel{2}}^{1}} \quad \textit{2 is a common factor.}$$

$$= \frac{1}{5}$$

Both methods have the same result.

You can use the same methods to multiply rational expressions.
Multiply $\dfrac{2x}{(x+1)^2}$ and $\dfrac{(x+1)}{x}$.

Method 1

Multiply, then simplify.

$$\frac{2x}{(x+1)^2} \cdot \frac{x+1}{x} = \frac{2x(x+1)}{x(x+1)^2}$$

$$= \frac{2\cancel{x}(\cancel{x+1})}{\cancel{x}(\cancel{x+1})(x+1)}$$

$$= \frac{2}{(x+1)}$$

Method 2

Simplify, then multiply.

$$\frac{2x}{(x+1)^2} \cdot \frac{x+1}{x} = \frac{2\cancel{x}}{(x+1)\cancel{(x+1)}} \cdot \frac{\cancel{(x+1)}}{\cancel{x}}$$

$$= \frac{2}{(x+1)}$$

Examples

1 Find each product.

$$\frac{6r^2}{5s^2} \cdot \frac{10rs}{6r^3}$$

Method 1 $\quad \dfrac{6r^2}{5s^2} \cdot \dfrac{10rs}{6r^3} = \dfrac{60r^3s}{30r^3s^2} \quad$ *Multiply.*

$$= \frac{\overset{1}{\cancel{2}} \cdot \overset{}{\cancel{2}} \cdot \overset{1}{\cancel{3}} \cdot \overset{1}{\cancel{5}} \cdot \overset{1}{\cancel{r}} \cdot \overset{1}{\cancel{r}} \cdot \overset{1}{\cancel{r}} \cdot \overset{1}{\cancel{s}}}{\underset{1}{\cancel{2}} \cdot \underset{1}{\cancel{3}} \cdot \underset{1}{\cancel{5}} \cdot \underset{1}{\cancel{r}} \cdot \underset{1}{\cancel{r}} \cdot \underset{1}{\cancel{r}} \cdot \underset{1}{\cancel{s}} \cdot s} \text{ or } \frac{2}{s} \quad \textit{Simplify.}$$

Method 2 $\quad \dfrac{6r^2}{5s^2} \cdot \dfrac{10rs}{6r^3} = \dfrac{\overset{1\,1}{\cancel{6r^2}}}{\underset{1\,s}{\cancel{5s^2}}} \cdot \dfrac{\overset{2\,1\,1}{\cancel{10rs}}}{\underset{1\,1}{\cancel{6r^3}}} \quad \textit{Simplify.}$

$$= \frac{2}{s} \qquad\qquad \textit{Multiply.}$$

❷ $\dfrac{3m}{n+2} \cdot \dfrac{n+2}{9m^2}$

$$\dfrac{3m}{n+2} \cdot \dfrac{n+2}{9m^2} = \dfrac{\overset{1}{\cancel{3m}}}{\cancel{n+2}} \cdot \dfrac{\overset{1}{\cancel{n+2}}}{\underset{3m}{\cancel{9m^2}}}$$ *3, m, and n + 2 are common factors.*

$$= \dfrac{1}{3m}$$ *Simplify.*

❸ $\dfrac{y-3}{y+5} \cdot \dfrac{2y^2 + 10y}{2y - 6}$

$$\dfrac{y-3}{y+5} \cdot \dfrac{2y^2 + 10y}{2y-6} = \dfrac{y-3}{y+5} \cdot \dfrac{2y(y+5)}{2(y-3)}$$ *Factor $2y^2 + 10y$ and $2y - 6$.*

$$= \dfrac{\overset{1}{\cancel{y-3}}}{\underset{1}{\cancel{y+5}}} \cdot \dfrac{\overset{1}{\cancel{2}}y(\overset{1}{\cancel{y+5}})}{\underset{1}{\cancel{2}}(\underset{1}{\cancel{y-3}})}$$ *2, y − 3, and y + 5 are common factors.*

$$= \dfrac{y}{1} \text{ or } y$$ *Simplify.*

❹ $\dfrac{x^2 - 25}{x^2 - 3x - 10} \cdot \dfrac{x+2}{x}$

$$\dfrac{x^2 - 25}{x^2 - 3x - 10} \cdot \dfrac{x+2}{x} = \dfrac{(x+5)(x-5)}{(x-5)(x+2)} \cdot \dfrac{x+2}{x}$$ *Factor $x^2 - 25$ and $x^2 - 3x - 10$.*

$$= \dfrac{(x+5)(\overset{1}{\cancel{x-5}})}{(\underset{1}{\cancel{x-5}})(\underset{1}{\cancel{x+2}})} \cdot \dfrac{\overset{1}{\cancel{x+2}}}{x}$$ *x − 5 and x + 2 are common factors.*

$$= \dfrac{x+5}{x}$$ *Simplify.*

Your Turn   **Find each product.**

a. $\dfrac{12x}{5y} \cdot \dfrac{20y^2}{36x^2}$

b. $\dfrac{x+2}{8} \cdot \dfrac{12x}{(x+2)(x-2)}$

c. $\dfrac{n-2}{6} \cdot \dfrac{3}{n^2 - 2n}$

d. $\dfrac{a^2 + 6a + 9}{a^2 - 4} \cdot \dfrac{a-2}{a+3}$

Look Back

Reciprocal:
Lesson 4–3

To divide a rational number by any nonzero number, multiply by its reciprocal. You can use the same method to multiply rational expressions.

$$\dfrac{2}{3} \div \dfrac{1}{2} = \dfrac{2}{3} \cdot \dfrac{2}{1}$$ *The reciprocal of*

$$= \dfrac{4}{3}$$ *$\dfrac{1}{2}$ is $\dfrac{2}{1}$.*

$$\dfrac{5}{x} \div \dfrac{y}{z} = \dfrac{5}{x} \cdot \dfrac{z}{y}$$ *The reciprocal of*

$$= \dfrac{5z}{xy}$$ *$\dfrac{y}{z}$ is $\dfrac{z}{y}$.*

Example

Sports Link

5 The Indianapolis 500 is a 500-mile automobile race. Each lap is $2\frac{1}{2}$ miles long. How many laps does a driver complete to race the entire 500 miles?

To find the number of laps, divide 500 by $2\frac{1}{2}$.

$$500 \div 2\frac{1}{2} = \frac{500}{1} \div \frac{5}{2} \qquad \textit{Express 500 and } 2\frac{1}{2} \textit{ as improper fractions.}$$

$$= \frac{500}{1} \cdot \frac{2}{5} \qquad \textit{The reciprocal of } \frac{5}{2} \textit{ is } \frac{2}{5}.$$

$$= \frac{\overset{100}{\cancel{500}}}{1} \cdot \frac{2}{\underset{1}{\cancel{5}}} \qquad \textit{5 is a common factor.}$$

$$= \frac{200}{1} \textit{ or } 200 \qquad \textit{Simplify.}$$

A driver completes 200 laps.

Examples

Find each quotient.

6 $\dfrac{6x^3}{y} \div \dfrac{2x}{y^2}$

$$\frac{6x^3}{y} \div \frac{2x}{y^2} = \frac{6x^3}{y} \cdot \frac{y^2}{2x} \qquad \textit{The reciprocal of } \frac{2x}{y^2} \textit{ is } \frac{y^2}{2x}.$$

$$= \frac{\overset{3\,x^2}{\cancel{6x^3}}}{\underset{1}{\cancel{y}}} \cdot \frac{\overset{y}{\cancel{y^2}}}{\underset{1\ 1}{\cancel{2x}}} \qquad \textit{2, x, and y are common factors.}$$

$$= \frac{3x^2y}{1} \textit{ or } 3x^2y \qquad \textit{Simplify.}$$

7 $\dfrac{2m+8}{m+5} \div (m+4)$

$$\frac{2m+8}{m+5} \div (m+4) = \frac{2m+8}{m+5} \cdot \frac{1}{m+4} \qquad \textit{The reciprocal of } (m+4) \textit{ is } \frac{1}{m+4}.$$

$$= \frac{2(m+4)}{m+5} \cdot \frac{1}{m+4} \qquad \textit{Factor } 2m+8.$$

$$= \frac{2\overset{1}{\cancel{(m+4)}}}{m+5} \cdot \frac{1}{\underset{1}{\cancel{m+4}}} \qquad (m+4) \textit{ is a common factor.}$$

$$= \frac{2}{m+5} \qquad \textit{Simplify.}$$

Your Turn

e. $\dfrac{12y}{7z^2} \div \dfrac{3xy}{2z}$

f. $\dfrac{x^2+4x+3}{x^2} \div (x+3)$

Sometimes it is necessary to factor -1 from one of the terms.

Example ⑧ Find $\dfrac{x^2 - 4}{2y} \div \dfrac{2 - x}{6xy}$.

$\dfrac{x^2 - 4}{2y} \div \dfrac{2 - x}{6xy} = \dfrac{x^2 - 4}{2y} \cdot \dfrac{6xy}{2 - x}$ *The reciprocal of $\dfrac{2 - x}{6xy}$ is $\dfrac{6xy}{2 - x}$.*

$= \dfrac{(x + 2)(x - 2)}{2y} \cdot \dfrac{6xy}{2 - x}$ *Factor $x^2 - 4$.*

$= \dfrac{(x + 2)(x - 2)}{2y} \cdot \dfrac{6xy}{-1(x - 2)}$ *Factor -1 from $2 - x$.*

$= \dfrac{(x + 2)\overset{1}{\cancel{(x - 2)}}}{\underset{1\,1}{\cancel{2y}}} \cdot \dfrac{\overset{3\;1}{\cancel{6xy}}}{-1\underset{1}{\cancel{(x - 2)}}}$ *$x - 2$, y, and 2 are common factors.*

$= \dfrac{3x(x + 2)}{-1}$ or $-3x(x + 2)$ *Simplify.*

Instead of factoring -1 from $2 - x$, you could have factored -1 from $x - 2$ and obtained the same results.

Your Turn

g. Find $\dfrac{x^2 - 9}{8x^3} \div \dfrac{6 - 2x}{4x}$

Check for Understanding

Communicating Mathematics

1. **Identify** and correct the error that was made while finding the quotient.

$\dfrac{a^2 - 25}{3a} \div \dfrac{a + 5}{15a^2} = \dfrac{3a}{(a + 5)(a - 5)} \cdot \dfrac{a + 5}{15a^2}$

$= \dfrac{\overset{1\,1}{\cancel{3a}}}{(\cancel{a + 5})(a - 5)} \cdot \dfrac{\overset{1}{\cancel{a + 5}}}{\underset{5\;a}{\cancel{15a^2}}}$ or $\dfrac{1}{5a(a - 5)}$

2. **Writing Math** Write a short paragraph explaining which method you prefer when finding the product of rational expressions: to simplify first and then multiply or to multiply first and then simplify. Include examples to support your point of view.

Guided Practice

⏱ Getting Ready Find the reciprocal of each expression.

Sample 1: $\dfrac{m}{2}$	**Sample 2:** $(a + b)$
Solution: The reciprocal is $\dfrac{2}{m}$.	**Solution:** The reciprocal is $\dfrac{1}{a + b}$.

3. $\dfrac{x^2}{4}$ 4. y 5. $\dfrac{x - 4}{x + 5}$ 6. $y + 6$

Find each product.

7. $\dfrac{a^2b}{b^2c} \cdot \dfrac{c}{d}$

8. $\dfrac{5(n-1)}{-3} \cdot \dfrac{9}{n-1}$

9. $\dfrac{2x-10}{x^2+x-12} \cdot \dfrac{x-3}{x-5}$

10. $\dfrac{y^2+3y-10}{2y} \cdot \dfrac{y^2-3y}{y^2-5y+6}$

Find each quotient.

11. $\dfrac{12a^3}{bc} \div \dfrac{3a^2}{bc}$

12. $\dfrac{9xy}{5} \div 3x^2y^2$

13. $\dfrac{3x^2+6x}{x} \div \dfrac{2x+4}{x^2}$

14. $\dfrac{x^2-5x+6}{2x^2} \div \dfrac{3-x}{4}$

Example 5

15. **Sewing** It takes $1\frac{1}{2}$ yards of fabric to make one flag for the school color guard. How many flags can be made from 18 yards of fabric?

Exercises

Practice

Find each product.

Homework Help	
For Exercises	See Examples
16–18	1
19–27, 40	2–4
28–39, 41	6–8
42–43	5

Extra Practice
See page 722.

16. $\dfrac{ab}{ac} \cdot \dfrac{c}{d}$

17. $\dfrac{3x}{2y} \cdot \dfrac{y^2}{6}$

18. $\dfrac{6a^2}{8n^2} \cdot \dfrac{12n}{9a}$

19. $\dfrac{3(a-b)}{a} \cdot \dfrac{a^2}{a-b}$

20. $\dfrac{2(a+2b)}{5} \cdot \dfrac{5}{3(a+2b)}$

21. $\dfrac{7s}{s+2} \cdot \dfrac{2s+4}{21}$

22. $\dfrac{3x+30}{2x} \cdot \dfrac{4x}{4x+40}$

23. $\dfrac{x+3}{x+4} \cdot \dfrac{x}{x^2+7x+12}$

24. $\dfrac{3a-6}{a^2-9} \cdot \dfrac{a+3}{a^2-2a}$

25. $\dfrac{x^2-9}{x+7} \cdot \dfrac{2x+14}{x^2+6x+9}$

26. $\dfrac{x}{x^2+8x+15} \cdot \dfrac{2x+10}{x^2}$

27. $\dfrac{n^2}{n^2-4} \cdot \dfrac{n^2-5n+6}{n^2-3n}$

Find each quotient.

28. $\dfrac{a^2}{b} \div \dfrac{a^2}{b^2}$

29. $\dfrac{7a^2b}{xy} \div \dfrac{7}{6xy}$

30. $2xz \div \dfrac{4xy}{z}$

31. $\dfrac{2a^3}{a+1} \div \dfrac{a^2}{a+1}$

32. $\dfrac{b^2-9}{4b} \div (b-3)$

33. $\dfrac{y^2+8y+16}{y^2} \div (y+4)$

34. $\dfrac{y^2}{y+2} \div \dfrac{y}{y+2}$

35. $\dfrac{m^2+2m+1}{2} \div \dfrac{m+1}{m-1}$

36. $\dfrac{x^2-4x+4}{3x} \div \dfrac{x^2-4}{6}$

37. $\dfrac{x^2+7x+10}{x-1} \div \dfrac{x^2+2x-15}{1-x}$

38. $\dfrac{x^2-16}{16-x^2} \div \dfrac{7}{x}$

39. $\dfrac{a^2-9}{9} \div \dfrac{6-2a}{27a^2}$

40. Find the product of $\dfrac{y-3}{8}$ and $\dfrac{12}{y-3}$.

41. What is the quotient when $\dfrac{x^2}{x^2-y^2}$ is divided by $\dfrac{x^2}{x+y}$?

42. Carpentry How many boards, each 2 feet 8 inches long, can be cut from a board 16 feet 6 inches long?

43. Probability Two darts are randomly thrown one at a time. Assume that both hit the target.
  a. What is the probability that both will hit the shaded region?
  b. Find the probability if $x = 3$.

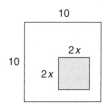

44. Critical Thinking Find two different pairs of rational expressions whose product is $\frac{5x^2}{8y^3}$.

Mixed Review

Simplify. *(Lesson 15–1)*

45. $\dfrac{14a^2b^4c}{2a^3bc}$

46. $\dfrac{m^2 - 16}{m^2 - 8m + 16}$

47. $\dfrac{8 - 4x}{x^2 - 4x + 4}$

48. Cycling You can use the formula $s = 4\sqrt{r}$ to find the fastest speed at which a cyclist can turn a corner safely and not tip over. In the formula, s is the speed in miles per hour, and r is the radius of the corner in feet. If $s = 8$ miles per hour, find r. *(Lesson 14–5)*

Use elimination to solve each system of equations. *(Lesson 13–5)*

49. $x + y = 4$
  $2x - 3y = -7$

50. $x + 3y = -4$
  $x - 2y = 6$

51. $3a + 4b = -25$
  $2a - 3b = 6$

52. Graph $2x + 3y \leq 12$. *(Lesson 12–7)*

Standardized Test Practice
Ⓐ Ⓑ Ⓒ Ⓓ

53. Extended Response Paul wants to start a small lawn-care company to earn money in the summer. He estimates that his expenses will be $200 for gasoline and equipment. He plans to charge $25 per lawn. The equation $P = 25n - 200$ can be used to find his profits. In the equation, n represents the number of lawns he cuts. *(Lesson 6–2)*
  a. Determine the ordered pairs that satisfy the equation if the domain is {5, 8, 10}.
  b. Graph the relation.

54. Multiple Choice A 60-kilogram mass is 140 centimeters from the fulcrum of a lever. How far from the fulcrum must an 80-kilogram mass be to balance the lever? *(Lesson 4–4)*
  Ⓐ 186.7 cm
  Ⓑ 34.3 cm
  Ⓒ 120 cm
  Ⓓ 105 cm

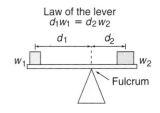

Law of the lever
$d_1 w_1 = d_2 w_2$

What You'll Learn

You'll learn to divide polynomials by binomials.

Why It's Important

Aviation You can use polynomials to determine distance, rate, and time. *See Exercise 32.*

In the previous lesson, you learned that some divisions can be performed using factoring.

$$(x^2 + 4x + 3) \div (x + 3) = \frac{x^2 + 4x + 3}{x + 3}$$

$$= \frac{(\cancel{x + 3})(x + 1)}{(\cancel{x + 3})} \quad \textit{Factor the dividend.}$$

$$= x + 1$$

Therefore, $(x^2 + 4x + 3) \div (x + 3) = x + 1$. Since the remainder is 0, the divisor is a factor of the dividend.

You can also use algebra tiles to model the division shown above.

Hands-On Algebra
Algebra Tiles

Materials: algebra tiles ▢ product mat

Step 1 Model the polynomial $x^2 + 4x + 3$. This represents the dividend.

Step 2 Place the x^2-tile at the corner of the mat. Arrange the three 1-tiles to make a length of $x + 3$. This represents the divisor.

Step 3 Use the remaining tiles to make a rectangle. The width of the rectangle, $x + 1$, is the quotient.

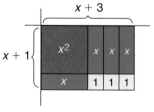

Try These

Use algebra tiles to find each quotient. Recall that you can add zero-pairs without changing the value of the polynomial.

1. $(x^2 + 6x + 9) \div (x + 3)$ **2.** $(x^2 - 4x + 4) \div (x - 2)$

3. $(x^2 + x - 6) \div (x + 3)$ **4.** $(x^2 - 9) \div (x + 3)$

5. What happens when you try to model $(x^2 + 5x + 9) \div (x + 3)$? What do you think your result means?

You can also divide polynomials using long division. Follow these steps to divide $2x^2 + 7x + 3$ by $2x + 1$.

Step 1 To find the first term of the quotient, divide the first term of the dividend, $2x^2$, by the first term of the divisor, $2x$.

$$
\begin{array}{r}
x \\
2x + 1 \overline{\smash{)}2x^2 + 7x + 3} \\
\underline{(-)\ 2x^2 + 1x} \\
6x
\end{array}
$$
$2x^2 \div 2x = x$
Multiply x and 2x + 1.
Subtract.

Step 2 To find the next term of the quotient, divide the first term of the partial dividend, $6x$, by the first term of the divisor, $2x$.

$$
\begin{array}{r}
x + 3 \\
2x + 1 \overline{\smash{)}2x^2 + 7x + 3} \\
\underline{(-)\ 2x^2 + 1x} \\
6x + 3 \\
\underline{(-)\ 6x + 3} \\
0
\end{array}
$$
Bring down 3; $6x \div 2x = 3$.
Multiply 3 and 2x + 1.
Subtract.

Therefore, $(2x^2 + 7x + 3) \div (2x + 1) = x + 3$.

Examples

Find each quotient.

1 $(6y - 3) \div (2y - 1)$

$$
\begin{array}{r}
3 \\
2y - 1 \overline{\smash{)}6y - 3} \\
\underline{(-)\ 6y - 3} \\
0
\end{array}
$$
$6y \div 2y = 3$
Multiply 3 and 2y − 1.
Subtract.

Therefore, $(6y - 3) \div (2y - 1) = 3$.

2 $(x^2 - 2x - 8) \div (x + 2)$

$$
\begin{array}{r}
x - 4 \\
x + 2 \overline{\smash{)}x^2 - 2x - 8} \\
\underline{(-)\ x^2 + 2x} \\
-4x - 8 \\
\underline{(-)\ -4x - 8} \\
0
\end{array}
$$
Multiply x and x + 2.
Subtract: $-2x - 2x = -4x$; bring down -8.
Multiply -4 and x + 2.
Subtract.

Therefore, $(x^2 - 2x - 8) \div (x + 2) = x - 4$.

Reading Algebra

Since $x - 4$ is a factor of $x^2 - 2x - 8$, you can check the quotient by simplifying $\dfrac{x^2 - 2x - 8}{x + 2}$ or by multiplying $(x + 2)$ by $(x - 4)$.

Your Turn

a. $(15a - 6) \div (5a - 2)$

b. $(x^2 - 7x + 10) \div (x - 2)$

If the divisor is *not* a factor of the dividend, the remainder will not be 0. The quotient can be expressed as follows.

$$\text{quotient} = \text{partial quotient} + \frac{\text{remainder}}{\text{divisor}}$$

Example ❸ Find $(8y^2 - 2y + 1) \div (2y - 1)$.

$$
\begin{array}{r}
4y + 1 \\
2y - 1 \overline{)8y^2 - 2y + 1} \\
(-)\ 8y^2 - 4y \\
\hline
2y + 1 \\
(-)\ 2y - 1 \\
\hline
2
\end{array}
$$

Multiply $4y$ and $2y - 1$.
Subtract. Then bring down 1.
Multiply 1 and $2y - 1$.
Subtract. The remainder is 2.

The quotient is $4y + 1$ with remainder 2.

So, $(8y^2 - 2y + 1) \div (2y - 1) = 4y + 1 + \dfrac{2}{2y - 1}$.

Your Turn

c. Find $(2a^2 + 7a + 3) \div (a + 2)$.

In an expression like $x^2 - 4$ there is no x term. In such situations, rename the dividend using zero as the coefficient of the missing term.

Example ❹ Find $(x^2 - 4) \div (x + 1)$.

$$
\begin{array}{r}
x - 1 \\
x + 1 \overline{)x^2 + 0x - 4} \\
(-)\ x^2 + 1x \\
\hline
-1x - 4 \\
-1x - 1 \\
\hline
-3
\end{array}
$$

Rename $x^2 - 4$ as $x^2 + 0x - 4$.
Multiply x and $x + 1$.
Subtract. Then bring down -4.
Multiply -1 and $x + 1$.
Subtract. The remainder is -3.

Therefore, $(x^2 - 4) \div (x + 1) = x - 1 + \dfrac{-3}{x + 1}$.

Your Turn

d. Find $(a^3 + 8a - 20) \div (a - 2)$.

If you know the area of a rectangle and the length of one side, you can find the width by dividing polynomials.

Example — **5** Find the width of a rectangle if its
Geometry Link area is $10x^2 + 29x + 21$ square units
and its length is $2x + 3$ units.

$2x + 3$

$10x^2 + 29x + 21$

To find the width, divide the area
$10x^2 + 29x + 21$ by the length $2x + 3$.

$$\begin{array}{r} 5x + 7 \\ 2x + 3\overline{)10x^2 + 29x + 21} \\ (-)\ 10x^2 + 15x \\ \hline 14x + 21 \\ (-)\ 14x + 21 \\ \hline 0 \end{array}$$

Multiply 5x and 2x + 3.
Subtract. Then bring down 21.
Multiply 7 and 2x + 3.
The remainder is 0.

Therefore, the width of the rectangle is $5x + 7$ units.
You can check your answer by multiplying $(2x + 3)$ and $(5x + 7)$.

Check for Understanding

Communicating
Mathematics

1. **Identify** the dividend, divisor, quotient, and remainder.
$(3k^2 - 7k - 5) \div (3k + 2) = k - 3 + \dfrac{1}{3k + 2}$

2. **Write** a division problem represented
by the model.

3. **Explain** how you know whether
the quotient is a factor of the
dividend.

Guided Practice

⏱ **Getting Ready** **Find each quotient.**

Sample 1: $x^2 \div x$	**Sample 2:** $4x^2 \div 2x$
Solution: x	**Solution:** $2x$

4. $3y^2 \div y$ 5. $6a^2 \div 2a$ 6. $10x^3 \div 5x$

Examples 1–4 **Find each quotient.**

7. $(x^2 + 4x) \div (x + 4)$ 8. $(a^2 + 3a + 2) \div (a + 1)$

9. $(2x^2 + 3x - 3) \div (2x - 1)$ 10. $(s^3 + 9) \div (s - 3)$

Example 5 11. **Geometry** Find the length of a
rectangle if its area is $2x^2 - 5x - 12$
square inches and its width is $x - 4$
inches.

$x - 4$ $2x^2 - 5x - 12$

Practice

Find each quotient.

12. $(12y - 4) \div (3y - 1)$

13. $(x^2 - 3x) \div (x - 3)$

14. $(8a^2 + 6a) \div (4a + 3)$

15. $(6r^3 - 15r^2) \div (2r - 5)$

16. $(a^2 + 6a + 5) \div (a + 5)$

17. $(x^2 + x - 12) \div (x - 3)$

18. $(s^2 + 11s + 18) \div (s + 2)$

19. $(a^2 - 2a - 35) \div (a - 7)$

20. $(c^2 + 12c + 36) \div (c + 9)$

21. $(3t^2 - 10t - 24) \div (3t - 4)$

22. $(2m^2 + 7m + 3) \div (m + 2)$

23. $(2b^2 + 3b - 6) \div (2b - 1)$

24. $(a^3 + 8a - 21) \div (a - 2)$

25. $(x^3 + 27) \div (x + 3)$

26. $(x^3 - 8) \div (x - 2)$

27. $(4x^4 - 2x^2 + x + 1) \div (x - 1)$

28. Find the quotient when $x^2 + 9x + 20$ is divided by $x + 4$.

29. What is the quotient when $2x^2 - 9x + 9$ is divided by $2x - 3$?

Applications and Problem Solving

30. **Geometry** The volume of a rectangular prism is $x^3 + 6x^2 + 8x$ cubic feet. If the height of the prism is $x + 4$ feet and the length is $x + 2$ feet, find the width of the prism.

31. **Transportation** The distance from San Francisco, CA, to New York, NY, is 2807 miles. To the nearest hour, find the number of hours it would take to travel this distance using each method of transportation at the given average speed. Use the formula $d = rt$.

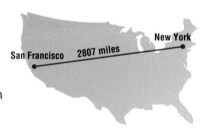

a. walking, 3 mph

b. stagecoach, 20 mph

c. automobile, 55 mph

d. jumbo jet, 608 mph

32. **Aviation** The distance d in miles flown by an airplane is given by the polynomial $x^2 + 501x + 500$. Suppose that $x + 500$ represents the speed r in miles per hour of the airplane. Use the formula $d = rt$ to find the following.

a. Find the polynomial that represents the time t in hours.

b. If $x = 5$, find the distance, rate, and time for the airplane.

33. **Critical Thinking** Find the value of k if the remainder is 15 when $x^3 - 7x^2 + 4x + k$ is divided by $x - 2$.

Mixed Review

Find each product or quotient. *(Lesson 15–2)*

34. $\dfrac{3x + 9}{x} \cdot \dfrac{x^2}{x^2 - 9}$

35. $\dfrac{a^2}{b^2} \div \dfrac{a^2}{b^2}$

36. **Sports** When a professional football team from the west coast plays a professional football team from the east coast on television on Monday nights, the west coast team wins 64% of the time. Express 64% as a fraction in simplest form. **Source:** Stanford University Sleep Disorders Clinic *(Lesson 15–1)*

Determine whether each system of equations has *one* solution, *no* solution, or *infinitely many* solutions. *(Lesson 13–2)*

37.

38.

39.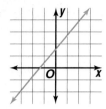

Solve each inequality. *(Lesson 12–4)*

40. $3x - 1 > 14$

41. $-7y + 6 \le 48$

42. **Multiple Choice** Which is the graph of $y = x^2 + 2$? *(Lesson 11–1)*

Ⓐ

Ⓑ

Ⓒ

Ⓓ

Quiz 1   Lessons 15–1 through 15–3

Find the excluded value(s) for each rational expression. Then simplify. *(Lesson 15–1)*

1. $\dfrac{m^2 - 2m}{m - 2}$

2. $\dfrac{x^2 - 4x + 4}{x^2 + 4x - 12}$

Find each product or quotient. *(Lesson 15–2)*

3. $\dfrac{y + 3}{2y + 8} \cdot \dfrac{y^2 + 5y + 4}{y^2 + 4y + 3}$

4. $\dfrac{a^2 b^2}{6a + 18} \div \dfrac{ab}{a^2 - 9}$

5. $\dfrac{6 - 2a}{a^2 + a - 2} \cdot \dfrac{a^2 + 3a + 2}{a - 3}$

6. $\dfrac{x^2}{x^2 - 16} \div \dfrac{x}{x^2 - 16}$

Find each quotient. *(Lesson 15–3)*

7. $(y^2 - 5y) \div (y - 5)$

8. $(n^2 + 2n - 10) \div (n - 2)$

9. $(x^3 - 1) \div (x - 1)$

10. **Fast Food** About 90% of Americans eat fast food in a given month. Of these, 60% use the drive-through window at least once. What percent of Americans use a fast-food drive-through at least once a month?
Source: Maritz Marketing Research Inc. *(Lesson 15–2)*

Rational expressions with like denominators are added or subtracted in the same way as number fractions with like denominators. In the following example, two steps are used to subtract $\frac{1}{10}$ from $\frac{4}{10}$.

Step 1 Add or subtract the numerators.

$$\frac{4}{10} - \frac{1}{10} = \frac{4-1}{10}$$

Step 2 Write the sum or difference over the common denominator.

$$= \frac{3}{10}$$

You can use this same method to add or subtract rational expressions with like denominators. Recall that in rational expressions, both the numerator and denominator can have variables.

Examples

Find each sum or difference.

① $\frac{2}{a} + \frac{4}{a}$

$\frac{2}{a} + \frac{4}{a} = \frac{2+4}{a}$ *The common denominator is a.*
Add the numerators.

$= \frac{6}{a}$ *Simplify.*

② $\frac{5x}{11} - \frac{2x}{11}$

$\frac{5x}{11} - \frac{2x}{11} = \frac{5x-2x}{11}$ *The common denominator is 11.*
Subtract the numerators.

$= \frac{3x}{11}$ *Simplify.*

③ $\frac{b}{b+3} - \frac{3}{b+3}$

$\frac{b}{b+3} - \frac{3}{b+3} = \frac{b-3}{b+3}$ *The common denominator is b + 3.*

Your Turn

a. $\frac{4m}{7} + \frac{m}{7}$ **b.** $\frac{7}{x} - \frac{3}{x}$ **c.** $\frac{n}{n-2} - \frac{1}{n-2}$

When adding or subtracting fractions, the result is not always in simplest form. To find the simplest form, divide the numerator and the denominator by the greatest common factor (GCF).

$$\frac{3}{8} + \frac{1}{8} = \frac{4}{8}$$

$$= \frac{\overset{1}{\cancel{4}}}{\underset{2}{\cancel{8}}} \quad \textit{The GCF of 4 and 8 is 4.}$$

$$= \frac{1}{2}$$

$$\frac{9}{16} - \frac{3}{16} = \frac{6}{16}$$

$$= \frac{\overset{3}{\cancel{6}}}{\underset{8}{\cancel{16}}} \quad \textit{The GCF of 6 and 16 is 2.}$$

$$= \frac{3}{8}$$

You can use similar steps to combine rational expressions.

Examples

Find each sum or difference. Write in simplest form.

4 $\dfrac{7}{10m} + \dfrac{1}{10m}$

$$\frac{7}{10m} + \frac{1}{10m} = \frac{7 + 1}{10m} \quad \textit{The common denominator is 10m. Add the numerators.}$$

$$= \frac{8}{10m} \quad \textit{Simplify.}$$

$$= \frac{\overset{4}{\cancel{8}}}{\underset{5}{\cancel{10}m}} \quad \textit{Divide by the GCF, 2.}$$

$$= \frac{4}{5m} \quad \textit{Simplify.}$$

5 $\dfrac{3}{3t} - \dfrac{9}{3t}$

$$\frac{3}{3t} - \frac{9}{3t} = \frac{3 - 9}{3t} \quad \textit{The common denominator is 3t. Subtract the numerators.}$$

$$= \frac{-6}{3t} \quad \textit{Simplify.}$$

$$= \frac{\overset{-2}{\cancel{-6}}}{\underset{1}{\cancel{3}t}} \quad \textit{Divide by the GCF, 3.}$$

$$= \frac{-2}{t} \text{ or } -\frac{2}{t} \quad \textit{Simplify.}$$

Your Turn

d. $\dfrac{a}{12} + \dfrac{2a}{12}$

e. $\dfrac{8}{9y} - \dfrac{5}{9y}$

Sometimes, the denominators of rational expressions are binomials.

Find each sum or difference. Write in simplest form.

6 $\dfrac{3}{x+2} + \dfrac{1}{x+2}$

$$\dfrac{3}{x+2} + \dfrac{1}{x+2} = \dfrac{3+1}{x+2} \qquad \textit{The common denominator is } x+2.$$
$$\textit{Add the numerators.}$$

$$= \dfrac{4}{x+2} \qquad \textit{Simplify.}$$

7 $\dfrac{y+2}{y-1} - \dfrac{8}{y-1}$

$$\dfrac{y+2}{y-1} - \dfrac{8}{y-1} = \dfrac{y+2-8}{y-1} \qquad \textit{The common denominator is } y-1.$$
$$\textit{Subtract the numerators.}$$

$$= \dfrac{y-6}{y-1} \qquad \textit{Simplify.}$$

Your Turn

f. $\dfrac{4m}{2m+3} + \dfrac{5}{2m+3}$

g. $\dfrac{5x}{x+2} - \dfrac{2x}{x+2}$

Sometimes you must factor in order to simplify the sum or difference of rational expressions.

Find each sum or difference. Write in simplest form.

8 $\dfrac{b}{b+4} + \dfrac{b+8}{b+4}$

$$\dfrac{b}{b+4} + \dfrac{b+8}{b+4} = \dfrac{b+(b+8)}{b+4} \qquad \textit{The common denominator is } b+4.$$
$$\textit{Add the numerators.}$$

$$= \dfrac{2b+8}{b+4} \qquad \textit{Simplify.}$$

$$= \dfrac{2(b+4)}{b+4} \qquad \textit{Factor the numerator.}$$

$$= \dfrac{2(\overset{1}{\cancel{b+4}})}{\underset{1}{\cancel{b+4}}} \qquad \textit{Divide by the GCF, } b+4.$$

$$= 2 \qquad \textit{Simplify.}$$

9 $\dfrac{15x}{4x-1} - \dfrac{3x+3}{4x-1}$

$$\dfrac{15x}{4x-1} - \dfrac{3x+3}{4x-1} = \dfrac{15x-(3x+3)}{4x-1} \qquad \textit{The common denominator is } 4x-1.$$
$$\textit{Subtract the numerators.}$$

$$= \dfrac{15x-3x-3}{4x-1} \qquad \textit{Distributive Property}$$

$$= \dfrac{12x-3}{4x-1} \qquad \textit{Simplify.}$$

$$= \frac{3(4x - 1)}{4x - 1} \qquad \textit{Factor the numerator.}$$

$$= \frac{3(4\cancel{x - 1})}{\cancel{4x - 1}} \qquad \textit{Divide by the GCF, } 4x - 1.$$

$$= 3 \qquad\qquad \textit{Simplify.}$$

Your Turn   **Find each sum or difference. Write in simplest form.**

h. $\dfrac{3r}{r + 5} + \dfrac{15}{r + 5}$

i. $\dfrac{m}{m + 3} - \dfrac{3m + 6}{m + 3}$

You can solve some geometry problems by adding rational expressions.

Example
Geometry Link

⑩ **Find the perimeter of rectangle *ABCD*.**

$$P = 2\ell + 2w$$

$$= 2\left(\frac{6}{x + 3}\right) + 2\left(\frac{2x}{x + 3}\right) \quad \textit{Replace } \ell \textit{ with } \frac{6}{x + 3} \textit{ and } w \textit{ with } \frac{2x}{x + 3}.$$

$$= \frac{2(6)}{x + 3} + \frac{2(2x)}{x + 3} \qquad \textit{Multiply.}$$

$$= \frac{12}{x + 3} + \frac{4x}{x + 3} \qquad \textit{Simplify.}$$

$$= \frac{12 + 4x}{x + 3} \qquad\qquad \textit{Add the numerators.}$$

$$= \frac{4(3 + x)}{x + 3} \qquad\qquad \textit{Factor the numerator.}$$

$$= \frac{4(\cancel{x + 3})}{\cancel{x + 3}} \qquad\qquad \textit{Divide by the GCF, } x + 3.$$

$$= 4 \qquad\qquad\qquad \textit{Simplify.}$$

The perimeter of rectangle *ABCD* is 4 centimeters.

Check for Understanding

Communicating Mathematics

1. **Explain** how the sum of two fractions with the same denominator can have a sum of zero.

2. **Identify** the mistake in the solution to the problem below. Then write the correct sum.

$$\frac{4x}{2x + 1} + \frac{2}{2x + 1} = \frac{4x + 2}{4x + 2} \text{ or } 1$$

3. Complete the table.

a	b	$a+b$	$a-b$	$a \cdot b$	$a \div b$
$\dfrac{2}{t}$	$\dfrac{1}{t}$	$\dfrac{3}{t}$	$\dfrac{1}{t}$	$\dfrac{2}{t^2}$	2
$\dfrac{12}{y}$	$\dfrac{4}{y}$				
$\dfrac{x}{x-1}$	$\dfrac{1}{x-1}$				

Guided Practice

 Getting Ready Find each sum or difference. Write in simplest form.

Sample 1: $\dfrac{4}{15} + \dfrac{3}{15}$

Solution: $\dfrac{4}{15} + \dfrac{3}{15} = \dfrac{4+3}{15}$

$\qquad\qquad\qquad = \dfrac{7}{15}$

Sample 2: $\dfrac{8}{9} - \dfrac{2}{9}$

Solution: $\dfrac{8}{9} - \dfrac{2}{9} = \dfrac{8-2}{9}$

$\qquad\qquad\qquad = \dfrac{6}{9}$ or $\dfrac{2}{3}$

4. $\dfrac{5}{7} + \dfrac{1}{7}$

5. $\dfrac{4}{12} + \dfrac{5}{12}$

6. $\dfrac{6}{10} - \dfrac{1}{10}$

Examples 1–9

Find each sum or difference. Write in simplest form.

7. $\dfrac{5}{x} + \dfrac{2}{x}$

8. $\dfrac{2t}{3} - \dfrac{t}{3}$

9. $\dfrac{7y}{y} - \dfrac{8y}{y}$

10. $\dfrac{a}{12} + \dfrac{2a}{12}$

11. $\dfrac{2x}{x-3} - \dfrac{6}{x-3}$

12. $\dfrac{2c+3}{c-4} - \dfrac{c-2}{c-4}$

Example 10

13. Geometry Find the perimeter of the rectangle.

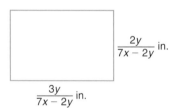

$\dfrac{2y}{7x-2y}$ in.

$\dfrac{3y}{7x-2y}$ in.

Exercises

• • • • • • • • • • • • • • • • • • •

Practice

Homework Help

For Exercises	See Examples
14–25, 35	1–5
26–33	6–9
34	10

Extra Practice
See page 723.

Find each sum or difference. Write in simplest form.

14. $\dfrac{7}{m} + \dfrac{4}{m}$

15. $\dfrac{x}{8} + \dfrac{6x}{8}$

16. $\dfrac{7}{15z} - \dfrac{3}{15z}$

17. $\dfrac{5t}{2} - \dfrac{4t}{2}$

18. $\dfrac{8p}{13} - \dfrac{3p}{13}$

19. $\dfrac{1}{6r} + \dfrac{6}{6r}$

20. $\dfrac{a}{2} + \dfrac{a}{2}$

21. $\dfrac{n}{3} + \dfrac{2n}{3}$

22. $\dfrac{5}{3y} - \dfrac{2}{3y}$

23. $\dfrac{5x}{24} - \dfrac{3x}{24}$

24. $\dfrac{8}{27m} + \dfrac{4}{27m}$

25. $\dfrac{5}{2z} + \dfrac{-7}{2z}$

26. $\dfrac{8}{y-2} - \dfrac{6}{y-2}$

27. $\dfrac{y}{a+1} - \dfrac{y}{a+1}$

28. $\dfrac{3}{x+2} - \dfrac{2}{x+2}$

29. $\dfrac{2x}{x+3} + \dfrac{6}{x+3}$

30. $\dfrac{8n+3}{3n+4} - \dfrac{2n-5}{3n+4}$

31. $\dfrac{10m-1}{4m-3} - \dfrac{8-2m}{4m-3}$

32. What is $\dfrac{3c-7}{2c-3}$ minus $\dfrac{c-4}{2c-3}$?

33. Find the sum of $\dfrac{x+y}{y-2}$ and $\dfrac{y-x}{y-2}$.

Applications and Problem Solving

34. Geometry Find the perimeter of the isosceles triangle.

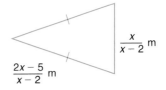

$\dfrac{x}{x-2}$ m

$\dfrac{2x-5}{x-2}$ m

35. Entertainment A musical group receives $15,000 for each concert. The five members receive equal shares of the money. They must pay 0.3 of their earnings for other salaries and expenses.

   a. Write a rational expression that represents how much money each member of the music group receives for a concert after expenses.

   b. Simplify the expression you wrote in part a to find how much each member of the group receives for a concert.

36. Critical Thinking Which rational expression is not equivalent to the others?

   a. $\dfrac{4}{x-5}$ **b.** $-\dfrac{4}{5-x}$ **c.** $\dfrac{-4}{x-5}$ **d.** $\dfrac{-4}{5-x}$

Mixed Review

Find each quotient. *(Lesson 15–3)*

37. $(a^2 + 9a + 20) \div (a + 5)$

38. $(x^2 - 2x - 35) \div (x - 7)$

39. $(2y^2 - 3y - 35) \div (2y + 7)$

40. $(t^2 + 12t + 36) \div (t + 9)$

41. Find the product of $\dfrac{5x^2y}{8ab}$ and $\dfrac{12a^2b}{25x}$. *(Lesson 15–2)*

42. Solve the system $y = x^2 - 2$ and $y = -x$ by graphing. *(Lesson 13–6)*

Standardized Test Practice
Ⓐ Ⓑ Ⓒ Ⓓ

43. Short Response Find $(3x + y)^2$. *(Lesson 9–5)*

44. Multiple Choice About 25% of all identical twins are "mirror" twins—a reflection of each other. In a group of identical twins, 35 pairs were found to be "mirror" twins. About how many pairs of identical twins would you expect to be in the group? *(Lesson 5–4)*

   Ⓐ 9   Ⓑ 140

   Ⓒ 70   Ⓓ 100

15-5 Combining Rational Expressions with Unlike Denominators

What You'll Learn
You'll learn to add and subtract rational expressions with unlike denominators.

Why It's Important
Teaching Teachers can use LCM when dividing their classes into groups.
See Example 3.

The **least common multiple (LCM)** is the least number that is a common multiple of two or more numbers. Suppose you want to find the LCM of 4, 6, and 9. Write multiples of each number until you find a common multiple.

multiples of 4: 0, 4, 8, 12, 16, 20, 24, 28, 32, **36**, . . .
multiples of 6: 0, 6, 12, 18, 24, 30, **36**, . . .
multiples of 9: 0, 9, 18, 27, **36**, . . .

Zero is a multiple of every number, but it cannot be the LCM. The LCM of 4, 6, and 9 is 36.

You can also use prime factorization to find the LCM.

$$4 = 2 \cdot 2 \qquad 6 = 2 \cdot 3 \qquad 9 = 3 \cdot 3$$

The prime factors are 2 and 3. The greatest number of times 2 appears is twice (in 4). The greatest number of times 3 appears is twice (in 9). So, the LCM of 4, 6, and 9 is $2 \cdot 2 \cdot 3 \cdot 3$ or 36. This is the same answer you got by listing the multiples.

Examples

Find the LCM for each pair of expressions.

1 $15a^2b, 18a^3$

$15a^2b = 3 \cdot 5 \cdot a \cdot a \cdot b$ *Factor each expression.*
$18a^3 = 2 \cdot 3 \cdot 3 \cdot a \cdot a \cdot a$

$\begin{aligned} LCM &= 2 \cdot 3 \cdot 3 \cdot 5 \cdot a \cdot a \cdot a \cdot b \\ &= 90a^3b \end{aligned}$ *Use each factor the greatest number of times it appears in either factorization.*

2 $x^2 + x - 6, 2x^2 - 3x - 2$

$x^2 + x - 6 = (x + 3)(x - 2)$ *Factor each expression.*
$2x^2 - 3x - 2 = (2x + 1)(x - 2)$

$LCM = (x + 3)(x - 2)(2x + 1)$ *Use each factor the greatest number of times it appears in either factorization.*

Your Turn

a. $12a, 8ab$

b. $x^2 - 9, x^2 - 2x - 3$

③ Mrs. Reimer wants student groups of 4, 6, or 8. What is the minimum number of desks needed?

Find the LCM of 4, 6, and 8.
$4 = 2 \cdot 2$
$6 = 2 \cdot 3$
$8 = 2 \cdot 2 \cdot 2$

LCM $= 2 \cdot 2 \cdot 2 \cdot 3$ or 24   *Use each factor the greatest number of times it appears in any of the factorizations.*

Mrs. Reimer must have at least 24 desks in her classroom.

To add or subtract fractions with unlike denominators, first rename the fractions so the denominators are alike. Any common denominator could be used. However, the computation is usually easier if you use the **least common denominator (LCD)**. Recall that the least common denominator is the LCM of the denominators.

Examples

Write each pair of rational expressions with the same LCD.

④ $\dfrac{9}{2m^2}, \dfrac{5}{6m}$

First find the LCD.
$2m^2 = 2 \cdot m \cdot m$ *Factor each expression.*
$6m = 2 \cdot 3 \cdot m$
LCD $= 2 \cdot 3 \cdot m \cdot m$ or $6m^2$

Then write each fraction with the same LCD.

$\dfrac{9}{2m^2} \cdot \dfrac{3}{3} = \dfrac{27}{6m^2}$ $\dfrac{5}{6m} \cdot \dfrac{m}{m} = \dfrac{5m}{6m^2}$

⑤ $\dfrac{1}{x+2}, \dfrac{3x}{4x+8}$

First find the LCD.
$x + 2 = x + 2$ *Factor each expression.*
$4x + 8 = 4(x + 2)$
LCD $= 4(x + 2)$

Then write each fraction with the same LCD.

$\dfrac{1}{x+2} \cdot \dfrac{4}{4} = \dfrac{4}{4(x+2)}$ $\dfrac{3x}{4x+8} = \dfrac{3x}{4(x+2)}$

Your Turn

c. $\dfrac{6}{b^3}, \dfrac{7}{ab}$

d. $\dfrac{x}{x-6}, \dfrac{x-3}{x-2}$

Use the following steps to add or subtract rational expressions with unlike denominators.

Step 1 Find the LCD.

Step 2 Change each rational expression into an equivalent expression using the LCD.

Step 3 Add or subtract as with rational expressions with like denominators.

Step 4 Simplify if necessary.

Find each sum or difference. Write in simplest form.

 $\dfrac{5}{6x} + \dfrac{7}{12x^2}$

Step 1 Find the LCD.

$$6x = 2 \cdot 3 \cdot x$$
$$12x^2 = 2 \cdot 2 \cdot 3 \cdot x \cdot x$$
$$\text{LCM} = 2 \cdot 2 \cdot 3 \cdot x \cdot x \text{ or } 12x^2$$
$$\text{LCD} = 12x^2$$

Step 2 Rename each expression with the LCD as the denominator. The denominator of $\dfrac{7}{12x^2}$ is already $12x^2$, so only $\dfrac{5}{6x}$ needs to be renamed.

$$\dfrac{5}{6x} \cdot \dfrac{2x}{2x} = \dfrac{10x}{12x^2} \qquad \textit{Why do you multiply } \dfrac{5}{6x} \textit{ by } \dfrac{2x}{2x}?$$

Step 3 Add.

$$\dfrac{5}{6x} + \dfrac{7}{12x^2} = \dfrac{10x}{12x^2} + \dfrac{7}{12x^2} \qquad \textit{Use the LCD.}$$

$$= \dfrac{10x + 7}{12x^2} \qquad \textit{This expression is in simplest form.}$$

 $\dfrac{m}{m^2 - 9} - \dfrac{3}{m - 3}$

$$m^2 - 9 = (m - 3)(m + 3) \qquad \textit{Factor each expression.}$$
$$m - 3 = m - 3$$
$$\text{LCM} = (m - 3)(m + 3)$$
$$\text{LCD} = (m - 3)(m + 3)$$

$$\dfrac{m}{m^2 - 9} - \dfrac{3}{m - 3}$$

$$= \dfrac{m}{(m - 3)(m + 3)} - \dfrac{3}{m - 3} \cdot \dfrac{m + 3}{m + 3} \qquad \textit{Multiply by } \dfrac{3}{m - 3} \textit{ by } \dfrac{m + 3}{m + 3}.$$

$$= \dfrac{m}{(m - 3)(m + 3)} - \dfrac{3m + 9}{(m - 3)(m + 3)} \qquad \textit{The LCD is } (m - 3)(m + 3).$$

$$= \dfrac{m - (3m + 9)}{(m - 3)(m + 3)} \qquad \textit{Subtract the numerators.}$$

$$= \dfrac{m - 3m - 9}{(m - 3)(m + 3)} \qquad \textit{Distributive Property}$$

$$= \dfrac{-2m - 9}{(m - 3)(m + 3)} \qquad \textit{Simplify.}$$

8 $\dfrac{5}{x-3} + \dfrac{7}{2x-6}$

$$\dfrac{5}{x-3} + \dfrac{7}{2x-6} = \dfrac{5}{x-3} + \dfrac{7}{2(x-3)} \qquad \textit{The LCD is } 2(x-3).$$

$$= \dfrac{5}{x-3} \cdot \dfrac{2}{2} + \dfrac{7}{2(x-3)} \qquad \textit{Multiply } \dfrac{5}{x-3} \textit{ by } \dfrac{2}{2}.$$

$$= \dfrac{10}{2(x-3)} + \dfrac{7}{2(x-3)} \qquad \textit{Simplify.}$$

$$= \dfrac{10+7}{2(x-3)} \qquad \textit{Add the numerators.}$$

$$= \dfrac{17}{2(x-3)} \qquad \textit{Simplify.}$$

Your Turn **Find each sum or difference. Write in simplest form.**

e. $\dfrac{3}{xy} + \dfrac{2}{y^2}$

f. $\dfrac{5a}{2a+6} - \dfrac{a}{a+3}$

Check for Understanding

Communicating Mathematics

1. **Write** two rational expressions with unlike denominators in which one of the denominators is the LCD of the rational expressions.

2. **Explain** the steps you would take to find the LCD for the expression $\dfrac{3}{8x^2} - \dfrac{5}{12xy} + \dfrac{7}{6y^2}$. Then simplify the expression.

3. **You Decide** Ashley found the sum of $\dfrac{4}{x-1}$ and $\dfrac{5}{x}$ to be $\dfrac{9}{2x-1}$. Malik found the sum to be $\dfrac{9x-5}{x^2-x}$. Who is correct? Explain why.

Guided Practice

Getting Ready **Find the LCM for each pair of numbers.**

Sample: 10, 12	**Solution:** $10 = 2 \cdot 5$ $\qquad$ $12 = 2 \cdot 2 \cdot 3$
	$\text{LCM} = 2 \cdot 2 \cdot 3 \cdot 5$ or 60

4. 6, 10 $\qquad\qquad$ **5.** 8, 18 $\qquad\qquad$ **6.** 12, 15

Examples 1 & 2 **Find the LCM for each pair of expressions.**

7. $6xy,\ 15y^2$ $\qquad\qquad\qquad$ **8.** $x+1,\ x^2+5x+4$

Examples 4 & 5 **Write each pair of rational expressions with the same LCD.**

9. $\dfrac{4}{a^2},\ \dfrac{5}{a}$ $\qquad\qquad\qquad$ **10.** $\dfrac{7}{t+3},\ \dfrac{13}{2t+6}$

Find each sum or difference. Write in simplest form.

11. $\dfrac{t}{6} + \dfrac{3t}{12}$

12. $\dfrac{2}{3a} - \dfrac{4}{9a}$

13. $\dfrac{2}{ab^2} + \dfrac{3}{ab}$

14. $\dfrac{6}{m} + \dfrac{5}{n}$

15. $\dfrac{x}{x^2 - 4} - \dfrac{4}{x + 2}$

16. $\dfrac{2}{3a - 1} - \dfrac{7}{15a - 5}$

17. Astronomy Earth, Jupiter, and Saturn revolve around the Sun about once every 1, 12, and 30 years, respectively. The last time Earth, Jupiter, and Saturn were lined up with the sun was in 1982, and Jupiter and Saturn could be seen close together, high in the sky at midnight. In what year will this happen again? **Example 3**

Exercises •

Practice

Homework Help

For Exercises	See Examples
18–23	1, 2
24–29	4, 5
30–51	6–8
52–54	3

Extra Practice

See page 723.

Find the LCM for each pair of expressions.

18. $3t, 9t^2$

19. $12a^2, 2ab^2$

20. $16m, 6mn$

21. $y + 2, y^2 - 4$

22. $x^2 + 9x + 14, x^2 + 3x + 2$

23. $x^2 - 9, 3x^2 - 8x - 3$

Write each pair of rational expressions with the same LCD.

24. $\dfrac{4}{m^3}, \dfrac{1}{m}$

25. $\dfrac{7}{2t}, \dfrac{8}{10t}$

26. $\dfrac{3}{6ab}, \dfrac{14}{4a^2}$

27. $\dfrac{5}{xy}, \dfrac{6}{yz}$

28. $\dfrac{x}{x^2 - 1}, \dfrac{-2}{x - 1}$

29. $\dfrac{10k}{3k + 1}, \dfrac{k}{3 + 9k}$

Find each sum or difference. Write in simplest form.

30. $\dfrac{a}{5} - \dfrac{a}{15}$

31. $\dfrac{d}{2} - \dfrac{d}{5}$

32. $\dfrac{t}{3} + \dfrac{2t}{7}$

33. $\dfrac{9}{4x} + \dfrac{3}{2x}$

34. $\dfrac{5}{2a} - \dfrac{3}{6a}$

35. $\dfrac{m}{3t} + \dfrac{1}{t}$

36. $\dfrac{4}{a^3} - \dfrac{2}{a}$

37. $\dfrac{7}{3x} - \dfrac{1}{6x^2}$

38. $\dfrac{2t}{5n^2} - \dfrac{1}{3n}$

39. $\dfrac{2}{x} + \dfrac{x + 3}{y}$

40. $\dfrac{6z}{7w} + \dfrac{2z}{w^3}$

41. $\dfrac{1}{4xy} + \dfrac{y}{10x^2}$

42. $\dfrac{9}{y + 1} - \dfrac{3}{4y + 4}$

43. $\dfrac{4a}{2a + 6} + \dfrac{3}{a + 3}$

44. $\dfrac{7}{x - 2} + \dfrac{3}{x}$

45. $\dfrac{y}{y^2 - 2y + 1} - \dfrac{1}{y - 1}$

46. $\dfrac{3a}{a^2 + 4a + 4} - \dfrac{2}{a + 2}$

47. $\dfrac{5x + 2}{2x - 1} + \dfrac{2x - 3}{8x - 4}$

48. $\dfrac{7}{a^2 + 2a - 3} - \dfrac{5}{a + 5a + 6}$

49. $\dfrac{x^2 + 4x - 5}{x^2 - 2x - 3} + \dfrac{2}{x + 1}$

50. Add $\dfrac{2x + 3}{x^2 - 4}$ and $\dfrac{6}{x + 2}$.

51. Simplify $\dfrac{m}{m - n} - \dfrac{5}{m}$.

52. **Entertainment**   The choreographer of a musical needs enough dancers so that they can be arranged in groups of 6, 9, and 12, with no one sitting out. What is the least number of dancers needed?

53. **Fitness**   Cynthia jogs around an oval track in 150 seconds. Minya walks around the track in 210 seconds. Suppose they start at the same time. After how many minutes will they meet back at their starting point?

54. **Critical Thinking**   Consider the numbers 18 and 20.
    a. Find the GCF and LCM of the numbers.
    b. What is the value of GCF · LCM?
    c. Describe the relationship between 18 · 20 and GCF · LCM.
    d. Describe how you could find the GCF of two numbers if you already know the LCM.

Mixed Review

Find each sum or difference. Write in simplest form.   *(Lesson 15–4)*

55. $\dfrac{9}{a} + \dfrac{7}{a}$ 56. $\dfrac{2}{5t} + \dfrac{3}{5t}$ 57. $\dfrac{4n}{n+1} - \dfrac{2n-2}{n+1}$

58. Find $(m^2 + 6m - 7) \div (m + 7)$.   *(Lesson 15–3)*

Simplify each expression.   *(Lesson 14–4)*

59. $7\sqrt{19} + 2\sqrt{19}$ 60. $6\sqrt{12} + 4\sqrt{3}$ 61. $8\sqrt{5} - 3\sqrt{2}$

62. **Geometry**   The length of the hypotenuse of a right triangle is $\sqrt{\dfrac{56}{25}}$ centimeters. Express the length as a radical in simplest form. *(Lesson 14–3)*

Standardized Test Practice

Ⓐ Ⓑ Ⓒ Ⓓ

63. **Short Response**   What is the solution of the system $y = 7 - x$ and $x - y = -3$? *(Lesson 13–3)*

64. **Multiple Choice**   Determine which polynomial can be the measure of the area of a square.   *(Lesson 10–5)*
    Ⓐ $2x^2 - 13x + 36$ Ⓑ $2n^2 + 20n - 100$
    Ⓒ $4a^2 + 12a + 9$ Ⓓ $4a^2 - 6a + 9$

Quiz 2   Lessons 15–3 through 15–5

Find each quotient.   *(Lesson 15–3)*

1. $(x^2 + 6x - 16) \div (x - 2)$ 2. $(2x^2 - 11x - 20) \div (2x + 3)$

Find each sum or difference. Write in simplest form.   *(Lesson 15–4)*

3. $\dfrac{a}{9} + \dfrac{7a}{9}$ 4. $\dfrac{5}{21t} + \dfrac{4}{21t}$

5. **Parades**   At the Veteran's Day parade, members of the Veterans of Foreign Wars (VFW) found that they could arrange themselves in rows of 6, 7, or 8, with no one left over. What is the least number of VFW members in the parade? *(Lesson 15–5)*

What You'll Learn

You'll learn to solve rational equations.

Why It's Important

Aviation Pilots can use rational equations to find wind speed. *See Exercise 9.*

Every week there is a 10-point quiz in math class. For the first five quizzes, Julia scored a total of 36 points, which gave her an average of 7.2. She is determined to get 10 points on each of the next quizzes until she brings her average up to an 8. On how many quizzes must she score 10 points in order to have an overall quiz average of 8 points?

Let x represent the number of quizzes on which she must score 10 points.

total number of quizzes = $5 + x$

sum of scores = $36 + 10x$

$$\text{average} = \frac{36 + 10x}{5 + x} \quad \begin{array}{l} \leftarrow \textit{sum of scores} \\ \leftarrow \textit{number of quizzes} \end{array}$$

$$8 = \frac{36 + 10x}{5 + x} \quad \textit{Julia wants to have an average of 8.}$$

The equation above is a **rational equation** because it contains at least one rational expression. *You will solve this problem in Exercise 39.*

There are three steps in solving rational equations.

Step 1 Find the LCD of each term.
Step 2 Multiply each side of the equation by the LCD.
Step 3 Use the Distributive Property to simplify.

Examples

Solve each equation. Check your solution.

1 $\dfrac{3a}{5} + \dfrac{3}{2} = \dfrac{7a}{10}$

$$\dfrac{3a}{5} + \dfrac{3}{2} = \dfrac{7a}{10} \qquad \textit{The LCD is 10.}$$

$$10\left(\dfrac{3a}{5} + \dfrac{3}{2}\right) = 10\left(\dfrac{7a}{10}\right) \qquad \textit{Multiply each side by the LCD.}$$

$$10\left(\dfrac{3a}{5}\right) + 10\left(\dfrac{3}{2}\right) = 10\left(\dfrac{7a}{10}\right) \qquad \textit{Distributive Property}$$

$$\overset{2}{10}\left(\dfrac{3a}{\underset{1}{5}}\right) + \overset{5}{10}\left(\dfrac{3}{\underset{1}{2}}\right) = \overset{1}{10}\left(\dfrac{7a}{\underset{1}{10}}\right) \qquad \textit{Divide each by the GCF.}$$

$$6a + 15 = 7a \qquad \textit{Simplify.}$$

$$6a + 15 - 6a = 7a - 6a \qquad \textit{Subtract 6a from each side.}$$

$$15 = a \qquad \textit{Simplify.}$$

Check: $\dfrac{3a}{5} + \dfrac{3}{2} = \dfrac{7a}{10}$ *Original equation*

$\dfrac{3(15)}{5} + \dfrac{3}{2} \overset{?}{=} \dfrac{7(15)}{10}$ *Replace a with 15.*

$9 + \dfrac{3}{2} \overset{?}{=} \dfrac{21}{2}$ *Simplify.*

$\dfrac{21}{2} = \dfrac{21}{2}$ ✓

2 $\dfrac{2}{3x} - \dfrac{1}{2x} = \dfrac{1}{6}$

$\dfrac{2}{3x} - \dfrac{1}{2x} = \dfrac{1}{6}$ *The LCD is 6x.*

$6x\left(\dfrac{2}{3x} - \dfrac{1}{2x}\right) = 6x\left(\dfrac{1}{6}\right)$ *Multiply each side by the LCD.*

$6x\left(\dfrac{2}{3x}\right) - 6x\left(\dfrac{1}{2x}\right) = 6x\left(\dfrac{1}{6}\right)$ *Distributive Property*

$\overset{2}{6x}\left(\dfrac{2}{\underset{1}{3x}}\right) - \overset{3}{6x}\left(\dfrac{1}{\underset{1}{2x}}\right) = \overset{1}{6x}\left(\dfrac{1}{\underset{1}{6}}\right)$ *Divide each by the GCF.*

$4 - 3 = x$ *Simplify.*

$1 = x$ *Subtract.*

Check: $\dfrac{2}{3x} - \dfrac{1}{2x} = \dfrac{1}{6}$ *Original equation*

$\dfrac{2}{3(1)} - \dfrac{1}{2(1)} \overset{?}{=} \dfrac{1}{6}$ *Replace x with 1.*

$\dfrac{2}{3} - \dfrac{1}{2} \overset{?}{=} \dfrac{1}{6}$ *The LCD is 6.*

$\dfrac{4}{6} - \dfrac{3}{6} \overset{?}{=} \dfrac{1}{6}$ $\dfrac{2}{3} = \dfrac{4}{6}$

$\dfrac{1}{6} = \dfrac{1}{6}$ ✓

3 $\dfrac{3x}{x+2} + \dfrac{1}{x+2} = 4$

$\dfrac{3x}{x+2} + \dfrac{1}{x+2} = 4$ *The LCD is x + 2.*

$(x+2)\left(\dfrac{3x}{x+2} + \dfrac{1}{x+2}\right) = (x+2)4$ *Multiply each side by the LCD.*

$(x+2)\left(\dfrac{3x}{x+2}\right) + (x+2)\left(\dfrac{1}{x+2}\right) = 4x + 8$ *Distributive Property*

$(\overset{1}{\cancel{x+2}})\left(\dfrac{3x}{\underset{1}{\cancel{x+2}}}\right) + (\overset{1}{\cancel{x+2}})\left(\dfrac{1}{\underset{1}{\cancel{x+2}}}\right) = 4x + 8$ *Divide each by the GCF.*

$3x + 1 = 4x + 8$ *Simplify.*

(continued on the next page)

$$3x + 1 - 3x = 4x + 8 - 3x \qquad \textit{Subtract 3x from each side.}$$
$$1 = x + 8 \qquad \textit{Simplify.}$$
$$1 - 8 = x + 8 - 8 \qquad \textit{Subtract 8 from each side.}$$
$$-7 = x \qquad \textit{Check the solution.}$$

④ $\dfrac{3}{r} - \dfrac{1}{r-1} = \dfrac{1}{r-1}$

$$\dfrac{3}{r} - \dfrac{1}{r-1} = \dfrac{1}{r-1} \qquad \textit{The LCD is } r(r-1).$$

$$r(r-1)\left(\dfrac{3}{r} - \dfrac{1}{r-1}\right) = r(r-1)\left(\dfrac{1}{r-1}\right) \qquad \textit{Multiply each side}$$
$$\textit{by the LCD.}$$

$$r(r-1)\left(\dfrac{3}{r}\right) - r(r-1)\left(\dfrac{1}{r-1}\right) = r(r-1)\left(\dfrac{1}{r-1}\right) \qquad \textit{Distributive Property}$$

$$\overset{1}{r}(r-1)\left(\dfrac{3}{r}\right) - r(\overset{1}{r-1})\left(\dfrac{1}{r-1}\right) = r(\overset{1}{r-1})\left(\dfrac{1}{r-1}\right) \qquad \textit{Divide by the GCF.}$$

$$(r-1)3 - r = r \qquad \textit{Simplify.}$$
$$3r - 3 - r = r \qquad \textit{Distributive Property}$$
$$2r - 3 = r \qquad \textit{Subtract.}$$
$$2r - 3 - r = r - r \qquad \textit{Subtract r from each side.}$$
$$r - 3 = 0 \qquad \textit{Simplify.}$$
$$r - 3 + 3 = 0 + 3 \qquad \textit{Add 3 to each side.}$$
$$r = 3 \qquad \textit{Check the solution.}$$

Your Turn **Solve each equation. Check your solution.**

a. $\dfrac{4x}{3} + \dfrac{7}{2} = \dfrac{9x}{12}$ **b.** $\dfrac{18}{b} = \dfrac{3}{b} + 3$ **c.** $\dfrac{4}{n+1} - \dfrac{2}{n} = \dfrac{5}{n}$

Recall that **uniform motion problems** can be solved by using the formula below.

$$\underbrace{distance}_{d} = \underbrace{rate}_{r} \cdot \underbrace{time}_{t}$$

Example **—⑤**
Travel Link

The Milbys rented a houseboat on the Sacramento River. The maximum speed of the boat in still water is 8 miles per hour. At this rate, a 30-mile trip downstream took the same amount of time as an 18-mile trip against the current. What was the rate of the current?

Explore Let c = the rate of the current.

Let $8 + c$ = the rate of the boat traveling downstream with the current. *When the boat is traveling with the current, you add the speed of the current to the speed of the boat.*

Let $8 - c$ = the rate of the boat traveling upstream against the current. *When the boat is traveling against the current, you subtract the speed of the current from the speed of the boat.*

Plan Since $d = rt$, then $t = \frac{d}{r}$.

	d	r	$t = \frac{d}{r}$
Downstream	30	$8 + c$	$\frac{30}{8 + c}$
Upstream	18	$8 - c$	$\frac{18}{8 - c}$

Solve

$$\frac{30}{8 + c} = \frac{18}{8 - c} \qquad \text{\textit{The time downstream equals the time upstream.}}$$

$$(8 + c)(8 - c)\left(\frac{30}{8 + c}\right) = (8 + c)(8 - c)\left(\frac{18}{8 - c}\right) \qquad \text{\textit{The LCD is}} \atop (8 + c)(8 - c).$$

$$(8 + c)(8 - c)\left(\frac{30}{8 + c}\right) = (8 + c)(8 - c)\left(\frac{18}{8 - c}\right) \qquad \text{\textit{Divide by GCF.}}$$

$$(8 - c)(30) = (8 + c)(18) \qquad \text{\textit{Simplify.}}$$
$$240 - 30c = 144 + 18c \qquad \text{\textit{Distributive Property}}$$
$$240 - 30c + 30c = 144 + 18c + 30c \qquad \text{\textit{Add 30c to each side.}}$$
$$240 = 144 + 48c \qquad \text{\textit{Simplify.}}$$
$$240 - 144 = 144 + 48c - 144 \qquad \text{\textit{Subtract 144 from each side.}}$$
$$96 = 48c \qquad \text{\textit{Simplify.}}$$
$$\frac{96}{48} = \frac{48c}{48} \qquad \text{\textit{Divide each side by 48.}}$$
$$2 = c \qquad \text{\textit{Simplify.}}$$

The rate of the current was 2 miles per hour.

Examine Check the solution to see if it makes sense. The houseboat goes downstream at $8 + 2$ or 10 miles per hour. A 30-mile trip would take $30 \div 10$ or 3 hours. The houseboat goes upstream at $8 - 2$ or 6 miles per hour. An 18-mile trip would take $18 \div 6$ or 3 hours. Both trips take the same amount of time, so the solution is correct.

A houseboat on a river

Check for Understanding

Communicating Mathematics

1. **List** two differences between linear equations and rational equations.

2. **Writing Math** Describe two different ways in which $\frac{4}{x + 1} = \frac{8}{x - 1}$ can be solved. Then determine whether there are other steps that you could use to solve the examples in this lesson.

Vocabulary

rational equation
uniform motion problems

Solve each equation. Check your solution.

3. $\dfrac{2x}{7} + \dfrac{1}{2} = \dfrac{x}{14}$ 4. $\dfrac{6}{c} = \dfrac{10}{c} + 4$ 5. $\dfrac{5}{m+3} - \dfrac{2}{m+3} = -9$

6. $\dfrac{a+1}{a} + \dfrac{a+4}{a} = 6$ 7. $\dfrac{2x}{x+3} + \dfrac{3}{x} = 2$ 8. $\dfrac{r-1}{r+1} - \dfrac{2r}{r-1} = -1$

Example 5

9. **Transportation**   An airplane can fly at a rate of 600 miles per hour in calm air. It can fly 2520 miles with the wind in the same time it can fly 2280 miles against the wind. Find the speed of the wind.

Exercises • • • • • • • • • • • • • • • • • •

Practice

Solve each equation. Check your solution.

10. $\dfrac{a}{6} + \dfrac{2a}{3} = -\dfrac{5}{2}$ 11. $\dfrac{1}{4} + \dfrac{5r}{8} = \dfrac{r}{4}$ 12. $\dfrac{b}{5} - \dfrac{2b}{15} = 1$

13. $\dfrac{x}{7} = \dfrac{x+3}{10}$ 14. $\dfrac{t-1}{4} = \dfrac{t}{3}$ 15. $\dfrac{2a-3}{6} = \dfrac{2a}{3} + \dfrac{1}{2}$

16. $\dfrac{4}{3y} + \dfrac{1}{y} = 7$ 17. $\dfrac{6}{x} - \dfrac{3}{2x} = \dfrac{1}{2}$ 18. $\dfrac{1}{2a} + \dfrac{3}{4a} = \dfrac{1}{4}$

19. $\dfrac{m+1}{m} + \dfrac{m+3}{m} = 5$ 20. $\dfrac{t+1}{t} + \dfrac{t+4}{t} = 6$ 21. $\dfrac{5}{5-p} - \dfrac{1}{5-p} = -2$

22. $\dfrac{5}{2x} - \dfrac{1}{6x} = 2$ 23. $\dfrac{1}{4x} + \dfrac{1}{6x} = 5$ 24. $\dfrac{3}{x} + \dfrac{4x}{x-3} = 4$

25. $\dfrac{n-3}{n} = \dfrac{n-3}{n-6}$ 26. $\dfrac{3}{r+4} - \dfrac{1}{r} = \dfrac{1}{r}$ 27. $\dfrac{5}{x+1} + \dfrac{1}{x} = \dfrac{2}{x^2+x}$

28. $\dfrac{5}{n-2} + \dfrac{2}{n} = \dfrac{1}{n}$ 29. $\dfrac{6}{t+1} - \dfrac{3}{4t+4} = \dfrac{3}{4}$ 30. $\dfrac{1}{2a+6} + \dfrac{3a}{a+3} = \dfrac{1}{2}$

31. $\dfrac{7}{a-1} = \dfrac{5}{a+3}$ 32. $\dfrac{1}{m+1} + \dfrac{5}{m-1} = \dfrac{1}{m^2-1}$

33. $\dfrac{j}{j+1} + \dfrac{5}{j-1} = 1$ 34. $\dfrac{6}{z+2} + \dfrac{3}{z^2-4} = \dfrac{7}{z+2}$

35. $\dfrac{1}{4m} - \dfrac{2}{m-3} = \dfrac{2}{m}$ 36. $\dfrac{x+2}{2x-1} + \dfrac{3x-6}{8x-4} = \dfrac{9}{4}$

37. What is the value of w in the equation $\dfrac{2w-3}{w-3} - 2 = \dfrac{12}{w+3}$?

38. Solve $\dfrac{3n}{n^2-5n+4} = \dfrac{2}{n-4} + \dfrac{3}{n-1}$.

Applications and Problem Solving

39. **Quizzes**   Refer to the application at the beginning of the lesson. Solve $8 = \dfrac{36+10x}{5+x}$ to find the number of quizzes on which Julia must score 10 points in order to have an overall quiz average of 8 points.

40. **Cycling**   A long-distance cyclist pedaling at a steady rate travels 30 miles with the wind. She can travel only 18 miles against the wind in the same amount of time. If the rate of the wind is 3 miles per hour, what is the cyclist's rate without the wind? (*Hint:* Use the formula $d = rt$.)

Homework Help

For Exercises	See Examples
10–20, 22, 23	1, 2
21, 24–39, 41	3, 4
40	5

Extra Practice
See page 723.

41. **Critical Thinking** What number would you add to the numerator and denominator of $\frac{2}{11}$ to make a fraction equivalent to $\frac{1}{2}$?

Mixed Review

42. **Space** Scientists can predict when an asteroid might hit Earth or come very close to Earth by studying their orbits. Toro is an asteroid that orbits the sun every 584 days. Earth orbits the sun every 365 days. Suppose Earth and Toro are lined up with the sun today. Find the LCM of 584 and 365 to determine how long before Earth and Toro arrive together back at that same point in their orbits. *(Lesson 15–5)*

An asteroid

43. Find the sum of $\frac{11}{4n^2}$ and $\frac{7}{4n^2}$. Write your answer in simplest form. *(Lesson 15–4)*

Name the set or sets of numbers to which each real number belongs. Let N = natural numbers, W = whole numbers, Z = integers, Q = rational numbers, and I = irrational numbers. *(Lesson 14–1)*

44. $\frac{16}{2}$ 45. -7 46. $\sqrt{15}$

47. **Farming** In order to have enough time in the growing season, a farmer has at most 16 days left to plant his corn and soybean crops. He can plant corn at a rate of 10 acres per day and soybeans at a rate of 15 acres per day. Suppose he has at most 200 acres available and he wants to plant both crops. *(Lesson 13–7)*

   a. Let c represent the number of days that corn will be planted and let s represent the number of days that soybeans will be planted. Write a system of inequalities to represent this situation.

   b. Graph the system of inequalities.

   c. List two ordered pairs that are possible solutions and explain what each ordered pair represents.

Solve each system of equations by graphing. *(Lesson 13–1)*

48. $y = 2x - 1$
    $y = -x + 5$

49. $x = 4$
    $x + y = 3$

Standardized Test Practice
Ⓐ Ⓑ Ⓒ Ⓓ

50. **Short Response** Write an irrational square root whose graph is between points A and B on the number line below. *(Lesson 8–6)*

Down the Drain

Materials

 calculator

Work Problems

Suppose you have a big job to do, like painting a house. How much faster could you get the job done if you had help? In this investigation, you will use algebraic techniques to solve **work problems**.

Investigate

1. A public swimming pool holds 100,000 gallons of water and has two drain outlets. If only outlet A is open, the pool will drain in 20 hours. If only outlet B is open, the pool will drain in 16 hours.

   a. The drainage rate of outlet A is 100,000 gallons ÷ 20 hours or 5000 gallons per hour. What is the drainage rate of outlet B?

   You could also use a spreadsheet for your calculations.

   b. Make a table like the one below and continue filling in columns 2, 3, and 4 until the total water drained in column 4 is more than 100,000 gallons.

Number of Hours	Water Drained by Outlet A (gal)	Water Drained by Outlet B (gal)	Total Water Drained (gal)
1	5000	6250	11,250
2	10,000	12,500	22,500

   c. Approximately how long will it take to drain the pool with both outlets open?

2. The problem above also can be solved using an algebraic equation.

   a. Since outlet A can drain the pool in 20 hours, it can drain $\frac{1}{20}$ of the pool in 1 hour. How much of the pool can outlet B drain in 1 hour?

   b. Use the table at the right and the following formula to write rational expressions for the work done w by outlets 1 and 2 in t hours.

	r	t	w
Outlet A	$\frac{1}{20}$	t	?
Outlet B	$\frac{1}{16}$	t	?

   $$\underbrace{\text{rate of work}}_{r} \cdot \underbrace{\text{time}}_{t} = \underbrace{\text{work done}}_{w}$$

c. If both outlets are open, then the following equation is true.

$$\frac{\text{work of}}{\text{outlet A}} + \frac{\text{work of}}{\text{outlet B}} = \frac{\text{the whole}}{\text{job}}$$

Write an equation to represent the time it takes the pool to drain if both outlets are open. Assume that 1 represents the whole job. *Why?*

d. Solve the equation in part b to find how long it takes to drain the pool if both outlets are open. Compare this answer to the estimate you made in Exercise 1d.

3. Micheal and Gus work for a ski resort. Their job is to prepare Bear Tooth Run for skiers. It takes Micheal 4 hours to prepare the run and it takes Gus 5 hours. How long would it take them to prepare the run if they work together?

a. In the swimming pool problem, you knew the total amount of gallons to be drained. In this problem since you do not know the size of the trail, write the work as a fraction done per hour. Micheal's rate is $\frac{1}{4}$ of the job per hour. What is Gus' rate?

b. Write an algebraic equation to represent the work done if Micheal and Gus work together.

c. How long would it take them both to prepare the run?

Extending the Investigation

Use a spreadsheet or write an algebraic equation to solve each problem.

- Ian and Mandy own a lawn care business. Ian can mow, trim, and fertilize one particular yard in 2 hours. Mandy can complete the same tasks for that yard in 4 hours. How long would it take them to complete this yard working together?

- In the summer, Sheila and her son, Cory, paint house exteriors. To paint a typical house, it takes Sheila about 45 hours. To paint the same size house, it takes Cory 55 hours. How long would it take them to paint one house working together?

Presenting Your Conclusions

Here are some ideas to help you present your conclusions to the class.

- Design a creative bulletin board displaying a problem from this investigation. Show two ways in which to solve the problem.

- Write a problem similar to those in this investigation. Solve the problem by using a spreadsheet and by using an equation. Write a one-page paper discussing the advantages and disadvantages of using each method to solve work problems.

 Investigation For more information on work problems visit: www.algconcepts.com

Understanding and Using the Vocabulary

After completing this chapter, you should be able to define each term, property, or phrase and give an example or two of each.

*inter*NET
CONNECTION **Review Activities**
For more review activities, visit:
www.algconcepts.com

excluded value *(p. 638)*
least common denominator *(p. 663)*
least common multiple *(p. 662)*
rational equation *(p. 668)*

rational expression *(p. 638)*
rational function *(p. 638)*
uniform motion problems *(p. 670)*
work problems *(p. 674)*

Choose the correct term or expression to complete each sentence.

1. An excluded value is any value assigned to a variable that results in a (denominator, numerator) of zero.

2. A rational expression is an algebraic fraction whose numerator and denominator are (rational numbers, polynomials).

3. To find $\dfrac{9}{x+y}$ divided by $\dfrac{3x+6y}{x}$, multiply $\dfrac{9}{x+y}$ and $\left(\dfrac{3x+6y}{x}, \dfrac{x}{3x+6y}\right)$.

4. The excluded values of $\dfrac{4x}{x^2-36}$ are (0 and 36, 6 and -6).

5. To simplify a rational expression, divide the numerator and denominator by their (greatest common factor, least common multiple).

6. The rational expression $\dfrac{x+2}{x+4}$ (is, is not) in simplest form.

7. The reciprocal of $a+b$ is $\left(-a-b, \dfrac{1}{a+b}\right)$.

8. To add or subtract rational expressions, first rename the expressions using the (greatest common factor, least common multiple) as the denominator.

9. The least common multiple of x^2 and $2x$ is $(2x^2, 2x^3)$.

10. The divisor in $(x^2+8x+15) \div (x+5) = (x+3)$ is $(x^2+8x+15, x+5)$.

Skills and Concepts

Objectives and Examples	Review Exercises

• **Lesson 15–1** Simplify rational expressions.

$$\dfrac{z^2-3z+2}{z^2-2z} = \dfrac{(z-1)(\overset{1}{\cancel{z-2}})}{z(\underset{1}{\cancel{z-2}})}$$

$$= \dfrac{z-1}{z}$$

Simplify each expression.

11. $\dfrac{4x^2yz}{16xy^3}$

12. $\dfrac{a^2-5a}{a-5}$

13. $\dfrac{n^2-16}{n^2+2n-8}$

14. $\dfrac{x^3-2x^2-3x}{x^2+x-12}$

 www.algconcepts.com/vocabulary_review

Objectives and Examples

• **Lesson 15–2** Multiply and divide rational expressions.

$$\frac{6x}{x^2 + x - 2} \cdot \frac{x^2 - 4}{2x^2}$$

$$= \frac{\overset{3\ 1}{6x}}{(x + 2)(x - 1)} \cdot \frac{(x - 2)(x + 2)}{\underset{1\ x}{2x^2}}$$

$$= \frac{3(x - 2)}{x(x - 1)}$$

$$\frac{a^3}{2c} \div \frac{a^2}{c^3} = \frac{a^3}{2c} \cdot \frac{c^3}{a^2} \qquad \textit{The reciprocal of } \frac{a^2}{c^3} \textit{ is } \frac{c^3}{a^2}.$$

$$= \frac{ac^2}{2}$$

Review Exercises

Find each product or quotient.

15. $\dfrac{4x^2 y}{y^2 z} \cdot \dfrac{z}{6y}$

16. $\dfrac{4a - 4b}{a} \cdot \dfrac{a^3}{8a - 8b}$

17. $\dfrac{x + 2}{x + 3} \cdot \dfrac{x}{x^2 - x - 6}$

18. $\dfrac{6m^2 n}{10p} \div 3m$

19. $\dfrac{y^2 - 4y + 4}{3} \div \dfrac{y^2 - 4}{9}$

20. $\dfrac{x^2 - 9}{2x} \div \dfrac{6 - 2x}{8x^2}$

• **Lesson 15–3** Divide polynomials by binomials.

$$
\begin{array}{r}
2x - 3 \\
x + 2\overline{\smash{)}2x^2 + x - 6} \\
\underline{(-)\ 2x^2 + 4x} \\
-3x - 6 \\
\underline{(-)\ -3x - 6} \\
0
\end{array}
$$

$2x^2 \div x = 2x$
Multiply $2x$ and $x + 2$.
Subtract.
$-3x \div x = -3$
Subtract.

Find each quotient.

21. $(a^2 + 6a) \div (a + 6)$

22. $(x^2 - 5x - 6) \div (x + 1)$

23. $(4y^2 + 8y + 5) \div (2y + 3)$

24. $(x^3 - 1) \div (x - 1)$

• **Lesson 15–4** Add and subtract rational expressions with like denominators.

$$\frac{x^2}{x + 3} - \frac{9}{x + 3} = \frac{x^2 - 9}{x + 3}$$

$$= \frac{(x + 3)(x - 3)}{x + 3}$$

$$= x - 3$$

Find each sum or difference. Write in simplest form.

25. $\dfrac{2x}{8} + \dfrac{4x}{8}$

26. $\dfrac{11}{15m} - \dfrac{2}{15m}$

27. $\dfrac{3x}{x + 4} + \dfrac{2x}{x + 4}$

28. $\dfrac{8x}{2x - 3} - \dfrac{2x + 9}{2x - 3}$

Mixed Problem Solving
See pages 724-731.

Objectives and Examples

Review Exercises

• **Lesson 15–5** Add and subtract rational expressions with unlike denominators.

$$\frac{y}{y + 2} - \frac{2}{y} \quad \textit{The LCD is } y(y + 2).$$

$$= \frac{y}{y + 2} \cdot \frac{y}{y} - \frac{2}{y} \cdot \frac{y + 2}{y + 2}$$

$$= \frac{y^2}{y(y + 2)} - \frac{2(y + 2)}{y(y + 2)}$$

$$= \frac{y^2 - 2(y + 2)}{y(y + 2)}$$

$$= \frac{y^2 - 2y - 4}{y(y + 2)}$$

Find each sum or difference. Write in simplest form.

29. $\dfrac{3}{4x} + \dfrac{5}{8x^2}$

30. $\dfrac{5}{x} - \dfrac{x + 1}{3x}$

31. $\dfrac{4}{a} + \dfrac{3}{a - 2}$

32. $\dfrac{y + 2}{y^2 - 4} - \dfrac{2}{y + 2}$

• **Lesson 15–6** Solve rational equations.

$$\frac{x}{4} - \frac{x}{6} = \frac{1}{4}$$

$$12\left(\frac{x}{4} - \frac{x}{6}\right) = 12\left(\frac{1}{4}\right)$$

$$\overset{3}{\cancel{12}}\left(\frac{x}{\cancel{4}}\right) - \overset{2}{\cancel{12}}\left(\frac{x}{\cancel{6}}\right) = \overset{3}{\cancel{12}}\left(\frac{1}{\cancel{4}}\right)$$

$$3x - 2x = 3$$

$$x = 3$$

Solve each equation. Check your solution.

33. $\dfrac{x}{3} - \dfrac{3x}{4} = \dfrac{1}{12}$

34. $\dfrac{6}{3x} - \dfrac{3}{x} = 1$

35. $\dfrac{x + 2}{x} + \dfrac{2}{3x} = \dfrac{1}{3}$

36. $\dfrac{m}{m + 1} + \dfrac{5}{m - 1} = 1$

Applications and Problem Solving

37. Carpentry Determine the number of pieces of $\frac{3}{8}$-inch plywood that are in a stack 30 inches high. *(Lesson 15–2)*

38. Boating The top speed of a boat in still water is 5 miles per hour. At this speed, a 21-mile trip downstream takes the same amount of time as a 9-mile trip upstream. Find the rate of the current. *(Lesson 15–6)*

39. Geometry The volume of a rectangular prism is $4x^3 + 14x^2 + 10x$ cubic inches. The height of the prism is $2x + 5$ inches, and the length of the prism is $2x$ inches. *(Lesson 15–3)*

a. Find the polynomial that represents the width.

b. If $x = 2$, find the length, width, height, and volume of the prism.

1. **Explain** why $x = -3$ is an excluded value for $\dfrac{2x}{x + 3}$.

2. **Identify** and correct the error that was made while finding the difference.

   $$\dfrac{4}{x + 1} - \dfrac{x - 3}{x + 1} = \dfrac{4 - x - 3}{x + 1}$$

   $$= \dfrac{1 - x}{x + 1}$$

3. **Find** the least common denominator of $\dfrac{5}{6a}$ and $\dfrac{2}{3a^2}$.

Find the excluded value(s) for each rational expression.

4. $\dfrac{8}{x}$

5. $\dfrac{6m}{m(m - 2)}$

6. $\dfrac{2n}{n^2 - 9}$

Simplify each expression.

7. $\dfrac{9a^3b^2}{15ab^5}$

8. $\dfrac{x^3 - x^2}{x - 1}$

9. $\dfrac{x - 2}{x^2 - 5x + 6}$

Find each sum, difference, product, or quotient. Write in simplest form.

10. $\dfrac{z^3}{8} \cdot \dfrac{10x^2}{z^3}$

11. $\dfrac{x^2 - 1}{3x} \div \dfrac{1 - x}{9x}$

12. $\dfrac{x^2 - x - 2}{x^2 - 4} \cdot \dfrac{x + 2}{x^2 + 4x + 3}$

13. $\dfrac{4a + 4b}{a} \div \dfrac{10a + 10b}{a^2}$

14. $\dfrac{5}{8x} + \dfrac{11}{8x}$

15. $\dfrac{t}{t + 5} - \dfrac{t - 6}{t + 5}$

16. $\dfrac{6}{5y} + \dfrac{7}{10y^2}$

17. $\dfrac{2}{x + 4} - \dfrac{x}{x^2 - 16}$

Find each quotient.

18. $(y^2 + 10y + 16) \div (y + 2)$

19. $(x^3 + x^2 - x - 1) \div (x - 1)$

20. $(a^2 - 10) \div (a + 3)$

21. $(x^2 - 5x + 8) \div (x - 2)$

Solve each equation. Check your solution.

22. $\dfrac{n + 2}{3} + \dfrac{n}{2} = \dfrac{1}{2}$

23. $\dfrac{5}{x + 2} - \dfrac{1}{4x + 8} = \dfrac{1}{4}$

24. **Geometry** Find the measure of the area of the rectangle in simplest form.

$\dfrac{x + 7}{x^2 - 25}$

$\dfrac{x^2 + 10x + 25}{x^2 - 49}$

25. **Transportation** A tugboat pushing a barge up the Mississippi River takes 1 hour longer to travel 36 miles up the river than to travel the same distance down the river. If the rate of the current is 3 miles per hour, find the speed of the tugboat and barge in still water.

Right Triangle Problems

You'll need to know how to identify right triangles and how to apply the Pythagorean Theorem.

If a and b are the lengths of the legs of a triangle and c is the length of the hypotenuse, then $a^2 + b^2 = c^2$.

Example 1

A 32-foot telephone pole is braced with a cable from the top of the pole to a point 7 feet from the base. What is the length of the cable, rounded to the nearest tenth?

Ⓐ 31.2 ft   Ⓑ 32.8 ft

Ⓒ 34.3 ft   Ⓓ 36.2 ft

Hint Since telephone poles are vertical and the ground is horizontal, the pole forms a right angle at the base.

Solution Draw a figure.

32 ft

7 ft

The cable forms the hypotenuse of a right triangle. Use the Pythagorean Theorem to find the length of the hypotenuse.

$a^2 + b^2 = c^2$ *Pythagorean Theorem*

$32^2 + 7^2 = c^2$ *Replace a with 32 and b with 7.*

$1024 + 49 = c^2$ *Multiply.*

$1073 = c^2$ *Add.*

$\sqrt{1073} = \sqrt{c^2}$ *Take the square root of each side.*

$32.8 \approx c$ *Use a calculator.*

To the nearest tenth, the length of the cable is 32.8 feet.

The answer is B.

Example 2

In the figure, $\overline{MO}$ is perpendicular to $\overline{LN}$, LO is equal to 4, MO is equal to ON, and LM is equal to 6. What is MN?

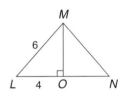

Ⓐ $2\sqrt{10}$ Ⓑ $3\sqrt{5}$ Ⓒ $4\sqrt{5}$

Ⓓ $3\sqrt{10}$ Ⓔ $6\sqrt{4}$

Hint Identify the types of triangles shown in the figure.

Solution $\triangle LMO$ is a right triangle. Use the Pythagorean Theorem to find MO.

$a^2 + b^2 = c^2$ *Pythagorean Theorem*

$a^2 + 4^2 = 6^2$ *Replace b with 4*

$a^2 + 16 = 36$ *and c with 6.*

$a^2 + 16 - 16 = 36 - 16$ *Subtract 16.*

$a^2 = 20$ *Simplify.*

$\sqrt{a^2} = \sqrt{20}$ *Take the square root.*

$a = \sqrt{4(5)}$ or $2\sqrt{5}$

$\triangle MON$ is an isosceles triangle. Use the Pythagorean Theorem to find MN.

$a^2 + b^2 = c^2$ *Pythagorean Theorem*

$\left(2\sqrt{5}\right)^2 + \left(2\sqrt{5}\right)^2 = c^2$ *a = 2√5, b = 2√5*

$20 + 20 = c^2$ *Simplify.*

$\sqrt{40} = \sqrt{c^2}$ *Take the square root*

$2\sqrt{10} = c$ *of each side.*

The answer is A.

After you work each problem, record your answer on the answer sheet provided or on a sheet of paper.

Multiple Choice

1. Quadrilateral *ABCD* is a rectangle. $\overline{AD}$ is 5 centimeters long, and $\overline{CD}$ is 12 centimeters long. What is length of $\overline{AC}$?   *(Lesson 8–7)*

   Ⓐ 13 cm
   Ⓑ 17 cm
   Ⓒ 30 cm
   Ⓓ 169 cm

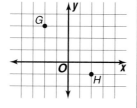

2. What is the distance between points *G* and *H* in the graph?
   (Lesson 14–2)

   Ⓐ 4
   Ⓑ $4\sqrt{2}$
   Ⓒ 8
   Ⓓ $8\sqrt{2}$

3. The hypotenuse of an isosceles right triangle has a length of 20 units. What is the length of one of the legs of the triangle?   *(Lesson 8–7)*

   Ⓐ 10   Ⓑ $10\sqrt{2}$ Ⓒ $10\sqrt{3}$
   Ⓓ 20   Ⓔ $20\sqrt{2}$

4. The streets on a jogging route form a right triangle. How long is the jogging route?
   (Lesson 8–7)

   Ⓐ 10 km
   Ⓑ 14 km
   Ⓒ 24 km
   Ⓓ 100 km

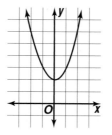

5. What is the equation of the graph?
   (Lesson 11–2)

   Ⓐ $y = -x^2 + 2$
   Ⓑ $y = x^2 - 2$
   Ⓒ $y = (x + 2)^2$
   Ⓓ $y = x^2 + 2$

6. What is the length of $\overline{BC}$?   *(Lesson 8–7)*

   Ⓐ 6
   Ⓑ $4\sqrt{3}$
   Ⓒ $2\sqrt{13}$
   Ⓓ 8
   Ⓔ $2\sqrt{38}$

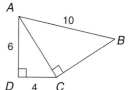

7. The base of a 25-foot ladder is placed 10 feet from the side of a house. How high does the ladder reach?   *(Lesson 8–7)*

   Ⓐ 12.5 ft   Ⓑ 15 ft
   Ⓒ 22.9 ft   Ⓓ 35 f

8. $\sqrt{3} + \sqrt{4}$ is between which pair of numbers?   *(Lesson 14–4)*

   Ⓐ 3 and 4
   Ⓑ 4 and 5
   Ⓒ 5 and 6
   Ⓓ 6 and 7

Grid In

9. Triangles *ABC* and *XYZ* are similar right triangles. If ∠*B* measures 55°, what is the degree measure of ∠*Z*?   *(Lesson 3–6)*

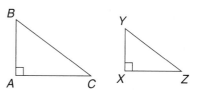

Extended Response

10. You want to fit a square cake inside a circular cake carrier that is 12 inches in diameter.   *(Lesson 8–7)*

    Part A   Draw a diagram showing side *s* of the *largest* square cake that will fit inside the carrier.

    Part B   Use your diagram to write an equation and then find the value of *s*.

Student Handbook

Prerequisite Skills Review

Operations with Decimals

- To add or subtract decimals, line up the decimal points. You may want to *annex*, or place zeros at the end of the decimals, to help align the columns. Then add or subtract.

Examples **1** Find 18.39 + 6.4.

$$
\begin{array}{r}
18.39 \\
+\ 6.4 \\
\end{array}
\quad \rightarrow \quad
\begin{array}{r}
18.39 \\
+\ 6.40 \\
\hline
24.79 \\
\end{array}
\quad \textit{Annex a zero to align the columns.}
$$

2 Find 5 − 2.17.

$$
\begin{array}{r}
5 \\
-\ 2.17 \\
\end{array}
\quad \rightarrow \quad
\begin{array}{r}
5.00 \\
-\ 2.17 \\
\hline
2.83 \\
\end{array}
\quad \textit{Annex two zeros to align the columns.}
$$

- To multiply decimals, multiply as with whole numbers. Then add the total number of decimal places in the factors. Place the same number of decimal places in the product, counting from right to left.
- To divide decimals, move the decimal point in the divisor to the right, and then move the decimal point in the dividend the same number of places. Align the decimal point in the quotient with the decimal point in the dividend.

Examples **3** Find 7.3(0.61).

$$
\begin{array}{r}
7.3 \quad \leftarrow \textit{1 decimal place} \\
\times\ 0.61 \quad \leftarrow \textit{2 decimal places} \\
\hline
73 \\
438 \\
\hline
4.453 \quad \leftarrow \textit{3 decimal places} \\
\end{array}
$$

4 Find 8.12 ÷ 5.8.

$$
\begin{array}{r}
1.4 \\
5.8\overline{)8.12} \\
5\,8 \\
\hline
2\,32 \\
2\,32 \\
\hline
0 \\
\end{array}
\quad \textit{Move each decimal point right 1 place.}
$$

Find each sum or difference.

1. $4.8 + 1.1$
2. $10.25 + 0.16$
3. $36.5 - 11.2$
4. $13.6 + 20.41$
5. $42.99 - 12.63$
6. $5.37 - 2.47$
7. $24 + 7.3$
8. $7.28 - 1.1$
9. $0.7 + 1.682$
10. $6.2 - 4.01$
11. $4.02 + 5.9 + 0.03$
12. $18 - 16.39$

Find each product or quotient.

13. $8(0.4)$
14. $5.72 \div 2$
15. $1.6(1.9)$
16. $7.2(3.5)$
17. $24(0.86)$
18. $4.86 \div 0.3$
19. $11.5 \div 4.6$
20. $9.3(30.23)$
21. $0.5(2.41)(6.7)$
22. $\dfrac{50.4}{36}$
23. $\dfrac{6.46}{0.68}$
24. $\dfrac{9.264}{0.24}$

Simplifying Fractions

- A fraction is in simplest form when the greatest common factor (GCF) of the numerator and the denominator is 1.
- To write a fraction in simplest form, divide both the numerator and the denominator by the GCF.

Examples   Write each fraction in simplest form.

❶ $\dfrac{6}{9}$

The GCF of 6 and 9 is 3.

$\dfrac{6}{9} = \dfrac{2}{3}$ *Divide 6 and 9 by 3.*

$\div 3$ $\div 3$

❷ $\dfrac{4}{20}$

The GCF of 4 and 20 is 4.

$\dfrac{4}{20} = \dfrac{1}{5}$ *Divide 4 and 20 by 4.*

$\div 4$ $\div 4$

- To change an improper fraction to a mixed number, divide the numerator by the denominator. Write the remainder as a fraction in simplest form.
- To change a mixed number to an improper fraction, multiply the whole number by the denominator and add the numerator. Write this sum over the denominator.

Examples   **❸** Write $\dfrac{8}{5}$ as a mixed number.

$\dfrac{8}{5}$ means $8 \div 5$.

$5\overline{)8}$ → 1 R3 or $1\dfrac{3}{5}$
$\dfrac{-5}{3}$

❹ Write $2\dfrac{3}{4}$ as an improper fraction.

$2\dfrac{3}{4} = \dfrac{(2 \times 4) + 3}{4}$

$= \dfrac{8 + 3}{4}$

$= \dfrac{11}{4}$

Write each fraction in simplest form.

1. $\dfrac{4}{8}$
2. $\dfrac{3}{9}$
3. $\dfrac{2}{12}$
4. $\dfrac{9}{15}$
5. $\dfrac{8}{20}$
6. $\dfrac{6}{21}$
7. $\dfrac{25}{30}$
8. $\dfrac{30}{40}$
9. $\dfrac{12}{16}$
10. $\dfrac{16}{36}$

Write each improper fraction as a mixed number.

11. $\dfrac{7}{6}$
12. $\dfrac{5}{3}$
13. $\dfrac{9}{2}$
14. $\dfrac{12}{5}$
15. $\dfrac{10}{4}$
16. $\dfrac{11}{2}$
17. $\dfrac{21}{10}$
18. $\dfrac{16}{6}$
19. $\dfrac{24}{14}$
20. $\dfrac{30}{8}$

Write each mixed number as an improper fraction.

21. $2\dfrac{1}{3}$
22. $6\dfrac{1}{2}$
23. $2\dfrac{4}{5}$
24. $4\dfrac{2}{5}$
25. $6\dfrac{1}{7}$
26. $3\dfrac{5}{8}$
27. $2\dfrac{7}{10}$
28. $3\dfrac{5}{12}$
29. $1\dfrac{10}{11}$
30. $5\dfrac{8}{9}$

Multiplying and Dividing Fractions

- To multiply fractions, multiply the numerators and multiply the denominators.
- When a numerator and denominator have a common factor, you can simplify before multiplying.

Examples   **Find each product.**

1 $\dfrac{1}{2} \cdot \dfrac{3}{5}$

$$\dfrac{1}{2} \cdot \dfrac{3}{5} = \dfrac{1 \cdot 3}{2 \cdot 5}$$
$$= \dfrac{3}{10}$$

2 $\dfrac{4}{9} \cdot \dfrac{3}{5}$

$$\dfrac{4}{9} \cdot \dfrac{3}{5} = \dfrac{4}{\overset{3}{9}} \cdot \dfrac{\overset{1}{3}}{5} \qquad \textit{Divide by the GCF, 3.}$$
$$= \dfrac{4 \cdot 1}{3 \cdot 5}$$
$$= \dfrac{4}{15}$$

- To divide fractions, multiply by the reciprocal of the divisor.

Examples   **Find each quotient.**

3 $\dfrac{5}{6} \div \dfrac{6}{7}$

$$\dfrac{5}{6} \div \dfrac{6}{7} = \dfrac{5}{6} \cdot \dfrac{7}{6} \qquad \textit{Multiply by the reciprocal.}$$
$$= \dfrac{5 \cdot 7}{6 \cdot 6}$$
$$= \dfrac{35}{36}$$

4 $\dfrac{2}{5} \div 4$

$$\dfrac{2}{5} \div 4 = \dfrac{2}{5} \div \dfrac{4}{1}$$
$$= \dfrac{2}{5} \cdot \dfrac{1}{4} \qquad \textit{Multiply by the reciprocal.}$$
$$= \dfrac{\overset{1}{2}}{5} \cdot \dfrac{1}{\underset{2}{4}} \qquad \textit{Divide by the GCF, 2.}$$
$$= \dfrac{1 \cdot 1}{5 \cdot 2}$$
$$= \dfrac{1}{10}$$

Find each product or quotient.

1. $\dfrac{1}{3} \cdot \dfrac{1}{2}$ 2. $\dfrac{2}{5} \div \dfrac{1}{2}$ 3. $\dfrac{1}{4} \div \dfrac{4}{5}$ 4. $\dfrac{2}{7} \cdot \dfrac{1}{3}$

5. $\dfrac{1}{8} \div \dfrac{1}{5}$ 6. $\dfrac{4}{9} \cdot \dfrac{2}{3}$ 7. $\dfrac{2}{7} \div 7$ 8. $\dfrac{6}{7} \cdot \dfrac{1}{5}$

9. $\dfrac{1}{2} \cdot \dfrac{2}{7}$ 10. $\dfrac{2}{3} \cdot \dfrac{1}{6}$ 11. $\dfrac{5}{9} \div \dfrac{2}{3}$ 12. $\dfrac{1}{8} \div \dfrac{1}{2}$

13. $\dfrac{2}{9} \div \dfrac{4}{7}$ 14. $\dfrac{3}{4} \cdot \dfrac{5}{6}$ 15. $\dfrac{5}{12} \cdot \dfrac{3}{7}$ 16. $\dfrac{3}{4} \div \dfrac{5}{6}$

17. $\dfrac{2}{9} \div \dfrac{2}{3}$ 18. $\dfrac{2}{3} \div 4$ 19. $\dfrac{7}{9} \cdot \dfrac{6}{7}$ 20. $\dfrac{3}{10} \cdot \dfrac{5}{18}$

21. $\dfrac{5}{12} \cdot \dfrac{8}{25}$ 22. $\dfrac{3}{8} \div 12$ 23. $\dfrac{9}{20} \cdot \dfrac{4}{15}$ 24. $\dfrac{8}{21} \div \dfrac{12}{15}$

Adding and Subtracting Fractions

- To add or subtract fractions with like denominators, add or subtract the numerators. Simplify if necessary.

Examples Find each sum or difference.

❶ $\dfrac{1}{5} + \dfrac{3}{5}$

$\dfrac{1}{5} + \dfrac{3}{5} = \dfrac{1+3}{5}$ *Add the numerators.*

$= \dfrac{4}{5}$

❷ $\dfrac{7}{12} - \dfrac{5}{12}$

$\dfrac{7}{12} - \dfrac{5}{12} = \dfrac{7-5}{12}$ *Subtract the numerators.*

$= \dfrac{2}{12}$

$= \dfrac{1}{6}$ *Simplify.*

- To add or subtract fractions with unlike denominators, first find the least common denominator (LCD). Rewrite each fraction with the LCD, and then add or subtract the numerators. Simplify if necessary.

Examples Find each sum or difference.

❸ $\dfrac{1}{3} - \dfrac{2}{9}$

The LCD of 3 and 9 is 9.

$\dfrac{1}{3} - \dfrac{2}{9} = \dfrac{3}{9} - \dfrac{2}{9}$ *Rewrite $\dfrac{1}{3}$ as $\dfrac{3}{9}$.*

$= \dfrac{3-2}{9}$ *Subtract the numerators.*

$= \dfrac{1}{9}$

❹ $\dfrac{3}{10} + \dfrac{1}{4}$

The LCD of 10 and 4 is 20.

$\dfrac{3}{10} + \dfrac{1}{4} = \dfrac{6}{20} + \dfrac{5}{20}$ *Rewrite the fractions using the LCD, 20.*

$= \dfrac{6+5}{20}$ *Add the numerators.*

$= \dfrac{11}{20}$

Find each sum or difference.

1. $\dfrac{1}{5} + \dfrac{2}{5}$

2. $\dfrac{7}{8} - \dfrac{2}{8}$

3. $\dfrac{1}{9} + \dfrac{4}{9}$

4. $\dfrac{6}{7} - \dfrac{2}{7}$

5. $\dfrac{1}{15} + \dfrac{3}{15}$

6. $\dfrac{4}{13} + \dfrac{1}{13}$

7. $\dfrac{5}{6} - \dfrac{4}{6}$

8. $\dfrac{9}{11} - \dfrac{3}{11}$

9. $\dfrac{3}{8} + \dfrac{1}{8}$

10. $\dfrac{5}{6} - \dfrac{1}{6}$

11. $\dfrac{7}{10} - \dfrac{3}{10}$

12. $\dfrac{1}{12} + \dfrac{7}{12}$

13. $\dfrac{1}{3} + \dfrac{1}{5}$

14. $\dfrac{5}{8} - \dfrac{1}{4}$

15. $\dfrac{4}{9} + \dfrac{1}{3}$

16. $\dfrac{9}{10} - \dfrac{3}{5}$

17. $\dfrac{3}{7} + \dfrac{1}{2}$

18. $\dfrac{3}{4} - \dfrac{2}{5}$

19. $\dfrac{2}{9} + \dfrac{2}{3}$

20. $\dfrac{9}{10} - \dfrac{3}{4}$

21. $\dfrac{2}{15} + \dfrac{7}{15}$

22. $\dfrac{1}{2} + \dfrac{1}{6}$

23. $\dfrac{5}{6} - \dfrac{3}{4}$

24. $\dfrac{3}{4} + \dfrac{1}{12}$

25. $\dfrac{4}{7} - \dfrac{1}{6}$

26. $\dfrac{2}{15} + \dfrac{2}{3}$

27. $\dfrac{5}{6} + \dfrac{1}{8}$

28. $\dfrac{7}{8} - \dfrac{3}{12}$

Fractions and Decimals

- To change a fraction to a decimal, divide the numerator by the denominator.
- A **terminating decimal** is a decimal like 0.75, in which the division ends, or terminates, when the remainder is zero.
- A **repeating decimal** is a decimal like 0.545454 . . . whose digits do not end. Since it is impossible to write all the digits, you can use bar notation to show that 54 repeats. We can write 0.545454 . . . as $0.\overline{54}$.

Examples Write each fraction as a decimal.

❶ $\frac{2}{5}$

$$\begin{array}{r} 0.4 \\ 5)\overline{2.0} \\ \underline{2\ 0} \\ 0 \end{array}$$ $\quad \frac{2}{5} = 0.4$

❷ $\frac{4}{9}$

$$\begin{array}{r} 0.44 \\ 9)\overline{4.00} \\ \underline{36} \\ 40 \\ \underline{36} \\ 4 \end{array}$$ *The pattern is repeating.*

$\quad \frac{4}{9} = 0.\overline{4}$

- Every terminating decimal can be expressed as a fraction with a denominator of 10, 100, and so on.
- Every repeating decimal can be expressed as a fraction.

Examples Write each decimal as a fraction.

❸ 0.8

$$0.8 = \frac{8}{10}$$
$$= \frac{4}{5}$$

❹ $0.\overline{1}$

Let $N = 0.\overline{1}$ or 0.111 . . . .
Then $10N = 1.\overline{1}$ or 1.111 . . . .

$$\begin{array}{r} 10N = 1.111 \ldots \quad \textit{Subtract.} \\ \underline{-1N = 0.111 \ldots} \quad \textit{N = 1N} \\ 9N = 1 \end{array}$$

$$N = \frac{1}{9} \quad \text{So, } 0.\overline{1} = \frac{1}{9}.$$

Write each fraction as a decimal.

1. $\frac{3}{4}$
2. $\frac{1}{5}$
3. $\frac{3}{8}$
4. $\frac{2}{9}$

5. $\frac{4}{11}$
6. $\frac{7}{10}$
7. $\frac{2}{15}$
8. $\frac{1}{6}$

9. $\frac{4}{15}$
10. $\frac{3}{20}$
11. $\frac{5}{6}$
12. $\frac{5}{8}$

Write each decimal as a fraction in simplest form.

13. 0.9
14. 0.6
15. 0.25
16. $0.\overline{3}$

17. $0.\overline{5}$
18. 0.4
19. 0.16
20. $0.\overline{6}$

21. 0.125
22. 0.35
23. $0.\overline{8}$
24. $0.\overline{7}$

Decimals and Percents

- To express a decimal as a percent, first express the decimal as a fraction with a denominator of 100. Then express the fraction as a percent.

Examples   Write each decimal as a percent.

1 0.38

$$0.38 = \frac{38}{100}$$
$$= 38\%$$

2 0.4

$$0.4 = 0.40$$
$$= \frac{40}{100}$$
$$= 40\%$$

Shortcut
To write a decimal as a percent, multiply by 100 and add the % symbol. $0.62 = 0.62 = 62\%$

3 0.07

$$0.07 = \frac{7}{100}$$
$$= 7\%$$

4 2.55

$$2.55 = \frac{255}{100}$$
$$= 255\%$$

- To express a percent as a decimal, rewrite the percent as a fraction with a denominator of 100. Then express the fraction as a decimal.

Examples   Write each percent as a decimal.

5 26%

$$26\% = \frac{26}{100}$$
$$= 0.26$$

6 9%

$$9\% = \frac{9}{100}$$
$$= 0.09$$

Shortcut
To write a percent as a decimal, divide by 100 and remove the % symbol. $62\% = 62\% = 0.62$

7 87.5%

$$87.5\% = \frac{87.5}{100}$$
$$= 0.875$$

8 125%

$$125\% = \frac{125}{100}$$
$$= 1.25$$

Write each decimal as a percent.

1. 0.62	**2.** 0.99	**3.** 0.14	**4.** 0.20
5. 0.15	**6.** 0.8	**7.** 0.5	**8.** 0.06
9. 0.42	**10.** 0.03	**11.** 0.1	**12.** 1.76
13. 0.08	**14.** 3.10	**15.** 1.05	**16.** 2.6

Write each percent as a decimal.

17. 12%	**18.** 37%	**19.** 86%	**20.** 51%
21. 30%	**22.** 90%	**23.** 2%	**24.** 55%
25. 5%	**26.** 12.5%	**27.** 73.6%	**28.** 134%
29. 208%	**30.** 120%	**31.** 60.5%	**32.** 200%

Fractions and Percents

- To express a percent as a fraction, write the percent as a fraction with a denominator of 100 and simplify.

Example **❶** **Express 24% as a fraction in simplest form.**

$$24\% = \frac{24}{100} \quad \textit{Write as a fraction with a denominator of 100.}$$

$$= \frac{\overset{6}{24}}{\underset{25}{100}} \quad \textit{Divide by the GCF, 4.}$$

$$= \frac{6}{25}$$

- To express a fraction as a percent, write a proportion and solve.

Example **❷** **Express $\frac{3}{5}$ as a percent.**

$$\frac{3}{5} = \frac{n}{100} \quad \textit{Write a proportion.}$$

$$3 \times 100 = 5 \times n \quad \textit{Find the cross products.}$$

$$300 = 5n$$

$$\frac{300}{5} = \frac{5n}{5} \quad \textit{Divide each side by 5.}$$

$$60 = n$$

So, $\frac{3}{5} = \frac{60}{100}$ or 60%.

Write each percent as a fraction in simplest form.

1. 25% 2. 80% 3. 70% 4. 15%

5. 6% 6. 56% 7. 40% 8. 98%

9. 19% 10. 35% 11. 22% 12. 64%

Write each fraction as a percent.

13. $\frac{3}{10}$ 14. $\frac{1}{5}$ 15. $\frac{3}{4}$ 16. $\frac{13}{25}$

17. $\frac{37}{50}$ 18. $\frac{1}{20}$ 19. $\frac{19}{20}$ 20. $\frac{42}{50}$

21. $\frac{2}{25}$ 22. $\frac{9}{10}$ 23. $\frac{8}{32}$ 24. $\frac{22}{40}$

Comparing and Ordering Rational Numbers

- To compare rational numbers, it is usually easier and faster if you write the numbers as decimals. In some cases, it may be easier if you write the numbers as fractions having the same denominator or as percents.

Examples   Replace each ● with <, >, or = to make a true sentence.

❶ $\frac{3}{5}$ ● 0.65

$\frac{3}{5} = 0.60$

Since $0.60 < 0.65$, $\frac{3}{5} < 0.65$.

❷ 0.18 ● 2%

$2\% = 0.02$

Since $0.18 > 0.02$, $0.18 > 2\%$.

- To order rational numbers, first write the numbers as decimals. Then write the decimals in order from least to greatest and write the corresponding rational numbers in the same order.

Example   **❸** Write $\frac{1}{2}$, 55%, and 0.20 in order from least to greatest.

Write each number as a decimal.

$\frac{1}{2} = 0.50 \qquad 55\% = 0.55 \qquad 0.20 = 0.20$

Write the decimals in order from least to greatest.

$$0.20 < 0.50 < 0.55$$
$$\downarrow \qquad \downarrow \qquad \downarrow$$

Write the corresponding rational numbers in the same order.

$$0.20 < \frac{1}{2} < 55\%$$

The numbers in order from least to greatest are 0.20, $\frac{1}{2}$, and 55%.

Replace each ● with <, >, or = to make a true sentence.

1. 0.35 ● $\frac{1}{4}$

2. 65% ● 0.7

3. 80% ● $\frac{4}{5}$

4. $\frac{1}{10}$ ● 1%

5. 12.5% ● $\frac{1}{8}$

6. 36% ● 3.6

7. $\frac{2}{3}$ ● 0.7

8. 0.2 ● 20%

9. $\frac{5}{6}$ ● $\frac{5}{7}$

10. $\frac{3}{8}$ ● 0.375

11. 0.9 ● 10%

12. 30% ● 3.9%

13. 0.9 ● $0.\overline{9}$

14. 0.15 ● $\frac{2}{15}$

15. 51% ● 51

16. 78% ● $\frac{7}{8}$

17. $\frac{6}{11}$ ● 50%

18. $0.\overline{4}$ ● $\frac{4}{9}$

Write the numbers in each set in order from least to greatest.

19. $0.52, 5\%, \frac{1}{2}$

20. $\frac{1}{3}, 40\%, \frac{1}{4}$

21. $20\%, 0.1, \frac{3}{10}$

22. $0.19, \frac{1}{5}, 15\%$

23. $63\%, \frac{2}{3}, 0.06$

24. $\frac{1}{9}, 0.\overline{2}, 19\%$

Extra Practice

Lesson 1–1 *(Pages 4–7)*   **Write an algebraic expression for each verbal expression.**

1. 4 less than x

2. the difference of 3 and m

3. 6 more than the quotient of b and 5

4. 2 less than 7 times k

Write a verbal expression for each algebraic expression.

5. $b + 4$

6. $12x$

7. $6 - k$

8. $5 - 2k$

9. $\dfrac{3}{x}$

10. $7 - \dfrac{m}{x}$

11. $\dfrac{6}{m} + 7$

12. $\dfrac{7}{2} + r$

Write an equation for each sentence.

13. Six times s plus seven equals fourteen.

14. Three minus k equals twelve plus two times j.

15. Five divided by m is equal to two times n minus three.

Lesson 1–2 *(Pages 8–13)*   **Find the value of each expression.**

1. $32 + 4 \cdot 2$

2. $6 - 3 \cdot 4 + 7$

3. $16 \div 4 + 3$

4. $6 \cdot 2 \cdot 4 - 1$

5. $25 \div 5 - (3 + 2)$

6. $\dfrac{3(7 + 1)}{3 \cdot 4}$

7. $[2(6 - 1)] \div 2$

8. $2(5 + 3) - 6$

Name the property of equality shown by each statement.

9. If $3y + 4 = 5x$ and $5x = 10$, then $3y + 4 = 10$.

10. If $13 + y = 18$, then $18 = 13 + y$.

11. $9 - 7x = 9 - 7x$

12. $2(6 + 4) = 2(10)$

Find the value of each expression. Identify the property used in each step.

13. $6 + 5 \cdot 2 - 1$

14. $9(8 \div 2 - 3)$

15. $24 - 1(3 + 15)$

16. $5(6) + 6(14 - 14)$

17. $11 + (7 - 28 \div 4)$

18. $\dfrac{7 + 21 \div 3}{3(2 - 1) + 4}$

Evaluate each algebraic expression if $m = 6$ and $b = 3$.

19. $8m + 2b$

20. $m(b - 3)$

21. $m - 3b + 4$

22. $\dfrac{6 \cdot 2 + b \cdot m}{m - 4}$

23. $\dfrac{m(6 - 2)}{b + 1}$

24. $(2m + 1)(6 - b)$

Lesson 1–3 *(Pages 14–18)*   **Name the property shown by each statement.**

1. $(7 + 7) + 2 = 7 + (7 + 2)$

2. $(3 \cdot m) \cdot 6 = 3 \cdot (m \cdot 6)$

3. $9 \cdot 13 = 13 \cdot 9$

4. $6 \cdot 3 \cdot k = k \cdot 6 \cdot 3$

5. $k + (p + q) = (k + p) + q$

6. $m + 5 = 5 + m$

Simplify each expression. Identify the properties used in each step.

7. $5(6r)$

8. $(k + 1) + 4$

9. $11 \cdot p \cdot 2$

10. $(s \cdot 4) \cdot 3$

11. $7 + y + 1$

12. $13 + w + 12$

Lesson 1-4 *(Pages 19–23)* **Simplify each expression.**

1. $6b + 3b$

2. $4(3k - 2k)$

3. $4a + 2b + 8a + 5b$

4. $9 + 12s - 2s$

5. $3 + 4p - 2p$

6. $8 + k - 7 + 6k$

7. $ab - b + 3ab$

8. $8(m + 3n) + 3m$

9. $6st - st + t$

10. $6(3 + 2m)$

11. $5(7 + 2r) + 2(r - 3)$

12. $4 + (7 + 5x)3$

13. Write $6m + (3k + m) + 2k + 4(m + 1)$ in simplest form. Indicate the property that justifies each step.

14. What is the value of $4x$ multiplied by the quantity $2x$ minus 3 if x equals 4?

Lesson 1-5 *(Pages 24–29)* **Solve each problem. Use any strategy.**

1. Simone has 16 players on her soccer team. There are 6 more boys than girls. How many girls and how many boys are there?

2. Dillon put $350 into a bank account that pays 6% simple interest. How much money will he have in four years?

3. How many ways can you make 50¢ using dimes, nickels, and pennies?

4. Kate lives 24 miles from school. Jake lives 6 miles closer to school than Kate does. How far does Jake live from school?

5. Nine paintings are for sale at an art gallery. Tara must choose two of the paintings.

    a. Make a chart or diagram to represent the problem.

    b. How many different combinations are there in all?

    c. How many different combinations would there be if there were eight paintings?

Lesson 1-6 *(Pages 32–37)* **Determine whether each is a good sample. Describe what caused the bias in each poor sample. Explain.**

1. Every other student in physical education class is surveyed to determine how many hours per week students exercise.

2. Every student at a school is asked to name their favorite subject in school.

3. Surveys are placed on every fifth car in a mall parking lot concerning the construction of a new lot.

4. Six people at a music concert are randomly chosen to find out their opinion of the concert.

Refer to the chart at the right.

5. Make a frequency table to organize the data.

6. What number of hours is most frequent?

7. How many students use the Internet for less than two hours per week?

8. How many more students use the Internet six or more hours than those who use it less than four hours?

Time on the Internet per Week (h)		
4	9	4
2	5	2
6	1	10
10	8	4
3	2	9
10	4	6
1	7	1

Lesson 1–7 *(Pages 38–43)* **People in three major cities were surveyed to see what transportation they used to get to work. The table shows the results.**

Type	New York, NY	Los Angeles, CA	Baltimore, MD
drive alone	24.0%	65.2%	75.5%
carpool	8.5%	15.4%	14.2%
public transit	53.4%	10.5%	2.7%
other	14.0%	8.9%	7.6%

Source: *Time Almanac 2000*

1. Make a histogram for each city that shows the results of the survey.
2. How do the histograms compare?
3. Make some assumptions to support your findings.

The record low temperatures for select states are shown.

4. How many states are represented?
5. What is the lowest record low temperature?
6. Which temperature occurs most frequently?
7. How many states have record low temperatures of less than $-50°C$?

Stem	Leaf
−6	2
−5	7 2 1 1
−4	6 4 4 3 0 0 0 0
−3	8 8 8 7 6 4 3
−2	8 7 7 7 6
−1	9 4

$-1|4 = -14°C$

Lesson 2–1 *(Pages 52–57)* **Name the coordinate of each point.**

1. A
2. B
3. C
4. D
5. E
6. F

Graph each set of numbers on a number line.

7. $\{-1, 0, 4\}$
8. $\{-3, -2, -1\}$
9. $\{0, 3, 5\}$
10. $\{-4, -2, 1, 2\}$
11. $\{-1, 0, 1, 2\}$
12. $\{-4, 2, 3, 4\}$

Replace each ● with < or > to make a true sentence.

13. $-3 ● 0$
14. $-4 ● -3$
15. $-8 ● 2$
16. $0 ● -1$
17. $|6| ● -2$
18. $-2 ● |-3|$
19. $|-7| ● |-4|$
20. $|16| ● -12$

Evaluate each expression.

21. $|7|$
22. $|-5|$
23. $|-7| - 5$
24. $-11 + |4|$
25. $|4| - |-12|$
26. $|-8| + |-3|$
27. $|5| - 11$
28. $|18| + |-21|$

Lesson 2–2 *(Pages 58–63)* **Write the ordered pair that names each point.**

1. G
2. H
3. I
4. J
5. K
6. L

Graph each point on a coordinate plane.

7. $P(5, -1)$
8. $Q(0, 0)$
9. $R(1, 2)$
10. $S(-2, 4)$
11. $T(-3, -2)$
12. $U(0, -3)$

Name the quadrant in which each point is located.

13. $(6, 4)$
14. $(-2, 1)$
15. $(-3, -3)$
16. $(0, 0)$

Lesson 2-3 *(Pages 64–69)*   **Find each sum.**

1. $2 + 14$
2. $8 + 4$
3. $-16 + (-7)$
4. $-13 + (-26)$
5. $(-21) + 14$
6. $-19 + 17$
7. $-5 + (-13) + (-24)$
8. $2 + 14 + (-12)$
9. $-16 + (-4) + (-11) + 5$
10. $12 + (-7) + 6 + (-18)$

Simplify each expression.

11. $6b + 2b$
12. $-8y + 23y$
13. $17k + (-18k)$
14. $7m - 7m$
15. $-8r + (-4r) + 12r$
16. $-6p + (-3p) + (-1p)$
17. $7c + 11c + (-12c)$
18. $-5x - 6x$

Lesson 2-4 *(Pages 70–74)*   **Find each difference.**

1. $16 - 7$
2. $16 - 21$
3. $-10 - (-9)$
4. $-4 - 6$
5. $3 - 9$
6. $7 - (-12)$
7. $-12 - (-4)$
8. $0 - 11$

Evaluate each expression if $x = 6$, $y = -8$, $z = -3$, and $w = 4$.

9. $y - z$
10. $3 - z$
11. $6 + y$
12. $w - y$
13. $14 - y - x$
14. $6 + x - z$
15. $y + z + w$
16. $w - z + 11$

Simplify each expression.

17. $7k - 4k$
18. $6m + (-3m)$
19. $3y - 7y$
20. $-14n - 17n$
21. $13s - (-2s)$
22. $9r - (-2r) - 4r$
23. $8c - 12c + 2c$
24. $b + 2b - 3b$

25. Corey spends $112 on movies each month. He spends $80 less on books. How much does he spend on books each month?

Lesson 2-5 *(Pages 75–79)*   **Find each product.**

1. $2(4)$
2. $3(-12)$
3. $-4(-13)$
4. $-5(8)$
5. $10(-3)(2)$
6. $-4(2)(-3)1$
7. $6(-1)(-1)(-2)$
8. $-8(-4)2$

Evaluate each expression if $x = -2$, $y = 4$, and $z = -6$.

9. $6x$
10. $2yz$
11. xyz
12. $-8xy$
13. $4x - 3z$
14. $6y - z$
15. $2x + z$
16. $10x + 5y$

Simplify each expression.

17. $-6(2b)$
18. $7(-4m)$
19. $-10(-3k)$
20. $4(5c)$
21. $(6g)(9h)$
22. $8m(-6n)$
23. $(-2r)(-3s)$
24. $(-5p)2q$

Lesson 2–6 *(Pages 82–85)* Find each quotient.

1. $49 \div 7$
2. $-12 \div 6$
3. $-36 \div (-12)$
4. $21 \div (-3)$
5. $36 \div (-6)$
6. $-20 \div 4$
7. $-16 \div (-4)$
8. $-32 \div (-4)$
9. $\frac{8}{2}$
10. $\frac{-54}{-9}$
11. $\frac{63}{-7}$
12. $\frac{-64}{8}$

Evaluate each expression if $x = -5$, $y = -3$, $z = 2$, and $w = 7$.

13. $25 \div x$
14. $-42 \div w$
15. $3 \div y$
16. $2x \div z$
17. $-3x \div y$
18. $x \div (-1)$
19. $xyz \div 10$
20. $yzw \div 2w$
21. $\frac{3y}{-3}$
22. $\frac{6-y}{y}$
23. $\frac{w}{-7}$
24. $\frac{w-x}{y}$

Lesson 3–1 *(Pages 94–99)* Replace each ● with <, >, or = to make a true sentence.

1. -2 ● -5
2. -3.4 ● -4
3. -0.32 ● -0.3
4. $4(-2)(-1)$ ● $2(-3)$
5. 0.6 ● $\frac{3}{5}$
6. $\frac{2}{3}$ ● 0.7
7. $\frac{7}{9}$ ● $\frac{7}{8}$
8. $-\frac{3}{18}$ ● $-\frac{2}{9}$

Write the numbers in each set from least to greatest.

9. $\frac{5}{6}, 0.3, \frac{1}{3}$
10. $0.4, -\frac{2}{3}, \frac{3}{4}$
11. $-\frac{4}{5}, -\frac{6}{7}, -0.7$
12. $\frac{1}{2}, -\frac{7}{10}, -\frac{2}{3}$
13. $0.\overline{4}, \frac{3}{8}, \frac{4}{6}$
14. $-\frac{1}{8}, \frac{2}{9}, -\frac{3}{12}$

15. Compare the numbers $\frac{6}{7}$ and $\frac{7}{8}$ using an inequality.

16. Using a number line, explain why $0.7 > 0.6$.

Lesson 3–2 *(Pages 100–103)* Find each sum or difference.

1. $-13.1 + (-21.4)$
2. $-8.14 - 0.13 + 1.11$
3. $6.2 - (-3.4)$
4. $-11.12 - 2.15 + 5.28 - 3.12$
5. $15.9 + 6.25 - 3.48 - 2.13$
6. $-17.6 + 0.3 - 3.7$
7. $-\frac{4}{5} + \frac{2}{3}$
8. $\frac{6}{7} - \frac{1}{8}$
9. $-\frac{3}{4} + \frac{2}{3}$
10. $3\frac{6}{7} - \left(-4\frac{5}{8}\right)$
11. $-6\frac{1}{6} + 6\frac{5}{6}$
12. $2\frac{8}{9} - 1\frac{1}{3}$

13. Find the value of $z - 0.25$ if $z = 0.5$.

14. Evaluate $m + 5\frac{3}{4}$ if $m = -3\frac{1}{2}$.

Lesson 3–3 (*Pages 104–109*) **Find the mean, median, mode, and range for each set of data.**

1. 8, 3, 2, 8, 1, 5, 8, 5

2. 11, 5, 18, 10, 14, 14, 18, 18, 14, 17, 15

3. 1.2, 5.4, 2.3, 3.2, 1.2, 5.3

4. 123, 153, 123, 114, 148, 114, 135

5. 47, 29, 77, 99, 50 47, 29

6. 6.5, 7.8, 5.7, 3.9, 9.9, 3.8, 5.5, 5.7

7.

8.

Stem	Leaf
11	0 1 7
12	3 4 4
13	5 5 6 7
14	2 2 *11|0 = 110*

Write a set of data with six numbers that satisfies each set of conditions.

9. The median is equal to the mean.

10. The median is less than the mean.

Lesson 3–4 (*Pages 112–116*) **Find the solution of each equation if the replacement sets are** $x = \{-2, -1, 0, 1\}$, $m = \{-1, 0, 1, 2\}$, **and** $d = \{6, 7, 8, 9\}$.

1. $x + 3 = 2$

2. $2d = 16$

3. $6m + 15 = 21$

4. $7m - 11 = \dfrac{3m}{2}$

5. $\dfrac{14 - 2}{x} = -6$

6. $\dfrac{2}{3(1 + 5)} = \dfrac{6 - 5}{d}$

Solve each equation.

7. $-17.8 - 12.2 = p$

8. $6.5 - 13.2 = y$

9. $x = 7 \cdot 2 + 1$

10. $j = 16 - 6 \cdot 2 \div 3$

11. $7 \cdot 4 - 4 \div 2 = r$

12. $w = 6.1 - 3.6 \cdot 4$

13. $n = 10.3 + 4.2 \div 3 - 1.2$

14. $[12 - (4 + 1)] = z$

15. $t = -4.2 - 3.3 \div 3$

16. $\dfrac{7 \cdot 3 + 3}{(3 - 1) \cdot 2} = k$

17. $\dfrac{-13 - 5}{3 \cdot 2} = s$

18. $\dfrac{6 - 3 \cdot 2}{12 \div 3 + 1} = m$

19. $x = \dfrac{5 \div (2 + 3)}{2 \cdot 2 - 3}$

20. $\dfrac{36 \div 9 + 1}{2 \cdot 5} = a$

21. $h = \dfrac{15 + 10 \div 2}{-3 \cdot 1 + 2}$

22. Find the solution of $6 - 3(8) = r$.

23. What is the value of k if $k = \dfrac{16 + 2}{-9}$?

Lesson 3–5 (*Pages 117–121*) **Solve each equation. Use algebra tiles if necessary.**

1. $g + 13 = 7$

2. $-8 = w + -6$

3. $m - 8 = -23$

4. $-3 + q = 9$

5. $-21 = a + 15$

6. $t - 12 = -3$

7. $d - (-6) = 14$

8. $s - 4 = 12$

9. $k - 9 = 18$

10. $c - (-3) = -8$

11. $13 + y = 4$

12. $h - 11 = 5$

13. $-24 = x + 7$

14. $16 = 9 + p$

15. $6 = -12 + b$

16. $-5 = -12 + b$

17. $m + 15 = 9$

18. $-8 = w + (-6)$

19. What is the value of y if $y - 4 = -15$?

20. When x is divided by 8, the result is 7. Find the value of x.

Lesson 3–6 *(Pages 122–127)* **Solve each equation. Check your solution.**

1. $-18 = w + 3$
2. $m + (-5) = 6$
3. $-16.4 = r - 7.4$
4. $m - 16 = -13$
5. $5 = x - (-3)$
6. $x + (-8) = -7$
7. $45 + j = -27$
8. $-56 = g - 32$
9. $n - (-26) = 41$
10. $6.4 + k = -3.2$
11. $-2.8 + k = 3.1$
12. $s + 2.5 = -1.3$
13. $13.4 + p = -2.4$
14. $17.3 = -2.7 + d$
15. $y + 2.17 = 5.67$
16. $\frac{5}{7} = z - \frac{6}{21}$
17. $n - \frac{5}{8} = -\frac{3}{4}$
18. $-\frac{2}{3} = j + \frac{7}{12}$

19. Solve for m if $-20 + m = -7$.
20. What is the solution of $h - 4 = 14$?

Lesson 3–7 *(Pages 128–131)* **Solve each equation. Check your solution.**

1. $|m| = -3$
2. $4 = |x|$
3. $|r| + 4 = 8$
4. $|y| - 8 = 11$
5. $|x + 5| = 10$
6. $|t + 8| = -2$
7. $|d - 4| = 10$
8. $6 = |-3 + p|$
9. $-4 = |5 + s|$
10. $|n - 11| = 15$
11. $-4 = |-2 + j|$
12. $|-2 + p| = 0$
13. $|k - (-2)| = 5$
14. $12 = |n + (-3)|$
15. $2 + |s + 3| = 8$
16. $5 = |-4 + g| + 17$
17. $16 = |z - 4| + 3$
18. $|s + 4| - 3 = 7$

19. How many solutions exist for $-3 = |4 + d|$?
20. How many solutions exist for $8 = |2 + k|$?

Lesson 4–1 *(Pages 140–145)* **Find each product.**

1. $3(8.2)$
2. $-7.3(3)$
3. $-6.2(3.5)$
4. $(2.1)(-1)$
5. $-3.2(-0.5)$
6. $-2.1(3)(-2)$
7. $16.2(0)$
8. $8.5(-1)(-2.2)$
9. $\frac{2}{5}\left(\frac{1}{9}\right)$
10. $\left(-\frac{1}{2}\right)\left(\frac{3}{8}\right)$
11. $-\frac{1}{4}\left(-\frac{6}{5}\right)$
12. $-1 \cdot \frac{1}{8}$
13. $-\frac{6}{7}(0)$
14. $4\left(-\frac{3}{4}\right)$
15. $\left(-\frac{3}{4}\right)\left(\frac{5}{6}\right)$
16. $\frac{1}{2}\left(-3\frac{1}{8}\right)$

Simplify each expression.

17. $-7.2(2p)$
18. $(-9y)(-4.1)$
19. $(6.2k)(-3)$
20. $-5(0.5j)$
21. $2x(-3.4y)$
22. $5.4m(-2n)$
23. $\frac{7}{13}p(26)$
24. $\left(\frac{3}{5}r\right)\left(-\frac{1}{3}\right)$
25. $\left(\frac{1}{2}x\right)\left(\frac{2}{3}y\right)$
26. $\frac{5}{6}k(6)$
27. $\left(-\frac{2}{3}s\right)\left(\frac{4}{5}r\right)$
28. $-\frac{1}{4}w(-3t)$

Lesson 4–2 *(Pages 146–151)*   **Find the number of possible outcomes by drawing a tree diagram.**

Men	Women
Troy	Ann
Malik	Lorena
Ben	Ellen
Aaron	

Exercise 2

1. tossing a coin twice
2. choosing different teams of one man and one woman
3. At a dinner party, you can choose either chicken or beef for your main dish, soda or juice for your drink, and pie, ice cream, or cake for your dessert. How many different meals are possible?

4. When choosing auto insurance, you have many choices, as shown in the table. How many different types of coverage are possible? Find the number of possible outcomes by using the Fundamental Counting Principle.

Liability	Underinsured Drivers	Collision	Comprehensive
$50,000	$50,000	$100	$100
$100,000	$100,000	$500	$500
$150,000	$150,000	$1000	$1000
	none		

5. choosing four cards from a standard 52-card deck
6. There are 5 multiple-choice problems on a quiz (with possible answers of a, b, c, and d). How many different combinations of answers are possible?

Lesson 4–3 *(Pages 154–159)*   **Find each quotient.**

1. $12 \div 1.5$
2. $3.1 \div (-3.1)$
3. $-8.2 \div (-4.1)$
4. $-18.6 \div 6.2$
5. $0 \div 7.3$
6. $-9.3 \div (-3.1)$
7. $1.6 \div (-16)$
8. $-0.4 \div 0.2$
9. $-\frac{1}{6} \div \left(-\frac{1}{5}\right)$
10. $\frac{1}{3} \div (-5)$
11. $-4 \div \frac{2}{3}$
12. $8 \div \left(-\frac{1}{4}\right)$
13. $6 \div \left(-\frac{3}{4}\right)$
14. $-\frac{2}{5} \div \left(-\frac{7}{8}\right)$
15. $4\frac{5}{8} \div \frac{5}{8}$
16. $-\frac{4}{3} \div \frac{5}{6}$

Evaluate each expression if $j = -\frac{1}{8}$, $k = \frac{3}{4}$, **and** $m = \frac{1}{6}$.

17. $\frac{8}{j}$
18. $\frac{k}{2}$
19. $\frac{m}{5}$
20. $\frac{4}{m}$
21. $\frac{j}{k}$
22. $\frac{2m}{k}$
23. $\frac{j}{8}$
24. $\frac{km}{j}$

Lesson 4–4 *(Pages 160–164)*   **Solve each equation.**

1. $6k = -54$
2. $0 = 8p$
3. $0.5y = 12$
4. $3.2m = -6.4$
5. $7.5 = -1.5w$
6. $2.1t = -6.3$
7. $-3n = 2$
8. $-7 = -3s$
9. $\frac{1}{3}x = -4$
10. $-3 = -\frac{5}{8}v$
11. $\frac{4}{5} = \frac{4}{9}r$
12. $-\frac{2}{3}z = \frac{1}{8}$
13. $-\frac{1}{2}a = 3$
14. $10 = \frac{c}{4}$
15. $\frac{21}{13} = -7j$
16. $-\frac{1}{18} = \frac{17}{18}b$

17. Solve $-1.5x = 18$. Then check your solution.

18. What is the solution of $\frac{s}{16} = \frac{1}{4}$?

Lesson 4–5 *(Pages 165–170)* **Solve each equation. Check your solution.**

1. $6 + 2y = 8$
2. $3 - 3r = 4$
3. $4 + (-0.5y) = 8$
4. $-7s - 3 = 18$
5. $0 = 1.5a + 3$
6. $3.1 = 0.7w + 1$
7. $6.3 = 0.4t - 2.1$
8. $6c + 1.5 = 3$
9. $\frac{r}{5} - 1 = 5$
10. $6 = -\frac{y}{3} + 3$
11. $10 = -\frac{x}{2} + 6$
12. $8 - \frac{b}{4} = 2$
13. $8 = -\frac{x}{3} + 2$
14. $\frac{6 - p}{2} = -3$
15. $-\frac{7}{8}c + 1 = 4$
16. $\frac{8 + m}{3} = -8$

17. What is the solution of $3 = \frac{r + 10}{-3}$?

18. Find the value of w in the equation $6 + \frac{2}{3}w = 12$.

Write an equation and solve each problem.

19. Three less than half a number x is 12. Find the number.

20. Start with a number k. If you add 3, multiply by 7 and subtract 4, you get 18. What is the value of k?

Lesson 4–6 *(Pages 171–175)* **Solve each equation. Check your solution.**

1. $6a = 3a + 3$
2. $16 + 5r = 3r$
3. $8x + 6 = 12 - 3x$
4. $7k + 3 = 2k$
5. $1.4y = -0.6y - 2$
6. $4 + 0.8p = -3.2p$
7. $6.4 + m = m - 2.3$
8. $\frac{1}{3}f = \frac{2}{3}f + 4$
9. $\frac{1}{8}t + 3 = -4 - \frac{1}{6}t$
10. $\frac{1}{2}w - 2 = \frac{2}{3}w$
11. $1 - \frac{7}{8}h = \frac{1}{4}h$
12. $\frac{2}{5}j + 3 = -\frac{3}{5}j$

13. Find the solution of $1.7x + 11 = -6.3x - 13$.

14. Six times a number k is 17 less than 2 times the number.
    a. Write an equation to represent the problem.
    b. Find the number.

Lesson 4–7 *(Pages 176–179)* **Solve each equation. Check your solution.**

1. $16 = -2(p - 1)$
2. $6(x + 5) = 12$
3. $8x + 6 = 3(4 - x)$
4. $8(6 - q) - 5 = -21$
5. $6t = 3(2t + 5)$
6. $-7 = 3k - 2(2k)$
7. $7(z + 3) = 2z + 4$
8. $-1(r + 1) - 2 = 2 + (r - 1)$
9. $5x = 3(1.5x - 3)$
10. $-4(m - 2.2) = -2(m + 1.4)$
11. $6 + \frac{1}{3}(9x + 3) = 6x + 2$
12. $4\left(\frac{1}{2}x + \frac{1}{2}\right) = 2x + 2$

13. Find the solution of $4[8 + 2(m - 2)] = 8$.

14. What is the value of r in $7(8 - r) + 3(2 + 2r) = 9$?

Lesson 5-1 *(Pages 188–193)* Solve each proportion.

1. $\dfrac{9}{8} = \dfrac{3}{10}$

2. $\dfrac{6}{4} = \dfrac{y}{10}$

3. $\dfrac{b}{12} = \dfrac{32}{8}$

4. $\dfrac{1}{3} = \dfrac{5}{r}$

5. $\dfrac{21}{28} = \dfrac{h}{7}$

6. $\dfrac{4}{5} = \dfrac{a}{100}$

7. $\dfrac{25}{3} = \dfrac{500}{x}$

8. $\dfrac{p}{24} = \dfrac{8}{4}$

9. $\dfrac{z}{6} = \dfrac{9}{2}$

10. $\dfrac{6.2}{s} = \dfrac{3.1}{2}$

11. $\dfrac{0.3}{0.4} = \dfrac{x}{8}$

12. $\dfrac{6+m}{m-3} = \dfrac{1}{2}$

13. $\dfrac{7+b}{4} = \dfrac{3}{5}$

14. $\dfrac{7-d}{8+d} = \dfrac{4}{9}$

15. $\dfrac{6}{d-2} = \dfrac{7}{d}$

16. $\dfrac{k+1}{3} = \dfrac{k}{6}$

17. Are $\dfrac{18}{4}$ and $\dfrac{9}{2}$ equivalent ratios? Explain your reasoning.

18. Find the value of x that makes $\dfrac{17}{3} = \dfrac{x}{9}$ a proportion.

Convert each measurement as indicated.

19. 7200 pounds to tons

20. 4.5 feet to inches

21. 2.1 quarts to pints

22. 460 meters to kilometers

23. 13 grams to milligrams

24. 16 milliliters to liters

Lesson 5-2 *(Pages 194–197)* On a map, the scale is 1.5 inches = 100 miles. Find the actual distance for each map distance.

	From	To	Map Distance
1.	Paris, France	Brussels, Belgium	3.15 inches
2.	Amsterdam, Holland	Frankfurt, Germany	4.5 inches
3.	Luxembourg, Luxembourg	Frankfurt, Germany	$2\frac{1}{4}$ inches
4.	Luxembourg, Luxembourg	Paris, France	$3\frac{3}{5}$ inches

5. **Puzzles** The picture on a puzzle box shows the puzzle to be 4 inches tall. The actual height is 12 inches. What is the scale of the picture?

Lesson 5-3 *(Pages 198–203)* Express each fraction or ratio as a percent.

1. 11 to 44

2. 36 out of 18

3. 8 out of 32

4. 12 to 24

5. $\dfrac{12}{20}$

6. $\dfrac{27}{9}$

7. $\dfrac{10}{8}$

8. $\dfrac{7}{21}$

9. 4 to 5

10. 3 out of 10

11. 6 to 8

12. 2 out of 20

13. $\dfrac{18}{27}$

14. 5 out of 25

15. 3 to 8

16. $\dfrac{13}{100}$

17. 12 to 3

18. 6 out of 12

19. 10 to 5

20. 5 out of 6

21. One out of three people at the conference agreed with the speaker.

22. Two fifths of the students passed the test.

Use the percent proportion to find each number.

23. What number is 30% of 120?

24. 125% of what number is 15?

25. What percent of 21 is 7?

26. Find 10% of 12.

27. 50 is what percent of 200?

28. 16 is 25% of what number?

Lesson 5-4 *(Pages 204–209)* **Use the percent equation to find each number.**

1. What number is 110% of 36?
2. 30 is 300% of what number?
3. Find 25% of 120.
4. What number is 30% of 200?
5. What number is 10% of 25?
6. Find 200% of 18.
7. 15 is 75% of what number?
8. 16 is 40% of what number?
9. Find 5% of 120.
10. What number is 500% of 5?
11. 35 is 20% of what number?
12. Find 150% of 80.
13. 45 is 80% of what number?
14. What number is 18% of 324?

15. **Banking** How long will it take Kristin to earn $180 if she invests $4000 at a rate of 6%?

16. **Banking** How much interest will Michael earn if he invests $575 at a rate of 7% for 3 years?

Lesson 5-5 *(Pages 212–217)* **Find the percent of increase or decrease. Round to the nearest percent.**

1. original: 20
   new: 22
2. original: 125
   new: 100
3. original: 18
   new: 9
4. original: 600
   new: 750
5. original: 30
   new: 10
6. original: 28
   new: 21
7. original: 12
   new: 15
8. original: 50
   new: 70

The cost of an item and a sales tax rate are given. Find the total price of each item to the nearest cent.

9. painting: $600; 6.5%
10. shoes: $85; 4%
11. book: $24.95; 3.5%
12. shirt: $29.99; 6%
13. piano: $1600; 5%
14. sweater: $45.99; 7%

The original cost of an item and a discount rate are given. Find the sale price of each item to the nearest cent.

15. jacket: $120; 40%
16. hat: $23.99; 30%
17. basketball: $25; 10%
18. video game: $49.99; 5%
19. bicycle: $425; 15%
20. snow skis: $225; 25%

21. What number is 20% less than $75?
22. Find the percent of increase from $75 to $85.

Lesson 5-6 *(Pages 219–223)* **Refer to the application on page 219. Find the probability of each outcome if a pair of dice are rolled.**

1. 2 odd numbers
2. a sum of 7
3. a sum of 4
4. even number on the first die
5. a sum greater than 11
6. a sum of less than 10

Find the odds of each outcome if the spinner at the right is spun.

7. greater than 10
8. blue
9. blue or yellow
10. not red
11. an even number
12. not a two
13. an odd number
14. less than 4

15. What is the probability that you select a seven at random from a standard deck of cards?

Lesson 5–7 *(Pages 224–229)*   **A card is drawn from a deck of ten cards numbered 1 through 10. The card is replaced in the deck and another card is drawn. Find the probability of each outcome.**

1. P(9 and then a 7)
2. P(6 and a 4)
3. P(an even and then a 2)
4. P(two numbers less than 6)
5. P(8 and then an odd number)
6. P(two odd numbers)

7. What is the probability of tossing a coin three times and getting two heads and a tail?

8. What is the probability of tossing a coin three times and getting a head, then a tail, then a head?

Determine whether each event is *mutually exclusive* or *inclusive*. Then find each probability.

9. There are 18 cars in a lot. There are 6 red cars, 8 blue cars, 3 white cars, and 1 purple car. What is the probability of randomly choosing a white or blue car?

10. In rolling a die, what is the probability that it is either an even or a four?

Lesson 6–1 *(Pages 238–243)*   **Express each relation as a table and as a graph. Then determine the domain and the range.**

1. $\{(6, 3), (2, 4), (3, 2), (5, 5)\}$
2. $\{(5.9, -3), (-2, 3.1), (0, 0), (-1, 7)\}$
3. $\{(-1, 6), (2, -5), (-2, 9), (0, 4)\}$
4. $\{(1, -3), (2.4, 6), (0, -5.1), (-4, 4)\}$
5. $\left\{\left(\frac{1}{4}, \frac{1}{6}\right), \left(-\frac{2}{3}, \frac{4}{5}\right), \left(-1, \frac{1}{3}\right)\right\}$
6. $\left\{(3, 1), \left(\frac{3}{4}, 2\right), (2, 0), \left(-\frac{2}{3}, -3\right)\right\}$

Express each relation as a set of ordered pairs.

7.

x	y
1	−0.5
2	0
3	1
4	3

8.

x	y
−3	4
−2	3
−1	2
0	1

9.

x	y
1	−3
2	−2
−3	−1
−1	0

Lesson 6–2 *(Pages 244–249)*   **Which ordered pairs are solutions of each equation?**

1. $6x + 2y = 12$ **a.** $(1, 3)$ **b.** $(1, 1)$ **c.** $(2, 0)$ **d.** $(2, 2)$
2. $3a + 6 = 7b$ **a.** $(4, 3)$ **b.** $(1, 0)$ **c.** $(2, 3)$ **d.** $(5, 3)$
3. $2m + n = 7$ **a.** $(1, 6)$ **b.** $(3, 1)$ **c.** $(0, 2)$ **d.** $(0, 6)$
4. $3r - 15 = -6s$ **a.** $\left(\frac{1}{3}, \frac{7}{3}\right)$ **b.** $(1, 2)$ **c.** $(2, 1)$ **d.** $\left(\frac{1}{3}, \frac{1}{2}\right)$

Solve each equation if the domain is $\{-2, -1, 0, 1, 2, 3\}$. Graph the solution set.

5. $y = 6x$
6. $y = -2x + 1$
7. $y = x - 8$
8. $x - y = 5$
9. $y = -3x - 4$
10. $3x - 2y = 6$
11. $4 - x = y$
12. $7 = 6x + 4y$

Find the domain of each equation if the range is $\{-1, 0, 1, 2\}$.

13. $y = 3x + 3$
14. $7 - 4x = y$
15. $6y = 2x$
16. $x = 5y - 2$
17. $2y = 3x + 1$
18. $5x - 15 = y$
19. $2x = y - 4$
20. $2y - 2x = 4$

Lesson 6-3 (Pages 250–255) Determine whether each equation is a linear equation. Explain. If an equation is linear, identify A, B, and C.

1. $4xy = 3$
2. $7x - 2y = 3$
3. $2xy + x = y$
4. $16y = 4x$
5. $x = 2y$
6. $3y = x + 4$
7. $y - 2 = x$
8. $x + 3y = 9$
9. $6x + 3 = y$
10. $2x^2 + y^2 = 8$
11. $4y + 8x = 0$
12. $y + xy = 1$

Graph each equation.

13. $x = y - 1$
14. $-3x + 2 = y$
15. $7y = 14x - 2$
16. $x + y = 4$
17. $2x + 3y = 0$
18. $x - y = 2$
19. $16 = 2x + 4y$
20. $5x - y = 15$

Lesson 6-4 (Pages 256–261) Determine whether each relation is a function.

1. $\{(3, 6), (4, -2), (-1, -3), (0, 6)\}$
2. $\{(-1, 8), (0, 3), (3, 3), (-2, 2)\}$
3. $\{(-1, 5), (3, 1), (5, 4), (-1, 3)\}$
4. $\{(-2, 4), (0, 0), (4, 2), (3, 0)\}$
5. $\{(0, 3), (1, 1), (1, -1), (2, 4)\}$
6. $\{(3, 3), (-3, 1), (4, 2), (-2, 3)\}$

7.

x	y
−4	2
−3	3
−2	2
−4	3
−5	4

8.

x	y
1	1
3	3
−1	1
2	2
0	0

9.

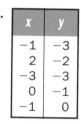

x	y
−1	−3
2	−2
−3	−3
0	−1
−1	0

Use the vertical line test to determine whether each relation is a function.

10.

11.

12.

If $f(x) = -3x - 1$ and $g(x) = 4x + 5$, find each value.

13. $f(3)$
14. $g(2)$
15. $g(-1)$
16. $f(-4)$
17. $g(-1.25)$
18. $f(5)$
19. $f(0.5)$
20. $g(0.2)$
21. $f(-3.1)$
22. $g(-1.5)$
23. $f(1.2)$
24. $g(2.1)$
25. $f\left(\frac{1}{8}\right)$
26. $g\left(\frac{2}{3}\right)$
27. $f\left(\frac{1}{2}\right)$
28. $g\left(-\frac{3}{5}\right)$
29. $f\left(-\frac{5}{8}\right)$
30. $g\left(\frac{1}{4}\right)$

Lesson 6-5 (Pages 264–269) Determine whether each equation is a direct variation. Verify the answer with a graph.

1. $x = y + 2$
2. $y = 3x$
3. $y = 4x + 2$
4. $y = 2x + 2$
5. $y = 2x$
6. $x = -2$

Solve. Assume that y varies directly as x.

7. If $y = -3$ when $x = -6$, find x when $y = 16$.

8. If $y = 9$ when $x = 5$, find y when $x = 10$.

9. Find x when $y = 24$ if $y = 16$ when $x = 6$.

10. Find x when $y = 3$ if $y = 15$ when $x = 4$.

11. If $x = 13$ when $y = 7$, find y when $x = 26$.

12. Find y when $x = 2.5$ if $y = 10$ when $x = 6$.

Lesson 6–6 *(Pages 270–275)* **Solve. Assume that *y* varies inversely as *x*.**

1. Suppose $y = 14$ when $x = 7$. Find x when $y = 18$.
2. Find y when $x = 8$ if $y = 15$ when $x = 12$.
3. If $y = 2.5$ when $x = 6.5$, find y when $x = 13$.
4. Suppose $y = \frac{5}{8}$ when $x = \frac{1}{3}$. Find y when $x = -\frac{1}{2}$.

Find the constant of variation. Then write an equation for each statement.

5. y varies inversely as x, and $y = 7$ when $x = 4$.
6. y varies directly as x, and $y = 8$ when $x = -1$.
7. y varies inversely as x, and $y = -3.5$ when $x = 7$.
8. y varies directly as x, and $y = 4$ when $x = -12$.

Lesson 7–1 *(Pages 284–289)* **Determine the slope of each line.**

1.
2.
3.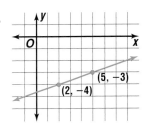

Determine the slope of the line passing through the points whose coordinates are listed in each table.

4.

x	y
−4	2
−2	4
0	6
2	8

5.

x	y
−1	12
0	6
1	0
2	−6

6.

x	y
0	1
5	2
10	3
15	4

Lesson 7–2 *(Pages 290–295)* **Write the point-slope form of an equation for each line passing through the given point and having the given slope.**

1. $(6, 4), m = -3$
2. $(-7, 3), m = 2$
3. $(2, 2), m = -1$
4. $(3, -2), m = 6$
5. $(1, 5), m = -5$
6. $(1, 2), m = -4$
7. $(-7, 2), m = \text{none}$
8. $(-3, -3), m = \frac{1}{2}$
9. $(-4, -2), m = \frac{4}{5}$
10. $(4, 0), m = \frac{1}{3}$
11. $\left(-\frac{1}{2}, -1\right), m = 4$
12. $\left(\frac{3}{4}, 4\right), m = -2$

Write the point-slope form of an equation for each line.

13. the line through points at $(8, 4)$ and $(7, 6)$
14. the line through points at $(0, -2)$ and $(-4, 2)$
15. the line through points at $(4, 7)$ and $(-1, 0)$
16. the line through points at $(2, 1)$ and $(1, 2)$
17. the line through points at $(4, -8)$ and $(6, -2)$
18. the line through points at $(-3, 1)$ and $(2, -3)$

19. Write an equation in point-slope form of a line that has a slope of 1.5 and passes through the point $(16, -5)$.

Lesson 7–3 *(Pages 296–301)* Write an equation in slope-intercept form of the line with each slope and *y*-intercept.

1. $m = -3, b = 8$
2. $m = 2, b = -3$
3. $m = -1, b = -6$
4. $m = 3, b = 2$
5. $m = 0, b = 4$
6. $m = -2, b = 1$
7. $m = 6, b = 4$
8. $m = -1.4, b = 7$
9. $m = \frac{5}{7}, b = -7$
10. $m = -\frac{2}{3}, b = 9$
11. $m = -\frac{4}{3}, b = -3$
12. $m = -\frac{1}{4}, b = 6$

Write an equation in slope-intercept form of the line having the given slope and passing through the given point.

13. $m = 4, (1, -2)$
14. $m = 3, (-5, -3)$
15. $m = -2, (-3, 0)$
16. $m = -4, (-2, 6)$
17. $m = -1, (0, 4)$
18. $m = 0, (7, 8)$
19. $m = 6, (12, -6)$
20. $m = 5, (-8, 3)$
21. $m = -\frac{4}{5}, (3, 6)$
22. $m = \frac{3}{8}, (-1, 1)$
23. $m = \frac{1}{3}, (0, 5)$
24. $m = -\frac{2}{3}, (-6, 1)$

Write an equation in slope-intercept form of the line passing through each pair of points.

25. $(2, 4)$ and $(3, 1)$
26. $(0, 1)$ and $(3, 0)$
27. $(-4, -3)$ and $(2, 1)$
28. $(5, 4)$ and $(7, -8)$
29. $(-2, 3)$ and $(9, -4)$
30. $(8, 12)$ and $(-2, -8)$
31. $(5, 3)$ and $(6, -8)$
32. $(3, 0)$ and $(8, -2)$
33. $(7, 4)$ and $(4, 7)$
34. $(-2, -2)$ and $(3, -3)$
35. $(8, 2)$ and $(-4, -1)$
36. $(6, 2)$ and $(3, -4)$

Lesson 7–4 *(Pages 302–307)* Determine whether each scatter plot has a *positive* relationship, *negative* relationship, or *no* relationship. If there is a relationship, describe it.

1.

2.

3.

Determine whether a scatter plot of the data for the following would show *positive*, *negative*, or *no* relationship between the variables.

4. study time and score on a test

5. your shoe size and your age

Lesson 7–5 *(Pages 310–315)* Determine the *x*-intercept and *y*-intercept of the graph of each equation. Then graph the equation.

1. $2x + 8y = 16$
2. $x + 2y = 2$
3. $x + y = 3$
4. $2x - 9y = 18$
5. $6x - y = 4$
6. $7y - x = -3$
7. $6x + y = -2$
8. $4y + 3x = -4$
9. $x + \frac{2}{3}y = 6$
10. $\frac{1}{2}x + 4y = 1$
11. $\frac{2}{3}x + \frac{1}{3}y = \frac{4}{3}$
12. $3y - 2x = -4$

Determine the slope and *y*-intercept of the graph of each equation. Then graph the equation.

13. $y = 8 - x$
14. $y = 5x - 2$
15. $y = 3x + 6$
16. $y = 4x + 8$
17. $y = -\frac{1}{2}x - 3$
18. $y = \frac{4}{3}x - 2$
19. $y = \frac{1}{6}x + 5$
20. $y = -\frac{3}{4}x + 6$
21. $-2y + 4x = 8$
22. $7x + 8y = -56$
23. $9y + 3x = 6$
24. $-5x + y = 6$

Lesson 7–6 *(Pages 316–321)* Graph each pair of equations. Describe any similarities or differences and explain why they are a family of graphs.

1. $y = 3x + 2$
$y = 3x$

2. $y = -2x + 1$
$y = 2x + 1$

3. $y = 3x$
$y = 4x$

4. $y = x + 4$
$y = 5x + 4$

5. $y = -4x + 1$
$y = -4x - 2$

6. $-\frac{3}{4}x + 2 = y$
$-x + 2 = y$

7. $y = \frac{1}{3}x$
$y = 3x$

8. $y = \frac{1}{2}x + \frac{1}{2}$
$y = 2x + 2$

Compare and contrast the graphs of each pair of equations. Verify by graphing the equations.

9. $y = 2x$
$y = 2x + 1$

10. $y = 0.25x + 1$
$y = 4x + 1$

11. $y = -3x - 2$
$y = 3x - 2$

12. $y = x - 4$
$y = x + 6$

13. $y = \frac{1}{2}x + 2$
$y = x + 2$

14. $y = -\frac{2}{3}x + 2$
$y = -\frac{2}{3}x$

15. $y = -5x + 4$
$y = -\frac{1}{2}x + 4$

16. $y = \frac{1}{3}x$
$5 + x = 3y$

Change $y = 0.5x - 5$ so that the graph of the new equation fits each description.

17. y-intercept is 2, same slope

18. negative slope, y-intercept is -3

19. shifted up 2 units, slope is 2

20. less steep positive slope, same y-intercept

Lesson 7–7 *(Pages 322–327)* Determine whether the graphs of each pair of equations are *parallel*, *perpendicular*, or *neither*.

1. $y = 3x + 2$
$y = -\frac{1}{3}x + 5$

2. $y = 3x + 4$
$y = 3x + 2$

3. $y = 3x + 2$
$y = \frac{1}{5}x - \frac{7}{5}$

4. $y = 2x + 4$
$3y = 6x + 6$

5. $y = \frac{1}{2}x + 4$
$4x - 3y = 12$

6. $2y = 10x - \frac{2}{5}$
$\frac{1}{5}x + y = 3$

7. $y = \frac{2}{3}x - 6$
$2x - 4 = 3y$

8. $2x + y = 6$
$y = \frac{1}{2}x + 4$

Write an equation in slope-intercept form of the line that is parallel to the graph of each equation and passes through the given point.

9. $y = 8x + 5$; $(0, 4)$

10. $y = 3$; $(-1, 2)$

11. $x = 2$; $(4, 3)$

Write an equation in slope-intercept form of the line that is perpendicular to the graph of each equation and passes through the given point.

12. $y = x + 5$; $(-2, 1)$

13. $y = 4x$; $(0, 0)$

14. $6x - 2y = 3$; $(0, 1)$

Lesson 8–1 *(Pages 336–340)* Write each expression using exponents.

1. $7 \cdot 7 \cdot 7$

2. $(-3)(-3)(-3)$

3. $5 \cdot 5 \cdot 5 \cdot 6 \cdot 4 \cdot 4$

4. 6 squared

5. $m \cdot m \cdot n \cdot n \cdot n$

6. 2 cubed

7. $4 \cdot r \cdot r \cdot r \cdot s \cdot s$

8. $(-2)(k)(k)(j)(j)$

Write each power as a multiplication expression.

9. 2^3

10. $(-4)^4$

11. $8^4 2^3$

12. k^5

13. $w^2 z^2$

14. $-6xy^2$

Evaluate each expression if $x = 4$, $y = -1$, $z = -2$, and $w = 1.5$.

15. y^3

16. $w(yz + 4)$

17. $z^3 + 2xy$

18. $2x^2 - z^5$

19. $-2(y^3 + w)$

20. wxy

Lesson 8–2 *(Pages 341–346)* **Simplify each expression.**

1. $2^2 \cdot 2^5$
2. $6^3 \cdot 6^4$
3. $x^4 \cdot x^4$
4. $y \cdot y^5$
5. $(a^3)(a^4)$
6. $(g^6h^3)(g^4h^3)$
7. $(r^2t^2)(rt^3)$
8. $(2k^2j)(-3kj)$
9. $(6m^3)(7m^2)$
10. $(2a^2b)(9ab^5)$
11. $(6a^3b)(4ac)$
12. $(8y^4)(3y^4)$
13. $(-3p^5q^2)(-pq^7)$
14. $(3xyz)(-4x^2y^3)$
15. $\dfrac{4^5}{4^2}$
16. $\dfrac{8^8}{8^7}$
17. $\dfrac{15m^{12}}{3m^5}$
18. $\dfrac{-4r^4s^5}{-rs^4}$
19. $\dfrac{-16m^5n^9p^{10}}{2m^2n^3p^6}$
20. $\left(\dfrac{2}{3}x^5y\right)(-12x^3y^2)$

21. Evaluate m^0.
22. Find the product of mn and $-3m^2n$.

Lesson 8–3 *(Pages 347–351)* **Write each expression using positive exponents. Then evaluate the expression.**

1. 8^{-3}
2. 2^{-4}
3. 10^{-3}
4. 5^{-1}
5. 9^{-3}
6. 7^{-2}
7. 3^{-5}
8. 2^{-6}

Simplify each expression.

9. $(m^6)(m^{-2})$
10. $r^0s^1t^4$
11. $6x^{-2}y^{-4}z$
12. $(a^{-3})(a^2)$
13. $\dfrac{1}{k^{-4}}$
14. $\dfrac{m}{m^{-5}}$
15. $\dfrac{6b^4}{3b^{-2}}$
16. $(r^{-3})(s^4)$
17. $-\dfrac{r^2s^5}{rs^6}$
18. $\dfrac{2a^2b^5}{ab^7}$
19. $\dfrac{-7cd^0}{14c}$
20. $-\dfrac{10a^7b^4}{2ab}$
21. $-\dfrac{4w^5z^4}{10z^7}$
22. $-\dfrac{2p^2q^7}{8p^4q}$
23. $\dfrac{56yz^4}{14y^5z^4}$
24. $\dfrac{x^{-3}y^4z^{-2}}{xy^2z^{-2}}$

25. Evaluate $7m^{-2}n^{-3}$ if $m = 4$ and $n = 2$.
26. Find the value of $5a^3b^{-1}$ if $a = -1$ and $b = 3$.

Lesson 8–4 *(Pages 352–356)* **Express each measure in standard form.**

1. 1.5 megaohms
2. 168 billion dollars
3. 2 megahertz
4. 76 milliamperes
5. 400 nanoseconds
6. 1.2 million dollars
7. 18 kilobytes
8. 93 micrograms

Express each number in scientific notation.

9. 178
10. 0.0098
11. 0.032
12. 106,000
13. 13.8
14. 269.3
15. 0.0000083
16. 100
17. 0.000016
18. 1.2
19. 0.3
20. 400,300
21. 17
22. 1852
23. 1900
24. 0.000000103

Evaluate each expression. Express each result in scientific notation and standard form.

25. $(6 \times 10^3)(4 \times 10^5)$
26. $(7 \times 10)(3.5 \times 10^2)$
27. $(2.1 \times 10^3)(1 \times 10^4)$
28. $(4 \times 10^{-3})(7 \times 10^4)$
29. $(1.5 \times 10^5)(6 \times 10^{-3})$
30. $(5 \times 10^{-1})(2.5 \times 10^{-4})$

Lesson 8–5 *(Pages 357–361)* **Simplify.**

1. $\sqrt{64}$
2. $-\sqrt{9}$
3. $\sqrt{121}$
4. $-\sqrt{225}$

5. $\sqrt{\dfrac{144}{81}}$
6. $-\sqrt{\dfrac{225}{100}}$
7. $\sqrt{\dfrac{256}{16}}$
8. $\sqrt{\dfrac{36}{64}}$

9. $-\sqrt{\dfrac{225}{441}}$
10. $-\sqrt{\dfrac{121}{289}}$
11. $-\sqrt{0.36}$
12. $\sqrt{0.81}$

13. $\sqrt{0.0025}$
14. $-\sqrt{0.0049}$
15. $\sqrt{0.0289}$
16. $-\sqrt{0.000196}$

17. Find the negative square root of 25.
18. If $x = \sqrt{1024}$, what is the value of x?

Lesson 8–6 *(Pages 362–365)* **Estimate each square root to the nearest whole number.**

1. $\sqrt{5}$
2. $\sqrt{10}$
3. $\sqrt{11}$
4. $\sqrt{15}$
5. $\sqrt{18}$
6. $\sqrt{24}$

7. $\sqrt{61}$
8. $\sqrt{126}$
9. $\sqrt{153}$
10. $\sqrt{412}$
11. $\sqrt{483}$
12. $\sqrt{504}$

13. $\sqrt{555}$
14. $\sqrt{621}$
15. $\sqrt{709}$
16. $\sqrt{981}$
17. $\sqrt{70.3}$
18. $\sqrt{81.4}$

19. $\sqrt{121.6}$
20. $\sqrt{153.2}$
21. $\sqrt{9.35}$
22. $\sqrt{13.6}$
23. $\sqrt{0.021}$
24. $\sqrt{0.29}$

25. Tell whether 11 is closer to $\sqrt{119}$ or $\sqrt{125}$.
26. Which is closer to $\sqrt{285}$, 16 or 17?

Lesson 8–7 *(Pages 366–371)* **If c is the measure of the hypotenuse and a and b are the measures of the legs, find each missing measure. Round to the nearest tenth if necessary.**

1.
2.
3.

4.
5.
6.

7. $a = 15, b = 17, c = ?$
8. $b = 12, c = 38, a = ?$
9. $a = 4, b = 5, c = ?$

10. $a = 10, c = 21, b = ?$
11. $a = 3, c = 8, b = ?$
12. $b = 20, c = 40, a = ?$

The lengths of three sides of a triangle are given. Determine whether each triangle is a right triangle.

13. 5 ft, 12 ft, 13 ft
14. 4 mi, 5 mi, 7 mi
15. 7 cm, 11 cm, 17 cm

16. 56 in., 70 in., 84 in.
17. 19 m, 24 m, 28 m
18. 31 mm, 37 mm, 49 mm

19. 3 ft, 6 ft, 12 ft
20. 2 in., 8 in., 10 in.
21. 6 mi, 8 mi, 10 mi

22. 10 cm, 24 cm, 26 cm
23. 20 m, 21 m, 29 m
24. 18 in., 26 in., 32 in.

Lesson 9–1 *(Pages 382–387)* Determine whether each expression is a monomial. Explain why or why not.

1. $6x$
2. $\dfrac{4}{k}$
3. $13b + 2a$
4. $7a^2b^{-1}$
5. m^2
6. $3x - y$
7. s^4
8. $-3x^2yz^3$

State whether each expression is a polynomial. If it is a polynomial, identify it as a *monomial, binomial,* or *trinomial.*

9. $11y$
10. $9xy^2 - y^3 + x^2$
11. $r^2 - r$
12. $\dfrac{1}{4}x - 2$
13. $3x - 11 + y - 2$
14. $6j^3k^2\ell - 7j^2$
15. $2.3mn^2$
16. $7r^2s^3t$

Find the degree of each polynomial.

17. y^2
18. $6x^3$
19. 4
20. $3j^2 - 2k + m$
21. $16m^2n + 14m^5n^3$
22. $11p^5q^2 - 6pq^7$
23. $x^2 + 2x + 3x^6$
24. $7a^2b^3 - 2ab$

Arrange the terms of each polynomial so that the powers of x are in descending order.

25. $4x^2 + x^3 - x$
26. $2x - 6x^5 - 3$
27. $5x^5 - 4x^2 + 3x - 2$
28. $6mx^5 - 4m^2x^6 + 4mx$
29. $3x^4y^5 + 2x^2y^3 + 6x^5y$
30. $2x - x^2 + 2$

Lesson 9–2 *(Pages 388–393)* Find each sum.

1. $\begin{array}{r} 7x - 2 \\ (+)\ x + 4 \\ \hline \end{array}$

2. $\begin{array}{r} 6x^2 - 2x - 1 \\ (+)\ 3x^2 - 4x - 7 \\ \hline \end{array}$

3. $\begin{array}{r} 2xy - 3x + 2 \\ (+)\ 4xy \quad\ \ - 7 \\ \hline \end{array}$

4. $(-3y + 7) + (4y - 2)$
5. $(7ab - 2a) + (2ab + 3b)$
6. $(4x^2 - 2xy + y^2) + (xy + 3y^2)$

Find each difference.

7. $\begin{array}{r} 3x - 4 \\ (-)\ 2x + 3 \\ \hline \end{array}$

8. $\begin{array}{r} 5m^2 -\ m + 2 \\ (-)\ 2m^2 + 2m + 5 \\ \hline \end{array}$

9. $\begin{array}{r} 11a^2 - 2a - 1 \\ (-)\ 5a^2 + 6a + 2 \\ \hline \end{array}$

10. $(13x - 2) - (9x + 4)$
11. $(7m + 4n) - (2m - n)$
12. $(6a + b) - (2b + 4)$

Find each sum or difference.

13. $(2x + 3) + (4x^2 + x - 7)$
14. $(mn + 2pn - mp) - (2pn - mp)$
15. $(y^2 + 2y - 3) + (2y^3 - y - 3)$
16. $(x^2y + 5xy - 9y^2) - (3x^2 - 2y^2)$
17. $(3x^2 + 2x - 4) + (5x^2 - 9)$
18. $(x^2y + 9xy - y^2) - (2x^2y - y^2)$

Lesson 9–3 *(Pages 394–398)* Find each product.

1. $4(3x + 2)$
2. $-3(7b - 4)$
3. $a(2a - 4)$
4. $2n(n - 5)$
5. $-y(10y - 5)$
6. $-3x(5x + 8)$
7. $7p(p^2 - 3)$
8. $-5m(3m^3 - 2m)$
9. $2r^3(7 - r + r^2)$
10. $-6x^3(x^2 - 4x)$
11. $0.75k(8k^3 - k^2)$
12. $2.4z^2(2z^2 - 3)$

Solve each equation.

13. $-3(4 - x) = 18$
14. $21 = 7(y - 11)$
15. $10x - 9 = 4(x - 1) + 1$
16. $7(r + 8) - 4 = -4(-7 - r)$
17. $6(s - 7) = 3(4s + 4)$
18. $-a + 6(a + 4) = 3(a + 5) + 1$

Lesson 9–4 *(Pages 399–404)* **Find each product. Use the Distributive Property or the FOIL method.**

1. $(m + 3)(m + 2)$
2. $(x - 4)(x + 6)$
3. $(y - 5)(y - 7)$
4. $(r + 9)(r - 3)$
5. $(2s - 3)(s + 5)$
6. $(a - 2)(4a + 8)$
7. $(7p - 3)(4p + 2)$
8. $(2k + 2)(2k + 1)$
9. $(d + 7)(3d - 5)$
10. $(8x + 2y)(4x + 3y)$
11. $(3m + n)(m - 3)$
12. $(7r + 8s)(5r - 3s)$
13. $(6x - 1)(x - 2)$
14. $(7p - 2)(3p + n)$
15. $(2p - 1)(p - 3)$
16. $(b + a)(a - b)$
17. $(5c + d)(3c - 2d)$
18. $(p + 1)(p + m)$
19. $(j^2 - 2)(j + 5)$
20. $(z^2 + r)(z^2 - r)$
21. $(n^2 + 1)(2n^2 - 3)$

Lesson 9–5 *(Pages 405–409)* **Find each product.**

1. $(m + 5)^2$
2. $(n - 3)^2$
3. $(2x + 3)^2$
4. $(3y - x)^2$
5. $(4a + b)(4a - b)$
6. $(2k + 2p)^2$
7. $(4x - 2)(4x + 2)$
8. $(1 + p)^2$
9. $(a - 2b)(a + 2b)$
10. $(6 + 3m)^2$
11. $(2 - 4t)^2$
12. $(r - 2s)(3r + 2s)$
13. $(2x - 7y)^2$
14. $(m + 3n)^2$
15. $2(p - q)^2$
16. $4(r + s)^2$
17. $k(2 + k)^2$
18. $4s(s - 1)^2$
19. $3r(r - 2)^2$
20. $y(y + 1)(y - 2)$
21. $p(p + 2)(2p - 3)$
22. $2(j + 3)(j - 3)$
23. $(x + 3)(x - 1)(x + 2)$
24. $6(m + 5)(m - 1)$

25. The area of a circle is given by the formula $A = \pi r^2$, where r is the radius of the circle. Suppose a circle has a radius of $k - 4$ inches.

   a. Write an equation to find the area of the circle.

   b. Find the area to the nearest hundredth if $k = 6$.

Lesson 10–1 *(Pages 420–425)* **Find the factors of each number. Then classify each number as *prime* or *composite*.**

1. 57
2. 22
3. 65
4. 17
5. 104
6. 18
7. 81
8. 73

Factor each monomial.

9. $12x^2$
10. $28m^2n$
11. $-33j^2k^2$
12. $54p^3$
13. $81ab^2$
14. $75xy$
15. $-13p^2q^2$
16. $105r^4$

Find the GCF of each set of numbers or monomials.

17. $13, 33$
18. $50, 75$
19. $32, 84, 144$
20. $32, 64, 96$
21. $-21, 15xy$
22. $4x^2, 2x, 8x$
23. $-3r^2s^2, -17rs$
24. $14rs, 12rst, 6t$
25. $7kr, 21k^2, 2kr$
26. $-16c^2d, -4cd, -8cd^2$
27. $24m^2n, 51m, 63m^2n^2$
28. $-1x^3yz, -7x^2y^2, -2x^3yz$

Lesson 10-2 *(Pages 428–433)* **Factor each polynomial. If the polynomial cannot be factored, write** *prime*.

1. $4m + 12$

2. $13n + n^2$

3. $2k^2 + 6k$

4. $7s^2t + 3$

5. $9pq^3 - 21pq^2$

6. $x^2y + 7y$

7. $3k - 4j$

8. $14m^2n - 18m$

9. $17cd - 14mn$

10. $16a^2b^2 - 3ab$

11. $16b^2c - 2ac + 4bc$

12. $12mn - 14m^2 + 16n$

13. $6ab - 7bc + 12ac$

14. $m^2n^3p + mn - mp$

15. $x^2y + 15xy + 5x$

Find each quotient.

16. $(16x + 4y^2) \div 4$

17. $(2rs + r) \div r$

18. $(7xy - 3x) \div x$

19. $(15a + 3b^2) \div 3$

20. $(6m^2n - 9m) \div 3m$

21. $(18cd^2 - 9cd) \div 9cd$

22. $(21a^2b - 14a^2) \div 7a^2$

23. $(16xy - 12xy) \div 4xy$

24. $(12a^2b + 4b) \div 4b$

25. $(36st^2 - 9st) \div 9st$

26. $(15rs^2 + 12r^2st) \div 3rs$

27. $(5xy^2z + 10x^2z) \div 5xz$

28. $(13r^2s - 26r^2) \div 13r^2$

29. $(32np + m^2np^2) \div np$

30. $(20r^2st + 15rs) \div 5rs$

Lesson 10-3 *(Pages 434–439)* **Factor each trinomial. If the trinomial cannot be factored, write** *prime*.

1. $x^2 + 4x + 4$

2. $n^2 + 6n + 9$

3. $t^2 - 8t + 16$

4. $w^2 + 5w - 2$

5. $r^2 + r - 12$

6. $p^2 - 8p + 4$

7. $s^2 - 7s - 8$

8. $d^2 + 21d + 110$

9. $y^2 - 10y + 3$

10. $q^2 - 3q - 28$

11. $4z^2 - 16z - 18$

12. $6c^2 - 3c - 3$

13. $2m^2 - 3m - 20$

14. $8y^2 + 6y - 2$

15. $3r^2 - 12r - 15$

Lesson 10-4 *(Pages 440–444)* **Factor each trinomial. If the trinomial cannot be factored, write** *prime*.

1. $5x^2 + 13x + 6$

2. $4w^2 + 7w + 3$

3. $2a^2 - 9a + 4$

4. $12t^2 + 18t + 2$

5. $4c^2 - 8c - 3$

6. $3r^2 + 15r + 12$

7. $8y^2 - 16y + 6$

8. $20n^2 - 40n + 20$

9. $10m + 3 + 3m^2$

10. $13 + 2m^2 + 14m$

11. $3q^2 - 4 - 4q$

12. $12d + 8 - 8d^2$

13. $2m^2 - 3mn - 2n^2$

14. $2a^2 + 4ab + 2b^2$

15. $12p^2 + 10mp + 2m^2$

16. $18x^2 + 15xy + 3y^2$

17. A rectangle has dimensions of $(y + 5)$ inches and $(y - 4)$ inches.
   a. Express the area as a trinomial.
   b. If y units are removed from the length, express the new area.

Lesson 10–5 *(Pages 445–449)* Determine whether each trinomial is a perfect square trinomial. If so, factor it.

1. $x^2 - 14x + 49$ **2.** $a^2 - 4a + 4$ **3.** $y^2 - 16y + 3$ **4.** $d^2 + 10d + 25$

5. $r^2 + 24r + 144$ **6.** $9c^2 + 12c + 4$ **7.** $4k^2 - 28k + 49$ **8.** $16m^2 - 8m + 1$

Determine whether each trinomial is the difference of squares. If so, factor it.

9. $x^2 - 9$ **10.** $4 - 49z^2$ **11.** $36x^2 - 4$ **12.** $17 - 3p^2$

13. $s^2 - 16r^2$ **14.** $9 - m^2n^2$ **15.** $32k^2 - 50$ **16.** $12b^2 - 48$

Factor each polynomial. If the polynomial cannot be factored, write *prime*.

17. $a^2 - a - 12$ **18.** $2s^2 - rs - r^2$ **19.** $2r^2 - 3$ **20.** $4m^2 + 48m + 144$

21. $6m^2 - 16m - 6$ **22.** $2xy - 8x$ **23.** $3a^2b^3 - 27b$ **24.** $16x^2 - 8xy + y^2$

Lesson 11–1 *(Pages 458–463)* Graph each quadratic function by making a table of values.

1. $y = 6x^2$ **2.** $y = 8x^2$ **3.** $y = -2x^2$ **4.** $y = -3x^2$

5. $y = 2x^2 + 1$ **6.** $y = x^2 - 3$ **7.** $y = -3x^2 + 5$ **8.** $y = -4x^2 - 3$

9. $y = x^2 + 3x - 6$ **10.** $y = -2x^2 - x + 5$ **11.** $y = 3x^2 - 2x$ **12.** $y = 0.25x^2 - 2x - 4$

Write the equation of the axis of symmetry and the coordinates of the vertex of the graph of each quadratic function. Then graph the function.

13. $y = 2x^2$ **14.** $y = 3x^2$ **15.** $y = 8x^2$ **16.** $y = x^2 + 2x$

17. $y = 3x^2 - 6x$ **18.** $y = -2x^2 + 1$ **19.** $y = 4x^2 - x + 2$ **20.** $y = -x^2 + 6x - 4$

21. $y = \frac{1}{2}x^2 - x + 1$ **22.** $y = -\frac{1}{2}x^2 - 2x + 2$ **23.** $y = 2x^2 + 3x - 1$ **24.** $y = -x^2 - 2x - 1$

Match each function with its graph.

25. $y = (x + 2)^2 - 2$ **26.** $y = -2x^2 - x - 1$ **27.** $y = -2x^2 + x + 2$

A.   **B.**   **C.**

Lesson 11–2 *(Pages 464–467)* Graph each group of equations on the same screen. Compare and contrast the graphs.

1. $y = x^2$
$y = 2x^2$
$y = 4x^2$

2. $y = x^2 + 1$
$y = x^2 - 4$
$y = x^2 - 8$

3. $y = (2x - 1)^2$
$y = (2x - 2)^2$
$y = (2x - 3)^2$

Describe how each graph changes from the parent graph of $y = x^2$. Then name the vertex of each graph.

4. $y = 8x^2$ **5.** $y = -2x^2$ **6.** $y = \frac{1}{2}x^2$ **7.** $y = (x - 1)^2$

8. $y = (x + 3)^2$ **9.** $y = (3x + 2)^2$ **10.** $y = -(x + 3)^2$ **11.** $y = -2x^2 - 3$

12. $y = 0.25x^2 + 0.5$ **13.** $y = -3x^2 + 8$ **14.** $y = (4x - 1)^2 + 3$ **15.** $y = (x + 2)^2 - 2$

Lesson 11-3 *(Pages 468–473)* **Solve each equation by graphing the related function. If exact roots cannot be found, state the consecutive integers between which the roots are located.**

1. $-x^2 - 2x + 24 = 0$ 2. $x^2 + 4x - 5 = 0$ 3. $x^2 - 4x - 2 = 0$ 4. $x^2 + 7x + 3 = 0$
5. $x^2 + 7x + 5 = 0$ 6. $x^2 - 6x + 6 = 0$ 7. $-x^2 - 3x - 2 = 0$ 8. $4x^2 + 2x + 1 = 0$
9. $-x^2 - 6x - 6 = 0$ 10. $x^2 - 11x + 4 = 0$ 11. $-2x^2 - 2x + 10 = 0$ 12. $x^2 + 12x + 20 = 0$
13. $-x^2 - 12x - 3 = 0$ 14. $-3x^2 - x + 8 = 0$ 15. $x^2 + 3x + 4 = 0$ 16. $x^2 + 5x - 9 = 0$

Use a quadratic equation to determine the two numbers that satisfy each situation.

17. Their sum is 21 and their product is 104.
18. Their difference is 8 and their product is 20.
19. Their sum is 13 and their product is 22.
20. Their sum is 32 and their product is 135.

Lesson 11-4 *(Pages 474–477)* **Solve each equation. Check your solution.**

1. $2r(r - 4) = 0$
2. $4k(k + 5) = 0$
3. $(s + 4)(s - 3) = 0$
4. $(m - 4)(m - 5) = 0$
5. $(3x - 4)(x - 2) = 0$
6. $(2y + 2)(2y - 4) = 0$
7. $(t + 2)(6t + 1) = 0$
8. $n^2 + n - 6 = 0$
9. $k^2 + 4k + 4 = 0$
10. $p^2 - 5p + 6 = 0$
11. $q^2 - 2q - 15 = 0$
12. $x^2 + 2x - 3 = 0$
13. $j^2 + 9j + 20 = 0$
14. $r^2 - 16r = 0$
15. $z^3 - 25z = 0$

For each problem, define a variable. Then use an equation to solve the problem.

16. Find two integers whose sum is 15 and whose product is 36.
17. The length of a swimming pool is 15 feet longer than it is wide. The area in square feet is 1350. Find the dimensions of the pool.
18. Find two integers whose difference is 12 and whose product is 13.

Lesson 11-5 *(Pages 478–482)* **Find the value of c that makes each trinomial a perfect square.**

1. $r^2 - 16r + c$
2. $k^2 + 12k + c$
3. $p^2 - 4p + c$
4. $n^2 + 2n + c$
5. $f^2 - 8f + c$
6. $s^2 + 18s + c$
7. $x^2 + 20x + c$
8. $r^2 + 14r + c$
9. $w^2 + 30w + c$
10. $h^2 + 10h + c$
11. $z^2 - 2z + c$
12. $m^2 - 6m + c$
13. $q^2 + 26q + c$
14. $t^2 + 28t + c$
15. $y^2 + 22y + c$
16. $z^2 + 24z + c$

Solve each equation by completing the square.

17. $z^2 + 10z + 12 = 0$
18. $h^2 - 8h - 15 = 0$
19. $y^2 + 3y + 1 = 0$
20. $w^2 + 15w = 5$
21. $m^2 + 2m = 0$
22. $t^2 + 2t = 18$
23. $r^2 - 20r + 24 = 0$
24. $p^2 - 2p = 32$
25. $q^2 - 7q + 12 = 0$
26. $n^2 - 4n - 16 = 0$
27. $x^2 + 10x = 12$
28. $r^2 + 12r = 0$

Lesson 11–6 *(Pages 483–488)* **Use the Quadratic Formula to solve each equation.**

1. $j^2 + 3j - 4 = 0$ **2.** $w^2 + 9w + 20 = 0$ **3.** $m^2 - 7m + 12 = 0$ **4.** $n^2 + 5n - 6 = 0$

5. $k^2 + 2k - 15 = 0$ **6.** $z^2 - 4z + 4 = 0$ **7.** $d^2 - 3d - 18 = 0$ **8.** $s^2 + 8s - 14 = 0$

9. $x^2 - 5x - 24 = 0$ **10.** $t^2 + 5t + 6 = 0$ **11.** $r^2 + 6r + 9 = 0$ **12.** $y^2 - 2y - 8 = 0$

13. $y^2 - y - 6 = 0$ **14.** $d^2 + 4d - 21 = 0$ **15.** $s^2 + 3s + 2 = 0$ **16.** $m^2 - 11m = -18$

17. $4k^2 - 4 = 8k$ **18.** $-3r^2 - 15r = -18$ **19.** $3p^2 + 5p + 2 = 0$ **20.** $6x^2 + 8x = 8$

21. $2p^2 + 5p = 3$ **22.** $4q^2 - 2 = -2q$ **23.** $-2c^2 - 4c - 2 = 0$ **24.** $-8m^2 - 10m = 2$

Lesson 11–7 *(Pages 489–493)* **Graph each exponential function. Then state the y-intercept.**

1. $y = 2^x$ **2.** $y = 5^x$ **3.** $y = 3^x - 1$ **4.** $y = 4^x + 4$

5. $y = 2^x - 3$ **6.** $y = 2^x + 3$ **7.** $y = 4^x + 3$ **8.** $y = 2^x - 6$

9. $y = 3^x - 3$ **10.** $y = 2^{4x} + 1$ **11.** $y = 4^{0.5x}$ **12.** $y = 5^{3x} + 1$

13. $y = 4^{3x} - 2$ **14.** $y = 3^{3x} - 2$ **15.** $y = 2^{0.5x} - 1$ **16.** $y = 2^{3x} - 4$

Find the amount of money in a bank account given the following conditions.

17. initial deposit = $6000, annual rate = 6.5%, time = 3 years

18. initial deposit = $1000, annual rate = 12%, time = 15 years

19. initial deposit = $2500, annual rate = 2%, time = 6 years

20. initial deposit = $5100, annual rate = 9%, time = 4 years

Lesson 12–1 *(Pages 504–508)* **Write an inequality to describe each number.**

1. a number less than 10

2. a number that is at least -4

3. a number greater than -2

4. a number less than or equal to 5

5. a number greater than 3

6. a number more than 12

7. a minimum number of 7

8. a number less than -1

9. a maximum number of 8

10. a number greater than -6

11. a minimum number of -8

12. a number more than -11

Graph each inequality on a number line.

13. $m < 8$ **14.** $n \le -4$ **15.** $x > 2$ **16.** $z < -7$

17. $r \ge 15$ **18.** $m \le -5$ **19.** $s > 8.4$ **20.** $y > 6.2$

21. $w \le 1.3$ **22.** $\ell \ge -2.4$ **23.** $p < -3.2$ **24.** $j > 4.3$

25. $t < \dfrac{1}{3}$ **26.** $r \le -2\dfrac{1}{4}$ **27.** $y \le \dfrac{1}{2}$ **28.** $q \ge 3\dfrac{5}{8}$

Write an inequality for each graph.

29.

30.

31.

32.

33.

34.

35.

36.

37.

Lesson 12–2 (*Pages 509–513*) Solve each inequality. Check your solution.

1. $r + 3 < 8$
2. $m + 4 > -2$
3. $j - 3 > 5$
4. $k - 6 < -13$
5. $-12 + w < 15$
6. $p + 11 \geq 5$
7. $0.4 + p \geq 1.2$
8. $x - 6.2 < 4$
9. $4.3 < 2.1 + y$
10. $\frac{1}{4} + r \leq 3\frac{3}{4}$
11. $k + 10.6 \geq -3.4$
12. $\frac{1}{6} + s \leq \frac{1}{3}$

Solve each inequality. Graph the solution.

13. $2n > n + 3$
14. $8 + 5x > 6x$
15. $3y - 2 < 4y$
16. $5d < 8 + 4d$
17. $9 \geq 3t - 4 - 2t$
18. $-2a < -3(a - 2)$
19. $6u \leq 5(u + 1)$
20. $11p \leq 2(5p + 4)$
21. $7s < 3(2s - 3)$
22. $3(x - 3) \leq 4x$
23. $7b < 4(2b - 3)$
24. $2c + 3 \leq 3(c - 5)$

Lesson 12–3 (*Pages 514–518*) Solve each inequality. Check your solution.

1. $-3k < 15$
2. $2r > -10$
3. $-4d \geq -12$
4. $\frac{x}{5} \geq 6$
5. $-\frac{y}{4} < 8$
6. $\frac{p}{7} \leq -3$
7. $-9\ell < 27$
8. $2s > 20$
9. $-t > 11$
10. $-\frac{d}{2} > 12$
11. $\frac{m}{6} \leq -5$
12. $-\frac{b}{4} < -9$
13. $-3p < 2$
14. $-8y \geq -4$
15. $7v \leq 3$
16. $2 < \frac{2}{3}j$
17. $\frac{5}{8}t \geq -5$
18. $-\frac{1}{4}w < 6$
19. $0.01r \leq 8$
20. $2.1m > 6.3$
21. $-2.25j \geq 9$
22. $\frac{k}{7.2} < -3$
23. $-\frac{r}{2.5} > 4$
24. $\frac{y}{0.4} \geq 2$

Lesson 12–4 (*Pages 519–523*) Solve each inequality. Check your solution.

1. $2a + 7 \leq 11$
2. $5r - 3 > 27$
3. $6 - 4m > 10$
4. $1 - 3s \geq 13$
5. $3 + 5d \leq -12$
6. $-8x - 1 < 15$
7. $6 + 2b \geq -2.4$
8. $5.1x - 2.4 < -7.5$
9. $3.3 + 4k > -8.7$
10. $\frac{6-r}{8} < 7$
11. $\frac{j}{3} + 7 > 9$
12. $\frac{4 + 3m}{7} \leq -5$
13. $2x - 1 < 8x + 2$
14. $11 \leq -(t + 4)$
15. $4(3 - 6r) > 18$
16. $\frac{2}{3}(k - 6) > 4$
17. $\frac{1}{3}(y + 3) < \frac{1}{2}(y - 2)$
18. $\frac{1}{8}(m + 3) \geq \frac{1}{4}(m - 3)$

Write and solve an inequality for each situation.

19. Three fifths times the sum of a number and 5 is greater than 15.
20. Four times the difference of a number and 3 is less than 24.

Lesson 12–5 *(Pages 524–529)* Write each compound inequality without using *and*.

1. $m > 2$ and $m < 6$ **2.** $j > -12$ and $j < 4$ **3.** $r < 6$ and $r \geq -1$

4. $x < 3$ and $x \geq 2$ **5.** $y \leq 5$ and $y \geq -3$ **6.** $s \geq -7$ and $s < -4$

Graph the solution of each compound inequality.

7. $w > 6$ or $w < 2$ **8.** $a < -4$ or $a \geq 4$ **9.** $z > 12$ and $z \leq 15$

10. $h \leq 20$ and $h \geq -3$ **11.** $s < -7$ or $s > -5$ **12.** $f \leq 8$ or $f > 9$

Solve each compound inequality. Graph the solution.

13. $4 < 2n < 10$ **14.** $-4 \leq x + 7 < 9$ **15.** $12 > v - 6 > -2$

16. $9 \leq 3q \leq 21$ **17.** $24 > b - 5 > -2$ **18.** $8 > c + 4 \geq 5$

19. $t + 3 > 15$ or $t - 5 < -12$ **20.** $-12 \leq -3d < 9$ **21.** $5.4 < 0.2k < 6$

22. $u - 12 > -3$ or $u + 11 < 2$ **23.** $-4 \leq g - \frac{1}{3} < 1$ **24.** $\frac{p}{3} > 3$ or $\frac{p}{3} \leq -2$

Lesson 12–6 *(Pages 530–534)* Solve each inequality. Graph the solution.

1. $|d + 7| > 15$ **2.** $|5c| > 30$ **3.** $|a - 2| < 17$ **4.** $|z + 3| > 12$

5. $|j - 12| < 10$ **6.** $|\ell + 8| \leq -14$ **7.** $|t - 6| < 9$ **8.** $|m - 4| \leq 3$

9. $|4x| < -20$ **10.** $|s + 5| < 8$ **11.** $|n + 1| \geq 7$ **12.** $|-2v| > 14$

13. $|y - 3| > 16$ **14.** $|r - 7| \leq -11$ **15.** $|-6b| > 18$ **16.** $|w - 4| \leq 1.5$

For each graph, write an inequality involving absolute value.

17.
$$-7\ -6\ -5\ -4\ -3\ -2\ -1\ \ 0\ \ 1\ \ 2\ \ 3$$

18.
$$-3\ -2\ -1\ \ 0\ \ 1\ \ 2\ \ 3\ \ 4\ \ 5\ \ 6\ \ 7$$

19.
$$-1\ \ 0\ \ 1\ \ 2\ \ 3\ \ 4\ \ 5\ \ 6\ \ 7\ \ 8\ \ 9$$

Write an inequality involving absolute value for each statement. Do not solve.

20. Alli's quiz score was within 5 points of her average of 85.

21. The 5-inch-wide picture frame was made with an accuracy of 0.1 inch.

Lesson 12–7 *(Pages 535–539)* Graph each inequality.

1. $x < -3$ **2.** $y > -6$ **3.** $y \geq x - 3$ **4.** $y < x + 2$

5. $y \geq -4x$ **6.** $6 \geq 2x + 4y$ **7.** $y > -2x + 6$ **8.** $x + 5y \geq 15$

9. $x + y \leq -2$ **10.** $-3x + 2 > y$ **11.** $y \leq -2(x - 1)$ **12.** $8 \geq y - 4x$

13. $y > -x - 8$ **14.** $7 + 3y \leq x$ **15.** $2x - y \geq -6$ **16.** $x - y \leq 5$

17. $3(6x + y) < 4$ **18.** $y \leq 4(x - 2)$ **19.** $-(4x - 3) \geq 2y$ **20.** $-2(3x + y) < 1$

For Exercises 21–24, write an inequality and graph the solution.

21. The difference of a number and three is less than or equal to eight.

22. Three times a number is greater than negative six.

23. One half the sum of a number and eight is greater than or equal to twelve.

24. A number minus three is less than another number.

Lesson 13–1 *(Pages 550–553)* **Solve each system of equations by graphing.**

1. $x = -2$
$y = 3$

2. $x = 5$
$y = x$

3. $x = -1$
$y = x - 2$

4. $x = -3$
$y = x + 4$

5. $y = -x - 4$
$y = x + 4$

6. $y = x + 2$
$y = 2x - 1$

7. $y = -x - 1$
$y = x - 1$

8. $x = 2$
$x + 2y = 4$

9. $y = 4x + 1$
$y = 3x$

10. $y = -3x - 2$
$2x - y = 2$

11. $y = 2$
$x - y = 3$

12. $x - y = 6$
$2x + y = 3$

13. $y = -\frac{1}{2}x$
$y = 3x + 7$

14. $y = \frac{1}{2}x - 3$
$x - y = 6$

15. $\frac{1}{4}x - 2 = y$
$x = -4y$

16. $2x + 3y = 12$
$4x + y = 4$

Lesson 13–2 *(Pages 554–559)* **State whether each system is** *consistent and independent, consistent and dependent,* **or** *inconsistent.*

1.

2.

3.

Determine whether each system of equations has *one* **solution,** *no* **solution, or** *infinitely many* **solutions by graphing. If the system has one solution, name it.**

4. $y = 2x + 1$
$y = -x - 2$

5. $x = 5$
$x - 4y = 1$

6. $y = 3$
$x - y = 2$

7. $y = x + 7$
$x = 7 - y$

8. $y = 2x - 6$
$y = 2x + 4$

9. $y = 10x - 16$
$y = 4x - 4$

10. $x + 2y = 5$
$x + y = 4$

11. $x + 4y = 5$
$2x + 6y = 6$

12. $y = \frac{1}{4}x - 1$
$y = -x + 4$

13. $\frac{1}{3}y = -x - \frac{1}{3}$
$y = -3x - 1$

14. $\frac{1}{3}x - 6y = 9$
$x - 18y = 27$

15. $y = \frac{3}{2}x$
$y = -\frac{2}{3}x + 13$

16. $x + y = 12$
$2x - y = 3$

17. $x - 2y = -3$
$-2x + 4y = 8$

18. $4x + 6y = 12$
$2x - 6 = -3y$

19. $3x + 3y = 6$
$4x - y = 3$

Lesson 13–3 *(Pages 560–565)* **Use substitution to solve each system of equations.**

1. $y = -x$
$x - y = -12$

2. $x = 5 - y$
$-3x + y = -3$

3. $y = x + 1$
$-x + 3y = -15$

4. $y = x + 2$
$2x - y = 1$

5. $y = 4x$
$x + y = 3$

6. $x = 3y$
$x + 3y = 4$

7. $y = 2x$
$2x - y = -1$

8. $y = 3x - 1$
$x + y = 4$

9. $y = 6x - 7$
$-2x - y = 3$

10. $x = 6 - y$
$x - 2y = 1$

11. $x + 4y = -8$
$3x - 6y = 0$

12. $x = 3 + y$
$3y + 2 = x$

13. $x - 7y = 0$
$2x + y = 0$

14. $x - 4y = 8$
$6y + 8 = 2x$

15. $6x - 3y = -2$
$y + 1 = x$

16. $x - \frac{1}{2}y = 14$
$x + \frac{1}{2}y = 2$

Lesson 13-4 (*Pages 566–571*)   Use elimination to solve each system of equations.

1. $x + y = 4$
$x - y = 16$

2. $x + 2y = 4$
$x + 3y = 6$

3. $x = 13 - y$
$x - y = -3$

4. $x - 6y = -1$
$x + 3y = -10$

5. $x - 3y = 7$
$x + 2y = 8$

6. $x - 4y = 5$
$-5x + 10y = -5$

7. $x - y = -2$
$2x + 2y = 16$

8. $x - y = 14$
$3x - 6y = 15$

9. $x - 13y = -2$
$7y - x = 5$

10. $x - 6y = 9$
$4x - 8y = 12$

11. $x + y = 8$
$16 + 3x = y$

12. $x - y = 9$
$4x - 2y = -8$

13. $x = 5 - y$
$6x - 4y = 0$

14. $x = 6 - y$
$2x - 8y = 2$

15. $4x + 3y = 12$
$8x - 4y = -12$

16. $6x - 3y = 18$
$6x + 3y = 12$

Lesson 13-5 (*Pages 572–577*)   Use elimination to solve each system of equations.

1. $x + 5y = 10$
$x + y = -6$

2. $x + 4y = 5$
$x - 2y = -7$

3. $x - 6y = 9$
$x + 3y = -9$

4. $-x - y = 4$
$x + y = -4$

5. $2x + y = -2$
$2x - y = 4$

6. $6x + y = -9$
$3x - y = 6$

7. $3x - y = 0$
$4x + 2y = 10$

8. $2x + 12y = 24$
$x + 7y = 18$

9. $4x - 8y = 0$
$x + 3y = -10$

10. $2x - 8y = 0$
$-x + 7y = -3$

11. $x - 8y = 5$
$-2x + 8y = -2$

12. $15x - y = 4$
$-6x + y = 5$

13. $4x - 16y = 24$
$2x + 2y = 12$

14. $x - 13y = 4$
$-2x + 10y = -4$

15. $-7x - y = 6$
$-3x + y = 4$

16. $6x - 3y = 18$
$10x + y = -12$

Lesson 13-6 (*Pages 580–585*)   Solve each system of equations by graphing.

1. $x = 4$
$y = x - 1$

2. $x = -2$
$y = x^2$

3. $x = -3$
$y = x^2 + 3$

4. $y = x^2$
$y = 3x$

5. $y = x^2$
$y = 2x$

6. $y = x^2 + 2$
$y = x + 4$

7. $y = x^2 + 1$
$y = -x + 1$

8. $y = x^2 + 3$
$y = 5x - 3$

9. $y = 4x^2$
$y = 8x$

10. $y = x^2 + 12$
$y = -x - 8$

11. $y = x^2 + 4$
$y = 3x + 2$

12. $y = -x^2 + 3$
$y = 3x + 5$

Use substitution to solve each system of equations.

13. $y = 2x^2$
$x = -2$

14. $y = -x^2$
$y = 5x + 6$

15. $y = -2x^2$
$y = 4x - 6$

16. $y = -3x^2$
$x = -5$

17. $y = \frac{1}{2}x^2 - 1$
$x = 4$

18. $y = -\frac{1}{2}$
$y = \frac{1}{2}x^2 - \frac{5}{2}$

19. $y = -\frac{1}{2}x^2 + \frac{7}{2}$
$x = 2x + 1$

20. $y = \frac{1}{4}x^2 - 2$
$y = 2$

21. $y = -x + 4$
$y = x^2 + 2$

22. $y = x^2 + 3$
$y = -7x^2 + 5$

23. $y = x^2 + x + 4$
$y = 6$

24. $y = 3x^2 + 6x - 9$
$y = -12$

Lesson 13–7 *(Pages 586–591)* Solve each system of inequalities by graphing.

1. $y > -2$
$y < 3$

2. $x \leq -4$
$y \geq -1$

3. $y < 0$
$x \geq -1$

4. $y \geq -2$
$x > 2$

5. $y > x$
$y > x + 1$

6. $y < x - 2$
$y > -1$

7. $y < -2$
$y > -2x$

8. $y \geq x$
$y \leq -3x$

9. $x < 3$
$y < x + 1$

10. $y \geq -2$
$y \leq -2x - 1$

11. $x + y > 0$
$2x + y \leq 1$

12. $y \geq x + 1$
$y < 2x + 2$

13. $y - 2 \leq x$
$y + 4 \geq x$

14. $y \leq 3x - 1$
$2x + 4 < y$

15. $2x + 4 > y$
$y - 2 \leq x$

16. $-3y - x > 2$
$2y + x < 0$

Lesson 14–1 *(Pages 600–605)* Name the set or sets of numbers to which each real number belongs. Let N = natural numbers, W = whole numbers, Z = integers, Q = rational numbers, and I = irrational numbers.

1. $\dfrac{21}{7}$

2. $\sqrt{13}$

3. $\sqrt{121}$

4. $47.13013001 \ldots$

5. $-\sqrt{38}$

6. 0.631

7. $\dfrac{1}{4}$

8. $-\dfrac{8}{2}$

9. -3

10. $-\dfrac{1}{6}$

11. $0.949949994 \ldots$

12. $-\sqrt{64}$

Find an approximation, to the nearest tenth, for each square root. Then graph the square root on a number line.

13. $-\sqrt{5}$

14. $\sqrt{14}$

15. $\sqrt{18}$

16. $\sqrt{29}$

17. $\sqrt{63}$

18. $-\sqrt{71}$

19. $\sqrt{82}$

20. $-\sqrt{93}$

21. $\sqrt{102}$

22. $\sqrt{145}$

23. $-\sqrt{201}$

24. $\sqrt{305}$

Determine whether each number is *rational* or *irrational*. If it is irrational, find two consecutive integers between which its graph lies on the number line.

25. $\sqrt{4}$

26. $-\sqrt{19}$

27. $-\sqrt{62}$

28. $\sqrt{33}$

29. $-\sqrt{54}$

30. $\sqrt{49}$

31. $-\sqrt{225}$

32. $-\sqrt{15}$

33. $\sqrt{196}$

34. $-\sqrt{152}$

35. $-\sqrt{181}$

36. $\sqrt{8}$

Lesson 14–2 *(Pages 606–611)* Find the distance between each pair of points. Round to the nearest tenth, if necessary.

1. $Q(2, 16), R(3, 15)$

2. $C(-4, -2), D(-6, -5)$

3. $P(-8, 1), Q(10, -7)$

4. $A(3, -9), B(1, -2)$

5. $G(-6, -14), H(-7, 2)$

6. $X(0, 0), Y(9, 9)$

7. $E(12, -2), F(-3, -4)$

8. $M(15, 3), N(8, -11)$

9. $V(-4, 4), W(6, -6)$

Find the value of a if the points are the indicated distance apart.

10. $A(a, 3), B(6, 5); d = 2$

11. $G(-1, 5), H(-8, a); d = \sqrt{85}$

12. $X(9, a), Y(5, -2); d = 4$

13. $P(6, 1), Q(a, -7); d = \sqrt{113}$

14. $C(-9, -2), D(0, a); d = \sqrt{90}$

15. $Q(a, -1), R(4, 5); d = 10$

16. $E(7, a), F(-2, 4); d = \sqrt{90}$

17. $M(a, 3), N(-1, 5); d = \sqrt{8}$

18. $V(-3, -3), W(a, 4); d = \sqrt{50}$

Lesson 14–3 *(Pages 614–619)* Simplify each expression. Leave in radical form.

1. $\sqrt{54}$
2. $\sqrt{80}$
3. $\sqrt{75}$
4. $\sqrt{300}$
5. $\sqrt{3} \cdot \sqrt{12}$
6. $\sqrt{4} \cdot \sqrt{8}$
7. $2\sqrt{6} \cdot \sqrt{18}$
8. $\sqrt{10} \cdot \sqrt{15}$
9. $\dfrac{\sqrt{18}}{\sqrt{9}}$
10. $\dfrac{\sqrt{12}}{\sqrt{4}}$
11. $\dfrac{\sqrt{20}}{\sqrt{8}}$
12. $\dfrac{\sqrt{75}}{\sqrt{5}}$
13. $\dfrac{4}{2 + \sqrt{12}}$
14. $\dfrac{6}{3 - \sqrt{8}}$
15. $\dfrac{3}{5 + \sqrt{3}}$
16. $\dfrac{5}{4 + \sqrt{10}}$

Simplify each expression. Use absolute value symbols if necessary.

17. $\sqrt{18a^2b}$
18. $\sqrt{21m^8}$
19. $\sqrt{120x^2y}$
20. $\sqrt{50jk}$
21. $\sqrt{16r^2s^3}$
22. $\sqrt{17n^3}$
23. $\sqrt{32c^4d^3}$
24. $\sqrt{12xy^2}$
25. $\sqrt{53m^2n^4}$
26. $\sqrt{51g^2h^2}$
27. $\sqrt{66mn^4}$
28. $\sqrt{48j^6k^2}$
29. $\sqrt{8a^2bc^2}$
30. $\sqrt{75x^6y^6}$
31. $\sqrt{30rs^2t^4}$
32. $\sqrt{196a^6}$

Lesson 14–4 *(Pages 620–623)* Simplify each expression.

1. $5\sqrt{3} + 7\sqrt{3}$
2. $4\sqrt{6} - 2\sqrt{6}$
3. $5\sqrt{7} + 2\sqrt{7}$
4. $13\sqrt{5} - 5\sqrt{5}$
5. $-11\sqrt{3} + 6\sqrt{2}$
6. $-6\sqrt{5} - 5\sqrt{5}$
7. $4\sqrt{2} + 3\sqrt{2} - 5\sqrt{2}$
8. $-3\sqrt{3} - 2\sqrt{3} - 4\sqrt{3}$
9. $2\sqrt{7} + 8\sqrt{7} - 14\sqrt{7}$
10. $4\sqrt{12} - 7\sqrt{3}$
11. $-2\sqrt{24} + 3\sqrt{6}$
12. $-6\sqrt{32} + 4\sqrt{8}$
13. $4\sqrt{2} - 3\sqrt{3} + 6\sqrt{2}$
14. $-7\sqrt{5} + 6\sqrt{3} - 2\sqrt{3}$
15. $\sqrt{105} - \sqrt{12} - \sqrt{18}$
16. $-8\sqrt{24} + 6\sqrt{12} - 3\sqrt{2}$
17. $4\sqrt{27} - 2\sqrt{48} + 3\sqrt{20}$
18. $\sqrt{52} - \sqrt{18} + \sqrt{120}$

19. If an equilateral triangle has a side of length $5\sqrt{6}$, what is the measure of the perimeter?

Lesson 14–5 *(Pages 624–629)* Solve each equation. Check your solution.

1. $\sqrt{m} = 4$
2. $\sqrt{y} = -2$
3. $\sqrt{2d} = 4$
4. $-\sqrt{5k} = 15$
5. $\sqrt{8t} = 4$
6. $\sqrt{r - 9} = 9$
7. $\sqrt{t + 5} = 7$
8. $\sqrt{n - 3} = 2$
9. $\sqrt{h + 5} = 0$
10. $\sqrt{2x + 6} = 6$
11. $\sqrt{4m + 5} - 6 = 2$
12. $\sqrt{3r - 2} + 5 = 4$
13. $r = \sqrt{r + 12}$
14. $k = \sqrt{3k + 10}$
15. $x - 10 = \sqrt{x + 2}$
16. $m = \sqrt{9m + 4} - 2$
17. $2 + \sqrt{4d + 4} = d$
18. $5 - \sqrt{5r - 6} = 9 - r$

Lesson 15–1 *(Pages 638–643)* **Find the excluded value(s) for each rational expression.**

1. $\dfrac{y}{y-6}$

2. $\dfrac{2m}{3m+6}$

3. $\dfrac{-n}{-8+4n}$

4. $\dfrac{j+1}{2j-6}$

5. $\dfrac{6}{r(r-2)}$

6. $\dfrac{8k}{k(k+1)}$

7. $\dfrac{2x}{(x-3)(x+1)}$

8. $\dfrac{7m}{(m+1)(m+1)}$

9. $\dfrac{2s}{s^2+2s}$

10. $\dfrac{q-2}{q^2-25}$

11. $\dfrac{3+p}{p^2-p-2}$

12. $\dfrac{x^2-2x+3}{x^2+x-6}$

Simplify each rational expression.

13. $\dfrac{8}{22}$

14. $\dfrac{6}{24}$

15. $\dfrac{30a}{36b}$

16. $\dfrac{2y^2}{4y}$

17. $\dfrac{2(x+1)}{x(x+1)}$

18. $\dfrac{(m-1)(m+2)}{(m+2)(m+3)}$

19. $\dfrac{2d^2-2d}{d^2-1}$

20. $\dfrac{-4j+12}{j^2-9}$

21. $\dfrac{n^2+2n-15}{n^2+3n-10}$

22. $\dfrac{2y^2+2y}{y^2-y-2}$

23. $\dfrac{x^2+2x-8}{x^2-7x+10}$

24. $\dfrac{r^2-r-6}{r^3-6r^2+9r}$

Lesson 15–2 *(Pages 644–649)* **Find each product.**

1. $\dfrac{xy}{x}\cdot\dfrac{z}{y}$

2. $\dfrac{2s}{3t}\cdot\dfrac{6}{4s^2}$

3. $\dfrac{7b^2}{c^2}\cdot\dfrac{3c}{b^2}$

4. $\dfrac{2p+1}{p^2}\cdot\dfrac{2p^2}{4p+2}$

5. $\dfrac{x(x+5)}{3}\cdot\dfrac{3}{x(2x+10)}$

6. $\dfrac{6r}{r+2}\cdot\dfrac{4r+8}{18}$

7. $\dfrac{3n+6}{n}\cdot\dfrac{n^2}{n^2+4n+4}$

8. $\dfrac{d^2+8d+16}{d^3}\cdot\dfrac{d^2}{d+4}$

9. $\dfrac{m}{m^2+4m+3}\cdot\dfrac{m+1}{m}$

Find each quotient.

10. $\dfrac{m}{n^2}\div\dfrac{2}{mn}$

11. $\dfrac{cd}{3a^2b}\div\dfrac{c^2}{3a^2}$

12. $6rs\div\dfrac{3r^2}{s}$

13. $\dfrac{2}{t-1}\div\dfrac{t}{2t-2}$

14. $\dfrac{y^2-2y+1}{2y}\div(y-1)$

15. $\dfrac{c^2-4}{3}\div\dfrac{c-2}{c+2}$

16. $\dfrac{x^2-9}{x^2}\div\dfrac{x+3}{x}$

17. $\dfrac{d}{d^2-16}\div\dfrac{6d}{d^2+2d-8}$

18. $\dfrac{b^2-4}{16b^2}\div\dfrac{2b+4}{4b}$

Lesson 15–3 *(Pages 650–655)* **Find each quotient.**

1. $(6x-3)\div(2x-1)$

2. $(m^2+4m)\div(m+4)$

3. $(16a^2+8a)\div(4a+2)$

4. $(k^2-5k-6)\div(k+1)$

5. $(s^2-5s+6)\div(s-3)$

6. $(2y^2-2y-24)\div(2y-8)$

7. $(m^2-6m+9)\div(m-3)$

8. $(r^2+3r-18)\div(r-3)$

9. $(s^2-3s-10)\div(s+2)$

10. $(4r^3-12r^2)\div(r-3)$

11. $(6p^3-10p^2)\div(3p-5)$

12. $(x^2+12)\div(x-3)$

13. $(t^2+8)\div(t-2)$

14. $(y^2+16)\div(y+4)$

15. $(4p^3-5p-3)\div(2p+3)$

Lesson 15-4 (Pages 656–661) Find each sum or difference. Write in simplest form.

1. $\dfrac{9}{r} + \dfrac{6}{r}$

2. $\dfrac{2a}{3} + \dfrac{5a}{3}$

3. $\dfrac{4}{j} - \dfrac{3}{j}$

4. $\dfrac{2b}{4} - \dfrac{4b}{4}$

5. $\dfrac{3}{2s} + \dfrac{1}{2s}$

6. $\dfrac{6}{3b} - \dfrac{9}{3b}$

7. $\dfrac{2a}{a} + \dfrac{7a}{a}$

8. $\dfrac{y}{16y} - \dfrac{y}{16y}$

9. $\dfrac{15}{8x} + \dfrac{1}{8x}$

10. $\dfrac{2}{9k} + \dfrac{5}{9k}$

11. $\dfrac{4}{6p} + \dfrac{10}{6p}$

12. $\dfrac{14}{25r} - \dfrac{19}{25r}$

13. $\dfrac{11}{10s} - \dfrac{12}{10s}$

14. $\dfrac{9}{2y} - \dfrac{3}{2y}$

15. $\dfrac{6}{6+m} + \dfrac{m}{6+m}$

16. $\dfrac{6s}{2-s} + \dfrac{3}{2-s}$

17. $\dfrac{2p}{3+j} - \dfrac{p}{3+j}$

18. $\dfrac{16}{m-5} - \dfrac{10}{m-5}$

19. $\dfrac{3r}{r+3} + \dfrac{9}{r+3}$

20. $\dfrac{-2x-4}{x+7} + \dfrac{2x+11}{x+7}$

21. $\dfrac{8}{6m-3} - \dfrac{4}{6m-3}$

22. $\dfrac{4x+5}{2x+6} + \dfrac{2-4x}{2x+6}$

23. $\dfrac{8-6r}{5r-2} + \dfrac{11r-10}{5r-2}$

24. $\dfrac{12t+1}{t+1} + \dfrac{3-8t}{t+1}$

Lesson 15-5 (Pages 662–667) Find the LCM for each pair of expressions.

1. $4mn, 6m^2$

2. $8x^2y, 20xy$

3. $4cd, 18d$

4. $d^2 - 9, d + 1$

5. $a^2 + a - 2, a^2 + 5a - 6$

6. $x^2 + x - 6, x^2 - x - 12$

Write each pair of rational expressions with the same LCD.

7. $\dfrac{2}{6mn}, \dfrac{1}{30}$

8. $\dfrac{5}{4p}, \dfrac{3}{2p}$

9. $-\dfrac{1}{6rs}, \dfrac{3}{4st}$

10. $\dfrac{5a}{3a^2b}, -\dfrac{4}{2a}$

11. $\dfrac{1}{y-2}, -\dfrac{3}{y^2-4}$

12. $\dfrac{m^2}{2m+4}, \dfrac{6m}{4m+8}$

Find each sum or difference in simplest form.

13. $\dfrac{r}{8} - \dfrac{r}{3}$

14. $\dfrac{2p}{6} - \dfrac{p}{5}$

15. $-\dfrac{r}{4} + \dfrac{3r}{7}$

16. $\dfrac{6t}{9} + \dfrac{t}{6}$

17. $\dfrac{2m}{5p^2} + \dfrac{2}{6p}$

18. $\dfrac{5x}{x-1} + \dfrac{7}{x}$

19. $\dfrac{6m}{j^2} + \dfrac{7m}{3j}$

20. $\dfrac{2t}{3+r} - \dfrac{9}{r^2}$

21. $\dfrac{4}{b^3} + \dfrac{3}{b}$

22. $\dfrac{6y+1}{2y+1} - \dfrac{1}{4y+2}$

23. $\dfrac{1}{x^2+x-2} + \dfrac{3}{x+2}$

24. $\dfrac{4}{b^2-16} + \dfrac{8}{b+4}$

Lesson 15-6 (Pages 668–673) Solve each equation. Check your solution.

1. $2 = \dfrac{2x}{5} + \dfrac{x}{10}$

2. $\dfrac{y}{5} + \dfrac{3y}{5} = \dfrac{4}{5}$

3. $\dfrac{m}{8} - \dfrac{3m}{4} = \dfrac{1}{2}$

4. $\dfrac{d}{5} = \dfrac{6d}{5} - \dfrac{3}{10}$

5. $\dfrac{8}{7v} + \dfrac{3}{4v} = -2$

6. $\dfrac{s+1}{2} + \dfrac{s}{4} = -2$

7. $\dfrac{2b-1}{2} = \dfrac{b}{3} - \dfrac{1}{4}$

8. $\dfrac{m-1}{3} + \dfrac{m+4}{3} = -5$

9. $\dfrac{4}{c+1} + \dfrac{2}{c+1} = -3$

10. $\dfrac{7}{3d} - \dfrac{5}{9d} = -3$

11. $\dfrac{1}{2m} + \dfrac{3}{m} = \dfrac{1}{4}$

12. $\dfrac{1}{r} + \dfrac{2}{r-5} = \dfrac{3}{r}$

13. $\dfrac{1}{3a+5} - \dfrac{4a}{6a+10} = \dfrac{1}{2}$

14. $\dfrac{1}{k-3} + \dfrac{k}{2} = \dfrac{k-2}{2}$

15. $\dfrac{r+2}{r^2-4} + \dfrac{1}{r-2} = \dfrac{1}{r}$

Mixed Problem Solving

Chapter 1 Writing Expressions and Equations

1. **Car Rental**   A rental company charges $120 dollars for a weekly car rental plus $0.25 per mile.   *(Lesson 1-1)*

   a. Write an expression for the total cost of driving 250 miles.

   b. Write an expression for the total cost of driving *m* miles.

2. **Entertainment**   The cost to go to the movies for *a* adults and *c* children is $8.50a + 5.50c$. If Mr. Washington takes his three children, what is the total cost? *(Lesson 1-2)*

3. **Fencing**   Emma wants to fence her horse pasture. Write and evaluate an expression for the amount of fence that she will need. *(Lesson 1-3)*

4. **Money**   Hats cost $12 each, and T-shirts cost $8 each. Use the Distributive Property to write an expression for the cost of 11 hats and shirts for a softball team. Then find the total cost.   *(Lesson 1-4)*

5. **Savings**   Marta deposited $300 into a savings account that pays 5% simple interest. How much more money will she have after 5 years than after 2 years? *(Lesson 1-5)*

6. **Education**   The table shows the number of hours that each student in the orchestra practiced in a week.

Hours of Practice							
14	3	2	15	4	14	13	3
13	17	5	4	12	3	13	15
16	21	12	13	3	14	5	23

   a. Make a frequency table of the data. *(Lesson 1-6)*

   b. Make a stem-and-leaf plot of the data. *(Lesson 1-7)*

Chapter 2 Integers

1. **Astronomy**   The table gives the surface temperature of the nine planets in our solar system. Order the temperatures from least to greatest.   *(Lesson 2-1)*

Planet Temperatures (°F)				
333	867	59	−81	−162
−323	−330	−369	−218	

   Source: *The World Almanac For Kids*

2. **Babysitting**   Ellie earns $6 per hour for babysitting. Find her total earnings for 1, 3, and 6 hours. Graph the ordered pairs (number of hours, earnings).   *(Lesson 2-2)*

3. **Money**   Nantan borrowed $30 from his parents. In the next two weeks, he paid them $40 and then borrowed $15 and $25. How much does he owe them? *(Lesson 2-3)*

4. **Geography**   The highest elevation in the world is Mt. Everest in Tibet, Nepal. It is 29,035 feet above sea level. The lowest elevation in the world is the Dead Sea in Israel and Jordan. It is 30,384 feet lower than Mr. Everest. What is the elevation of the lowest point in the world? *(Lesson 2-4)*

5. **Farming**   From 2000 to 2002, the number of farms in Missouri decreased by about 1000 per year. If there were 109,000 farms in 2000, how many farms were there in Missouri in 2002?   *(Lesson 2-5)*

6. **Education**   The table shows the cost of higher education in public institutions for several years. Find the average yearly change from 1999 to 2002.   *(Lesson 2-6)*

Year	Cost per year
1999	$7107
2000	$7310
2001	$7586
2002	$8046

   Source: U.S. Dept. of Education

Chapter 3 Addition and Subtraction Equations

1. **Shopping** B's Market sells oranges 10 for $3.99. QuickMart sells 8 oranges for $2.59. Which is the better buy? *(Lesson 3-1)*

2. **Population** The table shows the population for certain years in Pennsylvania. What is the change in population from 1999 to 2002? *(Lesson 3-2)*

Year	Population
1999	10,498,012
2000	11,881,643
2001	12,281,054
2002	12,335,091

Source: U.S. Bureau of the Census

3. **Parks** The table shows the acreage of five National Seashores. What is the mean acreage? *(Lesson 3-3)*

Seashore	Acreage
Assateague Island	39,733
Canaveral	57,662
Cape Cod	43,605
Cape Hatteras	30,321
Cape Lookout	28,243

Source: *Time Almanac*

4. **Health** The body mass index (BMI) is found by dividing weight in kilograms by height in meters squared or $BMI = \frac{w}{h^2}$. For a person that is 1.7 meters tall, what are the weights that correspond to a BMI of 18.5 and a BMI of 25? *(Lesson 3-4)*

5. **History** The *Mayflower* sailed for America in 1620. The Jamestown settlers had sailed for America 14 years earlier. Write an equation that represents these years. In what year did the Jamestown settlers sail for America? *(Lesson 3-6)*

6. **Games** To win a radio contest, Sam guesses the price of a car as $25,000. To win, he must be within $1000 of the actual price. Write and solve an absolute value equation to find the greatest and least values that the price could be so that Sam wins the car. *(Lesson 3-7)*

Chapter 4 Multiplication and Division Equations

1. **Track** For his training, Marcus runs 55 miles each week. If he ran 8 miles every day from Monday through Saturday, how many miles will he run on Sunday? *(Lesson 4-1)*

2. **School** If the morning consists of four class periods and Soledad needs to take Biology, Algebra, American History, and English 9, how many different morning schedules are possible? *(Lesson 4-2)*

3. **Food** A recipe for a fruit slush drink calls for $2\frac{1}{4}$ cups of sugar for each batch. How many batches can be made with 9 cups of sugar? *(Lesson 4-3)*

4. **Geometry** Find the height of the trapezoid if its area is 120 square inches. Use the formula $A = \frac{1}{2}h(b_1 + b_2)$. *(Lesson 4-4)*

5. **Money** Jessica received a cell phone bill for $43.99. If her phone plan charges $23.99 each month plus $0.10 per minute, write an equation for the number of minutes that Jessica talked. How many minutes did she talk? *(Lesson 4-5)*

6. **Car Rental** Able Car Rental charges $50 a day to rent a car and $0.25 per mile. Century Car Rental charges $30 a day and $0.35 per mile. The equations that represent the total amount A charged by each company for m miles driven are given. For how many miles will the companies charge the same amount? *(Lesson 4-6)*

   Able: $A = 50 + 0.25m$
   Century: $A = 30 + 0.35m$

7. **Gardening** Liam has 60 feet of rock to put a walkway around his garden. Write an equation to find the width w of the garden. Find the garden's dimensions. *(Lesson 4-7)*

MIXED PROBLEM SOLVING

Chapter 5 Proportional Reasoning and Probability

1. **Crafts**   Lanelle makes stuffed bears out of fabric for gifts. It takes $\frac{3}{4}$ of a yard for 1 bear. How many yards are needed for 8 bears?   *(Lesson 5-1)*

2. **Models**   Conor built a model of a stegosaurus with a length of 18 inches for his science project. The scale of the model is 3 inches = 5 feet. What is the actual length of a stegosaurus?   *(Lesson 5-2)*

3. **Money**   The table shows the number of each type of pizza sold at Pizza Palace over a lunch hour. What percent of each type of pizza were sold that hour?   *(Lesson 5-3)*

Type of Pizza	Number Sold
Pepperoni	15
Cheese	24
Other	11

4. **Fund-Raisers**   A charity sponsored a race to raise funds for cancer research. Adults paid $25 to participate and youths paid $10. There were 45 more youth participants than adults. They earned a total of $19,000. How many youth and adults participated?   *(Lesson 5-4)*

5. **Money**   Yuan bought a pair of shoes for $80 that were originally $95. Find the percent of increase or decrease in price. Round to the nearest percent.   *(Lesson 5-5)*

6. **Games**   To win a prize at a carnival, you must spin the spinner and land on a number greater than 5. What are the odds of spinning a number greater than 5?   *(Lesson 5-6)*

7. **Music**   Of the 16 CDs in Mariah's car, 5 are jazz, 3 are classical, 6 are soundtracks, and 2 are pop music. If Mariah chooses a CD at random, what is the probability that it is a soundtrack or a pop CD?   *(Lesson 5-7)*

Chapter 6 Functions and Graphs

1. **Olympics**   The table shows the winning times for the women's Olympics 200-meter dash since 1984.   *(Lesson 6-1)*

Year	Time (seconds)
1984	21.81
1988	21.34
1992	21.81
1996	22.12
2000	21.84
2004	22.05

Source: *Time Almanac*

a. Determine the domain and range of the relation.

b. Graph the relation.

c. Between which consecutive games was the decrease in time the greatest?

2. **Mail**   The equation $C = 1.42 + 0.42b$ gives the cost C to send a book in the mail for each pound b over 1 pound. Find the total cost for mailing books that weigh 3, 5, and 7 pounds.   *(Lesson 6-2)*

3. **Driving**   Raina drove at a constant rate of 70 miles per hour. The equation that can be used to find the distance d traveled in t hours is given by $d = 70t$. Graph the equation.   *(Lesson 6-3)*

4. **Money**   Michael's monthly cell phone bill is given by $C(x) = 24.99 + 0.15x$, where x is the number of minutes he talks. Suppose Michael talks for 184 minutes one month. What will be his phone bill?   *(Lesson 6-4)*

5. **Landscaping**   Ben charged Mr. Diaz $18 to landscape his yard for 3 hours. At this rate, what would Ben charge Ms. Lewis for 7 hours of landscaping?   *(Lesson 6-5)*

6. **Geometry**   The volume of a rectangular prism is 100 cubic meters. The area A of its base varies inversely as its height h. Write an equation that represents this inverse variation.   *(Lesson 6-6)*

Chapter 7 Linear Equations

1. **Entertainment** The linear function $C(x) = 19.50x$ represents the total cost of x tickets to an amusement park. Explain how changes in x affect $C(x)$. What is the slope of the graph of the equation? *(Lesson 7-1)*

2. **Elections** The table shows the national voter turnout in federal elections in 1996 and 2000. Write the point-slope form of an equation for the line passing through these points. *(Lesson 7-2)*

Year	Voters
1996	96,456,345
2000	105,586,274

Source: Federal Election Commission

3. **Business** Jamie holds scrapbooking parties. She charges $50 for supplies and $12 per person. Write an equation in slope-intercept form for the function if y is the total cost for x people. *(Lesson 7-3)*

4. **Driving** Determine whether a scatter plot for the distance traveled in a car for the time elapsed would show a *positive, negative,* or *no* relationship between the variables. Explain. *(Lesson 7-4)*

5. **Business** Denzel charges $400 for a landscape plan and $25 per hour for installation. Write a linear function to represent the total cost $f(x)$ as a function of the number of hours worked x. *(Lesson 7-5)*

6. **Track** Jason is running at a pace of 10 miles per hour while Cruz is running at a pace of 6 miles per hour. Make a conjecture about the similarities and differences in the graphs of the equations that represent the distance traveled by each runner d in time t. Verify by graphing the equations. *(Lesson 7-6)*

7. **Geometry** Determine whether $\triangle ABC$ is a right triangle if the vertices are $A(2, 1)$, $B(10, 15)$, and $C(3, 19)$. Justify your answer. *(Lesson 7-7)*

Chapter 8 Powers and Roots

1. **Sales** The Band Boosters will sell food at the basketball game. To determine how much popcorn to order, Monisha uses the formula $V = \pi r^2 h$ to find the volume of a cylindrical popcorn container. In the formula, $\pi \approx 3.14$, r is the radius of the base of the cylinder, and h is the height. Find the volume of the cylinder. *(Lesson 8-1)*

2. **Gardening** Hana is planting a vegetable garden. She wants it to be square as shown. Hana decides that the garden is not large enough so she doubles the length and width. Write an expression for the area of the new garden. *(Lesson 8-2)*

3. **Measurement** There are 10^{-3} grams in a milligram. Express 10^{-3} using positive exponents and evaluate. *(Lesson 8-3)*

4. **Astronomy** Mercury is 36 million miles from the Sun. Express the distance in scientific notation. *(Lesson 8-4)*

5. **Geometry** The area of a circle is given by the formula $A = \pi r^2$. If the area of a circle is 121π square feet, what is the radius? *(Lesson 8-5)*

6. **Decorating** Fiona's square room has an area of 154 square feet. What is a whole-number estimate of the length of a side? *(Lesson 8-6)*

7. **Skateboarding** How far from the base of the platform does the skateboarding ramp reach? *(Lesson 8-7)*

Chapter 9 Polynomials

1. **History**   Manassas National Battlefield Park in Virginia has an area of 5067 acres. A numeral in base 10 can be written in polynomial form such as $256 = 2(10)^2 + 5(10)^1 + 6(10)^0$. Write the number of acres in polynomial form.   *(Lesson 9-1)*

2. **Framing**   José designs picture frames in nonstandard sizes. The width of one rectangular frame is $2x + 5$ and the length is $5x - 9$.   *(Lesson 9-2)*

    a. What is the perimeter of this type of frame written as a polynomial?

    b. What is the perimeter of the frame if $x = 7$ inches?

3. **Gardening**   A garden located on the grounds of a new museum has width a and length $3a - 7$.   *(Lesson 9-3)*

    a. Write an expression in simplest form for the area of the garden.

    b. If a is 12 feet, find the area of the garden.

4. **Shipping**   Packages and More sells a variety of mailing boxes.   *(Lesson 9-4)*

    a. Find the length of box when $b = 10$ inches.

    b. Express the volume of the box shown as a polynomial.

    c. What is the volume of the box in cubic inches when $b = 1$ foot?

5. **Quilting**   Deidra is designing a quilt to use as a wall hanging in her room. She made the center portion of the quilt as a square with sides x inches long. She will surround the square with rectangles that have a width 2 inches smaller than the square and a length 2 inches longer than the square. Express the area of one rectangle as a polynomial in simplest form.   *(Lesson 9-5)*

Chapter 10 Factoring

1. **Parties**   Latisha has 90 candies and 54 cookies. If she divided the candies and cookies evenly among the bags, what is the greatest number of gift bags she can make so that there are no candies or cookies left over?   *(Lesson 10-1)*

2. **Home Improvement**   Carmen is installing hardwood floors in part of her house. Write an expression in factored form that represents the area that will be hardwood. *(Lesson 10-2)*

3. **Construction**   Kevin is going to put a hot tub that measures 8 feet by 6 feet within his new deck so that no decking will be installed below it. How many square feet of decking will Kevin need? Express the answer in factored form.   *(Lesson 10-3)*

4. **Decorating**   Cheyenne is painting a mural on one wall of her living room. The wall that she will paint has a window as shown. Write an expression for the area that will be painted in factored form. *(Lesson 10-4)*

5. **Construction**   Tavon drew plans for a square shed to put in his backyard. He then decided that he didn't want the shed to be square, so he reduced one dimension by a number and increased the other dimension by that same number. The new area of the shed floor is $x^2 - 16$. Factor this expression. *(Lesson 10-5)*

Chapter 11 Quadratic and Exponential Functions

1. **Sports** The height of a ball in feet is given by $h = -16t^2 + 64t + 5$ if the initial velocity is 64 feet per second and the ball is released 5 feet off the ground. What is the maximum height the ball reaches? *(Lesson 11-1)*

2. **Families of Graphs** Suppose the parent graph $y = x^2 + 7$ is shifted 4 units to the right. Write the new equation and name the vertex of the new graph. *(Lesson 11-2)*

3. **Construction** Lexi will surround a rectangular sitting area with rose bushes. She has enough bushes to extend 44 feet. *(Lesson 11-3)*
   a. Let w represent the width of the area. Write a quadratic equation for the area A of the sitting area. (*Hint:* Write an expression for the length in terms of the width.)
   b. Graph the equation you wrote. Can the width be 20 feet? Explain.

4. **Quilting** The length of a quilt is 3 feet more than the width. The area of the quilt is 40 square feet. Find the length and width of the quilt. *(Lesson 11-4)*

5. **Construction** The area of a daycare center's sandbox is 56 square feet. The walkway around it has equal width all the way around. What are the dimensions of the sandbox? *(Lesson 11-5)*

6. **Models** A model rocket is launched from a height of 2 feet off the ground. Use the function $h = -16t^2 + 155t + 2$ to find the approximate time that the rocket hits the ground after it is launched. *(Lesson 11-6)*

7. **Savings** Piper deposits $200 in a compound interest savings account with an annual interest rate of 6%. Use $A(t) = 200(1.06)^t$, where t is the number of years, to find how much will she have after 6 years. *(Lesson 11-7)*

Chapter 12 Inequalities

1. **Travel** On most airlines, a piece of luggage cannot weigh over 50 pounds. Write an inequality to represent this situation and graph it on a number line. *(Lesson 12-1)*

2. **Driving** Christopher drove from Bozeman, Montana, to Minneapolis, Minnesota, to visit his brother. He estimated that the trip was more than 1100 miles. On the first day he drove 540 miles. On the second day he had driven 280 miles when he stopped for lunch. How many more miles did he have to drive? *(Lesson 12-2)*

3. **School** Dewayne is preparing for a final exam. He can study 25 pages of his textbook per day. If the textbook has fewer than 310 pages, how many days will he need to study? *(Lesson 12-3)*

4. **Money** Ms. Parker is taking the French Club to a movie. The tickets cost $7.50 each, and there are 25 students in the club. If she wants to spend no more than $250, how much can she spend on refreshments? *(Lesson 12-4)*

5. **Track** If Jackson runs at a brisk pace, he needs to run between 50 and 60 minutes. If he runs at a slower pace, he needs to run between 55 and 75 minutes. Write an inequality that represents his possible running times. Graph the inequality. *(Lesson 12-5)*

6. **Sewing** Mira's job is to cut fabric at a fabric store. The fabric must be within $\frac{1}{8}$ yard of the desired length. The customer wants 2 yards of fabric. Write and solve an absolute value inequality that represents the possible lengths of the fabric. *(Lesson 12-6)*

7. **Fund-Raising** The band is selling small boxes of candy for $6 and large boxes for $10. They must raise at least $1170 for uniforms. Write and graph an inequality that describes the number of each size of box that could be sold to meet the goal. *(Lesson 12-7)*

Chapter 13 Systems of Equations and Inequalities

1. **Money**   A soccer team is celebrating the end of the season by going to a professional soccer game. Kara buys 40 tickets and spends $284. Youth tickets are $6 each and adult tickets are $8 each. Write and graph two equations that represent this situation. How many of each type of ticket did Kara buy?   *(Lesson 13-1)*

2. **Sports**   The 97th World Series was played in 2003. By 2003, an American League team had won the World Series 17 times more than a National League team. Write a system of equations to represent the situation. Find the solution of the system and explain what it means.   *(Lesson 13-3)*

3. **Food**   Inez is making a fruit punch by mixing a juice that is 25% water with a juice that is 40% water. How much of each juice should be used to make 5 gallons of a juice that is 35.5% water?   *(Lesson 13-3)*

4. **Movies**   Jamar bought 2 movie tickets and 1 bag of popcorn for $20. The next weekend, he bought 1 movie ticket and 1 bag of popcorn for $12.50. How much does a ticket cost?   *(Lesson 13-4)*

5. **Money**   The cost of 5 hamburgers and 4 hot dogs at Burger Joe's is $16. It costs $19 for 8 hamburgers and 2 hot dogs. Write a system of equations to represent this situation. What is the cost of a hamburger? a hot dog?   *(Lesson 13-5)*

6. **Geometry**   The area a of a rectangle that has a perimeter of 56 feet and a width of w feet is represented by $a = -w^2 + 28w$. Use a system of equations to find the width of a rectangle that has a perimeter of 56 feet and an area of 187 square feet.   *(Lesson 13-6)*

7. **Health**   Alvaro walks or jogs at least 3 miles a day. He walks at a rate of 4 miles per hour and jogs 8 miles per hour. He has a half hour to exercise today. How much time can he spend at both?   *(Lesson 13-7)*

Chapter 14 Radical Expressions

1. **Geometry**   The radius of a cylinder is given by $r = \sqrt{\dfrac{V}{\pi h}}$, where V is the volume of the cylinder, h is the height of the cylinder, and $\pi \approx 3.14$. What is the radius of a cylinder with a volume of 200 cubic feet and a height of 6 feet?   *(Lesson 14-1)*

2. **Walking**   Jariah's school is 10 miles north and 6 miles east of her house. Her workplace is 5 miles south and 15 miles west of her house.   *(Lesson 14-2)*

   a. Draw a diagram on a coordinate grid to represent this situation.

   b. How far is Jariah's school from her work? Round to the nearest tenth.

3. **Roller Coasters**   The first hill of a small roller coaster is 50 feet tall. If the coaster travels a horizontal distance of 90 feet as it climbs the hill, how long is the track up the hill?   *(Lesson 14-3)*

4. **Coordinate Geometry**   What is the perimeter of $\triangle ABC$?   *(Lesson 14-4)*

5. **Geometry**   The formula $A = \pi r^2$ gives the area A of a circle with radius r.   *(Lesson 14-5)*

   a. Write an equation for r in terms of A.

   b. What is the radius of a circle with an area of 52 square centimeters? Round to the nearest tenth.

   c. If the area of a circle is doubled, what is the change in the radius?

Chapter 15 Rational Expressions and Equations

1. **Elections** Chloe has a 75% chance of winning her student council election. Express 75% as a fraction in simplest form. *(Lesson 15-1)*

2. **Gift Wrapping** It takes $1\frac{2}{3}$ yards of ribbon to wrap a gift. How many gifts can be wrapped with a roll of ribbon that is 15 yards long? *(Lesson 15-2)*

3. **Home Design** Find the length of the kitchen if the area is $x^2 + 10x + 24$ square feet. *(Lesson 15-3)*

   $x + 4$ ft

4. **Sports** The time that it takes Mario to paddle his kayak 3 miles down a stream with a current of 15 miles per hour is given by $\dfrac{3}{x + 15}$, where x is Mario's paddling speed. The time that it would take Mario to paddle 5 miles down the same stream is given by $\dfrac{5}{x + 15}$. To find the time that it would take Mario to paddle 8 miles down the stream, add the two expressions. *(Lesson 15-4)*

5. **Geometry** Find an expression for the perimeter of the isosceles triangle. *(Lesson 15-5)*

   $\dfrac{6x}{x - 1}$
   $\dfrac{3}{x + 2}$

6. **School** Mandy's algebra teacher gives a 10-point quiz each week. After five weeks, Mandy has a total score of 36 for an average of 7.2 points on the quizzes. She would like to raise her average to 9 points. On how many quizzes must Mandy earn a 10 to reach her goal? *(Lesson 15-6)*

Becoming a Better Test-Taker

At some time in your life, you will probably have to take a standardized test. Sometimes this test may determine if you go on to the next grade or course, or even if you will graduate from high school. This section of your textbook is dedicated to making you a better test-taker.

TYPES OF TEST QUESTIONS   In the following pages, you will see examples of four types of questions commonly seen on standardized tests. A description of each type of question is shown in the table below.

Type of Question	Description	See Pages
multiple choice	4 or 5 possible answer choices are given from which you choose the best answer.	733–736
gridded response	You solve the problem. Then you enter the answer in aspecial grid and shade in the corresponding circles.	737–740
short response	You solve the problem, showing your work and/or explaning your reasoning.	741–744
extended response	You solve a multi-part problem, showing your work and/orexplaining your reasoning.	745–749

PRACTICE   After being introduced to each type of question, you can practice that type of question. Each set of practice questions is divided into five sections that represent the concepts most commonly assessed on standardized tests.

- Number and Operations
- Algebra
- Geometry
- Data Analysis and Probability
- Measurement

USING A CALCULATOR   On some tests, you are permitted to use a calculator. You should check with your teacher to determine if calculator use is permitted on the test you will be taking, and if so, what type of calculator can be used.

Test-Taking Tip

If you are allowed to use a calculator, make sure you are familiar with how it works so that you won't waste time trying to figure out the calculator when taking the test.

TEST-TAKING TIPS   In addition to Test-Taking Tips like the one shown at the right, here are some additional thoughts that might help you.

- Get a good night's rest before the test. Cramming the night before does not improve your results.

- Budget your time when taking a test. Don't dwell on problems that you cannot solve. Just leave that question blank on your answer sheet.

- Watch for key words like NOT and EXCEPT. Also look for order words like LEAST, GREATEST, FIRST, and LAST.

Multiple-Choice Questions

Multiple-choice questions are the most common type of questions on standardized tests. These questions are sometimes called *selected-response questions*. You are asked to choose the best answer from four or five possible answers.

To record a multiple-choice answer, you may be asked to shade in a bubble that is a circle or an oval, or to just write the letter of your choice. Always make sure that your shading is dark enough and completely covers the bubble.

The answer to a multiple-choice question is usually not immediately obvious from the choices, but you may be able to eliminate some of the possibilities by using your knowledge of mathematics. Another answer choice might be that the correct answer is not given.

Incomplete Shading
(A) (B) (*C*) (D)
Too light shading
(A) (B) (C) (D)
Correct shading
(A) (B) (C) (D)

Example

A storm signal flag is used to warn small craft of wind speeds that are greater than 38 miles per hour. The length of the square flag is always three times the length of the side of the black square. If y is the area of the black square and x is the length of the side of the flag, which equation describes the relationship between x and y?

(A) $y = \frac{1}{3}x^2$

(B) $y = \frac{1}{9}x^2$

(C) $y = x^2 - 1$

(D) $y = 3x$

(E) $y = 9x$

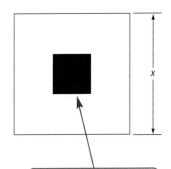

If x is 3 times the length of this side, then this side length is $\frac{1}{3}x$.

Strategy

Elimination
You can eliminate any obvious wrong answers.

For the area of a square, $A = s^2$. So, $A = x \cdot x$ or x^2.

The area of the black square is part of the area of the flag, which is x^2. Eliminate choices D and E because they do not include x^2.

$A = \left(\frac{1}{3}x\right)^2$ or $\frac{1}{9}x^2$ square units

So, $y = \frac{1}{9}x^2$. This is choice B.

Use some random numbers to check your choice.

Multiples of 3 make calculations easier.

Length of Flag (x)	Length of Black Square	Area of Black Square	Area $= \frac{1}{9}x^2$
12	4	16	$16 \stackrel{?}{=} \frac{1}{9}(12^2)$ ✓
27	9	81	$81 \stackrel{?}{=} \frac{1}{9}(27^2)$ ✓
60	20	400	$400 \stackrel{?}{=} \frac{1}{9}(60^2)$ ✓

Many multiple-choice questions are actually two- or three-step problems. If you do not read the question carefully, you may select a choice that is an intermediate step instead of the correct final answer.

Example 2

Strategy

Reread the Problem
Read the problem carefully to find what the question is asking.

Barrington can skateboard down a hill five times as fast as he can walk up the hill. If it takes 9 minutes to walk up the hill and skateboard back down, how many minutes does it take him to walk up the hill?

  Ⓕ **1.5 min**   Ⓖ **4.5 min**   Ⓗ **7.2 min**   Ⓙ **7.5 min**

Before involving any algebra, let's think about the problem using random numbers.

Skating is five times as fast as walking, so walking time equals 5 times the skate time. Use a table to find a pattern

Use the pattern to find a general expression for walk time given any skate time.

Skate Time	Skate Time × 5 = Walk Time
6 min	$6 \cdot 5 = 30$ min
3 min	$3 \cdot 5 = 15$ min
2 min	$2 \cdot 5 = 10$ min
x min	$x \cdot 5 = 5x$ min

The problem states that the walk time and the skate time total 9 minutes.

Use the expression to write an equation for the problem.

$x + 5x = 9$ skate time + walk time = 9 minutes
$6x = 9$ Add like terms.
$x = 1.5$ Divide each side by 6.

Looking at the choices, you might think that choice F is the correct answer. But what does x represent, and what is the problem asking?

The problem asks for the time it takes to walk up the hill, but the value of x is the time it takes to skateboard. So, the actual answer is found using $5x$ or $5(1.5)$, which is 7.5 minutes.

The correct choice is J.

Example 3

Strategy

Units of Measure
Make certain your answer reflects the correct unit of measure.

The Band Boosters are making ice cream to sell at an Open House. Each batch of ice cream calls for 5 cups of milk. They plan to make 20 batches. How many gallons of milk do they need?

  Ⓐ **800**   Ⓑ **100**   Ⓒ **25**   Ⓓ **12.5**   Ⓔ **6.25**

The Band Boosters need 5×20 or 100 cups of milk. However, choice B is not the correct answer. The question asks for *gallons* of milk.

4 cups = 1 quart and 4 quarts = 1 gallon, so 1 gallon = 4×4 or 16 cups.

$100 \text{ cups} \times \dfrac{1 \text{ gallon}}{16 \text{ cups}} = 6.25$ gallons, which is choice E.

Multiple Choice Practice

Choose the best answer.

Number and Operations

1. One mile on land is 5280 feet, while one nautical mile is 6076 feet. What is the ratio of the length of a nautical mile to the length of a land mile as a decimal rounded to the nearest hundredth?

    A 0.87   **B** 1.01   **C** 1.15   **D** 5.68

2. The star Proxima Centauri is 24,792,500 million miles from Earth. The star Epsilon Eridani is 6.345×10^{13} miles from Earth. In scientific notation, how much farther from Earth is Epsilon Eridani than Proxima Centauri?

    A 0.697×10^{14} mi   **B** 3.866×10^{13} mi

    C 6.097×10^{13} mi   **D** 38.658×10^{12} mi

3. In 1976, the cost per gallon for regular unleaded gasoline was 61 cents. In 2002, the cost was $1.29 per gallon. To the nearest percent, what was the percent of increase in the cost per gallon of gas from 1976 to 2002?

    A 1%   **B** 53%   **C** 95%   **D** 111%

4. The serial numbers on a particular model of personal data assistant (PDA) consist of two letters followed by five digits. How many serial numbers are possible if any letter of the alphabet and any digit 0–9 can be used in any position in the serial number?

    A 676,000,000   **B** 67,600,000

    C 6,760,000   **D** 676,000

Algebra

5. The graph shows the approximate relationship between the latitude of a location in the Northern Hemisphere and its distance in miles

from the equator. If y represents the distance of a location from the equator and x represents the measure of latitude, which equation describes the relationship between x and y?

    A $y = x + 69$ **B** $y = x + 690$

    C $y = 69x$ **D** $y = 10x$

6. A particular prepaid phone card can be used from a pay phone. The charge is 30 cents to connect and then 4.5 cents per minute. If y is the total cost of a call in cents where x is the number of minutes, which equation describes the relation between x and y?

    A $y = 4.5x + 30$ **B** $y = 30x + 4.5$

    C $y = 0.45x + 0.30$ **D** $y = 0.30x + 0.45$

7. Katie drove to the lake for a weekend outing. The lake is 100 miles from her home. On the trip back, she drove for an hour, stopped for lunch for an hour, and then finished the trip home. Which graph best represents her trip home and the distance from her home at various times?

   A

   B

   C

   D

8. Temperature can be given in degrees Fahrenheit or degrees Celsius. The formula $F = \dfrac{9}{5}C + 32$ can be used to change any temperature given in degrees Celsius to degrees Fahrenheit. Solve the formula for C.

    A $C = \dfrac{5}{9}(F - 32)$ **B** $C = F + 32 - \dfrac{9}{5}$

    C $C = \dfrac{5}{9}F - 32$ **D** $C = \dfrac{9}{5}(F - 32)$

9. Which of the following statements are true about the 4-inch quilt square?

R 2 in.

2 in.

_U_   _V_   _S_

_X_   _T_   _W_

Ⓐ _VSWT_ is a square.

Ⓑ _UVTX_ ≅ _VSWT_

Ⓒ Four right angles are formed at _V_.

Ⓓ Only A and B are true.

Ⓔ A, B, and C are true.

10. At the Daniels County Fair, the carnival rides are positioned as shown. What is the value of _x_?

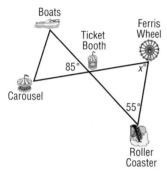

Boats

Ticket Booth

Ferris Wheel

85°

_x_°

Carousel

55°

Roller Coaster

Ⓐ 40   Ⓑ 47.5

Ⓒ 55   Ⓓ 70

Ⓔ 85

11. The diagram shows a map of the Clearwater Wilderness hiking area. To the nearest tenth of a mile, what is the distance from the Parking Lot to Bear Ridge using the most direct route?

Parking Lot

Crystal Spring 5 mi

4 mi

Double Falls

29 mi

20 mi

Bear Ridge   Lava Lake

Ⓐ 24 mi   Ⓑ 25.5 mi

Ⓒ 26 mi   Ⓓ 30.4 mi

12. Laura expects about 60 people to attend a party. She estimates that she will need one quart of punch for every two people. How many gallons of punch should she prepare?

Ⓐ 7.5

Ⓑ 15

Ⓒ 30

Ⓓ 34

13. Stone Mountain Manufacturers are designing two sizes of cylindrical cans below. What is the ratio of the volume of can A to the volume of can B?

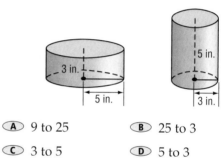

3 in.

5 in.

5 in.

3 in.

Ⓐ 9 to 25   Ⓑ 25 to 3

Ⓒ 3 to 5   Ⓓ 5 to 3

Data Analysis and Probability

14. The 2000 populations of the five least-populated U.S. states are shown in the table. Which statement is true about this set of data?

State	Population
Alaska	626,932
North Dakota	642,200
South Dakota	754,844
Vermont	608,827
Wyoming	493,782

Source: U.S. Census Bureau

Ⓐ The mode of the data set is 642,200.

Ⓑ The median of the data set is 626,932.

Ⓒ The mean of the data set is 625,317.

Ⓓ A and C are true.

Ⓔ B and C are true.

Gridded-Response Questions

Gridded-response questions are other types of questions on standardized tests. These questions are sometimes called *student-produced responses* or *grid-ins*, because you must create the answer yourself, not just choose from four or five possible answers.

For gridded response, you must mark your answer on a grid printed on an answer sheet. The grid contains a row of four or five boxes at the top, two rows of ovals or circles with decimal and fraction symbols, and four or five columns of ovals, numbered 0–9. Since there is no negative symbol on the grid, answers are never negative. At the right is an example of a grid from an answer sheet.

How do you correctly fill in the grid?

Example **①** **Diego drove 174 miles to his grandmother's house. He made the drive in 3 hours without any stops. At this rate, how far in miles can Diego drive in 5 hours?**

What value do you need to find?

You need to find the number of miles Diego can drive in 5 hours.

Write a proportion for the problem. Let m represent the number of miles.

$$\text{miles} \longrightarrow \frac{174}{3} = \frac{m}{5} \longleftarrow \text{miles}$$
$$\text{hours} \longrightarrow \qquad \qquad \longleftarrow \text{hours}$$

Solve the proportion.

$\dfrac{174}{3} = \dfrac{m}{5}$ Original proportion

$870 = 3m$ Find the cross products.

$290 = m$ Divide each side by 3.

How do you fill in the grid for the answer?

- Write your answer in the answer boxes.
- Write only one digit or symbol in each answer box.
- Do not write any digits or symbols outside the answer boxes.
- You may write your answer with the first digit in the left answer box, or with the last digit in the right answer box. You may leave blank any boxes you do not need on the right or the left side of your answer.
- Fill in only one bubble for every answer box that you have written in. Be sure not to fill in a bubble under a blank answer box

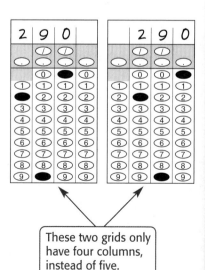

These two grids only have four columns, instead of five.

Many gridded response questions result in an answer that is a fraction or a decimal. These values can also be filled in on the grid.

How do you grid decimals and fractions?

Example ❷

What is the slope of the line that passes through (−2, 3) and (2, 4)?

Let $(-2, 3) = (x_1, y_1)$ and $(2., 4) = (x_2, y_2)$.

$$m = \frac{y_2 - y_1}{x_2 - x_1} \qquad \text{Slope formula}$$

$$= \frac{4 - 3}{2 - (-2)} \text{ or } \frac{1}{4} \qquad \text{Substitute and simplify.}$$

How do you grid the answer?

You can either grid the fraction $\frac{1}{4}$, or rewrite it as 0.25 and grid the decimal. Be sure to write the decimal point or fraction bar in the answer box. The following are acceptable answer responses that represent $\frac{1}{4}$ and 0.25.

Strategy

Decimals and Fractions
Fill in the grid with decimal and fraction answers.

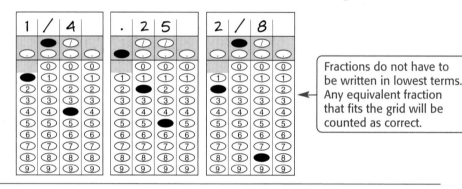

Fractions do not have to be written in lowest terms. Any equivalent fraction that fits the grid will be counted as correct.

Some problems may result in an answer that is a mixed number. Before filling in the grid, change the mixed number to an equivalent improper fraction or decimal. For example, if the answer is $1\frac{1}{2}$, do not enter 1 1/2 as this will be interpreted as $\frac{11}{2}$. Instead, either enter 3/2 or 1.5.

How do you grid mixed numbers?

Example ❸

Amber's cookie recipe calls for $1\frac{1}{3}$ cups of coconut. If Amber plans to make 4 batches of cookies, how much coconut does she need?

Find the amount of coconut needed using a proportion.

$$\frac{1\frac{1}{3}}{1} = \frac{x}{4}$$

$$4\left(1\frac{1}{3}\right) = 1x$$

$$4\left(\frac{4}{3}\right) = x$$

$$\frac{16}{3} = x$$

Leave the answer as the improper fraction $\frac{16}{3}$, as you cannot correctly grid $5\frac{1}{3}$.

Gridded-Response Practice

Solve each problem. Then copy and complete a grid like the one shown on page 738.

Number and Operations

1. China has the most days of school per year for children with 251 days. If there are 365 days in a year, what percent of the days of the year do Chinese students spend in school? Round to the nearest tenth of a percent.

2. Charles is building a deck and wants to buy some long boards that he can cut into various lengths without wasting any lumber. He would like to cut any board into all lengths of 24 inches, 48 inches, or 60 inches. In feet, what is the shortest length of boards that he can buy?

3. At a sale, an item was discounted 20%. After several weeks, the sale price was discounted an additional 25%. What was the total percent discount from the original price of the item?

4. The Andromeda Spiral galaxy is 2.2×10^6 light-years from Earth. The Ursa Minor dwarf is 2.5×10^5 light-years from Earth. How many times as far is the Andromeda Spiral as Ursa Minor dwarf from Earth?

5. Twenty students want to attend the World Language Convention. The school budget will only allow for four students to attend. In how many ways can five students be chosen from the twenty students to attend the convention?

Algebra

6. Find the y-intercept of the graph of the equation $3x + 4y - 5 = 0$.

7. Name the x-coordinate of the solution of the system of equations $2x - y = 7$ and $3x + 2y = 7$.

8. Solve $2b - 2(3b - 5) = 8(b - 7)$ for b.

9. Kersi read 36 pages of a novel in 2 hours. Find the number of hours it will take him to read the remaining 135 pages if he reads them at the same rate?

10. If $f(x) = x + 3$ and $g(x) = x^2 - 2x + 5$, find $6[f(g(1))]$.

11. The Donaldsons have a fish pond in their yard that measures 8 feet by 15 feet. They want to put a walkway around the pond that measures the same width on all sides of the pond, as shown in the diagram. They want the total area of the pond and walkway to be 294 square feet. What will be the width of the walkway in feet?

Geometry

12. $\triangle MNP$ is reflected over the x-axis. What is the x-coordinate of the image of point N?

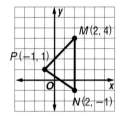

13. The pattern for the square tile shown in the diagram is to be enlarged so that it will measure 15 inches on a side. By what scale factor must the pattern be enlarged?

14. Find the measure of $\angle A$ to the nearest degree.

Test-Taking Tip ⒶⒷⒸⒹ

Question 14
Remember that the hypotenuse of a right triangle is always opposite the right angle.

15. Use the diagram for $\triangle ABO$ and $\triangle XBY$. Find the length of $\overline{BX}$.

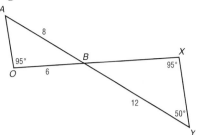

16. A triangle has a perimeter of 96 centimeters. The ratio of measures of its three sides is 6:8:10. Find the length of the longest side in centimeters.

17. The scale on a map of Texas is 0.75 inch = 5 miles. The distance on the map from San Antonio to Dallas is 8.25 inches. What is the actual distance from San Antonio to Dallas in miles?

Measurement

18. Pluto is the farthest planet from the Sun in this solar system at 2756 million miles. If light travels at 186,000 miles per second, how many minutes does it take for a particular ray of light to reach Pluto from the Sun? Round to the nearest minute.

19. Noah drove 342 miles and used 12 gallons of gas. At this same rate, how many gallons of gas will he use on his entire trip of 1140 miles?

20. The jumping surface of a trampoline is shaped like a circle with a diameter of 14 feet. Find the area of the jumping surface. Use 3.14 for π and round to the nearest square foot.

21. A cone is drilled out of a cylinder of wood. If the cone and cylinder have the same base and height, find the volume of the remaining wood. Use 3.14 for π and round to the nearest cubic inch.

14 in.

6 in.

22. The front of a storage building is shaped like a trapezoid as shown. Find the area in square feet of the front of the storage building.

Data Analysis and Probability

23. The table shows the average size in acres of farms in the six states with the largest farms. Find the median of the farm data in acres.

Average Size of Farms for 2001	
State	**Acres per Farm**
Alaska	1586
Arizona	3644
Montana	2124
Nevada	2267
New Mexico	2933
Wyoming	3761

24. The Lindley Park Pavilion is available to rent for parties. There is a fee to rent the pavilion and then a charge per hour. The graph shows the total amount you would pay to rent the pavilion for various numbers of hours. If a function is written to model the charge to rent the pavilion, where x is the number of hours and y is the total charge, what is the rate of change of the function?

25. A particular game is played by rolling three tetrahedral (4-sided) dice. The faces of each die are numbered with the digits 1–4. How many outcomes are in the sample space for the event of rolling the three dice once?

26. In a carnival game, the blindfolded contestant draws two toy ducks from a pond without replacement. The pond contains 2 yellow ducks, 10 black ducks, 22 white ducks, and 8 red ducks. The best prize is won by drawing two yellow ducks. What is the probability of drawing two yellow ducks? Write your answer as a percent rounded to the nearest tenth of a percent.

Short-Response Questions

Short-response questions require you to provide a solution to the problem, as well as any method, explanation, and/or justification you used to arrive at the solution. These are sometimes called *constructed-response, open-response, open-ended, free-response,* or *student-produced questions*. The following is a sample rubric, or scoring guide, for scoring short-response questions.

Credit	Score	Criteria
Full	2	Full credit: The answer is correct and a full explanation is provided that shows each step in arriving at the final answer.
Partial	1	Partial credit: There are two different ways to receive partial credit. • The answer is correct, but the explanation provided is incomplete or incorrect. • The answer is incorrect, but the explanation and method of solving the problem is correct.
None	0	No credit: Either an answer is not provided or the answer does not make sense.

On some standardized tests, no credit is given for a correct answer if your work is not shown.

Example ❶ Susana is painting two large rooms at her art studio. She has calculated that each room has 4000 square feet to be painted. It says on the can of paint that one gallon covers 300 square feet of smooth surface for one coat and that two coats should be applied for best results. What is the minimum number of 5-gallon cans of paint Susana needs to buy to apply two coats in the two rooms of her studio?

Full Credit Solution

First find the total number of square feet to be painted.
$$4000 \times 2 = 8000 \text{ ft}^2$$
Since 1 gallon covers 300 ft², multiply 8000 ft² by the unit rate $\dfrac{1 \text{ gal}}{300 \text{ ft}^2}$.

$$8000 \text{ ft}^2 \times \frac{1 \text{ gal}}{300 \text{ ft}^2} = \frac{8000}{300} \text{ gal}$$
$$= 26\frac{2}{3} \text{ gal.}$$

The steps, calculations, and reasoning are clearly stated.

Each can of paint contains 5 gallons, so divide $26\frac{2}{3}$ gallons by 5 gallons

$$26\frac{2}{3} \div 5 = \frac{80}{3} \div 5 = \frac{\overset{16}{\cancel{80}}}{3} \times \frac{1}{\underset{1}{\cancel{5}}} = \frac{16}{3} = 5\frac{1}{3}$$

The solution of the problem is clearly stated.

Since Susana cannot buy a fraction of a can of paint, she needs to but 6 cans of paint.

PREPARING FOR STANDARDIZED TESTS

Partial Credit Solution

In this sample solution, the answer is correct; however there is no justification for any of the calculations.

> There is no explanation of how $26\frac{2}{3}$ was obtained.

$$26\frac{2}{3} \div 5 = \frac{80}{3} \div 5$$

$$= \frac{\overset{16}{\cancel{80}}}{3} \times \frac{1}{\cancel{5}_1}$$

$$= \frac{16}{3}$$

$$= 5\frac{1}{3}$$

Susana will need to buy 6 cans of paint.

Partial Credit Solution

In this sample solution, the answer is incorrect. However, after the first statement, all of the calculations and reasoning are correct.

There are 4000 ft² to be painted and one gallon of paint covers 300 ft².

> The first step of doubling the square footage for painting the second room was left out.

$$4000\text{ft}^2 \times \frac{1\text{ gal}}{300\text{ ft}^2} = \frac{4000}{300}\text{ gal}$$

$$= 13\frac{1}{3}\text{ gal}$$

Each can of paint contains five gallons. So 2 cans would contain 10 gallons, which is not enough. Three cans of paint would contain 15 gallons which is enough.

Therefore, Susana will need to buy 3 cans of paint.

No Credit Solution

> The wrong operations are used, so the answer is incorrect. Also, there are no units of measure given with any of the calculations.

$$300 \times 2 = 600$$
$$600 \div 5 = 120$$
$$4000 \div 120 = 33\frac{1}{3}$$

Susana will need 34 cans of paint.

Short-Response Practice

Solve each problem. Show all your work.

Number and Operations

1. The world's slowest fish is the sea horse. The average speed of a sea horse is 0.001 mile per hour. What is the rate of speed of a sea horse in feet per minute?

2. Two buses arrive at the Central Avenue bus stop at 8 A.M. The route for the City Loop bus takes 35 minutes, while the route for the By-Pass bus takes 20 minutes. What is the next time that the two buses will both be at the Central Avenue bus stop?

3. Toya's Clothing World purchased some denim jackets for $35. The jackets are marked up 40%. Later in the season, the jackets are discounted 25%. How much does the store lose or gain on the sale of one jacket at the discounted price?

4. A femtosecond is 10^{-15} second, and a millisecond is 10^{-3} second. How many times faster is a millisecond than a femtosecond?

5. Find the next three terms in the sequence.

    $$1, 3, 9, 27, \ldots$$

Algebra

6. Find the slope of the graph of $5x - 2y + 1 = 0$.

7. Simplify $5 + x(1 - x) + 3x$. Write the result in the form $ax^2 + bx + c$.

8. Solve $17 - 3x \geq 23$.

9. The table shows what Gerardo charges in dollars for his consulting services for various numbers of hours. Write an equation that can be used to find the charge for any amount of time, where y is the total charge in dollars and x is the number of hours.

Hours	Charge	Hours	Charge
0	$25	2	$55
1	$40	3	$70

10. The population of Clark County, Nevada, was 1,375,765 in 2000 and 1,464,653 in 2001. Let x represent the years since 2000 and y represent the total population of Clark County. Suppose the county continues to increase at the same rate. Write an equation that represents the population of the county for any year after 2000.

Geometry

11. $\triangle ABC$ is dilated with scale factor 2.5. Find the coordinates of dilated $\triangle A'B'C'$?

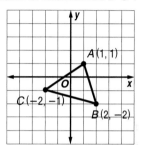

12. At a particular time in its flight, a plane is 10,000 feet above a lake. The distance from the lake to the airport is 5 miles. Find the distance in feet from the plane to the airport.

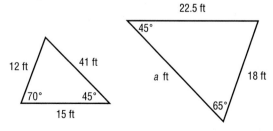

13. Refer to the diagram of the two similar triangles below. Find the value of a.

14. The vertices of two triangles are $P(3, 3)$, $Q(7, 3)$, $R(3, 10)$, and $S(-1, 4)$, $T(3, 4)$, $U(-1, 11)$. Which transformation moves $\triangle PQR$ to $\triangle STU$?

15. Aaron has a square garden in his front yard. The length of a side of the garden is 9 feet. Casey wants to plant 6 flowering bushes evenly spaced out along the diagonal. Approximately how many inches apart should he plant the bushes?

Measurement

16. One inch is equivalent to approximately 2.54 centimeters. Nikki is 61 inches tall. What is her height in centimeters?

17. During the holidays, Evan works at Cheese Haus. He packages gift baskets containing a variety of cheeses and sausages. During one four-hour shift, he packaged 20 baskets. At this rate, how many baskets will he package if he works 26 hours in one week?

18. Ms. Ortega built a box for her garden and placed a round barrel inside to be used for a fountain in the center of the box. The barrel touches the box at its sides as shown. She wants to put potting soil in the shaded corners of the box at a depth of 6 inches. How many cubic feet of soil will she need? Use 3.14 for π and round to the nearest tenth of a cubic foot.

4 ft
4 ft
4 ft

19. A line segment has its midpoint located at $(1, -5)$ and one endpoint at $(-2, -7)$. Find the length of the line to the nearest tenth.

Test-Taking Tip   Ⓐ Ⓑ Ⓒ Ⓓ

Question 20
Most standardized tests will include any commonly used formulas at the front of the test booklet. Quickly review the list before you begin so that you know what formulas are available.

20. A child's portable swimming pool is 6 feet across and is filled to a depth of 8 inches. One gallon of water is 231 cubic inches. What is the volume of water in the pool in gallons? Use 3.14 as an approximation for π and round to the nearest gallon.

6 ft
8 in.

Data Analysis and Probability

21. The table shows the five lowest recorded temperatures on Earth. Find the mean of the temperatures.

Location	Temperature (°F)
Vostok, Antarctica	−138.6
Plateau Station, Antarctica	−129.2
Oymyakon, Russia	−96.0
Verkhoyansk, Russia	−90.0
Northice, Greenland	−87.0

Source: *The World Almanac*

22. Two six-sided dice are rolled. The sum of the numbers of dots on the faces of the two dice is recorded. What is the probability that the sum is 10?

23. The table shows the amount of a particular chemical that is needed to treat various sizes of swimming pools. Write the equation for a line to model the data. Let x represent the capacity of the pool in gallons and y represent the amount of the chemical in ounces.

Pool Capacity (gal)	Amount of Chemical (oz)
5000	15
10,000	30
15,000	45
20,000	60
25,000	75

24. Fifty balls are labeled from 1 through 50. Two balls are drawn without replacement. What is the probability that both balls show an even number?

Extended-Response Questions

Extended-response questions are often called *open-ended* or *constructed-response questions*. Most extended-response questions have multiple parts. You must answer all parts correctly to receive full credit.

Extended-response questions are similar to short-response questions in that you must show all of your work in solving the problem, and a rubric is used to determine whether you receive full, partial, or no credit. The following is a sample rubric for scoring extended-response questions.

Credit	Score	Criteria
Full	4	A correct solution is given that is supported by well-developed, accurate explanations
Partial	3, 2, 1	A generally correct solution is given that may contain minor flaws in reasoning or computation or an incomplete solution. The more correct the solution, the greater the score.
None	0	An incorrect solution is given indicating no mathematical understanding of the concept, or no solution is given.

> On some standardized tests, no credit is given for a correct answer if your work is not shown.

Make sure that when the problem says to *Show your work*, show every aspect of your solution including figures, sketches of graphing calculator screens, or reasoning behind computations.

Example **The table shows the population density in the United States on April 1 in each decade of the 20th century.**

a. Make a scatter plot of the data.

b. Alaska and Hawaii became states in the same year. Between what two census dates do you think this happened. Why did you choose those years?

c. Use the data and your graph to predict the population density in 2010. Explain your reasoning

U.S. Population Density	
Year	People Per Square Mile
1910	31.0
1920	35.6
1930	41.2
1940	44.2
1950	50.7
1960	50.6
1970	57.4
1980	64.0
1990	70.3
2000	79.6

Source: *U.S. Census Bureau*

Full Credit Solution

Part a A complete scatter plot includes a title for the graph, appropriate scales and labels for the axes, and correctly graphed points.

- The student should determine that the year data should go on the *x*-axis while the people per square mile data should go on the *y*-axis.
- On the *x*-axis, each square should represent 10 years.
- The *y*-axis could start at 0, or it could show data starting at 30 with a broken line to indicate that some of the scale is missing.

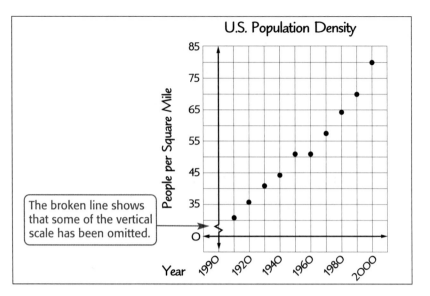

U.S. Population Density

The broken line shows that some of the vertical scale has been omitted.

Part b

1950–1960 because when Alaska became a state it added little population but a lot of land, which made the people per square mile ratio less.

Part c

About 85.0, because the population per square mile would probably get larger so I connected the first point and the last point. The rate of change for each year was $\frac{79.6 - 31.0}{2000 - 1910}$ or about 0.54. I added 10 × 0.54, or 5.4 to 79.6 to get the next 10-year point.

Actually, any estimate from 84 to 86 might be acceptable. You could also use different points to find the equation for a line of best fit for the data, and then find the corresponding y value for $x = 2010$.

Partial Credit Solution

Part a This sample answer includes no labels for the graph or the axes and one of the points is not graphed correctly.

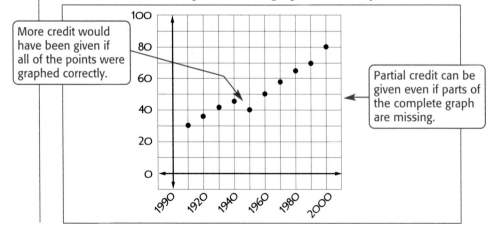

More credit would have been given if all of the points were graphed correctly.

Partial credit can be given even if parts of the complete graph are missing.

Part b Partial credit is given because the reasoning is correct, but the reasoning was based on the incorrect graph in Part a.

> 1940–1950, because when Alaska became a state it added little population but a lot of land, which made the people per square mile ratio less.

Part c Full credit is given for Part c.

> Suppose I draw a line of best fit through points (1910, 31.0) and (1990, 70.3). The slope would be $\dfrac{70.3 - 31.0}{1990 - 1910}$ or about 0.49. Now use the slope and one of the points to find the y-intercept.
>
> $$y = mx + b$$
> $$70.3 = 0.49(1990) + b$$
> $$-904.8 = b$$
>
> So an equation of my line of best fit is
> $$y = 0.49x - 904.8.$$
>
> If $x = 2010$, then $y = 0.49(2010) - 904.8$ or about 80.1 people per square mile in the year 2010.

This sample answer might have received a score of 2 or 1. Had the student graphed all points correctly and gotten Part B correct, the score would probably have been a 3.

No Credit Solution

Part a

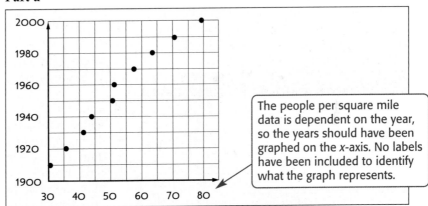

The people per square mile data is dependent on the year, so the years should have been graphed on the x-axis. No labels have been included to identify what the graph represents.

Part b

> I have no idea.

Part c

> 85, because it is the next grid line.

In this sample answer, the student does not understand how to represent data on a graph or how to interpret the data after the points are graphed.

Extended Response Practice

Solve each problem. Show all your work.

Number and Operations

1. The table shows what one dollar in U.S. money was worth in five countries in 1970 and in 2001.

Money Equivalent to One U.S. Dollar		
Country	1970 Value	2001 Value
France	5.5 francs	7 francs
Germany	3.6 marks	2 marks
Great Britain	0.4 pounds	0.67 pounds
Italy	623 lire	2040 lire
Japan	358 yen	117 yen

   a. For which country was the percent of increase or decrease in the number of units of currency that was equivalent to $1 the greatest from 1970 to 2001?

   b. Suppose a U.S. citizen traveled to Germany in 1970 and in 2001. In which year would the traveler receive a better value for their money? Explain.

   c. In 2001, what was the value of one franc in yen?

2. The table shows some data about the planets and the Sun. The radius is given in miles and the volume, mass, and gravity quantities are related to the volume and mass of Earth, which has a value of 1.

	Volume	Mass	Density	Radius	Gravity
Sun	1,304,000	332,950	0.26	434,474	28
Mercury	0.056	0.0553	0.98	1516	0.38
Venus	0.857	0.815	0.95	3760	0.91
Moon	0.0203	0.0123	0.61	1079	0.17
Mars	0.151	0.107	0.71	2106	0.38
Jupiter	1321	317.83	0.24	43,441	2.36
Saturn	764	95.16	0.12	36,184	0.92
Uranus	63	14.54	0.23	15,759	0.89
Neptune	58	17.15	0.30	15,301	1.12
Pluto	0.007	0.0021	0.32	743	0.06

   a. Make and test a conjecture relating volume, mass, and density.

   b. Describe the relationship between radius and gravity.

   c. Can you be sure that the relationship in part b holds true for all planets? Explain.

Algebra

3. The graph shows the altitude of a glider during various times of his flight after being released from a tow plane.

   a. What point on the graph represents the moment the glider was released from the tow plane? Explain the meaning of this point in terms of altitude.

   b. During which time period did the greatest rate of descent of the glider take place? Explain your reasoning.

   c. How long did it take the glider to reach an altitude of 0 feet? Where is this point on the graph?

   d. What is the equation of a line that represents the glider's altitude y as the time increased from 20 to 60 minutes?

   e. Explain what the slope of the line in part d represents?

4. John has just received his learner's permit which allows him to practice driving with a licensed driver. His mother has agreed to take him driving every day for two weeks. On the first day, John will drive for 20 minutes. Each day after that, John's mother has agreed he can drive 15 minutes more than the day before.

   a. Write a formula for the nth term of the sequence. Explain how you found the formula.

   b. For how many minutes will John drive on the last day? Show how you found the number of minutes.

   c. John's driver's education teacher requires that each student drive for 30 hours with an adult outside of class. Will John fulfill this requirement? Explain.

Geometry

5. Polygon *QUAD* is shown on a coordinate plane.

a. Find the coordinates of the vertices of *Q'U'A'D'*, which is the image of *QUAD* after a reflection over the *y*-axis. Explain.

b. Suppose point *M(a, b)* is reflected over the *y*-axis. What will be the coordinates of the image *M'*?

c. Describe a reflection that will make *QUAD* look "upside down." What will be the coordinates of the vertices of the image?

6. The diagram shows a sphere with a radius of *a* and a cylinder with a radius and a height of *a*.

a. What is the ratio of the volume of the sphere to the volume of the cylinder?

b. What is the ratio of the surface area of the sphere to the surface area of the cylinder?

c. The ratio of the volume of another cylinder is 3 times the volume of the sphere shown. Give one possible set of measures for the radius and height of the cylinder in terms of *a*.

Measurement

7. Alexis is using a map of the province of Saskatchewan in Canada. The scale for the map shows that 2 centimeters on the map is 30 kilometers in actual distance.

a. The distance on the map between two cities measures 7 centimeters. What is the actual distance between the two cities in kilometers? Show how you found the distance.

Test-Taking Tip ⒜ Ⓑ Ⓒ Ⓓ

Question 5
In a reflection over the *x*-axis, the *x*-coordinate remains the same, and the *y*-coordinate changes its sign. In a reflection over the *y*-axis, the *y*-coordinate remains the same, and the *x*-coordinate changes its sign.

b. Alexis is more familiar with distances in miles. The distance between two other cities is 8 centimeters. If one kilometer is about 0.62 mile, what is the distance in miles?

c. Alexis' entire trip measures 54 centimeters. If her car averages 25 miles per gallon of gasoline, how many gallons will she need to complete the trip? Round to the nearest gallon. Explain.

8. The diagram shows a pattern for a quilt square called Colorful Fan.

a. What is the area of region 6? Explain.

b. What is the area of region 1? Explain.

c. What is the ratio of the area of region 5 to the area of the entire square? Show how you found the ratio. Leave the ratio in terms of π.

Data Analysis and Probability

9. The table shows the Olympics winning times in the women's 1500-meter speed skating event. The times are to the nearest second.

Year	Time (s)	Year	Time (s)
1960	172	1984	124
1964	143	1988	121
1968	142	1992	126
1972	141	1994	122
1976	137	1998	118
1980	131	2002	114

a. Make a scatter plot of the data.

b. Use the data and your graph to predict the winning time in 2010.

10. There are 1320 ways for three students to win first, second, and third place during a debate.

a. How many students are on the debate team?

b. What is the probability that a student will come in one of the first three places if each student has an equal chance of succeeding?

c. If the teacher announces the third place winner, what is the probability that one particular other student will win first or second place?

Graphing Calculator Tutorial

General Information

- On a T1-84 Plus, any blue commands written above the calculator keys are accessed with the ⬚2nd⬚ key, which is also blue. Similarly, the yellow functions on a T1-83 Plus are accessed with the yellow ⬚2nd⬚ key. On both the T1-84 Plus and the T1-83 Plus, any green characters or commands above the keys are accessed with the ⬚ALPHA⬚ key which is also green. In this text, commands that are accessed by the ⬚2nd⬚ and ⬚ALPHA⬚ keys are shown in brackets. For example, ⬚2nd⬚ [QUIT] means to press the ⬚2nd⬚ key followed by the key below the yellow QUIT command.

- ⬚2nd⬚ [ENTRY] copies the previous calculation so it can be edited or reused.

- ⬚2nd⬚ [ANS] copies the previous answer so it can be used in another calculation.

- ⬚2nd⬚ [QUIT] will return you to the home (or text) screen.

- ⬚2nd⬚ [A-LOCK] allows you to use the green characters above the keys without pressing ⬚ALPHA⬚ before typing each letter. (This is handy for programming.)

- Negative numbers are entered using the ⬚(-)⬚ key, not the minus sign, ⬚-⬚.

- The variable x can be entered using the ⬚X,T,θ,n⬚ key, rather than using ⬚ALPHA⬚ [X].

- ⬚2nd⬚ [OFF] turns the calculator off.

Basic Keystrokes

Some commonly used mathematical functions are shown in the table below. As with any scientific calculator, graphing calculators observe the order of operations.

Mathematical Operation	Example	Keys	Display
evaluate expressions	Evaluate 2 + 5.	2 [+] 5 [ENTER]	$2 + 5$ 7
multiplication	Evaluate 3(9.1 + 0.8).	3 [(] 9.1 [+] .8 [)] [ENTER]	$3(9.1 + .8)$ 29.7
division	Evaluate $\frac{8-5}{4}$.	[(] 8 [−] 5 [)] [÷] 4 [ENTER]	$(8 − 5)/4$ .75
exponents	Find 3^5.	3 [∧] 5 [ENTER]	$3 \wedge 5$ 243
roots	Find $\sqrt{14}$.	[2nd] [√] 14 [ENTER]	$\sqrt{\ }$ (14 3.741657387
opposites	Enter −3.	[(−)] 3	-3
variable expressions	Enter $x^2 + 4x - 3$.	[X,T,θ,n] [x²] [+] 4 [X,T,θ,n] [−] 3	$X^2 + 4X − 3$

Key Skills

Each Graphing Calculator Exploration in the Student Edition requires the use of certain key skills. Use this section as a reference for further instruction.

A: Entering and Graphing Equations

Press [Y=]. Use the [X,T,θ,n] key to enter *any* variable for your equation. To see a graph of the equation, press [GRAPH].

B: Setting Your Viewing Window

Press [WINDOW]. Use the arrow or [ENTER] keys to move the cursor and edit the window settings. Xmin and Xmax represent the minimum and maximum values along the *x*-axis. Similarly, Ymin and Ymax represent the minimum and maximum values along the *y*-axis. Xscl and Yscl refer to the spacing between tick marks placed on the *x*- and *y*-axes. Suppose Xscl = 1. Then the numbers along the *x*-axis progress by 1 unit. Set Xres to 1.

C: The Standard Viewing Window

A good window to start with to graph an equation is the **standard viewing window.** It appears in the [WINDOW] screen as follows.

To easily set the values for the standard viewing window, press [ZOOM] 6.

D: Zoom Features

To easily access a viewing window that shows only integer coordinates, press [ZOOM] 8 [ENTER].

To easily access a viewing window for statistical graphs of data you have entered, press [ZOOM] 9.

E: Using the Trace Feature

To trace a graph, press TRACE . A flashing cursor appears on a point of your graph. At the bottom of the screen, x- and y-coordinates for the point are shown. At the top left of the screen, the equation of the graph is shown. Use the left and right arrow keys to move the cursor along the graph. Notice how the coordinates change as the cursor moves from one point to the next. If more than one equation is graphed, use the up and down arrow keys to move from one graph to another.

F: Setting or Making a Table

Press 2nd [TBLSET]. Use the arrow or ENTER keys to move the cursor and edit the table settings. Indpnt represents the x-variable in your equation. Set Indpnt to *Ask* so that you may enter any value for x into your table. Depend represents the y-variable in your equation. Set Depend to *Auto* so that the calculator will find y for any value of x.

G: Using the Table

Before using the table, you must enter at least one equation in the Y= screen. Then press 2nd [TABLE]. Enter any value for x as shown at the bottom of the screen. The function entered as Y_1 will be evaluated at this value for x. In the two columns labeled X and Y_1, you will see the values for x that you entered and the resulting y-values.

H: Entering Inequalities

Press 2nd [TEST]. From this menu, you can enter the $=$, $\neq$, $>$, $\geq$, $<$, and $\leq$ symbols.

Chapter	Page(s)	Key Skills
2	61	B
3	106	I
5	214	K
6	272	A, B, E
7	317	A, D, E
8	307	A, B, E
10	422	K
11	471, 491	A, D, E, F, G, I, J
12	521	A, D, E, H
13	551	A, D
14	625	A, B
15	638–639	A, B, E

I: Entering and Deleting Lists

Press STAT ENTER . Under L_1, enter your list of numerical data. To delete the data in the list, use your arrow keys to highlight L_1. Press CLEAR ENTER . Remember to clear all lists before entering a new set of data.

J: Plotting Statistical Data in Lists

Press Y= . If appropriate, clear equations. Use the arrow keys until Plot1 is highlighted. Plot1 represents a Stat Plot, which enables you to graph the numerical data in the lists. Press ENTER to turn the Stat Plot on and off. You may need to display different types of statistical graphs. To set the details of a Stat Plot, press 2nd [STAT PLOT] ENTER . A screen like the one below appears.

At the top of the screen, you can choose from one of three plots to store settings. The second line allows you to turn a Stat Plot on and off. Then you may select the type of plot: scatter plot, line plot, histogram, two types of box-and-whisker plots, or a normal probability plot. For this text, you will mainly use the scatter plot, line plot, and histogram. Next, choose which lists of data you would like to display along the x- and y-axes. Finally, choose the symbol that will represent each data point. To see a graph of the statistical data, press ZOOM 9.

K: Programming on the TI–83 Plus/TI–84 Plus

The TI–83 Plus and TI–84 Plus have programming features that allow you to write and execute a series of commands to perform tasks that may be too complex to perform otherwise. Each program is given a name. Commands begin with a colon (:), followed by an expression or an instruction. Most calculator features are accessible from the program mode.

When you press PRGM , you see three menus: EXEC, EDIT, and NEW. EXEC allows you to execute a stored program by selecting the name of the program from the menu. EDIT allows you to edit or change an existing program. NEW allows you to create a new program.

The following example illustrates how to create and execute a new program that stores an expression as Y and evaluates the expression for the designated value of X.

1. Press PRGM ► ► ENTER to create a new program.

2. Type EVAL ENTER to name the program. (Be sure that the A-LOCK is on.) You are now in the program editor, which allows you to enter commands. The colon (:) in the first column of the line indicates that it is the beginning of a command line.

3. The first command line will ask the user to choose a value for x. Press PRGM ► 3 2nd [A-LOCK] "ENTER X" ENTER . (To enter a space between words, press the 0 key when the A-LOCK is on.)

4. The second command line will allow the user to enter any value for x into the calculator. Press PRGM ► 1 X,T,θ,n ENTER .

5. The expression to be evaluated for the value of x is $x - 7$. To store the expression as Y, press X,T,θ,n − 7 STO► ALPHA [Y] ENTER .

6. Finally, we want to display the value for the expression. Press PRGM ► 3 ALPHA [Y] ENTER . At this point, you have completed writing the program. It should appear on your calculator like the screen shown below.

7. Now press 2nd [QUIT] to return to the home screen.

8. To execute the program, press PRGM . Then press the down arrow to locate the program name and press ENTER twice. The program asks for a value for x. Input any value for which the expression is defined and press ENTER . To immediately re-execute the program, simply press ENTER when the word *Done* appears on the screen. To break during program execution, press ON .

9. To delete a program, press 2nd [MEM] 2 7. Use the arrow and ENTER keys to select a program you wish to delete. Then press DEL 2.

While a graphing calculator cannot do everything, it can make some tasks easier. To prepare for whatever lies ahead, you should try to learn as much as you can. The future will definitely involve technology. Using a graphing calculator is a good start toward becoming familiar with technology.

Problem-Solving Strategy Workshops

Use a Table

Project

As a new DVD club member, you can choose seven movies. Each movie costs 1¢ plus $1.69 for shipping and handling. Within three years, you must order at least five more movies, each at the regular club price, plus the same shipping fee.

Type of Movie	Regular Club Price
Children's	$12.99
New Release	$24.99
All-Time Favorite	$16.99
Classic	$8.99

Suppose you buy a total of 12 movies. What is the lowest possible average cost per movie? the highest possible average cost per movie?

Working on the Project

Work with a partner and choose a strategy to help analyze and solve the problem. Here are some questions to help you get started.

- How much do you pay for the first shipment of seven 1¢ movies?

- What are the least and greatest amounts you can spend on the five required regular-priced movies?

Technology Tools

- Use a **spreadsheet** to calculate the average cost of the movies.

- Use **word processing software** to write your newspaper article.

 Research For more information about CD clubs, visit: www.algconcepts.com

Presenting the Project

Write an article for the school newspaper discussing the advantages and disadvantages of joining a DVD, book, or CD club.

- Research prices at retail or online stores. What would you pay for the same number of DVDs, books, or CDs required by the club?

- Show how the average cost per DVD, book, or CD changes as you buy more at the regular club price.

Draw a Diagram

Project

Sears Tower, the Magnificent Mile, Lake Shore Drive, Wrigley Field—you're in Chicago to see the sights! Suppose you are at the intersection of Illinois Street and Wells Street. You need to meet your group at the intersection of Ohio Street and Dearborn Street. How many ways can you walk there if you can only cross at intersections and can't backtrack?

Working on the Project

Work with a partner and choose a strategy. Develop a plan. Here are some suggestions to help you get started.

- Which intersections could you walk to from Illinois and Wells if you can't backtrack?

- Draw the map on grid paper and label the intersections with letters or numbers.

- Draw the different routes you can take on the grid paper.

Technology Tools

- Using **word processing software** to write an explanation of your solution.

- Use **drawing software** to draw your routes.

- Use **presentation software** to present your project.

 Research For more information about Chicago, visit:
www.algconcepts.com

Presenting the Project

Make a poster showing all of the routes. Include an explanation of your strategy for solving this problem. Make sure your explanation includes the following:

- a discussion of the number of possible routes if you have to meet your group one block past Ohio and Dearborn at Ontario and Dearborn, and

- a conjecture about the number of routes between any two intersections on the map.

Use an Equation

Project

At Funland, an amusement park, you can buy a 50-ticket pass to use at any time. The table shows the number of tickets needed for each ride.

You can use all 50 tickets by riding one 5-ticket ride 10 times. In what other ways can you use exactly 50 tickets?

Ride	Tickets Needed
The Mighty Axe	7
Log Chute	6
Kite-Eating Tree	5
Mystery Mine Ride	7
Ripsaw Roller Coaster	7
Screaming Yellow Eagle	6
Skyscraper Ferris Wheel	5
Tumbler	6

Working on the Project

Work with a partner to solve the problem.

- If you ride each ride once, can you use exactly 50 tickets?

- Suppose you ride six 5-ticket rides. Can you use all of the remaining tickets with none left over if you do not ride any more 5-ticket rides?

Technology Tools

- Use a **spreadsheet** to find the number of tickets a person can use for different combinations of rides.

- Use **word processing software** to design an ad campaign.

- Use **presentation software** to present your project.

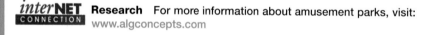 **Research** For more information about amusement parks, visit: www.algconcepts.com

Presenting the Project

Suppose the 50-ticket pass costs $25 and an all-day pass for one person costs $18.95. Design an ad campaign that answers the questions below.

- Suppose the amusement park is open from 10:00 A.M. to 9 P.M. About how many tickets could you use during that time period?

- What are the advantages and disadvantages of buying a 50-ticket pass? an all-day pass?

Look for a Pattern

Project

The United States government began issuing nine-digit Social Security numbers in 1935. An example is 123-45-6789. Each person can have only one number and the numbers are never reused. When do you think the government will run out of new numbers?

Year	Population	Births
1930	123,202,624	2,618,000
1940	132,164,569	2,559,000
1950	151,325,798	3,632,000
1960	179,323,175	4,258,000
1970	203,302,031	3,731,000
1980	226,542,203	3,612,000
1990	248,709,873	4,158,000
2000	281,421,906	4,059,000

Working on the Project

Work with a partner and choose a strategy to solve the problem.

- How many Social Security numbers are possible if each of the nine digits can be 0 through 9?

- Estimate the maximum number of Social Security numbers that could have been issued in 1935.

- Estimate the total number of people who had Social Security numbers by the year 2010 if every U.S. citizen had a Social Security number.

Technology Tools

- Use a **calculator** to find the total number of unique Social Security numbers.

- Use **spreadsheet software** to estimate the total number of Social Security numbers that will have been issued by the year 2010.

 Research For more information about Social Security numbers, visit: www.algconcepts.com

Presenting the Project

Write a one-page paper describing your results. Include the following:

- the total number of Social Security numbers possible,

- estimates for the total number of Social Security numbers issued by the year 2010 and the number issued each year after the year 2010, and

- the year you think that the government will run out of new Social Security numbers.

Guess and Check

Project

Congratulations! You have been selected to design the box for a new caramel corn snack. The manufacturers would like you to present three possible box designs. Each box must be a rectangular prism with a surface area of 150 square inches. Make a pattern for each of the three box designs. In a one-page paper, explain which box design you think will be best for the new product snack.

Working on the Project

Work with a partner and choose a strategy. Here are some suggestions to help you get started.

- Cut apart an empty cereal box along its edges to make a pattern for the surface area of a rectangular prism.

- Find the area of each of the six faces and the surface area of the prism.

- Suppose you don't know the measurements of the prism. Let ℓ represent the length, w represent the width, and h represent the height of the prism. Write a polynomial expression for the surface area of the prism.

Technology Tools

- Use a **calculator** to find the surface area of your patterns.

- Use **drawing software** to make patterns for your boxes.

*inter*NET
CONNECTION   **Research**   For more information about packaging, visit:
www.algconcepts.com

Presenting the Project

Prepare a portfolio of your box designs. Make sure your portfolio contains the following information:

- your calculations of the surface area of each box design,

- a one-page paper promoting the box design you think will be best, and

- a name for the new product and a logo or design for the box.

Use a Graph

Project

Do you want to be a millionaire? In this project, you will make a plan for saving money. Your goal is to reach $1,000,000 at the end of a 40-year period, following the guidelines listed below.

1. Your initial deposit is any amount of your choice.
2. The money is invested for exactly 40 years.
3. The annual growth rate of your investment is 5%.
4. Interest is calculated at the end of each year.
5. Choose an additional amount of money to deposit after interest is calculated each year. This amount is fixed, or the same, for all 40 years.
6. After 40 years, you should have as close to $1,000,000 as possible.

Working on the Project

Work with a partner and choose a strategy to help analyze and solve the problem. Here are some questions to help you get started.

- At the beginning of year 1, you decide to invest $5000. How much money will you have at the end of year 1, including interest?

- At the end of year 1, you invest an additional $2000. How much money will you have at the end of year 2, including interest?

Technology Tools

- Use a **spreadsheet** to calculate each year-end balance.
- Use **graphing software** to make a graph of your investment over 40 years.
- Use **presentation software** to present your project.

 Research For more information about investing, visit: www.algconcepts.com

Presenting the Project

Write a one-page paper describing your investment plan. Include:

- successful and unsuccessful strategies used to plan,
- a table or spreadsheet showing each year-end balance, and
- a graph of your savings over the 40-year period.

Check for Reasonableness

Project

Plan a one-week menu of foods that require little or no preparation.
1. Total Calories must be between 1200 and 1800 per day.
2. Calories from fat must be at most 30% of the total Calories. Note that 1 gram of fat has 9 Calories.
3. Total sodium must be less than 2400 milligrams.
4. Total protein must be at least 46 grams.

Working on the Project

Work with a partner to solve the problem.

- Write inequalities to represent the Calories, fat (g), sodium (mg), and protein (g) allowed per day.

- Determine whether the sample menu in the table below satisfies the nutritional guidelines.

Technology Tools

- Use a **spreadsheet** to help write a menu that meets all guidelines.

- Use **publishing software** to illustrate your menu.

Food (one serving)	Calories	Fat (g)	Sodium (mg)	Protein (g)
canned chicken noodle soup	90	2	940	6
granola cereal and skim milk	270	9	20	5
canned tuna in water (2 ounces)	60	0.5	250	13
frozen dinner	450	20	1530	10
white bread (3 slices)	240	3	480	6

 Research For more information about nutrition, visit: www.algconcepts.com

Presenting the Project

Compile a portfolio of your work for this project. Include the menu for each of seven days, the nutritional value of each food, and a list of inequalities to show that all guidelines were met.

Use a Formula

Working on the Project

Work with a partner and choose a strategy. Here are some suggestions to help you get started.

- Choose several values of r and solve the equation for t. Write the results as ordered pairs (r, t).

- Graph the ordered pairs on a coordinate plane.

- Research the average distance that four other animals could travel in one hour.

Technology Tools

- Use a **spreadsheet** or a **graphing calculator** to prepare your graphs.

- Use **presentation software** to prepare and give your presentation.

 Research For more information about animal speeds, visit: www.algconcepts.com

Presenting the Project

Prepare a presentation of your findings for your animals. Make sure that your presentation includes:

- a table of ordered pairs and a graph for each value of d,

- an explanation of your findings, including a comparison of the graphs, and

- a discussion of how these graphs differ from the graph of an equation such as $d = 20t$. Use the terms *direct variation* and *inverse variation* in your discussion.

A mathematics multilingual glossary is available at www.math.glencoe.com.

The glossary is available in the following languages.

Arabic	Haitian Creole	Russian	Urdu	Bengali	Hmong
Spanish	Vietnamese	Cantonese	Korean	Tagalog	

En www.math.glencoe.com encontrarás un glosario matemático plurilingüe.

Disponible en los siguientes idiomas:

Árabe	Criollo haitiano	Ruso	Urdu	Bengalí	Hmong
Español	Vietnamita	Cantonés	Coreano	Tagalo	

absolute value   The absolute value of a number *a* is its distance from zero on a number line and is represented by $|a|$. (p. 55)

The *absolute value* of -2 is 2, or $|-2| = 2$.

valor absoluto   El valor absoluto de un número *a* es la distancia que *a* dista de cero en la recta numérica, la que se denota por $|a|$. (pág. 55)

El *valor absoluto* de -2 es 2 ó $|-2| = 2$.

additive inverse   A number that when added to a given number results in a sum of zero. (p. 65)

The *additive inverse* of 4 is -4 because $4 + (-4) = 0$.

inverso aditivo   Número que sumado a un número dado da cero. (pág. 65)

El *inverso aditivo* de 4 es -4 porque $4 + (-4) = 0$.

algebraic expression   An expression consisting of one or more numbers and variables along with one or more arithmetic operations. (p. 4)

$4x$, ab, and $5x^2 - 7$ are *algebraic expressions*.

expresión algebraica   Expresión que consta de uno o más números y variables, junto con una o más operaciones matemáticas. (pág. 4)

$4x$, ab y $5x^2 - 7$ son todas *expresiones algebraicas*.

arithmetic sequence   A numerical pattern that increases or decreases at a constant rate or value. The difference between successive terms of the sequence is constant. (p. 110)

2, 5, 8, 11, 14, ... is an *arithmetic sequence*.

sucesión aritmética   Patrón numérico que aumenta o disminuye a una tasa o valor constante. La diferencia entre términos consecutivos de la sucesión es constante. (pág. 110)

2, 5, 8, 11, 14, ... es una *sucesión aritmética*.

base in a percent proportion   The number to which the percentage is compared. (p. 199)

$\text{percentage} \rightarrow \dfrac{1}{5} = \dfrac{20}{100} \leftarrow \text{percent}$
$\text{base} \rightarrow$

base de una proporción porcentual   Número al que se compara la parte. (pág. 199)

$\text{parte} \rightarrow \dfrac{1}{5} = \dfrac{20}{100} \leftarrow \text{porcentaje}$
$\text{base} \rightarrow$

base of an expression   In an expression of the form x^n, the base is *x*. (p. 336)

In 5^6, the base is 5.

base de una expresión   En una expresión de la forma x^n, la base es *x*. (pág. 336)

La base de 5^6 es 5.

best-fit line The line that most closely approximates the data in a scatter plot. (p. 308)

recta de óptimo ajuste Recta que aproxima más estrechamente los datos de una gráfica de dispersión. (pág. 308)

biased sample A sample in which one or more parts of the population are favored over others. (p. 32)

muestra sesgada Muestra en que se ha favorecido una o más partes de una población en lugar de otras. (pág. 32)

binomial An algebraic expression that has two terms. (p. 383)

binomio Expresión algebraica de dos términos. (pág. 383)

$x + 1, x^2 + y$, and $a - 2b$

$x + 1, x^2 + y$, y $a - 2b$

boundary A line or curve that separates the coordinate plane into regions. (p. 535)

frontera Recta o curva que divide el plano coordenado en regiones. (pág. 535)

box-and-whisker plot A diagram that divides a set of data into four parts using the median and quartiles. (p. 210)

diagrama de caja y patillas Diagrama que divide un conjunto de datos en cuatro partes usando la mediana y los cuartiles. (pág. 210)

circle graph A graph that compares parts of a set of data as a percent of the whole set. (p. 200)

gráfica circular Gráfica en que se comparan partes de un todo como porcentaje del todo. (pág. 200)

Employee Time Study
Estudio del tiempo de de empleados

coefficient The numerical factor of a term. (p. 20)

coeficiente El factor numérico de un término. (pág. 20)

The *coefficient* of $3x$ is 3.

El *coeficiente* de $3x$ es 3.

combination An arrangement or listing in which order is not important. (p. 152)

combinación Arreglo o lista en que el orden no es importante. (pág. 152)

Choosing toppings for a pizza is a *combination*.

Elegir los ingredientes para una pizza es una *combinación*.

common factor   A whole number that is a factor of each number in a set of numbers. (p. 422)

5 is a *common factor* of 10, 15, 25, and 100.

factor común   Un número que es un factor de cada número en un conjunto de números. (pág. 422)

5 es un *factor común* de 10, 15, 25 y 100.

common multiples   Multiples that are shared by two or more numbers. (p. 662)

Some *common multiples* of 2 and 3 are 6, 12, and 18.

múltiplos comunes   Múltiplos compartidos por dos o más números. (pág. 662)

Algunos *múltiplos comunes* de 2 y 3 son 6, 12 y 18.

complements   Two events are complements if the sum of their probabilities is 1. (p. 223)

When tossing a coin, heads and tails are *complements*.

complementarios   Dos eventos son complementarios si sus probabilidades suman 1. (pág. 223)

Al lanzar una moneda, cara y escudo son eventos *complementarios*.

composite number   A whole number greater than 1 that has more than two factors. (p. 420)

4 is a *composite number*. It has three factors: 1, 2, and 4.

número compuesto   Número entero mayor que 1 que posee más de dos factores. (pág. 420)

4 es un *número compuesto*. Tiene tres factores: 1, 2 y 4.

compound event   An event which consists of two or more simple events. (p. 224)

Drawing a card and then drawing a second card is a *compound event*.

evento compuesto   Evento que consta de dos o más eventos simples. (pág. 224)

Sacar una baraja y luego sacar una segunda baraja es un *evento compuesto*.

compound inequality   Two or more inequalities that are connected by the words *and* or *or*. (p. 524)

$x > 3$ and $x < 5$
or
$x < -3$ or $x > 7$

desigualdad compuesta   Dos o más desigualdades unidas por las palabras *y* u *o*. (pág. 524)

$x > 3$ y $x < 5$
ó
$x < -3$ ó $x > 7$

conjugate of a binomial   The conjugate of the binomial $a\sqrt{b} + c\sqrt{d}$ is $a\sqrt{b} - c\sqrt{d}$. (p. 616)

$6\sqrt{2} + 4\sqrt{3}$ and $6\sqrt{2} - 4\sqrt{3}$ are *conjugates*.

conjugado de un binomio   El conjugado de $a\sqrt{b} + c\sqrt{d}$ es $a\sqrt{b} - c\sqrt{d}$. (pág. 616)

$6\sqrt{2} + 4\sqrt{3}$ y $6\sqrt{2} - 4\sqrt{3}$ son *conjugados*.

consistent system   A system of equations that has at least one ordered pair that satisfies both equations. (p. 554)

sistema consistente   Sistema de ecuaciones para el que hay por lo menos un par ordenado que satisface ambas ecuaciones. (pág. 554)

constant of variation   The number k in an equation of the form $y = kx$. (p. 264)

In the equation of $y = 4x$, the *constant of variation* is 4.

constante de variación   El número k en una ecuación de la forma $y = kx$. (pág. 264)

En la equación, $y = 4x$, la *constante de variación* es 4.

coordinate   The number that corresponds to a point on a number line. (p. 53) **coordinate grid (see coordinate plane)**

coordenada   Número asociado con un punto de una recta numérica. (pág. 53) **cuadriculado coordenado (véase plano coordenado)**

The *coordinate* of point A is −4.
La *coordenada* de punto A es −4.

---

coordinate plane   A plane, also called a coordinate grid or coordinate system, in which a horizontal number line and a vertical number line intersect at their zero points. The two number lines are called *axes*. Their intersection point is called the *origin*. (p. 58)

coordinate system *See coordinate plane.*

plano coordenado   Plano, también llamado cuadriculado coordenado o sistema coordenado, en el que se han trazado una recta numérica horizontal y una vertical que se intersecan en sus puntos cero. Las rectas se llaman *ejes* y su punto de intersección se llama *origin*. (pág. 58)

sistema coordenado *Véase plano coordenado.*

---

counterexample   A specific case which proves a statement false. (p. 16)

contraejemplo   Ejemplo que prueba la falsedad de un enunciado. (pág. 16)

$1 \div 2 = \frac{1}{2}$ is a *counterexample* to "The quotient of two integers is always an integer."

$1 \div 2 = \frac{1}{2}$ es un *contraejemplo* a "El cociente de dos enteros es siempre un entero."

---

cross products   The product of the terms on the diagonals when two ratios are compared. (p. 95)

productos cruzados   Producto de los términos en diagonal de dos razones que se están comparando. (pág. 95)

In the proportion $\frac{2}{3} = \frac{8}{12}$, the *cross products* are 2×12 and 3×8.

En la proporción $\frac{2}{3} = \frac{8}{12}$, los *productos cruzados* son 2×12 y 3×8.

---

data   Numerical information gathered for statistical purposes. (p. 32)

datos   Información numérica que se reúne con fines estadísticos. (pág. 32)

---

deductive reasoning   The process of using facts, rules, definitions, or properties to reach a valid conclusion. (p. 30)

razonamiento deductivo   Proceso de usar hechos, reglas, definiciones o propiedades para sacar conclusiones válidas. (pág. 30)

---

degree of a monomial   The sum of the exponents of all its variables. (p. 384)

grado de un monomio   Suma de los exponentes de todas sus variables. (pág. 384)

The *degree* of the monomial $4x^3$ is 3.

El *grado* de un monomio $4x^3$ es 3.

degree of a polynomial   The greatest degree of any term in the polynomial. (p. 384)

grado de un polinomio   Grado mayor de los términos del polinomio. (pág. 384)

The *degree* of the polynomial $8x^4 + 2x^3 - 1$ is 4.

El *grado* de un polinomio $8x^4 + 2x^3 - 1$ es 4.

---

dependent system   A consistent system of equations that has an infinite number of solutions. (p. 554)

sistema dependiente   Sistema consistente de ecuaciones que posee un número infinito de soluciones. (pág. 554)

---

dependent variable   The variable in a relation whose value depends on the value of the independent variable. (p. 264)

variable dependiente   Variable de una relación cuyo valor depende del valor de la variable independiente. (pág. 264)

---

difference of squares   Two perfect squares separated by a subtraction sign. (p. 447)

diferencia de cuadrados   Cuadrados perfectos separados por el signo de sustracción. (pág. 447)

$a^2 - b^2 = (a + b)(a - b)$

---

dimensional analysis   The process of carrying units throughout a computation. (p. 190)

análisis dimensional   En un cálculo, proceso de tomar en cuenta las unidades. (pág. 190)

$\dfrac{55 \text{ miles}}{1 \text{ hour}} \cdot 3 \text{ hours} = 165 \text{ miles}$

$\dfrac{55 \text{ millas}}{1 \text{ hora}} \cdot 3 \text{ horas} = 165 \text{ millas}$

---

direct variation   An equation of the form $y = kx$, where $k = 0$. (p. 264)

variación directa   Ecuación de la forma $y = kx$, con $k = 0$. (pág. 264)

$y = 4x$
or
$y = \dfrac{1}{3}x$

$y = 4x$
o
$y = \dfrac{1}{3}x$

Distance Formula   The distance between two points, with coordinates (x_1, y_1) and (x_2, y_2), is given by $d = \sqrt{(x_2 - x_1)^2 + (y_2 - y_1)^2}$. (p. 607)

fórmula de la distancia   La distancia entre los puntos (x_1, y_1) y (x_2, y_2), viene dada por $d = \sqrt{(x_2 - x_1)^2 + (y_2 - y_1)^2}$. (pág. 607)

Distance between $(2, -1)$ and $(5, 3)$, is
$d = \sqrt{(5 - 2)^2 + (3 - (-1))^2}$
or 5.

La distancia entre los puntos $(2, -1)$ y $(5, 3)$, es
$d = \sqrt{(5 - 2)^2 + (3 - (-1))^2}$
ó 5.

---

domain   The set of the first numbers or abscissas of the ordered pairs in a relation. (p. 238)

dominio   Conjunto de los primeros números o abscisas de los pares ordenados de una relación. (pág. 238)

In the relation $\{(1, 2), (3, 5), (4, 0)\}$, the *domain* is $\{1, 3, 4\}$.

En la relación $\{(1, 2), (3, 5), (4, 0)\}$, el *dominio* es $\{1, 3, 4\}$.

empty set   A set with no elements shown by the symbol { } or ∅. A solution set with no members. Also called a null set. (p. 130)

conjunto vacío   Conjunto sin elementos, denotado por { } o ∅. Conjunto-solución sin elementos. (pág. 130)

---

equation   A mathematical sentence stating that two expressions are equal. (p. 5)

$3 \times (7 + 8) = 9 \times 5$

ecuación   Enunciado matemático que afirma la igualdad de dos expresiones. (pág. 5)

---

equivalent equations   Equations that have the same solution. (p. 122)

$2x + 3 = 9$ and $2x = 6$

ecuaciones equivalentes   Ecuaciones que tienen la misma solución. (pág. 122)

---

equivalent expressions   Expressions that have the same value or that have the same mathematical meaning for all replacement values of their variables. (p. 20)

$3 + 2 = 10 - 5$ and
$2x + 3x = 5x$

expresiones equivalentes   Expresiones que tienen el mismo valor o el mismo significado matemático para todos los valores de sustitución de sus variables. (pág. 20)

$3 + 2 = 10 - 5$ y
$2x + 3x = 5x$

---

evaluate   Find the value of an expression. (p. 10)

evaluar   Calcular el valor de una expresión. (pág. 10)

---

event   A specific outcome or type of outcome. (p. 147)

evento   Resultado específico o tipo de resultado. (pág. 147)

---

excluded values   Any values of a variable that result in a denominator of 0 must be excluded from the domain of that variable. (p. 638)

The *excluded value* of $y = \dfrac{1}{x - 1}$ is 1.

valores excluidos   Cualquier valor de una variable que dé un denominador nulo debe excluirse del dominio de dicha variable. (pág. 638)

El *valor excluido* de $y = \dfrac{1}{x - 1}$ es 1.

---

experimental probability   The ratio of the number of positive outcomes to the total number of events or trials in a probability experiment. (p. 220)

probabilidad experimental   En un experimento probabilístico, razón del número de resultados positivos al número total de eventos o pruebas. (pág. 220)

---

exponent   A number that indicates how many times a number or expression is to be multiplied by itself. (p. 336)

In the expression 5^3, the *exponent* is 3.

exponente   Número que indica el número de veces que un número o expresión se multiplica por sí misma. (pág. 336)

En la expresión 5^3, el *exponente* es 3.

---

exponential function   A function that can be described by an equation of the form $y = a^x$, where $a > 0$ and $a \neq 1$. (p. 489)

$y = 2^x$

función exponencial   Función de la forma $y = a^x$, con $a > 0$ y $a \neq 1$. (pág. 489)

expression A mathematical combination of numbers, variables, and operations. (p. 4)

$5 + 2^3, 4x, 6x + y^2$

expresión Combinación de números, variables y operaciones. (pág. 4)

F

---

factor In an algebraic or numerical expression, the quantities being multiplied are called factors. (p. 4)

3 and 11 are *factors* of 33.

factor En una expresión numérica o algebraica, se llaman factores las cantidades que se multiplican. (pág. 4)

3 y 11 son *factores* de 33.

---

factor a polynomial Express a polynomial as the product of monomials and polynomials. (p. 428)

$8y^2 + 10y$ in factored form is $2y(4y = 5)$.

factorizar un polinomio Escribir un polinomio como producto de monomios y polinomios. (pág. 428)

$8y^2 + 10y$ en forma factorizada es $2y(4y = 5)$.

---

factorial The expression of $n!$, read n factorial, where n is greater than zero, is the product of all positive integers beginning with n and counting backward to 1. (p. 153)

$5! = 5 \times 4 \times 3 \times 2 \times 1 = 120$

factorial La expresión $n!$, léase n factorial, con n un número natural, es el producto de todos los números naturales, empezando en n y contando hacia atrás hasta llegar a 1. (pág. 153)

---

family of graphs Graphs and equations of graphs that have at least one characteristic in common. (p. 316)

familia de gráficas Gráficas y sus ecuaciones que poseen por lo menos un rasgo común. (pág. 316)

---

formula An equation that states a rule for the relationship between certain quantities. (p. 24)

$A = \pi r^2$ is the *formula* for finding the area of a circle.

fórmula Ecuación que establece una relación entre ciertas cantidades. (pág. 24)

$A = \pi r^2$ es la *fórmula* para el área de un círculo.

---

frequency table A table of tally marks used to record and display how often events occur. (p. 33)

tabla de frecuencias Tabla en que se lleva la cuenta de la frecuencia de eventos. (pág. 33)

Age Edad	Tally Cuenta	Frequency Frecuencia
1-10	JHT III	8
11-20	JHT IIII	9
21-30	JHT	5
31-40	JHT JHT II	12
41-50	JHT III	8

---

function A relation in which exactly one element of the range is paired with each element of the domain. (p. 256)

$\{(5, -2), (3, 2), (4, -1), (-2, 2)\}$

función Relación en que a cada elemento del dominio le corresponde un único elemento del rango. (pág. 256)

---

function notation A way to name a function that is defined by an equation. (p. 258)

In *function notation*, the equation $y = 3x - 8$ is written as $f(x) = 3x - 8$.

notación funcional Forma de indicar una función definida por una ecuación. (pág. 258)

En *notación funcional*, la ecuación $y = 3x - 8$ se escribe $f(x) = 3x - 8$.

functional value   The element in the range that corresponds to a specific element in the domain. (p. 258)

If $f(x) = 3x - 8$, then $f(2) = 3(2) - 8$ or -2.

valor funcional   Elemento del rango que corresponde a un elemento específico del dominio. (pág. 258)

Si $f(x) = 3x - 8$, entonces $f(2) = 3(2) - 8$ ó -2.

 G

geometric sequence   A sequence in which each term after the nonzero first term is found by multiplying the previous term by a constant called the common ratio r, where $r \neq 0$ or 1. (p. 494)

1, 4, 16, 64, 256, …

sucesión geométrica   Sucesión en que cada término después del primero se obtiene del anterior multiplicándolo por una constante, la llamada razón común r, con $r \neq 0$ ó 1. (pág. 494)

graph   To draw, or plot, the points named by certain numbers or ordered pairs on a number line or coordinate plane. (p. 53)

graficar   Marcar en una recta numérica o plano coordenado los puntos dados por ciertos números o pares ordenados. (pág. 53)

greatest common factor (GCF)   The greatest number that is a factor of two or more numbers. (p. 422)

The *greatest common factor* of 30, 60, and 15 is 15.

máximo común divisor (MCD)   Número que es el factor común mayor de dos o más números. (pág. 422)

El *máximo común divisor* de 30, 60 y 15 es 15.

 H

half-plane   The two regions of the coordinate plane separated by the graph of a linear equation. (p. 535)

semiplano   Cualquiera de las dos regiones de un plano coordenado que divide la gráfica de una ecuación lineal. (pág. 535)

histogram   A bar graph in which the data are organized into equal intervals. The height of the bar represents the frequency in that interval. (p. 39)

histograma   Gráfica de barras en que los datos están organizados en intervalos iguales. La altura de cada barra viene dada por la frecuencia del intervalo correspondiente. (pág. 39)

hypotenuse   The side opposite the right angle in a right triangle. (p. 366)

hipotenusa   En un triángulo rectángulo, el lado opuesto al ángulo recto. (pág. 366)

identity   An equation that is true for every value of the variable. (p. 172)

identidad   Ecuación que se cumple para todo valor de la variable. (pág. 172)

$2x = 2x$

---

inclusive events   Two events that can occur at the same time and whose outcomes may be the same. (p. 227)

eventos inclusivos   Dos eventos que pueden ocurrir simultáneamente y cuyos resultados pueden ser los mismos. (pág. 227)

Rain on Saturday and rain on Sunday are *inclusive events.*

Lluvia el sábado y lluvia el domingo son *eventos inclusivos.*

---

inconsistent system   A system of equations with no ordered pair that satisfies both equations. (p. 555)

sistema inconsistente   Sistema de ecuaciones para el cual no existe un par ordenado que satisfaga ambas ecuaciones. (pág. 555)

---

independent events   Two or more events in which the outcome of one event does not affect the outcome of the other events. (p. 224)

eventos independientes   Dos o más eventos en los que el resultado de un evento no afecta el resultado del otro. (pág. 224)

Rolling a number cube and tossing a coin are *independent events.*

Lanzar un dado y lanzar una moneda son *eventos independientes.*

---

independent system   A system of equations with exactly one solution. (p. 554)

sistema independiente   Sistema de ecuaciones con una única solución. (pág. 554)

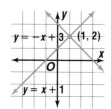

---

independent variable   The variable in a function whose value is subject to choice. (p. 264)

variable independiente   Variable de una función cuyo valor se puede elegir. (pág. 264)

---

inductive reasoning   A conclusion based on a pattern of examples. (p. 30)

razonamiento inductivo   Conclusión que se basa en un patrón de ejemplos. (pág. 30)

---

inequality   An open sentence that uses the symbol $<$, $\le$, $>$, or $\ge$ to compare two quantities. (p. 95)

desigualdad   Enunciado abierto que usa uno de los símbolos $<$, $\le$, $>$ o $\ge$ para comparar dos cantidades. (pág. 95)

$x < 9, y \ge -3$

---

integers   The set of whole numbers and their opposites $\{..., -2, -1, 0, 1, 2, ...\}$. (p. 52)

enteros   Conjunto de los números enteros junto con sus opuestos, $\{..., -2, -1, 0, 1, 2, ...\}$. (pág. 52)

intersection   The set of elements common to two or more sets as in compound inequalities and Venn diagrams. (p. 524)

intersección   Conjunto de elementos comunes a dos o más conjuntos, como en el caso de desigualdades compuestas o diagramas de Venn. (pág. 524)

---

inverse variation   An equation of the form $xy = k$, where $k \neq 0$. (p. 270)

$xy = 9$

variación inversa   Ecuación de la forma $xy = k$, con $k \neq 0$. (pág. 270)

---

irrational numbers   Numbers that cannot be expressed as fractions, terminating decimals, or repeating decimals. (p. 362)

$\pi, \sqrt{2}, 0.10110111\ldots$

números irracionales   Números que no pueden escribirse como fracciones, decimales exactos o decimales periódicos. (pág. 362)

L

least common denominator (LCD)   The least common multiple of the denominators of two or more fractions. (p. 663)

2 is the *least common denominator* of $\frac{1}{3}$, $\frac{3}{4}$, and $\frac{5}{6}$.

mínimo común denominador (mcd)   El múltiplo común menor de los denominadores de dos o más fracciones. (pág. 663)

2 es el *mínimo común denominador* de $\frac{1}{3}$, $\frac{3}{4}$ y $\frac{5}{6}$.

---

least common multiple (LCM)   The least nonzero number that is a common multiple of two or more numbers. (p. 662)

36 is the *least common multiple* of 3, 9, and 12.

mínimo común múltiplo (MCM)   Número no nulo que es el múltiplo común menor de dos o más números. (pág. 662)

36 es el *mínimo común múltiplo* de 3, 9 y 12.

---

legs   The sides of a right triangle that form the right angle. (p. 366)

catetos   En un triángulo rectángulo, los lados que forman el ángulo. (pág. 366)

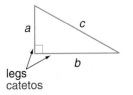

legs
catetos

---

like terms   Terms that contain the same variables raised to the same power. (p. 20)

$5x^2$ and $6x^2$ are *like terms*.

términos semejantes   Términos que tienen las mismas variables elevadas a los mismos exponentes. (pág. 20)

$5x^2$ y $6x^2$ son *términos semejantes*.

---

line graph   Numerical data displayed to show trends or changes over time. (p. 38)

gráfica lineal   Datos numéricos exhibidos de modo que muestren tendencias o cambios en el tiempo. (pág. 38)

Sales of Phone Cards
Venta de tarjetas de llamadas

Year (estimated)
Año (estimado)

linear equation An equation that can be written in the form $Ax + By = C$, where $A \neq 0$ and $B \neq 0$, whose graph is a straight line. (p. 250)

ecuación lineal Ecuación de la forma $Ax + By = C$, con $A \neq 0$ y $B \neq 0$. Su gráfica es una recta. (pág. 250)

mean The sum of the numbers in a set of data divided by the number of pieces of data in the set. (p. 104)

For $\{1, 2, 2, 6, 9\}$

$$mean = \frac{1 + 2 + 2 + 6 + 9}{5} \text{ or } 4$$

media Suma de los números de un conjunto de datos dividida entre el número de datos. (pág. 104)

Para $\{1, 2, 2, 6, 9\}$

$$media = \frac{1 + 2 + 2 + 6 + 9}{5} \text{ ó } 4$$

measure of central tendency A number or piece of data that represents the whole set of data. (p. 104)

medida de tendencia central Número o dato que representa todo un conjunto de datos. (pág. 104)

measure of variation A measure that describes the spread of the values in a set of data. (p. 106)

medida de variación Medida que describe la dispersión de los números de un conjunto de datos. (pág. 106)

median The middle number in a set of data when the data are arranged in numerical order. If the data has an even number, the median is the mean of the two middle numbers. (p. 104)

For $\{1, 2, 2, 6, 9\}$
↑
The *median* is 2.

mediana Número central de un conjunto de datos, cuando éstos se han ordenado numéricamente. Si el número de datos es par, la mediana es la media de los dos valores centrales. (pág. 104)

Para $\{1, 2, 2, 6, 9\}$
↑
La *mediana* es 2.

mode The number(s) or item(s) that appear most often in a set of data. (p. 104)

For $\{1, 2, 2, 6, 9\}$
↑
The *mode* is 2.

moda Dato(s) más frecuente(s) de un conjunto de datos. (pág. 104)

Para $\{1, 2, 2, 6, 9\}$
↑
la *moda* es 2.

monomial A number, a variable, or a product of a number and one or more variables. (p. 382)

$3, y, 2x, 5xy^2$

monomio Número, variable o producto de un número por una o más variables. (pág. 382)

multiplicative inverse The number that when multiplied by a given number results in a product of one. (p. 154)

The *multiplicative inverse* of 4 is $\frac{1}{4}$ because $4 \times \frac{1}{4} = 1$.

inverso multiplicativo Número que multiplicado por un número dado da uno. (pág. 154)

El *inverso multiplicativo* de 4 es $\frac{1}{4}$ porque $4 \times \frac{1}{4} = 1$.

multi-step equations   Equations with more than one operation. (p. 165) $3x - 4 = 11, \frac{x}{3} + 5 = 14$

ecuaciones de varios pasos   Ecuaciones con más de una operación. (pág. 165)

mutually exclusive events   Two or more events whose outcomes can never be the same. (p. 226)   Rolling a 2 and rolling a 4 on a dice are *mutually exclusive events.*

eventos mutuamente excluyentes   Dos o más eventos cuyos resultados no son jamás los mismos. (pág. 226)   Sacar un 2 y sacar un 4 en un dado son *eventos mutuamente excluyentes.*

natural numbers   The set {1, 2, 3, ... }. (p. 52)

números naturales   El conjunto {1, 2, 3, ... }. (pág. 52)

negative number   Any number that is less than zero. (p. 52) $-4, -\frac{1}{2}$

número negativo   Cualquier número menor que cero. (pág. 52)

number line   A line with equal distances marked off to represent numbers. (p. 52)

recta numérica   Recta en que se han marcado distancias a espacios iguales para representar números. (pág. 52)

numerical expression   A mathematical expression that has a combination of numbers and at least one operation. (p. 4) $4 + 2 \cdot 3$ is a *numerical expression.*

expresión numérica   Combinación de números y por lo menos menos una operación. (pág. 4) $4 + 2 \cdot 3$ es un *expresión numérica.*

odds   The ratio that compares the number of ways an event can occur (successes) to the number of ways the event cannot occur (failures). (p. 221)

posibilidades   Razón del número de maneras en que un evento puede ocurrir (éxitos) al número de maneras en que el evento no puede ocurrir (fracasos). (pág. 221)

open sentence   A mathematical statement with one or more variables. (p. 112) $n - 7 = 14$
$15 = m + 2$

enunciado abierto   Enunciado matemático con una o más variables. (pág. 112)

opposites   Every positive rational number and its negative pair. (p. 65) -1 and $1, \frac{1}{4}$ and $-\frac{1}{4}$

opuestos   Cada número racional positivo y su par negativo. (pág. 65) -1 y $1, \frac{1}{4}$ y $-\frac{1}{4}$

ordered pair   A pair of numbers used to locate a point in the coordinate plane or the solution of an equation in two variables. An ordered pair is written in the form (*x*-coordinate, *y*-coordinate). (p. 58)

par ordenado   Par de números que se usa para ubicar un punto en el plano coordenado o la solución de una ecuación de dos variables. Un par ordenado se escribe de la forma (coordenada *x*, coordenada *y*). (pág. 58)

origin   *See coordinate plane.*

origen   *Ver plano coordenado.*

outcome   One possible result of a probability event. (p. 146)

resultado   Resultado posible de un evento probabilístico. (pág. 146)

4 is an *outcome* when a number cube is rolled.

4 es un *resultado* cuando se lanza un cubo numérico.

parabola   The graph of a quadratic function. (p. 458)

parábola   Gráfica de una ecuación cuadrática. (pág. 458)

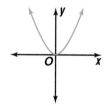

parallel lines   Lines in the same plane that never intersect and have the same slope. (p. 322)

rectas paralelas   Rectas en el plano que no se intersecan y que tienen la misma pendiente. (pág. 322)

parent graph   The simplest of the graphs or the anchor graph in a family of graphs. (p. 318)

gráfica madre   La más simple de las gráficas o gráfica áncora de una familia de gráficas. (pág. 318)

percent   A ratio that compares a number to 100. (p. 198)

porcentaje   Razón que compara un número con 100. (pág. 198)

76 out of 100 is 76 *percent* or 76%.

76 de 100 es 76 *por ciento* ó 76%.

percent of change   A ratio that compares the change in a quantity to the original amount. (p. 212)

porcentaje de cambio   Razón que compara el cambio de una cantidad con la cantidad original. (pág. 212)

perfect square   A number whose square root is a rational number. (p. 336)

25 is a *perfect square* since $25 = 5^2$.

cuadrado perfecto   Número cuya raíz cuadrada es un número racional. (pág. 336)

25 es un *cuadrado perfecto* porque $25 = 5^2$.

perfect square trinomial   Trinomials that are the square of a binomial. (p. 445)

$x^2 + 2x + 1$ is a *perfect square trinomial* since $x^2 + 2x + 1 = (x + 1)^2$.

trinomio cuadrado perfecto   Trinomio que es el cuadrado de un binomio. (pág. 445)

$x^2 + 2x + 1$ es un *trinomio cuadrado perfecto* porque $x^2 + 2x + 1 = (x + 1)^2$.

permutation   An arrangement or listing of a group of objects in which order is important. (p. 152)

Choosing a chairperson and a co-chair is a *permutation*.

permutación   Arreglo o lista en la que el orden es importante. (pág. 152)

Elegir a un director y a un subdirector es una *permutación*.

perpendicular lines   Lines which meet to form right angles and whose slopes have a product of -1. (p. 324)

rectas perpendiculares   Rectas que al intersecarse forman un ángulo recto y si sus pendientes existen, su producto es -1. (pág. 324)

point-slope form   An equation of the form $y - y_1 = m(x - x_1)$, where m is the slope and (x_1, y_1) is a given point on a nonvertical line. (p. 290)

In $y - 4 = 2(x - 3)$, the slope is 2 and (3, 4) is a point on the line.

forma punto-pendiente   Ecuación de la forma $y - y_1 = m(x - x_1)$, donde m es la pendiente y (x_1, y_1) un punto dado de una recta no vertical. (pág. 290)

En la ecuación $y - 4 = 2(x - 3)$, el pendiente es 2 y (3, 4) es un punto en la recta.

polynomial   A monomial or sum of monomials. (p. 383)

$5x$, $3x^2 + 2$

polinomio   Un monomio o una suma de monomios. (pág. 383)

power   An expression of the form x^n, read *x to the nth power*. (p. 336)

7^4 is 7 raised to the fourth *power*, or $7 \times 7 \times 7 \times 7$.

potencia   Expresión de la forma x^n, que se lee *x a la potencia enésima*. (pág. 336)

7^4 es 7 elevado a la cuarta *potencia*, ó $7 \times 7 \times 7 \times 7$.

prime factorization   Expressing a composite number as a product of its prime factors. (pp. 358, 421)

The *prime factorization* of 63 is $3 \times 3 \times 7$.

factorización prima   Número compuesto escrito como producto de sus factores primos. (ppág. 358, 421)

El *factorización prima* de 63 es $3 \times 3 \times 7$.

prime number   A whole number, greater than 1, whose only factors are 1 and itself. (p. 420)

2 is a prime number. It has two factors: 1 and 2.

número primo   Número entero mayor que 1 cuyos únicos factores son 1 y sí mismo. (pág. 420)

2 es un número primo. Tiene dos factores: 1 y 2.

prime polynomial   A polynomial that cannot be written as a product of two polynomials with integral coefficients. (p. 430)

$2x + 5y + 4$

polinomio primo   Polinomio que no puede escribirse como producto de dos polinomios con coeficientes enteros. (pág. 430)

---

probability   The ratio of the number of favorable outcomes for an event to the number of possible outcomes of the event. (p. 219)

$$P(a) = \frac{\text{number of favorable outcomes}}{\text{total number of possible outcomes}}$$

probabilidad   Razón del número de resultados favorables de un evento al número total de resultados posibles. (pág. 219)

$$P(a) = \frac{\text{número de resultados favorables}}{\text{número total de resultados posibles}}$$

---

product   The result obtained by multiplying two or more numbers or variables. (p. 4)

33 is the *product* of 3 and 11.

producto   El resultado de la multiplicación de dos o más números o variables. (pág. 4)

33 es el *producto* de 3 y 11.

---

proportion   An equation of the form $\frac{a}{b} = \frac{c}{d}$ stating that two ratios are equivalent. (p. 188)

$\frac{3}{12} = \frac{1}{4}$ or $\frac{x}{15} = \frac{2}{5}$

proporción   Ecuación de la forma $\frac{a}{b} = \frac{c}{d}$ en que se establece la equivalencia de dos razones. (pág. 188)

Pythagorean Theorem   If a and b are the measures of the legs of a right triangle and c is the measure of the hypotenuse, then $c^2 = a^2 + b^2$. (p. 366)

teorema de Pitágoras   Si a y b son las longitudes de los catetos de un triángulo rectángulo y c es la longitud de su hipotenusa, entonces $c^2 = a^2 + b^2$. (pág. 366)

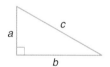

Q

quadrant   One of four regions into which the x- and y-axes separate the coordinate plane. (p. 60)

cuadrante   Cada una de las cuatro regiones en que los ejes x y y dividen el plano coordenado. (pág. 60)

Quadrant II / Cuadrante II $(-, +)$	Quadrant I / Cuadrante I $(+, +)$
Quadrant III / Cuadrante III $(-, -)$	Quadrant IV / Cuadrante IV $(+, -)$

---

quadratic equation   An equation of the form $ax^2 + bx + c = 0$, where $a \neq 0$. (p. 468)

$2x^2 + 5x + 18 = 0$

ecuación cuadrática   Ecuación de la forma $ax^2 + bx + c = 0$, donde $a \neq 0$. (pág. 468)

Quadratic Formula   The solutions of a quadratic equation in the form $ax^2 + bx + c = 0$, where $a \neq 0$, are given by the formula

$x = \dfrac{-b \pm \sqrt{b^2 - 4ac}}{2a}$. (p. 484)

fórmula cuadrática   Las soluciones de la ecuación cuadrática $ax^2 + bx + c = 0$, donde $a \neq 0$, vienen dadas por la fórmula

$x = \dfrac{-b \pm \sqrt{b^2 - 4ac}}{2a}$. (pág. 484)

quadratic function   A function described by an equation of the form $f(x) = ax^2 + bx + c$, where $a \neq 0$. (p. 458)

función cuadrática   Función de la forma $f(x) = ax^2 + bx + c$, donde $a \neq 0$. (pág. 458)

quadratic-linear system of equations   A system of equations involving a linear and a quadratic function. (p. 580)

sistema cuadrático-lineal de ecuaciones   Sistema con una ecuación lineal y una cuadrática. (pág. 580)

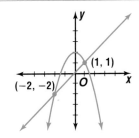

R

radical equation   An equation that contains a radical expression with the variable in the radicand. (p. 624)

$y = 2\sqrt{x} + 1$

ecuación radical   Ecuación que contiene una expresión radical con la variable en el radicando. (pág. 624)

radical expression   An expression that contains a square root. (p. 358)

$\sqrt{3}, \sqrt[4]{14}, 2\sqrt{x} + 1$

expresión radical   Expresión que contiene una raíz cuadrada. (pág. 358)

radical sign   The symbol $\sqrt{}$, used to indicate a root. (p. 357)

signo radical   El símbolo $\sqrt{}$, que indica una raíz. (pág. 357)

random   Outcomes occur at random if each outcome is equally likely to occur. (p. 220)

Tossing a coin and getting heads or tails is *random*.

aleatoriamente   Los resultados ocurren aleatoriamente si cada uno tiene la misma probabilidad de ocurrir. (pág. 220)

Lanzar una moneda y sacar cara o escudo es un evento *aleatorio*.

range of a data set   The difference between the greatest number and the least number in a set of data. (p. 106)

$\{1, 2, 2, 6, 9\}$
range = 9 − 1 or 8

rango de un conjunto de datos   Diferencia entre el dato mayor y el menor. (pág. 106)

rango = 9 − 1 ó 8

range of a relation   The set of second numbers in the ordered pairs of a relation. (p. 238)

In the relation $\{(1, 2), (3, 5), (4, 0)\}$, the *range* is $\{0, 2, 5\}$.

rango de una relación   Conjunto de los segundos números de los pares ordenados de la relación. (pág. 238)

En la relación $\{(1, 2), (3, 5), (4, 0)\}$, el *rango* es $\{0, 2, 5\}$.

rate   The ratio of two measurements having different units of measure. (p. 190)

60 miles per hour

tasa   Razón de dos medidas de unidades distintas. (pág. 190)

60 millas por hora

rate of change   The change in a quantity over time. (p. 285)

tasa de cambio   El cambio de una cantidad en el tiempo. (pág. 285)

ratio A comparison of two numbers by division. (p. 188)

razón Comparación de dos números mediante división. (pág. 188)

2 to 3; 1:2, $\frac{3}{5}$

---

rational equation An equation that contains rational expressions. (p. 668)

ecuación racional Ecuación que contiene expresiones racionales. (pág. 668)

$\frac{2x}{x+3} = 0$

---

rational expression An algebraic fraction whose numerator and denominator are polynomials. (p. 638)

expresión racional Fracción cuyo numerador y denominador son polinomios. (pág. 638)

$\frac{2x}{x+3}$

---

rational function A function that contains rational expressions represented as the quotient of two polynomials in the form $f(x) = \frac{g(x)}{h(x)}$, where $h(x) \neq 0$. (p. 638)

función racional Función que es el cociente de dos polinomios, de la forma $f(x) = \frac{g(x)}{h(x)}$, donde $h(x) \neq 0$. (pág. 638)

$f(x) = \frac{2x}{x+3}$

---

rational numbers The set of numbers that can be written in the form $\frac{a}{b}$, where a and b are integers and $b \neq 0$. (p. 94)

números racionales Conjunto de los números de la forma $\frac{a}{b}$, donde a y b son enteros y $b \neq 0$. (pág. 94)

$1 = \frac{1}{1}, \frac{2}{9}, -2.3 = -2\frac{3}{10}$

---

real numbers The set of rational numbers and irrational numbers together. (p. 600)

números reales Conjunto de los números racionales junto con el de los irracionales. (pág. 600)

$-1, 0, \sqrt{2}$

---

reciprocal The multiplicative inverse of a number. (p. 154)

recíproco El inverso multiplicativo de un número. (pág. 154)

$\frac{2}{3}$ and $\frac{3}{2}$

---

relation A set of ordered pairs. (p. 238)

relación Conjunto de pares ordenados. (pág. 238)

$\{(9, 3), (-1, 4), (4, 4), (-2, 0)\}$

---

rise The vertical change between any two points on a line on the coordinate plane. (p. 284)

elevación Cambio vertical entre dos puntos cualesquiera de una recta en el plano coordenado. (pág. 284)

---

root The solution of a quadratic equation. (p. 468)

raíz Solución de una ecuación cuadrática. (pág. 468)

---

run The horizontal change between any two points on a line on the coordinate plane. (p. 284)

carrera Cambio horizontal entre dos puntos cualesquiera de una recta en el plano coordenado. (pág. 284)

sample   Some portion of a larger group selected to represent that group. (p. 32)

muestra   Parte de un grupo más grande seleccionada para representarlo. (pág. 32)

sample space   The list of all possible outcomes for an event. (p. 146)

The *sample space* for rolling a number cube is (1, 2, 3, 4, 5, 6}.

espacio muestral   Lista de todos los resultados posibles de un evento. (pág. 146)

El *espacio muestral* de lanzar un dado es s (1, 2, 3, 4, 5, 6}.

scale   A ratio or rate between the actual size of an object and the size of its model. (p. 194)

escala   Razón o tasa entre el tamaño natural de un objeto y el de su modelo. (pág. 194)

scale drawing   A drawing that is similar to an actual object, but is either enlarged or reduced. (p. 194)

dibujo a escala   Dibujo semejante a un objeto real, pero más grande o más pequeño. (pág. 194)

scale model   A replica of an original object that is too large or too small to be built at actual size. (p. 194)

maqueta   Reproducción de un objeto real demasiado grande o demasiado pequeño como para ser construido de tamaño natural. (pág. 194)

scatter plot   A graph that shows the general relationship between two sets of data. (p. 302)

gráfica de dispersión   Gráfica que muestra la relación general entre dos conjuntos de datos. (pág. 302)

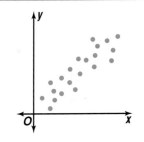

scientific notation   A system of writing numbers using multiplication and powers of ten. A number of the form $a \times 10^n$, where $1 < a < 10$ and n is an integer. (p. 353)

$95,700 = 9.57 \times 10^4$
$0.000024 = 2.4 \times 10^{-5}$

notación científica   Forma de escribir números usando multiplicación y potencias de diez. Un número de la forma $a \times 10^n$, donde $1 < a < 10$ y n un entero. (pág. 353)

set-builder notation   A concise way of writing a solution set. (p. 510)

$\{t|t < 17\}$ represents the set of all numbers t such that t is less than 17.

notación de conjuntos   Forma concisa de escribir un conjunto-solución. (pág. 510)

$\{t|t < 17\}$ representa el conjunto de todos los números t tal que t es menos que 17.

simplest form of an expression An expression is in simplest form when it is replaced by an equivalent expression having no like terms or parentheses. (p. 20)

forma reducida de una expresión Una expresión está reducida si no tiene términos semejantes o paréntesis. (pág. 20)

---

simplify Write an expression in simplest form. (p. 15)

reducir Simplificar una expresión. (pág. 15)

---

slope The ratio of the change in the y-coordinates (rise) to the corresponding change in the x-coordinates (run) as you move from one point to another along a line. (p. 284)

pendiente Razón del cambio en las coordenadas y (elevación) al cambio correspondiente en las coordenadas x (carrera) a medida que uno se desplaza de izquierda a derecha. (pág. 284)

---

slope formula The slope m of a line passing through two points is the ratio of the difference in the y-coordinates to the corresponding difference in the x-coordinates. (p. 284)

$$m = \frac{y_2 - y_1}{x_2 - x_1}$$

fórmula de la pendiente La pendiente m de una recta que pasa por dos puntos es la razón de la diferencia en las coordenadas y a la diferencia correspondiente en las coordenadas x. (pág. 284)

---

slope-intercept form An equation of the form $y = mx + b$, where m is the slope and b is the y-intercept. (p. 296)

forma pendiente-intersección Ecuación de la forma $y = mx + b$, donde m es la pendiente y b es la intersección y. (pág. 296)

---

solution A replacement value for the variable in an open sentence. A value for the variable that makes an equation true. (p. 112)

The *solution* of $12 = x + 7$ is 5.

solución Valor de sustitución de la variable de un enunciado abierto. Valor de la variable que hace verdadera una ecuación. (pág. 112)

La *solución* de $12 = x + 7$ es 5.

---

solution of a system of equations The ordered pair representing the solution common to both equations in a system of equations. (p. 550)

solución de un sistema de ecuaciones Pares ordenados cuyas coordenadas satisfacen ambas ecuaciones del sistema. (pág. 550)

---

solve an equation The process of finding all values of the variable that make an equation a true statement. (p. 112)

resolver una ecuación Proceso de hallar todos los valores de la variable que hacen verdadera una ecuación. (pág. 112)

---

square root One of the two equal factors of a number. If $a^2 = b$, then a is the square root of b. (p. 357)

12 is a *square root* of 144 since $12^2 = 144$.

raíz cuadrada Uno de dos factores iguales de un número. Si $a^2 = b$, a es una raíz cuadrada de b. (pág. 357)

12 es una *raíz cuadrada* de 144 porque $12^2 = 144$.

standard form of a linear equation   A linear equation of the form $Ax + By = C$, where $A \geq 0$, and A and B are real numbers and not both zero. (p. 250)

forma estándar de una ecuación lineal   Ecuación lineal de la forma $Ax + By = C$, donde $A \geq 0$ y A y B son números reales que no son ambos nulos. (pág. 250)

statement   A sentence that is either true or false, but not both. (p. 112)

enunciado   Frase que es verdadera o falsa, pero no ambas. (pág. 112)

stem-and-leaf plot   A display of data in which each piece of the data is separated into two numbers that are used to form a stem and a leaf. (p. 40)

diagrama de tallo y hojas   Presentación de datos en que cada uno está dividido en dos números que se usan para formar un tallo y una hoja. (pág. 40)

Stem Tallo	Leaf Hojas
0	7
1	4 7
2	4 7 7 7 9
3	5 7 8
4	
5	9

$1|4 = 14$

system of equations   A set of two or more equations with the same variables. (p. 550)

sistema de ecuaciones   Conjunto de dos o más ecuaciones con las mismas variables. (pág. 550)

$2x - 3 = y$
$3x + 5y = 7$

system of inequalities   A set of two or more inequalities with the same variables. (p. 586)

sistema de desigualdades   Conjunto de dos o más desigualdades con las mismas variables. (pág. 586)

$y > 2x + 6$
$4x - 3y < 12$

term in an expression   A number, a variable, or a product or quotient of numbers and variables. (p. 20)

término en una expresión   Número, variable, producto o cociente de números y variables. (pág. 20)

theoretical probability   The ratio of the number of favorable outcomes to the total number of possible outcomes. (p. 220)

probabilidad teórica   Razón del número de resultados favorables al número total de resultados posibles. (pág. 220)

The *theoretical probability* of rolling a 3 on a number cube is $\frac{1}{6}$.

La *probabilidad teórica* de sacar un 3 en un cubo numerado es $\frac{1}{6}$.

tree diagram   A diagram used to show the total number of possible outcomes of an event. (p. 146)

diagrama de árbol   Diagrama que se usa para mostrar todos los resultados posibles de un evento. (pág. 146)

bread   filling
white turkey / cheese
wheat turkey / cheese

pan   relleno
blanco pavo / queso
intebral pavo / queso

trinomial   An algebraic expression that has three terms. (p. 383)

trinomio   Expresión algebraica de tres términos. (pág. 383)

$x^2 + 2x + 1$
$a + b - c$

union   The graph of a compound inequality containing the word *or*; the solution is a solution of either inequality, not necessarily both. (p. 525)

unión   La gráfica de una desigualdad compuesta que contiene la palabra *o*; la solución es una solución de cualquiera de las desigualdades, no necesariamente de ambas. (pág. 525)

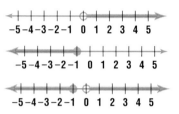

unit rate   A rate with a denominator of 1. (p. 190)

tasa unitaria   Tasa de denominador 1. (pág. 190)

55 miles per hour
99¢ per pound

55 millas por hora
99¢ por libra

variable   A letter or other symbol used to represent an unspecified number or value. (p. 4)

variable   Letra u otro símbolo que se usa para representar un número o valor general. (pág. 4)

Venn diagram   A diagram that uses circles to show relationships among sets of numbers or objects. (p. 52)

diagrama de Venn   Diagrama que usa círculos para mostrar relaciones entre conjuntos de números u objetos. (pág. 52)

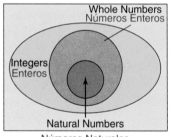

vertex of a parabola   The maximum or minimum point of a parabola. (p. 459)

vértice de una parábola   Su punto máximo o mínimo. (pág. 459)

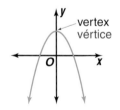

vertical line test   A test used to determine if a relation is a function. (p. 257)

prueba de la recta vertical   Prueba que se usa para determinar si una relación es una función. (pág. 257)

whole numbers   The set {0, 1, 2, 3, …}. (p. 16)

números enteros   El conjunto {0, 1, 2, 3, …}. (pág. 16)

x-axis   The horizontal number line on a coordinate plane. (p. 58)

eje x   La recta numérica horizontal de un plano coordenado. (pág. 58)

x-coordinate   The first number in an ordered pair. (pp. 59, 238)

coordenada x   Primer número de un par ordenado. (págs. 59, 238)

x-intercept   The x-coordinate at which a graph intersects the x-axis. (p. 296)

intersección x   La coordenada x donde una gráfica interseca el eje x. (pág. 296)

y-axis   The vertical number line on a coordinate plane. (p. 58)

eje y   La recta numérica vertical de un plano coordenado. (pág. 58)

y-coordinate   The second number in an ordered pair. (pp. 59, 253)

coordenada y   Segundo número de un par ordenado. (págs. 59, 253)

y-intercept   The y-coordinate at which a graph intersects the y-axis. (p. 296)

intersección y   La coordenada y donde una gráfica interseca el eje y. (pág. 296)

zero pair   The result of a positive counter paired with a negative counter. (p. 65)

par nulo   Ficha positiva junto con su ficha negativa. (pág. 65)

zeros   The roots, or x-intercepts, of a function. (p. 468)

ceros   Las raíces, o intersecciones x, de una función. (pág. 468)

Selected Answers

Chapter 1   The Language of Algebra

Page 3   Check Your Readiness
1. 18   **3.** 0   **5.** 64   **7.** 6   **9.** 120   **11.** 7   **13.** 6   **15.** 6
17. 0.99   **19.** 0.15   **21.** 0.55   **23.** 1   **25.** 17.5, 18,
18.5, 19   **27.** 14, 18, 20, 21, 26, 29, 30, 32, 37

Pages 6–7   Lesson 1–1
1. Sample answer: $3 + 7$, $10(6)$, $4 - 1$; $2x$, $5 + a$, gh
5. Sample answer: $7m$ **7.** Sample answer: 7 divided
by q **9.** $6 + m = 17$ **11.** Sample answer: 5 increased
by r is 15.   **13.** $9w + 3$ **15.** rs **17.** $5a + 3$
19. $1 - n$ **21.** $10 + h \cdot 1$ or $10 + h$ **23.** Sample
answer: the product of 9 and x **25.** Sample answer:
the difference of 6 and y **27.** Sample answer: 8 less
than the quotient of 3 and r **29.** $3 + w = 15$
31. $2 = \frac{7}{x}$ **33.** $3 - 5y = 2z$ **35.** Sample answer: 3
times r is 18.   **37.** Sample answer: h equals 10
minus i.   **39.** Sample answer: The quotient of t and
4 is 16.   **41.** Sample answer: Let $x =$ the variable;
$4x - 7 = 15 + c + 2x$ **43a.** $20 + 10(15 - 1)$
43b. $20 + 10(m - 1)$

Pages 11–13   Lesson 1–2
1. parentheses, brackets, fraction bar   **5.** Multiply 6
and 2.   **7.** Subtract 4 from 10.   **9.** 3   **11.** Transitive
13. $8(4 - 8 \div 2)$
    $= 8(4 - 4)$ *Substitution*
    $= 8(0)$ *Substitution*
    $= 0$ *Multiplicative Property of 0*
15. 8   **17.** 13   **19.** 14   **21.** 26   **23.** 40   **25.** 21   **27.** 8
29. Reflexive   **31.** Substitution   **33.** Transitive
35. $8(9 - 3 \cdot 2) = 8(9 - 6)$ *Substitution*
          $= 8(3)$ *Substitution*
          $= 24$ *Substitution*
37. $10(6 - 5) - (20 \div 2)$
    $= 10(1) - 10$ *Substitution*
    $= 10 - 10$ *Multiplicative Identity*
    $= 0$ *Substitution*
39. $6(12 - 48 \div 4) + 7 \cdot 1$
    $= 6(12 - 12) + 7$ *Substitution*
    $= 6(0) + 7$ *Substitution*
    $= 0 + 7$ *Multiplicative Property of 0*
    $= 7$ *Additive Identity*
41. 50   **43.** 23   **45.** 40   **47.** 5   **49.** 26   **51.** $311,392
53. 44 feet   **55.** Sample answer: Eight more than x
is 12.   **57.** Sample answer: The quotient of 25
and n is 5.   **59.** $3c = 27$ **61.** $b + 10 - 1 = 18$
63a. $4(20) + 7$ **63b.** 87

Page 13   Quiz 1
1. $2v - 5$ **3.** 13   **5.** 56 ft

Pages 17–18   Lesson 1–3
1. For any whole numbers that are multiplied,
the product is a whole number.   **3.** Jessie; the
Associative Property can be applied to addition
or to multiplication, but not to both at the same
time.   **5.** Associative $(+)$ **7.** $(3 + 47 + p)(7 - 6)$;
$50 + p$; Commutative $(+)$ **9a.** 188   **9b.** Use the
Commutative Property of Addition to change the
order of the numbers to $69 + 31 + 80 + 8$. Then
use the Substitution Property, adding 69 and 31
first.   **11.** Commutative $(+)$ **13.** Commutative
$(\times)$ **15.** Associative $(\times)$ **17.** $r \cdot (30 \cdot 5)$; $150r$;
Associative $(\times)$ **19.** $(6 + 3) + y$; $9 + y$;
Associative $(+)$ **21.** $2 \cdot 7 \cdot j$; $14j$; Commutative $(\times)$
23. false; sample counterexample:
    $(4 \div 2) \div 2 \overset{?}{=} 4 \div (2 \div 2)$
        $2 \div 2 \overset{?}{=} 4 \div 1$
           $1 \neq 4$
25. 100 board feet
27. Sample answer:
    $(8 - 5) - 3 \overset{?}{=} 8 - (5 - 3)$
        $3 - 3 \overset{?}{=} 8 - 2$
           $0 \neq 6$
29. 6   **31.** 7   **33.** A

Pages 21–23   Lesson 1–4
1. Sample answer: $1 + 2x + 3x + ab + 5ab$ **3.** b, e;
Write all the expressions in simplest form and find
two that are the same.   **5.** $8p, 9p$ **7.** $6bc, bc$ **9.** $y + 2$ **11.** $12 - 18m$ **13.** $7a + 7t$ **15.** $21f$ **17.** $3r$
19. $2g + 6$ **21.** $6x + 5y$ **23.** $15am - 12$ **25.** $7a + 5b$ **27.** $4y$ **29.** $46xy$ **31.** $3r + 4s$
33. $5(2n + 3r) + 4n + 3(r + 2)$
    $= 10n + 15r + 4n + 3r + 6$ *Dist. Prop.*
    $= 10n + 4n + 15r + 3r + 6$ *Commutative (+)*
    $= (10 + 4)n + (15 + 3)r + 6$ *Dist. Prop.*
    $= 14n + 18r + 6$ *Substitution*
35. 11   **37.** $52   **39a.** 179.8 lb   **39b.** 31 lb
41a. 96 ft^2 **41b.** $4(8 + 10 + 6)$ or 96
43. Commutative $(\times)$ **45.** Symmetric
47. 6   **49a.** $d + 30$ **49b.** 170

Pages 27–29   Lesson 1–5
1. Sample answer: You should look back to
determine whether the answer fits the problem,
whether there are other possible answers, or
whether there might be a better way to solve the
problem.   **3.** no   **5.** $n + 2$ **7a.** 25 $5 bills, 32
$1 bills   **7b.** Sample answer: Use an equation. Let
x represent the number of $1 bills. Then $x - 7$
represents the number of $5 bills. Write and solve
the equation $11(10) + 1x + 5(x - 7) = 267$.

9. Craig, 16; mother 40; sample explanation: Let m represent the age of the mother. Then $m - 24$ represents Craig's age. Solve the equation $m + (m - 24) = 56$. **11.** 7 craft books, 13 cookbooks

13a. Sample answer:

Person	1	2	3	4	5	6
Number of Handshakes	5	4	3	2	1	0

13b. 15   **13c.** 66   **15.** 9:00 P.M.   **17a.** $d = rt$
17b. 703 mi   **19.** 20 people   **21.** $8b$ **23.** $24a + 45$
25. A

Pages 34–37   Lesson 1–6
1. A frequency table shows the number of times a single event occurs. A cumulative frequency table shows the number of times an event occurs plus the previous events.   **3.** No, the sample is not large enough.

5.

Number of Goals	Tally	Frequency
1	III	3
2	IIII	5
3	I	1
4	IIII	4
5	III	3
6	III	3
7	I	1
8	I	1

7. 1 time   **9a.** 100 s

9b.

Cycle (s)	Frequency	Cumulative Frequency
80	33	33
90	42	75
100	60	135
110	25	160

9c. 75 times   **11.** Yes; the sample is random and representative of drivers in the area.   **13.** No; the sample is not representative of the entire country.
15. No; people might prefer a golfer who is from the same area.

17a.

Quiz Score	Tally	Frequency
6	II	2
7	IIII	4
8	IIII IIII I	11
9	IIII IIII	9
10	IIII	4

17b. 8   **17c.** Rather than adding all of the numbers, this formula uses multiplication to find the sum. Then the sum is divided by the number of scores, 30.
17d. 8.3   **19.** That is a typical breakfast time and they want the business.   **21.** Sample answers: a stadium, restaurant, school campus   **25.** 275 miles
27. nine more than x

Page 37   Quiz 2
1. $11 + 6 + 2a$, Commutative (+); $17 + 2a$
5.

Eye Color		
Color	Tally	Frequency
brown	IIII IIII	10
blue	IIII	4
green	IIII I	6
hazel	IIII	5

Blue occurs the least.

Pages 41–43   Lesson 1–7
1. Line graphs usually show change over time, while histograms compare quantities of similar nature.   **2.** A correct graph has a title, horizontal and vertical axes labeled, and equally spaced units on both axes.   **3.** Manuel; line graphs are usually used to show trends over time. Histograms are usually used to how frequency of items in sets of data.   **5.** 1997–1998   **7.** 14–15 cm

9.

Leaf Lengths in a Maple Tree Population

Length (cm)

11a. 81°F   **11b.** 12 days   **11c.** 63   **13.** 1993

15.

Men on Hold

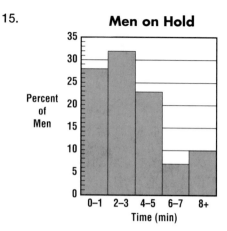

17. Sample answer: The histograms show that a greater percent of men than women are willing stay on hold for 0–1 minutes   **19.** 123   **21.** 71, 73, 92, and 100 occur three times.   **23.** Sample answer: More students in second period scored 80 or more, so second period did better   **25.** Yes; it is a random sample.   **27.** $8x - 3y$

Pages 44–46   Chapter 1 Study Guide and Assessment
1. term   **3.** algebraic expression   **5.** counterexample   **7.** sample   **9.** frequency tables   **11.** $5n$ **13.** $2y - 6 = 14$ **15.** 7   **17.** 14   **19.** 9   **21.** Commutative ($\times$), $20c$ **23.** Commutative ($\times$), $9x$ **25.** Associative ($+$), $g + 3$ **27.** $7v - 7$ **29.** $2h + ah$ **31.** $22 - 3d$
33a. 1200   **33b.** 80   **35.** 0

37.

Number	Cumulative Frequency
0	6
1	9
2	13
3	15
4	19

39. 10   **41.** 10–11

43.   **Cumulative Histogram**

45. 8   **47a.** 100; 62   **47b.** 88   **47c.** 10

Page 49   Preparing for Standardized Tests
1. C   **3.** D   **5.** C   **7.** A   **9.** 234

Chapter 2   Integers

Page 51   Check Your Readiness
1. $10x$ **3.** $19d$ **5.** $6y$ **7.** $6r$ **9.** $25p$ **11.** 19   **13.** 1
15. 54   **17.** 2   **19.** 42   **21.** 6   **23.** $32q$ **25.** $36x$
27. $26c$ **29.** $36st$

Pages 55–57   Lesson 2–1
1. Sample answer: temperature, elevation
3. Neither; 0 is neither negative nor positive.
5. -6 **7.** $+150$ **9.** 1
11. ![number line](−5 to 5)
13. $<$ **15.** 10   **17.** $-22, -17, -14, -7, -1, 9, 10, 13, 22, 26, 35$ **19.** 2   **21.** 0   **23.** -4
25. ![number line](−5 to 5)
27. ![number line](−5 to 5)
29. ![number line](−5 to 5)
31. $>$ **33.** $>$ **35.** $<$ **37.** $>$ **39.** 6   **41.** 2   **43.** 9
45. Aways; $|5| = 5$ and $|-5| = 5$, so $|5| = |-5|$.
47. 78, 14, $-14, -25, -36$ **49a.** $-36°$
49b. $-26°$ **49c.** $-36°$

51.   **Heights of U.S. Presidents**

53. 19 presidents   **55.** $6a$ **57.** $7m + 2n$
59. $7x + 11y$

Pages 61–63   Lesson 2–2
1. Start at the origin. Move 5 units to the left. Then move 1 unit up and draw a dot.

3.

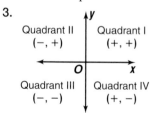

Quadrant II $(-, +)$ Quadrant I $(+, +)$
Quadrant III $(-, -)$ Quadrant IV $(+, -)$

5. $(3, -2)$

7.

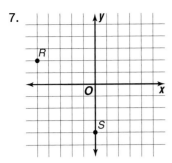

9. none   **11.** $(1, -1)$ **13.** $(0, -4)$ **15.** $(-3, -3)$

17–22.

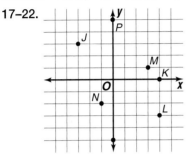

23. III   **25.** none   **27.** II   **29.** origin   **31.** I   **33.** IV
35a. $3, $9, $15

35b–c.

37a. $A(2, 1)$, $B(6, 1)$, $C(6, 5)$

37b.

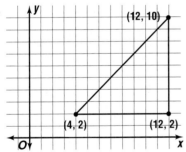

37c. The triangles have the same shape, but the second is larger.   **39.** -8685 **41.** Commutative $(+)$ **43.** Closure $(+)$ **45.** A

Page 63   Quiz 1
1. $>$ **3.** $<$ **5.** 11   **7.** III   **9.** II

Pages 68–69   Lesson 2–3
1.

$-5 + (-3) = -8$

3.

$-4 + 6$ $= 2$

5. $+$ **7.** $-$ **9.** $-$ **11.** 16   **13.** -1 **15.** -5 **17.** $2x$
19. $2a$ **21.** 12   **23.** 21   **25.** -11 **27.** -8 **29.** 9
31. -13 **33.** -8 **35.** -21 **37.** -55 **39.** 13
41. -22 **43.** -4 **45.** -1 **47.** -3 **49.** $-15x$
51. $3m$ **53.** 0   **55.** $8y$ **57.** -9 **59.** -1

61.

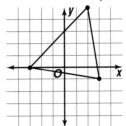

63. IV   **65.** II   **67.** -5 **69.** -10 **71.** 0   **73.** C

Pages 73–74   Lesson 2–4
1. To subtract an integer, add its additive inverse.
3. $10 + (-3)$ **5.** $-4 + (-8)$ **7.** -2 **9.** 11   **11.** 5
13. 8   **15.** 13   **17.** 7   **19.** -8 **21.** -13 **23.** -11
25. 16   **27.** -6 **29.** -9 **31.** 12   **33.** 7   **35.** -17
37. 3   **39.** -10 **41.** -12 **43.** $25n$ **45.** 50
47. $-34°F$ **49a.** false; $2 - 5 \neq 5 - 2$ **49b.** false;
$(6 - 2) - 3 \neq 6 - (2 - 3)$ **49c.** true   **51.** -20
53. -27 **55.** -8 **57.** $>$ **59.** $<$

Pages 77–79   Lesson 2–5
1. $3(-3) = -9$ **3.** -12 **5.** 80   **7.** 36   **9.** 28   **11.** $-18x$

13a.

13b. It was reflected over the x-axis.   **15.** -48
17. -9 **19.** 45   **21.** 0   **23.** -24 **25.** -39 **27.** 27
29. -84 **31.** -1 **33.** 99   **35.** -8 **37.** 30   **39.** 13
41. $-8a$ **43.** $32mn$ **45.** -22 **47.** $-100(5) = -500$

49a.

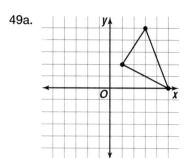

49b. It was reflected over both the x- and y-axis.
51. 4 **53.** -5 **55.** -8 **57.** -2 **59.** 0 **61.** C

Page 79 Quiz 2
1. -7 **3.** 32 **5.** 13 **7.** -45 **9.** -9

Pages 84–85 Lesson 2–6
1. $-10 \div 2 = -5$, $-10 \div (-5) = 2$ **3.** -5 **5.** -5
7. 4 **9.** -8 **11.** 12 **13.** 1 **15.** 6 **17.** -6 **19.** 5
21. -9 **23.** 10 **25.** -8 **27.** -5 **29.** -12 **31.** -6
33. -5 **35.** -3 **37.** 6 **39.** -7 **41.** -2
43a. -7000 farms **43b.** 10 to 49 and 2000 acres or
more **45.** 74 **47.** -54 **49.** 16 **51.** 80 **53.** -2
55. 4 **57.** B

**Pages 86–88 Chapter 2 Study Guide
and Assessment**
1. negative numbers **3.** opposites, additive inverses,
or zero pairs **5.** absolute value **7.** integers
9. ordered pair **11.** > **13.** < **15.** $-4, -3, -2, 0,$
$4, 7$ **17.** $(3, 2)$ **19.** $(-4, -3)$ **21.** I **23.** none
25. -6 **27.** -10 **29.** 5 **31.** -5 **33.** $2x$ **35.** $4m$
37. -8 **39.** 9 **41.** -4 **43.** -1 **45.** -1 **47.** 35
49. 36 **51.** 27 **53.** $-42m$ **55.** -7 **57.** -6
59. -8 **61.** $213 **63.** $-40(2) = -80$

Page 91 Preparing for Standardized Tests
1. A **3.** B **5.** B **7.** E **9.** -1

Chapter 3 Addition and Subtraction Equations

Page 93 Check Your Readiness
1. $\frac{1}{3}$ **3.** $\frac{1}{9}$ **5.** $\frac{2}{5}$ **7.** 5.6 **9.** 23.7 **11.** 33.1 **13.** 17.2
15. 5.2 **17.** 4.05 **19.** 37.6 **21.** $\frac{1}{2}$ **23.** $1\frac{3}{10}$ **25.** $\frac{1}{6}$

Pages 97–99 Lesson 3–1
1. Sample answer: $-\frac{1}{2}, \frac{4}{3}$, and 0.9 **5.** 0.75 **7.** 0.625

9.
```
←─┼──┼──┼──●──┼──┼──●──┼──┼──┼──┼─→
 -5 -4 -3 -2 -1  0  1  2  3  4  5
```
Since -3 is to the left of $-2.\overline{6}$ on the number line,
-3 is less than $-2.\overline{6}$. **11.** < **13.** = **15.** $\frac{7}{10}, \frac{6}{8}, \frac{4}{5}$
17. < **19.** < **21.** < **23.** > **25.** < **27.** >
29. $-2.002 > -2.02$

31.
```
     -5/6
←─┼──┼●●┼──┼──┼──┼──┼──┼──┼──┼──┼─→
 -1  -8/10                          0
```
Since $-\frac{5}{6}$ is to the left of $-\frac{8}{10}$ on the number line,
$-\frac{8}{10} > -\frac{5}{6}$. **33.** $\frac{1}{2}$, or 0.6, $\frac{5}{8}$ **35.** $-\frac{2}{6}, -\frac{1}{4}, -\frac{1}{8}$
37. $\frac{3}{8}, \frac{3}{5}, 0.\overline{6}$ **39.** a dozen eggs for $1.59;
$0.13 < $0.14 **41.** a 25-yd by 12-in. roll for $2.99;
$0.0004 and $0.0003 are both > $0.0002
43a. Firefly, Japanese Beetle, Mealworm
43b. Elm Leaf Beetle **45.** -7 **47.** -4 **49.** 48
51. 3822

Pages 102–103 Lesson 3–2
1. Add and subtract in the order in which they are
given; or group the positive numbers together and
the negative numbers together and then add.
3. 0.9 **5.** -15.05 **7.** 0 **9.** $\frac{1}{3}$ **11.** -13.3
13. -5.82 **15.** -1.26 **17.** -29.48 **19.** $-\frac{3}{8}$
21. $-5\frac{3}{10}$ **23.** $\frac{19}{35}$ **25.** $3\frac{1}{8}$ **27.** $-2\frac{3}{20}$ **29.** 6.9
31. $-6\frac{1}{6}$ **33.** $-$153.58 or a decrease of $153.58
35. $\frac{1}{4}$ foot above normal **37.** -20 **39.** $-12bc$
41. $11y + 14$

Pages 107–109 Lesson 3–3
1. When a set of data has one or more extremely
high or low values, the mean does not accurately
describe the values in the set of data. Consider
the set of data 3, 10, 5, 16, 148. The mean is 36.4.
3. Sonia; a few really high or really low values
could influence the mean so that it is not
representative of Eric's scores. **5.** 42 **7.** 0.151
9. 47.25 **11.** 15.5; 12.45; 12.5; 21.1 **13.** 62.3; 60;
72; 17 **15a.** Ms. Diaz: 201.625; 200.6; 210.0; 19.1;
Mr. Cruz: 201.625; 200.6; 210.0; 59.1 **15b.** Ms. Diaz;
her sales are more consistent **17.** 9; 8; none; 19
19. 0.25; 0.3; 0.3; 0.2 **21.** 60.8; 24; 20; 180
23. 10.375; 10.1; 10.2; 3.8 **25.** 10.8$\overline{3}$; 10.5; 10; 2.5
27. 19.5; 19.2; 18.3, 19.0, 21.0; 2.7 **29.** Sample
answer: 3.1, 1.2, 1.0, 1.5 **31.** Sample answer: 1, 7, 7,
7, 7, 7 **33.** 96 **35a.** $33,425; $29,050; none
35b. mean; highest value **35c.** median; lowest
value **37.** -9.9 **39.** -1
41.
```
←─┼──●──┼──●──●──┼──●──┼──●──┼──●──┼─→
 -5 -4 -3 -2 -1  0  1  2  3  4  5
```

Pages 114–116 Lesson 3–4
1. An open sentence is neither true nor false; a
statement is either true or false. **3.** Replace n with
each element of the replacement set to see which
replacement results in a true statement. **5.** 6

7. 11.7 **9.** $\frac{4}{5}$ **11.** 4 **13.** -3 **15.** 3 **17.** 4 **19.** -5
21. 2 **23.** 42 **25.** 11 **27.** 6 **29.** -1 **31.** 18
33. $5\frac{1}{4}$ **35.** 2 **37.** 6 h **39.** 65 days **41.** 41.2; 40;
40; 8 **43.** $2\frac{1}{3}$ **45.** $-4\frac{1}{12}$ **47.** D

Page 116   Quiz 1
1. < **3.** = **5.** 1.69 **7.** 2.8; 2.85; 3.0; 3.6 **9.** -2

Pages 120–121   Lesson 3–5
1.

3. $x + (-2) = 12$ **5.** $7 = -6 + x$ **7.** -4 **9.** -4
11. 6 **13.** 3 **15.** -6 **17.** -7 **19.** 8 **21.** 0
23. -3 **25.** 1 **27.** 2 **29.** 16 **31.** -7 **33.** -21
35. 11 **37a.** $t - 8 = -2$ **37b.** 6°F **39.** 2 **41.** 1
43. 54 **45.** Commutative (+)

Pages 125–127   Lesson 3–6
1. Replace m with -4 to see if $12 + m$ is equal to 8
3. 15; In the first equation, add 5 to each side to
get $x = 17$. By replacing x with 17 in the second
equation, the result is $17 - 2$ or 15. **5.** Sample
answer: -21 **7.** Sample answer: $+9$ **9.** Sample
answer: -9.3 **11.** -15 **13.** 1.1 **15.** $-\frac{1}{6}$
17a. Sample answer: $x - 1788 = 124$ **17b.** 1912
19. -4 **21.** 3 **23.** 13 **25.** -13 **27.** -48 **29.** 34
31. -3.8 **33.** $\frac{2}{3}$ **35.** $\frac{10}{9}$ or $1\frac{1}{9}$ **37.** 13
39. $m - (-15) = 20$; $m = 5$ **41a.** $c + 27.16 = 32.50$
41b. \$5.34 **43.** 0 **45.** 3.25 **47.** 6 **49.** \$6.76 **51.** D

Page 127   Quiz 2
1. -4 **3.** 23 **5a.** Let c = cost of Chicago meal;
$3.89 = c - 0.62$. **5b.** \$4.51

Pages 130–131   Lesson 3–7
1. Sample answer: $\left| a - 2 \right| = -5$ **3.** Sample
answer: The value inside the absolute value sign
can either be positive or negative. You have to
consider both cases, so there may be up to 2
solutions. **5.** $\{-3, 3\}$ **7.** $\varnothing$ **9.** $\{-7, 13\}$
11. $\left| s - 145 \right| = 6$; $\{139, 151\}$ **13.** $\{-2, 2\}$
15. $\{-6, 6\}$ **17.** $\varnothing$ **19.** $\{1\}$ **21.** $\varnothing$ **23.** $\{-21, 1\}$
25. $\{-19, 7\}$ **27.** $\{-9, 13\}$ **29.** $\varnothing$ **31.** 0
33. $\left| p - 165,000 \right| = 10,000$; $\{155,000, 175,000\}$
35. A **37.** 48 **39.** -14 **41.** B

Pages 132–134   Chapter 3 Study Guide and Assessment
1. median **3.** range **5.** rational number
7. inequality **9.** rational number **11.** > **13.** >

15. < **17.** $-1.1, 0, \frac{1}{8}, 0.25$ **19.** 10.5 **21.** -1
23. $5.\overline{72}$; 6; 5; 7 **25.** 12.92; 10; 10; 28 **27.** 0
29. -1 **31.** 45 **33.** -2 **35.** -4 **37.** 0
39. $x + 4 = 11$; $x = 7$ **41.** 12 **43.** 17 **45.** 9 **47.** $\frac{1}{2}$
49. $x + 12 = -108$; -120 **51.** $\{-8, 8\}$ **53.** $\{-4, 8\}$
55. $\varnothing$ **57.** \$70.625; \$72.5; \$50; \$75; \$45

Page 137   Preparing for Standardized Tests
1. E **3.** C **5.** B **7.** A **9.** 0.6

Chapter 4   Multiplication and Division Equations

Page 139   Check Your Readiness
1. 54 **3.** 0 **5.** 60 **7.** 20 **9.** 18 **11.** -7 **13.** -7
15. -7 **17.** -8 **19.** $\frac{1}{4}$ **21.** $\frac{2}{3}$ **23.** $\frac{3}{5}$ **25.** $7a$
27. t **29.** $4x - 8$ **31.** $2b - 3.2$ **33.** $32m - 12$

Pages 143–145   Lesson 4–1
1. Always, because rational numbers are closed
under multiplication. **3.** $\frac{1}{4} \cdot \frac{5}{6} = \frac{5}{24}$ **5.** -9
7. -14.7 **9.** -2.3 **11.** $-\frac{9}{4}$ or $-2\frac{1}{4}$ **13.** $\frac{3}{5}y$ **15.** $6cd$
17. 24.4 **19.** -7.1 **21.** -51 **23.** -29 **25.** 57.6
27. $\frac{5}{8}$ **29.** 0 **31.** -1 **33.** $\frac{21}{8}$ or $2\frac{5}{8}$ **35.** $-3x$
37. $8.8rs$ **39.** $\frac{6}{25}ab$ **41.** $\frac{5}{4}s$ or $1\frac{1}{4}s$ **43.** -4
45. \$135 **47.** $-\frac{2401}{40}$ m or $-60\frac{1}{4}$ m **49.** $\varnothing$
51. $\{-12, -6\}$ **53.** 1.4 **55.** $-\frac{5}{8}$ **57a.** 73 in.
57b. 64 in.

Pages 148–151   Lesson 4–2
1. Sample answer: the number of ways two or more
events can occur is the product of the number of
ways each event can occur. **3.** Ling; three
outcomes have 2 heads: HHT, HTH, THH. Three
outcomes have 2 tails: TTH, THT, HTT. **5.** sample
space **7.** outcome **9.** 36
11. 8 outcomes:

13. 15 outcomes:

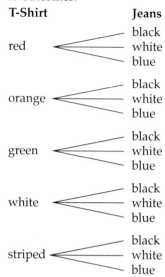

T-Shirt	**Jeans**
red	black / white / blue
orange	black / white / blue
green	black / white / blue
white	black / white / blue
striped	black / white / blue

15. 32   **17.** 30   **19.** Sample answer: choosing a red, green, or white car with or without a sunroof

21. 20   **23a.** 64   **23b.** 729   **25.** -16.2 **27.** $\frac{6}{35}$

29. $-\frac{5}{12}$ **31.** B

Pages 156–159   Lesson 4–3

1. No; the product of a negative number and a positive number can never equal 1.

5. $-\frac{3}{2}$ **7.** $-x$ **9.** 0.2   **11.** $-\frac{3}{14}$ **13.** $\frac{36}{5}$ or $7\frac{1}{5}$

15. $-\frac{15}{2}$ or $-7\frac{1}{2}$ **17.** 16 students   **19.** -2.5

21. -0.1 **23.** 12   **25.** 0   **27.** -1 **29.** $\frac{14}{15}$ **31.** 45

33. $-\frac{3}{32}$ **35.** -21 **37.** 12   **39.** $\frac{1}{10}$ **41.** $-\frac{2}{5}$

43. $-\frac{4}{5}$ **45.** 18   **47.** 2 min   **49.** \$12.50

51. 48   **53.** 6; 6; 8; 5   **55.** 20.5; 20; 20; 18   **57.** =

59. $\frac{1}{2}$-pound bag of cashews for \$3.15; \$6.30 < \$6.92

61. 22-ounce bottle for \$1.09; \$0.06 > \$0.05   **63.** A

Page 159   Quiz 1

1. -18.8 **3.** $\frac{20}{8}$ or $2\frac{1}{2}$ **5.** 27   **7.** 2.7   **9.** $\frac{28}{9}$ or $3\frac{1}{9}$

Pages 163–164   Lesson 4–4

1. Divide each side by 4.   **3.** Sample answer: $-8a = 56$ **5.** 8   **7.** 45   **9.** 4   **11.** 6   **13.** 15
15. -7 **17.** 28   **19.** 5   **21.** 64   **23.** -70 **25.** 27
27. $-\frac{5}{2}$ or $-2\frac{1}{2}$ **29.** 2   **31.** -8 **33.** $8x = 112$; 14

35. $\frac{3}{5}$ yr $= -9$; -15 **37a.** $6n = 288$ **37b.** 48

39. 110 mph   **41.** 20   **43.** $\frac{7}{12}$ **45.** 0   **47.** 57 ft

49. D

Pages 168–170   Lesson 4–5

1. Sample answer: If necessary, first undo addition and/or subtraction of constants, then undo multiplication or division of variables.   **3.** Jean; Soto did not multiply each side by 4 in the second step.
5. Add 5 to each side.   **7.** 2   **9.** -5 **11.** -20

13a. $y = 6(x - 1) + 21$ **13b.** $\frac{1}{3}$ yr or 4 months

15. -1 **17.** -7 **19.** 0   **21.** 6   **23.** 9.6   **25.** 8.1

27. 30   **29.** -14 **31.** 68   **33.** -7.2 **35.** $\frac{n-2}{-3} = \frac{1}{3}$;

1;   **37.** 3

-3 -2 -1 0 1 2 3 4

39a. $8c + 15 = 143$ **39b.** \$16   **41a.** $3x + 12 = 39$
41b. 9   **43.** 3   **45.** 6   **47.** 4   **49.** -81

Pages 173–175   Lesson 4–6

1. Add $2x$ to each side, subtract 5 from each side, divide each side by 5.   **3.** An equation that is an identity is true for every value of the variable; an equation with no solution is never true.   **5.** Subtract $0.5p$ from each side.   **7.** 2   **9.** identity   **11.** -5

13a. $3n = 5n - 24$ **13b.** 12   **15.** 19   **17.** 2   **19.** $\frac{5}{2}$

or $2\frac{1}{2}$ **21.** no solution   **23.** 0.7   **25.** -1 **27.** 3.4

29. -20 **31.** 60   **33.** 4.2   **35a.** $3w - 18 = 2w$
35b. 18 cm by 36 cm   **37.** Sample answer: $2x = x + 6$
39. -9 **41.** 12   **43.** 264 numbers   **45.** 31   **47.** 3.3

Page 175   Quiz 2

1. -6 **3.** 63   **5.** 72   **7.** 12   **9.** no solution

Pages 178–179   Lesson 4–7

1. Use the Distributive Property to remove the parentheses.   **5.** identity   **7.** -4 **9.** 8   **11.** 6

13. 5   **15.** -10 **17.** $\frac{1}{3}$ **19.** $\frac{7}{2}$ or $3\frac{1}{2}$ **21.** 2.5

23. no solution   **25.** 42   **27.** identity   **29.** -2
31. base 1 = 20 in.; base 2 = 8 in.   **33.** 17, 19   **35.** 1

37a. $\frac{1}{2}x - 4 = 7$ **37b.** 22   **39.** $-382, -369, -364,$
$-292, -229, 867$

Pages 180–182   Chapter 4 Study Guide and Assessment

1. event   **3.** tree diagram   **5.** identity
7. outcomes   **9.** outcomes   **11.** 22.8

13. $-\frac{3}{4}$ **15.** $-10.8y$ **17.** $-\frac{6}{7}s$

19. 6 outcomes:

Bread	**Meat**
wheat	ham / turkey / beef
white	ham / turkey / beef

21. -2 **23.** $\frac{8}{15}$ **25.** $-\frac{1}{8}$ **27.** $-\frac{3}{4}$ **29.** 8 **31.** -4
33. 12.3 **35.** 9 **37.** -2 **39.** 0 **41.** 2 **43.** -25
45. 6 **47.** 7.3 **49.** 1 **51.** 2 **53.** identity
55. 5 times **57.** 6 in.

Page 185 Preparing for Standardized Tests
1. A **3.** C **5.** E **7.** D **9.** 2

Chapter 5 Proportional Reasoning and Probability

Page 187 Check Your Readiness
1. 3 **3.** 9 **5.** 15 **7.** 0.25 **9.** 0.45 **11.** 1.25 **13.** $\frac{1}{3}$
15. $\frac{9}{10}$ **17.** $\frac{5}{6}$ **19.** $\frac{1}{100}$ **21.** $\frac{3}{20}$ **23.** $\frac{7}{25}$ **25.** 0.77
27. 0.62 **29.** 0.24

Pages 191–193 Lesson 5-1
1. Multiply 12 and 5. Then divide by 6. **3.** $75(4) =$ $100(3)$ **5.** $8(15) = 12(10)$ **7.** 60 **9.** 3.8 **11.** 5
13. 2 kg **15.** 4.5 gal **17.** 15 **19.** 4 **21.** 75 **23.** 9
25. 10.5 **27.** 3.5 **29.** 2.1 **31.** 5 **33.** 6 **35.** 37
37. Yes; the cross products are equal. **39.** 60 in.
41. 3000 mL **43.** 4 gal **45.** 0.52 km **47.** 9 pt
49. no **51.** 750 lb, 1050 lb **53.** It decreases.
55. -6 **57.** -2 **59.** $-\frac{1}{2}$ **61.** 12 **63.** $-\frac{2}{5}$

Pages 196–197 Lesson 5-2
1. Sample answer:

3. 150 mi **5.** 1 m = 35.2 m **7.** 120 mi
9. 100 mi **11.** 30 mi **13.** 1 in. = 15 in.
15. 0.5 c **17.** 1 in. = 30.2 mi **19.** $\frac{4}{3}$ **21.** -4

Pages 201–203 Lesson 5-3
1. P is the percentage, B is the base, and r is the percent or the rate per hundred.
3. $\frac{12}{24} = \frac{r}{100}$ **5.** $\frac{P}{60} = \frac{45}{100}$ **7.** 60% **9.** 70%
11. 25% **13.** 16.4 **15a.** 48%, 51%, 1%

15b. **Senators in 108th Congress**

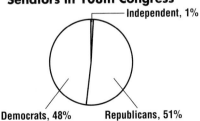

17. 62.5% **19.** 80% **21.** 75% **23.** 150% **25.** 60%
27. 6% **29.** 36% **31.** 190 **33.** 12.5 **35.** 60
37. 0.5% **39.** 9.3 **41.** 50%

43.

Mother's Day Flowers

45. 1 in. = 70 mi **47.** 2500 m **49.** -5.2 **51.** $\frac{2}{5}$

Page 203 Quiz 1
1. 27 **3.** 100 ft **5.** 950

Pages 207–209 Lesson 5-4
1. r represents the percent rate; R is the rate written in decimal form. **3.** Nikki; in the percent equation, the R is the decimal form of the percent. **5.** 41.8
7. 550 **9.** $75 **11.** 112 **13.** 117 **15.** 24 **17.** 625
19. 88 **21.** 64 **23.** 62.5 **25.** $50 **27a.** $P = 0.29(150)$
27b. 44 **29.** 7 adults **31.** $x = y$ **33a.** 180 mi
33b. 210 mi **33c.** 135 mi **35.** -31 **37.** 5

Pages 215–217 Lesson 5-5
1. First method: find the amount of the tax, then add the tax to the cost of the item. Second method: add the tax rate to 100%, then find the total cost.
5. $\frac{4}{14} = \frac{r}{100}$ **7.** 20% decrease **9.** $94.50 **11.** $1080
13. 83% **15.** 12% increase **17.** 16% decrease
19. 133% increase **21.** $42.80 **23.** $39.52
25. $26.64 **27.** $266 **29.** $29.75 **31.** $14.21
33. 100% **35.** 1999 and 2001; The amount of increase is larger and the base is smaller between these years. **37.** less than **39.** 62.5% **41.** 35%
43. 1.5 kg

Page 217 Quiz 2
1. 25 **3.** 20 **5.** $140.81

Pages 222–223 Lesson 5-6
1. Both theoretical and experimental probability are ratios that compare the number of favorable outcomes to the number of possible outcomes. Theoretical probability is based on known characteristics. It tells what should happen. Experimental probability is the result of an experiment or simulation. **3.** Yes; when the number of favorable outcomes is greater than the number of unfavorable outcomes, the odds of the outcome are greater than 1. **5.** $\frac{1}{36}$
7. 1:1 **9.** $\frac{1}{2}$ **11.** $\frac{1}{9}$ **13.** 0 **15.** 1:1 **17.** 7:1
19. 5:3 **21.** $\frac{1}{4}$ **23a.** 9:00, $\frac{1}{20}$; 1:00, $\frac{1}{40}$; 4:00, $\frac{1}{8}$
23b. 4:00 **25a.** $\frac{1}{2}$ **25b.** 9:11 **27.** $36 **29.** $180
31. A

Pages 227–229 Lesson 5–7
1. $P(A$ and $B)$ is the probability that both A and B occur; $P(A$ or $B)$ is the probability that either A or B occur. **3.** $\frac{1}{24}$ **5.** mutually exclusive, $\frac{2}{13}$

7. inclusive, $\frac{4}{13}$ **9.** 20.1% **11.** $\frac{1}{4}$ **13.** $\frac{1}{20}$ **15.** $\frac{3}{20}$

17. mutually exclusive, $\frac{2}{3}$ **19.** 0.013 **21a.** 34.2%

21b. 28% **21c.** 88.3% **23.** $\frac{1}{3}$ **25.** $\frac{3}{5}$ **27.** D

Pages 230–232 Chapter 5 Study Guide and Assessment
1. proportion **3.** impossible **5.** odds **7.** unit rate
9. Discount **11.** 45 **13.** 4 **15.** 10 mi **17.** 30 mi
19. 1 in. = 30.5 ft **21.** 60 **23.** 150% **25.** 50
27. 8.1 **29.** 150 **31.** 25% increase **33.** 33.$\overline{3}$%
decrease **35.** 100% **37.** $\frac{1}{11}$ **39.** $\frac{7}{11}$ **41.** $\frac{1}{2}$

43. mutually exclusive, $\frac{2}{3}$ **45.** $339.15

Page 235 Preparing for Standardized Tests
1. B **3.** C **5.** A **7.** C **9.** 2

Chapter 6 Functions and Graphs

Page 237 Check Your Readiness
1, 3, 5, 7, 9.

11. −6 **13.** 6 **15.** 10 **17.** −6 **19.** 10 **21.** 4
23. −15 **25.** 30 **27.** 4

Pages 241–243 Lesson 6–1
1. ordered pairs, table, graph **3.** The domain is the set of all first coordinates from the ordered pairs. The range is the set of all second coordinates from the ordered pairs.

5.

x	y
−3	−3.5
−1	3.9
0	2
5	2.5

domain: {−3, −1, 0, 5},
range: {−3.5, 3.9, 2, 2.5}

7. {(−3, −2), (1, 2), (−4, −2)}

x	y
−3	−2
1	2
−4	−2

domain: {−3, 1, −4},
range: {−2, 2}

9.

x	y
4	3
−2	3
−2	4
4	4

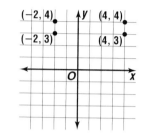

domain: {4, −2},
range: {3, 4}

11.

x	y
−2	0
3	−7
2	−5
−6	3
1	5

domain: {−2, 3, 2, −6, 1},
range: {0, −7, −5, 3, 5}

13.

x	y
$-\frac{1}{2}$	$\frac{1}{2}$
1	0
$-\frac{1}{2}$	−5

domain: $\left\{-\frac{1}{2}, 1\right\}$,

range: $\left\{\frac{1}{2}, 0, -5\right\}$

15. {(−3, 2), (−1, 1), (1, −2), (1, 3)}

x	y
−3	2
−1	1
1	−2
1	3

domain domain: {−3, −1, 1},
range: {2, 1, −2, 3}

17. {(−3, 2), (−2, 2), (−1, 4), (2, −2), (3, −2)}

x	y
−3	2
−2	2
−1	4
2	−2
3	−2

domain: {−3, −2, −1, 2, 3},
range: {2, 4, −2}

19. {(−3, −3), (−3, −1), (−1, 0), (2, −2), (3, 4)}

x	y
−3	−3
−3	−1
−1	0
2	−2
3	4

domain: {−3, −1, 2, 3},
range: {−3, −1, 0, −2, 4}

21. {(−1, 1), (0, 2), (1, 3), (2, 4)}

23. {(0, 0), (−1, −0.5), (1, 0.5), (−2, −1)}

25a. {(8, 4), (9, 3), (10, 2), (11, 1), (12, 0)}

25b.

x	y
8	4
9	3
10	2
11	1
12	0

27a. 3, 4, 5, 6, 7, 8, 9, 10, 11, 12, 13, 14, 15

27b. {(3, 1), (4, 1), (5, 2), (6, 2), (7, 3), (8, 3), (9, 4), (10, 3), (11, 3), (12, 2), (13, 2), (14, 1), (15, 1)}

29. 4:6 or 2:3   **31.** −24   **33.** B

Pages 247–249   Lesson 6–2
1. When the values are substituted in the equation, the equation is true.   **3.** Dan; when using variables other than x and y, it can be assumed that the domain goes with the variable that comes first alphabetically. Thus, m is the domain.   **5.** c, d

7. {(−2, 7), (−1, 2), (0, −3), (1, −8), (2, −13)}

9. {(−2, −5.5), (−1, −4), (0, −2.5), (1, −1), (2, 1/2)}

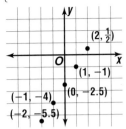

11. {(3, 25), (4, 24), (8, 20), (12, 16)}

13. d   **15.** b   **17.** c

19. {(−1, 3), (0, 0), (1, −3), (2, −6), (3, −9)}

21. {(−1, −7), (0, −6), (1, −5), (2, −4), (3, −3)}

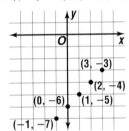

23. {(−1, 10), (0, 5), (1, 0), (2, −5), (3, −10)}

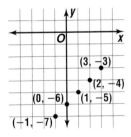

25. {(−1, −3), (0, −2), (1, −1), (2, 0), (3, 1)}

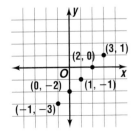

27. $\{(-1, 6), (0, 4), (1, 2), (2, 0), (3, -2)\}$

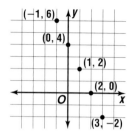

29. $\{(-1, -7), (0, -4), (1, -1), (2, 2), (3, 5)\}$

31. $\{0, 1, 2, 3\}$ **33.** $\{-1, 0, 1, 2\}$ **35.** 3

37a.

w	y
6.0	0.15
6.8	0.17
7.2	0.18
8.0	0.2
8.8	0.22

37b. $\{(6.0, 0.15), (6.8, 0.17), (7.2, 0.18), (8.0, 0.2), (8.8, 0.22)\}$

37c.

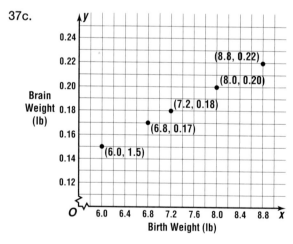

39a. $x + y = 90; x = 2y$ **39b.** $(60, 30)$
41. \$10.99   **43.** 5   **45.** -1 **47.** C

Pages 254–255   Lesson 6–3
1. The graph of $x = -8$ is a straight line parallel to the y-axis. For every value of y, $x = -8$.   **3.** no
5. yes; $A = 2, B = 0, C = 8$

7.   **9.**

11.

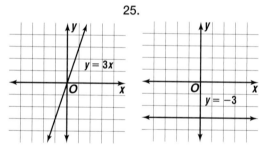

13. no   **15.** yes; $A = -6, B = 2, C = 0$
17. yes; $A = -1, B = 1, C = 0$ **19.** no
21. yes; $A = 5, B = 1, C = 9$

23.   **25.**

27.   **29.**

31.   **33.**

35.

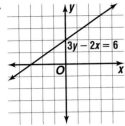

37.

x	y
0	-2
1	-1
2	0
3	1
4	2

39.

x	y
-2	-4
-1	-1
0	2
1	5
2	8

41a. $3x + 6y = 90$ **41b.** Sample answer: {(2, 14), (6, 12), (12, 9), (18, 6), (24, 3)}

41c.

43. {−8, −1, 0, 2, 4} **45.** 6 **47.** 4 **49.** $67 + 2h = 79$

Page 255   Quiz 1

1.

x	y
1	-1
3	5
0	-2
3	3

3. b, c

5a.

Total Savings vs. Number of Weeks

5b. 25 weeks   **5c.** $y = 500 + 80x$
$$y = 500 + 80(25)$$
$$y = 500 + 2000$$
$$y = 2500$$

Pages 259–261   Lesson 6–4
1. Sample answer: A function is a relation. However, in a relation, each member of the domain can be paired with more than one member of the range. In a function, each member of the domain is paired with exactly one member of the range.   **3.** Zina is correct. There are some graphs of straight lines that do not represent a function. The graph of $x = 2$ is a straight line. However, it is not a function since there are many y values for a single x value.   **5.** yes **7.** no **9.** −15
11. $4b - 3$ **13.** 3.25 s **15.** yes **17.** no
19. yes **21.** no **23.** no **25.** yes **27.** yes
29. −1 **31.** 13 **33.** 3.4 **35.** −1.7 **37.** 3
39. $-3\frac{1}{2}$ **41.** $-3h + 4$ **43.** $6a + 4$ **45.** −12
47a. $p(a) = 0.5a + 110$ **47b.** 128

47c.

Blood Pressure vs. Age

49a. −7

49b.

The lines intersect at $x = -7$.

51. {(2, 28), (4, 14), (7, 8)} **53.** 30% **55.** $\frac{3}{20}$ **57.** 1.2

Pages 267–269   Lesson 6–5
1. Divide each side of $y = kx$ by x, where $x \neq 0$
3. Sample answer: The length of the trip depends upon the amount of gasoline used. So, the length of the trip represents the dependent variable and the amount of gasoline used represents the independent variable.   **5.** 5 **7.** $\frac{2}{5}$

9. yes

11. 13 **13.** 4.7 yd
15. no

17. no

19. no

21. 25 **23.** 20 **25.** $26\frac{1}{4}$ **27.** $2\frac{2}{3}$ **29.** 0.6 bu
31. \$3.15 **33a.** $h = 7f$
33b.

33c. about 5.5 ft **35.** 11 **37a.** $x - 50 = 550$
37b. about 600 pounds **39.** 8 **41.** B

Page 269   Quiz 2
1. yes **3.** −7.4 **5.** 8 gal

Pages 273–275   Lesson 6–6
1. Multiply the value of x by the value of y.
3. Sample answer: In a direct variation, as x
increases, y increases. In an inverse variation, as x
increases, y decreases.   **5.** direct; 5   **7.** direct; $\frac{1}{3}$
9. direct; $\frac{1}{4}$ **11.** 6   **13.** 3.25 hours   **15.** 10
17. 0.09   **19.** $\frac{5}{8}$ **21.** $-\frac{1}{6}$; $y = -\frac{1}{6}x$ **23.** 0.3; $y = 0.3x$
25. a; inverse   **27.** c; direct   **29a.** 36   **29b.** 2.4 m
29c. $\ell w = 36$
29d.

31. 27   **33.** −14.7   **35.** 457

**Pages 276–278   Chapter 6 Study Guide
and Assessment**
1. linear equation   **3.** functional notation   **5.** inverse
7. $y = kx$ **9.** $y = 2x + 3$
11.

x	y
−1	0
0	0
2	3
4	1

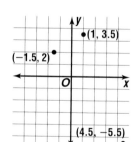

domain: {−1, 0, 2, 4},
range: {0, 3, 1}

13.

x	y
−1.5	2
1	3.5
4.5	−5.5

domain: {−1.5, 1, 4.5},
range: {2, 3.5, −5.5}

15. {(−4, 2), (−2, 1), (−2, −3), (0, −2), (1, 3)}

x	y
−4	2
−2	1
−2	−3
0	−2
1	3

domain: {−4, −2, 0, 1},
range: {2, 1, −3, −2, 3}

17. $\left\{\left(-1, \frac{1}{2}\right), (0, 0), (2, 1), (-4, 2)\right\}$

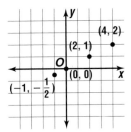

19. $\{(-1, 6.5), (0, 5), (2, 2), (4, -1)\}$

21. $\{-2, -1, 0, 1\}$ **23.** yes; $A = -1, B = -1, C = 0$
25. no
27.

29.

30.

33. no   **35.** yes   **37.** 4.5   **39.** 21   **41.** 12.5
43. 5   **45.** 8; $xy = 8$ **47.** $\{(5, 1), (10, 2), (15, 3),$
$(20, 4), (25, 5)\}$ domain: $\{5, 10, 15, 20, 25\}$, range:
$\{1, 2, 3, 4, 5\}$ **49.** 132 volts

1. A   **3.** E   **5.** E   **7.** B   **9.** 3/4

Chapter 7   Linear Equations
Page 283   Check Your Readiness
1. 4   **3.** -5 **5.** -1 **7.** -6 **9.** 18   **11.** 3
13. $-\frac{1}{2}$ **15.** -1 **17.** $\frac{6}{5}$ **19.** $(3, 4)$ **21.** $(3, 0)$
23. $(0, -3)$

Pages 287–289   Lesson 7–1
1a. Sample answer: **1b.** Sample answer:

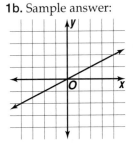

1c. Sample answer: **1d.** Sample answer:

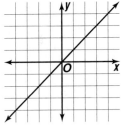

3. Percy; any two points on the line can be used to
find the slope.   **5.** 2   **7.** $\frac{3}{4}$ **9.** 4   **11.** $-\frac{6}{5}$ **13.** $-\frac{5}{3}$
15. $\frac{1}{3}$ **17.** -4 **19.** -2 **21.** $\frac{1}{3}$ **23.** $-\frac{2}{5}$ **25.** $-\frac{3}{8}$
27. $\frac{7}{100}$ **29a.** As x increases by 1, $f(x)$ increases by
0.25. The slope is 0.25.   **29b.** Sample answer: domain:
positive integers; range: positive multiples of 0.25
31. 12   **33.** no   **35.** -2

Pages 293–295   Lesson 7–2
1. the coordinates of a point on the line and the
slope of the line, or the coordinates of two points
on the line   **3.** $m = 2, (1, 6)$ **5.** $m = -3, (-9, -5)$
7. $y - 4 = \frac{3}{4}(x + 2)$ **9.** $y - 1 = 0$ or $y = 1$
11. $y - 3 = -\frac{5}{6}(x)$ or $y + 2 = -\frac{5}{6}(x - 6)$
13. $y - 5 = -2(x + 2)$ or $y + 7 = -2(x - 4)$
15. $y - 4 = -3(x + 1)$ **17.** $y - 1 = 6(x - 9)$
19. $y + 2 = \frac{4}{5}(x + 2)$ **21.** $x = 4$ **23.** $y + 4 =$
$-\frac{5}{2}\left(x - \frac{1}{2}\right)$ **25.** $y - 4 = \frac{1}{4}(x + 2)$ or $y - 5 = \frac{1}{4}(x - 2)$

27. $y = \frac{3}{4}(x + 2)$ or $y - 3 = \frac{3}{4}(x - 2)$ **29.** $y - 3 = -\frac{5}{2}(x + 3)$ or $y + 2 = -\frac{5}{2}(x + 1)$ **31.** $y - 5 = \frac{1}{2}(x - 4)$ or $y - 3 = \frac{1}{2}(x)$ **33.** $y + 4 = 3(x - 1)$ or $y - 5 = 3(x - 4)$ **35.** $y - 7 = -\frac{5}{8}(x + 3)$ or $y - 2 = -\frac{5}{8}(x - 5)$ **37.** $y - 6 = -\frac{1}{7}(x - 3)$ or $y - 7 = -\frac{1}{7}(x + 4)$ **39a.** Sample answer: $y - 5 = 5(x - 1)$ **39b.** Sample answer: domain: positive integers; range: positive multiples of 5   **41.** (4, 8)   **43.** {0, 1, 2}   **45.** {1, 2, 3}   **47.** 300% increase   **49.** 8.4

Pages 299–301   Lesson 7–3
1. Sample answer:

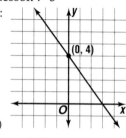

(3, 0)

3. Jacquie; the line $y = 2$ does not intersect the x-axis.   **5.** $y = \frac{1}{4}x - 3$ **7.** $y = 4x - 7$ **9.** $y = -\frac{1}{2}x - 3$ **11.** $y = 2x - 8$ **13a.** $y = 25x + 100$
13b. the pounds of cans he plans to collect each week
13c. 400 lb   **15.** $y = -2x + 5$ **17.** $y = \frac{3}{2}x + 7$
19. $y = -\frac{2}{5}x + 3$ **21.** $y = 4$ **23.** $y = 3x - 17$
25. $y = 5x$ **27.** $y = -2x + 13$ **29.** $y = -5x + 26$
31. $y = \frac{1}{2}x - \frac{9}{2}$ **33.** $y = -\frac{5}{4}x + 5$ **35.** $y = 4x - 12$
37. $y = -3$ **39.** $y = -\frac{4}{5}x + \frac{4}{5}$ **41.** $y = -\frac{3}{2}x + 6$; (4, 0), (0, 6)   **43.** $y = -\frac{2}{3}x + 2$; (3, 0), (0, 2)   **45.** $y = -3x + 11$ **47a.** $y = 8x$ **47b.** cost per person
47c. $320   **49.** $y - 5 = -2(x - 4)$

51.

x	y
−3	1
−1	5
1	3
2	6

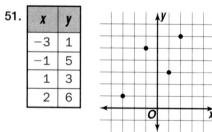

domain: {−3, −1, 1, 2}, range: {1, 5, 3, 6}
53. 9.144 km   **55.** −5   **57.** A

Page 301   Quiz 1
1. $-\frac{4}{3}$ **3.** $y - 4 = -2(x - 5)$ **5a.** 90

5b. The slope represents how much each problem is worth and the y-intercept represents the extra points students get with the curve.

Pages 305–307   Lesson 7–4
1. Graph the two sets of data as ordered pairs, where the x-coordinate is a member of one data set and the y-coordinate is a member of the other data set. The set of x-coordinates is the domain and the independent variable. The set of y-coordinates is the range and the dependent variable.

5a.

5b. Yes; as maximum temperature increases, so does the minimum temperature. There is a positive relationship.   **7.** Negative relationship; as speed increases, CO_2 decreases.

9.

There is a positive relationship; as the number of goals increases, the number of assists increases
11. Sample answer: Have higher maximum speed limits because CO_2 emissions decrease as speed increases.   **13.** no   **15.** negative

17a.

17b. No; there does not appear to be a relationship between elevation and precipitation.   **17c.** zero
19. $y = 5x + 2$ **21.** 1 cm = 5 m or 1:500   **23.** D

Pages 314–315   Lesson 7–5

1. two points, then draw a line through them

3. $-3, -3$

5. $2, 5$

7. $m = 1, b = 1$

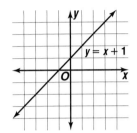

9. $m = \frac{1}{4}, b = -3$

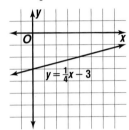

11a. $m = \frac{9}{5}, b = 32$ **11b.** $-16\frac{7}{9}$; It represents the temperature in degrees Celsius when the Fahrenheit temperature is $0°$.

13. $-5, -5$

15. $3, -1$

17. $-3, -6$

19. $-4, 3$

21. $2, -7$

23. $4, 8$

25. $m = -1, b = 2$

27. $m = 3, b = -1$

29. $m = 0, b = 4$

31. $m = \frac{3}{4}, b = -4$

33. $m = 2, b = 3$

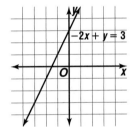

35. $m = -\frac{3}{5}, b = 2$

37a.

37b. $y = -\frac{7}{6}x + \frac{1}{3}$

37c. $m = -\frac{7}{6}, b = \frac{1}{3}$

39a. $f(x) = \frac{1}{4}x + 1$

39b.

39c. about $5\frac{1}{2}$ lb

41. Sample answer:

43. $\frac{1}{6}$ **45.** 0 **47.** 19 **49.** A

Pages 319–321 Lesson 7–6

1a. shifted down 4 units **1b.** less steep
1c. steeper **1d.** shifted up 3 units **3a.** Slope
represents rate. **3b.** A greater rate results in a
steeper positive slope.

5.

same y-intercept,
different slopes

7. same y-intercept, different slopes **9.** $y = x$
11.

same y-intercept,
different slopes

13.

same slope, different
y-intercepts

15.

same y-intercept,
different slopes

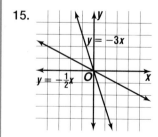

17. same y-intercept, different slopes **19.** same
slope, different y-intercepts **21.** same y-intercept,

different slopes **23.** $y = -\frac{4}{5}x$ **25.** $y = -\frac{4}{5}x + 2$
27. Sample answer: $y = -x + 6$ **29.** $y = 2x + 1$;
Sample answer: $y = 2x$ **31a.** Yes; they have the
same y-intercept, 0. **31b.** miles per gallon
31c. A barge gets 500 miles per gallon of fuel; its
graph has the steepest slope. **33.** Yes; they have
the same y-intercept, 7. **35.** $m = -2, b = 0$
37. 48 **39.** C

Page 321 Quiz 2
1. Positive; as air temperature increases, the
temperature of a glass of water would increase.
3. 4, 1

5a. Different slopes; they are a family because they
have the same y-intercept, 0. **5b.** cost per movie
5c. Store B

Pages 325–327 Lesson 7–7
1. Parallel lines have equal slopes. Perpendicular
lines have slopes that are negative reciprocals.
3. 4, $-\frac{1}{4}$ **5.** $\frac{2}{3}$, $-\frac{3}{2}$ **7.** neither **9.** The lines are
perpendicular. **11.** $y = \frac{3}{4}x - \frac{3}{2}$ **13.** $y = \frac{8}{3}x + 4$
15. perpendicular **17.** neither **19.** perpendicular
21. parallel **23.** perpendicular **25.** $y = -\frac{1}{2}x$
27. $y = -4x - 7$ **29.** $x = 2$ **31.** $y = \frac{1}{4}x + \frac{3}{4}$
33. $y = -\frac{1}{2}x + 1$ **35.** $y = -\frac{5}{3}x - \frac{5}{3}$ **37.** neither
39. Yes; the product of their slopes is -1.
41. 4 **43.** same y-intercept, different slopes
45. mutually exclusive **47.** mutually exclusive
49. 12.85

**Pages 328–330 Chapter 7 Study Guide and
Assessment**
1. false; y-intercept **3.** false; parallel **5.** true
7. false; horizontal **9.** true **11.** $\frac{2}{3}$ **13.** -2
15. $y - 5 = 3(x - 2)$ **17.** $y - 3 = -5(x)$
19. $y - 4 = -\frac{1}{6}(x - 10)$ or $y - 7 = -\frac{1}{6}(x + 8)$
21. $y = 3x - 9$ **23.** $y = 2x$ **25.** $y = 11x - 6$
27. no relationship

29. $2, -4$

31. $m = -2, b = 3$

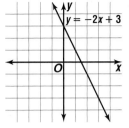

33. same y-intercept, different slope **35.** same slope, different y-intercepts **37.** neither **39.** parallel

41a.

41b. It would be steeper than graph A and less steep than graph B.

Page 333 Preparing for Standardized Tests
1. B **3.** E **5.** C **7.** A **9.** 4

Chapter 8 Powers and Roots

Page 335 Check Your Readiness
1. 36 **3.** 9 **5.** 625 **7.** 225 **9.** -48 **11.** $\frac{1}{16}$ **13.** 6
15. $\frac{1}{8}$ **17.** 3 **19.** 7.5 **21.** 6 **23.** 30.08 **25.** 4.5
27. 4.6

Pages 339–340 Lesson 8–1
1. A perfect square is the product of a number and itself. **3.** Becky; $(6n)(6n)(6n) = 6 \cdot 6 \cdot 6 \cdot n \cdot n \cdot n$, not $6n^3$. **5.** a^5 **7.** $12 \cdot 12 \cdot 12 \cdot 12$ **9.** $m \cdot m \cdot m \cdot m \cdot n \cdot n \cdot n$ **11.** 16^2 **13.** $2^3 3^2 5$ **15.** $(-2)^4$ **17.** 7^3
19. $2^2 3^3 5$ **21.** $3x^2 y^2$ **23.** $3 \cdot 3 \cdot 3 \cdot 3 \cdot 3$ **25.** $2 \cdot 2 \cdot 2 \cdot 2 \cdot 3 \cdot 3$ **27.** $y \cdot y \cdot y$ **29.** $6 \cdot a \cdot b \cdot b \cdot b \cdot b$
31. -32 **33.** -128 **35.** 24 **37.** 0.5 **39.** 4 **41.** 486
43a. 200.96 ft^2 **43b.** 21 bags **45.** $y = -2x + 6$
47.

49. 31.2 **51.** 50% **53.** B

Pages 344–345 Lesson 8–2
1. In the first case, the bases are the same, but in the second case, the bases are different. **3.** y^7 **5.** t^5
7. $12a^5$ **9.** n^3 **11.** xy^3 **13.** b^5 **15.** $10^3 = 1000$
17. 5^4 **19.** d^6 **21.** $a^3 b^5$ **23.** $r^7 t^8$ **25.** $15a^2$
27. $8x^3 y^5$ **29.** $m^7 b^2$ **31.** $m^3 n^3 a^2$ **33.** 9 **35.** k^5
37. $-12ab^4$ **39.** 4 **41.** $15a^3 c$ **43.** $2x^3$ **45.** $-16x^2$
47. 10^6 or 1,000,000 times **49a.** 10^{12} **49b.** 4^{15}
49c. x^8 **49d.** Multiply the exponents; $(x^a)^b = x^{ab}$.
51. 3 **53.** 15 **55.** 9.5 **57.** 3 **59.** D

Pages 349–351 Lesson 8–3
1. $\frac{1}{5^2} = \frac{1}{25}$ **3a.** Booker **3b.** Antonio **5.** $\frac{1}{3^3} = \frac{1}{27}$
7. $\frac{1}{n^3}$ **9.** $\frac{r^2}{pq^2}$ **11.** a^7 **13.** $\frac{c^3}{9a}$ **15.** $\frac{1}{2^5} = \frac{1}{32}$ **17.** $\frac{1}{4}$
19. $\frac{1}{r^{10}}$ **21.** a^2 **23.** $\frac{t^4}{s^2}$ **25.** $\frac{15r}{s^2}$ **27.** m^6 **29.** $\frac{1}{k^8}$
31. $\frac{a}{n^2}$ **33.** $\frac{1}{y^9}$ **35.** $3b^9$ **37.** $\frac{x^3}{4}$ **39.** $\frac{x^2}{7}$ **41.** $\frac{5}{8b^5 c}$
43. 18 **45.** 2^{-4} **47.** radar microwaves
49. $-10a^7$ **51.** n^6 **53.** z^4 **55.** 9^1 **57.** B

Page 351 Quiz 1
1. 6^3 **3.** 10^1 **5.** $2^5 3$ **7.** x^5 **9.** $-\frac{1}{3b^3}$

Pages 355–356 Lesson 8–4
1. no, 2.35×10^4 **5.** 68,000 **7.** 0.64 **9.** 0.0003
11. 0.0065 **13.** 5.69×10 **15.** 2.3×10^{-5}
17. $2.6 \times 10^3 = 2600$ **19.** 5,800,000,000
21. 0.0000000039 **23.** 0.009 **25.** 5.28×10^3
27. 2.683×10^2 **29.** 3.2×10^{-4} **31.** 4.296×10^{-3}
33. 1.2×10^{-2} **35.** 2.2×10^{-8} **37.** $6 \times 10^6 = 6,000,000$ **39.** $7.5 \times 10^2 = 750$ **41.** $2 \times 10^3 = 2000$
43. $5.5 \times 10^{-5} = 0.000055$ **45.** 5,000,000
47. Earth, Venus, Mars **49.** between 0.3 and 2 mm **51.** y^3 **53.** $-6c^6$ **55.** x^6 **57.** $a^3 b^3$
59. $-6x^2 y^6$ **61.** C

Pages 360–361 Lesson 8–5
1. 1, 4, 9, 16, 25, 36, 49, 64, 81, 100 **3.** $3^2 = 9$ and $\sqrt{9} = 3$ **5.** 11^2 **7.** $2^2 \cdot 7^2$ **9.** 2 **11.** -11 **13.** $\frac{1}{2}$
15. 52.5 mi **17.** 10 **19.** -14 **21.** 23 **23.** -26
25. -22 **27.** 17 **29.** $-\frac{3}{10}$ **31.** $-\frac{1}{4}$ **33.** $\frac{6}{7}$ **35.** $\frac{3}{4}$
37. -0.05 **39.** 0.03 **41.** 6 **43.** 52 m **45.** false; $-6(-6) \neq -36$ **47.** 6.3×10^4 **49.** 7.6×10^{-4}
51. $\frac{1}{x^4}$ **53.** $\frac{1}{x^2 y^3}$ **55.** Sample answer: $2x + 6 = x + 1$

Page 361 Quiz 2
1. 7×10^5 **3.** 21 **5.** square house, 30 ft $\times$ 30 ft

Pages 364–365 Lesson 8–6
1. Its decimal value does not terminate or repeat.
3. Sample answer: The area of a square garden is 200 square feet. Estimate the length of one side of the garden. **5.** 81, 100 **7.** 484, 529 **9.** 8 **11.** 16

13. 2 **15.** 4 **17.** 6 **19.** 11 **21.** 20 **23.** 24
25. 8 **27.** 11 **29.** 12 **31.** 0 **33.** $\sqrt{34}$ **35.** 17 h
37. 31 mi^2 **39.** 10, 11, 12, 13, 14, 15 **41.** 16
43. $\frac{13}{11}$ **45.** 0.000000004 **47.** $y = 5x - 17$
49. $y = -5x + 29$ **51.** C

Pages 369–371 Lesson 8–7
1. Sample answer:

3. false **5.** false **7.** 14.1 **9.** 8.1 **11.** yes **13.** 13
15. 8.2 **17.** 4.2 **19.** 6.7 **21.** 21 **23.** 11.4 **25.** no
27. no **29.** yes **31.** yes **33.** 9.4 m **35.** 6.9 ft
37. 8 **39.** 14 **41.** 15 **43.** 10 **45a.** $100
45b. $5 **47.** C

**Pages 374–376 Chapter 8 Study Guide
and Assessment**
1. prime factorization **3.** irrational numbers
5. exponent **7.** hypotenuse **9.** perfect squares
11. 9^5 **13.** $2^2 3^3$ **15.** $12 \cdot 12 \cdot 12$ **17.** $y \cdot y$
19. $5 \cdot x \cdot x \cdot y \cdot y \cdot y \cdot z \cdot z$ **21.** b^7 **23.** $a^4 b^3$
25. y^4 **27.** $3x^3$ **29.** $\frac{1}{2^3} = \frac{1}{8}$ **31.** $\frac{1}{x^4}$ **33.** y^7
35. $-5b^3$ **37.** 0.0000000015 **39.** 2.4×10^5
41. 4.88×10^9 **43.** 6×10^{11} **45.** 11 **47.** $\frac{2}{9}$
49. 4 **51.** 14 **53.** 7 **55.** 11.2 **57.** 10.4 **59.** 9 m

Page 379 Preparing for Standardized Tests
1. C **3.** B **5.** C **7.** D **9.** 25

Chapter 9 Polynomials

Page 381 Check Your Readiness
1. $7a$ **3.** $16g$ **5.** $8t$ **7.** $3m + 6$ **9.** $3p + 3q$
11. $ab + ac$ **13.** $6a + b$ **15.** $3y + 3$ **17.** n^6
19. $15d^2$ **21.** $12a^3$ **23.** $12jk^2$ **25.** $48nm^2$ **27.** $24g^4 h^4$

Pages 385–387 Lesson 9–1
1. Sample answer: $\frac{2}{x}$; it includes division.
3. $2y^2 + 5xy + 3x^2$ **5.** yes; product of numbers and
variables **7.** yes; binomial **9.** no **11.** 1 **13.** 2
15a. 20 **15b.** 9 **17.** no; includes addition
19. yes; product of numbers and variables
21. yes; product of numbers and variables
23. yes; trinomial **25.** no **27.** yes; binomial
29. yes; binomial **31.** yes; monomial **33.** yes;
monomial **35.** 2 **37.** 2 **39.** 1 **41.** 3 **43.** 6
45. 5 **47.** $-x^5 + x^2 - x + 25$ **49.** $5x^7 - 10x^6 +$
$3wx^2 + 6w^3x$ **51.** $7 + 2x - x^2 + 5x^3$ **53.** $y^6 +$
$5xy^4 - x^2y^3 + 3x^3y$ **55.** false **57a.** $-0.006t^4 +$
$0.14t^3 - 0.53t^2 + 1.79t$ **57b.** 4 **59.** yes **61.** yes
63. 7 **65.** 11 **67.** $\frac{x}{y^3}$ **69.** B

Pages 392–393 Lesson 9–2
1. Arrange like terms in column form.
3. $-2a - 9b$ **5.** $-x^2 - 8x - 5$ **7.** $-4xy^2 - 6x^2y + y^3$
9. $8y + 2$ **11.** $3x^2 - x + 2$ **13.** $2x + 2$ **15.** $x^2 + 3x$
$- 3$ **17.** $x^2 - x + 5$ **19.** $4x^2 + 3x + 3$ **21.** $14x + y$
23. $3n^2 + 13n + 11$ **25.** $7n^2 - 8n + 5$ **27.** $5x + 8$
29. $-x^2 + 3x - 3$ **31.** $-3x + y$ **33.** $a^2 + 3a - 1$
35. $5x^2 - xy - 2y^2$ **37.** $2pq + pr$ **39.** $-2x^2y - 7x^2y^2$
41. $-2x^2 + 6x - 7$ **43a.** $2w + 4$ in. **43b.** 34 in.
45. 2 **47.** 3 **49.** x^5y^3 **51.** $10xy^2$

Pages 396–398 Lesson 9–3
1. Distributive Property **3.** Consuelo; Shawn
forgot to multiply $2x$ and 4. **5.** $x^2 + 2x$ **7.** $-4x + 8$
9. $4z^3 - 8z^2$ **11.** $24y^2 + 9y - 15$ **13.** 4 **15.** 2
17. $3x^2$ **19.** $-3y - 9$ **21.** $x^2 - 5x$ **23.** $3z^2 - 2z$
25. $8x^2 - 12x$ **27.** $-2a^2 + 4a$ **29.** $-30y + 10y^2$
31. $5d^3 + 15d$ **33.** $-10a^3 + 14a^2 - 4a$ **35.** $40n^4 +$
$35n^3 - 15n^2$ **37.** $-14a^2 + 35a - 77$ **39.** $1.2c^2 - 12$
41. $x^2 - 2x$ **43.** $4y^2 - 6y + 2$ **45.** 5 **47.** -2
49. -1 **51.** 2 **53.** -3 **55.** 17 **57.** $4x^3 - 8x^2 + 4x$
59. $-4a^3 - 10a^2 - 10a + 66$ **61.** $21n^3 - 6n^2 -$
$46n + 28$ **63.** $8t^2 + t$ **65a.** $7y^2 + 35y$ units^2
65b. 1050 in^2 **67.** $7x - 2$ **69.** $3x^2 - 2x - 3$

71. 5 **73.** -11 **75.** $\frac{3}{10}$ **77.** 1:1

Page 398 Quiz 1
1. 4 **3.** $-2x^2 + 8x - 2$ **5a.** x ft, $2x + 40$ ft
5b. $2x^2 + 40x$ ft^2

Pages 402–404 Lesson 9–4
1.

x	3
x^2	$3x$
$-2x$	-6

$(x - 2)(x + 3) = x^2 + x - 6$

3. The Distributive Property is used to multiply
both two binomials and a binomial and a
monomial. But when you multiply two binomials,
there are four multiplications to perform. With a
binomial and a monomial there are only two
multiplications to perform. **5.** $6x$ **7.** $13m$ **9.** $14y$
11. $w^2 - 2w - 35$ **13.** $2y^2 + y - 3$ **15.** $6y^2 +$
$11y - 35$ **17.** $2a^2 - ab - 10b^2$ **19.** $m^4 + 3m^3 -$
$2m^2 - 6m$ **21.** $x^2 + 12x + 32$ **23.** $a^2 + 4a - 21$
25. $n^2 - 16n + 55$ **27.** $z^2 + 2z - 24$ **29.** $3x^2 -$
$10x - 8$ **31.** $3z^2 - 17z + 20$ **33.** $5n^2 - 13n - 6$
35. $9a^2 + 6a + 1$ **37.** $12h^2 - h - 6$ **39.** $2x^2 + x - 21$
41. $6y^2 - 24yz + 24z^2$ **43.** $6x^2 + 13xy - 5y^2$
45. $x^3 - 2x^2 + x - 2$ **47.** $3x^4 + 8x^2 - 3$ **49.** $x^3 -$
$x^2 - 6x$ **51.** $x^2 - 9$ **53a.** $3x + 2$ **53b.** $3x^2 + 4x - 1$
55a. $4x^3 + 2x^2 - 20x$ in. **55b.** 450 in^3 **57a.** $10 - 2t$
57b. 6 in. **59.** $2^3 x^3$ **61.** $3^2 (-2)^3$

63.

$y = -x + 2$

65. 84 students

Pages 408–409 Lesson 9–5
1a. ii **1b.** iii **1c.** i **3.** $y^2 + 4y + 4$ **5.** $m^2 - 18m + 81$ **7.** $x^2 - 49$ **9.** 90.25% **11.** $x^2 + 10x + 25$ **13.** $w^2 - 16w + 64$ **15.** $m^2 - 6mn + 9n^2$ **17.** $x^2 - 16y^2$ **19.** $25 + 10k + k^2$ **21.** $y^2 - 9z^2$ **23.** $9x^2 - 25$ **25.** $16 + 16x + 4x^2$ **27.** $a^2 - 6ab + 9b^2$ **29.** $x^3 + 2x^2 + x$ **31.** $x^2 - 4y^2$ **33.** $(40-1)(40+1) = 1600 - 1$ or 1599 **35a.** $\frac{1}{2}x^2 + \left(-\frac{9}{2}\right)$ **35b.** 8 square units **35c.** about 9.5 units **37.** $3x^2 - 4x - 3$ **39.** $y = 3x + 4$ **41.** $y = \frac{2}{5}x - 3$ **43.** B

Page 409 Quiz 2
1. $c^2 + 10c + 16$ **3.** $c^2 - 2c + 1$ **5.** $12x$ cm^2

Pages 412–414 Chapter 9 Study Guide and Assessment
1. i **3.** j **5.** c **7.** a **9.** e **11.** yes; monomial **13.** yes; trinomial **15.** 0 **17.** 4 **19.** $7cd - d^2 + 3d^3$ **21.** $9x^2 - 3x + 6$ **23.** $7s^2 - 6s - 2$ **25.** $4g^2 + 8g$ **27.** $-10s^2t^2 + 2st^2$ **29.** $3x - 15$ **31.** $2m^3 + 8m^2$ **33.** $10x^2 + 15x - 10$ **35.** -3 **37.** $y^2 + 4y - 12$ **39.** $x^2 - 4x + 3$ **41.** $2m^2 - 9m + 4$ **43.** $12y^2 + 10y + 2$ **45.** $10x^2 - 7x - 12$ **47.** $4x^2 + 12x + 9$ **49.** $x^2 - 4x + 4$ **51.** $25m^2 - 30mn + 9n^2$ **53.** $4a^2 - 9$ **55.** $25a^2 + 10a + 1$ **57.** $10\frac{1}{2}x^2 + 22x - 16$ **59a.** $64x^2 - 32x + 4$ **59b.** $(48x - 12)(48x - 12)$

Page 417 Preparing for Standardized Tests
1. C **3.** D **5.** B **7.** A **9.** 265

Chapter 10 Factoring

Page 419 Check Your Readiness
1. 3^4 **3.** 10^5 **5.** 2^25^4 **7.** p^3q^3 **9.** 5^2 **11.** k^9 **13.** $6m^2n$ **15.** $2x + 2y$ **17.** $48b + 16$ **19.** $14f - 8$

Pages 424–425 Lesson 10–1
1. 23, 29, 31, 37, 41, 43, 47 **3.** Write $8x^2$ and $16x$ as the product of prime numbers and variables where no variable has an exponent. Then multiply the common prime factors and variables. The GCF is $8x$. **5.** $3 \cdot 7$ **7.** $3 \cdot 17$ **9.** $2^2 \cdot 3^3$ **11.** 1, 2, 3, 6, 7, 14, 21, 42; composite **13.** $2 \cdot 2 \cdot 2 \cdot 3 \cdot x \cdot x \cdot y$ **15.** 5

17. 1 **19.** $7y^2$ **21.** 72 cm, 1 cm **23.** 1, 2, 4, 5, 10, 20; composite **25.** 1, 3, 5, 9, 15, 45; composite **27.** 1, 7, 13, 91; composite **29.** $-1 \cdot 3 \cdot 5 \cdot a \cdot a \cdot b$ **31.** $2 \cdot 5 \cdot 5 \cdot m \cdot m \cdot n \cdot n$ **33.** $2 \cdot 3 \cdot 3 \cdot 5 \cdot y \cdot z \cdot z$ **35.** 4 **37.** 18 **39.** 9 **41.** $3y$ **43.** 9 **45.** 1 **47.** 1 **49.** 5 **51.** 5 **53.** 5 **55.** 12 in. by 12 in. **57.** Every even number has a factor of 2. So, any even number greater than 2 has at least 3 factors—the number itself, 1, and 2—and is therefore composite. **59.** $4y^2 + 4y + 1$ **61.** $z^2 + 7z + 12$ **63.** $2n^2 + 9n + 4$ **65.** $6a - 2a^3$ **67.** 8.17×10^8

Pages 431–433 Lesson 10–2
1. Sample answer:

3. The Distributive Property is used to factor out a common factor from the terms of an expression. Examples: $5xy + 10x^2y^2 = 5xy(1 + 2xy)$, $2a^2 + 8a + 10 = 2(a^2 + 4a + 5)$ **5.** x **7.** $6y$ **9.** $5m^2n$ **11.** $2x(x + 2)$ **13.** prime **15.** $2ab(a^2b + 4 + 8ab^2)$ **17.** $c + 3$ **19.** $3(3x + 5)$ **21.** $2x(4 + xy)$ **23.** $3c^2d(1 - 2d)$ **25.** $mn(36 - 11n)$ **27.** prime **29.** $y^3(12x + y)$ **31.** $3y(8x + 6xy - 1)$ **33.** $x(1 + xy^3 + x^2y^2)$ **35.** $2a(6xy - 7y + 10x)$ **37.** $9x^2 - 7y^2$ **39.** $a + 2$ **41.** $2xy^2 + 3z$ **43.** $2x^2 + 3$ **45.** $(4x + y)$ ft **47a.** $8(a + b + 8)$ **47b.** $3(4a + b + 12)$ **49.** prime **51.** composite **53.** composite **55.** $3x^2 - 2x - 12$ **57.** Sample answer: $x^2 + 3x - 4$

Page 433 Quiz 1
1. $2^3 \cdot 3$ **3.** x^3; $x^3(a + 7b + 11c)$ **5.** $8(x + 1)$

Pages 438–439 Lesson 10–3
1.

$(x + 1)(x + 6)$

3. $-10, -3$ **5.** $-4, -3$ **7.** $-6, 1$ **9.** $(x + 3)(x + 2)$ **11.** $(a - 1)(a - 4)$ **13.** $(x + 5)(x - 2)$ **15.** prime **17.** $2(c - 7)(c + 1)$ **19.** $(b + 4)(b + 1)$ **21.** $(a + 3)(a + 4)$ **23.** $(y + 9)(y + 3)$ **25.** $(x - 5)(x - 3)$ **27.** $(c - 9)(c - 4)$ **29.** prime **31.** $(c + 3)(c - 1)$ **33.** $(r - 6)(r + 3)$ **35.** $(n + 15)(m - 2)$ **37.** $(x - 9)(x - 8)$ **39.** $(r + 24)(r - 2)$ **41.** $3(y - 4)(y - 3)$ **43.** $m(m + 1)(m + 2)$ **45.** $2a(a + 8)(a - 1)$ **47.** Sample answer: $x^2 + 7x - 9$ **49a.** $x^2 - 5x - 24$ **49b.** $(x - 8)(x + 3)$

51a.

51b.

C   c

c   Cc   cc

c   Cc   cc

53. $2x^2 + 5y^2$ **55.** $3a^2 + 4ab - 3b^2$ **57.** 4   **59.** 1
61a. $y = 0.2x - 398$ **61b.** the average increase each year   **61c.** \$3 billion

Pages 443–444   Lesson 10-4
1a. $2x^2 - 5x - 12 = (x - 4)(2x + 3)$ **1b.** $6x^2 - 11x + 3 = (2x - 3)(3x - 1)$ **1c.** $5x^2 + 28x + 15 = (5x + 3)(x + 5)$ **3.** $(2a + 3)(a + 1)$ **5.** $(5x + 3)(x + 2)$
7. $(2x + 7)(x - 3)$ **9.** $(5a - 2)(2a - 1)$
11. $2(3x + 5)(x + 1)$ **13.** $(2y + 1)(y + 3)$
15. $(2a + 1)(2a + 3)$ **17.** $(2q + 3)(q - 6)$
19. $(7a + 1)(a + 3)$ **21.** $(3x + 2)(x + 4)$
23. $(3x + 5)(x + 3)$ **25.** prime   **27.** $(7x - 1)(2x + 5)$
29. $(2m - 1)(4m - 3)$ **31.** $3(2x + 5)(x - 2)$
33. $x(2x - 3)(x + 4)$ **35.** $(6y - 1)(y + 2)$
37. $(2a - b)(a + 3b)$ **39.** $(3k + 5m)(3k + 5m)$
41. $3x, 2x - 1, x + 3$ **43a.** $2(y^2 - 7y - 2)$
43b. $2(y - 9)(y - 1)$ **43c.** 56 in^2, 18 in^3
45. $(x + 16)(x - 2)$ **47.** $3a(a^2 - 5a + 2)$ **49.** 25 ft
51. 10

Page 444   Quiz 2
1. $(x + 5)(x - 2)$ **3.** $(2x + 7)(x + 1)$
5. $(x + 3)(x - 2)$

Pages 448–449   Lesson 10-5
1. No; 7 is not a perfect square.   **3.** $(y + 7)^2$
5. $(a - 5)^2$ **7.** $2(2x - 5y)(2x + 5y)$ **9.** $3(x^2 + 5)$
11. $3(y - 1)(y + 8)$ **13.** $(r + 4)^2$ **15.** $(a + 1)^2$
17. $(2z - 5)^2$ **19.** $(3a + 4)^2$ **21.** $(7 + z)^2$
23. $(a - 6)(a + 6)$ **25.** $(1 - 3m)(1 + 3m)$ **27.** no
29. $2(z - 7)(z + 7)$ **31.** Sample answer: $25x^2 - 4$; $(5x - 4)(5x + 4)$ **33.** $5(x^2 + 5)$ **35.** $(y - 3)(y - 2)$
37. $2(x - 6)(x + 6)$ **39.** $xy(8y - 13x)$ **41.** prime
43. $2(5n + 1)(2n + 3)$ **45.** $(4w - 3)(2w + 5)$
47. $5(x + 1)(x + 2)$ **49.** $2x(x - 4)(x + 4)$ **51.** 2, 4
53. 4   **55.** $(4x - 1)(x + 3)$ **57.** $(m - 7)(m + 2)$
59. a^2 **61.** $\frac{d}{3c^2}$ **63.** C

Pages 450–452   Chapter 10 Study Guide and Assessment
1. false, $2^2 \cdot 3$ **3.** true   **5.** true   **7.** false, $x^2 - 9$
9. true   **11.** 5   **13.** 1   **15.** $9x$ **17.** $5(x + 6y)$
19. $6a(2b - 3a)$ **21.** $3xy(1 + 4xy)$ **23.** $5ab - 1$
25. $(x - 5)(x - 3)$ **27.** $(x - 4)(x + 2)$ **29.** $(x + 7)(x - 5)$ **31.** $2(n - 6)(n + 2)$ **33.** $(3x + 5)(x + 1)$
35. $(3a - 2)(2a + 1)$ **37.** $(2y + 3)(y - 6)$
39. $5(3a - 1)(a - 1)$ **41.** $(a - 6)^2$ **43.** $(5x + 2)^2$
45. $(2x + 3)(2x - 3)$ **47.** $12(c + 1)(c - 1)$

49.

Page 455   Preparing for Standardized Tests
1. C   **3.** D   **5.** D   **7.** B   **9.** 0.5

Chapter 11   Quadratic and Exponential Functions

Page 457   Check Your Readiness
1.   **3.**

5.

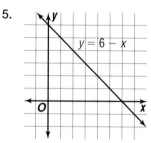

7. 18   **9.** 4   **11.** 2   **13.** $(x + 2)(x + 8)$
15. $(b - 5)(b + 3)$ **17.** $(2c - 1)(c + 3)$

Pages 461–463   Lesson 11-1
1. Sample answer: Quadratic functions have a degree of 2, but linear functions have a degree of 1. The graphs of quadratic functions are not straight lines as with linear functions.   **3.** the vertex
5. 1, 0, 4   **7.** 3, 4, 0

9.

11. $x = 0$; $(0, 2)$; D: {all reals}; R: {all reals greater than or equal to 2}

13. $x = 2$; $(2, 7)$; D: {all reals}; R: {all reals less than or equal to 7}

15a.

15b. 9 ft   **15c.** 6 ft

17.

19.

21.

23.

25. D: {all reals}; R: {all reals greater than or equal to 0}

27. D: {all reals}; R: {all reals greater than or equal to 0}

29. D: {all reals}; R: {all reals less than or equal to 1}

31. D: {all reals}; R: {all reals greater than or equal to -2}

33. D: {all reals}; R: {all reals less than or equal to 7}

35. D: {all reals}; R: {all reals greater than or equal to $-\dfrac{25}{4}$}

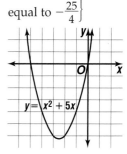

37. C   **39.** B   **41.** $x = -1$

43a. 150   **43b.** \$22,500   **43c.** The domain is all real numbers greater than or equal to 0. The range is all real numbers greater than or equal to 0 and less than or equal to 22,500.   **45.** $(x - 7)(x + 7)$ **47.** $2(g - 4)^2$ **49.** $-x^3 - 4x^2 - 8x + 5$; 3   **51.** 1.4 mi

Pages 466–467   Lesson 11–2
1. extremely narrow with vertex at $(0, 0)$, opening up
3. B, A, C   **5.** widens; $(0, 0)$ **7.** left 4 units; $(-4, 0)$
9. $y = 0.5(x - 4)^2 + 3$ **11.** shifts left, opens up
12. same axis of symmetry, shifts down, opens down
13. narrows; $(0, 0)$ **15.** right 7 units; $(7, 0)$ **17.** opens down, narrows; $(0, 0)$ **19.** narrows, up 1 unit; $(0, 1)$
21. widens, down 8 units; $(0, -8)$ **23.** left 1 unit, down 5 units; $(-1, -5)$ **25.** $y = (x + 8)^2$
27. $h(t) = -4.9(t - 34)^2 + 80$ **29.** $y = (x - 2)^2 - 4$
31. $(-1.5, -1.5)$ **33.** $a^2 + 3a - 28$ **35.** B

Pages 471–473   Lesson 11–3
1. Sample answer: The x-intercepts are the values for x where $f(x)$ equals 0. Before solving a quadratic function, you always set it equal to zero.
3. Sample answer:

5. $-1, 4$ **7.** 2, 3   **9.** between -2 and -1; between 2 and 3   **11.** -1 **13.** $-2, 7$ **15.** 5   **17.** between -5 and -4; between 0 and 1   **19.** between -1 and 0; between 2 and 3   **21.** $-4, -8$ or 4, 8

23.

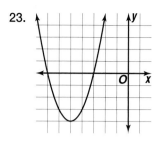

25. between 46 and 47 s **27.** The value of the function changes from negative when $x = 1$ to positive when $x = 2$. To do this, the value of the function would have to be 0 for some value of x between 1 and 2. Thus, the value of x represents the x-intercept of the function. This is the root of the related equation. **29.** up 4 units; $(0, 4)$ **31.** 4

Page 473   Quiz 1
1. $x = -3$; $(-3, -14)$ **3.** $x = -1$; $(-1, 7)$

5. right 6 units; $(6, 0)$ **7.** 2, 3   **9.** between -1 and 0; between 2 and 3

Pages 476–477   Lesson 11–4
1. Any number times 0 is 0. Therefore, if two or more numbers are multiplied and the result is 0, at least one of those numbers has to equal 0.
3. They are both correct, but factoring can only be used when the equation is not prime. **5.** 0, 5
7. 3 **9.** -2, 4 **11.** 0, -2 **13.** -7, 6 **15.** -1, 4
17. -8, -2 **19.** 3, 8 **21.** -3, 4 **23.** 0, 3 **25.** 2
27. -1, 5 **29.** 25 ft by 35 ft **31.** 8, 11; -11, -8
33. 3 ft by 1 ft by 1 ft **35a.** -5, 2 **35b.** Sample answer: $y = (x + 5)(x - 2)$ **37.** no real solution
39. The graph of $y = 0.25x^2$ is wider than the graph of $y = x^2$. The graph of $y = 4x^2$ is narrower than the graph of $y = x^2$.

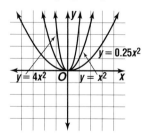

41. $(b + 7)(b - 6)$

Pages 481–482   Lesson 11–5
1. The equation is not factorable with integers and the coefficient of x^2 is 1. **3.** yes **5.** no **7.** 1
9. $\frac{9}{4}$ **11.** -3, -4 **13.** 9 ft by 11 ft **15.** 36 **17.** 1
19. 144 **21.** $\frac{25}{4}$ **23.** -3, 1 **25.** -5, -1 **27.** 0, 14
29. $1 \pm \sqrt{31}$ **31.** $3 \pm \sqrt{19}$ **33.** -5, 2 **35.** 3 s or 5 s **37.** -18 or 18 **39.** 2, 5 **41.** -3
43a. $2b(a + 4b)$ **43b.** $2a + 12b$ **43c.** 9 in. by 2 in., 18 in^2, 22 in.

Pages 486–487   Lesson 11–6
1. Sometimes; quadratic equations can have no real solutions. **3.** 48 **5.** 192 **7.** ± 5 **9.** -3, -1
11. -6, -1 **13.** 2, 3 **15.** no real solutions
17. 0, 8 **19.** $\frac{-1 \pm \sqrt{29}}{2}$ **21.** $\frac{-2 \pm \sqrt{7}}{2}$
23. $\frac{-7 \pm \sqrt{69}}{2}$ **25.** $\frac{5 \pm \sqrt{249}}{16}$ **27a.** 6.25 s
27b. about 1.02 s **29.** 4 **31.** 81 **33.** B

Page 487   Quiz 2
1. -6, 2 **3.** 3, 7 **5.** $3 \pm \sqrt{2}$ **7.** no real solutions
9. $\frac{-5 \pm \sqrt{13}}{2}$

Pages 491–493   Lesson 11–7
1. Sample answer: The graph begins almost flat, but then for increasing x-values, it becomes more and more steep. **3.** 256 **5.** 1.26
7. -5

9a.

Year	Balance	Pattern
0	500	$500(1.02)^0$
1	$500(1.02)1.02 = 510$	$500(1.02)^1$
2	$(500 \cdot 1.02)1.02 = 520.20$	$500(1.02)^2$
3	$(500 \cdot 1.02 \cdot 102)1.02 = 530.604$	$500(1.02)^3$

9b. $T(x) = 500(1.02)^x$ **9c.** \$714.12 **9d.** \$1428.25

11. 1   **13.** 2

15. −2   **17.** 0

19. $5304.50   **21.** $3920.88   **23.** $7986.70

25a.

Day	Cents	Pattern
1	1	2^0
2	$2 \cdot 1$	2^1
3	$2 \cdot 2 \cdot 1$	2^2
4	$2 \cdot 2 \cdot 2 \cdot 1$	2^3
5	$2 \cdot 2 \cdot 2 \cdot 2 \cdot 1$	2^4

25b. $y = 2^{x-1}$ **25c.** January 17   **27.** −2, 2
29a. $(2x + 4)x = 30$ **29b.** 3 yd by 10 yd
31. $y - 6 = 0$

Pages 496–498   Chapter 11 Study Guide and Assessment
1. f   **3.** c   **5.** j   **7.** a   **9.** d   **11.** $x = 0$; $(0, -3)$
13. $x = -1$; $(-1, 20)$

15.

17.

19. shifts right; $(8, 0)$ **21.** widens; $(0, 0)$
23. shifts right and up; $(1, 7)$ **25.** −3, 4
27. no real solutions   **29.** $-1 < x < 0, 3$
31. −4, 5   **33.** −8   **35.** −1, 1   **37.** 4   **39.** −3, 7
41. −1, −9   **43.** $-\frac{1}{3}, 1$ **45.** $\dfrac{-5 \pm \sqrt{13}}{2}$
47. 1   **49.** 4

51. $5.50

Page 501   Preparing for Standardized Tests
1. D   **3.** A   **5.** C   **7.** C   **9.** 2

Chapter 12   Inequalities

Page 503   Check Your Readiness
1. <   **3.** >   **5.** <

7.

9.

11. 5   **13.** 10   **15.** 39   **17.** −6, 6   **19.** −12, 16
21. −10, 10   **23.** 9   **25.** 15   **27.** 7   **29.** 4   **31.** 21

Pages 506–508   Lesson 12–1
1. i. b; ii. a; iii. d; iv. c   **3.** The graph of $x \le 4$ has a bullet at 4. The graph of $x < 4$ has a circle at 4.
5. $x < 14$

7.

9.

11. $x < -1$ **13.** $x > 4$ **15.** $x < 9$ **17.** $x \ge -8$

19.

21.

23.

25.

27.

29.

31. $x > -6$ **33.** $x \le 3$ **35.** $x \le 5$ **37.** $x \ge 2.5$
39. $f < 3$;

41. $w \le 3600$;

43. 7 yr **45.** no real solutions **47.** $2^3 \cdot 3$; two
49. $-6x^2y$

Pages 511–513   Lesson 12-2
1. They are the same. **5.** $8 + n \le 12$ **7.** $\{v \mid v \ge 17\}$
9. $\{h \mid h \le 21\}$ **11.** $\left\{y \mid y < 1\frac{5}{6}\right\}$

13. $\{x \mid x < -1\}$;

15. $\{n \mid n > 3\}$ **17.** $\{b \mid b < -5\}$ **19.** $\{r \mid r < 19\}$
21. $\{w \mid w > 22\}$ **23.** $\{t \mid t \le 0\}$ **25.** $\{c \mid c > -4\}$
27. $\{d \mid d < 5.4\}$ **29.** $\{z \mid z \le -0.5\}$ **31.** $\left\{v \mid v \le 2\frac{1}{2}\right\}$
33. $\{g \mid g < 13\}$;

35. $\{x \mid x \ge -16\}$;

37. $\{t \mid t \ge 9\}$;

39. $x - 8 \le 12$; $x \le 20$ **41.** $t \ge 99$ **43.** $p \ge 99$
45. Sometimes; examples are $x > x$ and $2x + 1 \le 2x$.
47. $x < -8$ **49.** $x \ge 5$ **51.** $(3x + 2)(x - 5)$ **53.** 32

Pages 516–518   Lesson 12-3
1. For both, reverse the sign when dividing or multiplying by a negative number.
3.

5. $\{x \mid x \ge -9\}$ **7.** $\{x \mid x \ge 20\}$ **9.** $\{a \mid a \le -30\}$
11. $\{h \mid h \ge -2\}$ **13.** $\{d \mid d \le -3\}$ **15.** $\{g \mid g > -36\}$
17. $\{x \mid x \ge 7\}$ **19.** $\{z \mid z > -3\}$ **21.** $\{c \mid c > -54\}$
23. $\left\{k \mid k > \frac{5}{3}\right\}$ **25.** $\left\{w \mid w > \frac{1}{2}\right\}$ **27.** $\{y \mid y \ge 6\}$
29. $\{v \mid v \ge 2\}$ **31.** $\{s \mid s \le 280\}$ **33.** $\{n \mid n < 105\}$
35. $x < 45$ **37.** at least 17 lawns **39.** $\{z \mid z \le 4\}$
41. $p > 600$ **43.** $4v^2 - 1$ **45.** $\frac{1}{400}$

Page 518   Quiz 1
1. $x < -1$
3. $\{x \mid x > 5\}$;

5. $\{z \mid z < -3\}$;

7. $\{x \mid x > -5\}$ **9.** $\left\{v \mid v \le \frac{15}{2}\right\}$

Pages 522–523   Lesson 12-4
1. subtraction, then division **3.** -7 **5.** -3.1
7. $\{y \mid y > 4\}$ **9.** $\{x \mid x < -1\}$ **11.** $\{w \mid w > -2\}$
13. $s \ge 92$ **15.** $\{b \mid b < 12\}$ **17.** $\{x \mid x \ge 4\}$
19. $\{g \mid g > 2\}$ **21.** $\{v \mid v \le 0.5\}$ **23.** $\{m \mid m \le -3\}$
25. $\{d \mid d < 1\}$ **27.** $\{x \mid x \le 2\}$ **29.** $\{y \mid y < 1\}$
31. $\{b \mid b < 11\}$ **33.** $\{x \mid x < 1\}$ **35.** at least \$260
37. $x > \$1484$ **39.** $\{p \mid p > 7\}$ **41.** $\{x \mid x \le -40\}$
43. $m \le \$19.50$ **45.** A

Pages 526–529   Lesson 12-5
1. an inequality made from two inequalities
3. intersection **5.** union **7.** $5 < \ell < 10$
9.

11. $\{c \mid 3 < c < 8\}$;

13. $\{v \mid 0 \le v \le 4\}$;

15. $\{j \mid j \le -1 \text{ or } j > 0\}$;

17. $9 \le t \le 11.5$;

19. $-8 < h < 8$ **21.** $2 \le g \le 5$ **23.** $6 < x \le 8$
25.

27.

29.

31. $\{x \mid 4 \le x \le 8\}$;

33. $\{w \mid 0 < w \le 3\}$;

35. $\{h \mid -3 \le h \le 3\}$;

37. $\{c \mid -5 < c < -4\}$;

39. $\{z \mid z \le -7 \text{ or } z > 4\}$;

41. $\{r \mid r \le -1 \text{ or } r > 0\}$;

43. $\{p \mid p \ge -1\}$;

45. $\{d \mid d > 1.5\}$;

47. $\{w \mid w \le -6 \text{ or } w > -2\}$;

49. $x < -3 \text{ or } x > 3$ **51.** $-3 \le x < 5$ **53.** $\{x \mid x \le 3 \text{ or } x > 9\}$ **55.** $\{k \mid 11 \le k \le 13\}$ **57a.** Sample answer: $11 \le h \le 14$

57b.

59. $6.6 < p < 9.4$ **61.** $\{y \mid y < 0\}$ **63.** $\{x \mid x \ge 5\}$
65. $\{b \mid b > 6\}$

67a.

67b. 34 in.   **69.** $3b - 18$ **71.** C

Pages 532–534   Lesson 12–6

1. $|x| < 7$ includes all numbers between -7 and 7. $|x| > 7$ includes all numbers to the left of -7 and to the right of 7.   **3.** Mia; $|x| \le 0$ includes only 0. $|x| \ge 0$ includes all real numbers.   **5.** $x \le 3$ and $x \ge -3$ **7.** $x > 8$ or $x < -8$
9. $\{x \mid -4 < x < 8\}$;

11. $\{t \mid t \ge 8 \text{ or } t \le 2\}$;

13. $\{s \mid s \ge -1 \text{ or } s \le -7\}$;

15. $\{m \mid -6 < m < 4\}$;

17. $\{z \mid -9 \le z \le -5\}$;

19. $\{p \mid -1 < p < 3\}$;

21. $\{t \mid -2 \le t \le 2\}$;

23. $\{k \mid -5.5 < k < 1.5\}$;

25. $\{y \mid y \le -9 \text{ or } y \ge 3\}$;

27. $\{x \mid x < -2 \text{ or } x > 2\}$;

29. $\{r \mid r \le 3 \text{ or } r \ge 5\}$;

31. $\{h \mid h < -6 \text{ or } h > 12\}$;

33. $|s - 90| < 4$ **35.** $|s - 65| \le 3$
37. $|x - 1| \ge 1$
39. $\{x \mid 0 < x < 8\}$;

41. $\{x \mid x < -10 \text{ or } x > 8\}$;

43. $|m - 3.25| \le 0.05; \ 3.20 \le m \le 3.30$
45.

47.

49. $\{y \mid y \le -1\}$ **51.** $-2, 6$

Page 534   Quiz 2

1. $\{x \mid x < 1\}$ **3.** $\{d \mid d \ge 1.2\}$
5. $\{n \mid -5 \le n < 1\}$;

7. $\{y \mid 5 < y < 9\}$;

9. $\{x \mid -4 < x < 4\}$;

Pages 538–539   Lesson 12–7

1. Graph a solid line for ≤ and ≥. Graph a dashed line for < and >.

3.

5.

7.

9.

11.

13.

15.

17.

19.

21.

23.

25.

27.

29.

31.

33a.

33b. Sample answer: {(4, 9), (2, 11), (3, 16)}
33c. positive, whole
35.

37. $\{h \mid 5 < h < 9\}$;

39. $\{b \mid b < -2 \text{ or } b > 1\}$;

41. 8.483×10^2

Pages 542–544   Chapter 12 Study Guide and Assessment

1. union   **3.** half-plane   **5.** boundary
7. half-plane   **9.** set-builder notation
11.

13.

15. $x \geq -5$ **17.** $\{x \mid x > 4\}$ **19.** $\left\{y \mid y \geq 1\frac{1}{6}\right\}$

21. $\{x \mid x \leq 3\}$ **23.** $\{n \mid n < -5\}$ **25.** $\{x \mid x > 15\}$

27. $\{w \mid w \leq 8\}$ **29.** $\{x \mid x \geq 4\}$ **31.** $\{t \mid t \leq 9\}$

33. $4x - 3 > 25; \{x \mid x > 7\}$

35. $\{y \mid -7 \leq y < -1\}$;

37. $\{a \mid a \leq 1 \text{ or } a \geq 2\}$;

39. $\{t \mid -4 \leq t \leq 6\}$;

41. $\{x \mid -4 < x < -2\}$;

43. $\{y \mid y \neq 2\}$;

45. $|s - 80| > 5$

47.   **49.**

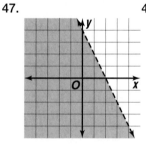

51. $\{x \mid 30 < x < 60\}$

Page 547   Preparing for Standardized Tests
1. A   **3.** C   **5.** A   **7.** B   **9.** 270

Chapter 13   Systems of Equations and Inequalities

Page 549   Check Your Readiness
1.

3.

5.

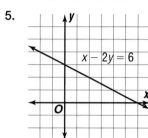

7. 3   **9.** 4   **11.** 2

13.

15.

17.

19.

21.

23.

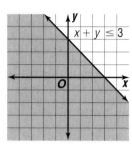

17. $(-2, -3)$ **19.** $(-6, 3)$

    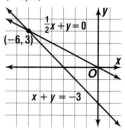

21. $(2, 6)$ **23.** $(10, -6)$
25. Sample answer: $y = x + 4$ and $y = -x + 4$
27. $\{x \mid 1 < x < 5\}$

29. $\{m \mid m > 2 \text{ or } m < -4\}$

31. $\frac{3}{2}$, 2

Pages 552–553   Lesson 13–1
1. the ordered pair for the point at which the graphs of the equations intersect   **3a.** $(-4, 2)$
3b. $(2, -2)$ **3c.** $(2, 4)$

5. $(-4, 1)$ **7.** $(1, -3)$

    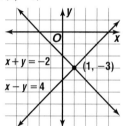

9. $(5, -4)$ **11.** $(0, 2)$

    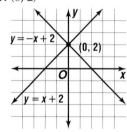

13. $(3, 6)$ **15.** $(5, 2)$

Pages 557–559   Lesson 13–2
1. two parallel lines, two intersecting lines, the same line   **5.** consistent and dependent
7. $(1, -1)$

9a. Consistent and independent; there is one solution, $(3, 150)$.   **9b.** After 3 seconds, both the cat and dog will be 150 feet from the dog's original starting position.   **11.** inconsistent   **13.** consistent and dependent   **15.** consistent and independent
17. infinitely many   **19.** $(5, -2)$

    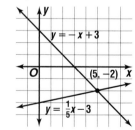

21. no solution   **23.** $(-1, 3)$

25. one; (5, 1)   **27.** There is no solution because the graphs of the equations are parallel lines. The second dog never catches up with the first.

29. Sample answer: $y = -\frac{1}{3}x$

31.

33. $4(x - 2)$ **35.** $3a(ab^2 + 2b - 3)$

Pages 564–565   Lesson 13–3

1. when the exact coordinates are not easily determined from the graph   **3.** Both are correct because you can substitute for either variable.
5. (4, 12)   **7.** (4, 0)   **9.** no solution   **11.** (1, 1)
13. (−3, −9)   **15.** no solution   **17.** (−4, 1)
19. (3, 4)   **21.** (2, 2)   **23.** (2, −10)   **25.** $\left(4, \frac{1}{3}\right)$
27. infinitely many   **29.** (13, 30)   **31a.** 5.5 hr
31b. 357.5 mi   **33.** No; the graphs of equations A, B, and C could form a triangle. Any two of the lines would be intersecting, but there is no point where all three lines intersect.
35. (−1, 4)

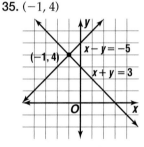

37. $-3 < x \le 2$ **39.** 460

Page 565   Quiz 1

1. (4, −1)   **3.** no solution

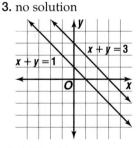

5. 12.5 lb of the $3.90, 37.5 lb of the $4.30

Pages 569–571   Lesson 13–4

1. when the coefficients of terms with the same variable are the same or additive inverses
3. addition   **5.** subtraction   **7.** (8, −1)   **9.** (2, 2)

11. $\left(\frac{16}{3}, 3\right)$ **13a.** $x + y = 42$ and $x = 2y + 3$
13b. 29, 13   **15.** (14, 5)   **17.** (−5, 15)   **19.** (0, 6)
21. (8, −1)   **23.** (−15, −22)   **25.** $\left(5, \frac{5}{2}\right)$
27. infinitely many   **29.** (−11, −4)   **31.** (2, 8)
33. $\left(2, \frac{1}{2}\right)$ **35a.** $13x + 2y = 137.50$ and
$9x + 2y = 103.50$ **35b.** $8.50 child, $13.50 adult
37a. $x + 2y = 29$ and $y = 3x - 10$ **37b.** 7 and 11
39. (11, −38)   **41.** consistent and dependent
43. $y < 4$ **45.** $x < -2$ **47.** $a^2 - 6a + 9$
49. $y - 1 = -5(x - 6)$ or $y + 4 = -5(x - 7)$

Pages 575–577   Lesson 13–5

1. when neither of the variables in a system of equations can be eliminated by simply adding or subtracting the equations   **3a.** substitution
3b. elimination ($\times$)   **3c.** elimination ($+$)
3d. elimination ($-$)   **5.** Multiply second equation by 2. Then add.   **7.** (2, 2)   **9.** (−3, 2)   **11.** no solution   **13a.** 2 mph   **13b.** 14 mph   **15.** (2, 0)
17. (−1, 3)   **19.** $\left(\frac{1}{2}, \frac{3}{4}\right)$ **21.** $\left(\frac{5}{2}, 1\right)$ **23.** (−6, −3)
25. infinitely many   **27.** (−2, −1)   **29.** $\left(\frac{7}{9}, 0\right)$
31. Sample answer: substitution or multiplication; (−2, 4)   **33.** Sample answer: graphing; no solution
35a. $6a + 10b = 108$ and $4a + 12b = 104$
35b. $8, $6   **37.** (7, −1)   **39.** $\left(\frac{3}{2}, 1\right)$ **41.** $m < -4$
43. $a \ge -4$ **45.** 5

Pages 583–585   Lesson 13–6

1. (−1, −7), (2, 0)   **3.** Sample answer: The following methods can be used to solve linear systems of equations: graphing, substitution, elimination using addition or subtraction, and elimination using multiplication. Graphing and substitution can be used to solve quadratic-linear systems of equations. Graphing is useful when you want to estimate the solution. The other methods are useful when you want to find the exact solution.
5. (0, 0), (2, 4)

7. no solution

9. (2, 0), (0, −4)   **11.** no solution

13. (−1, 4), (2, 1)

15. (−4, 7)   **17.** (8, 32), (−2, 2)   **19.** no solution
21. (3, 0), (0, −9)   **23a.** $A = 16$ and $A = 4x^2 - 28x +$
40   **23b.** $\ell = 8$ ft, $w = 2$ ft, $h = 1$ ft   **23c.** 16 ft³
25. Side length 4 units; solve the quadratic-linear
system of equations $p = s^2$ and $p = 4s$, where
p represents perimeter and s represents side length.
The solutions are (0, 0) and (4, 16).   **27.** (2, −1)
29. {−7, 7}   **31.** {4, −8}

Page 585   Quiz 2
1. (4, 2)   **3.** (6, 2)   **5a.** $x - y = 38$, $x = 3y - 2$
5b. 58, 20

Pages 589–590   Lesson 13–7
1. A system of linear equations may have at most
one solution if the equations are distinct. A system
of inequalities may have an infinite number of
solutions.   **3.** Kyle; all of the points in region B
satisfy both inequalities.   **5.** no

7.

9.

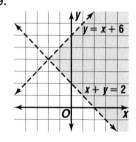

11. Sample answer: 2 dozen sugar, 4 dozen choc.
chip; 4 dozen sugar, 3 dozen choc. chip; 5 dozen
sugar, 2 dozen choc. chip

13.   **15.**

17.   **19.** no solution

21.   **23.**

25.

27. People who use the phone less than 75 minutes
a month should select the $0.15 per minute plan.
People who use the phone more than 75 minutes a
month should select the $0.05 per minute plan. The
middle plan is never the cheapest.
29. (2, 4), (−2, 4)   **31.** no solution

 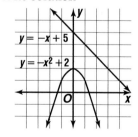

33. 4 lb of the $5.25, 1 lb of the $6.50   **35.** 5   **37.** A

Pages 592–594   Chapter 13 Study Guide and Assessment

1. true   **3.** false; all   **5.** true   **7.** true   **9.** false; elimination

11. $(1, 4)$

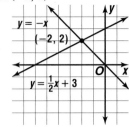

13. $(-2, 2)$

15. $(-6, 0)$ **17.** $(3, -2)$

19. infinitely many

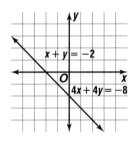

21. $\left(\frac{1}{2}, 1\right)$ **23.** no solution   **25.** infinitely many

27. $(4, -6)$ **29.** no solution   **31.** infinitely many

33. $(2, 0)$ **35.** $\left(\frac{6}{11}, \frac{1}{11}\right)$ **37.** $(22, 31)$

39. $(-2, -4), (1, -1)$

41. no solution

43. $(4, 18), (-1, 3)$ **45.** $(3, 3)$

47.

49.

51. 8 lb of almonds, 12 lb of cashews   **53.** 16 ft by 9 ft

Page 597   Preparing for Standardized Tests
1. D   **3.** A   **5.** B   **7.** D   **9.** 934

Chapter 14   Radical Expressions

Page 599   Check Your Readiness
1. $\frac{1}{2}$ **3.** 2   **5.** $-\frac{1}{11}$ **7.** 1   **9.** 4   **11.** 7   **13.** $13p$
15. $4v$ **17.** $6r + 3$ **19.** $y^2 - 9$ **21.** $z^2 - 25$
23. $9c^2 - 25$

Pages 603–605   Lesson 14–1
1. Sample answer: -4 **3.** Sample answer: $\sqrt{13}$
5. I   **7.** Q
9. -4.5

11. 7.3

13. rational   **15.** rational   **17.** Q   **19.** W, Z, Q
21. Z, Q   **23.** N, W, Z, Q   **25.** Z, Q   **27.** Q
29. 1.7

31. -3.2

33. -6.3

35. 7.2

37. 10.4

39. 15.8

41. rational   **43.** irrational; 6 and 7   **45.** rational
47. irrational; −2 and −3   **49.** irrational; 11 and 12
51. rational
53.

55. about 8.5 s   **57.** 5.2 ohms and 1200 watts or
4.6 ohms and 1500 watts   **59a.** $\frac{1}{2}$ **59b.** $\frac{1}{3}$ **59c.** $\frac{1}{4}$
61. (3, 11), (1, 3) **63.** $\{x \mid x < -7\}$ **65.** $\{m \mid m \le -1.8\}$
67. 0

69. $4(a + 3)(a - 3)$ **71.** 4   **73.** A

Pages 608–611   Lesson 14–2
1. Pythagorean Theorem   **3.** Both are correct since
it does not matter which ordered pair is first when
using the Distance Formula as long as the
coordinates are used in the same order.   **5.** 5
7. 9 or −3   **9.** 9.5 mi   **11.** 12   **13.** $\sqrt{58}$ or 7.6
15. $\sqrt{185}$ or 13.6   **17.** $\sqrt{18}$ or 4.2   **19.** 17 or −13
21. −12 or 6   **23.** 2 or −14   **25.** $\sqrt{194}$ or 13.9
27. 8 or 32   **29.** No; no two sides have the same
measure.   **31.** 16.3 mi
33. 3.5

35. 11.6

37. $\{k \mid k \ge 6\}$ **39.** $\{b \mid b \le -8\}$ **41.** $\varnothing$

Page 611   Quiz 1
1. W, Z, Q   **3.** I   **5.** 1.22, $1.\overline{2}$, $\sqrt{2}$ **7.** $\sqrt{98}$ or 9.9
9. 0 or 6

Pages 617–619   Lesson 14–3
1. to ensure nonnegative results   **3.** Greg is correct.
In its simplest form, the expression is written as
$\frac{3 - \sqrt{5}}{2}$ since the numbers 6, 2, and 4 in $\frac{6 - 2\sqrt{5}}{4}$

are all divisible by 2.   **5.** $2 - \sqrt{8}$; −4   **7.** $5\sqrt{3}$
9. $3\sqrt{10}$ **11.** $\frac{\sqrt{21}}{7}$ **13.** $6|x|\sqrt{y}$ **15.** $4\sqrt{2}$ units
17. $7\sqrt{2}$ **19.** $10\sqrt{5}$ **21.** 4   **23.** 15   **25.** 4
27. $2\sqrt{10}$ **29.** $\frac{2\sqrt{5}}{5}$ **31.** $\frac{8 + 2\sqrt{3}}{13}$ **33.** $\frac{24 + 4\sqrt{7}}{29}$
35. $2|m|\sqrt{10}$ **37.** $3|a|b\sqrt{6b}$ **39.** $6|x|y^2z^2\sqrt{xz}$
41a. $2\sqrt{3}$ mph   **41b.** 10 in.   **43.** 2 or −4   **45.** Z, Q
47. no solution   **49.** $\sqrt{79}$

Pages 621–623   Lesson 14–4
1. Sample answer: Add like terms.
3. The Distributive Property allows you to add
like terms. Radicals with like radicands can be
added or subtracted.   **5.** none   **7.** $8\sqrt{3}$, $-18\sqrt{3}$
9. $2\sqrt{5}$ **11.** $-5\sqrt{7}$ **13.** $-6\sqrt{2}$ **15.** $6\sqrt{3}$
17. $14\sqrt{5}$ **19.** $4\sqrt{7}$ **21.** $-2\sqrt{5}$ **23.** $-9\sqrt{2} + 8\sqrt{5}$ **25.** $7\sqrt{6}$ **27.** $14\sqrt{3}$ **29.** $-4\sqrt{2}$
31. $8\sqrt{2} + 6\sqrt{3}$ **33.** $-2\sqrt{5} - 6\sqrt{6}$ **35.** $5\sqrt{7}$
37a. $4\sqrt{3}$ **37b.** $6\sqrt{3} + 6$ **39.** $3n|p|\sqrt{5mn}$
41. $\sqrt{101}$ or 10.0   **43.** $-1 \cdot 2 \cdot 13 \cdot a \cdot a \cdot a \cdot b \cdot b$
44. $2 \cdot 2 \cdot 3 \cdot 3 \cdot x \cdot y \cdot y \cdot y \cdot y \cdot z$ **45.** $a^2 - 4c^2$
46. A

Pages 627–629   Lesson 14–5
1. Isolate the radical.   **5.** $a + 5 = 4$ **7.** 121   **9.** 4
11. 35.5 ft   **13.** no solution   **15.** −12   **17.** 2
19. 75   **21.** 63   **23.** no solution   **25.** 2 and 3
27. 6   **29.** 10   **31.** 140   **33a.** 1.6 m/s^2
33b. 9.8 m/s^2 **35.** $-5\sqrt{5}$ **37.** $-\sqrt{6} - 3\sqrt{2}$
39. $\sqrt{5}$
41.

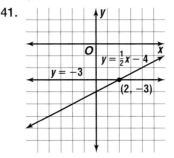

43. $x > 1$ **45.** 22.5

Page 629   Quiz 2
1. $3\sqrt{10}$ **3.** $\frac{7 - \sqrt{5}}{44}$ **5.** $12\sqrt{5}$ **7.** $7\sqrt{2}$
9. no solution

Pages 630–632   Study Guide and Assessment
1. $\sqrt{(x_2 - x_1)^2 + (y_2 - y_1)^2}$ **3.** radicals   **5.** $\sqrt{7ab}$
7. whole   **9.** no solution   **11.** Q   **13.** Z, Q

15. 2.2

17. 10.5

19. 5   **21.** $\sqrt{72}$ or 8.5   **23.** -1 or -5 **25.** $5\sqrt{2}$

27. $3\sqrt{3}$ **29.** $\dfrac{4+\sqrt{5}}{11}$ **31.** $10\,|a|\,c^2\sqrt{3b}$

33. $7\sqrt{5}$ **35.** 0 **37.** $22\sqrt{2}$ **39.** $-14\sqrt{5}$ **41.** 25

43. 9 **45.** no solution   **47.** 6 **49.** 150

51. integers, rational numbers   **53.** $10\sqrt{2}$ lb

55a. 3844 m   **55b.** 2601 m

Page 635   Preparing for Standardized Tests

1. A   **3.** A   **5.** A   **7.** C   **9.** 50

Chapter 15   Rational Expressions and Equations

Page 637   Check Your Readiness

1. 0, 1   **3.** 1, -2 **5.** 6, 2   **7.** k^7 **9.** $12d^5$ **11.** $30m^2n^2$
13. $w^2 + 4w$ **15.** $15c - 20$ **17.** $r^3 + 3r^2 - r$
19. $y^2 + 2y - 3$ **21.** $p^2 - p - 30$ **23.** $2s^2 + s - 10$

Pages 641–643   Lesson 15–1

1. The denominator is 0 when $x = 2$.   **3.** Darnell; x is not a factor of $x + 3$ and $x + 4$. Therefore, it cannot be divided out.   **5.** $4x(4x - 5)$ **7.** $(a - 5)(a + 4)$

9. $(x - 6)(x - 4)$ **11.** $0, -6$ **13.** $\dfrac{2}{5}$ **15.** $\dfrac{2y}{3z^3}$ **17.** $\dfrac{3}{4}$

19. $\dfrac{x - 2}{x + 1}$ **21.** $\dfrac{-(x + 5)}{x + 6}$ **23.** -5 **25.** 10 **27.** $-2, 3$

29. $0, -5$ **31.** $4, -4$ **33.** $\dfrac{1}{3}$ **35.** $\dfrac{2c}{3d}$ **37.** $-4bc$

39. $\dfrac{1}{5}$ **41.** $\dfrac{x + 3}{x + 1}$ **43.** $\dfrac{x}{2}$ **45.** $\dfrac{r + 2}{3}$ **47.** $\dfrac{y + 3}{y - 4}$

49. $\dfrac{y + 2}{y - 1}$ **51.** $\dfrac{z - 5}{z + 3}$ **53.** $\dfrac{8m}{m - 2}$ **55.** $\dfrac{c - 5}{c(c + 6)}$

57. $\dfrac{-x}{4 + x}$ **59.** $\dfrac{4}{3 + a}$ **61.** 9 yr   **63.** 8 lumens/m^2

65. 16   **67.** 19   **69.** $7\sqrt{3}$ **71.** $6\sqrt{3}$ **73.** 5   **75.** D

Pages 647–649   Lesson 15–2

1. The reciprocal of $\dfrac{a^2 - 25}{3a}$ was used instead of the reciprocal of $\dfrac{a + 5}{15a^2}$.   **3.** $\dfrac{4}{x^2}$ **5.** $\dfrac{x + 5}{x - 4}$ **7.** $\dfrac{a^2}{bd}$

9. $\dfrac{2}{x + 4}$ **11.** $4a$ **13.** $\dfrac{3x^2}{2}$ **15.** 12 flags   **17.** $\dfrac{xy}{4}$

19. $3a$ **21.** $\dfrac{2s}{3}$ **23.** $\dfrac{x}{(x + 4)^2}$ **25.** $\dfrac{2(x - 3)}{x + 3}$ **27.** $\dfrac{n}{n + 2}$

29. $6a^2b$ **31.** $2a$ **33.** $\dfrac{y + 4}{y^2}$ **35.** $\dfrac{(m + 1)(m - 1)}{2}$

37. $\dfrac{-(x + 2)}{x - 3}$ **39.** $\dfrac{-3a^2(a + 3)}{2}$ **41.** $\dfrac{1}{x - y}$ **43a.** $\dfrac{x^4}{625}$

43b. $\dfrac{81}{625}$ or about 13%   **45.** $\dfrac{7b^3}{a}$ **47.** $\dfrac{-4}{x - 2}$

49. $(1, 3)$ **51.** $(-3, -4)$ **53a.** $(5, -75)$, $(8, 0)$, $(10, 50)$

53b.

Pages 653–655   Lesson 15–3

1. dividend: $3k^2 - 7k - 5$; divisor: $3k + 2$; quotient: $k - 3$; remainder: $\dfrac{1}{3k + 2}$ **3.** The quotient is a factor of the dividend if the remainder is 0.   **5.** $3a$ **7.** x

9. $x + 2 - \dfrac{1}{2x - 1}$ **11.** $2x + 3$ in.   **13.** x **15.** $3r^2$

17. $x + 4$ **19.** $a + 5$ **21.** $t - 2 - \dfrac{32}{3t - 4}$

23. $b + 2 - \dfrac{4}{2b - 1}$ **25.** $x^2 - 3x + 9$

27. $4x^3 + 4x^2 + 2x + 3 + \dfrac{4}{x - 1}$ **29.** $x - 3$

31a. 936 h   **31b.** 140 h   **31c.** 51 h   **31d.** 5 h
33. 27   **35.** 1   **37.** one solution   **39.** infinitely many   **41.** $y \geq -6$

Page 655   Quiz 1

1. $2; m$ **3.** $\dfrac{1}{2}$ **5.** $\dfrac{-2(a + 1)}{a - 1}$ **7.** y **9.** $x^2 + x + 1$

Pages 659–661   Lesson 15–4

1. They are additive inverses.

3.

a	b	$a+b$	$a-b$	$a \cdot b$	$a \div b$
$\dfrac{2}{t}$	$\dfrac{1}{t}$	$\dfrac{3}{t}$	$\dfrac{1}{t}$	$\dfrac{2}{t^2}$	2
$\dfrac{12}{y}$	$\dfrac{4}{y}$	$\dfrac{16}{y}$	$\dfrac{8}{y}$	$\dfrac{48}{y^2}$	3
$\dfrac{x}{x - 1}$	$\dfrac{1}{x - 1}$	$\dfrac{x + 1}{x - 1}$	1	$\dfrac{x}{x^2 - 2x + 1}$	x

5. $\dfrac{3}{4}$ **7.** $\dfrac{7}{x}$ **9.** -1 **11.** 2   **13.** $\dfrac{10y}{7x - 2y}$ in.   **15.** $\dfrac{7x}{8}$

17. $\dfrac{t}{2}$ **19.** $\dfrac{7}{6r}$ **21.** n **23.** $\dfrac{x}{12}$ **25.** $-\dfrac{1}{z}$ **27.** 0

29. 2   **31.** 3   **33.** $\dfrac{2y}{y - 2}$ **35a.** Sample answer:

$\dfrac{15{,}000}{5} - \dfrac{0.3(15{,}000)}{5}$ **35b.** $2100   **37.** $a + 4$

39. $y - 5$ **41.** $\dfrac{3axy}{10}$ **43.** $9x^2 + 6xy + y^2$

Pages 665–667   Lesson 15–5

1. Sample answer: $\frac{1}{2x}, \frac{1}{4x}$ **3.** Malik; he found the LCD and then added the terms. Ashley incorrectly added the numerators and the denominators.

5. 72   **7.** $30xy^2$ **9.** $\frac{4}{a^2}, \frac{5a}{a^2}$ **11.** $\frac{5t}{12}$ **13.** $\frac{2+3b}{ab^2}$

15. $\frac{-3x+8}{x^2-4}$ **17.** 2042   **19.** $12a^2b^2$

21. $(y+2)(y-2)$ **23.** $(x+3)(x-3)(3x+1)$

25. $\frac{35}{10t}, \frac{8}{10t}$ **27.** $\frac{5z}{xyz}, \frac{6x}{xyz}$ **29.** $\frac{30k}{3(3k+1)}, \frac{k}{3(3k+1)}$

31. $\frac{3d}{10}$ **33.** $\frac{15}{4x}$ **35.** $\frac{m+3}{3t}$ **37.** $\frac{14x-1}{6x^2}$

39. $\frac{2y+x^2+3x}{xy}$ **41.** $\frac{5x+2y^2}{20x^2y}$ **43.** $\frac{2a+3}{a+3}$

45. $\frac{1}{(y-1)^2}$ **47.** $\frac{22x+5}{4(2x-1)}$ **49.** $\frac{x^2+6x-11}{(x+1)(x-3)}$

51. $\frac{m^2-5m+5n}{m(m-n)}$ **53.** 17.5 min   **55.** $\frac{16}{a}$ **57.** 2

59. $9\sqrt{19}$ **61.** in simplest form   **63.** (2, 5)

Page 667   Quiz 2

1. $x+8$ **3.** $\frac{8a}{9}$ **5.** 168

Pages 671–673   Lesson 15–6

1. Sample answer: In a linear equation, the variable cannot appear in the denominator and it cannot be raised to a power higher than 1.   **3.** $-\frac{7}{3}$ **5.** $-\frac{10}{3}$

7. 3   **9.** 30 mph   **11.** $-\frac{2}{3}$ **13.** 7   **15.** -3 **17.** 9

19. $\frac{4}{3}$ **21.** 7   **23.** $\frac{1}{12}$ **25.** 3   **27.** $\frac{1}{6}$ **29.** 6   **31.** -13

33. $-\frac{3}{2}$ **35.** $\frac{7}{5}$ **37.** 5   **39.** 2   **41.** 7   **43.** $\frac{9}{2n^2}$

45. Z, Q   **47a.** $c+s \le 16, 10c+15s \le 200$

47b.

47c. Sample answer: (7, 8), he could plant corn for 7 days and soybeans for 8 days; (2, 12), he could plant corn for 2 days and soybeans for 12 days.
49. (4, -1);

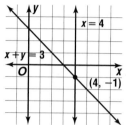

Pages 676–678   Study Guide and Assessment

1. denominator   **3.** $\frac{x}{3x+6y}$ **5.** greatest common factor   **7.** $\frac{1}{a+b}$ **9.** $2x^2$ **11.** $\frac{xz}{4y^2}$ **13.** $\frac{n-4}{n-2}$

15. $\frac{2x^2}{3y^2}$ **17.** $\frac{x}{(x+3)(x-3)}$ **19.** $\frac{3(y-2)}{y+2}$ **21.** a

23. $2y+1+\frac{2}{2y+3}$ **25.** $\frac{3x}{4}$ **27.** $\frac{5x}{x+4}$ **29.** $\frac{6x+5}{8x^2}$

31. $\frac{7a-8}{a(a-2)}$ **33.** $-\frac{1}{5}$ **35.** -4 **37.** 80 pieces

39a. $x+1$ **39b.** 4 in., 3 in., 9 in., 108 in^3

Page 681   Preparing for Standardized Tests
1. A   **3.** B   **5.** D   **7.** C   **9.** 35

Photo Credits

Index

INDEX

fraction bars, 8

fraction-decimal equivalencies, 97

fractions, 4, 48, 83, 90, 94, 170, 198, 234, 494, 534, 615, 617, 640, 642–643
  adding, 90, 203
  algebraic, 638, 676
  complex, 185
  equivalent, 673
  improper, 142, 153, 646
  least common denominators, 663
  multiplying, 90, 141–142
  in simplest form, 654
  simplifying, 185, 640, 654
  subtracting, 90, 656

frequency table, 32–37, 39, 46–47, 137, 175, 185

functional notation, 258, 276, 279

functional value, 258, 276

functions, 256–263, 269, 275–277, 289, 314–315, 393, 425, 454, 553
  cubic, 279
  dependent variable, 264, 267, 304–305, 309
  direct variation, 264–265, 267–268, 270, 273–274, 278, 356, 444, 637
  domain, 238–242, 245–249, 255–256, 276–277, 279, 289, 295, 301–302, 304–305, 308, 310, 371, 537, 649
  exponential, 379, 489–490, 496
  independent variable, 264, 267, 304–305, 309
  linear, 256, 283, 289, 330
  maximum points, 305
  minimum points, 305
  notation, 258, 276, 278
  parent graphs, 318, 464, 466–467, 473, 499, 585
  quadratic, 358, 471, 473, 496
  rational, 638
  relations, 238–242, 245, 254, 257, 259–260, 269, 275–277, 279, 289, 301
  vertical line test, 257, 259–260, 314

function table, 455

Fundamental Counting Principle, 146–147, 149, 180–181, 183

G

Gardening Link, 363, 437

GCF, 48, 139, 442, 450–451

Geography Link, 162

geometry,
  angles, 201, 223, 249, 255, 414, 512, 517, 546–547
  area of bases, 402
  area of circle, 340, 345, 487, 604
  area of circular regions, 336
  area of rectangle, 163, 307, 373, 414, 423–424, 432–433, 437, 476, 482, 605, 635, 679
  area of shaded regions, 396–397, 404, 409, 415, 433, 438–439, 444, 453
  area of square, 338, 345, 360–361, 364, 366, 371, 376, 446, 449, 667
  area of trapezoid, 26, 115, 178, 597
  area of triangle, 29, 164, 261, 365, 409, 414, 476, 597, 623
  coordinate planes, 58–60, 62, 89, 263, 488
  diagonals in polygons, 370, 372, 386, 610
  Distance Formula, 607–609, 613, 673
  graph systems of equations, 595
  height, 29, 596
  hexagon, 386
  hypotenuse, 366–370, 372, 374, 376, 379, 409, 619, 667, 681
  isosceles trapezoids, 610
  isosceles triangles, 546–547, 610, 680
  length, 345, 596, 680
  midpoint, 612–613
  nets, 25, 477, 585, 599
  octagon, 386
  parabolas, 458, 465–467, 496, 582
  parallel lines, 322, 325–328, 330–331, 454, 555, 562
  parallelograms, 323
  pentagons, 603
  perimeter of composite figures, 597
  perimeter of isosceles triangles, 661
  perimeter of quadrilateral, 620–622, 632
  perimeter of rectangle, 27, 279, 482, 659–660
  perimeter of square, 169, 339
  perimeter of triangle, 391–393, 610, 622, 632
  perpendicular lines, 322, 324–328, 330–331, 508
  Pythagorean Theorem, 366–369, 373–374, 376, 378, 379, 605, 609, 619, 623, 680

quadratic functions, 458, 471, 473, 496
quadrilaterals, 327, 546, 681
radius, 201, 596, 604, 618
rectangles, 69, 194, 246–248, 256, 275, 327, 681
rectangular solid, 27
right triangles, 249, 366, 368–372, 374, 379, 387, 409, 667, 680–681
similar triangles, 546–547
slope, 284–295, 298–301, 309–321, 323–324, 326, 328, 330–331, 340, 351, 365, 454–455, 493, 553–554, 557
surface area, 25, 27
trapezoids, 26, 177–178, 547
triangles, 63, 69, 77–79, 88, 164, 261, 546
volume of rectangular prisms, 403, 439, 442–443, 463, 654, 678

Geometry Link, 15, 77, 247, 323, 391, 396, 402, 423, 476, 621, 653, 659

graph
  compound inequalities, 529, 534
  equations, 251, 254–255, 265, 269, 279, 310, 313–315, 331, 340, 404, 501, 572, 584, 625
  exponential functions, 489–490, 492, 498–499
  functions, 459, 490, 499, 529, 639
  inequalities, 504–508, 535–536, 538–539, 542, 544–545, 559, 592, 649
  integers, 86
  linear equations, 251, 277, 307, 312, 329, 454, 550, 556
  on number lines, 89, 364
  ordered pairs, 60–63, 74, 237, 239, 254, 262, 310–311, 637
  pairs of equations, 319–320
  points, 60
  points on a coordinate plane, 61–62, 468
  profit functions, 472
  quadratic equations, 461, 467, 477, 482, 496–497, 540–541, 548
  quadratic inequalities, 541, 586
  related functions, 468–469, 472
  relations, 243, 248–249, 261, 458, 649
  sets of integers, 53
  sets of numbers on a number line, 56
  sets of points, 109

H

I

interest rates, 208

Internet Connection, 3, 31, 44, 51, 81, 86, 93, 111, 132, 139, 153, 180, 187, 211, 218, 230, 237, 263, 276, 283, 309, 328, 335, 346, 373, 374, 381, 411, 412, 419, 427, 450, 457, 488, 495, 496, 503, 541, 542, 549, 579, 591, 592, 599, 613, 630, 637, 675, 676

intersections, 524–525, 527, 542, 551, 554

intervals, 33, 38–39, 42, 46

inverse operations, 360

inverses, 124
  additive, 65–66, 68–69, 143, 155, 390–392, 412–413, 568, 572–573
  multiplicative, 154–155, 180

inverse variation, 270–276, 278, 289, 513, 629, 637

Investigations,
  Bits, Bytes and Bugs!, 80–81
  Box-and-Whisker Plots, 210–211
  Crazy Computers, 578–579
  A Cut-Out Caper, 612
  Dot-to-Dot, 30–31
  Down the Drain, 674
  How Many Centimeters in a Foot, 262–263
  Humidity Heats Things Up, 308–309
  It's Greek to Me!, 410–411
  Must Be TV, 372–373
  Parabolas and Pavilions, 540–541
  A Puzzling Perimeter Problem, 426–427
  Retail Rabbits, 152–153
  Sticky Note Sequences, 494–495
  Sugar & Pizza & Everything Nice!, 110–111

irrational numbers, 362, 364, 374, 485, 600–604, 611, 619, 623, 630, 633, 673

is, 48

isosceles trapezoids, 610

isosceles triangles, 546–547, 610, 680

keep an organized list, 442

kilo-, 352, 375

L

LCD, 100, 102, 664–665, 668, 670, 678–679

LCM, 667, 673

Landmark Link, 468

Landscaping Link, 431

last terms, 401

least common denominator, 663

least common multiple, 662, 676

least value, 106

leaves, 40

legs of triangles, 366, 369–370, 376, 378, 680

length, 596

length of hypotenuse, 680

length of rectangle, 345

less than, 12, 54, 363, 504–505

less than or equal to (≤) symbol, 504

Life Science Link, 5

like terms, 20–21, 47, 389–390, 412

linear equations, 250–251, 253–254, 276, 290, 328
  graphing, 251, 277, 307, 312, 329, 454, 550, 556
  point-slope form, 290–296, 298, 301, 323, 325
  slope-intercept form, 296–300, 307, 311–312, 322, 325, 327, 329, 331, 454, 534, 562
  standard form, 250, 253, 377

linear functions, 256, 283, 289, 330

linear graphs, 310

linear regression, 309

linear relationship, 449

line graph, 38, 41, 46–47, 136–137, 510

line plots, 134

lines, 250–251, 253–254, 276, 290, 328
  horizontal, 252, 313, 324, 328, 606
  number, 52–54, 58, 64, 75, 86, 94–95, 98, 109, 128, 170, 211, 604–605, 633, 673
  parallel, 322, 325–328, 330–331, 454, 555, 562
  perpendicular, 322, 324–328, 330–331, 508

slopes, 284–295, 298–301, 309–321, 323–324, 326, 328, 330–331, 340, 351, 363, 454–455, 493, 553–554, 557
  vertical, 257, 313, 322, 324, 328, 606

line segments, 69, 136

logical reasoning, 30

long division, 90

looking for patterns, 110, 285, 332, 491

lower quartiles, 210

M

Mail Link, 311

Manufacturing Link, 221, 442

map distance, 209, 231

Map Link, 195, 607

Marketing Link, 34

mathematical sentences, 112

Math In the Workplace
  Accountant, 218
  Broadcasting Technician, 346
  Car Dealer, 591
  Civil Engineer, 488

Math Online
  Personal Tutor, 5, 9, 15, 21, 54, 59, 67, 83, 101, 106, 124, 147, 162, 168, 172, 177, 191, 207, 215, 240, 247, 266, 272, 285, 292, 298, 325, 338, 343, 349, 354, 359, 363, 369, 382, 391, 396, 402, 407, 423, 476, 480, 485, 491, 510, 515, 522, 525, 531, 538, 552, 564, 569, 574, 587, 608, 616, 626, 641, 652, 659, 665, 671
  Data Update, 25, 42, 127, 151, 193, 240, 295, 361, 404, 420, 465, 512, 583, 614, 627, 671

matrices, 80, 578–579
  addition of, 80
  scalar multiplication of, 81

matrix subtraction, 81

maximum, 459, 496, 504

maximum point of function, 305

means, 104, 106–109, 116, 121, 131, 133–135, 158, 184, 209, 235, 281, 351, 544, 613

Measurement Link, 267, 532

measurements, 107

measurements, equivalent, 190

possible outcomes, 146, 149, 174, 181, 183, 219

powers, 336, 374, 383
of expressions, 336, 374
powers of powers, 345
powers of ten, 346, 352

predict, 38–39, 42, 46, 241

predictions, 38, 62, 290, 361, 404, 417

Preparing for Standardized Tests, 48–49, 90–91, 136–137, 184–185, 234–235, 280–281, 332–333, 378–379, 416–417, 454–455, 500–501, 546–547, 596–597, 634–635, 680–681, 732–749

prime, 77, 433, 451, 453, 605

prime factorization, 339–340, 351, 358, 360, 374, 421–422, 424, 433, 450, 508, 614–615, 617, 621, 662

prime factors, 421, 425

prime numbers, 358, 420–421, 424, 467

prime polynomials, 430, 436

principal, 205

principal square roots, 617

probability, 219–220, 222–223, 224–229, 230, 232, 233, 242–243, 279, 280–281, 315, 327, 333, 406–407, 467, 539, 649
of complements, 223
experimental, 223, 649
of independent events, 225, 406
of mutually exclusive events, 226
odds, 219, 221, 223, 230, 243, 398
predictions, 224

Problem-Solving Strategies,
check for reasonableness, 760
drawing diagrams, 28, 68, 73, 378, 610, 755
guess and check, 758
looking for patterns, 110, 285, 332, 491, 757
make a table, 110–111, 244–245, 252–253, 263, 458, 462, 468–471, 489, 501, 536, 612, 613, 674, 754
models, 119, 231, 388, 428, 460, 501
organizing data, 33, 47, 281
use a formula, 761
use a graph, 759
working backward, 165, 168, 519

writing equations, 57, 118, 120–121, 126–127, 133–134, 164, 168–170, 255, 269, 291–292, 333, 340, 477, 605, 681, 756
writing formulas, 27, 29, 43, 613

Problem-Solving Strategy Workshops, 754–761

product mat, 650

Product Property of Square Roots, 358, 614–615, 617

products, 4, 76, 89, 121, 141, 193, 518
of binomials, 408
of conjugates, 630
of diagonal terms, 96
of fractions, 324
of numbers and variables, 382
of powers, 341
of rational expressions, 647
of slopes, 324

Products of Powers Rule, 350

product of a sum and a difference, 407–408, 414, 447, 616

projections, 266

Properties, 9
Addition Property of Equality, 122, 124, 643
Addition Property of Inequalities, 509
Additive Identity, 9, 10
Additive Inverse, 66, 76, 85
Associative, 14–15, 17, 20, 23, 26, 29, 45, 47, 63, 74, 143, 337, 354
Closure, 16, 63, 74, 143, 158, 203
Commutative, 14–16, 18, 20, 26, 45, 47, 63, 67, 74–75, 77, 85, 143, 235, 321, 342, 354
Comparison, 95–96
Completeness, 601–602
Distributive, 19–23, 26, 45, 47, 67, 76, 85, 176–177, 182, 189, 206–207, 235, 298, 321, 323, 325, 389–390, 394–396, 399–404, 412, 429–431, 437, 451, 454, 470, 476, 480, 490, 520, 543, 562–563, 616, 620–621, 643, 658, 664, 668–670
Division Property of Equality, 160, 164, 643
Division Property of Inequalities, 514, 517
Identity, 10, 12, 15, 17, 26, 37, 121

Multiplication Property of Equality, 26, 161, 164
Multiplication Property of Inequalities, 515, 517
Multiplicative Identity, 10, 47
Multiplicative Inverse, 154–155
Multiplicative Property of –1, 143
Multiplicative Property of Zero, 10, 76
Product Property of Square Roots, 358, 614–615, 617
Quotient Property of Square Roots, 359, 615, 617
Reflexive, 9, 249
Substitution, 9, 10, 15, 19–20, 26, 166, 245, 247, 298, 312, 325
Subtraction Property of Equality, 341
Subtraction Property of Inequalities, 509
Symmetric, 9, 12, 20, 23, 47, 253
Transitive, 9, 12, 249
Zero Product, 26, 474–476, 582, 608, 638

Properties of Equalities, 9, 122, 134, 234

Properties of Inequalities, 643

Properties of Proportions, 288, 191

Properties of Square Roots, 358, 614–615, 617

proportions, 188, 191–192, 198, 201, 230, 270, 272
cross products, 96–97, 188–191, 195, 198–199, 212, 231–232, 267, 272, 278, 546
direct variations, 264–265, 267–268, 270, 273–274, 278, 356, 444, 637
percent, 198–199, 204, 213, 221, 230–231, 233, 261, 404, 417, 482
rate problems, 266, 575–576
similar triangles, 546

protractor, 201, 324

Punnett squares, 406, 434, 439, 452

Pythagorean Theorem, 366–369, 372–373, 374, 376, 378, 379, 605, 608, 619, 623, 680

Pythagorean triples,
3-4-5, 369, 378, 680
5-12-13, 378
7-24-25, 378

INDEX

Properties

Substitution (=)	If $a = b$, then a may be replaced by b.
Reflexive (=)	$a = a$
Symmetric (=)	If $a = b$, then $b = a$.
Transitive (=)	If $a = b$ and $b = c$, then $a = c$.
Additive Identity	For any number a, $a + 0 = 0 + a = a$.
Multiplicative Identity	For any number a, $a \cdot 1 = 1 \cdot a = a$.
Multiplicative (0)	For any number a, $a \cdot 0 = 0 \cdot a = 0$.
Multiplicative (−1)	For any number a, $-1 \cdot a = -a$.
Additive Inverse	For any number a, there is exactly one number $-a$ such that $a + (-a) = 0$.
Multiplicative Inverse	For any number $\frac{a}{b}$, where $a, b \neq 0$, there is exactly one number $\frac{b}{a}$ such that $\frac{a}{b} \cdot \frac{b}{a} = 1$.
Commutative (+)	For any numbers a and b, $a + b = b + a$.
Commutative (×)	For any numbers a and b, $a \cdot b = b \cdot a$.
Associative (+)	For any numbers a, b, and c, $(a + b) + c = a + (b + c)$.
Associative (×)	For any numbers a, b, and c, $(a \cdot b) \cdot c = a \cdot (b \cdot c)$.
Distributive	For any numbers a, b, and c, $a(b + c) = ab + ac$ and $a(b - c) = ab - ac$.
Comparison	For any numbers a and b, exactly one of the following sentences is true: $a < b, a > b$, or $a = b$.
Addition (=)	For any numbers a, b, and c, if $a = b$, then $a + c = b + c$.
Subtraction (=)	For any numbers a, b, and c, if $a = b$, then $a - c = b - c$.
Division and Multiplication (=)	For any numbers a, b, and c, with $c \neq 0$, if $a = b$, then $ac = bc$ and $\frac{a}{c} = \frac{b}{c}$.
Product Property of Square Roots	For any numbers a and b, with $a, b \geq 0$, $\sqrt{ab} = \sqrt{a} \cdot \sqrt{b}$.
Quotient Property of Square Roots	For any numbers a and b, with $a \geq 0$ and $b > 0$, $\sqrt{\frac{a}{b}} = \frac{\sqrt{a}}{\sqrt{b}}$.
Zero Product	For any numbers a and b, if $ab = 0$, then $a = 0$, $b = 0$, or both a and b equal 0.
Addition (>)*	For any numbers a, b, and c, if $a > b$, then $a + c > b + c$.
Subtraction (>)*	For any numbers a, b, and c, if $a > b$, then $a - c > b - c$.
Division and Multiplication (>)*	For any numbers a, b, and c, 1. if $a > b$ and $c > 0$, then $ac > bc$ and $\frac{a}{c} > \frac{b}{c}$. 2. if $a > b$ and $c < 0$, then $ac < bc$ and $\frac{a}{c} < \frac{b}{c}$.

* *These properties are also true for $<$, $\geq$, and $\leq$.*